THE GUINNESS ENCYCLOPEDIA OF THE HUMAN BEING

DR ROBERT M. YOUNGSON

GUINNESS PUBLISHING

First published 1994

Published in Great Britain by
Guinness Publishing Ltd., 33 London Road, Enfield,
Middlesex EN2 6DJ

Colour reproduction by
Bright Arts (HK) Ltd., Hong Kong

Printed and bound in Italy by New
Interlitho SpA, Milan

British Library Cataloguing in
Publication Data:
A catalogue record for this book is
available from the British Library.

ISBN 0-85112-565-4

Project Editor
Christine Winters

Additional Editing
Tina Persaud
Ian Crofton
Ben Dupré

Design
Amanda Sedge
Sarah Silvé

Systems Support
Alex Reid
Kathy Milligan

Picture Research
Image Select

Illustrators
Peter Harper
Pat Gibbon
David McCarthy
Robert Burns
Peters and Zabransky
Lesley Alexander
Amanda Williams

Art Director
David Roberts

Managing Editor
Stephen Adamson

CONSULTANT EDITORS

Dr Stephen Edgell.......Reader in Sociology, University of Salford
Professor Brian Gardiner......Professor of Vertebrate Palaeontology, King's College, University of London
Professor J.R. Hampton.......Professor of Cardiology, University Hospital, Nottingham
Dr Colin McEvedy.......Consultant Psychiatrist, West London Health Care
Dr Peter Moore.......Reader in Ecology, King's College, University of London
Dr Noel Smith.......Research Fellow in Biology, University of Sussex
Dr Jane Ussher.......Lecturer in Psychology, University College, London

Title page illustration: Group of children (Zefa)

PREFACE

Encyclopedias are usually deemed to require the separate input of many different people, each expert in his or her own particular field. This approach can be problematic, however, since methods and opinions are bound to conflict to some degree. There is another, and perhaps better, way to put an encyclopedia together. Find a harmless drudge with a reasonable general knowledge of the subject, and persuade him or her to spend a couple of years drafting an approved outline, and then writing the whole book. Then, separate the typescript into appropriate sections and send them to the experts. Next, set a team of experienced editors to review and correlate the experts' comments and to labour for months over the text until they achieve the greatest possible clarity and simplicity of expression. Finally, give the harmless drudge a last chance to protest at the mutilation of his or her deathless prose, but do not, of course, necessarily pay the smallest attention. The result, hopefully, is a book that expresses a broad spectrum of opinion, that is as up to date as possible, that has a uniform and readable style, and that is highly accurate and reliable.

This is how *The Guinness Encyclopedia of the Human Being* has been created. It will be clear that any faults it may contain are, in reality, the responsibility of the harmless drudge; any merits are to the credit of the experts and the editors. Even so, as the putative author, I am liable to be credited with any merit that may be found in the book. So I am more indebted than I can easily express to the panel of distinguished men and women, listed in these preliminaries, who have had to suffer the results of my lucubrations, and tactfully correct my many errors of conception. Even more am I indebted to the group of talented, hard-working and friendly people at Guinness Publishing who have worked with me on this project.

Specifically, I would express grateful thanks to Ian Crofton, who originally commissioned me to write the book, and who provided wise advice and guidance on resolving the detailed outline. Many thanks to Stephen Adamson, who succeeded him, and whose friendly encouragement, fresh outlook and many stimulating and acceptable suggestions, conspicuously helped the work along. The departure of Ben Dupré to Oxford University Press, early in the course of the work, was greatly regretted; this was a relaxed and productive association, illuminated by mutual respect. Thanks also to the patient and capable Tina Persaud for her valuable suggestions for additional text, her excellent editorial work, and her cheerful professionalism in the face of authorial asperity.

Most of all I must acknowledge with gratitude the work of my principal editor Christine Winters: whose energy and enthusiasm for the task greatly lightened my labours; whose editorial judgement I never found occasion to question; whose sensitive diplomacy averted many a crisis of offended *amour propre*; and whose support, encouragement and appreciation sustained my efforts through many months of toil.

HOW TO USE THIS BOOK

The Guinness Encyclopedia of the Human Being follows a **thematic approach**, and is arranged in a series of self-contained articles focusing on particular topics of interest. There is also an A-to-Z **Factfinder** section, providing quick access to information on key concepts, technical terms, major figures, and many other topics.

In the thematic section, the pages are colour coded to indicate the chapter to which they belong. Within each article there is a 'See Also' box referring to related spreads within the same section or elsewhere. There are also cross-references within the text itself to guide the reader to pages where further relevant details will be found.

The Factfinder consists of alphabetically arranged entries providing at-a-glance information on a range of matters and issues relating to the human being. Many of the entries apply to topics that receive some treatment in the main text; such entries tend to be brief, providing basic details such as a concise description of a particular disease or disorder, together with a page reference to the relevant article in the thematic section. The Factfinder also contains numerous entries for items not covered in the main text. In addition there are entries providing further details on concepts or technical terms to which only brief allusion is made in the thematic section.

The Factfinder is linked with the main thematic section by means of a simple cross-referencing and indexing system. The words in small capital letters refer the user to important related entries within the Factfinder itself.

Contents

1. HOW WE EVOLVED

Rocky coastline, Devon, England. (SPL)

2. HEREDITY

German family tree. Detail of painting by Daniel Bretschneider the younger (1623–58). (AKG)

3. THE BODY MACHINE

Rowing regatta on Lake Lucerne, Switzerland. (Gamma)

4. THE BRAIN AND THE SENSES

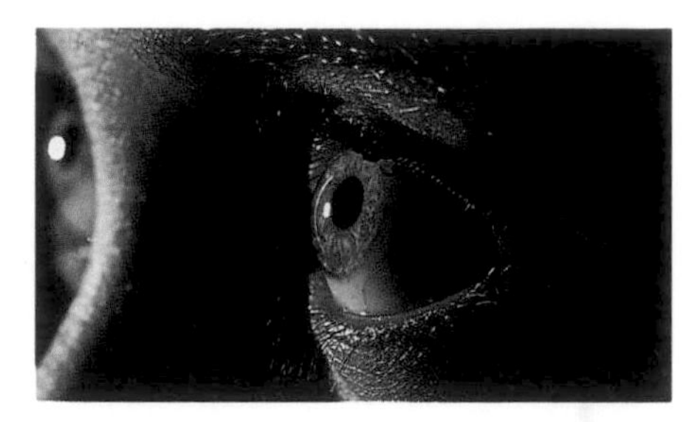

Close-up of a young woman's face. (Martin Dohrn, SPL)

5. THE CYCLE OF LIFE

Grandmother and grandson, the Philippines. (Explorer)

CONTENTS

6. HEALTH AND DISEASE

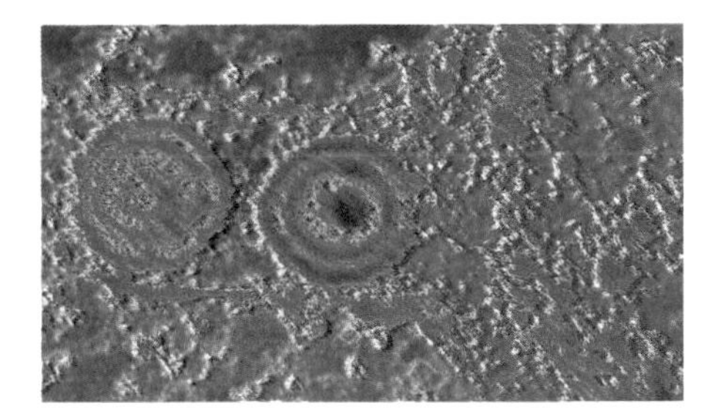

Herpes simplex, type II virus. (CNRI, SPL)

7. PSYCHOLOGY

At the Window (detail). Painting by Hans Heyerdahl (1857–1913). (BAL)

8. MENTAL DISORDERS

Cutting on the Stone of Madness. Painting by Pieter Brueghel the elder (c. 1525–69). (Giraudon, BAL)

9. OURSELVES AND OTHERS

Crowd at Federation Cup Match, Vancouver, Canada. (IB)

FACTFINDER

HOW WE EVOLVED

Our Place in Nature

As far as we know about two million different species of living things inhabit the earth today. All scientists now agree that these species, of which man (*Homo sapiens*) is one, arose through a long history of gradual change that can be traced back to very simple life forms.

These two million species account for only a tiny proportion – perhaps one thousandth – of the total number of species that have lived on earth since life began thousands of millions of years ago. Although modern humans, by virtue of their superior intelligence and ability to communicate highly complex information, have a uniquely dominant status in the animal kingdom, there is nothing biologically unique about the human species. The proposition that the human being is an animal is a statement of scientific fact – a reflection of the inescapable physical relationship between humans and the millions of other animal species.

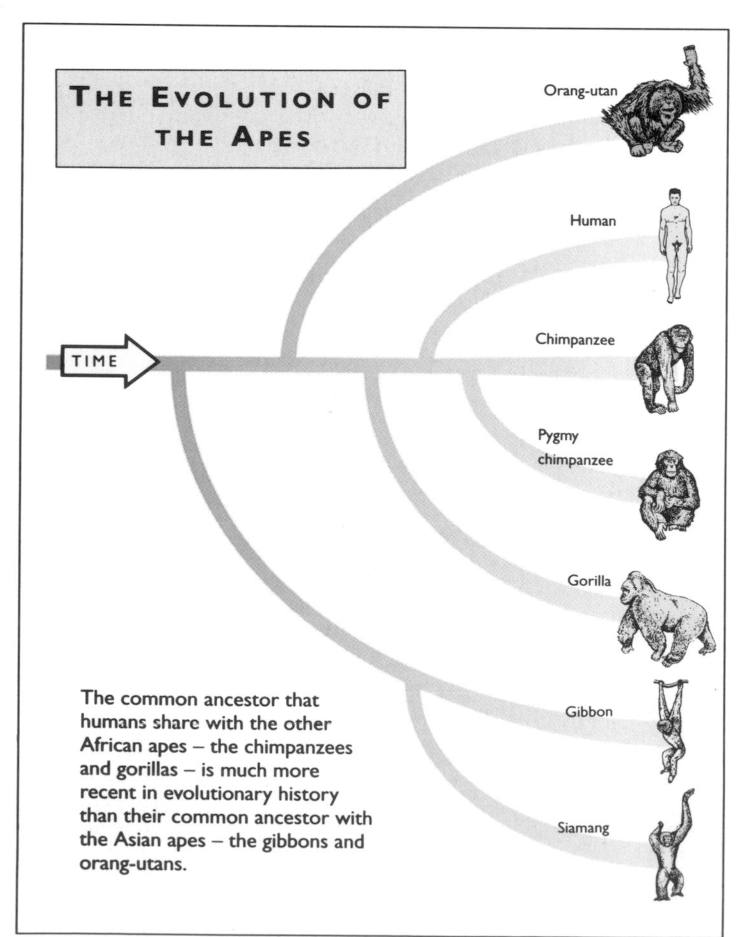

The common ancestor that humans share with the other African apes – the chimpanzees and gorillas – is much more recent in evolutionary history than their common ancestor with the Asian apes – the gibbons and orang-utans.

In comparing ourselves with closely allied species, it is hard to decide which are greater – the similarities or the differences. But if we exclude mental development and all its implications, there can be no doubt of the answer. On a purely physical level, the differences between humans and closely related species are so minor that, in some particulars, only an expert can distinguish one from another.

Comparative anatomy

Comparative anatomy is the study of the relationship between the body structure of different animals. Even an elementary knowledge of comparative anatomy immediately illustrates the remarkable similarities between humans and other animals.

Early studies and writings in this field were made by the Greek philosopher Aristotle (384–322 BC), who was concerned with the classification of all things and who dissected numerous animals so as to see how they were related. The Greek medical thinker and teacher Galen (c. AD 130–200) also took an interest in comparative anatomy, and dissected dogs, pigs, goats and monkeys, keeping meticulous records of his findings. However, the first major attempt at a systematized comparative anatomy was that of Georges Buffon (1707–88). This French nobleman spent 53 years compiling a monumental 44-volume work entitled *Natural History, General and Particular*, in which he presented to an appreciative public a great deal of information about the relationships between different species. A major step forward in setting the principles of the subject on a scientific basis was achieved by Baron Georges Cuvier (1769–1832). Inspired by Buffon, Cuvier became interested in comparisons between the anatomies of different animals. He was a true scientist, declining to form theories until he had sufficient observable facts on which to base them. Cuvier's 9-volume *The Animal Kingdom Arranged in Accordance with Structure* (1817–30) was widely influential, and he is regarded as the father of comparative anatomy.

A central element in comparative anatomy is the concept of *homology* – the relationship of body organs of different species that have the same evolutionary origin. In closely related species, corresponding structures are so similar as to leave no doubt about their common origin. But scientists such as Richard Owen (1804–92) also clearly showed how even less obviously related structures in remote species such as the wings of a bird and the arms of a human being, or the flipper of a seal and a person's hand, were built on the same plan or had a common pattern. By the middle of the 19th century the idea of homology was becoming widely appreciated.

The human being classified

Among the animals, a relatively small number of species possess backbones. These are known as the vertebrates, a category which includes fishes, birds,

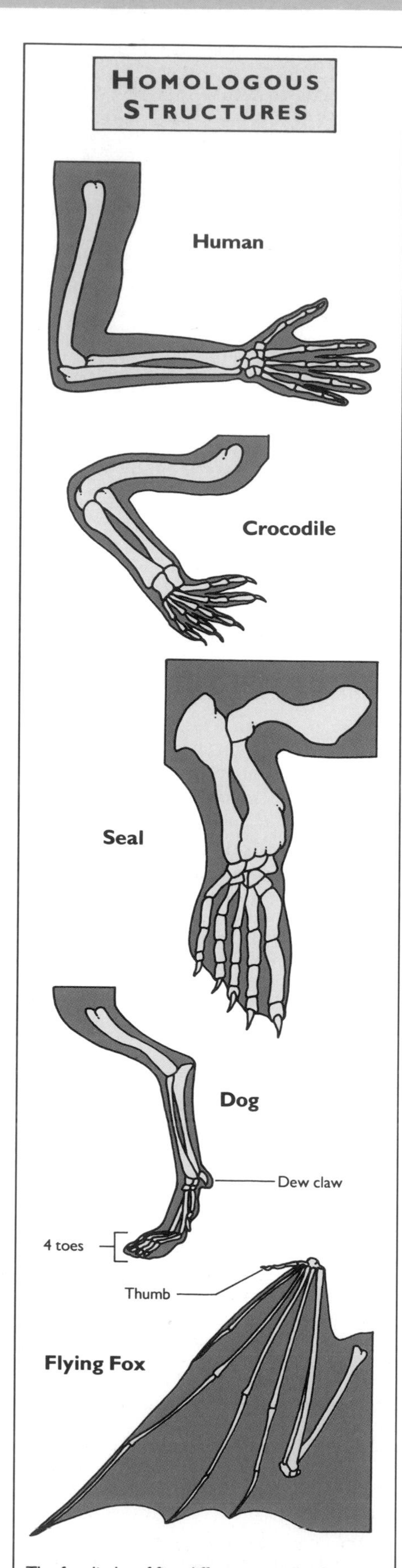

The forelimbs of five different animals, showing that the pentadactyl (five-digit) limb is common to them all. Such basically similar structures are described as *homologous*, and the existence of homologous structures in different organisms suggests that they have all evolved from a common ancestor.

The Classification of Man

RANK	NAME	MEMBERS	DISTINGUISHING FEATURES
KINGDOM	Animalia	All multicellular organisms except plants and fungi	Nervous system
PHYLUM	Chordata	Lancelet, sea squirts, all vertebrates	Notochord (central nerve cord)
SUBPHYLUM	Vertebrata	All vertebrates, i.e. fish, amphibians, reptiles, birds, mammals	Backbone (protecting notochord)
SUPER-DIVISION	Gnatho-stomata	All vertebrates except lampreys and hagfishes	Jaws
DIVISION	Osteichthyes	All vertebrates except lampreys, hagfishes and cartilaginous fishes	Principal component of skeleton is bone (not cartilage)
SUPERCLASS	Tetrapoda	All vertebrates except fish	Four limbs, the front pair being pentadactyl (i.e. having five digits)
CLASS	Mammalia	All mammals	Milk and sweat glands, hair
SUBCLASS	Eutheria	All mammals except monotremes and marsupials	Placenta
ORDER	Primates	Lemurs, lorises, pottos, galagos, tarsier, marmosets, tamarins, monkeys, apes	Fingers with sensitive pads and nails
SUPERFAMILY	Hominoidea	Apes (gibbons, orang-utan, gorilla, chimpanzees, humans)	No tail, broad chest, shoulder blades at back rather than sides
FAMILY	Hominidae	Gorilla, chimpanzees, humans	Upright posture, flat face, large brain; similar blood plasma protein
SUBFAMILY	Homininae	Chimpanzees, humans	Similar brain shape and anatomical and protein structures; tool and weapon users
GENUS	*Homo*	*Homo habilis*, *H. erectus* (both extinct), *H. sapiens*	Exclusively bipedal (walking on two feet), manual dexterity
SPECIES	*sapiens*	Neanderthal man (*H. sapiens neanderthalensis*, extinct), modern man	Double-curved spine
SUBSPECIES	*sapiens*	Modern man (*Homo sapiens sapiens*)	Well-developed chin

reptiles, amphibians, and mammals. Mammals are distinguished from other vertebrates by their ability to produce milk from special mammary glands, to suckle and nourish their offspring.

Among the mammals is a group called the primates. The principal defining feature of the primates is their possession of elongated fingers with sensitive pads and nails (rather than claws); most of them also have the ability to bend their thumbs across the palm so as to touch each of the other fingers. In addition, primates have large brains and well-developed eyes that are directed forward. The majority live mainly in trees, and all produce one or two young that are nurtured for a relatively long time after birth. The primates include lemurs, lorises, pottos, galagos, tarsiers, marmosets, tamarins, Old and New World monkeys, and apes (including human beings).

Narrowing the definition further, we can say that, in terms of strict classification, man is also an ape. Apes are primates with no visible tail, with shoulder blades at the back rather than at the sides, and with a Y-shaped pattern of grooves on the surface of the grinding teeth (the molars). The apes consist of the gibbons, the orang-utan, the gorilla, the chimpanzees and man. Chimpanzees are the apes most closely related to humans, and the two evolutionary lines are thought to have diverged as recently as about 5.5 million years ago.

All the apes sleep at night and all are predominantly vegetarian, apart from humans and the chimpanzees (the latter are known to hunt and kill small animals for food). Chimpanzees and gorillas spend most of their time on the ground. Apes usually have single young or sometimes twins and form families of varying size – up to as many as 50 or 60 individuals in the case of chimpanzees. Nonhuman apes have a life expectancy of about 30 years – much the same as that of early humans. The gorilla, orang-utan and gibbons are generally shy and retiring and will show aggressive responses only when threatened or when defending territory. However, chimpanzees are often inquisitive and extroverted, and are capable of considerable aggression.

In the entire animal kingdom the intelligence of nonhuman apes would appear to be second only to that of humans, though it is extremely difficult to set criteria for testing. Chimpanzees and gorillas are known to be able to learn a sign-language vocabulary of as many as a hundred words, and chimpanzees commonly use simple tools, especially when motivated by the desire for food. They use twigs to extract termites from nests, and in captivity pile up boxes to stand on to reach otherwise inaccessible food. They also use sticks and stones as weapons. It is possible that orang-utans are equally intelligent but are reluctant to display it in the presence of humans. None of the apes apart from man can speak, but this is probably only because their vocal equipment is not adapted to speech. Primates, like many other mammals, communicate comprehensively with each other by means of signs, gestures and calls.

The brain of a human being closely resembles that of the other apes. All the major features correspond in each case. The areas for voluntary movement, facial expression, bodily sensation, vision, hearing, smell, emotional responses, balance, motor coordination, and so on, are the same, qualitatively, in both. In all the apes, the sensory input to the brain areas concerned with emotion comes largely from the parts of the brain concerned with visual information processing. In less highly developed animals the connections come mainly from the parts concerned with smell appreciation.

The nonhuman apes suffer many of the same diseases as humans. Genetic disorders such as Down's syndrome are well known. Heart disease and strokes occur, and epilepsy has been observed. Infections that are virtually identical to those suffered by humans are commonplace.

Within the apes is a genus called *Homo*. Man, the only surviving species of this genus, has been given the specific name *sapiens*, meaning 'wise', 'judicious' or 'sensible'. Whether this is the most appropriate name must remain an open question, but no one questions that the species *Homo sapiens* is the dominant member of the group of apes.

Physical anthropology

Physical anthropology is the study of the evolutionary biology of the species *Homo sapiens*. Its main purpose is to try to elucidate the process by which present-day humans and their man-like ancestors evolved. Knowledge is constantly advancing as new fossil remains are discovered and improved techniques are applied. Scientists currently use carbon dating and other methods to establish ages; biochemical studies to compare the physiology of humans with that of other primates; detailed physical measurements of all parts of the body (anthropometry) to quantify comparisons; and genetic analysis – especially of mitochondrial DNA (⇨ p. 12), which is relatively unchanged through the generations – to help to determine human origins. Genetics has also been widely used in the studies of the more recent diversification of man. Various genetic markers, such as the range of blood groups, tissue types and various gene mutations, have proved valuable in tracing historic movements and interbreeding of human groups.

SEE ALSO
- HOW WE EVOLVED pp. 6–17
- HEREDITY pp. 20–31

The Concept of Race

Formerly the term 'race' was applied to different groups of humans in different parts of the world, who were distinguished from each other by highly visible variations such as skin colour, facial configuration, hair type, and so on. Thus anthropologists distinguished clearly defined 'races' such as Negroes, Pygmies, Bushmen, Caucasians, Mongoloids, Polynesians, and so on. These populations (as many scientists now prefer to call them) do, of course, have certain characteristics in common, but on a closer genetic analysis of many more factors it is seen that there is usually much more genetic variation within a so-called 'race' than between it and others. This is most markedly so in 'black' Africa, where there is by far the greatest genetic diversity to be found anywhere in the world.

Furthermore, generations of high population mobility and the increasing trend towards interbreeding between individuals formerly ascribed to different races have diluted and spread characteristics to a degree where, in many cases, 'race' as a genetic concept has become meaningless. The concept of 'ethnicity' – a common identity based on shared characteristics such as culture, language and religion – is now increasingly used by anthropologists and sociologists in place of race. This recognizes the differences between groups – and helps to explain conflict between them – without resorting to the kind of scientifically dubious concepts of genetic racial difference exploited by the Nazis.

The Story of Evolution

Until about the middle of the 19th century it was generally believed that all species were produced, fully developed, by a single act of creation. Some believed that species appeared by a process of 'spontaneous generation', others attributed the creation of species to God.

But as the arranging and classification of the multitude of different species proceeded – especially following the work of the Swedish naturalist Carl Linnaeus (1707–78) – people began to pay more attention to the close similarities between certain species. The more enquiring began to wonder whether there might be some kind of relationship between them. By the beginning of the 20th century most of their questions had been answered.

As interest in the structure of the earth developed and became formalized in the science of geology, two important facts emerged: that stratified layers of rock had been formed at different periods in the past, and that the fossils found in these different layers were the impressions of organisms that lived during these periods. From these observations developed the science of *palaeontology* – the study of prehistoric life based on the evidence of fossils.

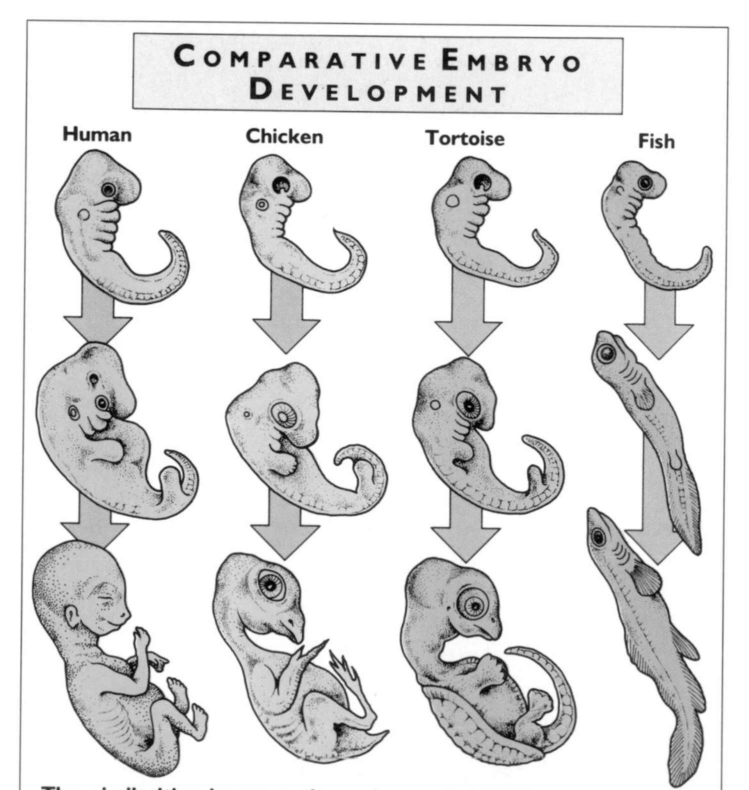

The similarities between the embryos of vertebrates at comparable stages of development have provided considerable support to the theory of evolution (⇨ text). Thus fishes, amphibians (not shown), reptiles, birds and mammals all start with a similar number of gill arches (the folds below the head) and a similar vertebral column. However, as the embryos develop, the similarities decrease and the individual species become more and more differentiated.

The record of the rocks

Fossils had, of course, been recognized for thousands of years. While Aristotle thought they were failed or abortive attempts at spontaneous generation from mud, Leonardo da Vinci and other Renaissance thinkers recognized them for what they were – the impressions left by once-living organisms. In the fifteenth century, however, it was not always safe to make too much of this now obvious fact for fear of reprisals from the Church.

The most impressive finding of the palaeontologists was the record of definite sequences of organisms, living at successive periods but showing general resemblances to their predecessors. As a rough generalization, the older the rocks in which they were found, the simpler were the life forms. This suggested the probability that modern living things came from earlier life forms by a gradual process of change – by evolution.

The earliest unmistakable fossils date back about 3400 million years. These have a vaguely cabbage-like structure and seem to be similar to modern aggregations of cyanobacteria (blue-green algae). In sediments laid down about 1500 million years ago, large numbers of protozoa (complex single-celled organisms) can be found. Sediments dating back about 590 million years contain the remains of multicellular soft-bodied animals such as segmented worms, and in deposits made about 570 million years ago we find fossils of molluscs, jellyfish, crustaceans and starfish similar to those existing today.

Fossils of numerous species of fish are found in deposits dating back 400 to 500 million years. Those of insects and amphibians appear slightly later. The first reptiles appeared about 340 million years ago, giving rise to the dinosaurs, which dominated life on earth from 230 to 65 million years ago. The earliest mammals emerged at about the same time as the dinosaurs, while by 150 million years ago (possibly much earlier) the birds evolved from one group of dinosaurs. Fossils of the first primates – the group of mammals that contains the living lemurs, monkeys and apes (including humans) – are found in deposits dating from about 65 million years ago. Remains of recognizably humanlike creatures occur only in deposits laid down in the past few million years.

The evidence of embryology

There is a striking parallel between many of the stages of evolution found in the fossil record and those occurring during the development of every human being as an embryo in the womb (⇨ p. 102). Each of us starts as a single fertilized cell. This divides repeatedly to form a hollow ball of cells. The ball then folds in on itself to form a two-layered structure called a *gastrula*, which is reminiscent of some of the simplest multicellular organisms. From this, a three-layered embryo develops. Each of the layers – the *ectoderm*, *endoderm*, and *mesoderm* – gives rise to particular organs and systems of the body. In the course of the development of these organs, the embryo passes through stages in which it clearly resembles, successively, the embryos of a range of organisms less complex than ourselves. It is as if each of us repeats in our bodily development the evolutionary history of our species.

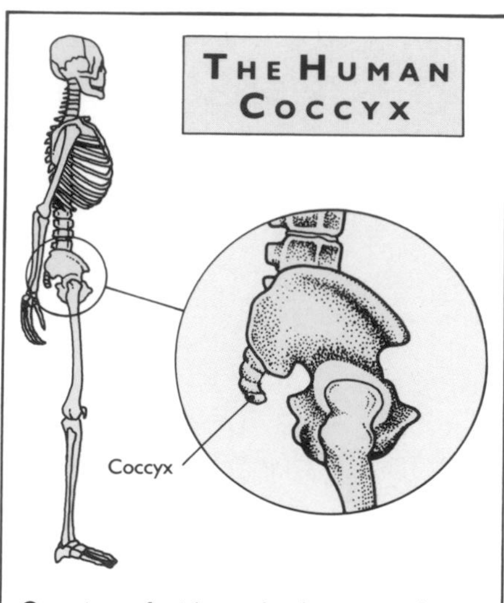

One piece of evidence that humans and monkeys had a common ancestor is the existence in humans of the coccyx – the bone at the base of the spine. The coccyx is the remains of a tail, and is an example of a vestigial structure, i.e. one that no longer functions. In very rare cases humans are actually born with a stump of a tail.

This concept was expressed by the German biologist Ernst Haeckel (1834–1919) in the maxim 'ontogeny recapitulates phylogeny'. Ontogeny is the developmental history of an individual from the time of conception; phylogeny is the sequence of events involved in the evolution of a species. Strictly speaking, this is not quite accurate. At no stage does a mammalian embryo resemble an adult fish, but at one stage it certainly does resemble the embryo of a fish.

If the embryos of different classes of animals are compared at certain stages, it is often very difficult to tell by inspection which is which. Haeckel was especially interested in this matter, and pointed out that the more closely two animals resemble one another in bodily structure, the longer their embryos remain indistinguishable. On the basis of this and similar evidence he, therefore, concluded that we had 'sufficient definite indications of our close genetic relationship with the primates'.

The evidence of body structure

As knowledge of the structure of the human body and of those of other species grew, it became apparent that the structural resemblances were much

Fossilized trilobite – *Arionellus ceticephalus.* Now extinct, trilobites were simple marine arthropods (the invertebrate group that includes crustaceans, insects, arachnids and centipedes), and were most abundant some 500 million years ago. The fossil record shows a large diversity in body shape and appendage structure, indicating a wide range of lifestyles. (Sinclair Stammers/SPL)

closer than had been thought. When the anatomies of the vertebrates (animals with backbones) were compared, it was seen that they differed essentially only in the shape and size of individual parts. The similarities between even remote vertebrate species were far greater than the differences. Many scientists contributed to this understanding, perhaps the most outstanding being the English comparative anatomist Richard Owen (1804–92), who showed the equivalence of many anatomical structures in terms not only of their position in the body but also of their developmental origins. Owen's concept of *homology* (⇨ p. 4) was illustrated by examples of organs in different species that not only shared the same essential anatomical structure (if not necessarily the same function) but also developed from the same germinal layer contained inside the embryo.

Equally striking was the finding that many animals possess *vestigial structures* that no longer serve any purpose for them but that are present, in functional form, in other species. Ostriches have rudimentary wings but cannot fly. Snakes have vestiges of hind-limb bones. Whales have a vestigial pelvic girdle. Humans have a vestigial tail in the form of the coccyx (⇨ illustration). Some marsupial mammals, which have been delivering live offspring for millions of years, still show indications of the egg tooth with which the young of various egg-laying species crack the eggshell on hatching. All these observations strongly suggested common ancestry and some kind of evolutionary process.

Pre-Darwinian theories

How were these suggestive facts to be explained? There were plenty of theories. For a time, one of the most influential was that of Jean Baptiste Lamarck (1744–1829), a French naturalist. Lamarck believed that characteristics acquired by an individual organism during the course of its life were passed on to its offspring. Thus giraffes acquired long necks by stretching up to reach the leaves at the top of trees. Tailless mice, he implied, could be produced by repeatedly cutting off the tails of mice and then letting them breed. Lamarck's ideas of evolution by use and disuse were incorporated in his *Zoological Philosophy* (1809) and were accepted by most of his contemporaries. A number of other theories along similar lines were put forward by other scientists.

A major difficulty for Lamarck's theory was to explain how acquired characteristics could be passed on. Lamarck suggested that bodily changes could somehow modify the sperm or the eggs (ova) in such a way that the new characteristics were passed on to the offspring. However, the theory did not represent observable fact. Cutting off the tails of mice was demonstrated to have no effect whatsoever on the length of the tails of their offspring, and Lamarck's theory was eventually abandoned. It was not long, however, until a much more credible theory of evolution was advanced, by Charles Darwin.

THE LYSENKO AFFAIR

In spite of the almost universal recognition by the beginning of the 20th century that all animals, including humans, acquire their inheritable characteristics at conception, a Russian agronomist (student of land cultivation), Trofim Denisovich Lysenko (1889–1976), produced a new theory of evolution, which was essentially that of Lamarck (⇨ main text). Working in the All-Union Genetics Institute at Odessa in the early 1930s, he claimed that he could alter the genetic constitution of strains of wheat by exposing them to various environmental factors.

Lysenko's ideas seemed to Stalin to fit in with Communist thinking: if wheat could be genetically improved by changing its environment, then so could people, given the correct political, social and economic system. In spite of the protests of more conventional colleagues, including N.I. Vavilov, Director of the Institute of Genetics of the Soviet Academy of Sciences, Lysenko's ideas were proclaimed official. In 1940 he was appointed in place of Vavilov, who was arrested and exiled. Lysenko remained the leading force in Soviet biology until well into the 1950s, thereby setting back genetics research and teaching in the USSR for many years. This odd sidelight in the history of evolutionary science had no effect on worldwide scientific opinion, for long before Lysenko's time the wider scientific community had accepted that acquired characteristics could not be genetically transmitted.

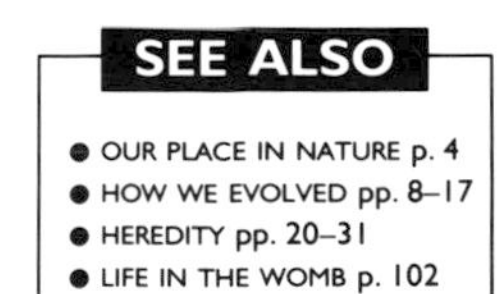
SEE ALSO
- OUR PLACE IN NATURE p. 4
- HOW WE EVOLVED pp. 8–17
- HEREDITY pp. 20–31
- LIFE IN THE WOMB p. 102

Natural Selection

The real breakthrough in the understanding of evolution was made by Charles Darwin (1809–82) and Alfred Russel Wallace (1823–1913). Their theories are fundamental to all modern thinking on the subject.

SEE ALSO

- OUR PLACE IN NATURE p. 4
- THE STORY OF EVOLUTION p. 6
- HUMAN EVOLUTION p. 10
- THE SPREAD OF HUMANS p. 12
- HUMANS AND THE ENVIRONMENT p. 14
- HEREDITY pp. 20–31

Darwin and Wallace first outlined the theory in a joint communication to the Linnaean Society in London in July 1858. However, the real impetus to the spread of the new ideas came with the publication in November 1859 of Darwin's book *On the Origin of Species by Means of Natural Selection, or the Preservation of Favoured Races in the Struggle for Life*. This book made Darwin famous – some at the time said infamous – and Wallace's contribution, which ranked with Darwin's, was largely ignored.

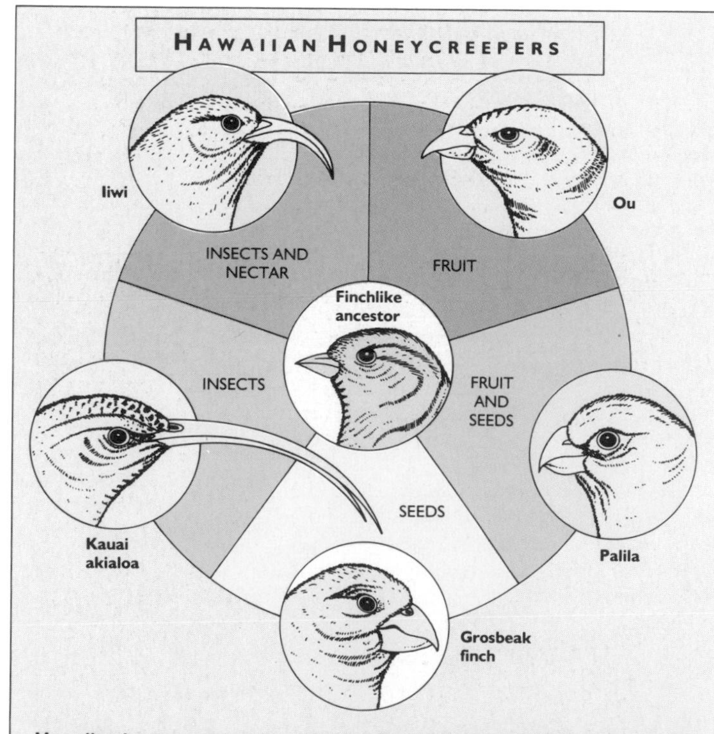

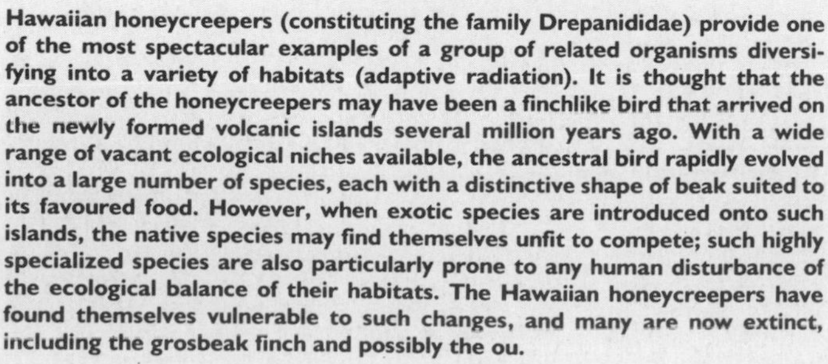

Hawaiian honeycreepers (constituting the family Drepanididae) provide one of the most spectacular examples of a group of related organisms diversifying into a variety of habitats (adaptive radiation). It is thought that the ancestor of the honeycreepers may have been a finchlike bird that arrived on the newly formed volcanic islands several million years ago. With a wide range of vacant ecological niches available, the ancestral bird rapidly evolved into a large number of species, each with a distinctive shape of beak suited to its favoured food. However, when exotic species are introduced onto such islands, the native species may find themselves unfit to compete; such highly specialized species are also particularly prone to any human disturbance of the ecological balance of their habitats. The Hawaiian honeycreepers have found themselves vulnerable to such changes, and many are now extinct, including the grosbeak finch and possibly the ou.

The voyage of the *Beagle*

While on a voyage of scientific exploration in HMS *Beagle*, during the 1830s, Darwin had visited the Galápagos Islands. There he had observed 13 species of finch, differing mainly in the shape and size of the bill, none of which was known to exist anywhere else in the world. Each species occupied its own island. Some of the finches were seed-eating, others insectivorous (insect-eating), and their beaks differed accordingly. It seemed highly probable to Darwin that a single species of finch – a seed-eating bird common on the nearby mainland – must have colonized the islands a very long time before and that the isolated descendants had evolved into different forms in an environment in which they were not in competition with other birds. These observations, and others made in the Galápagos and on the South American mainland, all provided clues to a new understanding of evolutionary processes.

In 1838 Darwin read *An Essay on the Principle of Population* by Thomas Robert Malthus (1766–1834), an English clergyman. In this book Malthus suggested that human populations invariably grow faster than the available food supply and that environmental factors such as starvation, disease or war are necessary to limit them. Darwin's thoughts returned to the finches. It occurred to him that the first occupiers of the islands would have multiplied unchecked until they outstripped the available seed supply. Many would have died, but those that happened to be able to adapt to a different diet, for instance insects – perhaps by a naturally occurring variation such as a differently shaped beak – could then flourish and multiply, until they in turn outstripped their food supply. Purely by chance, some variations might prove to be well adapted to the current environment, so the birds would survive to breed, while others might prove poorly adaptive, and disappear. It was a matter of the survival of those best fitted to the environment. Over the course of millions of years this process, operating very slowly, could be seen to give rise to radical alterations in the characteristics of organisms that shared common ancestors in the past.

New species could evolve by the splitting of one species into two or more different ones, largely as a result of geographical isolation of populations. Such isolated populations would experience different environmental pressures, undergoing different spontaneous changes, and so would evolve along different lines. If isolation prevented interbreeding with other stock derived from the same ancestors, these differences might become great enough for interbreeding to be no longer possible, and thus for a new species to be established.

This was the great idea of how changes in species occurred by 'natural selection', and Darwin saw that this principle alone was sufficient to explain evolution. Darwin never understood how inheritable changes took place or how the necessary spontaneous variations in

Charles Robert Darwin, English naturalist whose theories on the evolution of species according to the law of the 'survival of the fittest' led to heated debate between evolutionists and religious fundamentalists. Watercolour by George Richmond (1809–96). (AKG)

species (mutations) occurred (⇨ p. 26), but this did not in the least detract from the power or persuasiveness of his theory.

Objections to Darwin's theory

One of the difficulties faced, even in Darwin's time, by those who supported his theory was the problem of sterility between different species. It is part of the definition of a species that its members are incapable of breeding successfully with members of other species. Mating and fertilization can, of course, occur between closely related species and new individuals can occasionally result but – at least in the case of animals – these are usually sterile or even more seriously defective. If species evolve by the selection of small chance changes, why should not close species be able to interbreed?

With our advantages in understanding genetics we now know that different species have incompatible genes and often even different numbers of chromosomes. As a result, the accurate matching up of chromosomes from the father with those from the mother cannot take place. Fertilized cells require matched pairs of chromosomes if they are to function normally (⇨ p. 22).

Another criticism of Darwin's theory – first mooted in the early 1970s – is that evolution is not a steady, gradual process of change from one species to another as Darwin and his contemporaries believed. These critics have detected in the fossil record long periods of equilibrium (balance) or stasis (stagnation), interrupted by sudden jumps: new species appear to enter the fossil record suddenly, and change little during their term of existence. This theory is called *punctuated equilibrium*. Darwin believed this to be a false impression arising from the incompleteness of the fossil record, but his critics have suggested that it shows that natural selection acts on species rather than individuals. The fierce debate aroused by these criticisms has prompted a great deal of new research into the fossil record. The theory that is now most generally accepted, known as *punctuated gradualism*, maintains that evolutionary change is slow, but at certain times the tempo of evolutionary change is punctuated by a fairly rapid speeding up. However, this is still achieved by the normal process – as described by Darwin – of natural selection acting on individuals.

Variation and mutation

We now understand, in considerable detail, how it happens that individuals within a species commonly differ in many features from one another and from their parents. We also know how it is that mutations – sudden, seemingly inexplicable changes in inheritable characteristics – occur in individuals (⇨ pp. 22, 26 and 28).

Variation within a species is commonplace. Humans provide one of the most striking examples: differences in skin and hair colour, height, weight, body type, intelligence, and so on, abound. Humans may now have enormous power over their environment, and are able to interfere with many of the natural selective processes, but these processes continue to occur nevertheless. The lifetime of any individual (or even of many generations) is an insignificantly short period in the context of evolution, and changes are not apparent. This makes it difficult for us to perceive that we are subject to the same evolutionary forces we readily accept as acting on other species.

Every group of organisms, humans included, produces over the course of very long periods a large number of mutant types. Some of these changes are so damaging that the affected organisms do not survive long enough to breed and the mutation is lost. Many are of little value in assisting in the process of adaptation to changing environments. A few, however, so change the individual that the power of survival is enhanced. If the characteristic caused by the mutation offers an advantage to the individual, a single such individual can, in the course of a number of generations, give rise to a very large number of individuals possessing the desirable new characteristic.

Major changes of this kind are very rare, and it is not surprising that we have recognized none in ourselves in the course of recorded history. Notwithstanding our ability to insulate ourselves from the effects of natural selection, and to modify its operation by improvements in hygiene, nutrition, medical treatment and contraception, there is no reason to believe that we are immune to the same evolutionary processes that affect all other living organisms.

CREATION AND SCIENCE

Popular interest in palaeontology (the study of prehistoric life) was intense in the latter half of the 19th century. The Churches had little choice at the time but to counter, with all their power and authority, the inferences that were being made by many about the story of creation. This was a formidable task, especially for those with some idea of scientific thinking.

One idea put forward to account for the evidence was that successive creations of living forms had succumbed to successive catastrophes. It was still widely held at that time that the world had been created in the year 4004 BC – a date worked out by Bishop James Ussher (1581–1656) from evidence in the Bible. One difficulty with this chronology was the inescapable evidence of the time necessary for the formation of sedimentary rocks. Another was to account for the presence of fossils. An ingenious, but not very plausible, suggestion put forward by some churchmen was that when God created the world he also created the fossils to provide a test for people's faith.

James Ussher, Bishop of Armagh. (ME)

After the publication of the *Origin of Species*, argument rose to a climax. At the meeting of the British Association for the Advancement of Science at Oxford in June 1860 Samuel Wilberforce (1805–73), Bishop of Oxford, rose to his feet to smite the evolutionists. Professor Thomas Henry Huxley (1825–95), who had appointed himself official apologist for his friend Darwin, listened intently and soon found that the bishop, for all his bluster, had failed to grasp the essential point of the matter. The bishop's discourse was witty, sarcastic and deeply hostile to Darwin, but contained no real substance. Finally, flushed with the success of his speech, the bishop turned to Huxley and with mock politeness asked whether it was through his grandfather or his grandmother that he claimed to be descended from a monkey. He then sat down to enthusiastic applause.

Huxley's manner was as grave as the bishop's had been boisterous. He outlined Darwin's ideas, referred to the bishop's obvious ignorance of them, and ended by remarking quietly that while he would not be ashamed to have a monkey as an ancestor, he would certainly be ashamed to be connected with a man who used great and obvious gifts to obscure and conceal the truth. In the sensation that followed, Huxley was congratulated and complimented, even by the clergy, with a frankness that surprised him. Huxley did not glory in his triumph, but in private correspondence referred to himself, jokingly, as *episcopophagus* – a 'bishop eater'. To others, he was afterwards always known as 'Darwin's bulldog'.

HMS *Beagle* off the coast of South America during the voyage in which Darwin first began to formulate a theory of evolution. (ME)

Human Evolution

The idea that humans evolved from a nonhuman species was widely rejected until the end of the 19th century. But as the evidence of comparative anatomy and of the fossil record accumulated, this conclusion gradually became irresistible. It soon became apparent that both humans and the other apes had developed from a common ancestor.

The evidence of human evolution is open to various interpretations, and many claims by some experts have been challenged by others. There is, however, a general consensus about the broad lines of the evolution of human beings.

The rise of the mammals

The earliest mammals on earth date back some 220 million years. They are believed to have evolved from mammal-like reptiles, the synapsids, which had begun to develop the important characteristic of *thermoregulation*, the internal regulation and maintenance of body temperature. They also showed signs of differentiating teeth into front teeth for grasping and tearing prey and back teeth for chewing and grinding food. The synapsids became extinct during the age of the dinosaurs, but their inconspicuous descendants – small mouse-like creatures, the first mammals – survived when the dinosaurs fell. From these lowly beginnings, the mammals rose to dominance during the Cenozoic era, which began about 65 million years ago.

The reason for the success of the mammals is unclear. The extinction of the dinosaurs may have allowed both birds and mammals to proliferate. Mammal structure was potentially very adaptable. Limbs could be modified into flippers, wings and digging organs, so mammals could adapt to very different habitats – the surface of the land, the sea, the air or the subterranean environment. Thus mammals were able to diversify into land animals, into marine forms like dolphins, whales, seals and walruses, into underground diggers such as moles, and into flying and gliding animals like bats and flying squirrels.

Primate evolution

The order of mammals to which humans belong is called the primates (⇨ p. 4). The primates today consist of around 180 species, most of whom are primarily tree-dwellers. By about 45 million years ago primates similar to modern lemurs had developed. They had relatively large brains, forward-pointing eyes, and nails instead of claws. Ten million years later, a number of monkey-like primates appeared. This line included *Aegyptopithecus*, a small creature thought to be in the direct ancestral line of the apes (including humans). By about 20 million years ago the first true monkeys and apes had appeared. Fossils of these creatures have been found in large numbers in Africa, where it is thought that the changes leading to humans occurred. There is now general agreement that the evolutionary line that led to humans probably diverged from the line leading to our closest ape relatives, the chimpanzees, around 5.5 million years ago.

One of these early apes was *Dryopithecus*, thought by some to be a common ancestor of the modern apes and ourselves. The identity and timing of the separation of the lines leading to the other apes and ourselves is in dispute. One descendant of *Dryopithecus* is believed by some to be the earliest man-like creature, although others believe it to be an ancestor of the orang-utan. This is *Ramapithecus*, an ape-like animal that abandoned the trees as its main habitat. During this period, which may be as recent as 6 million years ago, climates were warm, mild and rainy, and supported many species of plant and animal life. In the grasslands of Africa some primates, including *Ramapithecus*, became adapted to life on the ground, feeding mainly on roots, seeds and berries.

A more popular ancestral candidate is the later genus *Australopithecus*, many fossils of which have been found in Africa. This creature lived between about 4 million and 1 million years ago in what are now Kenya, Tanzania, Ethiopia and the Transvaal. *Australopithecus* walked upright and had short canine teeth, while its molar teeth were adapted for hard grinding of roots and berries. Its brain was larger than that of other apes and some specimens were about as tall as modern humans. There is reason to believe that it could defend itself with simple weapons, although it probably did not use tools. There are four recognized species – *Australopithecus africanus*, *A. afarensis*, *A. robustus*, and *A. boisei*. The latter two species were contemporary with early members of the genus *Homo*. Early tools found in conjunction with *Australopithecus* remains are probably the work of humans who shared the same sites.

The first humans

The evidence provided by fossil remains suggests that the first human species (i.e. members of our own genus, *Homo*) appeared near the end of the Pliocene epoch – between 2 and 1.5 million years ago. The period since then has seen enormous climatic variations. There has been

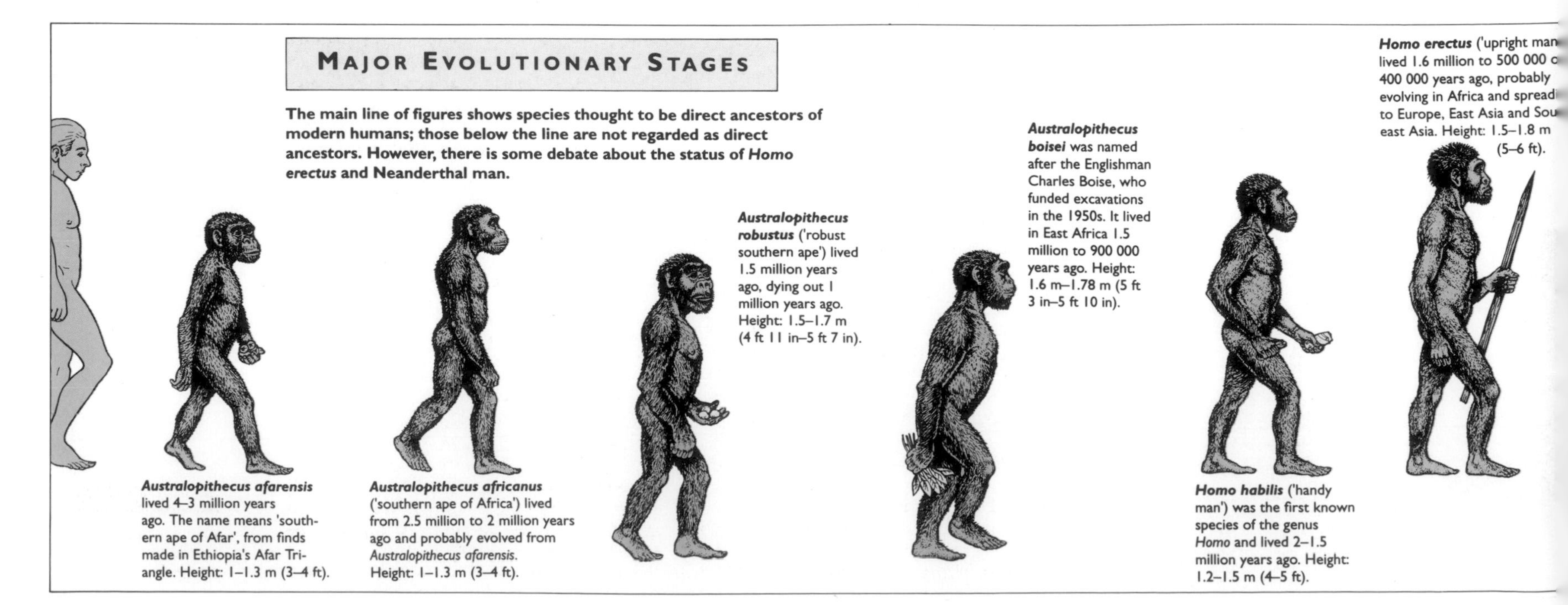

MAJOR EVOLUTIONARY STAGES

The main line of figures shows species thought to be direct ancestors of modern humans; those below the line are not regarded as direct ancestors. However, there is some debate about the status of *Homo erectus* and Neanderthal man.

Australopithecus afarensis lived 4–3 million years ago. The name means 'southern ape of Afar', from finds made in Ethiopia's Afar Triangle. Height: 1–1.3 m (3–4 ft).

Australopithecus africanus ('southern ape of Africa') lived from 2.5 million to 2 million years ago and probably evolved from *Australopithecus afarensis*. Height: 1–1.3 m (3–4 ft).

Australopithecus robustus ('robust southern ape') lived 1.5 million years ago, dying out 1 million years ago. Height: 1.5–1.7 m (4 ft 11 in–5 ft 7 in).

Australopithecus boisei was named after the Englishman Charles Boise, who funded excavations in the 1950s. It lived in East Africa 1.5 million to 900 000 years ago. Height: 1.6 m–1.78 m (5 ft 3 in–5 ft 10 in).

Homo habilis ('handy man') was the first known species of the genus *Homo* and lived 2–1.5 million years ago. Height: 1.2–1.5 m (4–5 ft).

Homo erectus ('upright man') lived 1.6 million to 500 000 or 400 000 years ago, probably evolving in Africa and spreading to Europe, East Asia and South-east Asia. Height: 1.5–1.8 m (5–6 ft).

a succession of ice ages in which the valleys of much of Europe and North America were covered with great sheets of thick ice. With so much of the earth's water frozen, the seas shrank and the ice and exposed land provided bridges between the continents, allowing wider movement of populations. During these periods plants and animals adapted to arctic conditions, and the northern plains supported large herds of caribou and other large grazing animals on which early humans, who had developed skills as hunters, were able to prey.

Between these long glacial periods there were comparatively short periods – perhaps about 10 000 years – of warmth. Farther south, especially in Africa, cool rainy periods, during which forests spread widely, alternated with dry periods, leading to the formation of deserts and grasslands.

The first evidence of the development of behaviour distinguishing the genus *Homo* from other apes appears in indications of the extensive use of tools and implements made of stone and bone. They were used for hunting and for gathering food, and as containers. There is also abundant later evidence of tribal or family life and of the division of labour in the endless task of finding enough to eat.

The earliest human species is known as *Homo habilis* – the specific name meaning 'handy, able or skilful'. More than fifty specimens of this type were found at Olduvai Gorge in East Africa between 1964 and 1981. This species was physically similar to *Australopithecus* but had a larger brain. *H. habilis* made and used simple tools, including stone choppers and sharp-edged cutting implements made of flakes of flint. Some sites show evidence of butchery of quite large animals – up to the size of hippopotamus – but it is not clear from this whether they had been hunted and killed or merely scavenged after death.

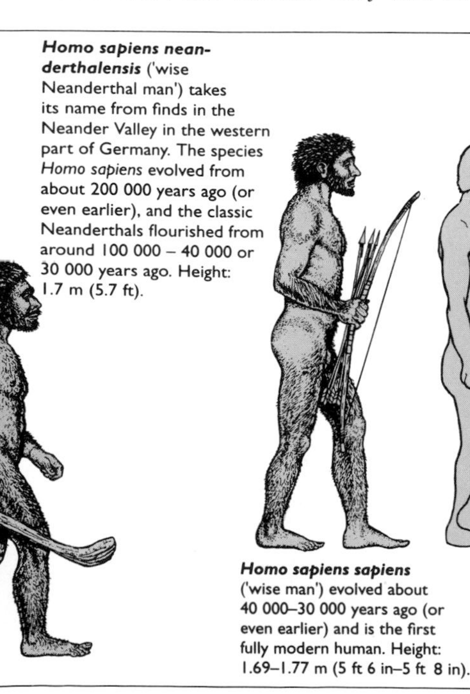

Homo sapiens neanderthalensis ('wise Neanderthal man') takes its name from finds in the Neander Valley in the western part of Germany. The species *Homo sapiens* evolved from about 200 000 years ago (or even earlier), and the classic Neanderthals flourished from around 100 000 – 40 000 or 30 000 years ago. Height: 1.7 m (5.7 ft).

Homo sapiens sapiens ('wise man') evolved about 40 000–30 000 years ago (or even earlier) and is the first fully modern human. Height: 1.69–1.77 m (5 ft 6 in–5 ft 8 in).

THE THREE AGES OF NEAR-EASTERN AND EUROPEAN PREHISTORY

AGE	STONE AGE					BRONZE AGE	IRON AGE
	PALAEOLITHIC			MESOLITHIC	NEOLITHIC		
	LOWER	MIDDLE	UPPER				
PERIOD BEGAN (approximately)	1 million years ago	100 000 years ago	30 000 BC	10 000 BC	9000-4000 BC (spreading from Near East to W Europe)	3000-2000 BC	1200-500 BC (both periods earliest in Near East and SE Europe, spreading to W and N Europe)
DOMINANT HOMINID	*Homo erectus*	Neanderthal man (*H. sapiens neanderthalensis*)	Modern man (*Homo sapiens sapiens*)				
TECHNOLOGY	Simple stone tools, e.g. hand axes Use of fire	More specialized stone tools, e.g. spear heads, knives	Development of stone blades and bone tools Beginnings of art	Use of bow and arrow	Beginnings of agriculture First towns in Near East	Bronze artefacts First cities	Iron artefacts

Homo erectus

In 1891 a Dutch army surgeon, Eugène Dubois, found some fossilized bones at Trinil in central Java. These included the vault of a skull and a very human-like thigh bone. Dubois was convinced that the creature whose remains he had found walked erect and had characteristics intermediate between those of apes and humans. Adopting a term first coined by Ernst Haeckel, he called it *Pithecanthropus erectus*. Subsequent discoveries of similar remains in other parts of Indonesia (Java man) and China (Peking man), and later in Africa and Europe (Heidelberg man), suggested that here was a species more advanced than *H. habilis*. Now known as *H. erectus* ('upright man'), this species lived some 1.6 million to 500 000 or 400 000 years ago. Some specimens have anatomical features close to those of *H. habilis*, others closely resemble our own species, *H. sapiens*.

The only major anatomical differences between *H. erectus* and *H. sapiens* are in the skull and the teeth. The limb bones are very similar, although those of *H. erectus* are usually more robust. The differences in brain size – derived from measuring cranial volume – are significant. The cranial volume of *H. erectus* ranges from 750 cm^3 to 1225 cm^3, the average of 14 skulls from Java, China and Africa being 940 cm^3. The range for modern humans is from about 1000 to 2000 cm^3, with an average of about 1450 cm^3. The largest brains of *H. erectus* were therefore larger than the smallest brains of modern humans. The cranial volumes of the known samples of *Australopithecus* range from 440 cm^3 to 520 cm^3. Those of *H. habilis* average 640 cm^3.

H. erectus was an active big-game hunter, and was intelligent enough to be able to adapt to a range of habitats. Around 500 000 years ago they developed the use of fire for warmth, cooking, protection and in hunting. As well as eating meat, eggs, fish and small rodents, their diet consisted of roots, nuts, fruit and fleshy leaves. They preferred to live in open or lightly wooded areas, and used two-edged hand axes and various pounders and cutting flakes. There is no evidence that they buried their dead.

Neanderthal man

Homo sapiens ('wise man') probably shared a common ancestor with *H. erectus*, although whether *H. sapiens* actually evolved directly from *H. erectus* or not is the subject of debate (➪ p. 12). In either case, *H. sapiens* appears to have evolved gradually from about 200 000 years ago, or even earlier. At that stage, however, *H. sapiens* did not particularly resemble modern human beings. They had heavy ridges above their eyes and a skull that sloped sharply back with little or no forehead. The teeth were large and protuberant. The size of the brain, however, was well within the modern range.

Some time between about 100 000 and 40 000 years ago *H. sapiens* developed a tool culture, evidence of which has been found in the Neander Valley (*Neandertal*) of Western Germany, Le Moustier in France, Gibraltar, Italy and many other sites. Mousterian tools were made from large thin flakes struck from flint cores and used as blanks from which a range of cutting and scraping implements could be fashioned. These were made by the Neanderthals (often regarded as the subspecies *H. sapiens neanderthalensis*). Mousterian tools allowed these people to cut and carve wood, to cut meat, to scrape hides and to make thongs.

The Neanderthals were advanced peoples who lived in caves or huts or even in rough stone-constructed dwellings, sometimes forming substantial societies. They regularly used fire, and killed and ate a wide variety of animals. They were probably capable of trapping game, and they may have been cannibals. They were capable of looking after the aged, whose knowledge and experience they may have valued, and they performed rituals such as the formal burial of the dead. It is likely that they communicated through speech.

The Neanderthals seem to have disappeared by about 40 000 or 30 000 years ago, when they were replaced by humans of modern type (*H. sapiens sapiens*), exemplified by Cro-Magnon man, whose remains were first found in a rock shelter at Cro-Magnon in the Dordogne, France, in 1868. There is considerable controversy as to whether the Neanderthals evolved gradually into modern humans, or whether they were displaced or killed by emigrating groups of moderns from other areas (➪ p. 12).

SEE ALSO
- HOW WE EVOLVED pp. 4–9 and 12–17
- HEREDITY pp. 20–31

The Spread of Humans

There is considerable controversy among scientists concerning the geographic origins of modern humans. Two main hypotheses dominate the argument: that modern humans evolved at about the same time from different pre-modern ancestors in different parts of the world; or that we are all descended from modern people who evolved in Africa. Both sides base their opinions on genetic and fossil evidence.

Unfortunately, the available evidence is of necessity flimsy, and is open to various interpretations. The essence of the current argument is whether modern humans originated in one place and migrated to displace earlier species or subspecies, or whether they appeared in many different parts of the world at about the same time.

The African genesis and the Eve hypothesis

The concept known as the 'Eve hypothesis' has attracted much interest and attention. It was originated by Allan Wilson, a molecular biology anthropologist at the University of California at Berkeley, and two of his graduate students. According to this hypothesis, first published in *Nature* in 1987, modern humans are a new species of primate that arose in Africa. The implication of this is that humans then spread from Africa to all parts of the inhabited world where they replaced the indigenous populations, such as the Neanderthals, probably by conquest and slaughter. Although some supporters of this view cite fossil evidence, the Eve hypothesis is based mainly on genetic data, which, it is claimed, prove that we are all descended from a single woman, 'Eve', who lived in Africa.

In addition to the DNA found in the chromosomes in the nucleus of every cell of our bodies (➪ p. 22), each cell also contains hundreds of energy-producing mitochondria (➪ p. 38). In each mitochondrion there are many copies of a quite separate and distinct gene sequence known as the *mitochondrial DNA*. This is a circular body containing a series of genes whose function is known, together with some 'junk' DNA. Unlike chromosomal (nuclear) DNA, which is inherited from both parents, mitochondrial DNA is derived from the ovum and thus comes only from the mother. Whether we are male or female, therefore, our mitochondrial DNA comes to us in a direct maternal line from the past. There is no arguing the fact that all modern humans inherited their mitochondrial DNA from a common female ancestor. But the degree of its divergence gives rise to the question of where, and how long ago, she lived.

By comparing the mitochondrial DNAs of people of different ethnic groups, scientists can determine how closely they are related to one another. It is also possible to map a tree of their evolution. The production of ancestral trees based on these relationships allows us – in theory – to work backwards until we reach the common female ancestor. When scientists first made these comparisons they were struck by the divergence of African mitochondrial DNA. This pattern of DNA, with large numbers of mutations, was far more widespread than that of any other group. This suggested that modern humans have inhabited Africa much longer than anywhere else.

To establish a time scale for human history it is necessary to assume that the

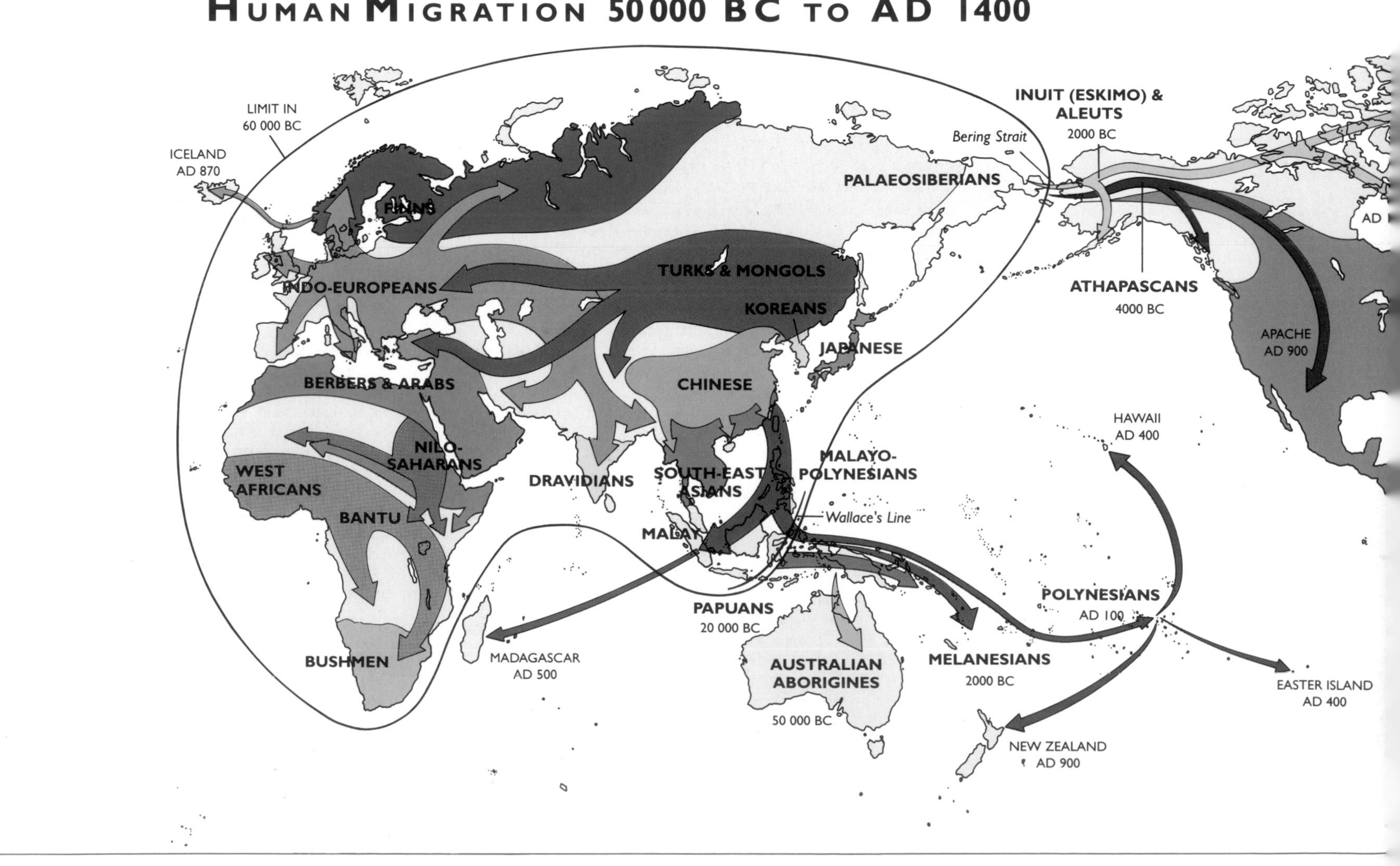

average rate of mutation of DNA is steady and that this rate can be correctly assessed. It has been calculated that, on average, about 3 in every 100 bases in the DNA (⇨ p. 26) are changed every million years. This figure, initially questioned, is now generally accepted. The most recent calculations, then, suggest that Eve lived in Africa around 600 000 years ago.

Multiregional evolution

Those who reject the Eve hypothesis base their arguments both on criticism of the Berkeley work and on what they hold to be strong fossil evidence. A study of the phylogenetic (evolutionary) tree produced by the Berkeley team showed that some of the data were clearly wrong. For instance, a group of African !Kung bushmen were shown to be split on the deepest branches of the tree although they are all known to be closely related. Criticisms have also been made of the way the DNA had been analysed and of the fact that samples had been taken from American Africans instead of native Africans. Computer analysis of the data and other methods of tree construction showed less suggestive results than in the original work, and even produced some trees with non-African roots. Furthermore, there are some genetic analyses that support the view that humans arose in different geographic areas at about the same time. There is little or no argument among anthropologists that the fossil record of the prior species *Homo erectus* (⇨ p. 11) starts in Africa and that *Homo erectus* migrated to Europe and Asia about 1 million years ago. So far as they go, these facts support the Eve hypothesis. But those who subscribe to the multiregional evolution hypothesis argue that once *Homo erectus* had migrated throughout the Old World the species continued to evolve, firstly into Neanderthal man (*Homo sapiens neanderthalensis*), who, in turn, gradually evolved into modern humans (*Homo sapiens sapiens*). Supporters point out that a series of *Homo erectus* skulls have been found in various regions that show distinct similarities to modern skulls from the same regions.

Disinterested scientists believe that neither case has been proven, and maintain that the evidence remains ambiguous. There is, however, a lively expectation that the development of new and more convincing methods of genetic and phylogenetic analysis will settle the matter one way or the other.

The spread of populations

Until about 60 000 years ago *Homo sapiens* was confined to the same area of the globe as the preceding species *Homo erectus*: Europe, Asia and Africa, the three continents that together constitute the Old World. Then, starting about 50 000 years ago, humans began to probe the barriers that had previously restrained them: first Wallace's line – the boundary in the Indonesian Archipelago that separates the fauna of Asia and Australasia; then the Bering Strait between Asia and the Americas. Wallace's line represents the point where the Indonesian islands start to get so far apart that people without boats could not get from one to another. By 50 000 BC the ancestors of today's Australian Aborigines must have had some serviceable craft for they reached the north of Australia at about this time. Later, the Papuans crossed the line too. Most Papuans got no further than New Guinea, but one group colonized the islands further east and developed the characteristics we term Melanesian. Around 1000 BC these Melanesians were overtaken by an offshoot of the Malayo-Polynesian population that had meanwhile displaced the Papuans from most of Indonesia. These new arrivals, the Polynesians, proved the most intrepid explorers of all. They colonized nearly all the worthwhile island groups of the Pacific from the Hawaiian chain in the north to New Zealand in the south. Eastward they reached as far as Easter Island. During the same period – the middle centuries of the 1st millennium AD – their cousins, the Malay, crossed the Indian Ocean and settled Madagascar.

Expansion into the Americas started much later than the movement into Australasia and was less complicated. Most Amerindians are descendants of a single group of mammoth hunters that made its way over what is now the Bering Strait but was then, around 10 000 BC, thanks to the lowered sea level characteristic of the last Ice Age, a land bridge. This stock spread rapidly through North, Central and South America (by 9000 BC), then, more slowly, through the Caribbean Islands (500 BC–AD 500). In 4000 BC another group, the Athapascans, arrived in the northwest: most of them stayed there, but the ancestors of the Apache moved into the North American southwest in the course of the 10th century AD. The third and final set of immigrants consisted of the Arctic hunters and fishermen from whom the present-day Aleuts and Inuit (Eskimo) are descended. The Inuit spread right across the north of Canada to reach Greenland around AD 1000. By the 15th century, they had eliminated the tiny colonies established there by the Indo-European Norse, who had settled Iceland in around AD 870 and had reached Greenland about a century later.

While all this was going on movements of great significance were taking place within the Old World. Among the most important were the expansion of the Indo-Europeans into Western and Northern Europe, Central Asia, Iran and India; the later countermovement by the Turks of Central Asia that created the Turkey of today; and the expansion of the Bantu through sub-Saharan Africa. These three were the main winners in the competition for space: the losers were groups who were slow to develop their material culture – peoples such as the Bushmen in Africa and the Papuans in Indonesia. However, space is not the only indicator of success – the Chinese did not increase their territory much but by diligently developing their agricultural base they became the most populous nation on earth. The world's population in AD 1400 was about 350 million of which around 80 million – nearly one in four – were Chinese. By contrast the people who had remained hunters and gatherers, the Bushmen and Australian Aborigines and, by climatic necessity, the Finns and Palaeosiberian peoples, numbered only a few hundred thousand.

This was the scene prior to the last major act in the story of human migrations – the European colonial era. This got under way at the end of the 15th century, initially with just a trickle of adventurers into the Americas, but by the later 19th century this trickle had turned into a flood, with tens of millions of people migrating from Europe into the United States, and smaller numbers into other parts of the Americas, southern Africa, and Australasia. Colonization was accompanied in some areas by the virtual elimination of the native peoples, and also by the forced importation into the Americas of as many as 20 million slaves from Africa. However, this is by no means the end of the story, and populations today – with the increased opportunities offered by modern transport – are perhaps more mobile than ever before. Significant movements in the last few decades include the migration of people from former colonies into Western Europe, and of eastern Asians and Latin Americans into North America.

SEE ALSO
- HOW WE EVOLVED pp. 4–11 and 14–17
- HEREDITY pp. 20–31

Humans and the Environment

No living thing on earth, including the human organism, is unaffected by its environment. We human beings are no exception. We depend for our very survival on interactions not only with other human beings, but also with other animals and plants, and other aspects of our physical surroundings.

Even though we have an unsurpassed ability to alter and manipulate our environment, this is not unique to our species; beavers build dams, for example, and ants 'farm' aphids. Nor does our power over nature set us apart from it or above it; we are as much a part of nature and as dependent on it as any other form of life. The environment is made up of both non-living and living elements. The non-living environment consists of natural factors such as climate, landscape and soil, as well as man-made components such as buildings, cities and pollution. The living environment comprises the interacting *community* of plants and animals found in any one area. Taken together, the non-living and living environments in an area make up an *ecosystem*, and the totality of ecosystems around the world makes up the *biosphere*, the layer around our planet where all life forms exist.

Energy flow and the food chain

Virtually all the energy necessary to maintain life on earth is ultimately derived from the sun. This energy is not directly accessible to animals, but it can be used by plants. Plants (together with certain bacteria) have the unique ability to use solar energy to convert simple inorganic molecules into the complex organic compounds that provide all animals, including humans, with their nutritional requirements.

Among these groups of complex organic compounds are the carbohydrates, such as sugars and starch. Carbohydrates are made by plants via *photosynthesis*, a chemical process in which sunlight is used to convert atmospheric carbon dioxide and water into glucose (a simple sugar) and oxygen. Similarly, all animals ultimately derive their protein (➪ p. 36) from plants: protein contains nitrogen, which plants generally obtain from the soil in the form of nitrates, and carbon, which is obtained from the carbohydrates created by photosynthesis.

This ability of plants to utilize solar energy forms the basis of all food chains. Plants, known as *primary producers*, are eaten by herbivores (*primary consumers*), which in turn are eaten by carnivores (*secondary consumers*) – which may themselves even be eaten by bigger, more powerful carnivores. Human beings are mixed feeders, with a diet including both plant and animal constituents.

Adapting to different habitats

Our earliest human ancestors, *Homo habilis* (➪ p. 10), developed the use of simple tools, but were restricted to the tropical regions of Africa. The successor to *Homo habilis* was *Homo erectus*, whose ability to use fire helped the species to adapt to the cooler climates of Europe and Asia. With the arrival of *Homo sapiens* the pace of technological development quickened, and the ability to make more effective weapons, clothes, shelters and boats enabled humans to colonize every continent except Antarctica (➪ p. 12).

These early humans were all hunter-gatherers, a way of life that is very much in tune with the environment. Once the game and berries and so on in one area were exhausted, the small group would move on to another, giving the first area a chance to recover. Populations were never big enough to damage the environment permanently, and humans were much like other large mammals – whether carnivores or herbivores – in the need to have a fairly large territory to exploit. If the population grew, then there was always uninhabited territory elsewhere.

One of the first signs of humans having a permanent impact on the environment came around the end of the last Ice Age, 10 000 years ago, with the extinction of a number of large mammal species, such as the mammoth and woolly rhinoceros. This is thought to have been due to the increasing sophistication of Palaeolithic hunting techniques, although it is possible that climatic change may have played a part.

Changing the environment

For most of their history, humans have been hunter-gatherers, reliant on harvesting wild animals and plants. Things began to change around the end of the last Ice Age, when humans first started to manage their environment in order to ensure a constant food supply. This was the beginning of agriculture – the domestication of wild animals and the planting and tending of edible plants. Although dogs were probably tamed earlier, the first agricultural animal to be domesticated was the sheep, in the Middle East around 9000 BC. The domestication of cattle, goats and pigs followed shortly afterwards. Livestock farming started to spread to Europe around 6000 BC. Arable agriculture also first emerged in the Middle East, in the period 9000–7000 BC, and the crops grown were primitive forms of wheat and barley. Between 6000 and 4000 BC agricultural societies based on

Energy Flow in the Biosphere

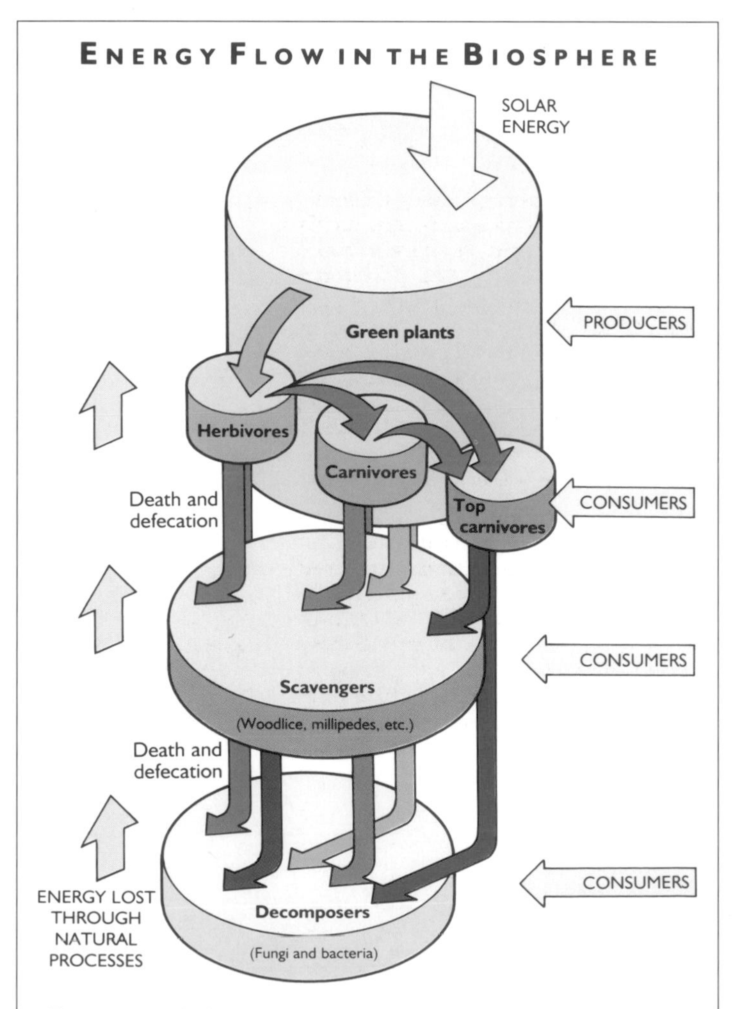

Energy enters the biosphere in the form of solar energy, which is used by the primary producers – predominantly green plants – to fuel the photosynthetic reactions by which atmospheric carbon dioxide is converted into organic sugars. Virtually all other organisms – animals, fungi and most bacteria – are consumers, ultimately relying on the organic matter of plants for their support.

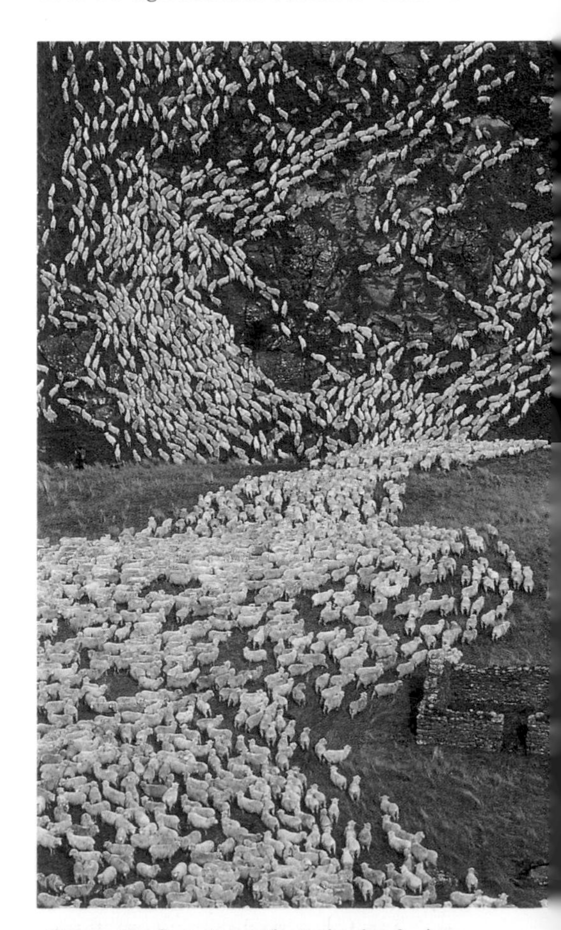

Sheep, the first domesticated animal – here being driven in huge numbers – have served human communities for nearly 11 000 years.

(Popperfoto)

rice, wheat and maize (corn) developed in many other parts of the world, in the fertile river valleys of China, India, Egypt and Central and South America.

Initially agriculture depended entirely on human labour, the first 'plough' being a simple digging stick. However, with the development of the hard wooden plough (later tipped with bronze, then iron), and the domestication of draught animals (such as ox, horse and camel), the mechanization of agriculture began. With this mechanization, fewer hands were needed to produce adequate amounts of food, freeing some members of the settlement to follow different occupations – for example, as craftsmen.

This process gradually brought about major changes in the way that human beings lived, changes that first emerged in the Middle East between the Tigris and Euphrates rivers. This area, known as Mesopotamia (Greek for 'the land between two rivers'), was flooded every spring, the rivers covering the plain with rich alluvial soil. The local farmers developed an elaborate system of dykes and reservoirs to control the floods, a system that must have needed a considerable amount of organization and cooperation to develop. Variations in the fertility of the soil meant that some farmers became richer than others, and were able to employ landless labourers – the beginnings of a class system. Food surpluses also enabled many more people to make a living away from the land – as artisans and merchants, and even as administrators. All this contributed to the development of a more complex, hierarchical society, a society that needed a centre for buying and selling goods and a centre from which control could be administered: the beginnings of urban civilization.

These first cities were built in Mesopotamia between 4000 and 3000 BC by the people known as the Sumerians, and in the following millennium cities also began to appear in the valleys of the Nile, Indus and Yellow River (Huang He).

The agricultural impact

Although irrigation lay behind the success of the Sumerians, it was also the cause of their decline. Wherever intensive irrigation is practised on poorly drained semi-arid land, the salt level in the soil starts to rise, until eventually no crops can be grown. This is what happened in Sumer, whose agricultural power declined, making it a prey to outside invaders. Irrigation is just one of the ways in which human alteration of the environment can have irreversibly damaging effects. Various other forms of poor land management can also lead to desertification. Clearing forests – whether for firewood or to make space for agriculture – can result in the topsoil being washed off, leaving the land eroded and barren. Overgrazing – particularly in drier areas – can bring about semi-desert conditions in a matter of a few generations.

Careful land management – such as building cultivation terraces to prevent erosion, protecting trees and controlling grazing – can avoid these problems. But where there is a breakdown of the social fabric – through war, disease, or rapid population growth – controlled land management becomes impossible, and environmental degradation sets in. History is strewn with examples of this process, from the semi-desertification of Mediterranean lands following the fall of the Roman Empire to the recent famines in the Horn of Africa following years of warfare in the region.

The industrial impact

If the first major change in the relationship of humans with the environment came with the beginnings of agriculture, the second came with largescale industrialization, a process that began in the 18th century. The impact of industrialization on the environment has been in two major areas.

Firstly, there have been the ever-growing requirements of industry for non-renewable natural resources – either in the form of fuels for energy, or raw material for manufacturing processes. This has resulted in scarring and despoilment of the environment by such activities as quarrying and mining, activities that often mean the area concerned cannot be used again for productive purposes. Secondly, industrialization has resulted in the polluting of the environment, largely with the chemical by-products of industrial processes. Pollutants can have a direct effect on the health of humans and animals – for example, many forms of air pollution can cause respiratory diseases, and thousands of man-made chemicals have been implicated as carcinogenic (cancer-causing) agents. But pollutants can also have more indirect effects, which in the end may prove harmful to life on earth. Examples of this include the possibilities of global warming through excessive burning of fossil fuels, and the destruction of the ozone layer – which screens out much of the sun's harmful ultraviolet rays – by chlorofluorocarbons.

Most people in the developed world are now aware of these problems, and governments and international agencies are coming under increasing pressure to control harmful industrial activities. There is usually an economic price to pay, but it is becoming increasingly necessary to weigh up short-term gain against the long-term health of the planet as a whole.

Sixty per cent of the world's population, mainly in Asia and in Central and South America, depends on rice for sustenance. Only 2% of the total annual rice production enters the international commodities market. One way to increase crop yield is to cut horizontal terraces out of previously inaccessible slopes and use these for planting, as here in Bali. Being level, terraces also help prevent soil erosion by gravity – an important benefit in areas of poor soil fertility. (Popperfoto)

Modern Food Production

The amount of energy expended on the production of food is an important factor in our interaction with the environment. In traditional arable agriculture, yields were very much dependent on the natural fertility of the soil. This could be enhanced by techniques such as weeding, irrigation, manure application and crop rotation, but generally the energy invested by humans was considerably less than that yielded. This is because plants are very efficient at converting solar energy into food energy.

This is not the case in livestock farming. Large herbivores are not at all efficient (less than 4% in the case of rangeland cattle) at converting plant material into meat for human consumption. The remaining energy is expended on keeping warm, moving, breathing and digesting. This inefficiency is why meat is too expensive for the majority of the world's population.

Modern farming techniques have boosted productivity in terms of yields, but, in terms of energy efficiency, we apply very heavy subsidies. In arable farming, this energy subsidy comes in terms of the fossil fuels required to run farm machinery and to manufacture artificial fertilizers and pesticides. The result is that every unit of energy we consume in the form of potatoes, for example, has required the application of a similar amount of energy in the form of fossil fuels used to grow the potatoes. Even greater energy subsidies are applied in intensive livestock rearing – factory farming. The animals are kept in heated buildings and their movements restricted to reduce energy loss. Most significantly, they are fed energy-intensive food (grain in the case of cattle), which they can convert to meat much more efficiently than if they have to browse and digest vast amounts of leafy food. However, large energy subsidies are applied in constructing and heating the buildings, and in growing the grain and transporting it to the animals. In the USA, more than half of the grain produced is fed to cattle to make meat – a very inefficient use of food resources, and a luxury that only wealthy societies can afford.

SEE ALSO

- HOW WE EVOLVED pp. 4–13
- THE TALKING ANIMAL p. 16
- WHAT THE BODY IS MADE OF p. 36
- FUEL AND MAINTENANCE 2: DIET p. 46

Medieval field systems, still visible from the air, Laxton in Nottinghamshire, England. (Aerofilms)

The Talking Animal

The ability to speak distinguishes the human being from all other animals, and the acquisition of this facility must have given humans a commanding advantage over all other forms of life. The power of speech made possible a much more complex level of communication – not only of needs, warnings and emotions but also, for instance, concerning technological innovation and in social organization.

The development of language and the discovery that language could be recorded in permanent form as written or drawn symbols gave us humans another unique advantage – the ability to preserve records of experience and to transmit these to others we may never meet. We have no way of knowing when people began to use language. The oldest written records date back about 5000 years, but this is a negligible period in the context of the prehistory of humans. It is almost certain that language was being developed for a long period before any written records were made. We cannot be sure that the first humans were able to communicate by speech. Indeed it seems probable, from the analogy with the methods used by other animals, that they originally engaged in vocal but nonverbal communication, and by gestures and other visual signs.

Theories of the origins of language

Body language has long been a form of communication, and remains an important means of expression for many animals, including ourselves. Some scientists believe that since the significance of certain bodily movements and gestures would have been well understood, speech may have originated in verbal imitation of such gestures – the jaw, lip and tongue unconsciously mimicking them, much in the way that a child will move its mouth during early attempts to write. This is the 'mouth-gesture' or 'Ta-ta' theory.

Proponents of the 'Bow-wow' theory hold that speech is purely imitative of natural sounds and arose when humans discovered that they could make noises similar to those in nature. Onomatopoeic (words sounding like their meaning) origins are, of course, common and familiar, as when children call dogs 'bow-wows', cows 'moo-moos' and steam trains 'choo-choos'. A similar theory, the 'Pooh-pooh' or 'Ouch' theory, suggests that language originated from the 'instinctive' cries made in response to emotions of various kinds. The 'Yo-heave-ho' theory would have it that language developed from the grunts and exclamations inseparable from hard labour. The trouble with these theories, though, is the severe limitation they place on vocabulary.

One of the most fanciful ideas was the 'Ding-dong' theory of the German philologist Friedrich Max Müller (1823–1900). Müller was impressed with the concept of resonance and came to believe that there was a mystical harmony between sound and sense. The idea seems to have originated with the Greek philosopher Plato (c. 427–c. 347 BC), who also believed that there was a correspondence between names and their objects. None of these theories carries much weight nowadays, however.

It has been suggested that we might discover something about the origins of language by observing how children learn to speak. Some trials are said to have been based on this idea. The Greek historian Herodotus (c. 485–c. 425 BC) records that the Egyptian king Psammatichos isolated two infants in a mountain hut in the hope that they would begin to speak spontaneously. The king decided that the sounds they uttered were Phrygian – then believed to be the original language of humanity. King James IV of Scotland is said to have conducted a similar experiment in the expectation that the children would speak good Hebrew – then thought to be the root of all modern languages. The naive notion, implicit in these beliefs, is that language is inborn rather than environmentally acquired.

Even studies of the languages of contemporary 'primitive' societies tell us nothing of origins. Far from being basic and providing clues to what original languages may have been like, such languages are often rich in vocabulary and complex in grammar and syntax.

Can other primates use language?

Although apes do not appear to have the necessary neurological equipment or physical apparatus for the full articulation of language, some, especially chimpanzees, can certainly understand and use it. A chimp named Washoe was taught a vocabulary of 130 words in American Sign Language by the researchers Allan and Beatrice Gardner. Another chimp, Sarah, was able to converse with researcher David Premack using symbols consisting of pieces of plastic of different shape, size, colour and texture. These chimps went further and were able, without instruction, to combine symbols to produce new words and to construct sentences of a kind. A refrigerator, for instance, was designated 'open-eat-drink'. Sarah was found capable of accurate performance of logical operations of the type, 'if this, then that' (for instance, 'If I am hungry, then I must eat.')

There are good physical reasons why animals other than humans cannot use speech. Vocal sounds made by chimpanzees and other animals appear to be controlled mainly by centres in the limbic area of the brain (⇨ p. 74) – the part responsible for emotional response rather than reason. Stimulation of these centres produces the whole range of utterances of which the animal is capable. In humans, vocal articulation is under the control of the cortex of the brain (⇨ p. 76). Destruction by disease of the cortical speech areas in humans – a common consequence of a stroke – will abolish speech. Cortical destruction has no effect on vocalization in nonhuman primates. It thus appears that, by an evolutionary step, the connections for speech have been transferred from the more primitive instinctive or reflex centres in the other animals to the cerebral cortex in humans, making possible the voluntary and considered use of language. It is a moot point when this transfer occurred. We do not know, for instance, whether or not Neanderthal man was capable of articulate language. There is evidence that the vocal tract of the Neanderthals was unsuitable for speech. Some scientists go so far as to suggest that the appearance of language may have been the defining characteristic of modern man – *Homo sapiens sapiens*.

Language and the brain

The American linguist Noam Chomsky (1928–) believes that the brain is structurally programmed for the organization of language. He does not think it possible that children could learn a language without such programming, especially its complex grammatical and syntactical elements, as quickly as they do. He holds that the rapidly acquired competence to formulate an endless variety of sentences could never be explained by mere learning. There must therefore be some kind of 'mental organ' operating on built-in sets of linguistic rules dictating what is possible and what is not. Chomsky's ideas, with their emphasis on the existence of linguistic universals at a physiological level, have been controversial but very influential.

These ideas have gained support from the study of *creoles* by the American linguist Derek Bickerton. Creoles are a class of languages that arise spontaneously when people who speak mutually unknown languages are thrown into prolonged close contact. More than a hundred creoles are known, and their grammatical structures have been studied. The first step in the development of a creole is the use of a *pidgin* – a simple and inefficient language in which a submissive group (often, in the past, slaves) adopts some of the vocabulary of a dominant group and uses it with some of the grammar of its own language. The children of pidgin speakers, however, typically speak a creole. In this they use the same vocabulary as the pidgin, but with a particular grammar that differs from anything of which they can have had experience. Extraordinarily, all creoles, whether old or recent, share a common, or at least similar, grammar. For instance, they all deal with tense in the same way.

When children are learning their native language, they make characteristic errors or nonconformities. A study of children's mistakes in different languages shows that they bear a strong resemblance to creole grammar. For instance, they do not use a change in word order to indicate interrogation, but rely solely on a rising intonation. Also like creole speakers, they often use a negative subject with a negative verb: 'Nobody don't love me.' Many experts believe that the language of early humans may have resembled a creole. Cultural forces would, of course, have imposed large changes on languages, which would gradually have evolved to meet contemporary needs.

A common source?

In the words of the New English Bible translation of Genesis, Chapter 11: 'Now the whole earth had one language and few words.' The story of the Tower of Babel, which follows, describes how God confused the speech of humans 'that they may not understand one another's speech' and 'scattered them abroad from there over the face of all the earth'. This first language was long believed to be Hebrew, and all the languages of the world were thought to derive from it.

Recently a more scientific version of this idea has been getting a lot of attention. The new hypothesis stems from the linguists' success in classifying languages and grouping them into families. This process began in the late 18th century when people realized that languages that were geographically widely separated, as for example English and Iranian, often had a surprising number of features in common. Subsequent work made it apparent that most languages could be placed in one of a dozen or so groupings, like Indo-European (the one that contains English and Iranian) or Afro-Asiatic (containing, among others, Berber, Hebrew and Arabic). But the process of classification was also one of historical reconstruction, for following the Indo-European tree to its roots took the linguists back in time to an undifferentiated parent language spoken by the ancestral Indo-Europeans. We can now be reasonably sure that in 6000 BC these ancestral Indo-Europeans all lived in the area between the Balkans and south Russia: their spread to all the other parts of the world where Indo-European languages are spoken today can be accounted for by subsequent migrations (➪ pp. 12–13), the vast majority of which are well documented. The differences between the individual languages can be related to the time that has elapsed since they split off from the main language stem (➪ illustration).

What is true for an individual language family such as Indo-European is also true for the 'superfamilies' that have been set up by the current generation of linguists. The Nostratic superfamily, for example, links together Indo-European, Afro-Asiatic, Elamo-Dravidian, Uralian (Finn) and Altaic (Turko-Mongol). It also takes us back to the Upper Palaeolithic (c. 30 000–10 000 BC) when there were probably no more than a million human beings in the entire Eurasian land mass. According to the linguists they would have spoken either Nostratic or one of the other two Eurasian languages of the era, Eurasiatic and Sino-Caucasian.

This brings us within striking distance of a single ancestral language for all humanity, a sort of linguistic equivalent of the 'Eve hypothesis' (➪ p. 12). Many, perhaps a majority, of linguists are reluctant to go so far at this stage, but the tree builders have won many converts in the last decade. It would be surprising if they were to slow down now they are within sight of their goal.

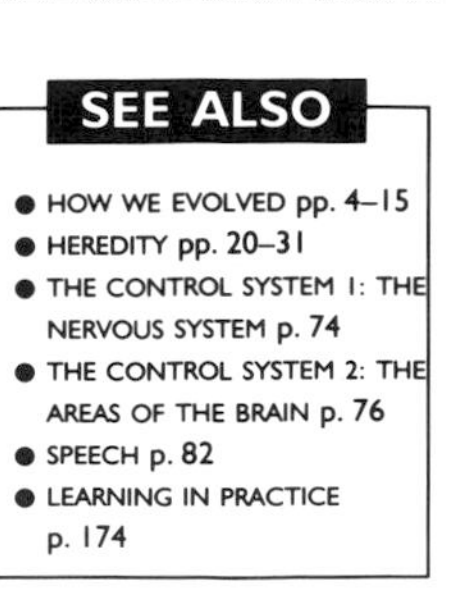
SEE ALSO
- HOW WE EVOLVED pp. 4–15
- HEREDITY pp. 20–31
- THE CONTROL SYSTEM 1: THE NERVOUS SYSTEM p. 74
- THE CONTROL SYSTEM 2: THE AREAS OF THE BRAIN p. 76
- SPEECH p. 82
- LEARNING IN PRACTICE p. 174

MIGRATION AND LANGUAGE EVOLUTION

This diagram represents one way of putting together what is known about the spread of modern humans, *Homo sapiens*, and the evolution of their languages. The division marked number 1 indicates the break-out of modern humans from sub-Saharan Africa, their original homeland, to North Africa and Eurasia. In making this move, some time around 100 000 BC, *Homo sapiens* was merely following in the footsteps of predecessor *Homo erectus*, who had done the same thing a million years earlier. The same is true of the next move, into the Indonesian archipelago, 2. But 3, the colonization of Australia, represents a new departure, as does 4, the colonization of the Americas. Movement number 5, taking the Malayo-Polynesians into Indonesia, does not involve new territory but is interesting because it pushed the original Indopacific inhabitants, the ancestors of the present-day Andamanese and Papuans, into opposite ends of the island arc.

The colours filling the right-hand branches of the tree represent the languages spoken by these pioneering groups and their descendants up to AD 1. There is little dispute about those to the right of the 5000 BC line, but the ones that lie between 5000 and 15 000 BC are still controversial. The uncoloured connections to the left of the 15 000 BC line are based on migration theory and as yet have no linguistic support at all.

The colours used in this diagram correspond to those used on the human migration map (➪ pp. 12–13).

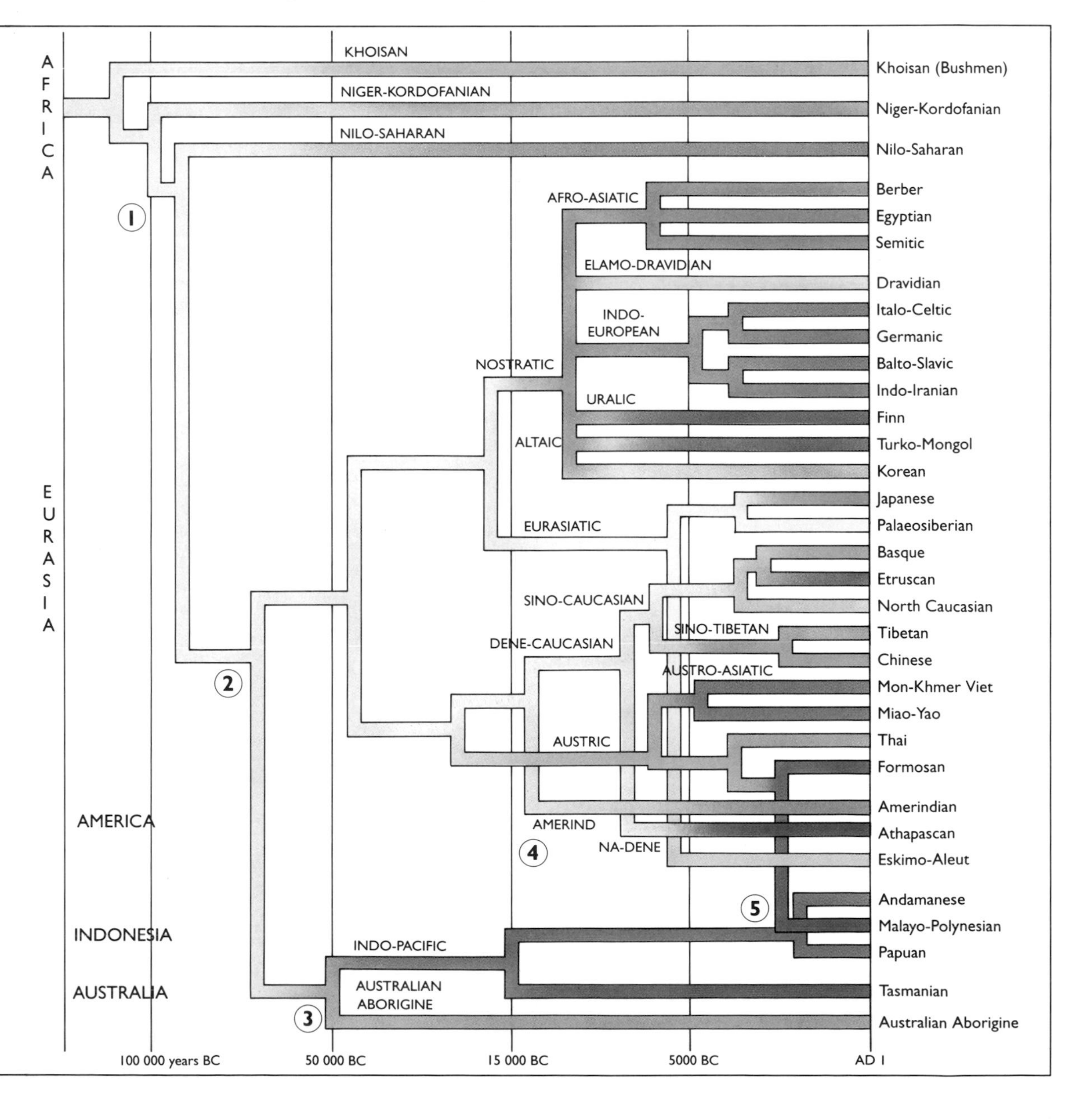

Frz
Katharina
12 sept: 1655
Scopus vitæ meæ Christus

HEREDITY

The Pattern of Inheritance

The science of genetics derives from a paper delivered in 1865 by an Austrian monk called Gregor Johann Mendel (1822–84). He laid the foundations for a science that during the 20th century was to lead to the greatest advances ever made in biology and in the understanding of the human organism.

Mendel, who had spent seven years experimenting with the breeding of dwarf and tall pea plants, read a paper outlining his results to the Natural History Society of Brünn (now Brno in the Czech Republic). His lecture aroused no discussion and no questions were asked. Nevertheless the paper was published in the Transactions of the Society but was otherwise ignored. Following this discouraging reception of his work Mendel abandoned attempts to publicize his researches, and when he died in 1884 his paper had long been forgotten. Mendel's paper was rediscovered in 1900 and its significance began to be appreciated. Six years later the term 'genetics' was introduced. Heredity is the transmission from one generation to the next of genetic factors that determine the characteristics of the individual. For centuries, humans had successfully bred domestic animals and plants by selecting those with certain desired features. But how these characteristics were passed on in reproduction remained unknown.

Early views

The earliest thinkers put forward a number of suggestions as to how heredity might work. The Greek philosopher Pythagoras (582–509 BC) taught that all hereditary features were passed on in the father's seminal fluid. Two centuries later Aristotle (384–322 BC) came nearer the truth. He suggested that the mother also produced semen, and that the two fluids fused in the womb to produce a new individual whose characteristics therefore derived from both parents.

Little progress in understanding was made until the 17th century, when the Dutchman Anton van Leeuwenhoek (1632–1723) – the first scientist to put the microscope to serious use – discovered what he called 'animalcules' in human semen. These were spermatozoa (⇨ p. 96), and eventually it became widely accepted that they were the carriers of the hereditary factors from the father to the child. Later, when animal ovaries were studied, eggs (ova) were discovered and their significance was understood as the vehicles of hereditary factors from the maternal line.

In spite of these advances, ideas on the transmission of individual characteristics remained confused. As late as the 18th century some scientists believed that they could see miniature human beings in sperms and ova, and postulated that individuals were fully formed in the testes and ovaries. According to their theory, development in the womb consisted simply of growth from microscopic to visible dimensions – the doctrine of 'preformation'.

By the end of the 18th century microscopic examination and dissection of animal embryos had begun to show that this idea was wrong. The adult pattern of organs was not present at conception but developed during the early weeks of embryonic life. This principle, known as

Gregor Johann Mendel, monk and botanist. (AR)

epigenesis, was soon confirmed by countless observations. Notions of heredity were, however, advanced no further by this development. Scientists of the time believed every separate structure or organ of the body gave off tiny particles that somehow collected in the sperms or ova. After conception these particles moved apart to form the appropriate organ or part in the new individual, thereby determining resemblances to the parents.

In a monastery garden

Mendel's work on the artificial fertilization of peas, from which he derived a completely new theory of heredity, was a model of experimental science, almost unique for its time. In the course of seven years, Mendel grew some 28 000 garden pea plants in beds and pots and studied seven of their characteristics. One of these – the seventh in his list – was plant height. To study height he produced a large number of apparently pure-bred peas by careful self-pollination, guarding them against insect pollination. The peas were of two types, tall and dwarf. The dwarf plants' seed produced only dwarf plants. In the case of the seeds from the tall plants, however, some produced tall plants while others produced dwarf ones. Evidently there were two kinds of tall pea plants – those that bred true (i.e. passed on their own characteristics) and those that did not. Mendel then crossbred true-breeding tall plants with dwarfs. The outcome, in every case, was a tall plant. He then self-pollinated these hybrid plants. One quarter of the offspring were true-breeding dwarfs, one quarter were true-breeding tall plants, and the remainder were of the non-true-breeding tall variety.

Mendel's findings for the other characteristics, such as colour, were the same. His conclusion was a very important one. When plants with different characteristics were crossbred, the outcome was not a blend of features. Instead, characteristics distributed themselves among the offspring in accordance with simple arithmetical ratios. Mendel was also able to

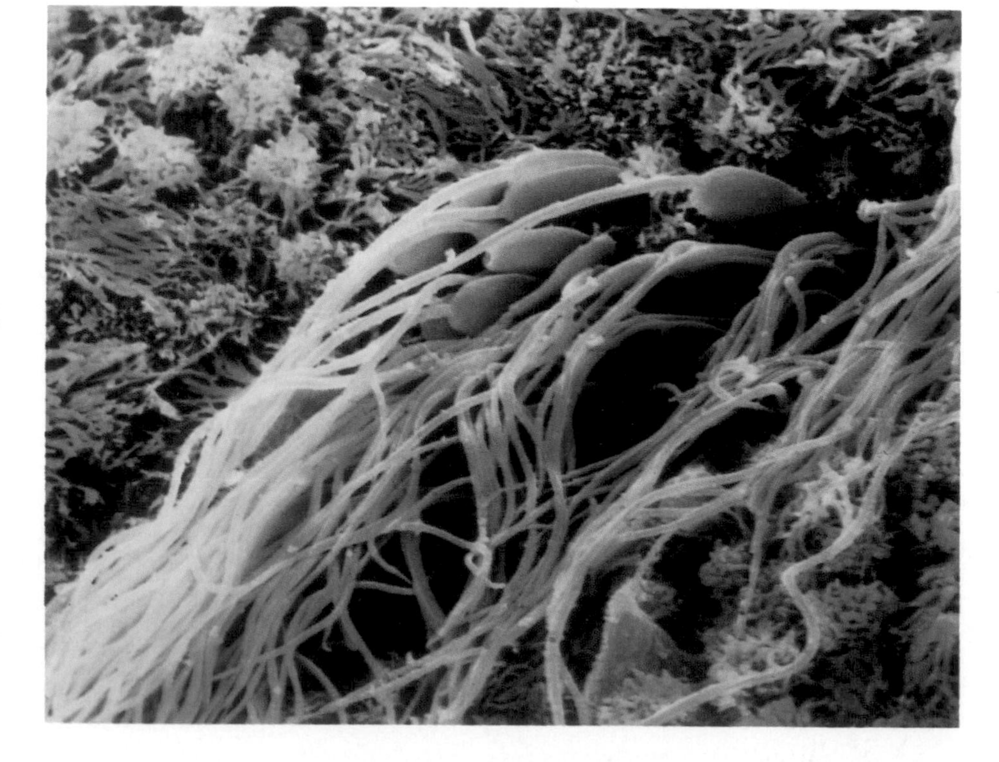

Human beings are diploid organisms – they have two copies of each chromosome in every cell. The exceptions, however, are the sex cells, spermatozoa (right) and ova. These cells, the gametes, are haploid – that is they contain one of each chromosome. As sex cells fuse at fertilization, a single diploid cell, the zygote, is produced, with the full complement of genetic information.
(Prof. P. Motta, Dept. of Anatomy, University 'La Sapienza', Rome, SPL)

show that different pairs of contrasted characteristics could be inherited independently of each other without interacting in any way. Tallness was a dominant characteristic, dwarfness was recessive. Mendel used these terms (➪ below) and showed how the characteristics were related mathematically. It seems likely that the apparent complexity of his mathematics repelled the ordinary reader and this may account for the scant attention his paper received.

Mendel had little appreciation of the fundamental importance of his work. Although he had read Darwin's *Origin of Species* he seems not to have grasped the relevance of his work to Darwin's theory. His paper was concerned only with the rules for the hybridization of plants and said nothing of the wider implications.

Mendel rediscovered

By 1900, Hugo Marie de Vries (1848–1935), Professor of Botany at the University of Amsterdam, had effectively worked out the laws of inheritance by research on plants. Before publishing, he reviewed the literature to see if anything had been done on the subject. It was with mixed feelings that he discovered Mendel's paper, and realized that the laws had been discovered a quarter of a century before. De Vries was not the only scientist to be thus disappointed. Two others, Erich von Tschermak and Carl Correns, unknown to him or to each other, had independently worked out the laws and then had found Mendel's paper. In the best tradition of science, all three immediately acknowledged Mendel's priority and referred to their own work only in confirmation of it.

But de Vries was able to go further. In 1901 he proposed a new theory – the mutation theory. Aware that new varieties of plants and animals that bred true occasionally occurred, he suggested that such alterations in hereditary characteristics were the result of actual permanent changes – *mutations* – that had taken place in the hereditary material of the parent. This theory has been universally accepted and remains the explanation for the one remaining gap in Darwin's theory – the origin of the spontaneous changes in species.

The role of the chromosomes

In 1882 Walther Flemming (1843–1905), Professor of Anatomy at Kiel University in Germany, published a book called *Cell Substance, Nucleus, and Cell Division.* Flemming had been using stains to show up the details of normally transparent cells and had discovered thread-like bodies in cell nuclei that stained deeply. These he called *chromosomes* (literally, 'coloured bodies'), because of the ease with which they were stained. No sooner had Mendel's work been publicized by de Vries than biologists began to realize that the chromosomes, as described by Flemming, could account for the transmission of hereditary characteristics.

By 1902 it had been recognized that chromosomes were present in pairs in normal body cells, but that in the single cells constituting sperms and eggs only one member of each pair was present. During fertilization, the fusion of sperm and egg restored the number of chromosomes to normal, half of them coming from the father and half from the mother. This observation was immediately seen to support Mendel's ideas of the movement of 'hereditary units'.

THE LAW OF DOMINANCE

Mendel's experiment on inherited characteristics using red and white pea flowers

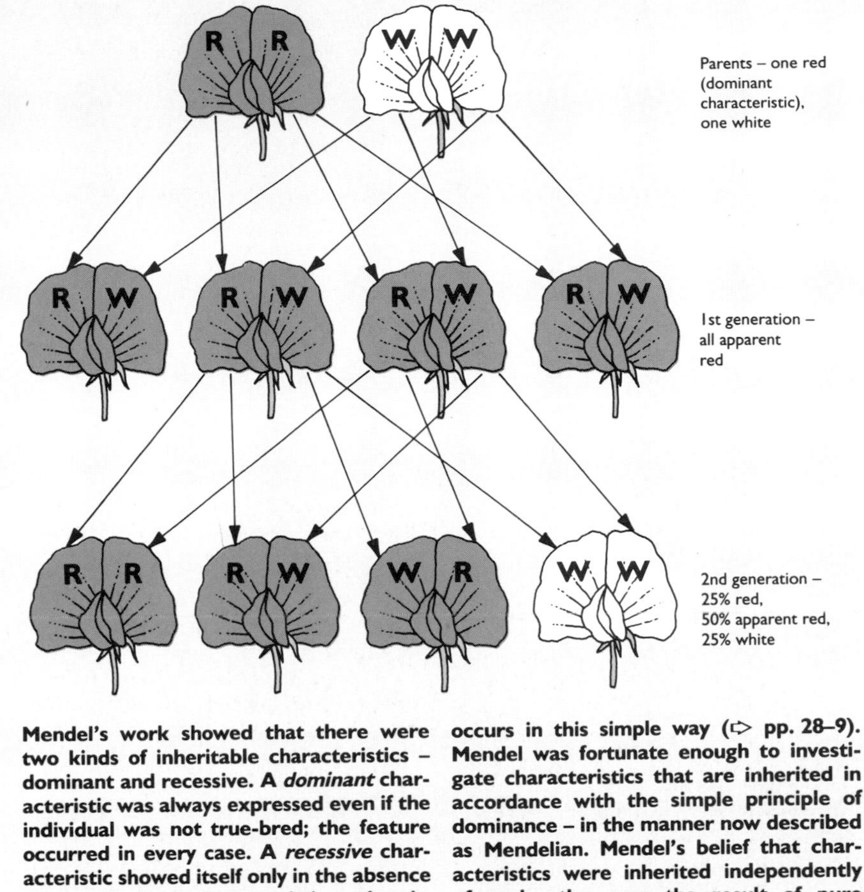

Mendel's work showed that there were two kinds of inheritable characteristics – dominant and recessive. A *dominant* characteristic was always expressed even if the individual was not true-bred; the feature occurred in every case. A *recessive* characteristic showed itself only in the absence of the dominant characteristic – that is, only if the individual had received the recessive feature from both parents. This is a central principle of genetics, although we now know that not all inheritance occurs in this simple way (➪ pp. 28–9). Mendel was fortunate enough to investigate characteristics that are inherited in accordance with the simple principle of dominance – in the manner now described as Mendelian. Mendel's belief that characteristics were inherited independently of each other was the result of pure coincidence. The seven characteristics he studied happened, as we now know, to be controlled by single genes on separate chromosomes.

Since there were thousands of hereditable characteristics and only 23 pairs of chromosomes in human beings, this could mean only that each chromosome must carry the genetic material for a large number of features. So it was postulated that there were many separate units on each chromosome, each one possibly controlling an individual characteristic. These units were called *genes*.

The most important work at this stage was done by the American biologist Thomas Hunt Morgan (1866–1945), who selected for study the tiny fruit flies of the genus *Drosophila*. These flies bred quickly and easily and their cells contained only four chromosomes. By studying successive generations of *Drosophila*, Morgan was able to show that de Vries was right about spontaneous mutations, and also that various groups of characteristics were often inherited together. This was called *linkage*, and it showed that particular sets of genes always occurred on a particular chromosome. Morgan, however, found an important exception to the rule. It became apparent that during the process of cell division pairs of chromosomes occasionally exchanged segments with each other. This was called *crossing-over* (and is in fact usual in larger animals, including humans; ➪ p. 23). Morgan's work progressed to the point where he was able to show the approximate position of certain genes on a chromosome, and he showed that the further apart two genes were located on a chromosome, the less likely were the characteristics they caused to be linked.

Morgan's work on *Drosophila* culminated in his publication in 1926 of *The Theory of the Gene*, which contained gene maps for *Drosophila* and placed genetics on a new and sound foundation.

SEE ALSO

- THE STORY OF EVOLUTION p. 6
- HEREDITY pp. 20–31
- CELLS, THE STUFF OF LIFE p. 38
- CHILDBIRTH THROUGHOUT HISTORY p. 94
- PASSING ON LIFE p. 96
- HEREDITARY DISEASES p. 148

Chromosomes and Cell Division

The advance of biological science, especially medical science, has brought with it an increased awareness of the fundamental importance of genetics. Recent years have seen a dramatic enlargement of our understanding of the role of heredity in determining not only the physical characteristics of organisms, but also the way in which organisms function and in many cases 'malfunction'.

Genetics is the study of the way in which inheritable factors are coded in the cells of an organism, how this coding is maintained during cell reproduction, how it is passed on from one generation to the next, and how it is expressed in physical features.

The chromosomes

The physical basis of heredity is essentially the same in plants and animals, including humans. The cell is the basic building structure of almost all living things (➪ p. 38) and, apart from the bacteria, almost every cell contains a central part – the nucleus – which consists largely of a mass of thread-like material called *chromatin*. The chromatin strands are remarkably long in relation to their thickness. Each one is a protein-coated molecule of DNA (➪ p. 24) containing several thousands of the chemical groups we call genes (➪ pp. 21 and 26) strung along its length, in single file. Most of the time the chromatin remains bundled up like a loose ball of wool, but when the cell is about to divide it becomes coiled up in an extremely compact way to form a number of separate bodies called *chromosomes*. Loose chromatin does not stain well so it is almost impossible to see by normal light microscopy. But when it is coiled up to form chromosomes it stains readily – hence the name (literally 'coloured bodies').

With the exception of sperms and eggs and their immediate parent cells, each human cell contains 46 chromosomes, arranged as 23 pairs. If cells are photographed as they are dividing, pictures of the 46 chromosomes can be obtained and then enlarged. The photographic print can then be cut up to separate them and they can be arranged in a standard, numbered order called the *karyotype* (➪ illustration). Whole chromosome abnormalities can thus immediately be detected. When photographed for a karyotype, all 46 chromosomes are dividing and are seen duplicated into two arms still stuck together to form a roughly X-shaped body. The slight constriction at the crossing of the arms is called the *centromere*. The position of the centromere, the resulting long and short arms, the size of the chromosome and the banding pattern allow each one to be identified and numbered.

Autosomes and sex chromosomes

In women the members of each pair of chromosomes appear identical, but in men this applies only to 22 of the 23 pairs. The two chromosomes numbered 23 are called the *sex chromosomes*. In women they are also called the X chromosomes. But in men, the sex chromosomes consist of one X (which looks just like a female X chromosome) and one Y, which is much smaller. It is this XY configuration that determines maleness, while the XX configuration determines femaleness. The chromosomes other than the sex ones are called *autosomes*. So a characteristic caused by a gene on an autosome is called an *autosomal characteristic* and one caused by a gene on a sex chromosome is called a *sex-linked characteristic*.

Normal male karyotype shown here in false-colour light micrograph. The XY chromosomal configuration that determines maleness is visible in the bottom right-hand corner. (CNRI, SPL)

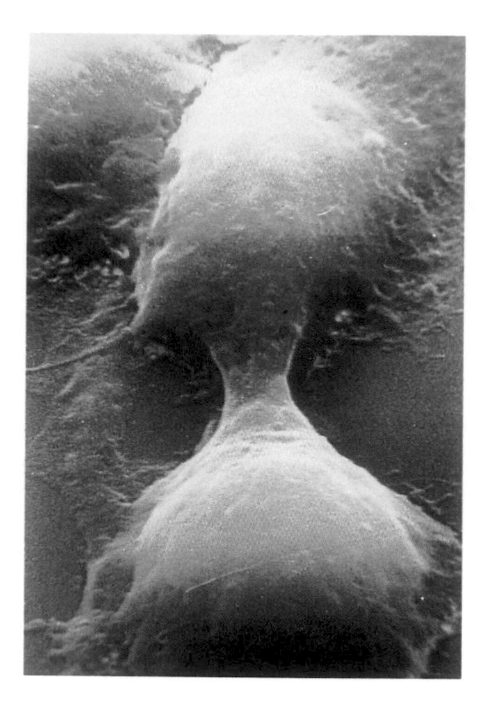

Cultured human cells in telophase. This is the last stage of mitosis, in which the chromosomes of the two daughter cells are grouped together at each separating pole to form new nuclei. (RB)

If reproductive cells – sperms and eggs (ova) – contained the normal number of 46 chromosomes, the penetration of an ovum by a sperm would result in a rather crowded cell with 92 chromosomes and a very abnormal individual. For this reason, in the course of their maturation, the reproductive cells undergo a special series of divisions, the net result of which is to reduce each pair of chromosomes to one. This is called reduction division or *meiosis* (➪ below). So sperms and ova each contain only 23 chromosomes. And this is called the *haploid* number of chromosomes.

As soon as a sperm penetrates an egg (➪ pp. 97 and 102), a dense membrane forms around the ovum to prevent the entry of any others. Thus the fertilized ovum has 46 chromosomes (the *diploid* number), one member of each pair of chromosomes being derived from the mother and the other member from the father. This applies to the sex chromosomes as well as to the autosomes. One X chromosome comes from the mother's ovum, the other sex chromosome (an X in the case of a girl, a Y in the case of a boy) comes from the father's sperm. During meiosis in the production of sperms, the separation of the sex chromosomes means that half of them will contain a Y chromosome and half an X chromosome. If a sperm with an X chromosome is the first to penetrate the ovum, the result will be a girl. If a sperm with a Y chromosome is first, the result will be a boy.

How Cells Reproduce

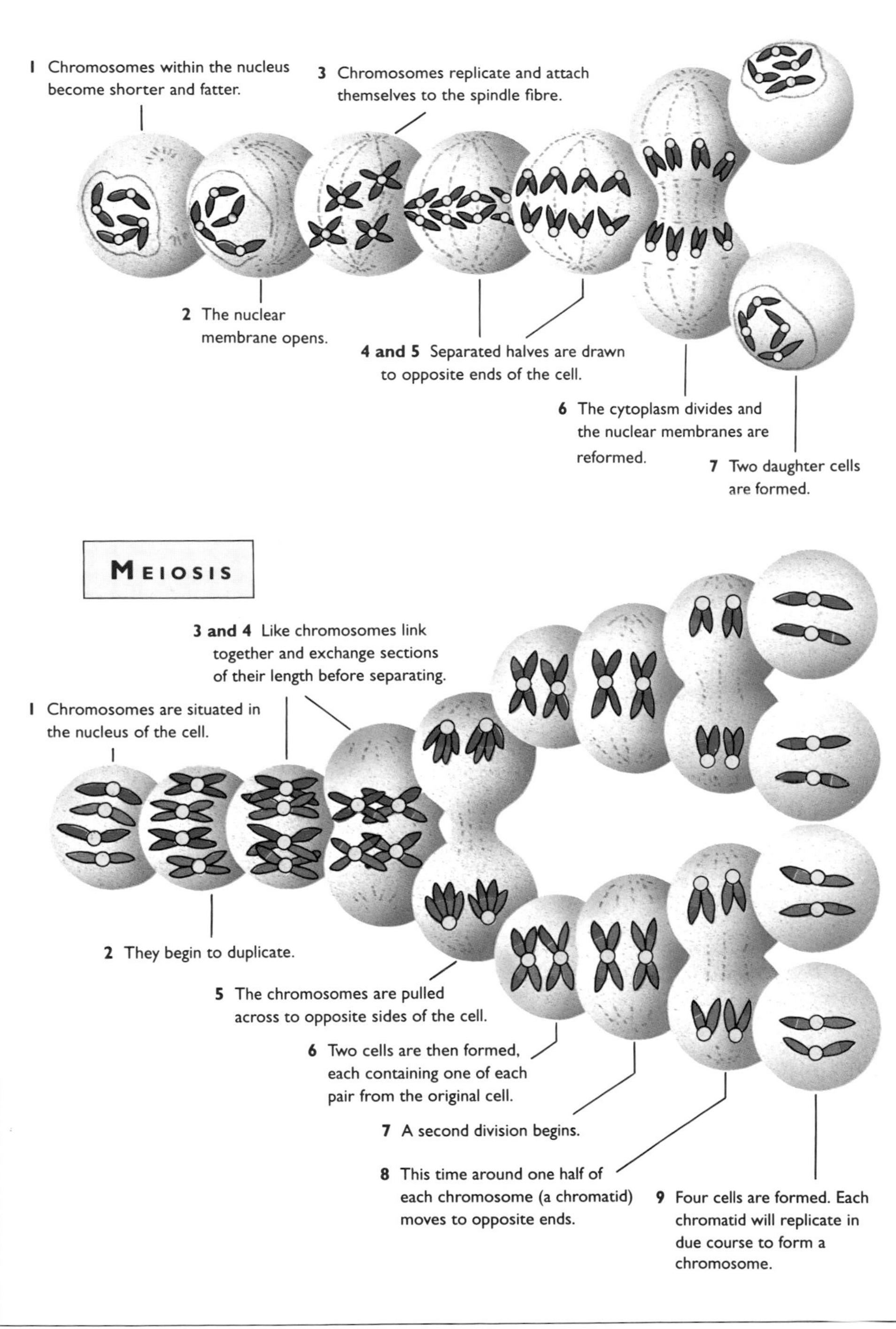

MITOSIS Normal cell division, or *mitosis*, results in two 'daughter' cells genetically identical to the cell from which they are derived. Each of the threads of chromatin – the DNA strands – is exactly duplicated and the two sets of pairs of new threads separate to form the chromosomes of the daughter cells. This separation involves a temporary reorganization of the cell and much mechanical movement of chromosomes.

The first sign that a cell is about to divide is the appearance within the nucleus of long chromatin threads which divide longitudinally to form new strands. Following this division, the new paired chromatin threads coil up to form pairs of new chromosomes. At first, these are stuck together at the centromere to produce the characteristic X-shaped double or replicated chromosomes seen in the karyotype and known as *chromatids*. These double chromosomes now form up in a rough circle around the equator of the cell, and a delicate structure of fine strands or fibres, radiating to the equator from each end of the cell, appears. This is called the *spindle apparatus* and its function is to pull apart the two chromosome rods that form the arms of the X of each double chromosome. As the strands of the spindle contract, the two halves of the replicated chromosome behave as if they are being pulled apart, separating at the centromere, and are dragged to opposite ends of the cell.

In this way each end of the cell contains 46 single chromosomes. The chromosomes now uncoil and a nuclear membrane forms around each of the two bundles of chromatin. The spindle disappears. The centre of the cell begins to narrow until an hour-glass shape is formed, and soon the cell divides into two daughter cells, each containing the full complement of 46 chromosomes. Barring accidents, the chromosomes in each of the daughter cells are identical in every way to the chromosomes in the mother cell. This remarkable process of duplication is going on all the time in millions of cells in all our bodies.

MEIOSIS A different form of cell division is necessary when the cells in our ovaries and testes divide to form ova and sperms. This kind of division is called *meiosis*, and in addition to producing the haploid number of chromosomes in the ova and sperms, it has another important feature – it ensures that when we pass on to our offspring the characteristics we inherited from our parents, our offspring (other than identical twins) will differ considerably from each other.

The first stage of meiosis is similar to that of mitosis. The chromatin strands, each representing a single chromosome, replicate longitudinally, and the resulting pairs of daughter strands coil up and stick together to form X-shaped double chromosomes (chromatids). These now congregate in pairs with their original partners. The duplicated chromosome I that came originally from the individual's father aligns itself with the duplicated chromosome I from that person's mother, and so on for all 23 pairs. The partners in each pair of double chromosomes now twist intimately together and become closely aligned along their entire length. While in this relationship they exchange several short corresponding segments with each other – a process known as *crossing-over*. This occurs in a random manner so that the genetic material inherited from the individual's father and the mother becomes thoroughly mixed and new chromosomes are formed, each with a unique blend of the genes from both parents.

The cell now divides, but in this division the X-shaped chromatids do not have their arms pulled apart as in mitosis. Instead, one of each of the pairs of the double chromosomes goes intact to each daughter cell. So far as we know, it is a matter of pure chance which of the pairs goes to which daughter cell. In this way, the random redistribution of genes already effected by crossing-over is further increased.

Each daughter cell now has 23 chromatids. A second division occurs, but this time the halves of the chromatids are pulled apart by a spindle, as in mitosis, and each daughter cell receives a single chromosome from each pair – a total of 23. A third division, producing a total of four sperms or ova from the original parent cell, does not alter the fact that each germ cell (sperm or ovum) now has half the normal number of chromosomes – the haploid number.

Homologous chromosomes

Of the 23 pairs of chromosomes in normal body cells, one member of each pair is a maternal chromosome and is matched with its fellow, which is a closely similar paternal chromosome. These pairs of chromosomes are called *homologous chromosomes*. The corresponding genes at any particular site (*locus*) in a pair of homologous chromosomes also form a pair. These paired genes may be identical, in which case the individual concerned is said to be *homozygous* for that gene, or they may be slightly different, in which case the individual is *heterozygous* for that gene. In this case the dominant gene (➪ p. 21) will be expressed.

Of the huge number of gene loci on our chromosomes, all of us are homozygous for some, heterozygous for others. In most cases, the distinction between homozygous and heterozygous genes is a matter of relative indifference to our health – they merely determine some characteristic such as our skin colour. But many loci can have genes that are far from unimportant and in these cases the difference between the homozygous and the heterozygous condition may make all the difference in the world (➪ p. 28).

- HEREDITY pp. 20–31
- CELLS, THE STUFF OF LIFE p. 38
- PASSING ON LIFE p. 96
- LIFE IN THE WOMB p. 102
- HEREDITARY DISEASES p. 148

DNA, the Blueprint of Life

One of the greatest advances in science in the 20th century – indeed of all time – was the elucidation of the physical basis of inheritance. This chapter in the history of science has already had an explosive impact on biology, and its effect on the future of humanity will certainly be as great.

The scientists who took the final step in the elucidation of the structure of the DNA molecule based their research on a body of knowledge that went back for almost a century. Like most science, the final crystallization of truth was possible only as a culmination of the work of scores of other patient scientists. But to an unprepared world, the news came as a bolt from the blue. On 25 April 1953 a short paper appeared in the scientific journal *Nature*. It started with the words 'We wish to suggest a structure for the salt of deoxyribose nucleic acid (D.N.A.). This structure has novel features which are of considerable biological interest.' Near the end of the article the authors modestly stated: 'It has not escaped our notice that the specific pairing we have postulated immediately suggests a possible copying mechanism for the genetic material.'

The authors, James Watson and Francis Crick (⇨ boxes), had been working in the Cavendish Laboratory, Cambridge, which was headed at that time by Sir William Lawrence Bragg (1890–1971). Aware that several other scientists, especially the world-renowned chemist Linus Pauling, were actively investigating the structure of DNA, they showed a draft of the paper to Bragg. Bragg, whose earlier pioneering work on x-ray crystallography had provided an essential technique and a source of data for Watson and Crick, saw at once that the molecular structure they were now suggesting answered many questions with convincing elegance. He agreed to send the paper to the editor of *Nature* with a strong recommendation for early publication.

Throughout the decade that followed, many biology laboratories were engaged in an intense experimental study of the Watson-and-Crick model, and as the evidence in favour of it mounted and nothing was found to disprove it, the now-celebrated double-helix structure of the molecule gradually became accepted beyond dispute. This structure could reproduce itself, could provide a code for the formation of proteins, and contained sufficient capacity for the estimated requirements of the whole human genetic pool. In 900 words, Watson, a 28-year-old American ornithologist-turned-biochemist, and Crick, a 37-year-old English physicist-turned-molecular biologist, had revolutionized scientific thinking.

Francis Crick

Francis Harry Compton Crick was born in 1916 and received his basic scientific education as a student of physics at University College, London. During the Second World War he worked for the British Admiralty on the development of magnetic mines. His interests later turned to biology, and in 1947 he moved to Cambridge, where for two years he worked at Strangeways Research Laboratory in the University. He then moved to the Medical Research Unit at the Cavendish Laboratory. There he worked under Max Perutz and Lawrence Bragg using x-ray-diffraction methods to study the structure of proteins. James Watson joined him there in 1951 and an unofficial collaboration arose between the two men on the structure of DNA.

After the publication of the *Nature* paper in 1953, and the award of a PhD, Crick spent the next twenty years in Cambridge working on the problem of how DNA carries the genetic information and how this is translated into proteins (⇨ p. 26). Crick was the first to suggest the role of transfer RNA in the assembly of -amino acids into proteins. The basis of the genetic code, as triplets of bases, was finally established by Crick in 1961, and he went on to determine the code for some of the 20 amino acids occurring in proteins.

In 1977 Crick moved from Cambridge to take up a position in the Salk Institute in San Diego, California, where he embarked on a third scientific career in research into brain function.

The raw materials

Well before this time it had been appreciated that it was the nucleic acids in cells, rather than proteins, that were the essential elements in chromosomes (⇨ p. 22). It had also been known for seventy years that chromosomes replicated themselves during cell division (⇨ p. 23). The basic chemistry of the giant nucleic acid molecules had been worked out, and the exact nature of the chemical groups from which they were constructed – nitrogenous bases and sugar–phosphate groups – was known. Much work had also been done, mainly by Maurice Wilkins and his assistant Rosalind Franklin using x-ray diffraction methods at King's College, London, to demonstrate certain regularities in the way these groups occurred. These x-rays did not show every repetition of the chemical units in the molecule, but they were consistent with the idea that the patterns were showing every tenth chemical repetition from the same angle. If the chain were coiled into a helix (or spiral), this is exactly what would happen. The idea of a helical molecule was not new. Linus Pauling had already shown that some collagen protein molecules took this form. There were also good reasons to suppose that the molecule now known as deoxyribonucleic acid (DNA) might have a helical shape.

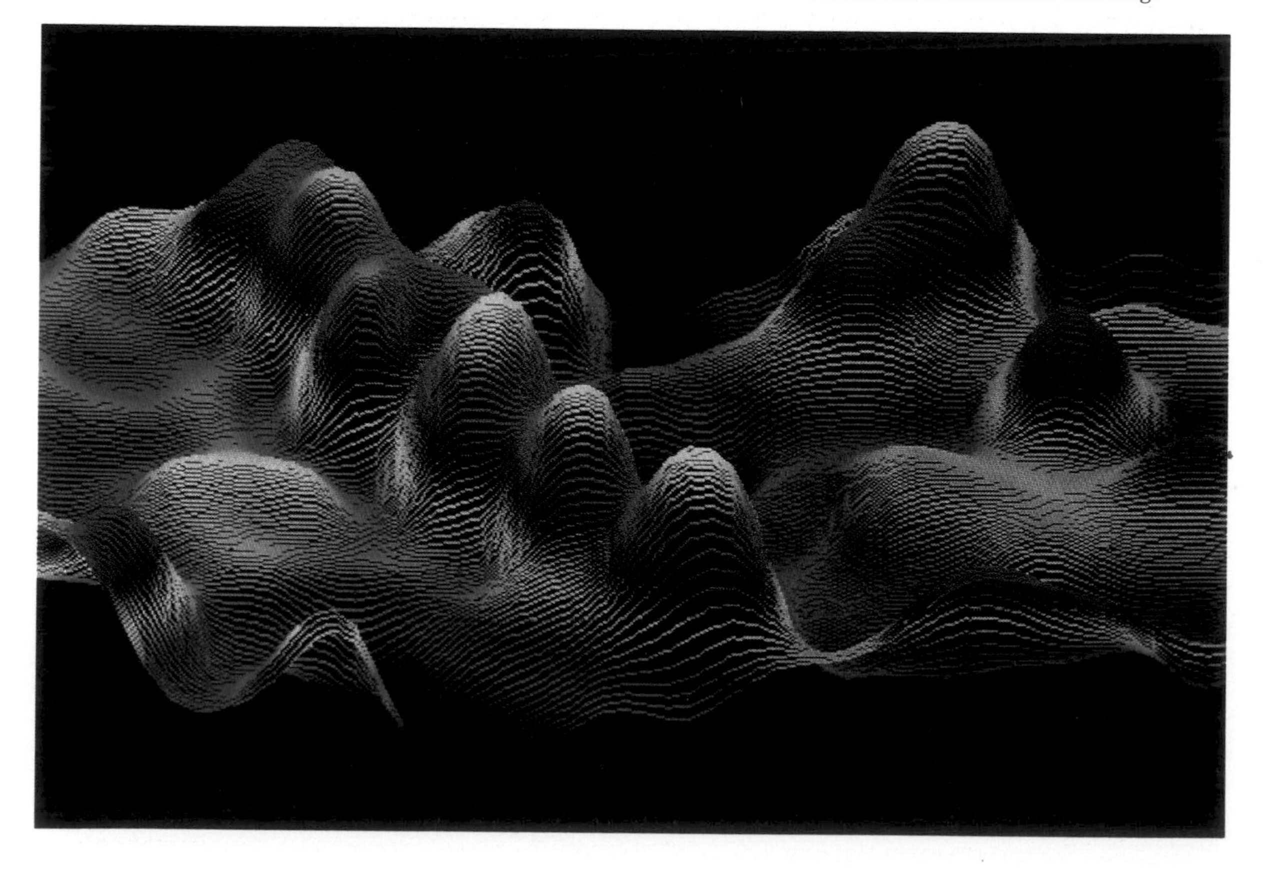

The DNA molecule represented in false-colour scanning, tunnelling micrograph. The orange/yellow peaks (centre-left) correspond to the ridges of the double-helical 'backbone'. (Magnification x 1 600 000 at 6 x 7cm size.) (Lawrence Livermore Laboratory, SPL)

In December 1952 news reached Watson and Crick, by way of Linus Pauling's son Peter, who was working at Cambridge, that Pauling had worked out a structure for DNA. When they saw a draft of Pauling's paper, however, they realized to their great relief that the idea was flawed, and they resumed their work with renewed energy.

'A structure this pretty just had to exist'

Like other groups of scientists working on the structure of the DNA molecule, Watson and Crick were essentially engaged in solving a jigsaw puzzle. They were trying to fit known pieces together in a way that would account for a number of known facts and that would, somehow, produce a structure that could reproduce itself automatically. They knew that DNA contained four nitrogenous bases – adenine, guanine, cytosine and thymine. They also knew that the number of adenine bases was the same as that of thymine bases, and that the number of guanines was the same as that of cytosines. There was no fixed ratio, however, between adenine and guanine or thymine and cytosine. They had accepted that the 'backbone' of the molecule was a helically shaped sugar–phosphate strip and that the bases were attached to this. The question was, how?

One major problem was that all four bases were of a different size, and it was hard to see how they could be attached to each other and to the backbone without producing unacceptable irregularity. Watson was working on the idea of two helical backbones with pairs of identical bases connected between them. The different sizes could be accommodated if the two helical strips twisted in opposite directions. Bitter disappointment followed, though, when the idea was shown to conflict with chemical principles. But Watson persisted in trying various arrangements of bases between two backbones and suddenly realized that an adenine base linked to a thymine had the same shape and size as a guanine base linked to a cytosine. This was chemically impeccable and also explained the riddle of why the number of adenines and thymines was the same and the number of guanines and cytosines was the same. The way in which the complementary pairs bonded together meant that adenine could pair only with thymine, and guanine only with cytosine.

Watson at once realized that, given one longitudinal half of the molecule and a nearby supply of the necessary bases, the other half could automatically be constructed. Crick saw the promise in this idea and for the next few days the two men worked frantically to build a model of the new molecule. It was a double helix with the sugar–phosphate backbones on the outside and the base pairs joining them like the steps on a spiral staircase. When it was completed, they checked every detail to ensure that it satisfied both the x-ray data and the rules of stereochemistry (which is the study of the spatial arrangement of atoms in molecules). Everything fitted. As Watson said, 'A structure this pretty just had to exist.'

In 1962 Crick, Watson and Wilkins were jointly awarded the Nobel prize for Physiology or Medicine.

— James Watson —

James Dewey Watson was born in 1928 in Chicago and enrolled in the University of Chicago at the age of 15. He graduated in 1947 and proceeded to do postgraduate work on viruses at Indiana University, where he was awarded a PhD in 1950. His initial interests had been in ornithology, but from 1950 to 1951 he continued working on virus research at the University of Copenhagen. While there, he became strengthened in his conviction that an elucidation of the chemical structure of DNA was necessary if heredity was to be understood, and he resolved to work on the problem. Accordingly, at the age of 23, he moved to the Cavendish Laboratory at Cambridge, where he began to study the possible structure of the DNA molecule.

Soon after the publication of the *Nature* paper in 1953 Watson moved to the California Institute of Technology, where he worked until 1955. He then moved to Harvard, where he did research on molecular biology and where, in 1961, he was appointed Professor of Biology. In 1965 his book *Molecular Biology of the Gene* appeared, and in 1968 he published *The Double Helix*, a personal account of the discovery of the structure of DNA and of the other people involved.

From 1988 to 1992 Watson was Director of the Human Genome project at the National Institutes of Health, Washington – the largest single endeavour in biological science – which aims to map the entire genetic chromosomal structure of the human being (⇨ p. 26).

The Structure of DNA

Deoxyribonucleic acid (DNA) is the basic genetic material of most living organisms. Although a large and apparently complex molecule, the structure of DNA is in fact astonishingly simple.

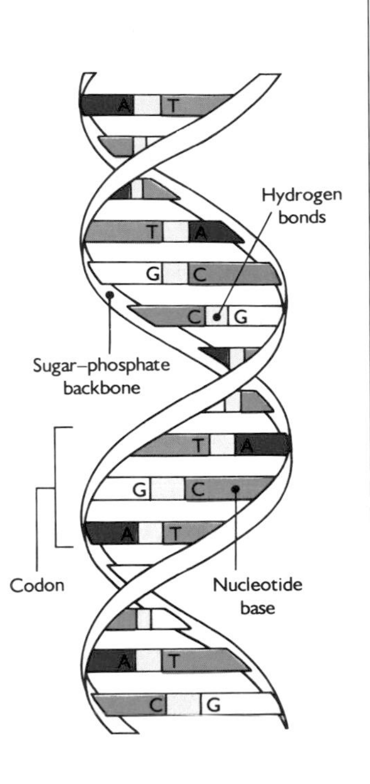

A single DNA molecule consists of two separate strands wound around each other to form a double-helical (spiral) structure. Each strand is made up of a combination of just four chemical components known as *nucleotides*, all of which have the same basic composition: each nucleotide consists of a sugar molecule (deoxyribose) linked to a phosphate group to form the helical backbone; different nucleotides are distinguished only by the identity of the nitrogen-based unit (called a *nucleotide base*) bonded to the sugar molecule. The four bases – adenine (A), cytosine (C), guanine (G) and thymine (T) – lie in the central region of the double helix, with each base linked by hydrogen bonds to a specific complementary base on the partner strand: A pairs only with T, and G only with C.

This simple structure explains the two key properties of DNA – how it codes for the manufacture of amino acids (from which proteins are formed) and its capacity to replicate itself. Each combination of three bases (known as a *triplet* or a *codon*) within a DNA molecule codes for a particular amino acid, while the specific pairing of bases explains how two identical DNA molecules can be produced by the separation of the two strands of the parent molecule.

Winners of the Nobel prize, 1962. From left to right: Maurice Wilkins, John Steinbeck (for literature with *The Grapes of Wrath*), John Kendrew and Max Perutz (jointly for chemistry with their discovery of myoglobin and work on haemoglobin), Francis Crick and James Watson.
(Keystone)

SEE ALSO
- HEREDITY pp. 20–31
- WHAT THE BODY IS MADE OF p. 36
- CELLS, THE STUFF OF LIFE p. 38

The Genetic Code

The great sequence of genes strung out along our chromosomes may be considered a 'blueprint' or programme for the construction of the human body. But this blueprint is unusual because it is essentially self-reading, whose operation determines the workings of proteins, the principal structural elements of the body.

SEE ALSO

- DNA, THE BLUEPRINT OF LIFE p. 24
- WHAT THE BODY IS MADE OF p. 36
- CELLS, THE STUFF OF LIFE p. 38
- FUEL AND MAINTENANCE 2: DIET p. 46

Cracking the genetic code, that triumph of scientific deduction (⇨ p. 24), revolutionized the understanding of genetics and led to the foundation of a new science – molecular biology. The code is concerned with one thing only – the order of the sequence of the constituents that make up proteins. The human *genome* – the entire gene collection – contains the codes for about 100 000 different proteins. A genome works like a complex parallel-processing computer, different parts of which are operating at different times, regulating each other's activity in generating the appropriate proteins – mostly enzymes (⇨ below) – to cause cells to function and, during body growth and repair, to take the appropriate form. This behaviour of the genome is the basis of cell differentiation (⇨ pp. 38 and 102) and is the ultimate explanation of how our bodies are put together.

Proteins and amino acids

Proteins, from which the body is largely made, are very elaborate molecules consisting often of hundreds of amino acids strung together. *Amino acids* are fairly simple organic compounds, all of which contain an amino group ($-NH_2$) in the molecule. The acidic feature is the carboxyl group $-COOH$ (⇨ p. 36). Their most important property is the ability readily to form a bond between the amino group of one amino acid and the carboxyl of another by eliminating one molecule of water (H_2O). This is called a *peptide bond* and it allows amino acids to link together in long chains.

Although at least 26 different amino acids have been found in nature, there are only 20 different ones in the body. The same 20 are found in all animal tissues. Each of the large number of different proteins is formed by stringing these amino acids together in a particular, specific order. A choice of two amino acids from 20 gives 400 permutations. A string of 10 from 20 can be arranged in 100 000 000 000 000 different ways.

The process of linking amino acids together to form proteins is called *polymerization*. A *polymer* is a long molecule consisting of smaller units (*monomers*) joined together in a chain. For example, the plastic polyethylene (more commonly known as polythene) is a polymer of ethylene monomers. Protein polymers contain varying numbers of amino acids – sometimes as many as hundreds of thousands. When proteins in food are eaten, the reaction that occurs, under the influence of specific digestive enzymes, is a depolymerization, a breakdown of the molecule into shorter lengths called *polypeptides*. These are then broken down into separate amino acids, which are small enough to be absorbed into the bloodstream and carried to the cells, where they are available for resynthesis into new proteins.

Proteins are not only structural materials. Many proteins are *enzymes* – specific substances that act as catalysts, enabling particular chemical reactions to occur very quickly, including the reactions that result in the body being put together in a particular way from raw materials supplied in the food. Once we know how cells produce appropriate enzymes we have gone a long way to explaining how living organisms – including human beings – grow and develop. By the early 1960s Francis Crick (⇨ pp. 24–5) and other scientists had largely solved this problem. The secret lay in the sequence of bases (chemical components that are the functional opposites of acids) in the DNA molecule. It was necessary to have a code that could specify the 20 different amino acids present in the body. The order of these acids in a particular protein could, they found, be represented by the order in which the code for each amino acid occurred along the DNA molecule. The simplest code would be one in which a short sequence of bases would stand for a particular amino acid. A code consisting of two out of the four bases would not be enough as this would give only 16 possible combinations and 20 are needed. A permutation of three out of the four bases gives 64 combinations, so a triplet of bases could be the answer. Sequences of four or more bases would work equally well, but as three would do the trick it seemed best to settle for the most economical solution. In the end it was found to be the case that a single triplet of bases did indeed specify a single amino acid.

Putting it all together

Genes are lengths of DNA. Information is stored in the DNA as sequences of three bases, taken from a possible four, which are read consecutively. Of all the possible unique triplets of bases, 20 stand for the 20 amino acids, and the sequence of triplets of bases in the DNA thus determines the order of amino acids in the protein that is to be formed. The order of the bases along the DNA molecule is unique for any gene, and each gene forms a single kind of protein.

The order of bases in the DNA is determined by patterns that arose in the remote past – sometimes millions of years ago. That that order has usually been accurately preserved is one of the astonishing facts of biology. Changes that have occurred in the base sequence for any reason (mutations) are also copied with the same degree of accuracy.

DNA replication

DNA replication can occur only after the chromosomes – which are coiled coils – have unwound to form extended lengths of helical DNA (chromatin; ⇨ p. 22). They do this as a result of the action of an enzyme called helicase, which unwinds about 100 revolutions each second. When a double-helix DNA molecule replicates, the first step is for the two strands to separate. When they do so the stage is then set for each of the single separated strands to act as a template for the production of a new strand.

Contrary to expectation, separation does not start at one end and proceed along to the other. Because of the great length of the strand, this would take far too long. The double helix actually separates locally at many points to form open loops in which replication occurs. As this happens, each of these loops then opens progressively in both directions until the whole length of DNA has split. The copying of DNA is a largely automatic process

WHY ARE PROTEINS SO IMPORTANT?

The 'building blocks' of proteins, the amino acids, are widespread in nature. Of the 20 amino acids from which we are largely constructed, 12 can be synthesized in the body from other food materials. The remaining eight cannot be formed in the body and must be provided in the diet. For this reason they are known as the *essential amino acids*. The absence of essential amino acids from the diet, although rare, is a serious nutritional deficiency that leads to structural failure because, although the code for the construction of proteins is present, not all the raw materials are. Amino-acid chains can be put together in a bewildering variety of arrangements to form a vast number of different proteins. In practice, only a small proportion of all the possible ways are used, and proteins of similar function from different animals bear a close resemblance to one another. The protein insulin, for instance, involves a sequence of 51 amino acids and differences in its form occur in a maximum of only three places in pigs, sheep, horses and humans. Pig insulin differs from human insulin in only one place and has been widely used in the management of human diabetes.

Most of the proteins formed by the ribosomes are enzymes, which may immediately begin to operate within the cell. Some act to regulate the functioning of other genes. During body development, this switching-on of different genes, under the influence of the genetic code, results in the production of enzymes that determine whether a particular cell should become, for example, a cell of the brain, kidney or liver, as required.

Of those proteins that are not enzymes, some are muscle proteins, others structural connective tissue. Skin, hair, antibodies, haemoglobin and many hormones are proteins. Of the structural proteins of the body, the most important is *collagen*. This is a family of fibrous proteins that form into bundles with high tensile strength, that is they are capable of withstanding high stresses without breaking. Collagen is the most abundant protein in mammals, and makes up about a quarter of the total body weight. It is the main structural element in bones, teeth, skin, tendons, blood vessels and cartilage. Collagen not only holds cells together to form organs, it also has a directive role in the growth and development of tissue. The collagen molecule is a triple-stranded helical rod and is one of the longest proteins known. The proper formation of these three-dimensional strands from the basic procollagen (precursor) molecules requires the action of an enzyme that is activated by vitamin C. Severe deficiency of vitamin C causes scurvy, a disease whose symptoms include poor healing of wounds and weakness of the small blood vessels, and which is due to defective collagen.

in which the base adenine always links with thymine and the base guanine always links with cytosine (⇨ pp. 24–5), the new bases being attracted from the immediate cell environment. In this way, a double helix becomes two identical double helices, one of which goes to each of the new cells that are forming (⇨ box, p. 23). As this process gets under way, the helical chromatin undergoes a further coiling into the compact form we recognize as chromosomes.

The role of RNA

Proteins are not made directly on the DNA molecule itself but in special cell bodies (organelles) known as *ribosomes*. These lie within the cytoplasm, the part of the cell outside the nucleus (⇨ p. 38). First, a length of double helix separates longitudinally to form a loop, exposing the sequence of single bases that together constitute the gene and specify the protein. Base triplets different from those coding for amino acids indicate where the gene starts and where it ends. A new complementary strand is now made on the exposed bases. This takes place because the exposed bases of DNA attract the constituent elements of *RNA* (*ribonucleic acid*) from the cell cytoplasm where they float freely. The RNA bases in turn attract the sugar-phosphate molecules, which form its backbone. These strands form at a rate of about 50 bases per second. RNA production is rather slapdash and the accuracy of copying is poor, about one mistake occurring every 100 000 bases. But many copies are made and the occasional wrong one does not matter. RNA uses three of the same bases as DNA, the exception being thymine. Another base called uracil is substituted for it, and this links with the DNA adenine.

The RNA chains formed in this way are called *messenger RNA* (mRNA) because they carry the code of the gene out through pores in the nucleus of the cell to the ribosomes in the cytoplasm. Before the codes are used to select the amino acids, each mRNA length is 'edited' to get rid of sections that are not needed, and the remaining coding sequences are spliced together. It is this edited version of the mRNA that is read by the ribosomes.

The cell fluid contains millions of samples of the 20 amino acids, most derived from food but some synthesized in the body. Before they can be joined together in the right order to form new protein molecules, however, they have to be brought to the ribosomes in the correct sequence. This is done by yet another kind of RNA, *transfer RNA* (tRNA), which moves around in the cell fluid picking up the 20 different amino acids and carrying them to the ribosome site. There, the mRNA, the tRNA and the ribosome all work together to form the chain of protein. The ribosome is itself a tiny RNA/protein body that moves along the strand of mRNA checking the sequence of bases, selecting amino acids from the tRNA in the right order, and linking them together to form proteins.

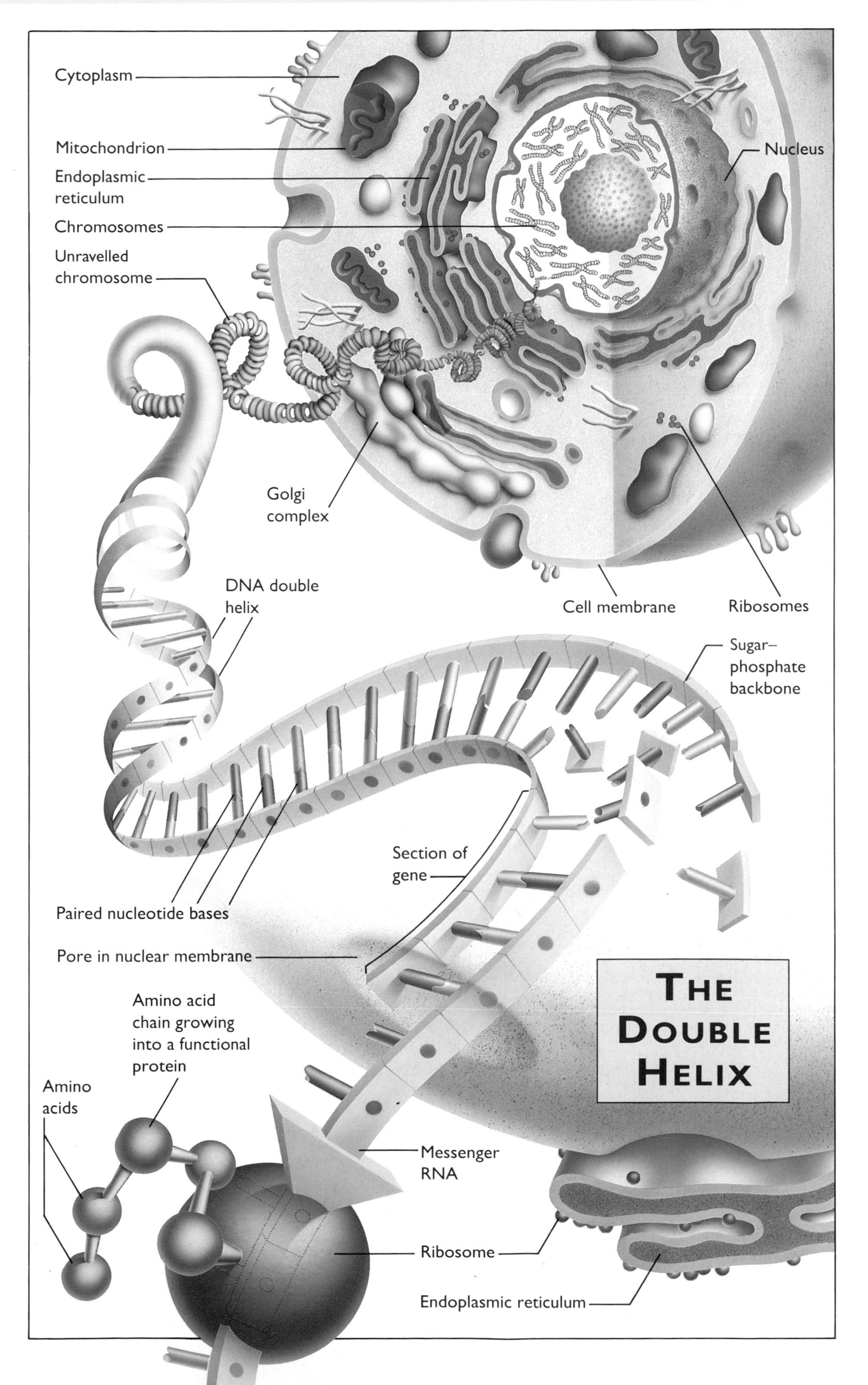

THE DOUBLE HELIX

The Transmission of Genetic Characteristics

Each human being is the unique outcome of an immensely complex inheritance acted upon by a no less complicated environment. Our inheritance comes to us, through our parents, from many generations of forebears, but the gene total possessed by each of us differs significantly from that of our parents and even more from that of earlier ancestors. Through the generations it has been progressively altered by mixing and by substitution. Most of these changes are harmless but a few are beneficial or damaging.

SEE ALSO

- HEREDITY pp. 20–31
- PASSING ON LIFE p. 96
- HEREDITARY DISEASES p. 148
- WHAT IS INTELLIGENCE? p. 168

Today we understand, as never before, the part played by genetics in determining our bodily characteristics and disorders. Most of what we consider normal qualities are in fact highly complex, involving the interplay of many genetic factors and environmental influences. Because of their complexity, these are difficult to study. To see genetics operating more obviously it is necessary to consider those conditions caused by abnormal single genes. In general, the effects of these are to cause bodily abnormalities.

Single-factor inheritance

Many disorders are due to defects in a single gene – an error in the genetic code sequence of bases in the DNA (➪ p. 26). Such disorders are inherited in a simple manner in accordance with the laws discovered by Mendel (➪ pp. 20–1). It is thus possible to predict with accuracy the risk of such a disorder recurring in a family in which it is known already to have occurred. If the gene concerned is situated on any but the two sex chromosomes (➪ p. 22), the effect it produces is called an *autosomal trait*. If, though, the gene is on one or other of the sex chromosomes, the effect is called a *sex-linked trait*. In either case, the trait may be dominant or recessive.

Dominant genes

An autosomal dominant trait is one in which the condition will appear even if the defect-producing gene is present at only one of the two corresponding positions (*loci*) on the autosomal chromosome pair, the other gene being normal. This is known as the *heterozygous* state and is what usually happens where defect-producing genes occur since most of these single-gene conditions are rare and the chances of both parents carrying the gene are small. A few autosomal dominant conditions are common, however, and some people suffering from them will have the gene at both loci. These people are said to be *homozygous*. The effect is, however, the same as if only one affected gene is present.

Most people with an autosomal dominant disorder will have an affected parent. If not, the condition is probably due to a new mutation that has altered one of the gene pairs in a sperm or egg cell of one of the parents. This is a fairly common way for such conditions to arise. If a person with a heterozygous autosomal dominant condition has a family with a normal person, there is a fifty-fifty chance with each new child that the condition will be inherited. This is because only one gene is needed to produce the disorder and, on average, half the sperms or ova of the affected person will contain that gene. If a person is homozygous for an autosomal dominant gene, all their offspring will suffer from the condition.

The extent to which autosomal dominant genes produce their effect can vary considerably. The resulting variation in the severity of the condition concerned is described as its *expressivity*. Similarly, the effect of some dominant genes can be modified by other genes and by environmental effects; the result is also a variability in the effect of the gene. This variability is called *penetrance*.

Diseases caused by dominant genes include Marfan's syndrome (where the body's structural collagen is weakened), polycystic kidney disease, Huntington's chorea (a rare brain disorder involving loss of nerve cells and progressive dementia), familial high cholesterol levels, achondroplasia (a form of dwarfism), brittle-bone disease, and hereditary spherocytosis (a form of anaemia involving abnormal red blood cells).

Deformed red blood cell (far left) in sickle-cell anaemia. Abnormal haemoglobin is produced, which is insoluble when the blood is deprived of oxygen. It then precipitates, forming elongated crystals that distort the cell. Sickle cells are rapidly removed from the circulation, leading to anaemia. To the right of the sickle cell is a normal, biconcave cell, an echinocyte (literally 'spiky cell'), displaying signs of damage, and a red cell beginning to sickle. (False-colour scanning electron micrograph, magnification x 1655 at 35mm across.) (Jackie Lewin, Royal Free Hospital, SPL)

Recessive genes

Autosomal recessive traits appear only if the gene concerned is present at both loci – the homozygous state. A single recessive mutant gene coupled with a normal gene – the heterozygous state – produces no effect, or, occasionally, a very minor effect. Many of the disorders caused by recessive genes involve the defective production of an enzyme needed for the normal functioning of one of the numerous biochemical processes of the body. Almost all human physiology functions under the influence of enzymes – proteins that act as catalysts to speed up chemical processes (➪ p. 26). The great majority of genes specify enzymes, and a defective gene will result in an enzyme that is partly or wholly ineffective.

In heterozygous people (heterozygotes), one of the pair of genes specifies the normal enzyme, so body cells produce about half the normal amount. Regulatory mechanisms, however, ensure that the biochemical processes operate adequately and it is rare for any discernible effect to occur. If both genes are defective (that is, the individual is homozygous for that defect), none of the normal enzyme is produced and the effects are often very serious, producing conditions such as cystic fibrosis (which produces disorders in the glandular tissue), albinism (absence of the pigment melanin ➪ p. 56), sickle-cell anaemia and beta-thalassaemia (disorders of the blood), various types of congenital blindness, deaf-mutism, and many other metabolic disorders due to defective or absent enzymes.

Recessive conditions are rare. Most of the offspring of people with autosomal recessive traits do not develop the condition, because such traits can be inherited only if the affected individual produces offspring with a person who either has the condition or is heterozygous for it. If two people with the condition pair up, all their children will have the condition, because all four genes concerned are affected and any combination will involve two affected genes. If a person who is homozygous for a recessive condition has children with a heterozygous person, there will be a fifty-fifty chance with each child that the condition will appear. If two heterozygous people have a family there will be a fifty-fifty chance, both for the sperm and for the egg, that it will carry the gene. As Mendel appreciated, a quarter of the children will, on average, inherit two normal genes and will be normal, a half will inherit one defective gene and will be normal, and a quarter will inherit two defective genes and will be affected by the condition (➪ p. 20).

Following on from all this, it is often because their parents are related that individuals have such conditions. This is because the parents have inherited the same defective gene from a common ancestor. First cousins, for instance, have a one-in-eight chance of being heterozygous for the same recessive gene. So if there is a family history of a recessive

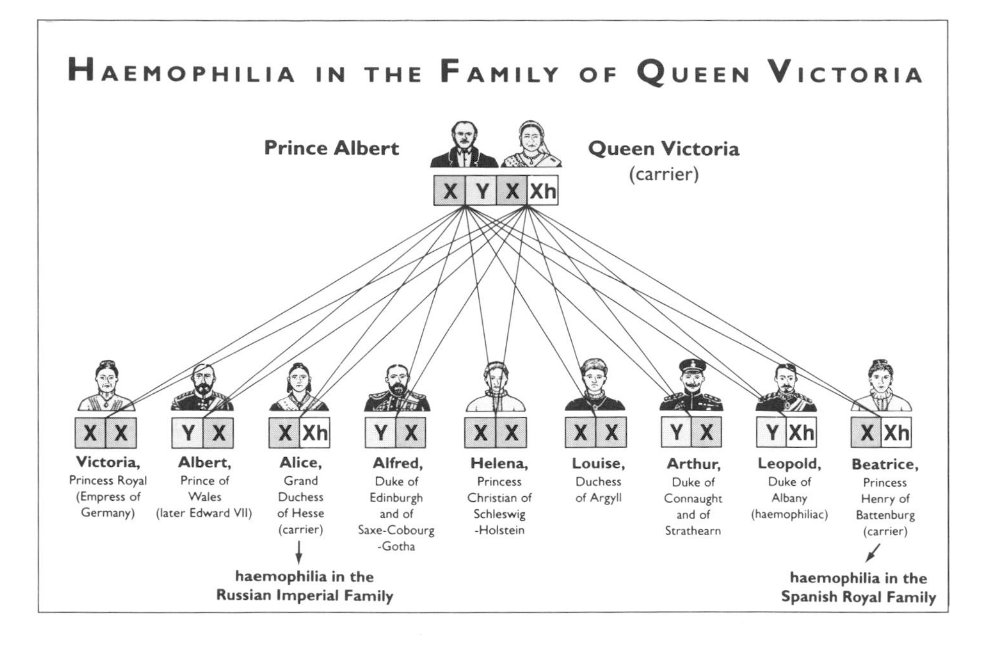

condition, producing children with a cousin greatly increases the risk of their developing the disease.

Sex-linked conditions

Sex-linked disorders are caused by genes situated on one of the sex chromosomes – the X or the Y chromosome (➪ p. 22). Genes on the X chromosome produce X-linked characteristics or disorders, while those on the Y chromosome produce Y-linked features. Since only males have Y chromosomes, Y-linked features occur only in males and all the sons of a man with a Y-linked condition would inherit it. The Y chromosome is, however, very small and there are no positively proved single-gene Y-linked disorders. Since all well-known sex-linked conditions arise from genes on the X chromosome, the terms 'sex-linked' and 'X-linked', in practice, mean the same thing.

X-linked conditions may be dominant or recessive. So far, only a few dominant X-linked conditions have been found. One is a form of rickets (a bone disorder caused by vitamin D deficiency) that resists the usual treatment with vitamin D. An affected male will have the gene on his solitary X chromosome and will therefore pass on the condition to all his daughters but to none of his sons (to whom he passes on only Y chromosomes). An affected female will have the gene on one of her two X chromosomes. Offspring of hers, of either sex, will therefore have a fifty-fifty chance of acquiring the trait.

X-linked recessive inheritance is especially interesting. Because women have two X chromosomes, they may carry either one defective gene (the heterozygous state) or two (homozygous). If they are homozygous, they themselves will suffer from the condition, but this is very rare, as it implies that the father had the condition and the mother also had at least one gene for it. So X-linked recessive conditions are almost always heterozygous in women, who remain unaffected but are carriers of the gene. Men, on the other hand, have only one X chromosome and show the full effects of any X-linked mutant gene – the Y chromosome, being different, has little effect. In short, X-linked recessive conditions predominantly affect males and are transmitted by healthy females, who are carriers of the gene, to their sons. An affected man cannot transmit the disease to his sons, because he passes on only Y chromosomes to sons. His X chromosomes, carrying the gene, are passed on only to daughters.

If a woman carrying an X-linked recessive gene for a disease on one of her X chromosomes produces offspring with a normal man, each child will have a fifty-fifty chance of acquiring the mutant gene. Thus, on average, half her sons will have the disease and half her daughters will be carriers of the gene. X-linked recessive diseases include haemophilia (a disorder of the blood's ability to clot), Duchenne muscular dystrophy, colour blindness, the testicular feminization syndrome, and Fabry's disease (a degenerative disorder of the brain).

Multifactorial inheritance

The majority of human characteristics, such as intelligence, height, weight and family resemblance, are the result of multifactorial inheritance (the effect of many different genes working together) and environmental influences. We cannot, therefore, apply the simple rules of Mendelian inheritance to them. A predisposition or liability to many common body disorders is also inherited in a similar way as a so-called *polygene trait*, that is, via several genes. Such conditions are not directly caused by the genes, but people who inherit them show a higher than average tendency to develop the conditions, which often 'run in the family'. Disorders occurring as a result of multifactorial causes include asthma, schizophrenia, cleft palate, high blood pressure, coronary artery disease, and duodenal and stomach ulcers.

Queen Victoria and her family, photographed in the late 1880s. (Popperfoto)

WHOLE CHROMOSOME DISORDERS

Some of the most serious genetic disorders occur as a result of gross abnormalities affecting the number or structure of whole chromosomes. Such disorders are much commoner than is generally supposed. About half of all fetuses that suffer spontaneous abortion (miscarriage) show abnormal chromosomes, and about one live-born baby in 200 has a chromosome abnormality. Many abnormalities of this kind occur at or around the time of fertilization. The absence of a whole autosomal chromosome in the fetus is invariably fatal.

The commonest whole-chromosome disorder is Down's syndrome. This is caused by an extra chromosome number 21 – a state known as trisomy-21. Trisomy-13 and trisomy-18 also occur, and these are even more severe in their effects and are usually fatal early in life. Gross abnormalities of the sex chromosomes are also common. Klinefelter's syndrome, which is caused by one or more extra X chromosomes in males, features sterility, a eunuch-like body with underdeveloped genitals and female breasts, and often mental underdevelopment. The XYY configuration, on the other hand, has effects ranging from complete normality to mental disability, possibly combined with criminal tendencies. Turner's syndrome occurs in females who have only one X chromosome; they are of short stature and do not develop normal secondary sexual characteristics (➪ p.108); they do not menstruate and are sterile. In addition, they often have other features such as webbing of the sides of the neck and various congenital abnormalities, such as eye and bone disorders.

Young girl with Down's syndrome. Caused by the presence of an extra chromosome number 21, the physical signs are slightly slanted eyes, a round head, flat nasal bridge, fissured tongue and generally short stature. People with Down's syndrome experience certain learning difficulties, but their problems need not be severe – many are able to hold down jobs and maintain a degree of social independence. (Hattie Young, SPL)

Nature Versus Nurture

The 'nature versus nurture' argument is of far more than purely academic importance. If the variations in the complex of human qualities generally understood by the term intelligence are entirely due to heredity, then many of us are condemned from birth to a lifetime of inferiority. But is this premise true?

The relative importance of genetic endowment ('nature') and environmental influences ('nurture') in determining the human characteristics we call intelligence, personality, abilities, academic performance and so on has been argued for centuries. In spite of advances both in genetics and in our understanding of psychology, those who would seem to be best qualified to pronounce on the matter still appear to be as divided in their opinions as ever.

Heredity versus environment

Historically, most people have plumped for heredity as the main determinant of human behaviour and have cited families in which succeeding generations have shown high levels of achievement in particular spheres. The English scientist Francis Galton wrote a book called *Hereditary Genius* in which he recorded numerous families of poets, scientists, judges, statesmen, literary figures, musicians, painters, and even oarsmen and wrestlers. He concluded that outstanding abilities of these kinds were transmitted in proportion to closeness of descent. Galton's work raised many challenging questions and greatly stimulated the nature–nurture argument. He examined, for instance, the way in which environmental influences might affect fertility and breeding patterns, such as the selection of mates.

Ignatius Loyola, founder of the Society of Jesus. His *Spiritual Exercises* (1548) remains the basic training manual for Jesuits. (ME)

Identical twins are ideal subjects for studies on the influence of both heredity and environment on the individual. Even though they share the same genetic make-up, and may have been reared in very close proximity, their separate environments will be subtly different, and consequent differences in their personalities will become apparent. (Frajnich)

The view that intelligence is mainly a matter of genetics has had support from some psychologists, but their work has not always been beyond reproach (➪ below). Opinions of this kind have often been formed with scant regard to the effects of the environment on the growing child. In the past a few discerning individuals – such as Ignatius Loyola (1491–1556), founder of the Jesuits – clearly understood the enormous effect environmental influences could have on the young. He famously said, 'Let me have the child until it is seven and I care not who has it after.'

In the early part of the 20th century scientists quickly ranged themselves on one side or the other. The eminent geneticist William Bateson (1861–1926) held that there were genetic grounds for people becoming artists, poets, scientists, actors, servants or farmers. On the other hand, John B. Watson (➪ pp. 162–3), the American behaviourist psychologist, took an extreme position. He claimed that any healthy baby, taken at random without regard to the parents' own abilities, could be trained to become a specialist of any type: a doctor, lawyer, artist – or even a genius.

INBREEDING

There would appear to be a conflict between the aims of positive eugenics and the almost universal prohibition of incest (the incest taboo). It was once widely held that human inbreeding inevitably gave rise to 'degenerative' characteristics. This view probably arose from false generalization from particular cases. Inbreeding will, of course, greatly increase the chances that offspring will be homozygous for recessive characteristics (➪ pp. 28–9), but to argue from this that *all* inbreeding will necessarily produce adverse results is fallacious.

Few social scientists, however, now believe that this idea was the origin of the incest taboo, and it is generally accepted that the real purpose of the prohibition of sex between closely related people was to avoid the often violent consequences of sexual rivalry and jealousy, with parent-child and brother-sister relationships becoming exploitative rather than protective.

Some societies have allowed exceptions to be made to the incest taboo for cultural reasons. Examples are the frequent marriages between brother and sister in ancient Egyptian royal families, among the Incas, and in historical Hawaiian society. In almost all modern societies, incest is proscribed by law, and most individuals react 'instinctively' against the idea. Canon law lays down detailed prescriptions of the permitted relationships, and these differ in extent in different legal systems.

Avoidance of incest is commonly associated with rules providing for marriage outside the family. This is a genetically healthy situation leading to the importation into families of new and useful genes as well as new and valuable cultural elements.

Twin and adoption studies

Since identical twins have identical genes, they offer ideal experimental subjects for studies into the nature–nurture problem. Such twins normally share the same environment (so far as different individuals can be said to do so), so the chief interest is in twins who have grown up separately. However, identical twins seldom are reared apart, and even if they are, the environments are usually not sufficiently different to make the observations worthwhile.

Studies have been made of children of financially deprived families who were adopted by well-to-do families soon after birth. When the adopted children were compared with their biological siblings who had not been adopted, they showed consistently higher IQ scores. The adopted children averaged an IQ of 111; their non-adopted siblings averaged 95. It can of course be argued, though, that variations of this kind might be due to factors such as differences in nutrition,

SEE ALSO
- THE STORY OF EVOLUTION p. 6
- NATURAL SELECTION p. 8
- HEREDITY pp. 20–9
- HEREDITARY DISEASES p. 148
- WHAT IS INTELLIGENCE? p. 168

EUGENICS

Nazi propaganda poster exhorting the public to support the party's relief work for 'Aryan' mothers and children. Symbolism glorifying the Nazi belief in the superiority of the 'master race' is evident. The sun behind the woman's head resembles a halo and casts her as a madonna figure. Hers and her baby's fair hair and skin epitomize the Nazi ideal of a race of genetically pure, strong and healthy individuals; and the rural scene in the background represents the idyllic future that National Socialism purported to offer – at the expense of those who were genetically 'inferior'. (AKG)

For centuries livestock farmers have known that they could alter the qualities of their animals by selective breeding. Inevitably, the question has arisen whether similar methods could, or should, be applied to humans. The ancient Greeks held that men distinguished in war or in other valued activities should be encouraged to breed, while socially undesirable people should not. Such deliberate control of human genetic traits is called *eugenics*, a term coined by scientist and explorer Francis Galton in 1883. Positive eugenics implies active encouragement of mating to promote supposedly desirable physical or mental qualities; negative eugenics implies its discouragement so as to eliminate those qualities that are considered to be undesirable.

Unfortunately, the need is first to identify good and bad qualities – on which there can be no firm consensus – and then to decide whether they are genetically determined. One putative example of positive eugenics is the selection of sperm donors for artificial insemination of women whose partners are sterile (⇨ pp. 98–9). Doctors who perform this procedure try to select donors, usually from the ranks of medical students, whose mental and physical abilities are well above average. Such donors are, of course, anonymous. The Repository for Germinal Choice, an institution founded in 1979, maintains stores of sperm from a number of Nobel Prize winners for the use of those women who desire it. Again, this idea is based on possibly defective premises. It seems unlikely that the qualities that lead to a person winning a Nobel prize are determined mainly by his or her chromosomes; environmental influences and chance will also play a significant part. Another attempt at positive eugenics is a measure adopted by the Singapore government in 1983, when it legislated for the provision of educational advantages for children born to educated women.

Early this century eugenicists were strongly advocating the sterilization of classes of people thought to be of defective stock – people with psychiatric disorders, epilepsy, mental disability, or a history of criminal activity or sexual 'deviation'. Several European countries and 27 states in America enacted laws providing for voluntary or compulsory sterilization of such people. The culmination of this sinister side of eugenics was, of course, Hitler's attempt to produce a 'master race' by encouraging mating between pure 'Aryans' and by the murder of millions of people whom he claimed to have inferior genes.

rather than to those in education, as there is some evidence that nutritional deprivation in children has an effect on their intelligence levels.

The compromise position

The strength of opposed opinion on the nature–nurture question has tended to deflect attention from the essential interdependence of the two major elements. Most authorities now agree that the truth of the matter must lie somewhere between the two extremes. Few genetic traits of importance can operate without some environmental contribution, and this contribution may be highly significant. Consider, as one example out of many that could be cited, a purely genetic inborn metabolic error in which a gene necessary for the normal breakdown of the amino acid phenylalanine is deficient. If a child with this error consumes a normal diet containing this amino acid, a toxic product will accumulate in the blood and the brain; the child will have the disease known as phenylketonuria and will suffer mental underdevelopment. But if the condition is detected early and the child given a phenylalanine-free diet, he or she will grow up normally.

A helpful, if limited, analogy is that of the personal computer. One might consider the role of heredity as providing the hardware and the role of the environment as providing the software. A person could inherit hardware of high or low quality, but the achievements realized would depend equally on the quality of the input (software).

THE BODY MACHINE

The Emergence of Medical Science

Until about the beginning of the 19th century very few people had any real understanding of how the human body works. Before then some very unscientific ideas prevailed, which were derived from philosophical theories rather than from observation and testing. Although these were often unsupported by evidence, they exercised considerable influence, both on medical practice and on the way people looked at themselves.

Every age has had its authorities and, throughout history, medicine has had more than its share. Until the dawn of rational enquiry, most medical 'experts' did more harm than good. The most notable of these were so powerful that to question their pronouncements was considered tantamount to heresy. As late as the 18th century, serious medical men who, in the interests of their patients, tried to substitute observation and reason for the arbitrary dogma of the contemporary Establishment, were hounded out of their appointments and professionally ruined. Probably the most successful medical pundit of all – so successful, indeed, that his influence held up the advance of medical knowledge for the next fifteen hundred years – was the Greek physician Galen.

Galen of Pergamum (Latin name *Claudius Galenus*), AD 130–201, physician, anatomist and physiologist. (AKG)

Galen and the four humours

Claudius Galen (c. AD 130–201) was born in Pergamum in Asia Minor (modern Turkey) and studied the works of Hippocrates and Aristotle at Corinth and Alexandria. He was surgeon to the gladiators at Pergamum and personal physician to Marcus Aurelius and other Roman emperors. Galen was a highly intelligent and successful man who based his system of body function on Aristotle's idea that everything in nature was designed to fulfil God's purpose. He taught that the body contained three separate 'spirits': a 'vital' spirit in the heart; a 'natural' spirit in the liver; and an 'animal' spirit in the brain. He was also much influenced by the extensive Egyptian and other texts on human dissection, subsequently lost when the great library of Alexandria was sacked and burned by the Arabs in the 7th century AD.

Galen's chief dogma was that the body, like all nature, was composed of four elements – earth, air, fire and water – and that everything that happened in the body, from illness to the manifestations of the personality, were governed by the relative proportions of the four humours derived from them – blood, phlegm, black bile and yellow bile (⇨ p. 116). These Galen elaborated into a vast and fantastic system that was to dominate medical thinking for centuries. An important consequence of the entirely imaginary humoral theory was the idea that bloodletting – cutting into a vein and allowing blood to flow out – was a valuable therapeutic measure for almost all diseases. This was one of Galen's central dogmas and was responsible for the deaths of countless millions of people weakened by blood loss right up until the 19th century.

Although Galen derived considerable anatomical information from his own dissections of dogs, pigs and goats, and was able to show by experiment that cutting the spinal cord caused paralysis, many of the ideas about the workings of the body that he promulgated in his voluminous writings were seriously mistaken. He taught, for instance, that blood originated in the liver and was consumed in the other organs. Then, he asserted, it passed through tiny pores in the wall between the two sides of the heart and was mixed with air on the left side. Such pores do not exist. He taught that the pulsation of the arteries served the same purpose as breathing, and he seemed to be confused as to whether or not arteries contained air – as the name (Latin: *arteria*, a 'windpipe') implies and the Greek philosophers believed. On the grounds that the action of the heart was involuntary, he denied, wrongly, that it was a muscle. He insisted, contrary to the visible evidence, that the heart lay in the exact centre of the body. Furthermore, he believed that the brain generated a 'vital spirit' that passed through hollow nerves to the muscles, which it then activated. This particular idea was still universally credited as late as the end of the 18th century. He taught that phlegm originated in the brain rather than in the linings of the nose and throat. In addition, he stated that cataracts were caused by a 'humour' from the brain that solidified behind the lens, and he insisted, against all evidence, that the lens occupied the centre of the eyeball.

Galen repeatedly claimed to respect, even revere, observed and demonstrable fact. But, in truth, his ideas were always dominated by his preconceived beliefs. His writings were so powerful, so explanatory, so plausible and, above all, so authoritative, that they remained the standard texts until the 16th century, when they were forced to give way under the logical assaults of the Flemish anatomist Andreas Vesalius and, later, the English physician William Harvey (⇨ below).

Andreas Vesalius and human anatomy

Andreas Vesalius (1514–64), the foremost 16th-century anatomist and professor at the University of Padua, did not, at first, dare to challenge the writings of the great authority. But when artists, such as Leonardo da Vinci, who sketched accurately from human dissections and had no interest in medical dogma, showed that Galen was frequently wrong, Vesalius took heart. In 1543 he published his masterpiece, *De Humani Corporis Fabrica*, a beautifully illustrated text on human anatomy, based on meticulous dissections and observations of dead bodies. This work directly confronted much of the error in Galen's teachings, and showed that many of Galen's ideas had been derived from observations on animals and had been applied, without reservation, to human beings.

Vesalius' book contained exact and carefully accurate descriptions of the visible anatomy of the whole body – the skeleton, muscles, heart and blood vessels, nervous system and organs. The book quickly became famous. Ambroise Paré (1510–90), the outstanding surgeon of the 16th century, incorporated Vesalius' work into the anatomical section of his own classic 1564 textbook on surgery, and translated much of the book into French, thus greatly extending its influence. By the end of the century, Vesalius had a monopoly on human anatomy, and his influence on other medical matters was also great. His book contained nothing that was not already accessible to any patient dissector, but it marked a radical change in method. From then on, speculative philosophy, dogmatic assertion and traditional authority had to give way to empirical observation.

Vesalius was succeeded at Padua and elsewhere by a number of distinguished men whose contributions to human anatomy and physiology remain essential to this day. Among these was Gabriello Fallopio (1523–62), who enlarged on Vesalius' work, and is remembered eponymously for

Harvey's Demonstration of the Blood's Circulation

The veins of the forearm can be made prominent by briefly tying a moderately tight band around the upper arm. The effect of the valves in these veins, in allowing the blood to flow only in the direction of the heart, can then readily be demonstrated. Harvey himself explained this very clearly. In his own words: 'But so that this truth may be more openly manifest, let an arm be ligated [tied] above the elbow in a living human subject as if for a blood-letting. At intervals there will appear, especially in country folk and those with varicosis, certain so to speak nodes and swellings (Fig. 1, B, C, D, E, F) not only where there is a point of division, but even where none such exists, and these nodes are produced by valves. If by milking the vein downwards with the thumb or a finger you try to draw the blood away from the node or valve (Fig. 2, O), you will see that none can follow your lead because of the complete obstacle provided by the valve; you will also see that the portion of the vein (Fig. 2, OH) between the swelling and the drawn-back finger has been blotted out, though the portion above the swelling or valve is fairly distended (Fig. 2, OG). If you keep the vein thus emptied and with your other hand exert a pressure downwards towards the distended upper part of the valves, you will see the blood completely resistant to being forcibly driven beyond the valve.'

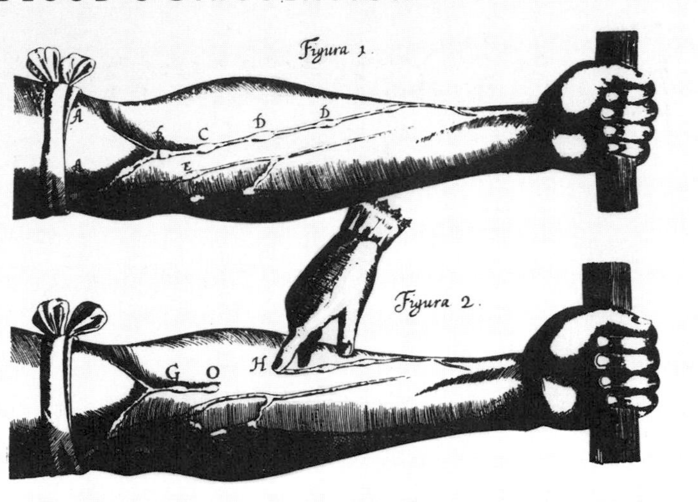

Engraving from Harvey's *De Motu Cordis* (1628), in which he demonstrated that valves in the arteries and veins created a one-way circulation. (LSI)

his description of what is now called the uterine or fallopian tube (⇨ pp. 96–7), and Hieronymus Fabricius (1537–1619), who published a detailed description of veins and their valves. Fabricius contented himself with pure description, drawing no conclusions from his findings, but his work was to have momentous consequences.

William Harvey and the circulation of the blood

The fame of the Paduan school (⇨ above) spread widely and attracted foreign scholars such as William Harvey (1578–1657), who was studying there when Fabricius produced his book on veins. Harvey returned to London where he soon became a successful medical practitioner, but never ceased to ponder on the anomalies between Galen's teaching and what he had learned at Padua. Finally, nearly thirty years later, and after much thought and experimentation based on close observation, he decided to publish his findings. In 1628, under the title *Exercitatio Anatomica de Motu Cordis et Sanguinis* (*Movement of the Heart and Blood: an Anatomical Essay*), he outlined for the first time, in terms that would brook no argument, the way in which the blood was driven, by the pumping action of the heart, through the arteries to all parts of the body, and was returned, by way of the veins, to the heart. The calculated volume moved in this way was such that Galen's idea – of blood being continuously manufactured in the liver from ingested food – was seen to be manifest nonsense.

One difficulty with Harvey's theory, however, was that there appeared to be no channel in which the blood could flow from the smallest visible branches of the arteries supplying organs to the smallest branches of the veins draining blood from them. Harvey therefore postulated large numbers of fine connections. Eventually, in 1661, he was shown to be correct when Marcello Malpighi (1628–94), using the recently developed microscope (⇨ pp. 128–9), demonstrated the system of capillaries that links up all the body's arteries and veins.

This idea of a continuous circulation was revolutionary and marked the beginning of modern physiology. Harvey saw clearly that the same blood was moved round the body continuously and that the nutrients in it were replenished as it passed through the intestines. Harvey's idea, soon generally accepted as obviously correct, was, in its way, as important as Darwin's theory of evolution.

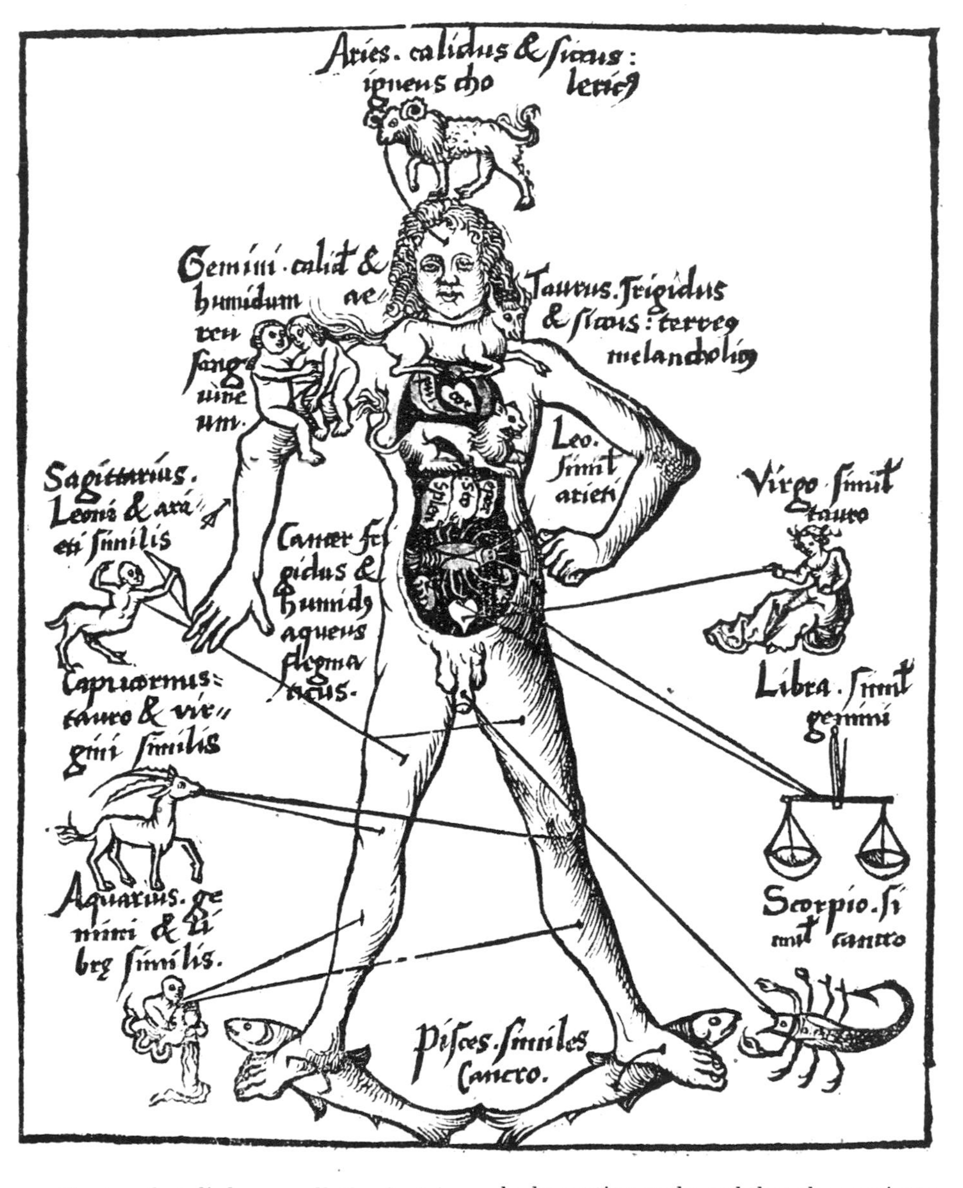

Blood-letting points, corresponding to the parts of the body associated with the signs of the zodiac and the four humours. From Gregor Reisch, *Margarita Philosophica*, 1508. (AR)

SEE ALSO

- BODY AND SOUL p. 70
- CHILDBIRTH IN HISTORY p.94
- CONCEPTS OF DISEASE p.116
- DISEASE IN HISTORY p. 118

What the Body is Made of

Chemistry is the branch of science concerned with the composition, properties and reactions of substances. Biochemistry is the chemistry of living organisms and is the real basis of an understanding of how the body works. It is a dynamic science, concerned with the thousands of chemical reactions that underlie everything that happens in the body.

The discovery in 1953 that DNA was a sequence of codes that selected amino acids (⇨ p. 26), and that the function of almost all genes was to produce enzymes (⇨ p. 76), put biochemistry firmly in the forefront of medical interest and research. Questions about what the body is made of can no longer be answered simply in terms of the chemical composition of the visible structure. Any reasonable answer must now also include reference to the thousands of enzymes present in every cell, busily engaged in accelerating the chemical reactions vital for life. The body is not a fixed physical structure; everything in it is constantly changing and interchanging. Calcium in a bone at breakfast time may be circulating in the bloodstream by mid-morning, and deposited in another bone or excreted in the urine before lunch time. Amino acids, derived from the breakdown of dietary protein, and carried in the blood through the liver soon after lunch, may be incorporated into a leg muscle before evening, and released, to be used as fuel, the next day. This dynamic action is performed under the influence of specific enzymes. All enzymes are proteins, so, to understand how the body is put together, it is necessary to have some idea of the structure of these important molecules and how they function. Also essential for the proper functioning of the human body, however, are fats, carbohydrates and trace elements (⇨ below).

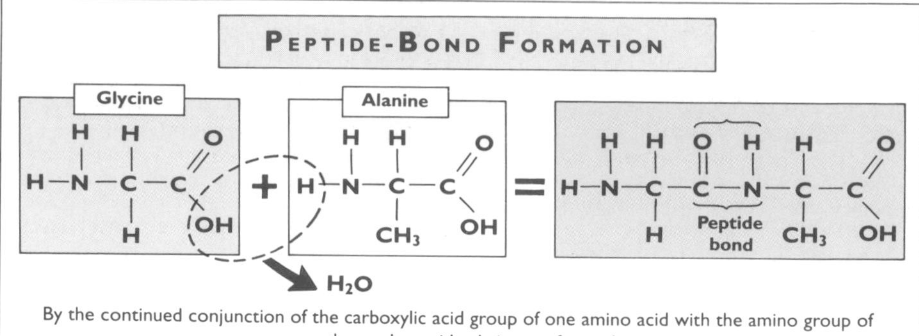

By the continued conjunction of the carboxylic acid group of one amino acid with the amino group of another, polypeptide chains are formed.

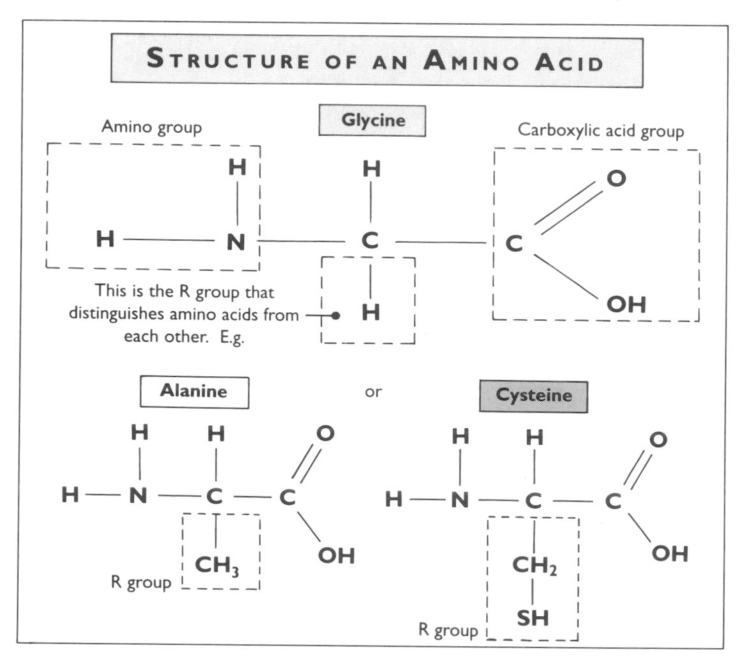

The central role of proteins

The whole structure of the body is based on protein. About half the dried weight of the average cells consists of it. This central importance is reflected in the derivation of the term: the Greek word *protos* means 'first' or 'earliest'. The bones are made of protein impregnated with minerals. The muscles are almost pure protein, as are the tendons and ligaments. The skin and its appendages – the hair and nails – are mainly composed of protein. All the connective tissue of the body, the structural material that forms the scaffolding of the organs, is also made of protein.

The blood contains a considerable quantity of protein. Some of this is nutritional; some maintains the necessary viscosity and water-retaining power (osmotic pressure) of the blood; some is needed for blood clotting; and some constitutes the large range of antibodies that provide us with protection against infection. The blood also contains large numbers of enzymes (⇨ below), all of which are proteins. Every cell in the body contains thousands of different enzymes and other proteins.

Proteins are large, complex molecules. They are made up of chains of up to 20 different amino acids linked together in various ways. Chemically, the amino acids are fairly simple. Each contains two functional groups: an amino group (one nitrogen atom linked to two hydrogen atoms – NH_2); and a carboxylic acid group (one carbon atom linked to two oxygen atoms and one hydrogen atom – COOH). Both of these groups are connected to the same central carbon atom. The amino acids differ from each other by virtue of another group – the R (radical) group – which is also connected to the same central carbon atom.

Protein molecules are built up by virtue of the ease with which the carboxyl group of one amino acid becomes attached to the amino group of another. When this happens, a molecule of water (H_2O) is eliminated. This is called a *peptide bond* and is one of the most important chemical linkages in the whole of science as it allows any amino acid to link with any other. Two or three amino acids linked together by peptide bonds constitute a dipeptide and a tripeptide respectively. Many amino acids linked together make up a polypeptide. When a polypeptide forms it has a free amino group at one end and a free carboxyl group at the other. One or more polypeptide chains constitute a protein molecule. The difference in the two ends of the polypeptide chain allows the chain to be incorporated in the protein molecule in a particular direction.

Protein molecules also have three-dimensional structures that determine their biological properties. These arise because the long polypeptide chain becomes spontaneously folded or twisted in a particular way as a result of the interaction between amino acids that may be

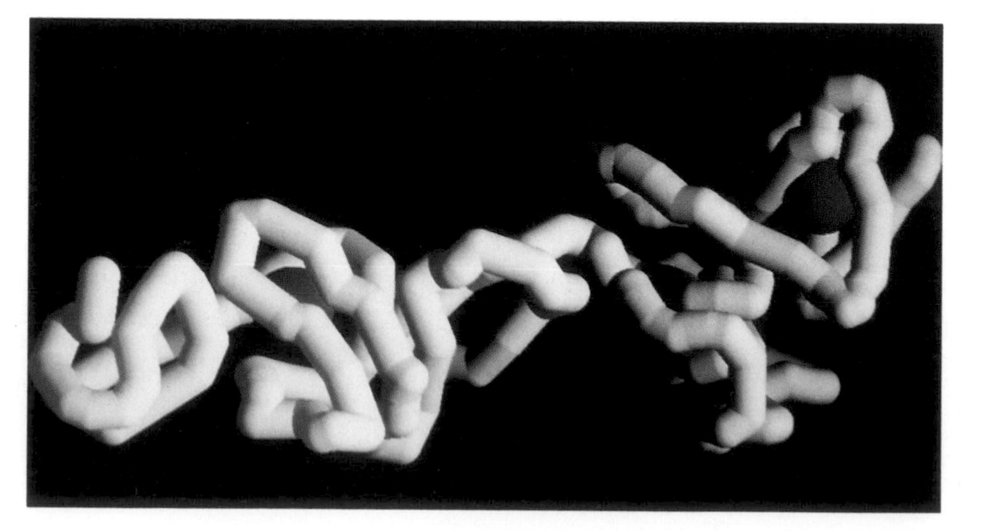

The most abundant protein in the circulation, Human Serum Albumin is active in maintaining the volume of blood plasma and the transport of a number of substances. This computer model demonstrates the protein's three-dimensional structure, knowledge of which will allow pharmaceutical chemists to design drugs to be carried by the protein more efficiently through the body. The red spheres represent the binding points of aspirin (right) and diazapine (left). (NASA, SPL)

The Chemical Elements of the Body

MAIN CONSTITUENTS

Hydrogen (63%)
Oxygen (26%)
Carbon (9%)
Nitrogen (1%)

These make up over 98% of the total body weight. Of this, hydrogen and oxygen, combined as water (H_2O), account for 60%.

MINERAL ELEMENTS

Calcium
Phosphorus
Potassium
Sulphur
Sodium
Chlorine
Magnesium
Iron

Together, these total only 0.7% of the total body weight.

TRACE ELEMENTS

Copper
Manganese
Iodine
Zinc
Selenium

quite distant from each other along the chain. Often the chain twists into a regular helix. Helical molecules may, in turn, twist into a tertiary, or even quaternary, configuration, mainly under the influence of the R groups.

Structural proteins, such as collagen, of which the skeleton, the connective tissue, and the keratin of the hair and nails are largely made, need be designed only for strength and stability. These are fibrous proteins, consisting of bundles of helical molecules. The keratin of hair and the myosin of muscles have a regularly repeating single helix. The strong collagen of bones and tendons is a triple helix. Proteins with a more active function, such as enzymes and antibodies, require a specific, three-dimensional shape to perform their actions. The remarkable specificity of enzymes is the result of the unique three-dimensional configuration of each – allowing the right enzyme to slot into the right place to promote a chemical reaction.

Disruption of the three-dimensional shape of a protein molecule occurs when it is heated or acted on by strong reagents. This is called 'denaturation'. A denatured protein can never recover its original shape or properties. So the transparency of egg white or of the crystalline lens of the eye cannot be restored if the former has been boiled or the latter is suffering from a cataract.

Enzymes and their functions

Enzymes are organic catalysts – molecules that can enormously accelerate chemical reactions without themselves being changed. Whereas the energy contained in fossil fuels can be released instantly at very high temperatures by burning, body fuels have, somehow, to be oxidized at low temperatures (about 37°C/98° F). This can be done only by enzyme action. Thousands of similar chemical reactions are going on all the time in our bodies. Among other things these reactions are concerned with the breakdown of food to a form that can be absorbed and utilized (digestion); with repair and replenishment of body tissues; with transport of vital materials such as oxygen and fuels throughout the body; with synthesis of new organic molecules and their storage and release; with body growth; with the control of gene action to generate new enzymes; and with the repair of damaged DNA. The list is endless.

Enzymes bind to the molecules whose action they control, causing the formation of weak bonds that impose strains on these molecules and allow chemical breaks or linkages that could not otherwise occur. Enzymes also position the substances about to react with each other in correct orientation so as to promote linkages.

Fats (lipids)

Enzymes would not be able to coordinate their activities were it not for the membranes that delineate a cell from its surroundings and divide a cell into functional compartments. Cell membranes also provide anchorage points for some important proteins. Fats, especially cholesterol, which is a steroid (➪ pp. 80–1), form the main elements of these vital and highly active biological membranes. Other important steroid fats include a number of hormones (➪ pp. 78–9), such as those produced by the adrenal glands and the sex glands.

Ideally, fats should not form important visible structural elements in the body, but they are important as energy stores. Fats are oily or semisolid substances, and are found, in variable quantity, in bag-like cells under the skin and within the abdomen. Most body fats are known as *triglycerides* and are made up of three fatty acids linked to a backbone of *glycerol* (glycerine). These fatty acids may be the same or a mixture of different kinds and may be *saturated* or *unsaturated*. A saturated fatty acid has only single bonds between carbon atoms; an unsaturated fatty acid has double bonds between some of the carbon atoms. The degree of saturation of a fat determines the melting point so that many unsaturated fats are liquid at room temperature while saturated fats, in contrast, tend to be solid at such temperatures.

Carbohydrates

Carbohydrates, as the name implies, are made from carbon and the constituents of water – hydrogen and oxygen. Carbohydrates are the main structural elements in plants, but in humans their greatest importance is as a source of energy – a fuel. All sugars are carbohydrates. Carbohydrates do have minor, but important, structural functions in the body, especially the deoxyribose sugar molecules that link with phosphates to form the 'backbones' of the double helix of DNA. They also link with protein to form substances, such as *mucopolysaccharides*, which help form the structure of cartilage and occur elsewhere in the body.

The principal fuel of the body – glucose – is a carbohydrate, derived mainly from food. Glucose is a simple sugar, or monosaccharide, and can be derived from the breakdown, by digestion, of more complex carbohydrates such as the disaccharide table sugar (sucrose) or the polysaccharide starch. Polymerization (chain formation) of simple sugars to form polysaccharides is analogous to the linkage of amino acids to form polypeptides and proteins (➪ above).

Glucose is stored in the liver in the compact form of the polysaccharide *glycogen*. This consists of a long, branching chain of linked glucose molecules. When glucose is needed for fuel, glycogen is broken down by enzyme action to release glucose molecules into the blood.

How Enzymes Work

The catalytic action of enzymes is three-fold. An enzyme may bring about just a small modification to the structure of the *substrate* (body chemical) upon which it is acting; it may split a substrate; or it may join two together.

Substrate
Enzyme

The specific shape of an enzyme means that it will combine only with a particular substrate of complementary dimensions.

Combined substrate and enzyme

The substrate combines with the enzyme at its active site and a chemical change is brought about in the substrate. Here it splits into two. The enzyme itself is unchanged by this process and can go on to combine with another suitable substrate.

Products
Enzyme

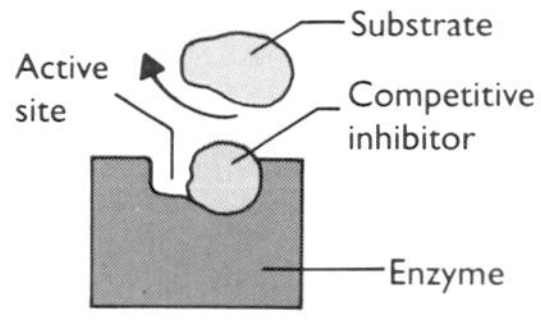

Enzymic action can be blocked in two ways. A competitive inhibitor (a substance very similar to the substrate) will block the possibility of enzyme and substrate combining.

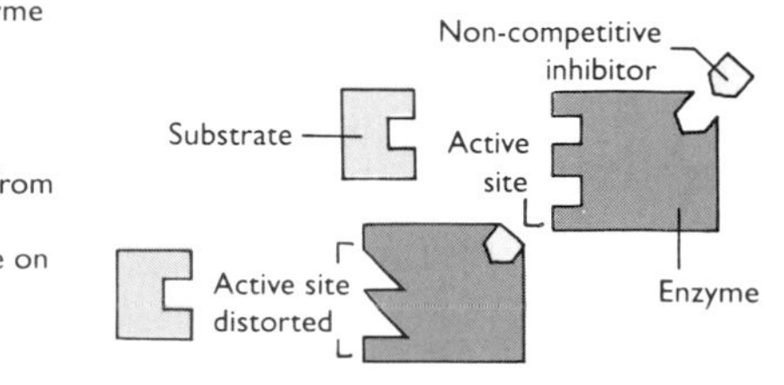

A non-competitive inhibitor, which may be quite different from the substrate, will distort the active site by acting elsewhere on the enzyme.

SEE ALSO

- CHROMOSOMES AND CELL DIVISION p. 22
- DNA, THE BLUEPRINT OF LIFE p. 24
- THE GENETIC CODE p. 26
- CELLS, THE STUFF OF LIFE p. 38

DNA and Amino Acids

DNA consists of a double-helical backbone of a polymerized sugar-phosphate compound, linked by pairs of bases (➪ pp. 24–5). These bases are: adenine, thymine, guanine and cytosine. Adenine always links to thymine, and guanine always links to cytosine. The order of these base pairs along the molecule is the genetic code. In RNA (➪ pp. 26–7), thymine is replaced by a different base, uracil.

The 20 amino acids from which all body proteins are constructed are:

Alanine	**Proline**	**Glutamine**	**Arginine**
Isoleucine	**Tryptophan**	**Glycine**	**Histidine**
Leucine	**Valine**	**Serine**	**Lysine**
Methionine	**Asparagine**	**Threonine**	**Aspartic acid**
Phenylalanine	**Cysteine**	**Tyrosine**	**Glutamic acid**

Cells, the Stuff of Life

The word 'cell' comes from the Latin *cella*, meaning 'store' or 'larder'. *Cellula* means 'little cell', and gives rise to the word 'cellulose' – the main structural material of plants. The word 'cell' was first used in a biological context in the 17th century but the full implications of cell studies were not appreciated until the middle of the 20th century with the development of the science of molecular cell biology.

It was the 17th-century English physicist and microscopist Robert Hooke (1635–1703) (⇨ box) who first used the term in this sense. Although he did not recognize the importance of his discovery, the name caught on and subsequent research by such men as the Scottish botanist Robert Brown (1774–1858), the German botanist Matthias Schleiden (1804–81) and the German physiologist Theodor Schwann (1810–82) established that all living things, whether plant or animal, were made of cells or the products of cells.

Every human being starts life as a single cell – the fertilized ovum (or zygote; ⇨ p. 96). This cell divides into two; each of these divides into two, and so on. If this were all that happened, however, the result would simply be a large mass of identical cells. Early in development, cells begin to specialize, acquiring different functions and properties (⇨ p. 102). In this way, four broad categories of cell are formed. These are connective tissue cells, surface-covering (*epithelial*) cells, muscle cells and nerve cells. Within these categories are some 200 distinguishable types of cell, all with different properties. Cells join together to form tissues, and different tissues join to form organs. Organs, in turn, join up to form systems, and the sum of these is the human body.

The latter part of the 20th century has seen an explosive growth of knowledge about cell structure and function – a growth that has revolutionized biology and medicine, and will have an incalculable effect on the future of humanity. Molecular cell biology has become the most active of the sciences, and now occupies a much larger proportion of the space of general scientific journals than any other discipline. An understanding of the cell is the beginning of an understanding of life.

Rough endoplasmic reticulum, coloured orange in this electron micrograph. Clearly visible attached to its walls are the ribosomes, which appear as black discs. (RB)

What are cells?

Although cells are immensely complex entities, engaged in constant physical and biochemical activity, each one represents the simplest structural and functional unit from which an organism, such as the human being, can be composed. Cells vary greatly in size, from a thousandth of a millimetre across or smaller, in the case of small microorganisms, to several centimetres in the case of large avian and reptilian eggs. The cells that make up the human body, however, lie in the range between about one hundredth of a millimetre and about a tenth of a millimetre. The largest human cell is the female's ovum, which is about a tenth of a millimetre in diameter. So almost all of them are of microscopic dimensions.

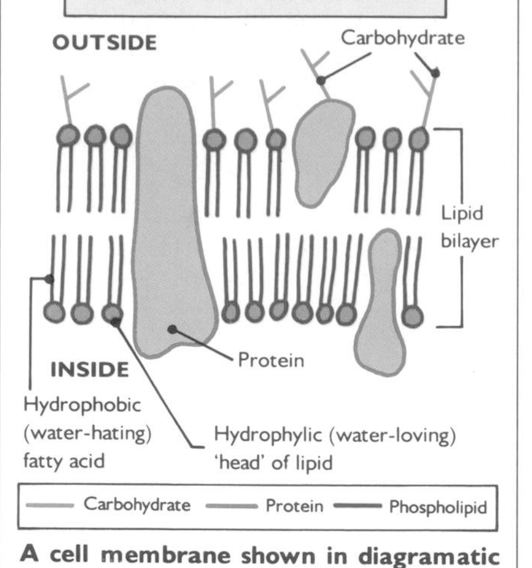

A cell membrane shown in diagramatic cross-section. A membrane is a highly dynamic structure composed of two layers of phospholipids arranged with their hydrophobic (water-hating) 'tails' inwards and their hydrophilic (water-loving) 'heads' facing outwards. Protein molecules are interposed in lipid layers, some being in either the outer or inner lipid layer whilst others traverse the whole membrane. These molecules are mobile and can be moved within the membrane. Carbohydrate molecules are attached to the outer surface of the membrane and are involved with molecular recognition. This total structure is called a unit membrane and is essentially typical in its general form for all membranes in living organisms.

The outer cell wall – the cell membrane – is far more than simply a bag for the cell contents. It is a complex fat and protein structure that contains many specialized sites for the receipt of information from the external environment and from adjacent cells (receptor sites), and sites at which chemical substances of many kinds can be pumped into and out of the cell.

When a dead tissue is cut into a thin slice that is then mounted onto a microscope slide and stained, the part of the cell that appears central and densely stained is called the *nucleus*. The term comes from the Latin *nux*, meaning 'nut'. The nucleus contains the chromosomes – a metre-long delicate strand of DNA that contains the unique sequence of bases that specify the proteins that control our physical construction (⇨ pp. 24–5). The nucleus is enclosed in a nuclear envelope that contains pores through which information from the DNA can be carried by messenger RNA (⇨ pp. 26–7). Surrounding the nucleus, within the cell, is the *cytoplasm*. This contains many important structures suspended in a fluid called the *cytosol*. These structures are called cell *organelles* (miniature organs).

The cell organelles

The little organs of the cell are all essential for life, and each kind serves a separate purpose. They were discovered and elucidated in various ways – by the use of electron microscopy, by separation by powerful gravitational forces using ultracentrifuges, by chemical analysis and by biochemical experiments to determine their function.

Mitochondria are tiny oval or rod-shaped bags, each with a smooth outer membrane and a much-folded inner membrane. Their name comes from the Greek *mitos*, 'thread', and *kondrion*, 'small grain'. The cell cytoplasm contains hundreds of mitochondria, the number depending on the amount of energy required by the cell. Highly active cells, such as those in the liver, may contain as many as 1000. Mitochondria contain enzymes that accelerate the chemical processes by which energy is released through the combination of glucose and oxygen, with the production of carbon dioxide. This process is mediated by the chemical substance *adenosine triphosphate* (ATP) (⇨ pp. 46–7). Mitochondria also contain a circular genome of DNA (that is, their own complete set of genetic material) whose sequence was ascertained in the early 1990s. Mitochondrial DNA contains genes connected with muscle tissue, and defects in these genes can cause certain forms of muscular dystrophy. This genome is unique in that it is inherited from the mother only. This fact has been of great interest to archaeologists and evolutionists (⇨ pp. 10–11).

The *endoplasmic reticulum* is the largest organelle, and extends throughout most of the cell cytoplasm. It takes two forms – a series of flattened sacs covered with tiny granules, and a network of smooth, non-granular, tubules of different diameter. These are called, respectively, the rough (granular) and the smooth (agranular) reticula. So far as can be judged from electron-microscopic studies, the interiors of the two types are continuous with each other. The granules on the rough reticulum are much more important than their tiny size might suggest; these are the vital ribosomes (⇨ below). The smooth reticulum has a number of functions, the most important being the synthesis of the various fatty acids and phospholipids needed to form the cell membrane. In liver cells the smooth reticulum contains many enzymes that detoxify dangerous chemicals, converting them into safer conjugated (chemically linked) and soluble forms that can easily be excreted from the body.

As well as studding the endoplasmic reticulum, the *ribosomes* also exist as free bodies in the cell cytoplasm. Ribosomes are protein bodies, and they contain nucleic acids (⇨ pp. 22–3) It is in the ribosomes that amino acids, free in solution in the cytosol, are selected in the right order, in accordance with the code carried by messenger RNA from the DNA, and are strung together to form new proteins, mostly enzymes. If the new proteins have been synthesized in the ribosomes attached to the endoplasmic reticulum, they pass into the interior of the reticulum and are delivered to another group of organelles known as the *Golgi apparatus*, named after the Italian histologist (microscopic anatomist) Camillo Golgi (1843–1926). Those formed by the free ribosomes are released into the cytosol, ready to act immediately.

Several minutes after they are synthesized, proteins formed in the rough endoplasmic reticulum are transferred into the Golgi apparatus. This is situated near the cell nucleus, and consists of a series of flattened, membranous sacs surrounded by a number of spherical bubbles, or vesicles. These vesicles form initially on the surface of the rough reticulum in areas not coated with ribosomes. Proteins within the rough endoplasmic reticulum pass into these vesicles, which then travel though the cytosol and fuse on to the surface of the Golgi complex, transferring their contents into the Golgi sacs. Secondary 'transfer' vesicles are now formed on the surface of the Golgi sacs, and these are 'tagged' by the addition of a carbohydrate or phosphate group to indicate where they should go. Golgi vesicles have been called the 'traffic police' of the cell as they play a key role in directing the many proteins that are formed within the cell to their required destination.

The *lysosomes* are tiny oval or spherical bodies surrounded by a single membrane that encloses digestive enzymes and acids. A typical cell may contain several hundred lysosomes. Their function is to scavenge and remove cell debris; this includes pieces of DNA and RNA, and organelles that are damaged and no longer functioning. They do this by engulfing the material, breaking it down by means of digestive enzymes, which work only in the highly acid environment of the inside of the lysosomes, and carrying it to the outer cell membrane where it is ejected from the cell. Some cells, for example the *macrophages*, are equipped to take in large particles of unwanted material such as bacteria and foreign (nonbody) protein. They do this by forming vesicles which then fuse with the lysosomes and transfer their contents into the lysosomes, where they are digested and destroyed.

Most cells contain a protein filamentous structure, known as the *cytoskeleton*, that maintains the cell shape and allows cell movement, as in the case of the amoeboid immune system cells or muscle cells. The cell filaments are of different diameters and are solid or tubular. All cells contain fine filaments made of the protein *actin*, which has the power to contract. The smallest filaments – the actin microfilaments – can be assembled and taken down rapidly, allowing the cell to change shape when necessary. The larger filaments are, however, more permanent. Muscle cells are largely composed of the contractile protein myosin.

Inside The Human Cell

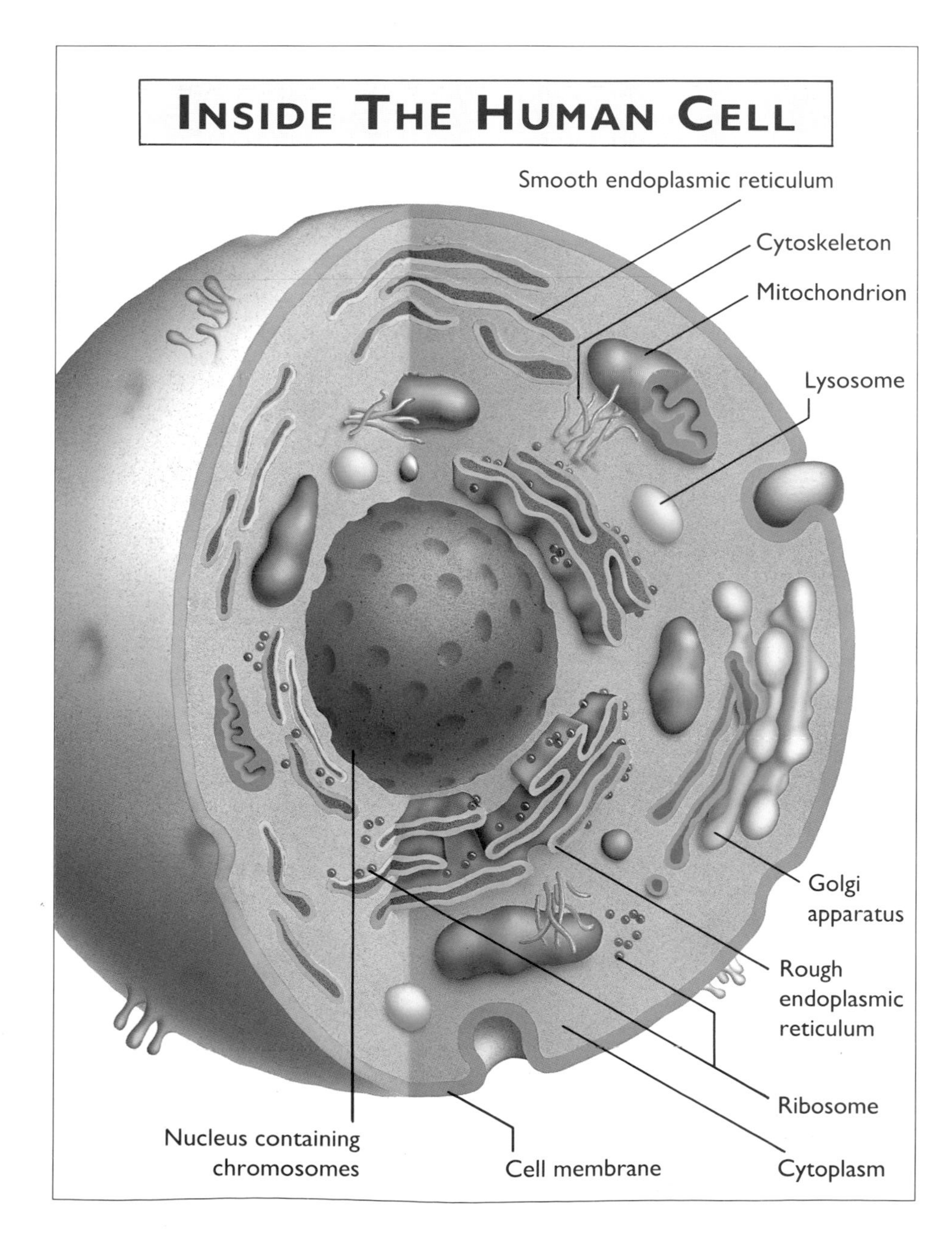

Robert Hooke and the First Description of Cells

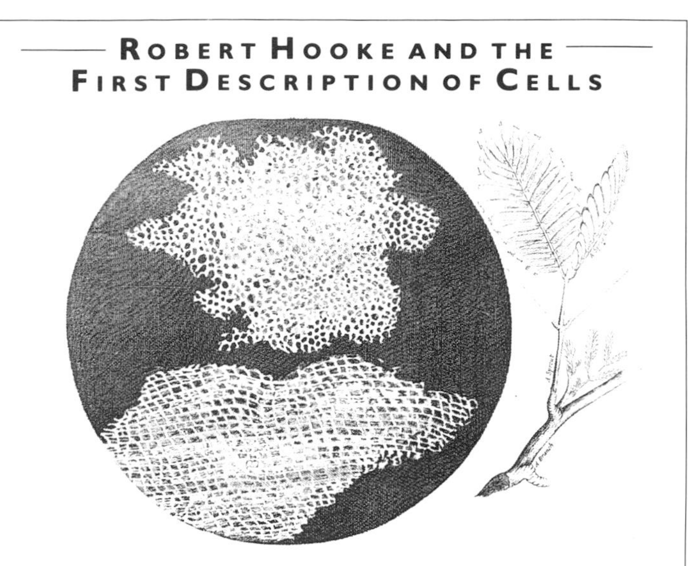

The cellular structure of cork and a sprig of sensitive plant, from the genus *Mimosa*. From Robert Hooke's *Micrographia*, 1665. (AR)

Hooke was born on the Isle of Wight in 1635. The son of a poor clergyman, he had to work as a waiter to pay his fees at Oxford University. There he met the great chemist and physicist Robert Boyle (1627–91), who at once recognized and encouraged his mechanical genius. Hooke became interested in every aspect of science, and conducted many important and brilliant studies in physics (especially light and gravitation), astronomy, mechanical engineering, geology and biology. He invented the hair spring that made watches possible. In 1663 he was elected a Fellow of the Royal Society and became Secretary in 1677. Unfortunately, he was a quarrelsome, jealous, mean-minded and censorious man whose greatest satisfaction was to prove others wrong. His position in the Royal Society provided him with plenty of scope for his malice, and he became notorious for his sustained attacks on other scientists. His unremitting persecution of Isaac Newton, for instance, may well have contributed to the mental breakdown of the great mathematician and physicist.

The microscope, made famous by the Dutch biologist Anton van Leeuwenhoek (1632–1723; ⇨ p. 94), was further developed by Hooke, who produced a compound instrument and used it to make many observations of nature. These, he published in his book *Micrographia* – a collection of exceptionally beautiful drawings with a text in English. Among many other things, he described the microscopic appearances of thin slices of cork that showed regular rows of geometric spaces. These spaces Hooke called 'cells'. Later studies showed that these were actually empty spaces – the dead remnants of living cells – but the observation was of fundamental importance and led others to look for, and find, a cellular structure in all living organisms.

SEE ALSO

- HUMAN EVOLUTION p. 10
- CHROMOSOMES AND CELL DIVISION p. 22
- DNA, THE BLUEPRINT OF LIFE p. 24
- THE GENETIC CODE p. 26

Movement 1: the Skeleton

The reason why bones appear so different from the other body tissues is that the protein (collagen) framework of which they are made is densely impregnated with mineral salts, mainly complex phosphates of calcium. In health, these salts account for about 60% of the weight of the bone. If the minerals are removed from bones they retain their general shape but are as flexible as rubber.

The collagen scaffolding (matrix) in a bone is laid down in well-organized strands to form struts and girders, disposed in such a way as to withstand the normal stresses of standing and walking. Indeed, if an engineer were to design a bone, it would have to be made very much as evolution has accomplished it. The strength of bones depends as much on the collagen matrix as on the mineral salts. Unfortunately, as the body gets older there is a progressive loss of both collagen and minerals.

The skeleton of the living body bears little resemblance to the dried bones of a skeleton in a museum. Living bone is in a dynamic stage of flux, there being a constant interchange of its constituent materials, such as calcium, phosphates, amino acids and fatty acids. There is also a constant consumption of glucose, oxygen and other materials. Calcium has many bodily functions in addition to the mineralization of bone, and there is a constant interchange of calcium between the bones and the blood. This is strictly regulated by hormones (➪ p. 78), and by the action of the kidneys (➪ p. 64). Maintaining the correct blood calcium levels is vital to life and is, in fact, more important than calcification (deposition of calcium salts) of the bones. If there is a shortage of calcium it is always the bones that are the first to suffer. Inadequate calcium leads to softening and distortion of the bones – rickets in children and osteomalacia in adults (➪ pp. 126–7).

SEE ALSO

- WHAT THE BODY IS MADE OF p. 36
- CELLS, THE STUFF OF LIFE p. 38
- MOVEMENT 2: MUSCLES p. 42
- HEALING AND REPAIR p. 62

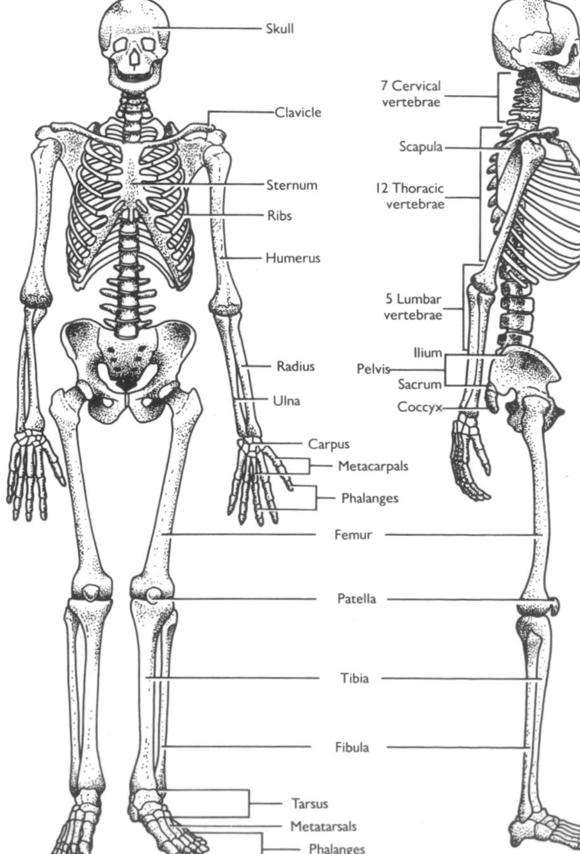

Osteoblasts and osteoclasts

Embedded in the collagen matrix of bones are cells called *osteoblasts*. Their purpose is to manufacture collagen to maintain and repair bone structure. The osteoblasts are especially active during the period of body growth in childhood and adolescence (➪ pp. 106–9) and after bone fractures. Throughout life, their activity is also greatly influenced by sex hormones (➪ pp. 108–9), both male and female, which act on them to promote collagen formation and calcification. This is why women suffer more than men from osteoporosis – a general loss of bone bulk and strength – since women lose hormones after the menopause while men continue to secrete them throughout their lives (➪ pp. 110–11). Another important stimulus to osteoblast activity and resultant bone bulk is the physical loading of the bones – especially the weightbearing involved in standing and walking. Astronauts living in conditions of zero gravity quickly lose bone mass, as do people lying in bed. Weight-bearing exercise in youth promotes strength and thickness in the bones, and such exercise throughout life minimizes the natural rate of decline in bone bulk. Men, on average, naturally have heavier bones than women, and are less likely to suffer in old age from dangerous weakening of their bones.

Bone growth, remodelling and fracture repair involve reabsorption as well as construction, which are performed by a different group of cells known as *osteoclasts*. These secrete digestive enzymes that break down collagen polymers to

Sheets of compact bone in the femur, magnified 1200 times in this false-colour scanning electron micrograph. During bone formation, osteoblasts lay down a bone matrix in a lamellar (sheet-like) form, which is subsequently mineralized by the formation of calcium phosphate crystals. (Prof. P. Motta, Dept. of Anatomy, University 'La Sapienza', Rome, SPL)

release amino acids (➪ pp. 36–7) and calcium salts. Osteoclasts are especially active after fractures, removing splinters of bone, cleaning up the break, preparing it for osteoblastic new bone formation, and even, over the years, gradually remodelling and realigning bones that have healed askew after fractures.

Bone marrow

In addition to its structural function and its function as a calcium depot, bone is also the place where red and white blood cells are produced. This occurs in the marrow – a soft tissue, well suppliedwith blood vessels, that occupies the space within bones. The blood cells develop from primitive *stem* cells (cells from which more mature cells form) in the red marrow of the flat bones – the ribs, breastbone (sternum), shoulder blades (scapulae), pelvic bones and skull – and in the bodies of the bones of the spine (vertebrae; ➪ below). This process of blood-cell production is continuous, and is necessary to make up for the steady losses of cells in the circulation. (Red cells live for only 120 days.) The bone marrow therefore has a rich blood supply into which large quantities of red cells, white cells and platelets (➪ pp. 50–1) are poured. The marrow of the long bones does not normally produce blood cells (but may do so under certain conditions, such as leukaemia) and is filled mainly with yellow fat-storage cells.

The skeleton as a body framework

The general shape of the body is conferred by the 206 bones of the skeleton, which

Cartilage viewed in polarized light. (RB)

provide attachments for the muscles, and support and protection for the internal organs (⇨ illustration). The central axis of the skeleton consists of the skull and the spine (vertebral column). This is a curved column of individual bones, called vertebrae, all of the same general shape but enlarging progressively in size and altering in proportion, from the top to the bottom. Each vertebra has a stout, roughly circular body in front and an arch behind that encloses a wide opening. The bones are secured neatly together by fibro-cartilaginous intervertebral discs and longitudinal ligaments, and the sequence of arches forms a long, flexible tube in which lies the spinal cord (⇨ pp. 74–5). There are 7 vertebrae in the neck (cervical vertebrae), 12 in the chest (thoracic or dorsal vertebrae) and 5 in the lumbar region. The bottom end of the vertebral column is firmly attached to the central bone of the pelvis – the sacrum. The ring-shaped pelvis has a deep hollow, the acetabulum, on either side, into which articulates the almost spherical head of the thigh bone (femur). The acetabulum and the head of the femur together form the ball-and-socket hip joint.

The chest vertebrae provide attachment for the 12 pairs of ribs. The front ends of most of these connect, by flexible cartilages, either to the breastbone or to cartilages attached to it. This structure is called the thoracic cage. Lying almost free over the back of the upper ribs are the two flat shoulder blades (scapulae). Running outwards and upwards from the upper corners of the sternum lie the two collar bones (clavicles), the outer ends of which are attached by ligaments to bony projections on each scapula. The scapulae and clavicles together form the shoulder girdle. The head of the upper arm bone (humerus) moves in a shallow hollow on the outer surface of the shoulder blade.

There is an analogy between the bones of the arm and of the leg, in that the humerus of the upper arm corresponds to the femur of the thigh. The radius and ulna of the forearm correspond to the tibia and fibula of the lower leg. The 8 carpal bones of the wrist correspond to the 7 tarsal bones of the foot. The 5 metacarpals of the hand correspond to the 5 metatarsals of the foot and the 14 phalanges of the fingers correspond to the 14 phalanges of the toes.

How Bones Grow

During childhood and adolescence all the bones of the body increase greatly in size. Individuals will reach their genetically determined height and size as long as adequate quantities of building materials – amino acids, fatty acids, vitamins and minerals – are provided by their diet. Body height increases mainly by growth of the long limb bones – the femur, tibia and fibula in the leg. This occurs by virtue of special growth plates of cartilage (gristle), called the *epiphyseal plates*, which are situated near both ends of the bones. Throughout the whole growth period, the cartilage is the site of a cooperative osteoclast and osteoblast activity. In each epiphyseal plate, the edge of the cartilage nearer the centre of the bone is gradually converted to bone while, at the same time, new cartilage grows outward at the edge further from the centre. In this way the bone length progressively increases. Inadequate nutrition will stunt growth; excessive nutritional intake cannot, however, increase height – only width. Once body growth is complete – at the age of about 25 – the epiphyseal plates are converted wholly to bone and no further growth of the long bones is possible.

The activity of the cells in the epiphyses is under the control of the growth hormone *somatotrophin*, produced by the pituitary gland (⇨ p. 78). The amount secreted determines the extent of growth and the ultimate height of the individual. If, for any reason, the hormone is not produced during the growth period, the individual will suffer from dwarfism; if excess hormone is produced during childhood or adolescence, however, the individual becomes a giant. Absence of somatotropin is rare, and can be remedied artificially by injections of growth hormone. Excess growth hormone is usually the consequence of a tumour of the hormone-secreting cells of the pituitary gland. The production of excess hormone after the epiphyses have fused causes the condition of *acromegaly* – a disorder in which those bones not formed from epiphyseal plates, such as the jaw, skull, spine and hand and foot bones, continue to enlarge until the cause is removed.

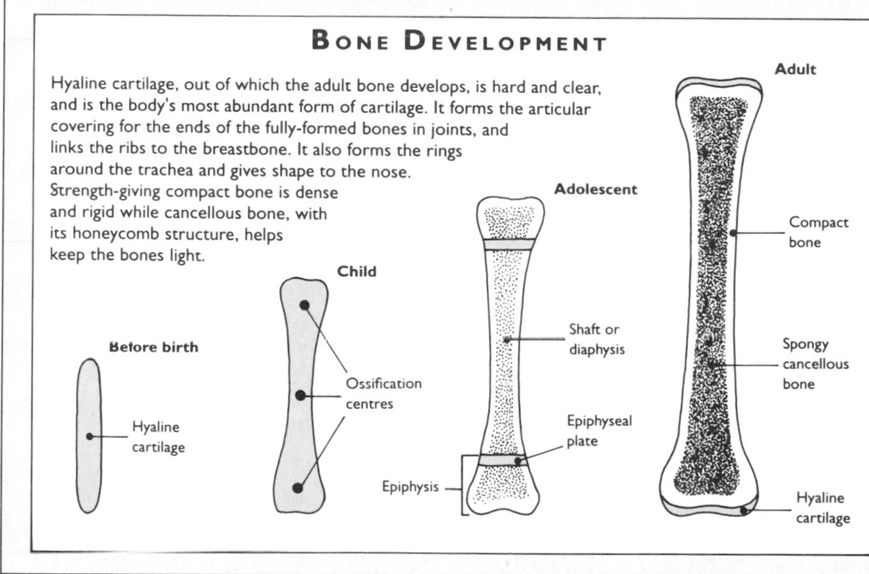

The skull

The primary purpose of the skull is to accommodate and protect the most important part of the body – the brain – and to provide support and protection for the main sensory organs – the eyes, ears and nose. The floor of the interior of the skull is moulded to the exact shape of the surface of the brain, and is provided with suitably placed holes (*foramina*) through which the nerves emerging directly from the brain pass. The most conspicuous of these holes is the *foramen magnum* in the centre of the floor. The spinal cord runs through this to enter the longitudinal channel in the vertebral column. The front part of the skull – the facial skeleton – contains the eye sockets (the *orbits*), and the double-sided air passage for the nose. It also contains four pairs of bony *sinuses*. These are hollows in the bone that lighten the skull and provide resonance to the voice. The lower jaw, or *mandible*, articulates with the skull high up on each side, just in front of the ear openings. The vault of the skull is made up of six double-layered, thin, curved bones that are fixed together by complex, jigsaw-like joints (*sutures*).

The joints

Joints are the junctions between bones, and may be movable or not. There are three kinds – *fibrous*, *cartilaginous*, and *synovial*. Fibrous joints occur where bones are held firmly together with ligaments, and allow little or no movement. Such joints occur in the pelvis and the vertebral column. Cartilaginous joints are somewhat more mobile because of the flexibility of gristle (cartilage). The most conspicuous cartilaginous joints are those between the ribs and the breastbone. Their flexibility allows the rib cage to move during breathing. Synovial joints, such as those at the shoulder, elbow, hip and knee, are freely mobile. The bearing surfaces of synovial joints are covered with a thin layer of cartilage and this is lubricated with synovial fluid that exudes from the cartilage under pressure. Such joints are enclosed in capsules of tough, fibrous tissue lined with a membrane, the synovial membrane, that secretes the synovial fluid. Synovial joints are of several types. They include hinge joints (knee and finger), ball-and-socket joints (shoulder and hip), pivotal joints (upper end of radius bone and between the upper two cervical vertebrae), saddle joints (elbow), condyloid (between the skull and first vertebra), and plane or sliding joints (wrist and feet). Movement of synovial joints is restricted by ligaments, cords of tough, elastic protein (collagen; ⇨ pp. 36–7) whose function is to support the joints. Usually these are external, but in some cases – as in the knees – are internal. Any inflammation of a joint is called *arthritis* (⇨ pp. 126–7).

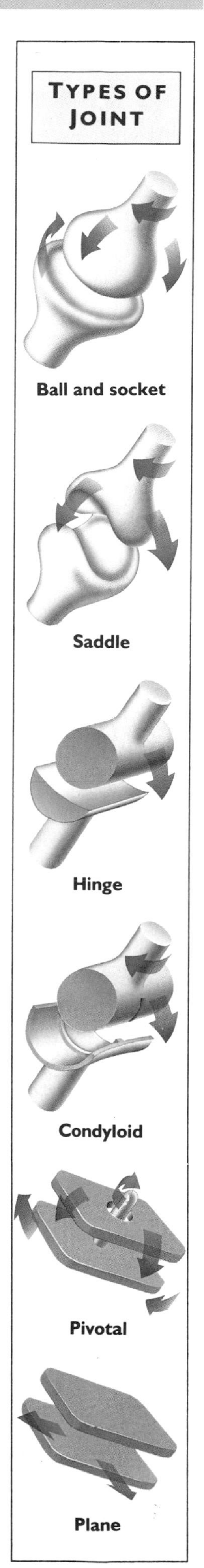

Movement 2: Muscles

Movement of the body, or of any part of it, is possible only as a result of the action of muscles. This movement is the result of the muscle becoming shorter, a process called 'contraction'.

SEE ALSO

- WHAT THE BODY IS MADE OF p. 36
- CELLS, THE STUFF OF LIFE p. 38
- MOVEMENT 1: THE SKELETON p. 40
- FUEL AND MAINTENANCE pp. 44–52
- EXERCISE AND BODY UNITY p. 54
- THE CONTROL SYSTEM 2: THE AREAS OF THE BRAIN p. 76

A contracting muscle changes its length only, not its volume, so that when it shortens it becomes thicker. Muscles can work only by shortening and exerting a pull or, if circularly arranged, a squeeze. Even when we seem to be pushing something, the muscle action involved is a pulling one – often to straighten a joint. Contraction involves the conversion of chemical energy into movement (*kinetic*) energy. Many muscles act across joints, so that when the muscle shortens the joint bends (➪ below). Others change the diameter of body cavities or tubes, so that the contents are moved. The most striking example of this kind of muscle action is demonstrated by the heart as it pumps blood. Some other muscles, such as the multiple muscles of the tongue, cause movement by pulling on each other.

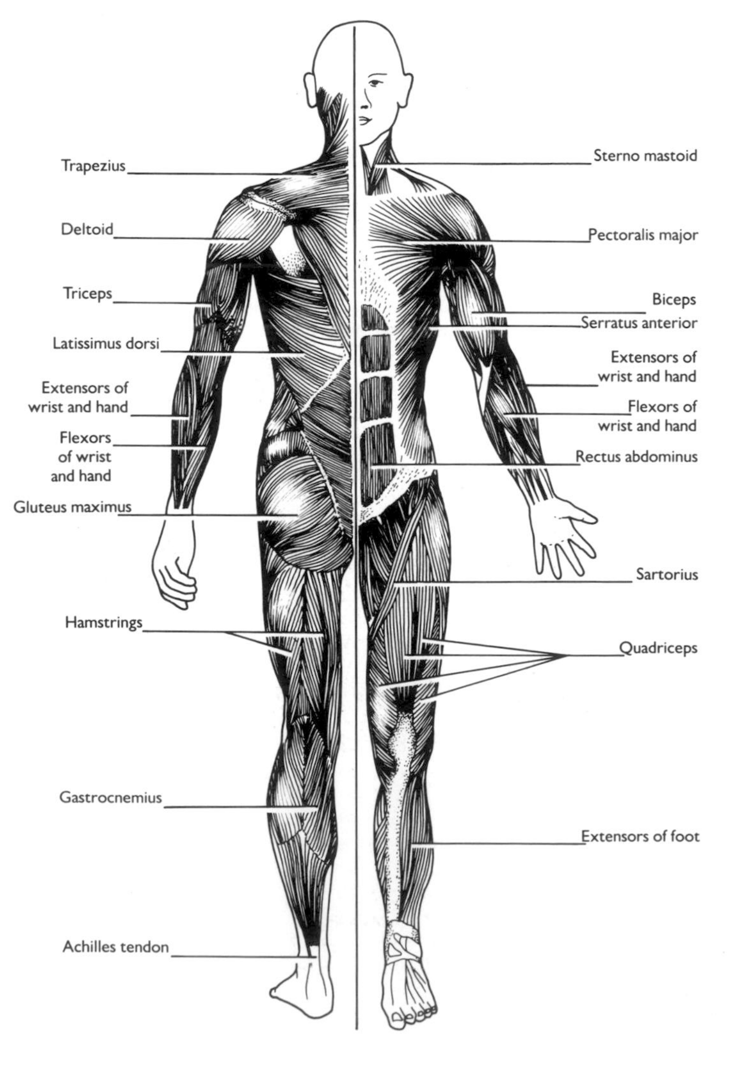

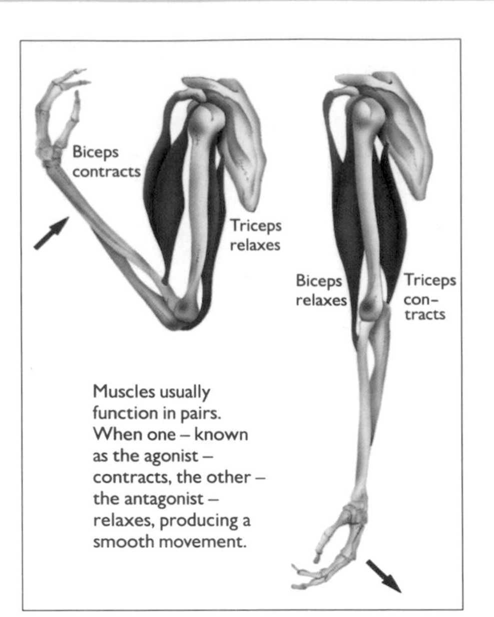

Muscles usually function in pairs. When one – known as the agonist – contracts, the other – the antagonist – relaxes, producing a smooth movement.

On average, muscle constitutes about 40% of the weight of adult males and about 36% of the weight of adult females. Because of minor anatomical variations, the exact number of voluntary muscles (those causing willed movement) in the body is uncertain, but is in the region of 660. Muscle is about as efficient as an internal combustion engine (around 25%) in converting chemical energy into work. A reasonably fit man can exert about half a horsepower (373 joules per second) – at least for a few minutes.

The most conspicuous muscles of the body are those of the limbs. Limb muscles work in opposing groups, those on one side of a joint being balanced by a comparable group working on the other side. As one group contracts, the opponents relax, without, however, losing tension. The coordinated action of these opposing groups allows stability and smoothly controlled movement. While movement is taking place in a limb, the rest of the body must be appropriately braced, and this may involve simultaneous contraction of muscles on either side of joints so as to stabilize them. The control of this complex and ever-changing interaction is provided by a compact, real-time 'computer' called the cerebellum (➪ pp. 72–3), which is connected to the voluntary centres in the brain and to the muscles by way of the peripheral nerves.

The main voluntary muscle groups

Several muscles act on the shoulder joint. The powerful 'shoulder-pad' muscle, the *deltoid* (➪ illustration), raises the whole arm while other muscles simultaneously act to fix the shoulder blade, on which the upper arm bone articulates. The elbow joint is flexed by the *biceps*, which crosses the front of it, and extended by the *triceps*, which crosses behind. The biceps also rotates the forearm. Among the most impressive examples of muscle control and interaction are those of the forearm that cause the complex movements of the hand. Almost all the movements of the fingers are mediated by contraction of forearm muscles, operating through long, fine tendons (cords of collagen) that run across the front and the back of the wrist into the hand. There are a few fine muscles in the hand itself.

The powerful hip and buttock muscles are concerned mainly with moving and stabilizing the hip joints in walking, running and climbing. The large muscle group that originates from the front of the thigh (the *quadriceps* muscles) has a common insertion in a large tendon that encloses the kneecap, and is firmly fixed to a bump on the top of the front of the tibia, the main lower leg bone. This group straightens the knee as in kicking. The opposing group, the *hams*, on the back of the thigh, are attached to the back of the tibia by the *hamstrings*, and bend (flex) the knee. Two of the main calf muscles are inserted into the back of the heel bone (calcaneum) by way of the *Achilles tendon*. Others act on the foot, by way of long, fine tendons, analogous to those of the forearm muscles that act on the hand.

The most important of our voluntary muscles are those that maintain breathing. These are the *intercostal* muscles that pull the ribs upwards and outwards, so increasing the side-to-side dimension of the chest, and the *diaphragm*, a dome-shaped sheet of muscle situated under the lungs, that simultaneously flattens (➪ pp. 52–3). The resulting increase in the volume of the chest causes air to be sucked into the lungs from outside. When these muscles relax, the lungs collapse by elastic recoil.

The abdominal contents are less well protected than those of the chest, since the chest is encased in the ribcage, but the abdominal wall contains strong sheets of muscle that can contract quickly to form a firm, protective barrier. The same muscles can strongly compress the contents of the abdomen, helping us to empty the bladder and rectum (➪ pp. 64–5), and, in childbirth, to help to push out the baby (➪ pp. 104–5). The spine (vertebral column; ➪ pp. 40–1) is surrounded by a strong mass of longitudinal muscle that supports it, maintaining the proper relationship of the bones, and allowing bending and twisting in any direction. Most back trouble is caused by weakness of these paravertebral muscles.

Much of our muscle action is somewhat less energetic, however. All the subtleties of facial expression and the indications of the emotions are achieved by the contraction of muscles under the skin, acting with great precision to alter the configuration of the mouth, the eyebrows, the eyelids, the nostrils and the cheeks. Delicate and finely controlled muscles in the voice box (larynx; ➪ pp. 82–3) act through levers on the vocal cords, tensing and loosening them so as to change the pitch of the voice, and prolong or interrupt the sounds produced. Tiny, but rapidly acting, muscles in the middle ear

Voluntary Muscle Contraction in Action

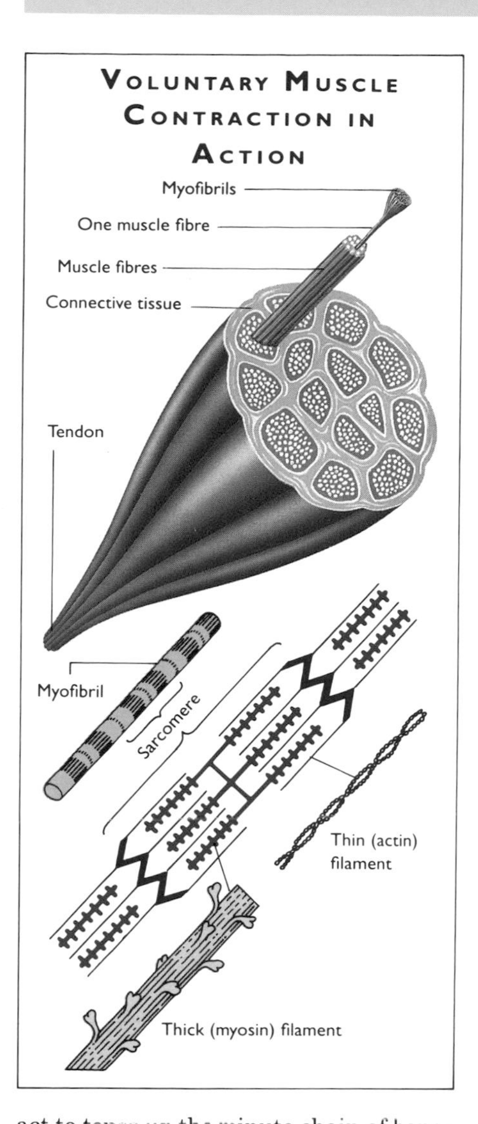

act to tense up the minute chain of bones that connect the ear drums to the inner ear (⇨ pp. 86–7), so as to protect us from unduly loud noises. Tiny, circular muscles constrict the pupils (⇨ pp. 84–5) of the eyes; radially placed muscles work to widen them.

Biting and chewing are made possible by muscles arising from the base of the skull and the outside of the temple bones, and attached to the jawbone. These muscles of mastication (chewing) are capable of pulling the jawbone powerfully upwards, and rocking it sideways, so that food can be cut and torn by the front teeth, and ground by the molars.

As well as the voluntary muscles, there exists another class – the involuntary or smooth muscles. These are not involved in active bodily movement, being involuntarily controlled by the autonomic nervous system (⇨ pp. 72–3). Involuntary muscles are found, for instance, in the arteries, which they constrict. And the process of peristalsis in digestion (⇨ pp. 44–5) is effected by the involuntary muscles. Microscopically, they do not show the characteristic transverse stripes that give voluntary muscle its appearance. The cardiac muscle, also in a class of its own, differs from other muscle in that it has the property of spontaneous contraction and in that its fibres are arranged in a network, so that a contraction in one sets off a contraction in the next, and so on (⇨ pp. 50–1).

The fine structure of muscle

Muscles consist of bundles of thousands of individual muscle fibres, bound together by collagen, the same protein that provides the framework of bones (⇨ p. 40). Each fibre is a single, cylindrical cell, up to 30 cm (1 ft) long and up to $^{1}/_{10}$ mm ($^{1}/_{250}$ in) in diameter, containing a bundle of several hundred smaller subfibres called *myofibrils*. Myofibrils, in turn, contain bundles of smaller protein filaments arranged along them in a repeating pattern. Each unit of this repeating pattern is called a *sarcomere*, and it is in the sarcomeres that the real action occurs. Sarcomeres in adjacent fibres lie accurately in line; this is why voluntary muscle fibres have a striped appearance.

The structure of sarcomeres is far too fine to be seen with anything less than a powerful electron microscope. They are made up of thick filaments of the protein myosin, between which run thin filaments of another protein, actin (⇨ pp. 36–7), like interlacing fingers (⇨ illustration). The thick filaments are linked together in groups so as to leave spaces between into which the free ends of the thin filaments, which are also connected together in groups, can move.

What happens during contraction?

Muscle contraction requires a lot of energy, and this is provided by the many mitochondria (enzyme-containing cell constituents; ⇨ pp. 38–9) lying within the muscle fibre alongside the sarcomeres. Contraction of the muscle fibre does not occur by shortening of the myosin and actin filaments as might be supposed. In fact, the contraction results from a sliding of the thin filaments more deeply between the thick filaments. This is brought about by the action of short protrusions on the myosin filaments, called cross bridges, which bind, briefly, to the thin filaments and, acting much in the manner of the oars of a rowing boat, pull the two sets of filaments together. The links between the myosin and the actin filaments then break. This action is performed repeatedly so as to pull the thick and thin filaments as far as possible into each other. The minute shortening movement achieved by a single sarcomere is multiplied many times by the number of sarcomeres in each fibre, and the muscle fibre, as a whole, shortens to about two thirds of its resting length. Fibres cannot contract less than fully; they do so completely or not at all.

What causes muscles to contract?

Muscles are made to contract by electrical nerve impulses entering them from motor nerves (⇨ pp. 76–7). Like the muscles, these nerves are also made up of bundles of fibres. One motor nerve fibre is connected to a group of muscle fibres, and each time an electrical pulse (nerve impulse) reaches the end of the nerve fibre, all the connected muscle fibres contract fully, albeit briefly. Fibres can contract as frequently as 50 times a second, and by varying the frequency of contraction and the number of fibres involved, the nervous system can effect the most delicate gradations of muscle power.

Muscles and Exercise

Intense, sustained muscular work builds up the bulk of muscles, and disuse soon leads to wasting (*atrophy*). When muscles are built up by exercise, the number of muscle fibres does not, as one might expect, increase, but each fibre enlarges by an increase in the number of its myofibrils (⇨ main text). This *hypertrophy* increases the power of the muscle. In addition, the blood supply to exercised muscles increases, so that the fuels from which energy is derived can be more efficiently delivered. The build-up of simple substances, such as amino acids (⇨ pp. 36–7), into complex proteins, such as myosin or actin, is called *anabolism* (from Greek *ana-*, 'upwards', and *ballein*, 'to throw').

Different kinds of exercise affect muscles in different ways, because muscles contain two kinds of fibre that can be selectively improved. Periods of intense, maximum-effort exercise increase the number of myofibrils in the fast-acting fibres that are used only during such work. By contrast, prolonged exercise of moderate intensity, such as jogging, increases the number of mitochondria in the other muscle fibres. It follows, then, that athletes must work hardest at the kind of activity in which they wish to excel.

Galvani and the Twitching Frog Legs

Luigi Galvani (1737–98) was a lecturer in medicine at the University of Bologna who, in 1775, became Professor of Anatomy there. In 1771, in the course of his scientific work, he discovered, by accident, that the muscles in frogs' legs twitched strongly whenever they were brought into contact with the free ends of two joined but dissimilar metals. Galvani had already demonstrated that the muscles would twitch when connected to an electrical generating machine or to an electrically charged Leyden jar (a primitive capacitor), and was convinced, therefore, that electricity was involved in muscle contraction. But from this new observation he concluded that the electricity came from the muscle rather than from the metals. On the basis of this, he formulated a theory of animal electricity to which he clung tenaciously for years. Ironically, Galvani was wrong about the source of electricity in this case. In 1794 the Italian physicist Alessandro Volta (1745–1827), after whom the unit of electrical potential, the volt, is named, demonstrated that connected dissimilar metals produced an electric current. Galvani died a disappointed man. Later research, however, showed that there was much more in his idea of animal electricity than anyone at the time could possibly foresee; nerve impulses, which cause muscles to contract, are essentially electrical in nature.

Luigi Galvani demonstrating the behaviour of the muscles in a frog's legs when stimulated by electricity. Artist's impression from *Le Journal de la Jeunesse*, 1880. (AR)

Fuel and Maintenance 1: Digestion

The human digestive system is essentially a fuel and materials processing factory: the input is a wide range of edible materials, and the product is a comparatively small list of absorbable materials – mainly glucose, amino acids, fatty acids, vitamins and minerals – that are needed by the body. Foodstuffs are chemically complex, and considerable physical and biochemical processing is needed before they are reduced to the state in which they can be absorbed and assimilated by the body.

SEE ALSO
- WHAT THE BODY IS MADE OF p. 36
- DIET p. 46
- WASTE DISPOSAL p. 64

In a sense, the main part of the digestive system – the intestinal tract – can be regarded as lying outside the body. The inner surface of the gut, from the mouth to the anus, is continuous with the surface of the skin. This means that unabsorbable objects and materials can pass through the intestinal tract without ever entering the body proper.

Mixing and Moving

The commonest kind of muscle action in the bowel produces an effect known as *segmentation*, the purpose of which is to mix the bowel contents with enzymes and to promote absorption. In this process, short segments of circular wall muscle contract and squeeze the contents, while the segments between these contracting zones relax. The relaxed zones then contract and the contracted zones relax. Segmentation ensures thorough mixing but has no effect on the net movement of the bowel contents.

When digestion and absorption are largely complete, the bowel contents must be moved on. In *peristalsis*, a number of short separate segments of bowel wall contract, while the segments lying on the downward side of the contracting segments relax; this process is then repeated again and again, progressing a little further along the intestine on each occasion. The effect of this is to squeeze the bowel contents in the direction of the anus. If peristalsis fails, as in certain severe intestinal diseases, obstruction of the bowel occurs.

Mechanical processing

The first stage in the processing of food is to reduce it to a *bolus* – a soft, chewed mass of suitable size and consistency for swallowing. This is the function of the teeth, the muscles of mastication, the tongue, and the salivary glands. The teeth are of different shapes and have different functions (⇨ diagram). The eight central sharp-edged teeth, the *incisors*, are cutters and are used to sever food by biting. On either side of these are the pointed *canines* and *premolars*, which are adapted for tearing food. Food broken up in these ways is then shifted by the tongue to lie between the *molars*; these have irregular upper surfaces of much greater area than the other teeth and are used as crushers and grinders.

Lubrication is provided by three pairs of salivary glands – the *parotid* glands in the cheeks, the *sublingual* glands immediately under the tongue, and the *submandibular* glands set deep within the curve of the jawbone. The presence of food in the mouth (or even the contemplation of food; ⇨ pp. 88–9) prompts these glands to produce an alkaline fluid known as saliva, which runs through short ducts into the mouth. Saliva also contains an enzyme, *amylase*, that quickly promotes the breakdown of the carbohydrate polysaccharide starch to simpler (and sweeter) sugars – a piece of chewed bread or potato held in the mouth soon becomes sweet.

Swallowing

Once the bolus of food is in a suitable state, it is shifted to the centre of the mouth by the tongue. There follows a complex sequence of muscular actions, which are controlled by nerve centres in the brain stem (⇨ pp. 76–7). First the bolus is pushed to the back of the mouth by the middle part of the tongue; at the same time the rear part of the tongue flattens, and the soft palate presses firmly upwards to seal off the mouth from the cavity of the nose. Once the bolus enters the upper part of the throat (the *pharynx*), swallowing becomes largely involuntary, but it can be inhibited by violent action. The muscles surrounding the pharynx relax, and at the same time the entrance to the *larynx* (voice box) is closed off by a leaf-shaped cartilage (the *epiglottis*), so preventing food from entering the trachea (windpipe). Breathing is temporarily stopped and the vocal cords are pressed tightly together. Next, the muscle ring at the upper end of the *oesophagus* (gullet) relaxes and food passes down. As this happens the vocal cords relax and breathing is resumed. The bolus is now carried rapidly down the oesophagus by the action of *peristalsis* (⇨ box).

The stomach

The stomach is more than simply a bag-like reservoir for food – it also has important digestive functions. Its outlet to the small intestine is narrow, and in this region the three muscular coats of the stomach wall thicken to form a strong muscular ring called the *pylorus*, which controls emptying. The stomach lining is thick and deeply folded, and contains many glands. The secretions from these include an important protective mucus, which prevents the stomach from digesting itself, a digestive enzyme (*pepsin*) that can break down protein, and a strong acid (hydrochloric acid).

The stomach is highly active, constantly churning the contents to ensure thorough mixing. About 2 litres (3½ pints) of hydrochloric acid are secreted each day. This has various functions: it kills germs in food and prevents it from putrefying in the intestine; it activates the production of pepsin; and it acts on the cell walls of food to release the cell contents, so that the protein can be acted on by pepsin.

The small intestine

The adjective 'small' refers to the diameter of the small intestine, not to its length, which is some 6 m (20 ft). The first short part, the *duodenum*, is so called (from the Latin) because it is about 12 finger-breadths long. The other parts are, successively, the *jejunum*, where most digested material is absorbed, and the *ileum*, where a considerable amount of water is reabsorbed. The *bile duct* from the liver and the *gall bladder*, and the duct from a large gland – the *pancreas* – enter the duodenum at the same point.

Teeth

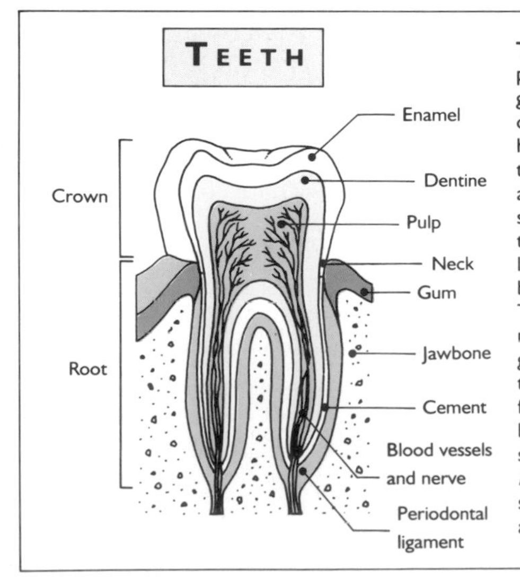

Tooth structure. The part of the tooth above the gumline (the *crown*) is covered by a layer of very hard *enamel*. Beneath this there is a layer of *dentine*, a less hard bone-like substance that is nourished by the innermost *pulp*; the latter is supplied with blood vessels and nerves. The *roots*, one to three in number, extend below the gum and are covered by a thin layer of *cement*. They fit into a socket in the jawbone, where they are secured by the *periodontal ligament*, which absorbs shock as the tooth impacts against food.

The 20 *primary teeth* (also known as the *deciduous* or *milk teeth*) normally begin to erupt at around 6 months and are gradually replaced by the *permanent teeth* from about the age of 6 years. The last of the 32 permanent teeth – the four third molars or wisdom teeth – usually appear between the ages of 18 and 21.

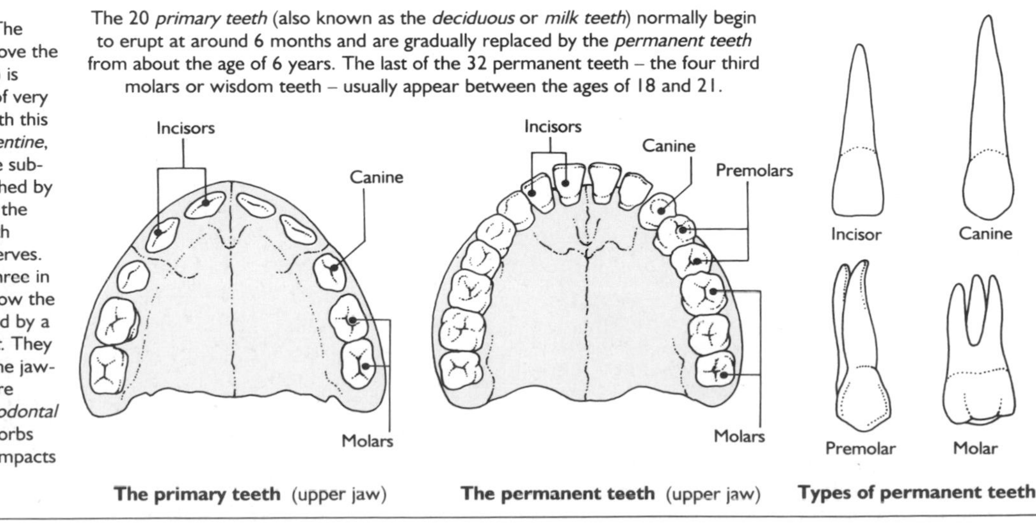

The primary teeth (upper jaw) **The permanent teeth** (upper jaw) **Types of permanent teeth**

The lining of the small intestine also contains glands that secrete digestive enzymes.

By the time food leaves the stomach, the pepsin has already acted on the protein, breaking it down to polypeptides (⇨ pp. 36–7). In the duodenum the polypeptides meet enzymes from the pancreas, mainly *trypsin*, that complete the breakdown to amino acids, while further enzymes in the pancreatic juice break down complex carbohydrates and fats. Not all carbohydrates can be split by human digestive enzymes. Polysaccharides such as cellulose, and pectin and various other fibres are not digested and pass on through the gut. Soluble fibre, however, serves a useful purpose in binding cholesterol in the bile and in reducing the incidence of various diseases of the large intestine. The other carbohydrates in the diet are easily broken down to monosaccharides by the pancreatic enzymes. Pancreatic amylase splits starch to dextrin and then to maltose; this is then split into molecules of the monosaccharide glucose by maltase, also from the pancreas. The disaccharide lactose is split to the monosaccharides glucose and galactose, and the disaccharide sucrose is split into glucose and fructose. These monosaccharides are easily absorbed.

The most important function of *bile* is to emulsify (reduce to tiny globules) fats so that they can more readily be acted on by the fat-splitting enzymes. They are broken down to glycerol and fatty acids. The latter react with the alkaline intestinal juice to form soaps, and these too have an emulsifying action on other fats.

The lining of the small intestine is covered with millions of fine, finger-like projections called *villi*, which greatly increase the total surface area. Each villus contains fine blood vessels and lymph vessels (⇨ pp. 58–9), and the wall is so thin that small molecules can easily pass through into the bloodstream or, in the case of fats, into the lymph vessels. This movement is called *absorption*, and after a meal a large quantity of simple sugars, amino acids and fatty acids, together with minerals and vitamins, passes into the bloodstream. Most of the blood returning from the intestine then passes directly to the liver (⇨ p. 48), where further processing occurs and many of the absorbed materials are temporarily stored.

The colon

The colon or large intestine is shorter than the small intestine – about 1.5 m (5 ft) long – but is much wider. Its lining is smooth and free from villi. It secretes no enzymes and produces only mucus. The main function of the colon is to absorb water and concentrate and firm up the faeces. Towards its lower end is a wider segment, called the *rectum*. When colon contents move into the rectum, the individual experiences a conscious desire to defecate. The contents of the lower colon consist largely of bacteria, but also include cells cast off from the intestinal lining and a small amount of cellulose residue from the diet. Their brown colour is derived from the bile.

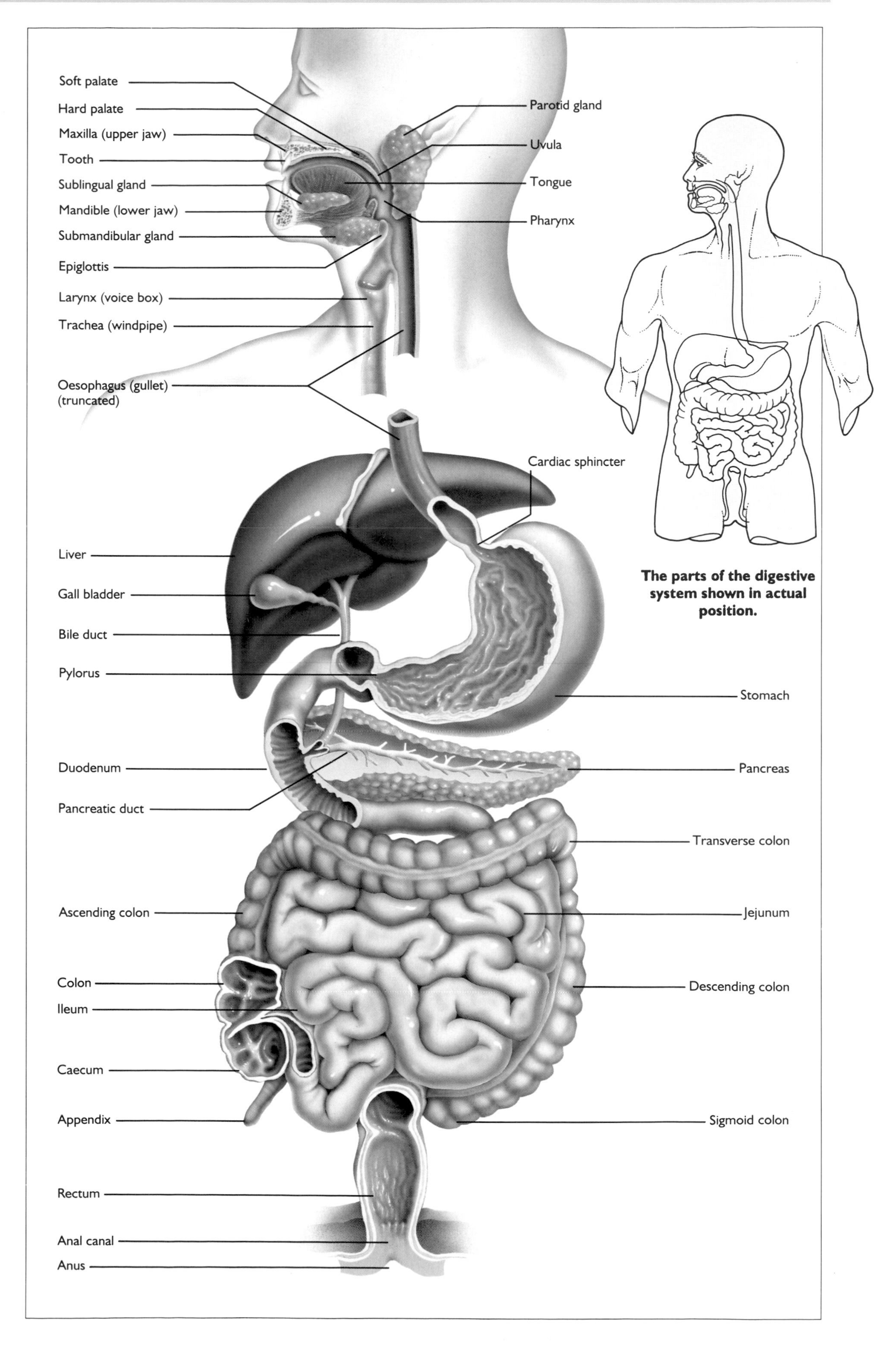

The parts of the digestive system shown in actual position.

Fuel and Maintenance 2: Diet

No one questions that the size of our bodies can be affected by the amount we eat, but there is also a widespread belief that the state of our bodies can be significantly affected, in a qualitative sense, by the nature of our diet.

SEE ALSO

- WHAT THE BODY IS MADE OF p. 36
- CELLS, THE STUFF OF LIFE p. 38
- MOVEMENT pp. 40–2
- FUEL AND MAINTENANCE pp. 44–51
- WASTE DISPOSAL p. 64
- TASTE AND SMELL p. 88

With the exception of the original single fertilized cell from which we developed, every molecule in our bodies has been acquired from outside, initially, in the womb, via the mother, but thereafter mostly by way of the mouth. In a quite literal sense our bodies are made of what we eat and drink.

An adequate water intake is essential as everything that happens in the human body takes place in a watery environment, both inside and outside the cell. It is important to bear in mind what happens to the food we eat before it is assimilated into our bodies (➪ pp. 44–5). A particular amino acid is exactly the same as any other sample of the same amino acid, whether it is derived from an expensively cooked Chateaubriand steak or from a tin of cat meat, and its effect on the body will be the same. Vitamin C derived from a fresh orange is identical in every way to vitamin C derived from a cheap multivitamin pill, and the effect of the two molecules on the body is identical. Since nearly all the food we eat is broken down, in the process of digestion, to simple dietary constituents – sugars, fatty acids, amino acids, minerals and vitamins – the source of these constituents is, physiologically, a matter of indifference. There is, however, as we shall see, another sense in which the quality of our diet is very important.

What is food?

Food is the source of three groups of substances required by the body – fuel to be burnt (oxidized; ➪ pp. 42–3) to provide energy for all the physiological functions; constructional materials for growth, maintenance and repair (➪ pp. 40–1); and atoms and molecules needed for biochemical cell functions (➪ pp. 38–9).

The standard body fuels are glucose and fatty acids. These are derived from the carbohydrates in the diet – the sugars and starches – and from the fats (triglycerides; ➪ pp. 36–7). Although glucose is the principal fuel, comparatively little of it is stored in the body – and that mainly in the liver as a polymer called glycogen that can break down easily to release free glucose molecules. Long-term energy stores are in the form of fats.

The glycerine (glycerol) and fatty acids absorbed after a meal are carried to the fat cells, mainly under the skin, where they are resynthesized to triglycerides, which consist of three fatty acids linked to a 'backbone' molecule of glycerol. Between meals, some of the fat in these stores is broken down to glycerol and fatty acids, and these are released into the blood for use as fuel in the cells. Amino acids serve as fuel only when the other fuel sources are almost exhausted; this use of them can cause severe muscle wasting.

The main body-building 'bricks' are the amino acids (➪ p. 36). They are derived from protein in the diet by the action of the protein-splitting (*proteolytic*) enzymes of the stomach, pancreas and small intestine (➪ pp. 44–5). Some of the 20 necessary amino acids can also be synthesized by the body. Those that cannot are called essential amino acids because they must be provided in the diet. Amino acids are small, fairly simple molecules that readily link together to form polypeptides, which, in turn, link up to form proteins (➪ p. 36). The order determines the type of protein produced, and is prescribed in the DNA by

VITAMINS AND THEIR FUNCTION

Vitamins are organic substances that are required in very small amounts for specific chemical reactions in cells. These reactions cannot use more than the quantity needed; an excess of vitamins has no useful effect, and can be dangerous, especially in the case of vitamins A and D. However, a normal, well-balanced diet will provide all the vitamins needed. Some vitamins, such as vitamin C, are stored in large quantity in the body; others are hardly stored at all. Some, such as vitamin K, are even manufactured in the body. These facts have a bearing on how readily nutritional deficiency can occur.

Vitamins have a wide range of functions. Vitamin A promotes normal production of a substance called *retinol-binding protein*, and is needed for growth, maintenance of surface tissues (epithelia; ➪ pp. 62–3), normal vision and reproduction. All the members of the large B group are *co-enzymes* – chemicals necessary to allow particular enzymes to work so that cell function can proceed normally. Vitamin C is necessary for the proper synthesis of collagen (➪ p. 40) and thus for healthy body structure. Vitamin D is a hormone that regulates calcium deposition in bone. Vitamin E stabilizes cell membranes, preventing them from being damaged, while vitamin K is necessary for the normal production of the blood-clotting factors (Factors II, VII, IX and X) by the liver.

***Free radicals* are highly active chemical groups capable of damaging body tissue by breaking down the surrounding membranes of cells. They are produced by disease, radiation and smoking, amongst other things. Vitamins C and E can 'mop up' free radicals, and their role in a wide range of conditions associated with free-radical damage is currently the subject of much research.**

Vitamin B12 (Cyanocobalamin) is essential for the healthy growth of red blood cells in the bone marrow. Good sources include liver, fish, eggs and meat. (RB)

A wholefood diet, free from preservatives and added sugar, is the best way to ensure a balanced intake of nutients. Particularly important are foods that are high in fibre – unrefined cereals, vegetables, fruits and pulses – as these bind to bile cholesterol and prevent it being reabsorbed. Lower cholesterol means a reduced risk of coronary artery disease, the West's number one killer. (RB)

How is Energy Released from Body Fuels?

A remarkable 'cyclical' biochemical process goes on continuously in every cell in the body. The purpose of this is to release the energy needed for all the numerous cell activities, such as the formation of proteins, muscle contraction and movement of materials within and between cells. This process is called the *Krebs cycle* after Sir Hans Adolf Krebs (1900–81) who first described it in 1937. Krebs had to flee Nazi Germany in 1933, and moved to the UK, first to Cambridge and then to the University of Sheffield where he taught for 20 years. He was awarded the Nobel Prize for Physiology or Medicine in 1953.

The Krebs cycle, or citric acid cycle, is very complicated. In simplified form it involves a sequence of enzymes found in the mitochondria (➪ pp. 38–9) that act on the three fuel molecules – glucose, fatty acids and amino acids – to produce carbon dioxide, and to build up a substance of central importance called *adenosine triphosphate* (ATP) from a simpler form *adenosine diphosphate* (ADP). ATP, which has three phosphorus atoms in the molecule, and ADP, which has two, are present in every living cell, from bacteria to humans. In the synthesis of a molecule of ATP from a molecule of ADP, a large amount of energy, derived from the oxidation of the food fuels, must be added. This energy, stored in the ATP, is soon released where it is needed in the cell, when ATP breaks down to ADP. With each turn of the Krebs cycle, an ADP molecule has a phosphorus and some oxygen atoms added (*oxidative phosphorylation*), plus energy, to change it back to a molecule of ATP.

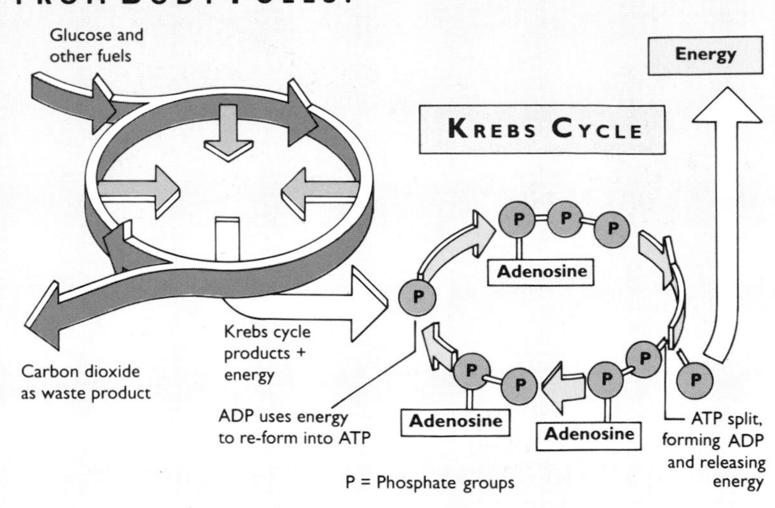

the order of the sequence of bases (➪ pp. 36–7).

Amino acids are present in all cells, both free and linked to form proteins. Other structural elements derived from the food include the minerals calcium and phosphorus, which contribute rigidity to the protein scaffolding of the bones.

The biochemicals that are necessary for the proper structure and function of the body include the vitamins, the essential amino acids, the minerals calcium, phosphorus, magnesium, sodium, potassium, fluorine, and some metallic elements such as iron, copper, zinc, selenium and manganese. Apart from some vitamins that can be synthesized in the body, we have no other source of any of these substances, and they must therefore be present in the diet.

Diet and disease

The coronary arteries supply the heart muscle with blood. If they are narrowed by plaques of cholesterol-containing material (*atheroma*), the supply of blood to the heart muscle may be insufficient to allow normal functioning under increased demand, as during exercise. This is called *ischaemia*, and it causes *angina pectoris* and severe limitation of activity. If a blood clot forms on top of a plaque and obstructs the artery completely (*coronary thrombosis*), or if the narrowed artery goes into spasm, the result is a heart attack with grave or even fatal results.

Scientific trials have repeatedly shown that men with a high proportion of saturated fats (➪ pp. 36–7) in their diet are more likely than average to suffer heart disease from narrowing of the coronary arteries. Protein intake, per se, appears to have no effect on coronary artery disease, but it is difficult to avoid saturated fats when consuming a high-protein diet, because animal protein is associated with fats. Surveys have also shown that those with a high intake of vegetables, starch (potatoes and cereal-based foods) and other complex carbohydrates are less likely than average to develop coronary disease. The kind of fat laid down in the body deposits reflects the type of the dietary fat intake. It is now well established that a diet high in polyunsaturated fats reduces body cholesterol. There is clear evidence also that a high intake of fibre reduces the incidence of intestinal disorder and can reduce levels of blood cholesterol.

Certain polyunsaturated fatty acids, especially linoleic acid, are believed to reduce the tendency of blood to clot. If this is so, an increase in linoleic acid in the blood may reduce the tendency to coronary thrombosis. For example, the proportion of linoleic acid in the tissue fats of men in Scotland is significantly lower than in those of men in Sweden, and the incidence of coronary heart disease is one third higher in Scottish men. Coronary heart disease was formerly very uncommon in black people in eastern and southern Africa, but is being increasingly diagnosed among the more affluent sections of the population as they increasingly adopt a Western dietary style.

Cholesterol is an essential ingredient in the cell structure and is needed to form certain steroid hormones (➪ pp. 78–9). Excess body cholesterol, however, is dangerous. Blood cholesterol varies more, from person to person, than any other blood constituent, and appears to be the most important of the known risk factors for coronary artery disease. In more than twenty studies in different countries, total blood cholesterol was found to be related directly to the development of coronary disease. The higher the blood cholesterol, the more likely the development of coronary problems. This association held for both sexes, and in every population studied the risk of disease was higher for people with higher levels of blood cholesterol.

High total blood cholesterol implies high levels of the cholesterol carriers in the blood. These are called *low-density lipoproteins* (LDLs), and the risk is associated with the level of LDLs in the blood, rather than with the absolute amounts of cholesterol. A high intake of polyunsaturated fats, relative to saturated fats, lowers the levels of low-density lipoproteins. At more than 10% of the total calorie intake, the level of saturated fats (animal and dairy product fats that are usually solid at low room temperatures) is unhealthily high.

Effects of Starvation

Starvation is the result of insufficient food or of the body's failure to absorb or utilize food. Its chief effect is weight loss and wasting of the body. First, in order to provide energy for vital functions, the fat stores are consumed, then the muscles, including the heart muscle, and other organs, especially the liver. The starving person also suffers vitamin deficiencies, especially those of vitamin A – dryness of the eyes (*xerophthalmia*) and blindness from corneal melting. Unrelieved starvation leads to uncontrollable diarrhoea brought about by changes in the intestines and proneness to infection because of deficiency in the body's immune system from shortage of antibody protein. It inevitably leads to death from heart failure as a result of loss of heart muscle, though death from starvation usually takes many weeks. In the Western world the most common cause is anorexia nervosa, the result of a defect in perception of body image on the part of the affected person, who undergoes strenuous efforts to avoid eating. Unfortunately, in the less developed countries, famines still occur frequently.

Drought in Ethiopia in 1984 led to widespread famine and death from starvation. An estimated 800 000 people died. The sight of suffering such as this prompted pop singer Bob Geldof to found the high-profile Band Aid charity for famine relief in Africa. (WHO)

Fuel and Maintenance 3: the Liver

Systems that use fuel require a storage arrangement, and in the human being – indeed in all vertebrates – this function is performed by the liver. The liver, which is the largest organ in the body, weighing about 1.5 kg, is, however, far more than just a fuel store. It is also the major processing plant of the body, carrying out a multiplicity of different metabolic functions, many of them essential to life.

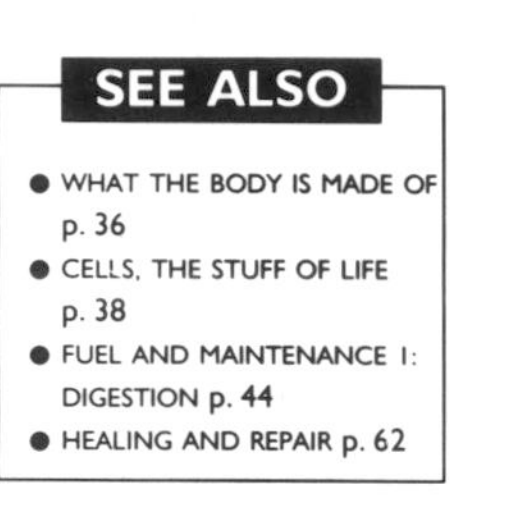
SEE ALSO
- WHAT THE BODY IS MADE OF p. 36
- CELLS, THE STUFF OF LIFE p. 38
- FUEL AND MAINTENANCE 1: DIGESTION p. 44
- HEALING AND REPAIR p. 62

This important organ occupies the upper-right-hand corner of the abdomen, and extends across the midline to the left side. It is a spongy, reddish-brown structure, moulded to fit high under the upwardly domed diaphragm, so that most of it lies behind the ribs. It is roughly wedge-shaped, with the narrower edge pointing across to the left. On the right side, the liver extends down to the level of the lower ribs. The lower edge slopes upwards at an angle, across the front of the stomach, and the wedge-shaped smaller left lobe of the liver lies mainly under the lower part of the breastbone (sternum) and the cartilages for the ribs on the left side. On the underside of the liver lies the gall bladder.

The liver and the digestive system

The intestines have a profuse blood supply carried by short trunks coming off the main artery of the body (the aorta; ➪ p. 50). The main purpose of this massive blood supply is not to nourish the tissues of the intestines but to carry away from the bowel digested food substances that have been absorbed through its inner lining (➪ pp. 44–5). Almost all the blood returning from the intestine, the stomach and the spleen (➪ pp. 58–9) passes to the liver by way of a short, wide vein (the portal vein; ➪ pp. 50–1), which enters it on the underside. The liver is so active in its chemical processes (metabolism) that it also needs a large, freshly oxygenated blood supply. This is provided by the hepatic artery, a branch from the aorta, which enters the liver at about the same point as does the portal vein. When the body is at rest, almost a quarter of the entire blood flow is contained in the liver. But during exercise the flow into the liver is greatly reduced, and more blood becomes available to take oxygen and nutrients to the muscles.

Within the liver, the portal vein and the hepatic artery each divide into tree-like structures, the smallest branches of which end in millions of tiny liver lobules (➪ illustration). Here the blood from these vessels comes into intimate contact with the liver cells. These lobules are the functional units of the liver, and it is in the liver cells within them that the many complex biochemical processes go on. In parallel with the blood system of the liver is a network of fine, branching drainage tubules – the *biliary system* – running in the opposite direction. These branches join to form the *bile duct*, which enters the small intestine at the duodenum (➪ pp. 44–5) and which has a short side branch to the gall bladder where the bile is stored and concentrated until a meal containing fats is taken. Bile acts as an emulsifier for dietary fats, turning them into a kind of milk of tiny globules that can easily be absorbed. In the absence of bile, as in the case of a bile duct obstruction, much of the dietary fat is lost in the stools. Most people over 40 have gall stones in the gall bladder, made mainly of cholesterol (➪ pp. 36–7) – which is a major constituent of bile. Usually these stones are harmless, but if large enough they may obstruct the outlet of the gall bladder or may impact in the bile duct and thus cause trouble. The idea of 'biliousness' (congestion caused by a build-up of bile) is a fiction of 19th-century medicine.

The liver works constantly to process nutrient substances from the intestines. The raw materials – glucose, amino acids, fats, minerals and vitamins – enter it in the nutrient-rich blood coming from the

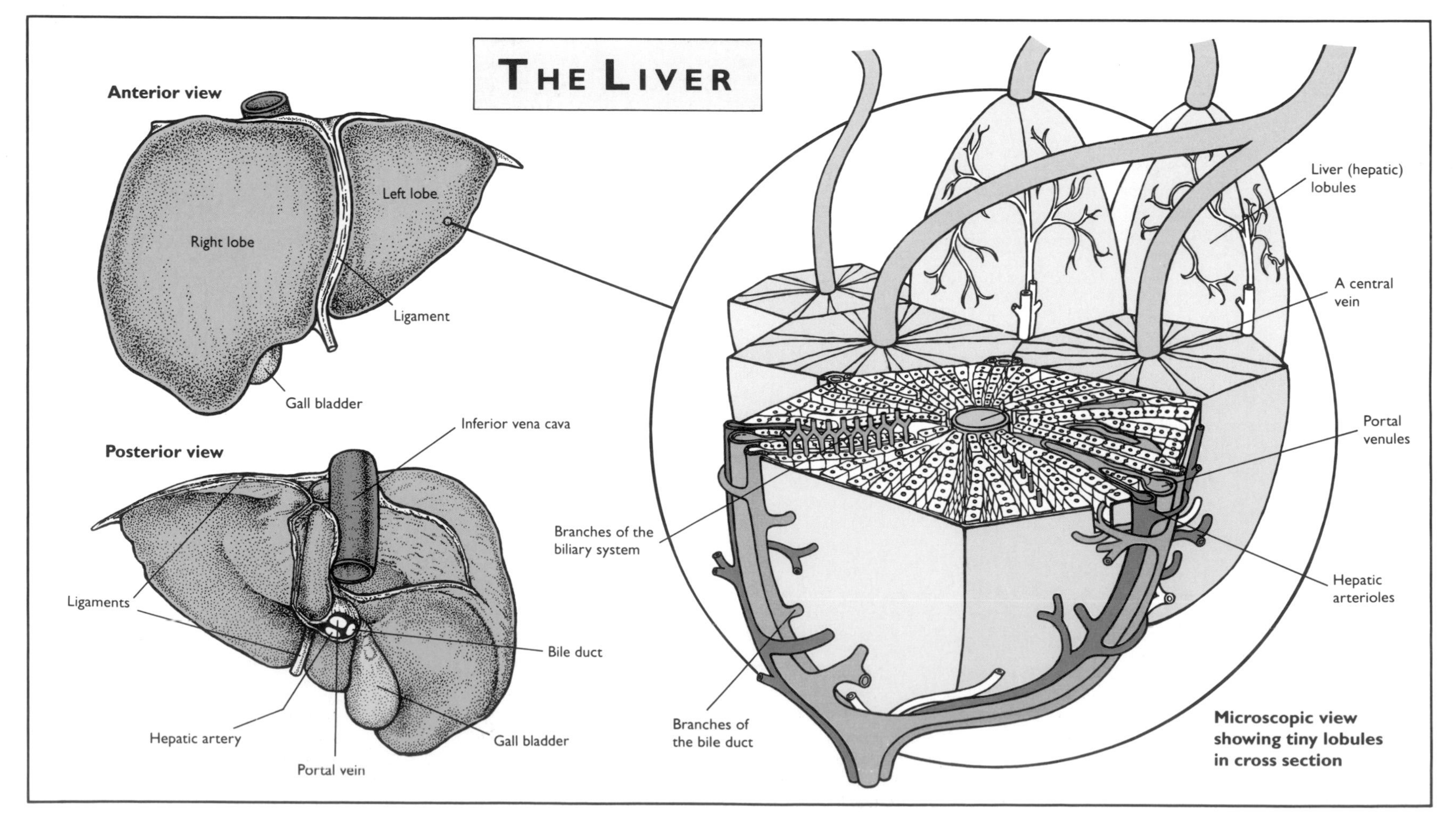

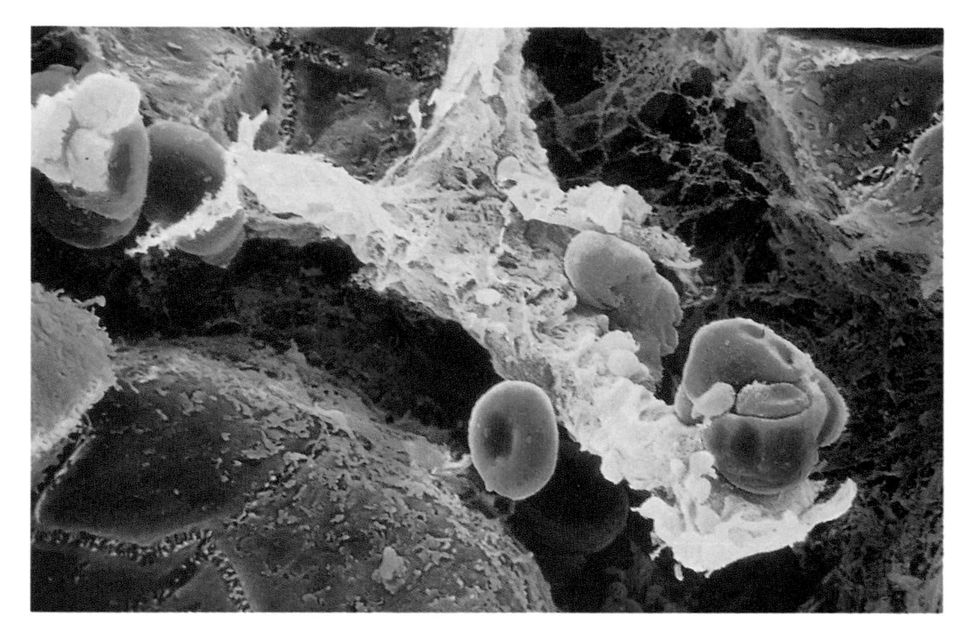

Removal of dead red blood cells in the liver. The irregularly shaped Kupffer cells (yellow) serve to purge blood of old and redundant cells, bacteria and other foreign material by engulfing and breaking them down in a process called phagocytosis. False-colour scanning electron micrograph (magnification x1970 at 6x7 cm).
(Prof. P. Motta, Dept. of Anatomy, University 'La Sapienza', Rome, SPL)

intestine by way of the portal vein (➪ above). From this blood, the liver takes up glucose and synthesizes from it a highly concentrated polymer storage form of carbohydrate called glycogen (➪ pp. 36–7). On demand, glucose can immediately be released from this material. The liver also deals with fats and proteins, converting them into the simpler constituents and into glucose, as required by the body, and, when necessary, converting one into the other. Amino acids are built up into the complex proteins required by the blood and the immune system, or broken down and converted to carbohydrate or fat, as the need dictates.

Other liver functions

Protein is constantly being broken down in the body – especially in the muscles – and new protein formed. Protein breakdown releases ammonia, which is poisonous in excess. The liver, however, prevents accumulation by constantly synthesizing a safe ammonia-containing compound called *urea*, which is excreted as one of the main constituents of urine. Another important function of the liver cells is to deal with any toxic materials that may have been ingested. Among hundreds of other potentially dangerous substances, alcohol and other drugs are broken down to safer forms. Although the liver has a remarkable capacity to detoxify many substances, it may suffer damage in so doing (➪ below).

The liver also takes up the products of exhausted and dead red blood cells – red cells have a lifespan of only about 120 days – and converts these into a pigment known as *bilirubin* that, together with other substances, forms the bile. If, for any reason, such as hepatitis (liver inflammation) or obstruction to the bile duct, the bilirubin is unable to escape from the liver, it accumulates in the blood and stains the skin and the whites of the eyes yellow. This yellowing – known as jaundice – is one of the major indications that something is seriously wrong with the liver or the bile-drainage system.

Cholesterol, which has several important functions in the body, is formed in the liver in large quantities; much of this passes down the bile duct into the intestine to be absorbed into the blood along with the food (➪ p. 44). The liver is, however, sensitive to the levels of cholesterol in the blood. Whenever the blood cholesterol rises, it reduces its rate of cholesterol production. This happens because cholesterol itself inactivates the enzyme necessary for its own synthesis in the liver. So, when blood cholesterol tends to rise, as after a fatty meal, liver production of cholesterol drops; and when blood cholesterol drops, liver production rises. By such a *homeostatic* (equilibrium-maintaining) mechanism, the blood levels are kept remarkably constant even if the dietary intake of cholesterol is high. Saturated fats in the diet (➪ pp. 46–7), however, raise blood cholesterol levels by stimulating cholesterol synthesis, while polyunsaturated fats tend to lower blood cholesterol. The level of these substances in the diet is much more relevant to the health of the individual than the actual level of dietary cholesterol.

Regeneration, scarring and failure

The liver has remarkable powers of regeneration after disease, poisoning or physical injury. There is, however, a limit to the abuse that the liver will tolerate and there are some poisons, such as a single overdose of paracetamol (Panadol) or years of overdosage with alcohol, that can overcome its regenerative capacity. When this happens, functional liver cells are replaced by scar tissue (fibrosis; ➪ pp. 62–3). Because of the architecture of the liver lobule, this scar tissue is laid down in a characteristic manner, around the lobules, to produce the condition of cirrhosis of the liver (from the Greek *kirrhos*, meaning 'orange-coloured'). Some functional liver tissue remains, but this may not be enough to fulfil the metabolic needs of the body. This, if it happens, is called liver failure. There will be inadequate nutrition, jaundice, and accumulation in the body of ammonia and many other toxic substances. The affected person will gradually be poisoned and will sink into a coma. Liver failure is invariably fatal unless a liver transplant can be performed.

Because so much of the blood flows through the liver on its way back to the heart, it is the commonest site for the development of secondary cancer that has spread from a primary site elsewhere in the body. Enlargement of the liver, along with irregularity in shape, easily felt through the skin along its lower edge, is a common clinical sign of spread (metastatic) cancer. In the West, primary liver cancer is, however, rare, being caused mainly by a combination of *aflatoxin* (a poison produced by a fungus that grows on certain foods in damp conditions) with hepatitis B, or by liver parasites. Secondary cancer is commonplace everywhere.

Liver Transplant

The first liver transplant in Britain was performed in 1968 in Cambridge by Professor Roy Calne. Many thousands have since been done, mainly on people with long-term (chronic) terminal liver disease, especially cirrhosis. A smaller number of transplants have been done on people who have suffered acute liver failure from poisoning and primary liver cancer.

Most liver transplants have been performed with donated organs from people, unrelated to the recipient, who have been accidentally killed. Increasingly, however, partial liver transplants from closely related donors, often parents, are being done. These provide a better tissue match and suffer fewer rejection problems (➪ pp. 58–9). The regenerative power of the liver is so great that it is possible to transplant a large enough part of the liver from one person to another to provide the needs of the host without putting the donor at significant risk. After the operation new liver tissue regenerates in both. The rate of regeneration of the liver can be studied in both donor and host by CT scanning (➪ pp. 128–9). This shows that grafted partial livers increase in size by anything from 60 to 200% within a month. For reasons that are not clear, regeneration in the donors is slower and may take three times as long to complete.

The decision to perform a liver transplant is not lightly made as the risks are by no means negligible, and these must be weighed against the patient's chances of surviving with the disease. Risks include: failure of the donated liver to function; acute rejection of the donated liver; the development of hepatitis in the new liver, mainly from one of a variety of infections; and the development of a severe form of bile obstruction. Strenuous immunological and other measures are necessary to try to prevent or treat these complications.

The results of liver transplantation are improving steadily. Most surgeons expect a one-year survival rate of at least 75% and a five-year survival rate (which is deemed to be a cure) of 60% or more. Most failures resulting in death occur early – usually within a month of operation. Patients who survive, however, are able to lead a normal life. Some long-term liver transplant recipients can, after a number of years, safely be weaned off immunosuppressive (antirejection) drugs. Females who receive male livers may be found, after several years, to have male cells in other parts of the body, such as the skin, the lymph nodes (➪ p. 50) and the blood vessels. This is called *chimerism* and appears to have no undesirable effects.

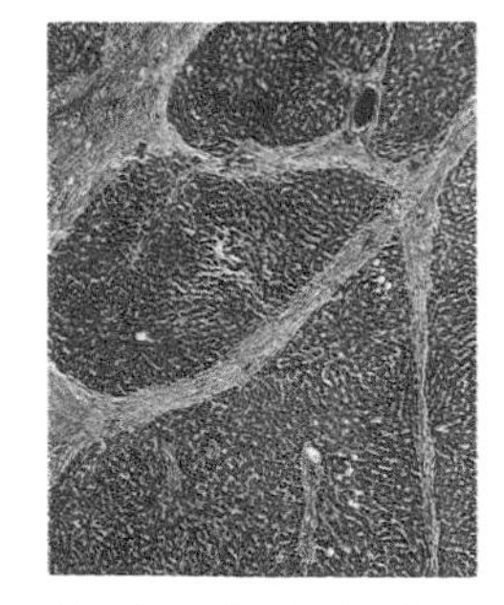

Section of a cirrhotic liver. Cirrhosis is characterized by the formation of fibrous tissue (here stained lighter) that disturbs the functioning of the liver lobules. These structural changes disrupt the flow of blood through the organ and result in large, dilated veins that are liable to rupture. Causes include chronic alcohol abuse and chronic hepatitis, but in cases of cryptogenic cirrhosis, as shown in this light micrograph, there is no known cause. (Biophoto Associates, SPL)

Fuel and Maintenance 4: the Circulatory System

Blood is a complex fluid, partly liquid, partly cellular, that circulates around the body by the pumping action of the heart. The red cells carry vital oxygen to every part of the body, while the white cells play an important role in the body's immune system (➪ pp. 58–9).

The blood also supplies glucose fuel to the tissues and organs, and carries away waste products. Lack of blood to any part of the body, with the failure to supply the vital materials it transports, is the most serious assault possible on the tissues, and by far the most frequent cause of serious disease and death. Failure of blood passage, even for a very few minutes, through organs such as the brain or the heart is seriously damaging, and often fatal. An adequate idea of the blood circulation (➪ pp. 52–3) is fundamental, therefore, to the understanding of the functioning of the human body.

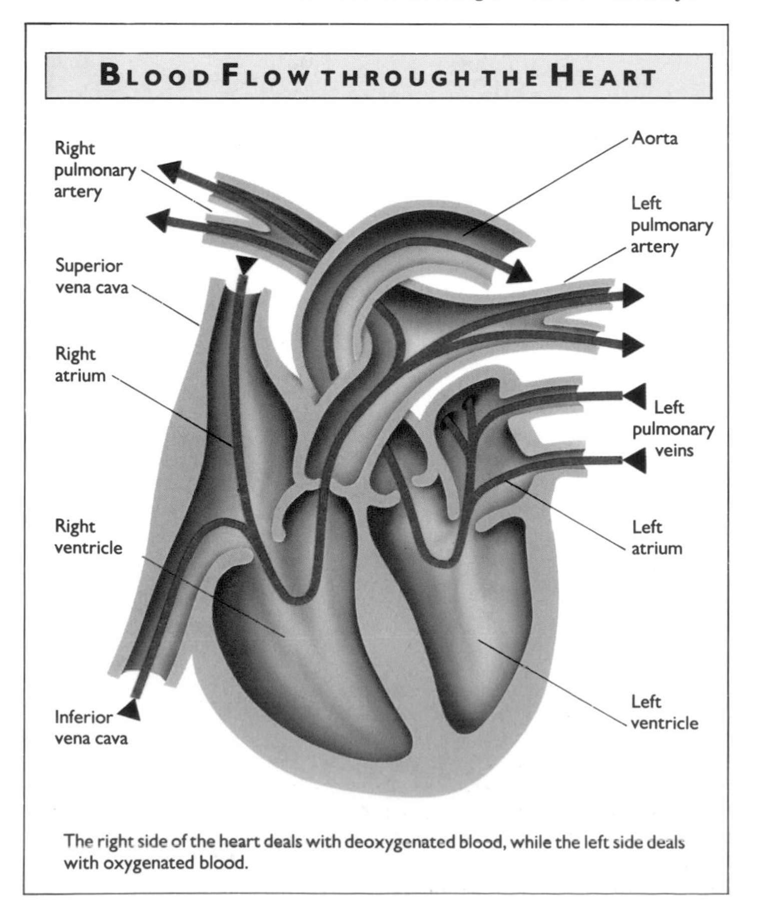

The right side of the heart deals with deoxygenated blood, while the left side deals with oxygenated blood.

Until the time of William Harvey (➪ p. 35) it could hardly have been said that there was any scientific basis for the practice of medicine. Indeed, apart from some traditional methods derived from observation, medical diagnosis and treatment were based largely on systems of thought that were little more than fanciful nonsense. But once Harvey had shown that blood moves in a continuous circulation through all the tissues of the body, and had shown it in such a way as to demonstrate the power of the experimental and observational method, advances were rapid. It was not long before the detailed structure of the heart, and the function of its various parts, were universally understood. These were major advances that, over the years, were powerfully to widen the scope and usefulness of medicine.

The structure of the heart

The heart is a controlled pump, immediately responsive to every change in the body's needs for fuel. It maintains two separate but interconnected circulations and does so, in many people, without flagging, for upwards of eighty years. It is constructed almost entirely of muscle, and also contains blood vessels and lining and covering membranes. It has four chambers, and is equipped with valves that ensure that the blood can move only in one particular direction. The two upper chambers are called *atria*; the lower chambers are called *ventricles*. The walls of the lower chambers are much thicker than those of the upper chambers because it is these ventricles that do the real work.

Functionally, there are two hearts, so separated in their purposes that doctors commonly refer to 'the right heart' or 'the left heart'. In the event of dysfunction, it is usual for one side of the heart to fail while the other side remains normal. The right side of the heart is less powerful than the left because it pumps blood only through the lungs. The left pumps blood around all the rest of the body.

The heart muscle is of a kind almost unique in the body in that it has the property of spontaneous rhythmical contraction at a steady rate without any external stimulus. The muscle fibres of which the heart is constructed are joined together in a branching network (a *syncytium*) so that the contraction of one fibre sets off a contraction in adjoining fibres. In addition, the heart muscle contains bundles of specialized muscle fibres that convey the electrical impulses associated with contraction in a systematic manner from the top of the organ to the bottom. As a result, the contraction starts at the top and proceeds downwards.

The heart muscle works very hard and itself needs an excellent blood supply. Because the chambers have an impervious lining, blood must be supplied to the muscle from outside. This is the function of the *coronary arteries*. These are two small arteries that branch out from the main outflow artery of the body (the *aorta*) immediately after it leaves the heart. The left coronary artery divides near its origin into two large branches and the three vessels course over the surface of the heart like a crown – hence 'coronary'. From these many smaller branches penetrate the muscle to supply it with blood. The coronary veins then drain this blood into the right upper chamber.

The circulation

Blood pumped away from the heart travels, at first, in *arteries*. These are strong, elastic-walled vessels, able to withstand the pressure. Blood from the tissues is returned to the heart by *veins*, which are less strong and have thinner walls. Blood in veins is at a much lower pressure than blood in arteries (➪ pp. 52–3).

From the powerful left ventricle, blood is pumped into the aorta through a one-way valve, called the *aortic valve*. Just above this valve are the openings of the coronary arteries. The blood is immediately distributed, by way of a number of large arteries, branching off the aorta, to the head (*carotid* and *vertebral*) and the arms (*brachial*). The aorta then arches through 180° and runs down behind the heart, giving off many small branches to supply various tissues in the chest and the respiratory muscles of the chest wall. After passing through a hole in the back of the diaphragm into the abdomen, the aorta gives off branches to supply all the organs of the abdomen, including the intestines, where blood picks up absorbed nutrients (➪ pp. 44–5). The branches to the kidneys are especially prominent as all the blood in the body must pass though the kidneys many times a day (➪ pp. 64–5). In the lower abdomen the aorta divides into the two large *iliac* arteries, one for each leg. All these arteries branch repeatedly to supply the organs and the limbs as well as the bones, tendons and skin.

Blood Groups and Blood Transfusion

Pre-20th-century attempts to restore health by transfusing blood from animals, such as dogs, usually ended in the death of the patient. Transfusion of human blood was sometimes wonderfully effective, but this too was often fatal. Transfusion was actually prohibited in England, France and Italy as being too dangerous. But in 1900 there came a remarkable breakthrough. By checking the effect of mixing blood serum (the fluid part without the red cells) from one person with the red blood cells from another, Karl Landsteiner (1868–1943), an Austrian bacteriologist and immunologist, made a remarkable and important discovery. A particular serum would cause the red cells from one person to clump together but would have no effect on those from another person. Landsteiner decided to check all the possible combinations and was soon able to show that human red blood cells fell into four groups, which he arbitrarily named A, B, AB and O. Human blood serum from any person with type A red blood cells could be mixed with type A red cells and no clumping would occur. If the serum were mixed with cells from a person with type B cells, however, clumping always occurred.

We now know that people with A cells have antibodies (⇨ pp. 58–9) in their serum to B red cells; people with B cells have antibodies to A cells; people with AB cells have no antibodies to red cells; and those with O cells have serum antibodies to both A and B. This means that group AB people can receive blood from anyone ('universal recipients'; ⇨ table) and group O people can donate blood to anyone ('universal donors'). Transfused serum antibodies are so quickly diluted that they have no effect. It is the clumping of the transfused red cells that matters.

Donor	Recipient	Compatibility
O, A	A	YES
O, B	B	YES
O, A, B, AB	AB	YES
O	O	YES
AB, B	A	NO
AB, A	B	NO
AB, A, B	O	NO

After types O and AB, the next most important blood group is that involving the rhesus factor. This is of importance only if a person without the factor (rhesus negative) receives multiple blood transfusions with rhesus positive blood, or, more commonly, if a rhesus negative mother carries a rhesus positive fetus (the father being rhesus positive). In such a case the mother may produce antibodies against the fetal red cells. This seldom causes any problems in the first pregnancy but, as antibody levels rise, may have very serious effects on the fetus in later pregnancies, causing massive blood breakown and brain damage. Rhesus disease can be prevented by giving the mother gamma globulin (⇨ pp. 58–9).

At each branch of an artery the total cross-sectional area of the two branches exceeds that of the parent trunk. Therefore, because the same volume of blood has more space, there is a progressive drop in blood pressure, and a slowing in the rate of flow, as arteries form smaller branches. Small arterial branches are called *arterioles*, and, as the size of the branches decreases so does the thickness of the walls. The smallest, and most profuse, branches have very thin walls and merge imperceptibly into the smallest blood vessels of all – the *capillaries*. At any time, about 5% of the blood is in the capillaries, and it is in these that the real function of the whole circulatory system – the exchange of nutrients and waste products – is performed.

Capillaries are neither arteries nor veins but form an important intermediate class. Capillaries are everywhere in the body. If every other structure were removed, its shape would still be easily recognizable by the pattern of the capillaries. Capillary walls consist of little more than single layers of flattened cells, so loosely fixed together by their edges that small pores are left between them, through which water and mobile white cells can pass. Some capillaries, especially in the brain and kidneys, have a second inner layer of cells. These are less permeable than the majority, and form the so-called 'blood–brain barrier' that keeps many organisms out of the brain. Nutrients and other vital substances from the blood – oxygen, glucose, fatty acids, amino acids, minerals and vitamins – are able to pass easily through the capillary walls or through the pores between the cells. Waste products from cell metabolism, especially carbon dioxide, can as easily diffuse into the blood. Capillaries are surrounded by tissue fluid in which the cells of the body are bathed; the initial interchange is between the blood and this fluid. At the same time, an interchange is going on between the tissue fluid and the cells. There is also a constant interchange of water between the blood in the capillaries and the tissue fluid.

On the opposite side of each capillary mass from the arterioles lies the system of tiny veins (the *venules*) by which the blood is carried away. Venules join up to form larger veins and these, in turn, run into the main veins returning the blood to the heart. Running parallel to each of the major arteries is a major vein in which the blood runs in the opposite direction.

The whole of the vein drainage system ends up in two massive veins, each called a *vena cava* (from the Latin *vena*, 'vein' and *cavus*, 'channel'). The upper (*superior*) vena cava drains the head, neck and arms; the lower (*inferior*) vena cava drains the lower part of the body. These direct the blood into the upper chamber on the right side of the heart – the right atrium. From there the blood passes down through a valve into the right ventricle, which pumps it to the lungs by way of the lung (*pulmonary*) arteries. The lung circulation takes the same general form as that of the rest of the body except that the capillaries lie in intimate contact with the lung air sacs (alveoli; ⇨ pp. 52–3). This allows blood that is low in oxygen to be reoxygenated, and allows carbon dioxide carried by the blood from the tissues to be passed out into the air sacs to be expired. Refreshed blood returns to the heart via four large pulmonary veins, but this time it is directed to the upper chamber on the *left* side – the left atrium. From here it passes down through a valve to the left ventricle to be pumped around the body.

Blood Circulation

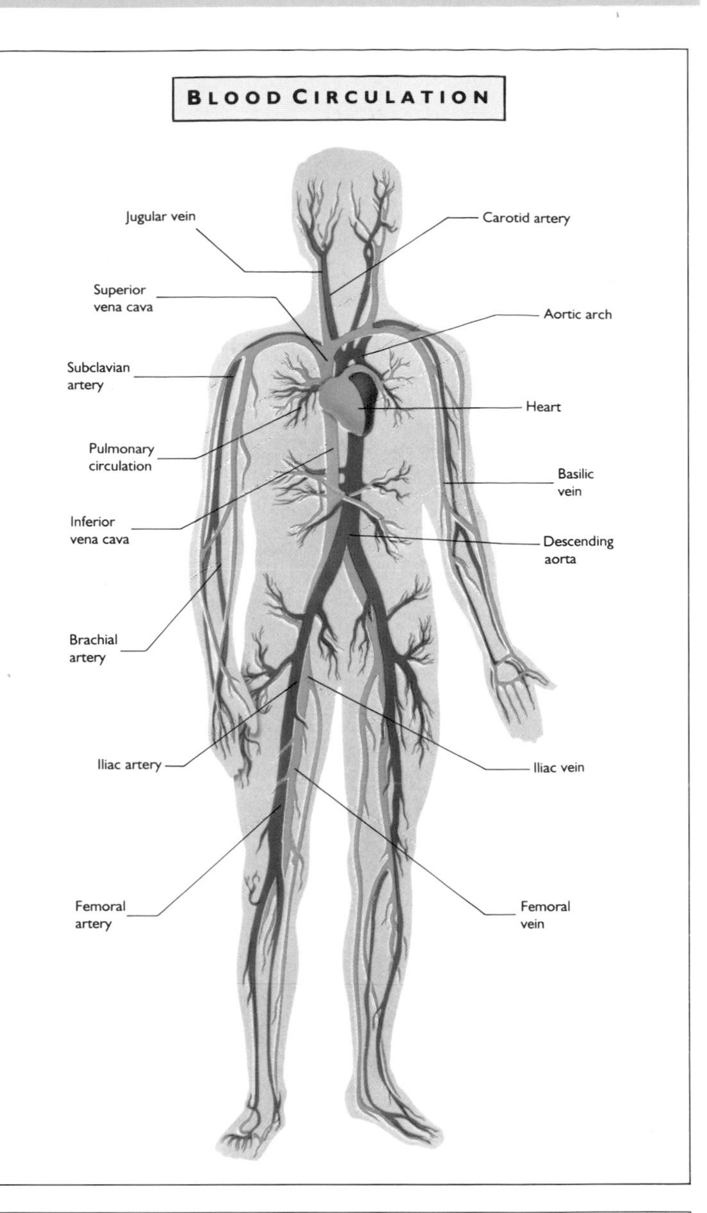

The Lymphatic System

In the capillary beds there is a net outflow of fluid from the capillaries into the tissue fluid of some 3 litres (6 pints) per day. Because blood in the arteries is at a higher pressure than blood in the veins, more fluid passes out at the arteriolar side than returns at the venous side. Excess tissue fluid formed in this and other ways would soon cause water-logging (*oedema*) of the tissue were it not for the *lymphatic* drainage system, which returns it to the circulation. Lymph vessels have one-way valves, and external pressure on them by muscles maintains the flow back into veins at the root of the neck. These lymph vessels drain fluid through lymph nodes, which contain large collections of *lymphocytes* – cells of the immune system that can usually deal with infective organisms or other unwanted material. Any rise in the pressure in the veins, for any reason, such as, for instance failure of the right side of the heart to pump adequately, will cause an increase in the tissue fluid. This may exceed the draining ability of the lymphatic system and oedema may result.

SEE ALSO

- WHAT THE BODY IS MADE OF p. 36
- CELLS, THE STUFF OF LIFE p. 38
- FUEL AND MAINTENANCE 1: DIGESTION p. 44
- OXYGEN: THE VITAL ELEMENT p. 52
- PROTECTION pp. 58–61
- WASTE DISPOSAL p. 64

Oxygen: the Vital Element

One can survive for weeks without food and for days without water, but even a few minutes without oxygen is likely to be fatal. *Oxidation* – the process whereby substances are chemically combined with oxygen – is probably the most important of all chemical reactions. Without it, animal and plant life is impossible.

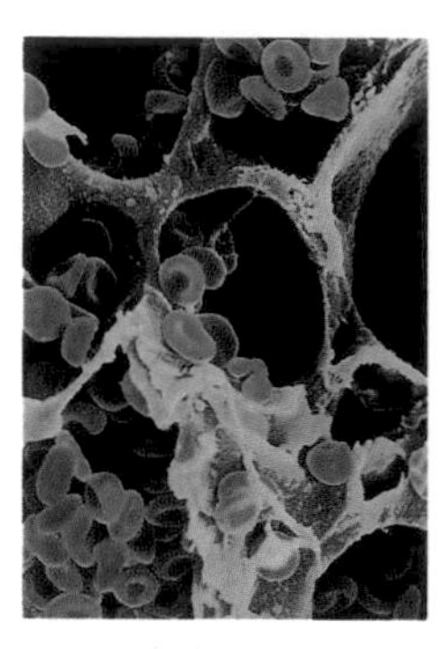

Section through the lung (scanning electron micrograph) showing the thin-walled capillaries together with red blood cells (erythrocytes). (RB)

Oxidation is the ultimate source of energy for nearly all living organisms, and is the reaction involved in the natural decomposition of organic matter. It can be rapid, as in burning, when energy is quickly released, or slow as in the reactions in the cells of the human body or in decomposition. Slow organic oxidation reactions are usually brought about by the catalytic action of enzymes (➪ p. 36). The oxidation of body fuels – glucose and fatty acids – ends in the production of carbon dioxide and water during the Krebs cycle (➪ pp. 51–2).

In order to survive, the human body therefore requires a constant supply of oxygen. To be able to make use of it, the body has a system in which oxygen can be taken from the atmosphere and rapidly distributed to every cell. This process is effected by close cooperation between the circulatory system (➪ p. 50) and the respiratory system. The air we breathe consists of about 20% oxygen and 78% nitrogen, with traces of rare gases and about 0.03% of carbon dioxide. Nitrogen is an inert gas, and is largely ignored by the body; it becomes important only when the body is exposed to such high pressures (such as in deep-sea diving) that the gas dissolves in the body fluids in large quantity.

The respiratory system

Air is drawn into the lungs by the continuous, and largely unconscious, action of the muscles of respiration. These are under the control of the respiratory centre in the brainstem (➪ pp. 74–5). These muscles are, however, voluntary, and, unlike the heart, have no inherent tendency to contract. They do so only when stimulated by nerve impulses from a controlled source in the respiratory centre. If the respiratory centre is destroyed by disease or injury, breathing stops, and only artificial respiration by mechanical means can preserve life.

The ribs are pulled upwards by the muscles lying between them (the intercostal muscles; ➪ pp. 42–3), in a manner resembling the movement of Venetian blinds; in so doing they also rotate outwards. Because of their curved shape, this movement widens the chest and increases its volume. At the same time, the muscles of the upwardly domed diaphragm contract, pulling down its central fibrous portion, and causing the whole structure to flatten. This, too, adds to the internal volume of the chest. The consequent reduction in air pressure in the chest immediately results in air being pushed into it from outside by the higher atmospheric pressure.

This air passes through the nose or mouth, down the throat, through the opening into the voice box (larynx; ➪ pp. 82–3), between the vocal cords in the larynx, down the windpipe (or *trachea*) and the branching air tubes (*bronchi*), down the smaller air tubes (*bronchioles*) to end up in the 300 million or so tiny air sacs of the lungs (the *alveoli*). This causes the lungs, which are elastic, to expand. When inspiration (breathing in, as opposed to expiration, breathing out) is complete, the ribcage falls, the diaphragm returns to its domed shape and the lungs deflate. This whole cycle is repeated at a rate that varies, with the needs of the body or the state of the emotions, from about 12 breaths per minute at rest rising to as high as 180 per minute during strenuous exercise or extreme anxiety.

Gas exchange between the atmosphere and the blood occurs in the alveoli of the

The Discovery of Oxygen

In 1771 the Swedish apothecary and chemist Karl Wilhelm Scheele (1742–86), while heating mercuric oxide, discovered a new gas with remarkable properties. It was colourless and odourless, and small animals, such as mice, kept in an atmosphere of this gas became frisky. A glowing wood splint plunged into it would burst into flame. Scheele called the gas 'fire air', and wrote a book about it in which he described his experiments. Unfortunately, his dilatory publisher did not get the book out until 1777, by which time the English scientist Joseph Priestley (1733–1804) had reported his own similar experiments, and had taken the credit – still generally acknowledged – as the discoverer of the element oxygen.

Scheele was a remarkable man, the extent of whose chemical discoveries is almost unrivalled in science. His achievements, though, remain little known. In a life devoted exclusively to the pursuit of science – and probably shortened by his habit of tasting all the new substances he came across – he isolated the elements arsenic, barium, chlorine, manganese, molybdenum, nitrogen and oxygen, and produced scores of important new compounds, including hydrogen sulphide and copper arsenite ('Scheele's green'). In fact, his contributions to chemistry probably exceeded those of any other scientist of the time. In 1775 he was elected to the Royal Academy of Sciences of Sweden – a unique honour for an apothecary's assistant.

Neither Scheele nor Priestley fully grasped the role of oxygen in combustion and respiration, however. In 1774 Priestley visited the French scientist Antoine Laurent Lavoisier (1743–94), and told him about his experiments. Lavoisier immediately repeated these and soon saw the importance of the newly discovered gas and its relation to air. He gave oxygen its current name (mistakenly derived from Greek roots meaning 'acid maker'), and showed that air contained two main gases, one that supported combustion (oxygen) and one that did not (nitrogen). He studied the heat produced by animals breathing oxygen, and demonstrated the relationship of respiration to combustion. Lavoisier, anxious to be known as the discoverer of an element, did not acknowledge the help of Priestley, whom he regarded as an amateur. Scheele, the first discoverer of oxygen, was ignored and later forgotten.

Antoine Laurent Lavoisier (centre) demonstrating his discovery of oxygen. From Louis Figuier, *Vies des Savants Illustres du XVIIIe Siècle*, 1874. (AR)

SEE ALSO

- WHAT THE BODY IS MADE OF p. 36
- CELLS, THE STUFF OF LIFE p. 38
- MOVEMENT pp. 40–3
- FUEL AND MAINTENANCE pp. 44–51
- EXERCISE AND BODY UNITY p. 54

Respiration and Exercise

Strenuous exercise demands a greatly increased supply of oxygen, to oxidize sugar and release energy, and a greatly increased rate of disposal of carbon dioxide. During exercise, the ventilation of the lungs may increase up to twenty-fold. But this, in itself may not be sufficient to supply the extra oxygen needed by the muscles. During sedentary breathing many of the capillaries surrounding the alveoli are closed simply because the blood pressure in the lungs, caused by the contraction of the heart's right ventricle (➪ pp. 50–1), is insufficient to open them. During exercise, however, the heart rate and force increase, and the pressure in the lung (pulmonary) circulation rises. This causes more of the capillaries to open so that there is a greater surface area available for the passage of oxygen from the alveoli into the blood, and similarly for the passage of carbon dioxide in the opposite direction.

The increased efficiency of gas exchange during exercise prevents any marked change occurring in the levels of oxygen and carbon dioxide in the blood. This poses the question: what is the stimulus for the greatly increased respiratory rate during exercise? Surprisingly, in spite of much research, this remains unexplained. It seems probable, although unproved, that the muscles and joints send stimulating messages to the respiratory centre. Possibly, the abrupt increase in ventilation during exercise is the result of a conditioned reflex (➪ pp. 172–3).

lungs. Each of these air sacs is surrounded by a network of blood capillaries (➪ pp. 51–2); these are part of the blood circulation system of the lungs. The walls of both the alveoli and the surrounding capillaries are so thin that the blood, spread out in the fine capillaries, is brought into contact with the air in the alveoli.

Oxygen transport

Each millilitre (cubic centimetre) of blood contains about 5 billion oxygen-carrying red blood cells (*erythrocytes*). These tiny disc-shaped cells are concave on both sides so as to increase their surface area, and each is filled with about 300 million molecules of the iron-containing protein *haemoglobin*. Globins are a class of proteins and every haemoglobin molecule consists of four such protein chains folded together, each being wrapped round a complex, called a *haem*, that consists of iron and the coloured compound protoporphyrin. The iron atoms in the haem can readily form loose bonds with oxygen, each atom linking to one molecule of oxygen (O_2). This is the secret of oxygen transportation. Whenever haemoglobin finds itself in an environment of high oxygen content (as in the alveoli) the iron atoms link up with oxygen molecules. In an environment of low oxygen content, the haemoglobin gives up oxygen.

The demand for oxygen in the tissues is high, and oxygen is rapidly taken up. At the end of the metabolic processes (➪ pp. 50–1) some of the oxygen comes into combination with carbon from food to make carbon dioxide (CO_2) and with hydrogen to make water (H_2O). The result of this is that the environment through which the blood in the tissue capillaries passes is low in oxygen. Any oxygen linked to haemoglobin is thus given up to the surrounding tissues. Therefore, all the blood returning to the heart and pumped to the lungs is low in oxygen. Then, as it passes through the capillaries surrounding the alveoli it enters an environment of high oxygen content, and is immediately, and automatically, replenished, every haemoglobin molecule picking up four molecules of oxygen. Haemoglobin linked with oxygen is a bright red colour; haemoglobin without oxygen is a dull purple. These colour differences readily distinguish arterial blood, which has just come from the lungs, from venous blood, which has just come from the tissues.

The carbon dioxide is carried mainly in the form of bicarbonate (HCO_3) dissolved in the blood serum (the fluid part) and leaves the blood in the alveoli simply by diffusing through the alveolar walls into an atmosphere of lower carbon dioxide content. So the expired air always contains substantially more carbon dioxide than the inspired air.

The control of respiration

The maintenance of an adequate supply of oxygen to the body is so important that without a feedback control mechanism to correct oxygen lack, we would be unlikely to survive. Danger arises when there is either a drop in the level of oxygen or a rise in the level of carbon dioxide in the blood. The amounts of both gases are automatically regulated. The main artery of the body (the aorta; ➪ p. 50) and two of its branches to the head (the carotid arteries) carry in their walls collections of monitoring cells, the *carotid bodies*, that can detect changes in the amounts of oxygen carried by the blood. When a drop is detected, messages are sent to the respiratory centre in the brainstem to cause it to increase the rate of respiration by sending impulses more rapidly to the respiratory muscles.

Any rise in the level of carbon dioxide in the tissues is at once reflected in a rise in the amount of carbon dioxide carried by the blood. This blood passes through all parts of the body, including the respiratory centre. As it does so it is monitored, and, if a rise in carbon dioxide is detected, the centre immediately increases the rate of respiration. A high rate of respiration often occurs as a response to acute anxiety. This is helpful if the individual is able to make a physical response (the so-called 'fight or flight' response), but if the anxiety occurs in a situation in which exertion is not needed, overbreathing (*hyperventilation*) merely results in the abnormal loss of carbon dioxide, and an increase in the alkalinity of the blood, leading to a drop in calcium and possibly to spasms of the muscles (*tetany*) and dizziness.

Breathing

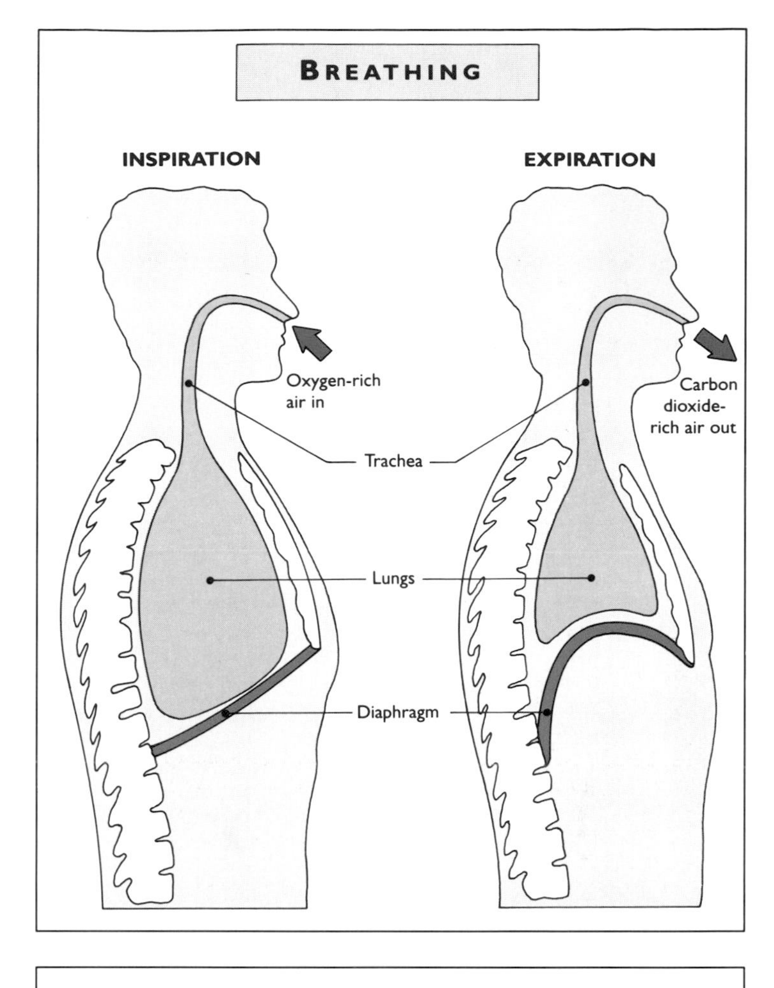

Capillaries surrounding the Alveoli

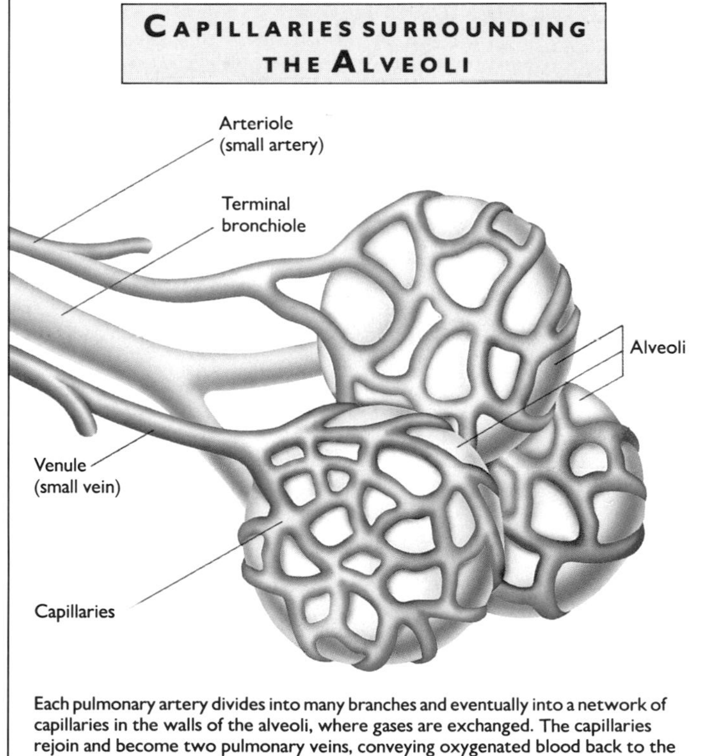

Each pulmonary artery divides into many branches and eventually into a network of capillaries in the walls of the alveoli, where gases are exchanged. The capillaries rejoin and become two pulmonary veins, conveying oxygenated blood back to the heart.

Exercise and Body Unity

There is more to the advantages of exercise than meets the eye in the shape of bulkier muscles. A study of the effects of exercise is a study in the unity and the dynamic interrelationships of all the major systems of the body – the skeleton, muscles, heart and circulation, respiratory system, the brain and nervous system, and the endocrine system.

For early humans physical fitness was often a matter of life or death. As the need to be fit is deeply ingrained in us it is not surprising that exercise is important in promoting health, and that lack of exercise is damaging. During adolescence and early adult life, the quantity and type of exercise taken has a permanent, lifelong bearing on the physical state and quality of operation of most of the body systems. This is because any part of the body grows and develops in proportion to the amount it is used, and this is particularly so during the period of maximal body growth. Exercise causes growth hormone to be released. It is difficult, later, to compensate for lack of exercise during this period. But whatever the early experience, exercise remains important throughout the whole of life. Physical improvement is still possible after a person has reached the mid-life point.

Exercise and the heart

Exercise causes a large increase in the pumped output of the heart. Heart output is the amount pumped with each heartbeat multiplied by the number of beats in a given time – such as a minute. This averages about 5 litres (10 pints) per minute during rest, and increases by several times during exertion. A reasonably fit person may, during exercise, have an output of 15–20 litres (30–40 pints) per minute, and a trained athlete can achieve an output as high as 35 litres (70 pints) per minute. Such an increase implies that a major improvement in heart performance and efficiency can be achieved simply by sustained exertion. During exercise, a great increase in heart output is necessary to provide the markedly increased blood flow though the muscles without which exercise would be impossible. With training, the main improvement is in the volume of blood ejected with each beat of the heart (the *stroke volume*), but the efficiency of pumping must also be maintained up to a high rate.

For such improvement in heart efficiency to take place, actual structural changes have to occur. These are considerable. The heart's muscle fibres become enlarged by an increase in the number of sarcomeres (➪ pp. 42–3), the number of mitochondria in each muscle fibre also increases, the branches of the coronary arteries enlarge, and the ratio of capillaries to muscle fibres becomes greater by the budding out of new vessels. As a result of these changes, the resting heart rate becomes lower, the blood pressure drops, there is an increase in the total blood volume, and the recovery to normal after exertion is more rapid. All these effects lessen the risks of coronary heart disease – the number one killer of the Western world, and an increasing health risk in less well-developed nations.

Exercise and the respiratory system

Exercise takes two forms, *aerobic* and *anaerobic*. During aerobic exercise the body is able to take in as much oxygen as the muscles use up. Anaerobic exercise, on the other hand, consists of short bursts of intense activity in which the blood supply delivers less oxygen than the muscles need. As a consequence, lactic acid builds up in the muscles causing fatigue, aches and cramp. This type of exercise cannot be sustained for long and probably does not improve the overall fitness of the heart and lungs. With aerobic exercise changes similar to those in the heart muscle are observed in the muscles used in respiration. Useful exercise is necessarily associated with breathlessness, which is simply a more forceful action of the muscles of respiration (➪ pp. 40–1) – between the ribs (the intercostal muscles) and the diaphragm. In strenuous exercise, the prominent muscles of the neck that are able to pull up the whole ribcage are also brought into action. These are called the accessory muscles of respiration. Lung expansion is greater because, with long-term aerobic exercise, the ability to increase the maximum volume of the chest improves. The effect of these changes, together with the changes in the heart, is an increase in the amount of oxygen (➪ p. 52) that can be supplied in a given time to the tissues. An improved oxygen supply increases the individual's stamina – strength, energy levels and staying power.

Exercise and the Elderly

Controlled studies of the effect of exercise on people in their seventies and eighties have shown that, after 10 months of moderate but regular exercise, oxygen consumption per unit heartbeat increases by up to 30%. Increased oxygen consumption is an indication of an increase in cell metabolic processes (➪ pp. 46–7), but also reflects an improvement in heart function, especially in the output per beat. The same studies showed that the *vital capacity* of the lungs – the maximum amount of air that can be exhaled after a maximal inspiration – could be improved by 20%. In experiments, the average person showed a 36% increase in the rate of ventilation that their lungs could achieve. At the end of the trials, the aerobic capacity of most of the men who were studied was about the same as that of the average nonexercising 40-year-old man.

Exercise and fuel supply

The energy required by the muscles during exercise derives from the mobilization of large quantities of fuel in the form of glucose and fatty acids (➪ pp. 42–3). The body's glucose is compacted into long-chain glucose polymers called glycogen. Glycogen is stored in the muscles and liver, and can readily

Polarized light micrograph of crystals of glucose. Glucose is a simple, monosaccharide sugar and is essential for energy, and indeed for life. A severe, untreated drop in blood levels rapidly leads to coma and death. (Magnification x 80 at 35 mm across.) (John Walsh, SPL)

release single molecules of glucose. Muscle glycogen is the first to be mobilized, and, for the first 10 minutes or so of exertion, this is the main source of fuel. Not all the muscle glycogen is consumed at this stage, however. After about 10 minutes the muscles begin to use a higher proportion of the glucose and fatty acids brought to them in the bloodstream by the increased blood flow associated with exercise. To prevent the blood glucose from dropping to a dangerously low level, a condition known as *hypoglycaemia*, glucose is released from glycogen in the liver. But liver stores are limited, and, as exerise continues, a larger amount of fatty acid is consumed.

When liver glycogen is in danger of being exhausted, the liver makes up the deficit in glucose by converting amino acids (➪ p. 36), glycerine (glycerol) and lactic acid into glucose. The glycerol comes from the breakdown of fats (triglycerides), which also releases fatty acids to be used as fuel. Lactic acid is produced in large quantities in the muscles during prolonged exercise, released into the blood and carried to the liver where it can be converted into glucose. So efficient is the mechanism for the maintenance of fuel supplies that strenuous exercise causes only a moderate drop in the levels of glucose in the blood, along with a drop in the secretion of insulin (➪ pp. 78–9). Cortisol from the adrenal glands and growth hormone from the pituitary gland are secreted in increased amounts during exercise. Fatigue occurs when the glycogen stores in the muscles become depleted.

Bone bulk and exercise

Bone bulk is an asset, as is well known to the 25% or so of elderly people, mostly women, who suffer hip and other fractures during minor falls, as a result of general bone loss of bulk and weakening (osteoporosis; ➪ p. 110). Both men and women suffer progressive loss of bone mass with age, but this is accelerated by inactivity and can be minimized by regular exercise. The compressive forces produced by walking or running, and the effect of muscle tension on tendons, provide a stimulus to osteoblast activity (➪ pp. 40–1) to maintain the bulk of bone. Athletes have significantly increased bulk in the bones involved in their particular activity, while people who are bedridden for long periods, or astronauts in conditions of zero gravity, can lose 1% of bone volume per week.

Morale and exercise

It is an almost universal experience that exercise induces a feeling of wellbeing. One of the effects of depression is to make the person concerned disinclined to engage in strenuous physical activity. If people with mild to moderate depression can somehow be persuaded to undertake exercise, however, the results are excellent. Trials have shown that such depression can be cured by exercise alone, even in depressives who have failed to respond to drug treatment.

Two views of exercise. Strenuous cardiovascular exercise such as that displayed here by Kriss Akabusi and Antonio Pettigrew in the 4 x 400 m relay (1991 World Athletics Championships) increases muscle bulk and improves heart and lung capacity. The gentler practice of yoga, which makes use of particular exercises and postures, has a beneficial effect on breathing, body alignment and suppleness without necessarily increasing the body's aerobic capacity. (Allsport/Colorific!)

Several factors may account for the effect of exercise on the mind. These are most readily comprehensible if the body–mind relationship is seen as inextricable, rather than as some kind of duality. In this view, it is hardly surprising that an increase in the general efficiency and ease of function of one, either mind or body, should be associated with a healthier function of the other. Exercise produces: an improved oxygen supply to the brain, leading to an increase in the capacity for action; an awareness of one's growing physical powers; the sense of physical wellbeing; pride in performance; and relief of embarrassment over physical deficiencies.

There is a further factor in the pleasure produced by exercise. *Endorphins* and *enkephalins* are natural morphine-like substances (*opioids*), each consisting of a sequence of five amino acids (➪ p. 36). They were discovered after it was found that the brain has specific receptors capable of responding only to morphine-like substances. Endorphins and enkephalins are produced by the brain and pituitary gland during, or even in anticipation of, painful or severely stressful experiences. It is these substances that prevent us from feeling pain at the time of severe injury, and they have all the pleasure-producing properties of the opium alkaloids, most especially morphine. There is evidence to suggest that taxing physical exercise is a sufficient stimulus for the production of these natural opioids. There is no doubt that many runners, in particular, experience a 'high' while exercising and some appear to have withdrawal symptoms when deprived of their exercise. Whether this is a major factor in the mood-elevating effect of exercise, however, has not been established.

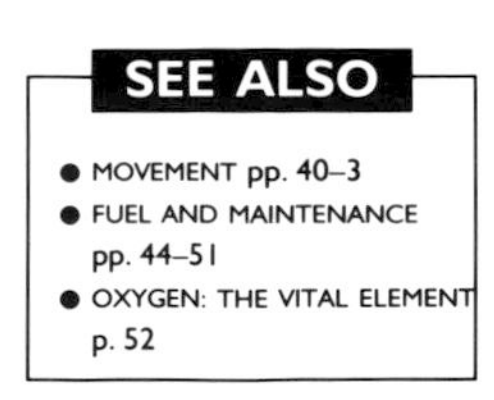
SEE ALSO
- MOVEMENT pp. 40–3
- FUEL AND MAINTENANCE pp. 44–51
- OXYGEN: THE VITAL ELEMENT p. 52

The Body Covering

The internal body environment is largely fluid. All the inner cells are bathed in tissue fluid (➪ pp. 50–1) and most of the body substance is infiltrated by fluid-containing blood and lymph vessels. The body therefore needs a waterproof outer covering to prevent the loss of this fluid. The skin, however, is much more than a simple sealant. It is a major organ in its own right with several important functions.

Skin provides a remarkable degree of protection from the many hazards of the outside world, and is self-renewing and self-repairing. It is also the main effector of body temperature regulation. Being exquisitely sensitive to touch, pressure, pain, itching, and heat and cold, it provides a major sensory interface between the body and the outer world, endlessly receiving environmental information and sending it to the brain.

The skin has a total area of 1.5–2 m² and varies considerably in thickness, from the semitransparent fineness of the eyelids, at one extreme, to the protective solidity of the soles of the feet, at the other. It affords an excellent shield against various forms of radiation, especially light and heat. Solar radiation, otherwise severely damaging because of its high ultraviolet content, is largely absorbed by the skin pigment *melanin*, and exposure to increased solar radiation results, most noticably in white people, in an automatic protective increase in melanin production (tanning). Excessive exposure, however, increases the risk of contracting skin cancer.

A healthy, intact skin offers remarkable resistance to bacterial attack, and accumulations of bacteria in the outer layer are disposed of by the constant process of shedding of this layer. Dry skin has high electrical resistance. Considerable biochemical activity occurs in the cells of the deeper layers of the skin, a notable feature being the synthesis of vitamin D from body steroids under the influence of sunlight. This is sufficient to prevent rickets in children or osteomalacia (➪ pp. 126–7) in adults adequately exposed to the sun, even if the diet is deficient in the vitamin.

The structure of the skin

The skin has two layers. The outer *epidermis*, as the name implies (*epi* is Greek for 'on' or 'above'), lies upon the inner layer, the *dermis*. The epidermis has two main functions. It generates an outermost dead layer of flat scales of keratin (➪ pp. 36–7) that is constantly being shed and replaced by cells from below, and it synthesizes the pigment melanin. The epidermis is structurally simple and has no nerves or blood vessels. It acts as a rapidly replaceable surface, resistant to, or capable of tolerating, much abrasion and wear.

Between the epidermis and the dermis is a collagen membrane called the *basal lamina*. Attached to the outer side of this is the *basal cell layer*, the deepest layer of the epidermis. It is the basal cells that give rise to all the cells nearer the surface – the *prickle cell* layer and the flattened outer layer – and it is the basal cells also that grow abnormally in the ultraviolet light-induced skin cancer *basal cell carcinoma* or 'rodent ulcer'. Most of the basal cells are called *keratinocytes* because they produce the prickle cells. The other basal cells are pigment-generating cells called *melanocytes*. The ratio of melanocytes to keratinocytes determines the colour of the skin, dark-coloured people having a high proportion of melanocytes, which produce the yellow-to-brown pigment melanin, made from the amino acid tyrosine (➪ pp. 36–7) linked to a protein. Prickle cells are the cells that grow abnormally in common warts, while, when affected by rare malignant change, melanocytes form *malignant melanomas*.

The dermis, or 'true' skin (also called the *corium*), is thicker than the epidermis, and is less uniform in structure. It has a connective tissue framework, largely made of collagen, which enmeshes blood vessels, lymph vessels, nerve endings, glands and various types of cell. The finest blood vessels are arranged in numerous finger-like structures (*papillae*) that push upwards, indenting the epidermis. A transverse cut made at about the level of the interface between dermis and epidermis – as in skin grafting – shows countless bright red points where the papillae are cut through. The papillary vessels are the capillaries (➪ p. 51) of the larger skin arterioles. The latter have an important layer of circular muscles in their walls that can greatly narrow them, so limiting the rate of blood flow through the skin. This is part of the mechanism of temperature regulation (➪ box).

The nerve endings in the dermis are of a variety of types, specialized to respond to different stimuli and to provide different modalities of sensation. They respond to deformation, such as that caused by light touch, to pricking, to strong pressure, to painful and itching stimuli and to cold, and send nerve impulses to the brain (➪ pp. 90–1).

Sweat glands

Sweat production occurs as part of the body's temperature-control and excretory systems. The sweat glands each have a deep secreting part, lying under the skin, and a duct that runs up through both layers of the skin to reach the surface. The sweat passes from these ducts through the small openings in the skin known as pores. Most of the sweat glands are *eccrine glands*, which produce a fluid that is 99% water, containing a dozen dissolved substances, mainly salt and urea. But

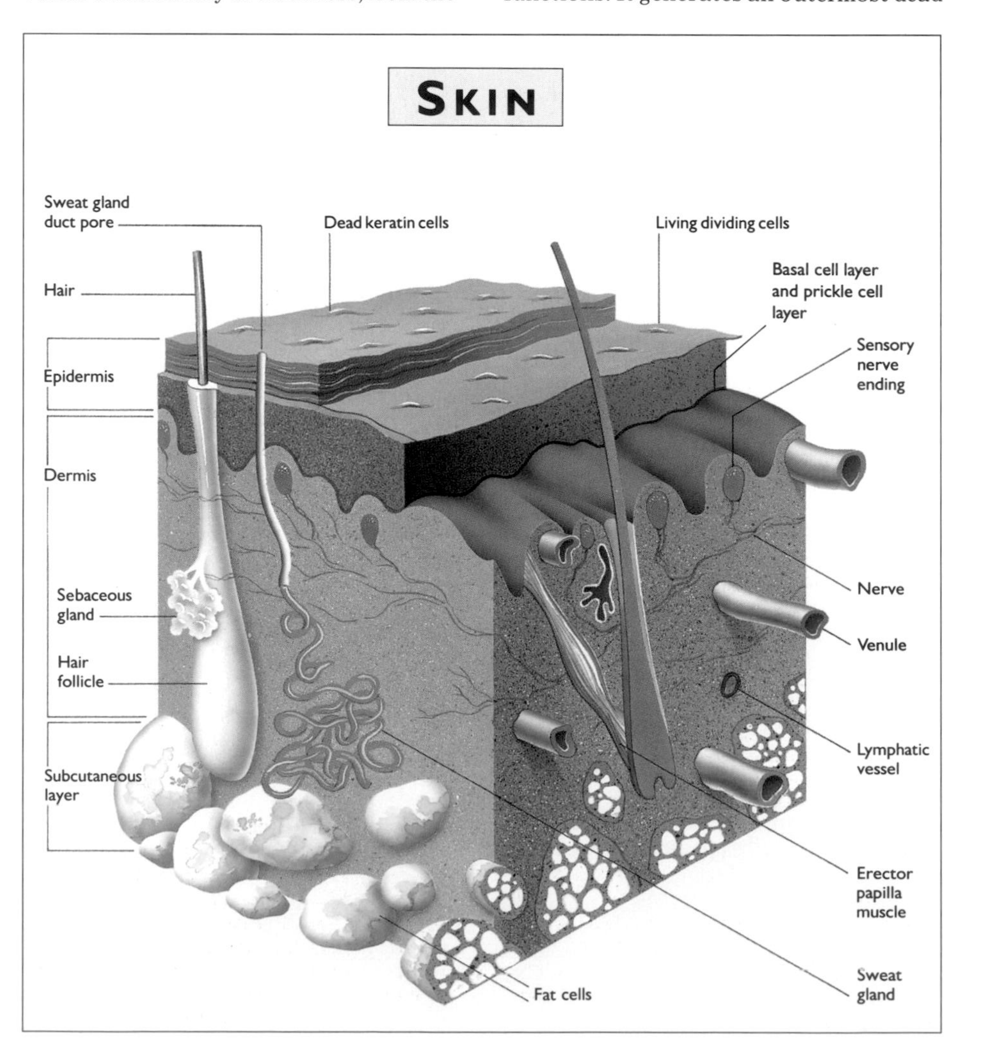

SEE ALSO
- WHAT THE BODY IS MADE OF p. 36
- CELLS, THE STUFF OF LIFE p. 38
- HEALING AND REPAIR p. 62
- WASTE DISPOSAL p. 64

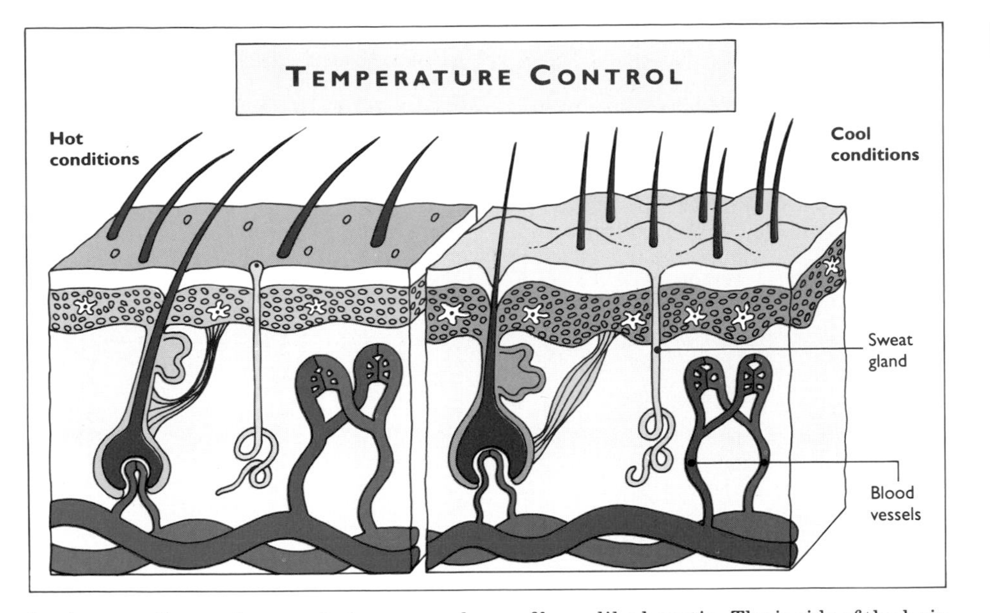

in the armpits, groins, genital areas and around the nipples, the sweat glands are of a special type called *apocrine glands*. These are larger than the eccrine sweat glands, and many of them open into the hair follicles rather than onto the skin surface. The main difference, however, from the eccrine sweat glands is that apocrine glands also throw off solid organic material from their linings. This is initially odourless, but is quickly acted on by skin bacteria to produce a characteristic body odour.

Hair

Hairs are threadlike filaments of the protein keratin, which are secreted by the hair *follicles* in the skin. These follicles are tube-like structures, wider at the bottom (the bulb) to accommodate the growing cell mass (the *papilla*). The papilla and the bulb are liberally supplied with blood vessels to bring in the enormous numbers of amino acids and the large quantity of energy fuel needed to synthesize hair protein. The papilla also contains nerve endings. Above the bulb, the tube, or sheath, of the follicle is lubricated by fatty sebaceous material (*sebum*) secreted into it by small adjoining sebaceous glands that open into the sheath. This material facilitates the slide of the hair outwards as it is continuously created at its lower end. In some areas, such as the outside of the nose, the hairs in the follicles are so small in comparison with the sebaceous glands that the pores appear to be concerned solely with sebum production.

Straight hair is produced by follicles that are circular in cross section; curly hair comes from follicles that are elliptical in cross section. Follicles that are curved along their longitudinal axis produce very curly hair. Each hair follicle has a tiny muscle, the *erector pili* muscle, lying between it and the surrounding connective tissue. When this muscle contracts, as it does under the influence of intense emotion, the hair literally 'stands on end'.

Individual hairs each have an outer layer (the *cuticle*) of overlapping flat cells. Under this is the thick *cortex*, consisting of horn-like keratin. The inside of the hair is made of softer, rectangular cells. Hair colour is determined by the concentration and depth of the melanin pigment, which produces the whole spectrum of hair colour, from black to blond. Very fair people and albinos have no melanocytes. When the melanocytes die, the hair turns grey and then white, but hair cannot spontaneously lose its colour, so stories of hair turning white overnight are apocryphal.

So far as head hair is concerned, production is periodic. For two to five years at a stretch, most of the cells of the bulb are actively reproducing and secreting the protein keratin from which the hairs are made. This growth phase is followed by a resting spell of about three months during which no growth occurs. The hairs separate from their papillae and form clubbed ends, but remain in the follicle. When growth resumes, the old hairs are pushed out by the new growing hairs.

Nails

The nails are protective plates that cover the vulnerable finger and toe ends. They are invaluable tools for many manipulative purposes, such as picking, scratching and unravelling. They also provide counter-pressure when feeling with the finger-pads, stopping the finger-ends splaying out. The nail is a curved plate of keratin – the same protein from which hair and skin is made – that rests on the *nail bed*. Nails grow continuously from the *nail matrix*, the growing zone of the bed at the root of the nail that extends about one third of the distance from the root towards the tip of the nail, taking some four to five months, in the case of fingernails, to reach the tip of the finger. Toenails grow more slowly, taking about three times as long. Damage to the matrix invariably distorts, or prevents growth of, the nail. The nail separates from the bed just short of the tip of the digit. The inturned skin edge around the nail is called the *nail fold*, and the *cuticle* is the free skin edge over the pale half moon-shaped area (*luna*). The purpose and cause of the luna have not been established.

The Skin and Temperature Control

It is essential that the temperature of the body should be kept within narrow limits in spite of wide variations in the environmental temperature. Most body functions can continue normally up to about 40° C (104° F) body temperature, but children commonly suffer convulsions at temperatures above 40.5° C (104.9° F), and permanent brain damage is likely in anyone whose temperature rises above 42° C (107.6° F). Lowered temperature is less dangerous, causing a general slowing of all metabolic processes, but, because of the retardation of vital biochemical processes, the heart stops beating (*cardiac arrest*) at very low temperatures.

Blood temperature is monitored constantly by regulating centres in the hypothalamus (➪ pp. 78–9), which is situated on the underside of the brain, just above the pituitary gland. Any variations from normal 37° C (98.6° F) automatically activate mechanisms to compensate. Extra heat is produced, when needed, by means of shivering – rapid, repetitive muscle contraction – and heat loss is prevented by contraction of the muscles in the walls of the skin arterioles so that the vessels are narrowed and the flow of blood through the skin is minimized. In fever, loss of excessive heat occurs almost entirely via the skin. Under the control of the hypothalamus, the arteriolar wall muscles relax so that the blood flow through the skin increases, and heat is lost by radiation. If this is insufficient, the sweat glands are stimulated to secrete sweat. Evaporation of sweat from the skin surface results in considerable heat loss from the latent heat of evaporation. Evaporation of 1 g (0.03 oz) of water requires 539 calories and much of this heat is taken from the body.

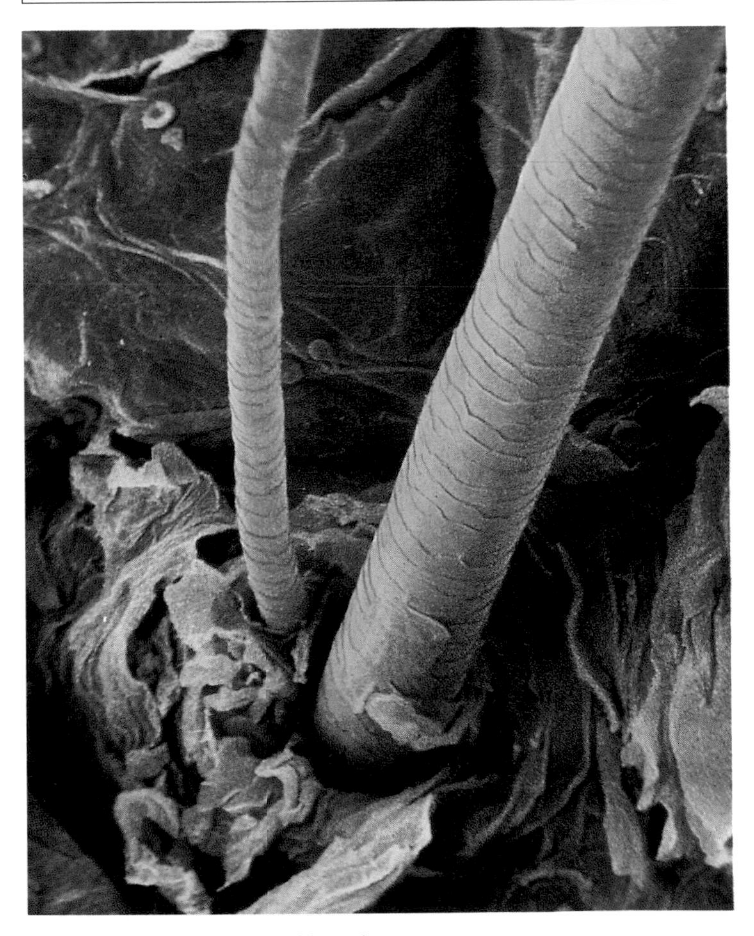

Human hairs, shown in this false-colour scanning electron micrograph, penetrating the outer layer of squamous (scaly) cells that comprise the epidermis. The overlapping cuticle cells are clearly visible. In damaged head hair, as after repeated chemical treatments, this regular overlapping pattern is disturbed. One result of this is that light is reflected less uniformly and the hair loses its shine. (CNRI, SPL)

Protection Against Attack 1

Organisms that cause human diseases have, by evolution, adapted almost perfectly to the conditions that prevail within the body – the temperature, supply of nutrients and other chemicals, the moistness and the darkness. If it were not for the operation of an elaborate system of defences, the interior of the living human body would come close to being an ideal culture medium for infecting organisms. This system of defences is known as the *immune system*.

SEE ALSO

- WHAT THE BODY IS MADE OF p. 36
- CELLS, THE STUFF OF LIFE p. 38
- PROTECTION AGAINST ATTACK 2 p. 60
- HEALING AND REPAIR p. 62
- HEALTH AND DISEASE pp. 116–59

Despite the use of the word 'immune' to describe this mechanism, we are clearly *not* immune from infection. In only a few instances is there an approximation to absolute immunity. The term arose from knowledge of an atypical disease – smallpox – in which almost complete immunity followed a single attack or followed vaccination with a similar organism – a characteristic that made possible the complete eradication of the disease (⇨ box).

In general, the body's ability to resist infection is determined, on the one hand, by the virulence of and size of the dose of organisms to which it is exposed, and, on the other, by the effectiveness of its resistance to attack. The outcome will depend on the relative strength of the opposing forces. In a healthy person, a minor assault by a small dose of organisms of low virulence is easily repulsed, while a large dose of highly virulent organisms might be overwhelming. In a person with impairment of the immune system a small dose of an organism of low virulence may lead to infection. In severe immune deficiency, the body becomes vulnerable to organisms that do not normally cause infection at all – as in AIDS (Acquired Immune Deficiency Syndrome; ⇨ pp. 60–1).

The nature of the body's defences

A primary requirement of an active defence system – a system that defends by attack – is that it should be able to distinguish friend from foe. As the immune system consists of no more than a large collection of separate and often free-moving cells of different kinds, the use of the term 'recognition' in this context is somewhat metaphorical. Immunological recognition is a chemical process concerned with the relationship of the molecular shape of 'recognition sites' (receptors) on the surface of the defending cells to molecular structures on the surface of invading organisms and other foreign material – such as a donated graft. This relationship becomes effective when the defender and the invader come into contact so that the fit between the surface molecules can be tried. The specifically but variably shaped molecules on the surface of invaders are called *antigens*. For historic reasons, arising from a less complete medical understanding than we have today, the term 'antigen' is still often applied to the whole invader. Immunological recognition displays remarkable specificity. The immune response can, for instance, distinguish very small differences between antigens on closely similar strains of influenza virus. The immune system also has a demonstrable memory. Once contact with a particular invader has been made, the response to subsequent contact with the same invader will be stronger, more rapid and more effective. This process is known as *amplification* of the response.

The class of cells responsible for the final destruction of invaders, whether organisms or other foreign material such as a transplanted donor organ, are called *phagocytes* – from the Greek *phago* meaning 'eater' and *cyte* meaning 'cell'. There are two main varieties – the *macrophages* ('big eaters') and the *microphages*, ('small eaters', also called *polymorphs*). Macrophages are large cells that live for years in most of the tissues of the body and accumulate gradually at sites of infection. They are attracted to these sites by substances released when body tissues are injured. They can engulf invaders, flowing around them by amoebic action, and contain protein-splitting and other enzymes (⇨ pp. 36–7), which they use to destroy the invaders. Polymorphs are smaller, short-lived cells that constitute the majority of the white cells of the blood, and are attracted to the site of infection in enormous numbers. They, too, actively engulf bacteria and digest them. In the process, they are often killed by bacterial toxins, and dead polymorphs are a large constituent of pus, together with bacteria and tissue debris.

Phagocytes have limited inherent power of recognition of invaders but recognition is greatly enhanced when certain other substances produced by the body have become attached to the surfaces of the invaders. These substances are called *antibodies*.

The antibodies

Apart from the phagocytes, the most important cells of the immune system are the *lymphocytes*. These small round cells are found free in the blood circulation and tissue fluids, and are also packed tightly in millions in the lymph nodes found in the neck, armpits, groins and around the main blood vessels of the chest and abdomen (⇨ pp. 50–1). They also occur in enormous numbers in the spleen. They fall into two large classes, the B lymphocytes (B cells) and the T lymphocytes (T cells).

The B cells produce antibodies with the help of the T cells. The latter also directly attack certain invaders, such as viruses. Antibodies, also known as *immunoglobulins*, are proteins that attach themselves to invading organisms or foreign substances and neutralize them so that they can be recognized and destroyed by the phagocytes.

Antibody production is a remarkable process. The B cells exist in millions of different genetically determined types. When an invader enters the body, helper T cells identify its antigens and select, from the range of B cells, the one or two

EDWARD JENNER AND THE ORIGINS OF IMMUNOLOGY

Smallpox, now one of the few diseases to have been totally eliminated, was once the scourge of nations. Fifty million people died from the disease in Europe in the 18th century and countless others were disfigured by the deep, pitting scars left by it. Since it was known that an attack conferred immunity, attempts had been made to achieve this by exposing people to material from smallpox scabs. Many of these people had serious attacks of smallpox and a large number of them died.

As a young assistant medical practitioner, Edward Jenner (1749–1823) had heard a milkmaid tell his principal that she could never get smallpox because she had contracted the mild disorder, cowpox, from a cow's udder while milking. Many years later, while in practice in Gloucestershire, it occurred to Jenner to make practical use of this suggestion. In 1796 he scratched some material from cowpox pustules on the hand of a milkmaid into the arm of a healthy young boy. The boy developed cowpox. Two months later, Jenner, with what many thought criminal foolhardiness, inoculated the boy with smallpox. Nothing happened. In 1798 he was able to repeat the experiment, and, in the same year, he published his book *An Inquiry into the Causes and Effects of the Variolae Vaccinae.* Although his temerity was criticized, others tried and proved the method, and Jenner's fame soon spread around the world. It was not long before thousands of people, including the Royal Family, had been vaccinated. In recognition of his achievement, Parliament voted Jenner £30,000.

The medical Establishment, however, was less enthusiastic than the lay public about Jenner's success. When he applied for a fellowship of the London College of Physicians he was told he would have to be examined in his knowledge of the works of Galen (⇨ p. 34). Quite rightly, Jenner refused and his application was rejected.

Edward Jenner vaccinating a small boy. From Edwin Hodder, *Heroes of Britain in Peace and War* (c. 1880). (AR)

capable of producing antibodies that will best fit the invader. Once these B cells have been selected they rapidly start to divide, and produce large collections of identical cells (clones) called *plasma cells*. These are the actual antibody 'factories'. Plasma cells pour out antibodies at the rate of many thousands per second. Meanwhile, B cells produce clonal *memory cells*. If subsequent invasion, by the same organism or material, occurs, the response is much more rapid because the memory cells can immediately clone plasma cells to produce large quantities of the appropriate antibodies, by-passing the original selection process.

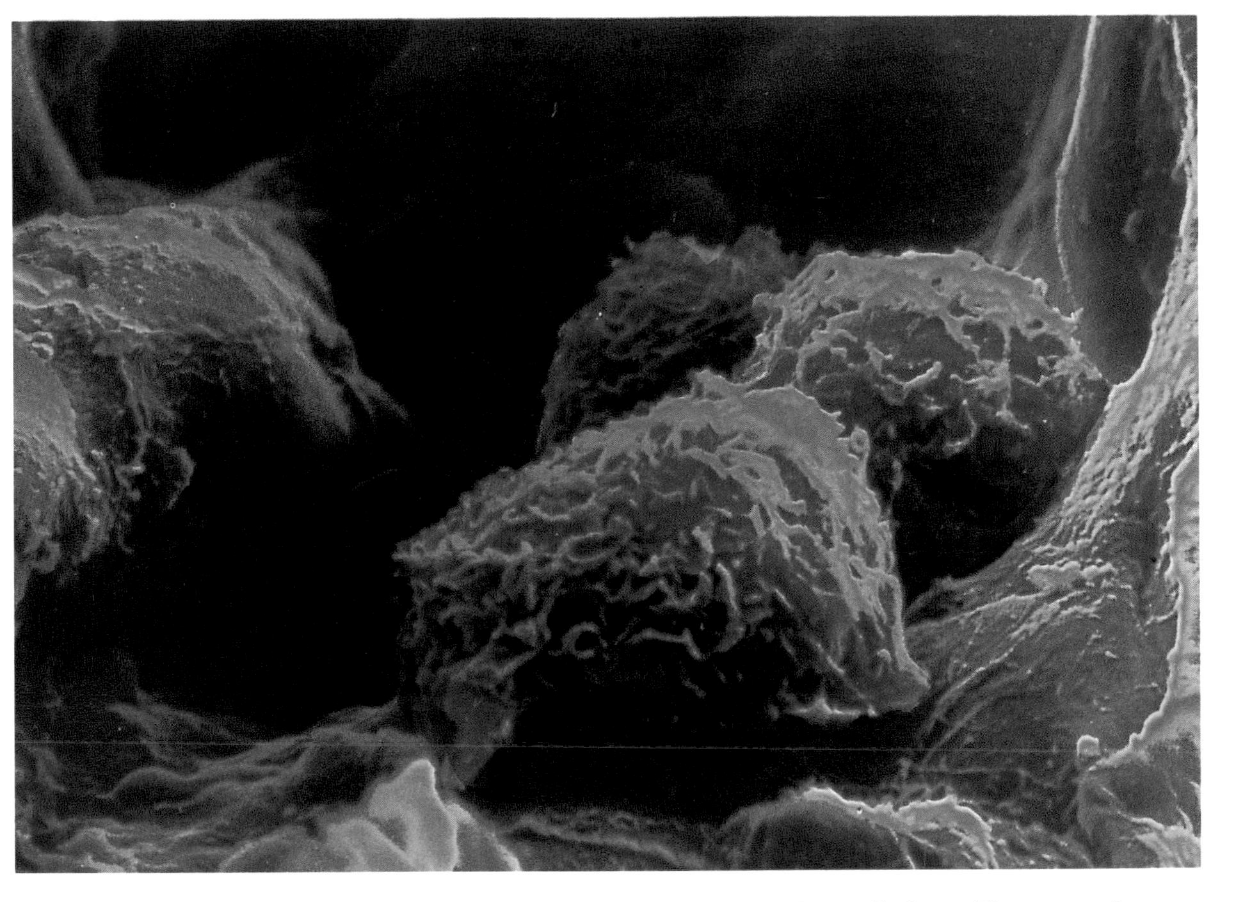

Three macrophages located in an alveolus (air sac) of the human lung. Macrophages are relatively large scavenger cells. Here their job is to clear the lungs of dust, pollen, bacteria and some components of tobacco smoke through phagocytosis (engulfing). Macrophages are, however, vulnerable to some toxic pollutants and their destruction can give rise to pulmonary disease. In cases of asbestosis and mesothelioma (a type of lung cancer) the macrophages are impaled and destroyed by needle-like asbestos particles. (False-colour scanning electron micrograph, magnification x2000 at 6 x 7 cm.) (Secchi-Lecaque, Roussel-UCLAF, CNRI, SPL)

Passive immunity

There are five classes of immunoglobulins (Ig) – IgG, IgA, IgM, IgI and IgE. Each class has a particular general function, and each may consist of any one of thousands of different specific antibodies. The most plentiful immunoglobulin is IgG (gamma globulin), which is the class involved in defence against most bacterial infections and against many viruses and toxins. Human gamma globulin and other immunoglobulins, from pooled human plasma, can be used therapeutically to help people already suffering from infections. This is called *passive immunity* and is valuable in providing protection against many infections, such as hepatitis, chickenpox, measles, rubella, tetanus and poliomyelitis. It is especially useful for people with an inherent or acquired immune deficiency. One form of such deficiency is called *agammaglobulinaemia* – a congenital absence of gamma globulin.

Active immunity

A more generally helpful and widely used method of conferring relative immunity is known as *active immunization*. To achieve this, organisms are either killed or treated in such a way as to be harmless while still carrying the specific antigens that provoke an immune response. If such organisms are suspended in an injectable fluid and introduced into the body, the T and B cell responses result in the production of large quantities of immunoglobulins and memory cells. Such a response may be effective for a lifetime but, more often, 'booster' doses are required from time to time to maintain the antibody levels (*titres*).

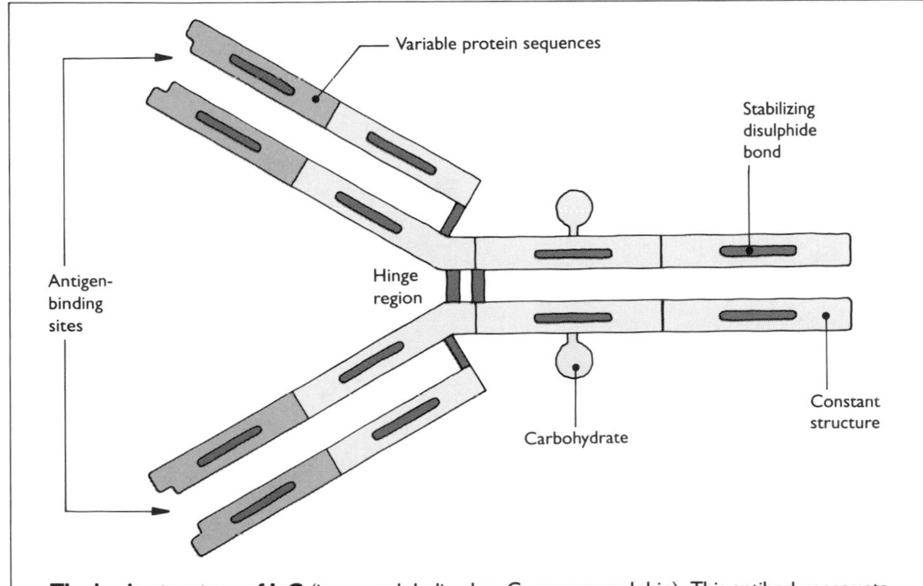

The basic structure of IgG (immunoglobulin class G, or gammaglobin). This antibody accounts for 75% of the immunoglobulins in the blood of healthy people. It is widely distributed in the body tissues and protects against a wide range of infecting organisms. Flexibility in the hinge region permits variation in the distance between the two antigen binding sites, allowing them to operate independently.

The terms 'immunization', 'inoculation' and 'vaccination' are interchangeable. The latter is a historic curiosity arising from the central part the cow (Latin, *vacca*) played in the origins of immunization (⇨ box). Immunization against infectious disease has been one of the most successful enterprises in medicine, and has been responsible for an enormous reduction in the incidence of human suffering and tragedy. In countries where acceptance of immunization is high, crippling and fatal diseases like poliomyelitis and diphtheria have been virtually eliminated, and the incidence of many other less serious conditions has been greatly reduced. Immunization has been so successful in the Western world that generations have grown up never knowing the terrors of some of these diseases. Public health authorities are seriously concerned that, for this reason, parents may neglect the immunization of their children.

Immunization is readily available against diphtheria, whooping cough, tetanus, poliomyelitis, measles, tuberculosis, mumps, rubella (German measles), influenza, hepatitis A and B, rabies, cholera, typhoid, anthrax and yellow fever. Medical opinion is that every child should be protected against the first five of these as a matter of routine policy, and, to prevent the ravages of rubella on the early fetus – blindness, heart defects and mental retardation – all girls should be vaccinated against rubella between their tenth and fourteenth birthdays.

Protection Against Attack 2

Lymphocytes (white blood cells) adhering to the interior surface of an artery in the human liver. (False-colour scanning electron micrograph, magnification x 800 at 6 x 7 cm.) (Secchi-Lecaque, Roussel-UCLAF, CNRI, SPL)

Phagocytes – literally 'eating cells' – whether large or small, are the ultimate scavengers of the body, engaged in an unremitting clean-up operation. Much of their work is concerned with picking up and digesting the bodies of foreign invaders – bacteria, viruses and other unwanted matter. But the large phagocytes (macrophages) are also concerned with the scavenging of some of our own body cells that have become dangerous to us and have had to be destroyed. The task of recognizing cells that must be destroyed, whether foreign invaders or infected or cancerous body cells, is the principal function of the immune system. To understand how this is done it is necessary to look a little more deeply into this most complex of all the body systems.

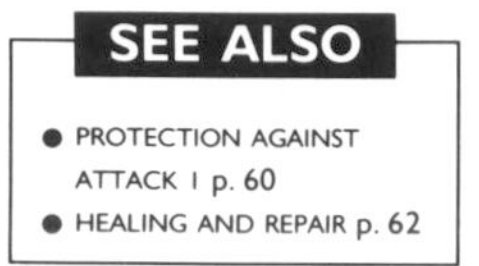

SEE ALSO

- PROTECTION AGAINST ATTACK I p. 60
- HEALING AND REPAIR p. 62

The end product of the activity of the immune system is the production of specific antibodies which latch on to invaders, such as bacteria, and inactivate them so that they can be destroyed by the phagocytes. The most numerous cells of the immune system, the lymphocytes, all look very much alike – small round cells with large nuclei – but in fact they fall into a number of radically differing groups. The production of antibodies is the function of one of the two classes of lymphocytes – the B cells (➪ pp. 58–9). The other class is made up of the T cells. In this usage, the term 'cell' is synonymous with the term 'lymphocyte'.

The T cells and their interactions

The T cells – so called because their early processing occurs in the thymus gland in the upper chest – fall into a number of distinct classes, the most important of which are the *killer* cells and the *helper* cells. Killer cells, often called *cytotoxic* cells, are aggressive lymphocytes that roam the body searching for, and destroying, cells bearing 'suicide notes'. These cells, asking to be killed, have been invaded by viruses or have developed cancerous tendencies. Their 'suicide notes' are short lengths of linked amino acids (➪ pp. 36–7) called peptides. These are formed within the cell by breakdown of the proteins of, for instance, invading viruses, and are then transported to the cell surface to be displayed, like recognizable flags, on the outside. Such peptides are antigens (➪ pp. 58–9) and occur in an enormous variety, depending on what has gone wrong in the cell.

Killer and helper T cells

Killer T cells arise from an unactivated form of T lymphocyte, the *pre-killer*. These carry antigen receptors on their surfaces that are able to 'lock on' to the peptides. When this happens the now activated T cell, able to recognize the specific peptide

WHO DISCOVERED THE AIDS VIRUS?

AIDS was first reported in 1981, in Los Angeles and New York. As soon as its importance was appreciated, an immense research effort was mounted all over the world to discover the cause. In September 1983, Luc Montagnier, head of the Pasteur Institute in Paris, sent to Robert Gallo, head of the tumour cell biology laboratory at the American National Cancer Institute, samples of a virus which had been isolated in his laboratory from an AIDS patient by F. Barré-Sinoussi, C. Cherman, F. Rey and others. They called the virus LAV. Seven months later Gallo held a press conference at which he announced he had discovered the cause of AIDS – a virus he called HTLV-III. He also announced that he had developed a blood test – an antigen–antibody reaction using the new virus as an antigen. Gallo was advised to patent the test and did so. The test proved to be a major money-spinner although Gallo, as a government employee, has a strict limit placed on his own earnings from it.

In 1985 France sued the United States government, implying that Gallo had used the French virus for the test. For years, the Gallo team had been working on this class of viruses (human retroviruses) and had described a similar virus, HTLV-I, in 1976. Workers in Gallo's laboratory, notably Mikulas Popovic, had discovered how to culture HTLV-III and to produce the large quantities needed for commercial exploitation of the test. Much other work of fundamental importance in immunology had come out of Gallo's laboratory, but efforts to trace the real origins of the virus now claimed to be the cause of AIDS and now called HIV have, so far, proved fruitless. Gallo has acknowledged the possibility that his virus may have arisen from contamination with LAV, so perhaps the truth of the matter will never be known. Eventually, the French and American governments agreed that Luc Montagnier and Robert Gallo should share the credit.

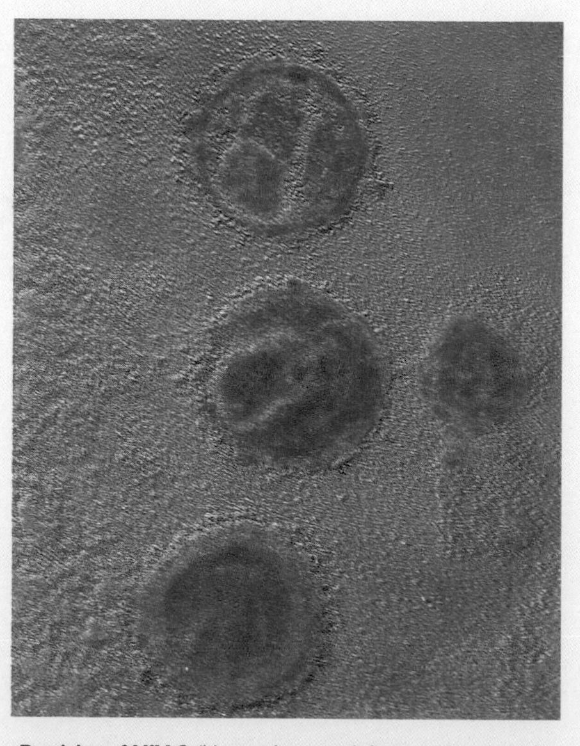

Particles of HIV-2 (Human Immunodeficiency Virus type 2). HIV-2 is more closely related to the Simian Immunodeficiency Virus (SIV) that affects monkeys than is HIV-1. However, like HIV-1, it attacks the body's white blood cells, exposing the infected person to a range of opportunistic infections. (Magnification x 57 000 at 6 x 7 cm size.) (Institut Pasteur, CNRI, SPL)

THE IMMUNE SYSTEM

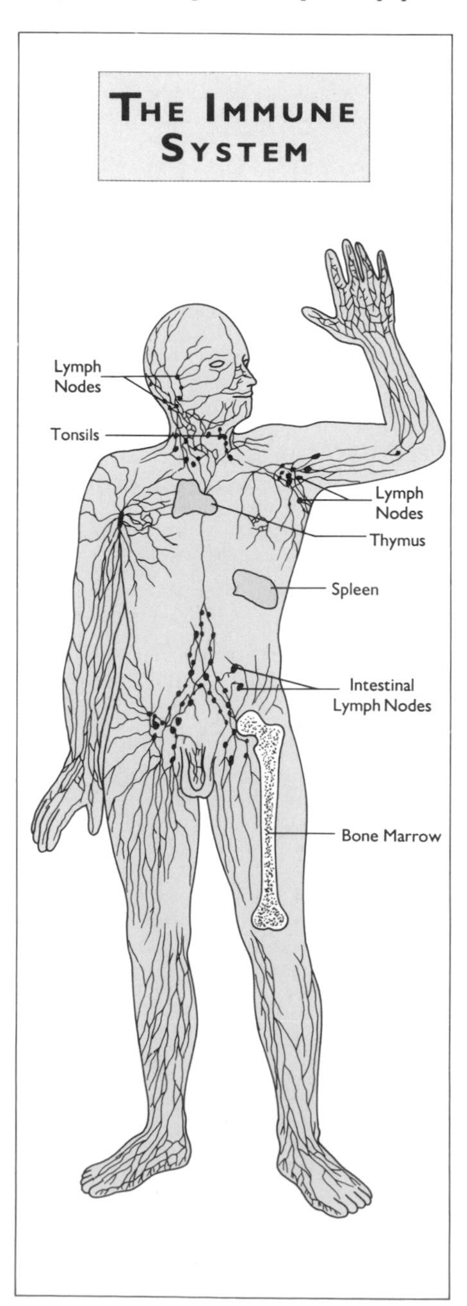

displayed by the damaged body cell, divides repeatedly. All the daughter cells (the members of the same clone) are now active killers able to deal with other cells that are putting out the same flags. They destroy the abnormal cells by the use of enzymes (➪ pp. 36–7).

The helper T cells also carry receptors for peptide flags, but these have a different, though equally important, function. Again, the unactivated T cell – the pre-helper cell – has to be converted, this time into a helper cell. This occurs when macrophages that have taken up antigens, especially bacterial toxins, put out specific flags on their surfaces. The unactivated T cells lock onto these, and are converted to helper cells. These then divide to form large families of identical helper cells (clones), which select and latch on to the correct B cells. The helper cells then secrete powerful stimulating substances called *interleukins*, which prompt the B cells to multiply and clone the plasma cells. The latter then secrete the specific antibodies to tackle the antigens.

The diversity of T-cell receptors

The antigen receptors on the surfaces of the T cells have two functions – to recognize 'self', and to identify the peptide 'flags' put out by abnormal cells. The structures of both types of antigen are determined (coded for) for by genes (➪ pp. 20–1), which we inherit from our parents. But, although the self-recognizing receptors come from only two genes in each individual, the genes for the flags on abnormal cells are inherited as many small and separate segments of DNA that undergo random recombination in the developing lymphocytes. The result is a huge collection of some 100 million genes, producing an equal number of different T-cell receptors. Many of these are exactly what are needed and do the job perfectly, binding to nonself-peptides presented by own-body cells. Some, unfortunately, are of a type that can lock on to self-peptides. Lymphocytes that can do this are able to clone gangs of cells that attack normal body cells in the process known as autoimmune disease.

Various theories have been put forward to explain why such an attack on normal body cells does not usually happen. The main argument among the experts has been about whether or not clones of T cells capable of attacking the body are automatically removed at an early stage. This has now been proved. We now know that the discrimination between helpful and dangerous T lymphocytes is made in the thymus, and that all the antigens that might later be encountered on body cells ought to be present in the thymus. T cells that do not have the receptors to bind to any sites in the thymus are useless, and die within about three days of their creation. T cells that are found capable of attacking normal cells are destroyed. Only those that can form useful killer and helper lymphocytes are allowed to proceed into the blood circulation to perform their function.

DISORDERS OF THE IMMUNE SYSTEM

Many different diseases are caused by the phenomenon known as *autoimmunity* in which cells of the immune system appear to regard certain body cells as foreign, and wrongly attack and destroy them. Such autoimmune diseases include diabetes mellitus, the muscle weakness known as myasthenia gravis, certain thyroid diseases, rheumatoid arthritis, and possibly multiple sclerosis. The precise way in which autoimmunity occurs has been unclear for many years, but it seems most likely that it is due to the survival of the dangerous T cells that are normally destroyed by the thymus. This probably occurs because of the absence from the thymus of the particular antigens that are present on the cells attacked in autoimmune disease.

Another form of immune disorder occurs when the body is infected by organisms whose antigens closely resemble those of certain body tissues. There is a particular strain of streptococcus bacterium that commonly causes severe throat infections. This organism carries antigens closely similar to those occurring on the heart lining, the lining of the joints and in the kidneys. Infection with this organism promotes the production of antibodies in the normal way, but these antibodies not only attack the organisms but also lock on to the normal peptides in the heart, brain, joints and kidneys. The result may be rheumatic fever, rheumatic heart disease, St Vitus' dance (rheumatic chorea) or glomerulonephritis (inflammation of the kidneys).

IMMUNE DEFICIENCY

This has been known for many years – long before the AIDS epidemic (➪ pp. 118–19). Several forms are present from birth and some of these are due to a genetic inability to manufacture T cells. In other less severe forms, the genes for the globulin protein from which antibodies are made are missing or defective. One such is the disorder *hypogammaglobulinaemia*, in which there is a severe deficiency of gamma globulin (IgG; ➪ pp. 58–9). Failure of normal development of the thymus can cause severe early immune deficiency, as in the DiGeorge syndrome. Infants affected in this way show many of the features of AIDS – including widespread thrush infections, *Pneumocystis carinii* pneumonia and chronic diarrhoea. Vaccination with live vaccines is usually fatal. Various other specific T-cell disorders can occur, the best known being the acquired immunodeficiency syndrome (AIDS).

AIDS

The acquired immune deficiency syndrome is caused by a virus, the human immunodeficiency virus (HIV), and is spread mainly by sexual contact but also by contact, through any breach in the skin, with infected blood or other body fluids. Intravenous drug abusers are commonly infected. The virus carries an enzyme (➪ pp. 36–7) that enables it to make DNA copies of its own genetic material in the helper T cells. After binding to the T-cell membrane the virus injects its own genes into the cell. These genes replicate rapidly inside the lymphocyte to form thousands of copies of the virus, which then burst out of the cell and spread to, and kill, other helper cells. The result is a profound deficiency of helper T cells. Some shortage of killer cells also occurs with reduced production of interleukin. B cells are not affected, and some antibody production occurs, although the only value of this appears to be to enable tests for HIV infection to be performed. The antibodies do not have any significant effect on the virus.

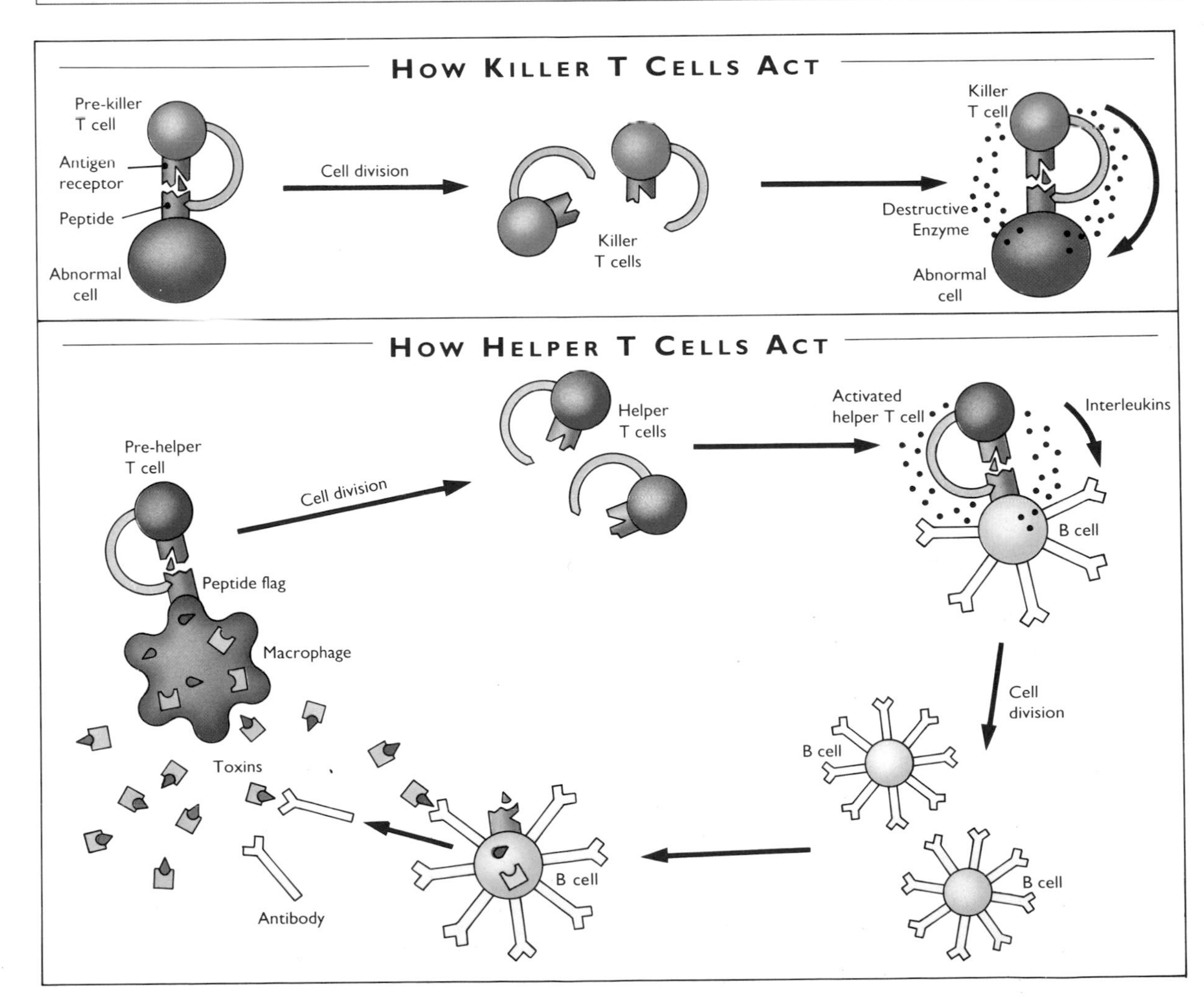

Healing and Repair

The body is exposed to so much physical injury that without a built-in system of healing and repair, few of us could survive long. Anything as mobile as the human body is bound to suffer injury of all kinds, and, without automatic repair, the accumulation of physical damage, major and minor, would be disastrous.

SEE ALSO

- WHAT THE BODY IS MADE OF p. 36
- CELLS, THE STUFF OF LIFE p. 38
- MOVEMENT pp. 40–3
- THE BODY COVERING p. 56
- PROTECTION AGAINST ATTACK pp. 58–61

Surgery is implicitly based on this fundamental healing response of tissues, and would be impossible without it. Structural damage from infecting organisms, although less visible, would, on aggregate, be equally dangerous, and would often be more rapidly fatal than apparently more serious injuries. Failure to repair such microinjuries would soon lead to widespread and fatal tissue destruction. But, even in cases of severe immune deficiency (⇨ pp. 60–1), the processes of repair continue as normal. Healing, whether of obvious wounds or of microscopic tissue damage, is a dynamic series of events that involves remarkable action on the part of many body cells.

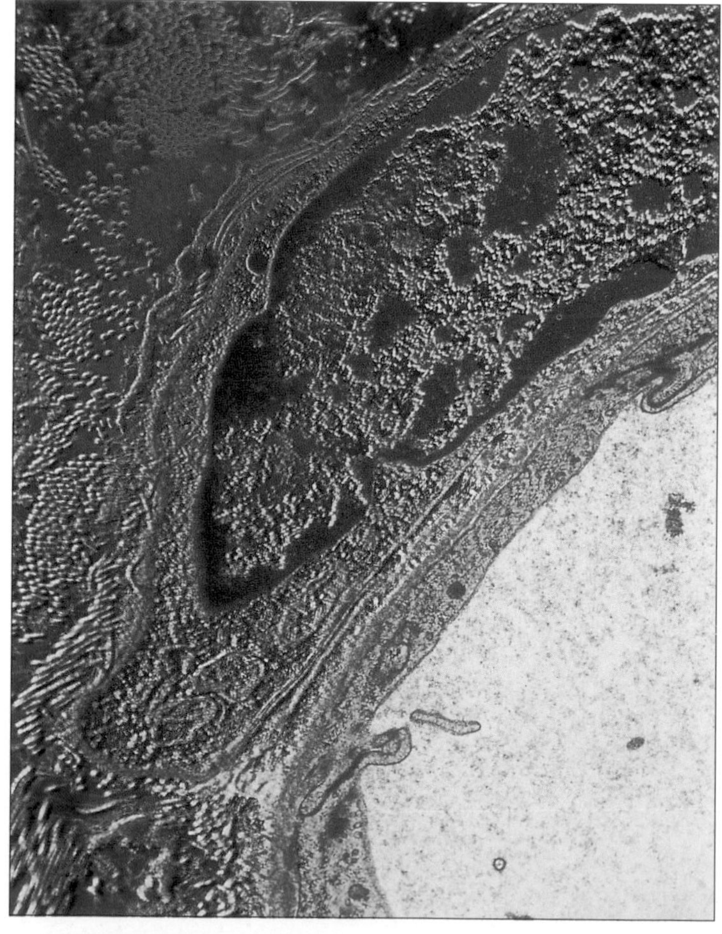

A fibroblast cell (green and red at centre top) adjacent to a blood capillary (white structure at bottom right). The lining of the capillary, made of a layer of single flattened cells (endothelium), appears yellow. Fibroblasts are responsible for the synthesis and secretion of collagen (pink-red dots at top and left), the protein that in large part makes up connective tissue and bone. (False-colour transmission electron micrograph, magnification x 600 at 35 mm across.) (CNRI, SPL)

The area of the body most exposed to environmental injury is, as one would expect, the skin surface. Healing and repair are so frequently necessary here that we have evolved a system of continuous replacement of the surface layer – a system that continues to operate whether injury occurs or not. As the cells of the outer layer, the epidermis, are pushed out towards the surface, they die, become flattened and hardened and are eventually cast off (⇨ pp. 56–7). The effects of minor surface, injuries, such as abrasions and shallow cuts, are thus eliminated, and a normal surface restored. When necessary, the skin even responds to the threat of injury, or to undue sustained local pressure, by an automatic protective thickening of this layer – the formation of calluses. Deeper injuries, however, which involve tissues below the level of the epidermis, need a more elaborate repair process.

Closed-wound healing

Disruptive injury to deep tissues immediately brings into action an impressive series of changes. At first, all the small blood vessels around the injury constrict, but then almost immediately widen to a more-than-normal degree. As they do so the smallest vessels – the capillaries – become abnormally permeable, and proteins from the blood pass into the injury site. Large numbers of white blood cells of the immune system migrate through the walls of the capillaries and congregate on and around the injured surfaces. This process is called inflammation, and it is the body's basic response to injury of any kind. The inflammatory reaction can often, by itself, cope with the problem. Torn blood vessels leak blood into the wound space, and this soon clots so that the whole area between the surfaces becomes filled with semisolid material. If the wound is clean and relatively free from organisms and foreign material, the white blood cells will be able to cope with the organisms and minor tissue debris present. The area will rapidly become sterile, and the healing processes may proceed. If there is much contamination or foreign material, however, the inflammation may persist for months as the immune system tries to dispose of the irritants, and healing may be delayed. The basic surgical management of a wound is to cut it wide open, remove all foreign matter and dead or dying tissue, and bring the edges closely together with deep stitches.

If the edges of a wound are kept together, surface healing is rapid. Within 24 hours the edges of the wound begin to thicken, and the living cells of the epidermis enlarge and migrate across the wound plane. This is a rapid process, and, in about 48 hours, the whole wound surface will have acquired a new covering. This is called *epithelialization* (⇨ below). The healing of the deeper part of the wound still has a long way to go, however. At this stage the edges are held together only by delicate strands of the protein fibrin, which forms the main part of the blood clot. It is now that the definitive healing cells, the *fibroblasts*, come into action. Fibroblasts are highly active cells that might be considered as mobile protein 'factories'. They have oval nuclei and are of varying contour, but, when active, they are usually star- or spindle-shaped. Within two or three days of wounding they appear in their millions, probably from the surrounding tissues, where they have been lying in a resting state. By about the tenth day, fibroblasts are the most numerous of the cells present. Each one contains a large, rough endoplasmic reticulum (⇨ pp. 38–9) covered with enormous numbers of ribosomes – the cell organs responsible for protein synthesis under the control of DNA (⇨ pp. 24–5). Fibroblasts are programmed to synthesize the structural protein collagen (⇨ pp. 36–7), and each one busily pours out new, long, delicate collagen fibrils into the wound. Fibroblasts use the fibrin strands in the blood clot as scaffolding, and actively move along them from one side of the wound to the other, generating collagen as they go. They do not move, as many cells do, by putting out protrusions and flowing into them (amoeboid action), but by way of tiny, adhesive 'feet' that move from front to back, thus propelling them forwards.

The stage of collagen formation lasts for several weeks. During this time the fibroblasts produce a quantity of the protein that is enormous in relation to their size. The collagen fibres form bundles of ever-increasing thickness, that are randomly oriented between the tissue surfaces, and eventually build up into a massive collagen structure that binds the edges of the wound together. This structure is called a *scar*. Over the course of subsequent years the scar gradually changes in bulk and appearance, usually remodelling itself so that it becomes stronger and less conspicuous. Occasionally, collagen production is excessive and results in an ugly, heaped-up scar known as a *keloid*.

Open wounds

The way that wounds heal that are left open for want of medical attention, or that cannot be closed because the skin edges are too far apart, is even more remarkable. Soon after the injury, the surface of the wound becomes covered with fluid that oozes from the edges. This fluid, called plasma, is invaded with fibroblasts and with large numbers of buds of tiny blood vessels that grow out from the raw surfaces of the wound. These growing vessel buds produce a rough, granular appearance, and were long ago given the name of *granulation tissue*. At the same time, epidermis from the wound edges

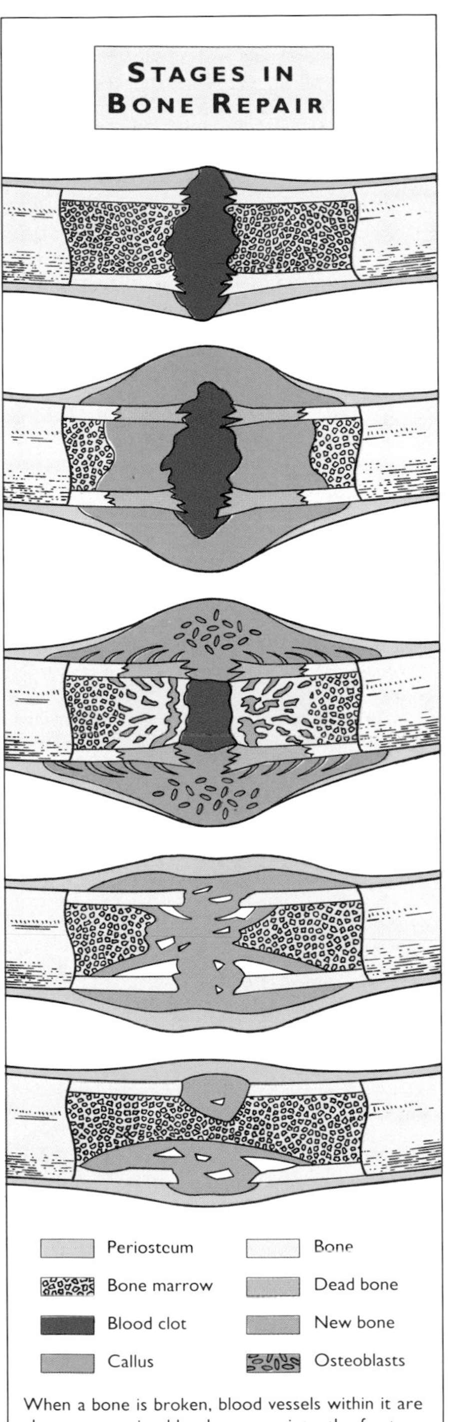

When a bone is broken, blood vessels within it are also torn causing blood to pour into the fracture where it forms a clot. The broken arteries go into spasm causing bone at the fracture site to die. Soon, however, fibroblasts appear on the site where they differentiate into bone-forming osteoblasts and cells that make up the tissue that surrounds the bone (periosteum). Dead bone is absorbed and replaced by the action of the osteoblasts in the callus.

begins to grow down the raw surfaces. After two or three days the edges of the wound begin to be pulled towards each other, so that the size of the wound is greatly reduced. No new skin is formed and wound contraction involves stretching of the surrounding skin.

Why open wounds contract remains obscure. Although a lot of structural collagen is formed by the fibroblasts in the wound, collagen is not a contractile protein, and does not cause wound contraction. It is possible that granulation tissue contains specially modified fibroblasts capable of shortening in a manner similar to muscle cells (➪ pp. 42–3), and pulling the edges of the wound together. Although almost all wounds that are left open will eventually contract and heal over, the process often has undesirable side-effects, such as limitation of joint movement by shortening of the surrounding skin, or the production of ugly, wide scars. For these reasons, the medical treatment of wounds that cannot be closed is almost always by skin, or skin and muscle, graft from another part of the body. Application of a graft prevents wound contraction, as does the application of steroid drugs, which simulate the action of some of the adrenal gland hormones (➪ pp. 80–1).

Bone healing

The structure of bone (➪ pp. 40–1) is a scaffolding of collagen impregnated with mineral salts, especially salts of calcium and phosphorus. Bone contains many blood vessels, which allow a constant interchange of materials. When a bone is broken a blood clot forms between the broken ends, and, as in a soft-tissue wound, large numbers of white cells collect in this clot. Fibrin that has formed in the clot connects the broken ends of the bone, but, in relation to the forces acting on a fracture site by the weight of the parts concerned, provides no structural strength. Again, as in soft-tissue wounds, large numbers of fibroblasts invade the site, using the fibrin strands as pathways. Fibroblasts in bone are called osteoblasts and are either identical to or similar to the fibroblasts in soft-tissue wounds.

The process of laying down collagen to re-establish the shape of a bone requires a higher level of organization than the formation of scar tissue elsewhere, and another kind of cell, the osteoclast, is involved. Osteoclasts are large cells with multiple nuclei, and are thought by some to develop from fibroblasts. They are, however, destructive rather than constructive, and more closely resemble *macrophages* (➪ pp. 58–9). Osteoclasts secrete enzymes (*collagenases*) that break down collagen and release minerals. These enzymes allow the osteoclasts to remove unwanted pieces and spicules of bone, and prepare the site for new collagen deposition in an orderly manner. The cooperative action of these two cells, the osteoblasts and the osteoclasts, will lead, in the course of a few weeks, to the replacement of the blood clot by a mass of semisolid tissue called *callus* that bridges the gap between the bone ends. Within this mass, new bone is being formed by a simultaneous process of collagen synthesis and mineralization. Inadequate immobilization will damage the callus bridge, and may lead to the production of a false joint of gristle (cartilage). Immobilization methods include: plaster of Paris or cold-setting plastic splints, which take various forms depending on the nature of the fracture; external fixators consisting of a steel rod to which strong pins, screwed into the bone, are attached; traction, usually applied by weights attached to cords, often secured to steel pins transfixing the bone; internal fixation (open operation) using screws, bolts, metal plates, or metal rods within the bone.

Surfaces and adhesions

All body surfaces, inside and out, are covered with a nonstick layer. This is called an *epithelium*, and it is an essential guard against the body's tendency to heal surfaces together. If the pads of a finger and thumb were to be sewn together, no healing would occur even if left for several weeks. But if, before stitching, the skin were removed over the areas to be brought in contact, healing would occur rapidly and would be secure, without stitches, in about two weeks. Local loss of the epithelial covering of internal structures can cause much trouble because of the tendency for raw surfaces to heal together (*adhesions*). Such loss may result from surgical intervention or from certain uncommon disease processes that damage surface cells. Adhesions between abdominal organs can cause distortion and other problems such as dangerous intestinal obstruction. Fortunately regeneration of damaged epithelium is normally rapid and effective, and, in the absence of persisting disease, will usually restore the normal nonstick condition of surfaces.

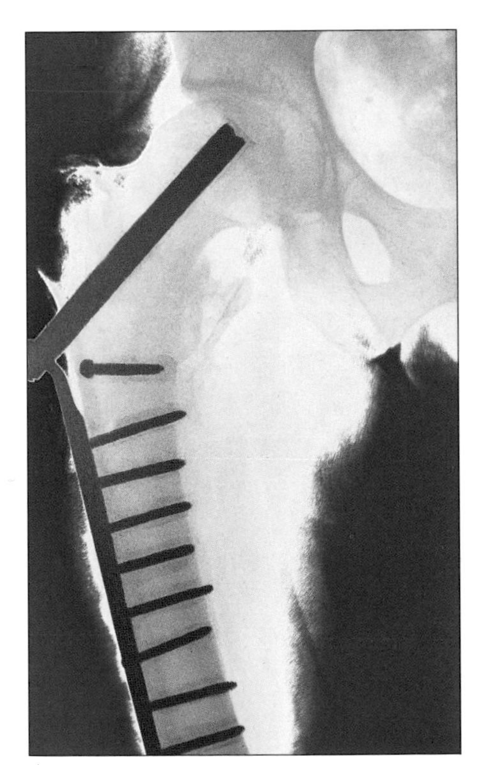

Broken femur (thigh bone) pinned into position, shown in false-colour x-ray. The fracture occurs at the neck of the femur near to where the head fits into the acetabulum in the pelvis to form the ball-and-socket hip joint. Provided that the bone is not deprived of an adequate blood supply, healing can be expected to take place within two to three months. (SPL)

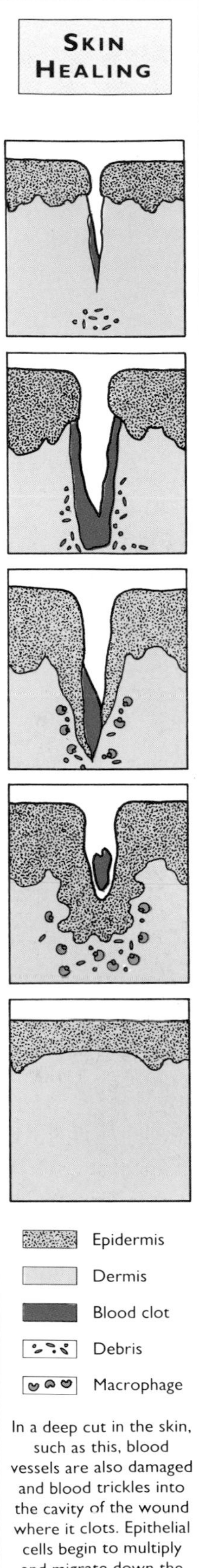

In a deep cut in the skin, such as this, blood vessels are also damaged and blood trickles into the cavity of the wound where it clots. Epithelial cells begin to multiply and migrate down the edges of the wound. Simultaneously, macrophages remove dead cells, bacteria and any other debris, while capillaries grow into the blood clot to supply blood to the multiplying dermis and epidermis cells.

Waste Disposal

Like any other chemical engineering plant, the human body continuously produces unwanted waste material, which must be disposed of.

Much of our energy is derived from carbohydrate, which leaves an effluent consisting of carbon dioxide (CO_2) and water. Carbon dioxide is the major waste product of body chemistry (metabolism) and is disposed of, along with water vapour, with every expired breath. The amount of water lost in the breath is roughly equal to that produced by the process of metabolism (➪ pp. 46–7).

Sir William Bowman and his Capsule

William Bowman was born in the English west Midlands in 1816. His medical training began with an apprenticeship to a Birmingham surgeon, after which he studied at King's College Hospital, London, and in Europe. At the age of 23 he decided to devote himself exclusively to scientific research, in the course of which he was the first to describe the hollow capsule that surrounds each of the millions of glomeruli of the kidney (➪ main text). A year later, in 1840, he was appointed Professor of Physiology and General and Morbid Anatomy at King's. There he did not waste his time. In 1841 he became a Fellow of the Royal Society, in 1844 a Fellow of the Royal College of Surgeons, and in 1845 he published the first of five volumes of the major work *Physiological Anatomy and Physiology of Man* (1845–56). He was made a Baronet in 1884.

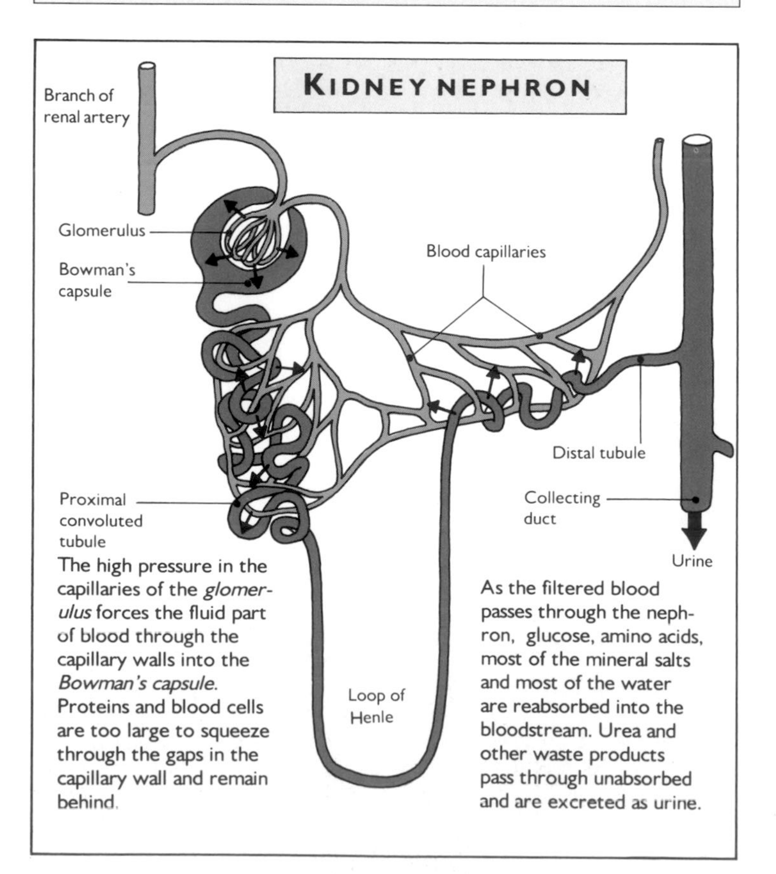

The high pressure in the capillaries of the *glomerulus* forces the fluid part of blood through the capillary walls into the *Bowman's capsule*. Proteins and blood cells are too large to squeeze through the gaps in the capillary wall and remain behind.

As the filtered blood passes through the nephron, glucose, amino acids, most of the mineral salts and most of the water are reabsorbed into the bloodstream. Urea and other waste products pass through unabsorbed and are excreted as urine.

The breakdown of the protein fragments, amino acids (➪ pp. 36–7), to form carbohydrates or fats involves a process known as *deamination* – removal of amino chemical groups that contain ammonia (NH_3). In any quantity, this ammonia would be poisonous, so it is converted in the liver to a stable and inert substance called urea (➪ pp. 48–9). About 30 g (1 oz) of urea is produced each day. This is the main waste product that arises from protein breakdown. A principal function of the body's urinary excretory system is to dispose of urea.

The two kidneys – reddish brown, bean-shaped structures, each about 11 cm (4½ in) long – lie in pads of fat high up on the inside of the back wall of the abdomen, one on each side of the spine. Kidney function, in disposing of urea and in maintaining correct levels of other substances in the blood, is essential to life. Its importance is reflected in the location of the kidneys close to, and on either side of, the main artery and vein of the body – the aorta and the inferior vena cava (➪ pp. 50–1). The kidneys are connected to these large blood vessels by short, wide arteries and veins – the *renal* vessels – and they receive blood under high pressure from the aorta. Approximately 125 ml (4 fl oz) of blood is filtered by the kidneys every minute – a total of some 180 litres (317 pints) per day.

How do the kidneys work?

The functional unit of the kidney is called the *nephron*. This is a microscopic structure of which there are about one million in each kidney. The nephron consists of a filter and a complex system of tubules. The filter is simply a small tuft of blood capillaries called the *glomerulus*, surrounded by a cup-shaped, double-walled structure known as *Bowman's capsule*, which is best regarded as a thin-walled, hollow sphere pushed in (invaginated) at one side by the glomerulus. The hollow interior of the capsule leads into the start of the tubule system of the nephron. From the capsule the tubule takes a few twists and turns and then runs straight for a time before looping back to the region of Bowman's capsule, where it has a few more twists, before ending in a urine-collecting duct (➪ below). The tiny blood vessels that form the glomerulus are derived from branches of the renal artery. On exiting from the glomerulus they form a dense network of capillaries, surrounding the entire tubular system of the nephron. The capillaries drain into tiny veins that join to form the draining veins of the kidney (renal veins). These empty into the inferior vena cava. All the blood entering each nephron thus passes first to the glomerular capillary tuft inside Bowman's capsule, then passes on to come into intimate contact with the whole of the nephron tubule system before returning to the circulation.

The formation of urine starts with a massive filtration of blood plasma (blood from which the cells have been removed) through the glomerulus into Bowman's capsule. This filtrate, however, contains a lot of water and much low-molecular-weight material that is essential to the body; large-molecule substances, such as proteins (➪ pp. 36–7) and fats, are too large to pass through the gaps in the capsule wall. The final composition of the urine is determined by the remarkable powers of selective reabsorption and secretion of the tubule system, because if all the filtrate were lost as urine we would be dead from dehydration in a matter of hours. As the crude filtrate passes down the tubules from Bowman's capsule, much of the water, together with any substances needed by the body, is reabsorbed into the blood. At the same time, certain unwanted substances that have not passed into Bowman's capsule are actively secreted into the urine by the tubule. Sodium, potassium, calcium, chloride, bicarbonate, phosphate, glucose, amino acids, vitamins and many other substances are returned to the blood and conserved. Selective reabsorption is a highly complicated process under the control of various hormones from the adrenal gland (➪ box), the pituitary gland (➪ pp. 78–9) and the parathyroid glands.

Urine

In health, urine is a sterile (organism-free) solution containing water, urea, uric acids and several inorganic salts in varying concentrations. By adjusting the amount of water and acidic substances reabsorbed, and hence the composition of the urine, the kidneys are largely responsible for ensuring that the body contains the right amount of water and that the blood is of the correct degree of acidity. Although a large fluid intake is desirable, we almost always drink more water than is physiologically necessary, so the kidneys regularly excrete urine of appropriate concentration to maintain the body's water balance. On average, about 1 ml (0.04 fl oz) of urine is formed per minute – a daily output of 1200–2000 ml (48–80 fl oz). The volume produced varies greatly with variations in fluid intake and the amount of fluid lost in sweat.

The urine is usually acid and contains various products from blood and body-cell breakdown. The yellow colour comes from the bile. Most drugs or their breakdown products are also eliminated through the kidney. Many diseases, but especially those of the kidneys, cause characteristic variations in the constitution of the urine, and, because of this, laboratory examination of the urine can often provide valuable diagnostic information.

The urinary drainage system

The urine-collecting tubules form a separate branching system of ever-widening ducts that end in a conical drain called the *renal pelvis* (Latin, *renes*, 'kidneys' and *pelvis*, 'basin'). This runs into a hollow tube, the *ureter*, 40–5 cm (16–18 in) long, which descends to enter the lower rear wall of the *urinary bladder*. The ureter is not a passive tube. Its wall contains circular muscle fibres capable of peristalsis (➪ pp. 44–5) so that the urine can be 'milked' down to the bladder. The power of these muscles is well demonstrated by the agonizing pain caused when they attempt to move a urinary stone downwards (*renal colic*).

THE URINARY SYSTEM

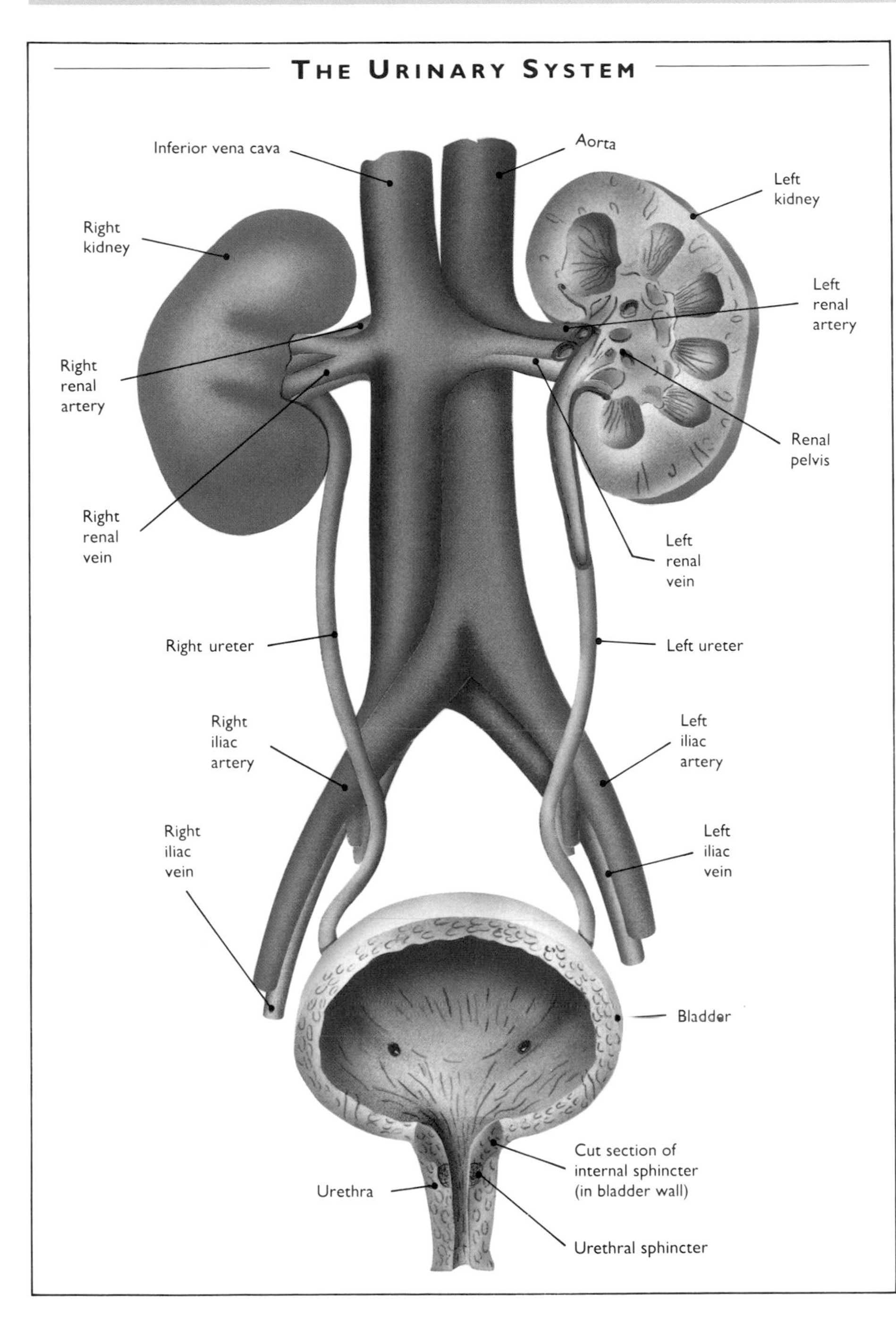

The bladder is a muscular bag lying low in the pelvis, immediately behind the pubic bone. In the male, the outlet drainage tube, the *urethra*, occupies the underside of the penis. In the female, it runs down just in front of the front wall of the vagina, opening between the vagina and the clitoris (➪ pp. 96–7). As urine enters the bladder from the kidneys it relaxes progressively to accommodate the increasing volume, storing it until it may conveniently be disposed of. When full, the adult bladder contains about 350 ml (¾ pint) of urine, and this volume is retained by the tight contraction of muscles surrounding the first part of the urethra. In health, we are unconscious of the accumulation of urine until the pressure in the bladder reaches about 20 cm of water (that of the weight of a column of water 20 cm high). At this point there is a conscious desire to urinate, and if the inclination is gratified, voluntary relaxation of the urethral sphincter (the muscular ring around the urethra just below the bladder) allows the urine to pass to the outside by way of the urethra. If the desire to urinate is repressed, the bladder relaxes a little, and the impulse ceases. But as pressure rises higher, the impulse returns. This sequence may be repeated several times, but occurs at ever shorter intervals until there is a continuous and urgent demand for release, accompanied by rhythmical contraction of the bladder muscle. At this extremity, control can be exercised only by a major effort of will on the part of the now wholly preoccupied subject who is tensely contracting all the muscles in the floor of the pelvis. Eventually, at a pressure of about 100 cm of water, regardless of the circumstances, the sphincter relaxes and the bladder spontaneously empties.

OTHER KIDNEY FUNCTIONS

The kidneys have functions other than excretion. When the blood pressure falls below normal or the blood volume drops, the kidneys automatically release an enzyme called *renin* into the blood. Renin acts on a large blood protein to split off a small polypeptide (➪ pp. 36–7) called *angiotensin I*. As the blood containing angiotensin I passes through the lungs it encounters a converting enzyme that changes angiotensin I to *angiotensin II*. The latter hormone acts on the muscle in the walls of blood vessels causing them to constrict. This at once raises the blood pressure. Angiotensin II is also a powerful stimulator of the hormone aldosterone from the adrenal glands, and a stimulus to thirst and sodium appetite. Aldosterone acts on the kidney tubules, causing them to reabsorb sodium and retain water. This kind of automatic feedback control mechanism (servomechanism) is typical of the way the body works. An important group of drugs for controlling blood pressure, the *angiotensin-converting enzyme inhibitors* (ACE inhibitors) work by this principle. The kidneys also produce a substance, *erythropoietin*, which stimulates the rate of formation of blood cells in the bone marrow.

Intestinal waste disposal

Food residue, mainly cellulose, generally forms only a small proportion of the faeces (excrement). Only people on a very high-fibre diet have much food residue from cellulose (for which we have no digestive enzymes). The faeces consist of about two thirds water and one third solid matter. The latter consists largely of harmless bacteria that have multiplied in the colon, some cellular debris scraped from the lining of the bowel (epithelium; ➪ pp. 62–3), food cellulose, bile pigments and small amounts of salts.

Defecation, the process of emptying the lower bowel, is initiated by the stretching of the wall of the part of the bowel immediately above the anal canal (the *rectum*) by the mass movement of faeces into it from the colon. Stretching causes a conscious desire to defecate, but this is readily inhibited by voluntary decision, and is not experienced again until the next movement of faeces from the colon. This can occur repeatedly. Undue deliberate inhibition leads to constipation and hardening of the stools (faeces) by progressive water removal by absorption through the bowel wall. Voluntary defecation involves relaxing the two anal sphincters, taking a deep breath, suppressing the breathing and contracting the abdominal muscles so that pressure is applied to the outside of the rectum.

SEE ALSO
- WHAT THE BODY IS MADE OF p. 36
- FUEL AND MAINTENANCE I: DIGESTION p. 44

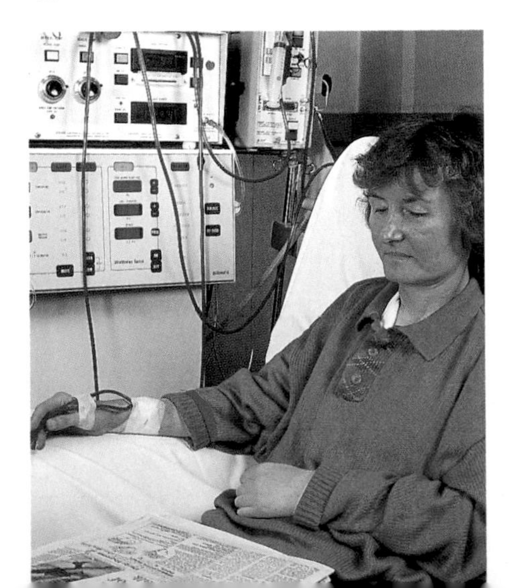

After kidney failure the sufferer relies on renal dialysis to replace the lost function of the kidneys. A stream of blood from one of the patient's arteries is circulated through the dialysis machine, which removes waste materials and returns the purified blood to the patient through a vein. (Simon Fraser, Royal Victoria Infirmary, Newcastle upon Tyne, SPL)

Biorhythms

Periodicity is very much a feature of the human organism, which is locked into a number of body functional cycles (biorhythms), especially the diurnal 24-hour (circadian) clock rhythm.

SEE ALSO

- HEALING AND REPAIR p. 62
- THE CONTROL SYSTEM pp. 74–81
- VISION p. 84
- DREAMS p. 178

Other biorhythms include the regular patterns of brain waves revealed by the electroencephalogram (⇨ pp. 128–9), the roughly monthly occurrence of the menstrual cycle in women (⇨ pp. 96–7), sleep cycles (⇨ pp. 178–9), and the cyclical release of hormones in pulses at intervals of minutes or hours. These rhythms are controlled by biological clocks – self-sustaining timing mechanisms that control the occurrence or varying intensity of physiological events. These clocks are essentially free-running, but some, if not all, of them are maintained at a fixed frequency by synchronization with some external influence. At present, we know most about physiological events that have a circadian periodicity. It is not surprising that human functioning is regulated by 'clocks' with a periodicity linked to natural phenomena, such as the cycle of the day. Throughout the course of evolution, natural events, especially the sequence of day and night, must have had a profound effect on all organisms. These effects are demonstrable in many species today. The fiddler crab, for instance, even under laboratory conditions, shows a resting–activity cycle of 24 hours, 50 minutes – the length of the lunar (tide) day. This activity is synchronous with low tides caused by the gravitational effect of the Moon.

Children need more sleep than adults. Babies sleep for up to 16 hours in every 24, adults only about 7 or 8. (Popperfoto)

Dangerous Circadian Patterns

Circadian patterns occur in several common disorders that frequently have a fatal outcome. It has been known for some time that the incidence of heart attacks, of sudden death from heart disease, and of strokes is significantly higher during the early morning hours than at other times of the day. Scientists have identified a number of circadian mechanisms that contribute to this state of affairs. The ability of the blood to break down clots that have formed (*fibrinolytic activity*), and so prevent thrombosis (⇨ pp. 126–7), is lowest in the early morning. The tendency for mutual adhesion of *blood platelets* (the cell fragments that initiate clotting by clumping together) is greatest at that time. Adrenaline (⇨ pp. 80–1) and cortisol levels (⇨ above) and blood pressure are highest at that time, as is the activity of the enzyme renin (⇨ pp. 64–5). Perhaps most important of all, the tendency for arteries, such as the coronary arteries of the heart, to narrow (by tightening of the muscles in their walls) is significantly greater in the morning than later in the day. This results in reduced blood flow to the parts supplied by the arteries, and is partly or wholly due to the effect of adrenaline on the arteries. Studies of the efficiency of small doses of aspirin in preventing heart attacks by reducing platelet stickiness have shown that they are most effective in the mornings.

Circadian rhythms in humans

Many important physiological processes have a circadian periodicity. The natural adrenal steroid hormone *cortisol* controls many vital body processes and is essential for life. Production of this hormone rises at the time of waking, peaking at about 9 a.m., and is at its lowest level at about midnight. This is the result of an increased output of a stimulating hormone, adrenocorticotropic hormone (ACTH), from the pituitary gland (⇨ pp. 78–9), prompted by the suprachiasmatic nucleus (⇨ below). This periodicity can be destroyed by stress, depression and heart failure. The circadian rhythm also has a link with the immune system or at least with our resistance to infection, as the peak of cortisol production reduces the efficacy of the immune system. Resistance to the effects of many drugs is similarly dependent on the time we take them.

Thyroid-stimulating hormone from the pituitary is also released in accordance with the circadian rhythm. The peak output is at around 11 p.m., and the lowest output around 11 a.m. Thyroid hormones act directly on almost all body cells to control their rate of metabolism. Endorphin (⇨ pp. 54–5) and sex-hormone production vary in a circadian manner. Blood pressure varies characteristically over the 24-hour cycle, as do blood sugar levels. For people who are fasting, the secretion of acid by the stomach is maximal at 10 p.m. and minimal at 9 in the morning. The tension in the muscles of the air tubes of the lungs (bronchi; ⇨ pp. 52–3) is greatest at 4 a.m. and least at 4 p.m., a pattern that is exaggerated in people with asthma.

Light and the body clock

Body temperature varies slightly but predictably during a 24-hour cycle. There is a reduction in heat production and increased heat loss in the evening, while in the early morning heat production rises. Body temperature is at its lowest in the middle of the night, and peaks around mid-afternoon.

Light input variations are undeniably important in bringing all the behaviour

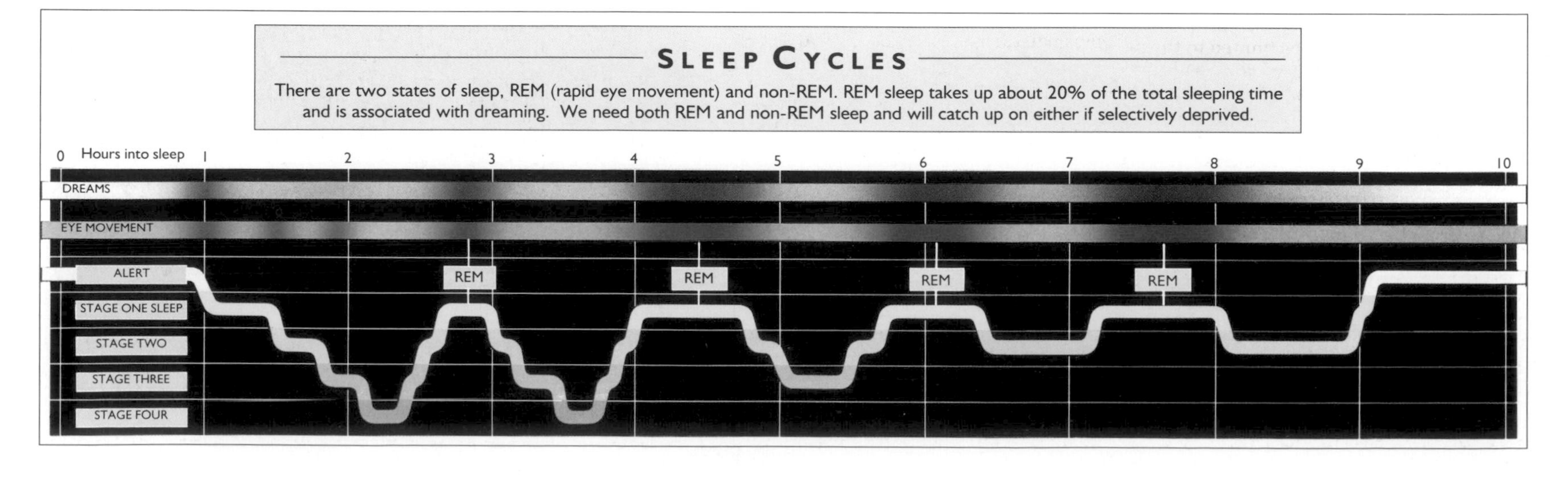

patterns and other physiological functions into synchronization with the day-and-night rhythm, but scientists have wondered whether such patterns are anything more than a simple response to light variations. Trials suggest that we do have an internal circadian-rhythm clock. People kept in isolation chambers, with no indication of time, and no light clues to day or night, continue to follow a regular sleep-and-wake cycle. This cycle does not, however, have a 24-hour periodicity but is nearer to 25 hours. Over the course of a month or so, people kept in such conditions get completely out of synchronization with day and night. Body temperature averages 36.5° C (97.7° F) at 4 a.m. and 37.5° (99.5° F) at 4 p.m. If these subjects are exposed to artificial light–dark cycles of a periodicity of less or more than 24 hours, and light is of the intensity of morning daylight, the characteristic day–night body temperature swings soon fall into step with the new cycles. These may have a periodicity of anything from 21 to 28 hours. Exposure to bright light after the normal time of the lowest body temperature causes the cycle to lengthen, while exposure to bright light before the usual time shortens it. These findings show that the natural period of the body clock is not caused by the day–night cycle but is simply synchronized by it.

Synchronization of a free-running periodic system by another of fixed but slightly different frequency occurs very easily, in engineering as well as in nature, and the linkage between the two systems need not be strong. A very small stimulus, repeatedly applied, is usually all that is necessary.

Melatonin and the suprachiasmatic nuclei

The small, oval, flattened *pineal gland* is centrally placed in the brain immediately above the top of the brainstem. In some simple animals, such as lampreys (jawless fish), the pineal acts as a kind of primitive eye, sensitive to light and dark. Descartes, the 17th-century French philosopher, taught that the pineal was the channel by which the mind interacted with the body, and for a long time the pineal was regarded by philosophers as the 'seat of the soul'. Until recently, physiologists knew little or nothing of its function, and its interest in medicine was limited to the fact that it often acquired a few grains of calcium that showed up on x-rays and could be used to demonstrate shift of the midline of the brain to one side as a result of tumours. We now know that the pineal produces a hormone called *melatonin*, which is released during darkness. The function of melatonin in human beings has not been fully established but there is little doubt that it is connected with body rhythms. Scientists are still studying its relationship to the seasonal affective disorder (SAD), to jet lag (⇨ box) and to its role in the onset of puberty.

About one million nerve fibres run back from the retina of each eye in the optic nerve to form the neurological visual 'pathways'. The large majority of these fibres pass right back through the brain to end in the visual cortex – the part of the brain in which the processes of bringing vision into consciousness occur. A substantial number of them, however, run to other regions of the brain. Among these are two tiny nerve cell nuclei, the *suprachiasmatic nuclei*, lying in an area near the midline of the lower surface of the brain (the hypothalamus; ⇨ pp. 78–9). The suprachiasmatic nuclei (so called because they lie immediately above the chiasma, or partial crossing, of the optic nerves) seem to be the biological clocks for circadian rhythm. If they are destroyed, as by a nearby brain tumour, circadian rhythm is lost. Disease of the hypothalamus has long been known to cause disturbances of the normal pattern of sleep, often with the desire to sleep during the day and to remain awake at night. The suprachiasmatic nuclei respond to the presence of melatonin by reducing their nervous activity.

Neurophysiologists are still arguing about the details of this mechanism, but the evidence adduced so far strongly suggests that body synchronization with the day–night cycle involves the action of the pineal gland in turning off the clock, and that of the connections from the retinas in turning it on. There is growing conviction that the suprachiasmatic nuclei, acting through the hypothalamic connections, organize the timing of such events as sleeping and waking, body temperature variations, times of eating and so on.

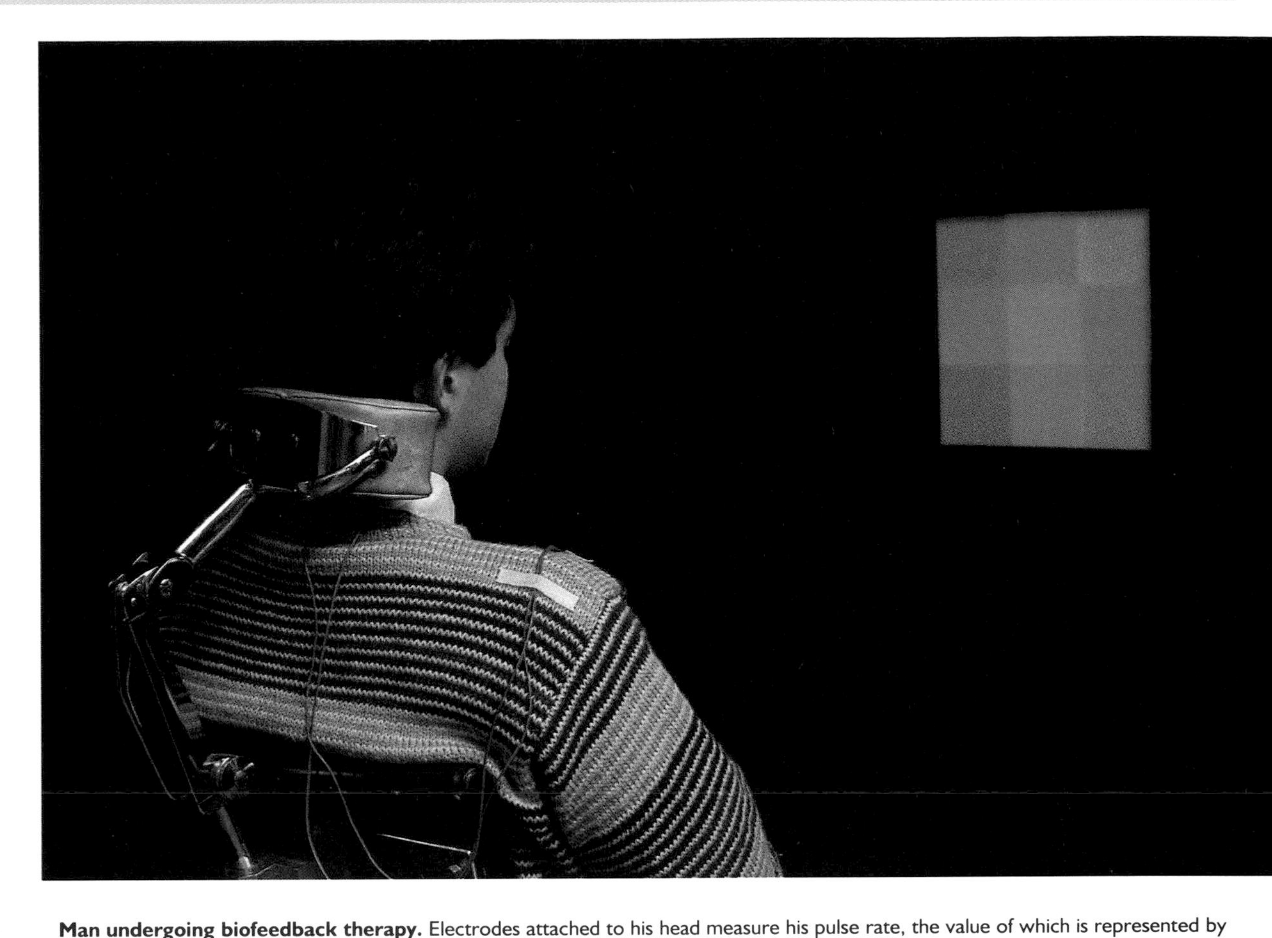

Man undergoing biofeedback therapy. Electrodes attached to his head measure his pulse rate, the value of which is represented by the squares of coloured light. By distinguishing between different colours, subjects may become able voluntarily to control their heart rates. Exactly how the technique works is unclear but therapists claim that it can also control blood pressure and migraine attacks, and relax spastic muscles. (Martin Dohrn, SPL)

Jet Lag

Rapid travel across time zones causes physiological disturbances because of the change in the experienced periodicity of light and dark, which quickly falls out of synchronization with the various periodic body functions. The result may be insomnia, tiredness during the day, a feeling of light-headedness, a sense of ill-being, and reduced mental and physical performance. These effects are not restricted to air travel but are experienced by people, usually research scientists, who have recently moved to the Arctic or Antarctic, astronauts travelling in space, or by people who have recently lost all vision – people, in short, who are deprived of their normal day–night light cycles. Jet lag tends to be worse when travelling east because of the difficulty in shortening the day; since the innate circadian period is longer than 24 hours, lengthening the day, when travelling west, is easier.

When travelling by air there are some useful steps that can be taken to prevent, or at least reduce, the effects of jet lag. These include:

- **Limiting alcohol intake during the flight to a moderate level.**
- **Drinking plenty of water so as to avoid the effects of dehydration during the flight.**
- **Avoidance of insomnia during the flight by the use of short-acting sedatives, such as Diazepam.**
- **On arrival, deliberate and immediate adjustment of sleeping and eating times to those of the new time zone.**
- **The use of short-acting sedatives to produce sleep during the normal sleeping times of the new time zone.**
- **Avoidance of eating during the normal sleeping times in the new time zone since eating tends to delay the adjustment of the body to the new circadian patterns.**

Trials of melatonin to effect resynchronization of the circadian rhythm have suggested that this might prove useful in the management of jet lag. The quality of the research has, however, been strongly criticized.

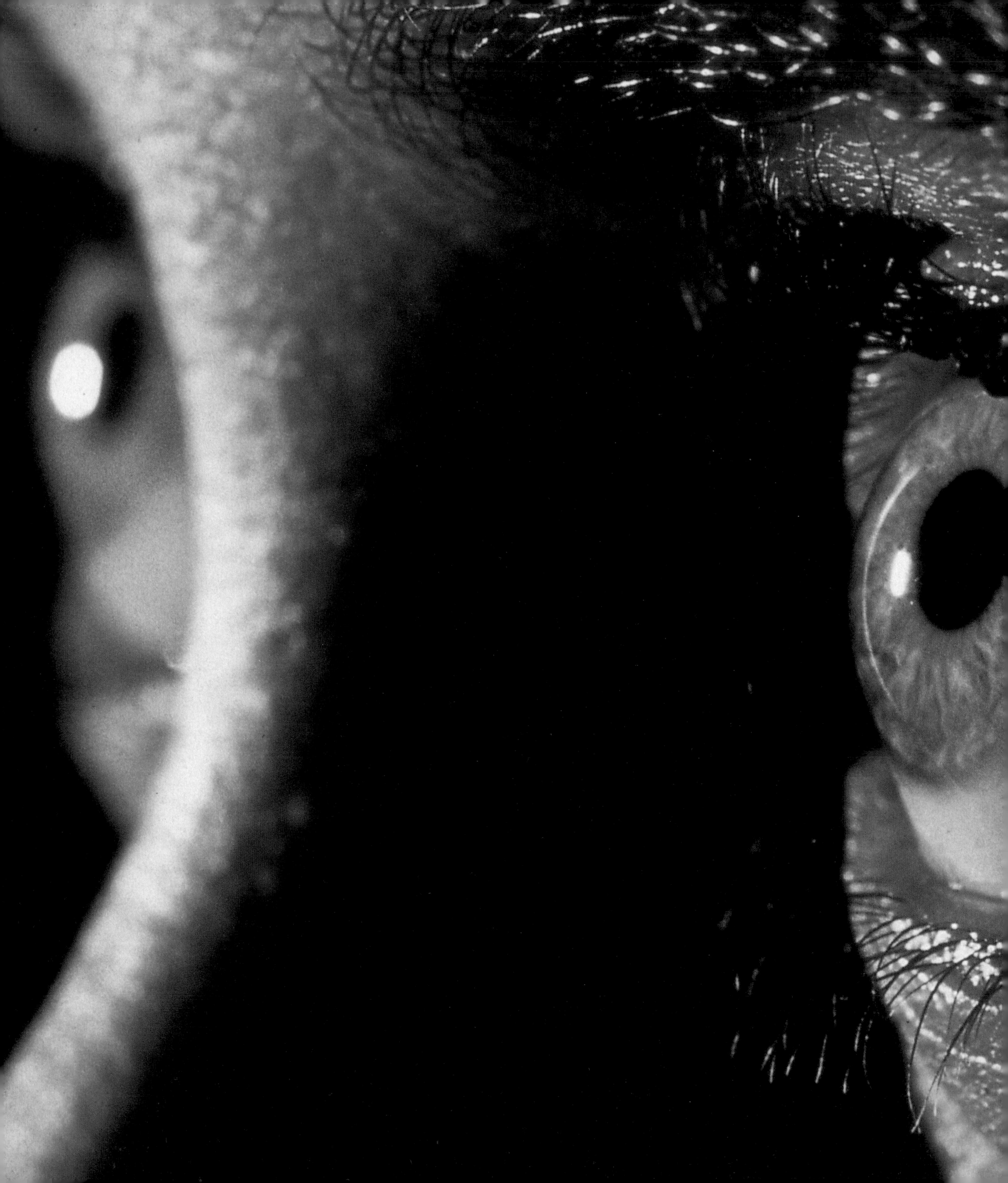

THE BRAIN AND THE SENSES

Body and Soul

Many people clearly distinguish three separate entities – body, mind and soul – and the interrelationship of these has exercised philosophers from the beginning. To most people today, the notion of the mind and the body as distinct entities seems self-evident, but there are increasing numbers, among both philosophers and physiologists, to whom such dualism is no longer acceptable.

The soul has usually been defined, by philosophers, theologians as well as by ordinary people, as being distinct from, but somehow linked to, the body. The body is manifestly temporary but the soul, to most of those who believe in its existence, is permanent and persists after the death of the body. It is said to be the spiritual part of the human being and is often thought of as animating the body. The term 'spiritual' has meaning only in a theological context and conveys as little to the materialistic scientist as does the term 'soul'. The idea that the conscious part of the human being (whether described as soul or mind, according to belief) somehow brings about all bodily action seems to be common to much human experience. But even this is now being questioned.

Baron Gottfried Wilhelm von Leibnitz (1646–1716). (AKG)

Ancient ideas

Early religions identified the soul as the vitalizing force of the body and, because of the psychological need to relate ideas to something concrete, often located the soul in particular parts of the body, such as the heart, the kidneys or the pineal gland (➪ pp. 162–3). Sometimes the soul was identified with the breath or the pulse. In early Jewish theology the soul (*nefesh*, which originally meant 'neck' or 'throat') was not thought of as existing apart from the body. As the writer of Ecclesiastes (9:5) puts it: 'For the living know that they will die, but the dead know nothing, and they have no more reward.'

Greek and, in particular, Platonic thought was quite clear on the matter: humans were divided into two parts – body and soul. The soul, or *psyche*, was both pre-existent and immortal. Plato thought of the body as the prison of the soul or mind and regarded philosophy as the preparation for the separation of the two in death. Aristotle took a more sophisticated view, which has echoes in present thought. He regarded the soul as being inseparable from the body and defined it as a 'capacity' of the body, just as vision is a 'capacity' of the eye.

Greek ideas about the body and soul had great influence on early Christian ideas, but the central belief in the resurrection of Jesus Christ imposed on these ideas new concepts about the nature of the soul. Even so, throughout the history of the Christian Church, there has never been a universally accepted definition of the soul. The notion of some form of personal survival after death is, however, basic to Christian theology, and, implicit in this, is the idea that the soul has independence of the body, even if the body is held to be resurrected with the soul.

Saint Thomas Aquinas – faith and reason

Thomas Aquinas (c. 1225–74) was the son of an Italian count. Well versed in Jewish, Greek and Islamic philosophic thought, he confidently tackled their challenges to Christianity, firm in his conviction of the unity of truth because of its origin in God. Thomas rejected Plato's idea of humans as rational souls inhabiting helpless bodies. With Aristotle, he conceived the human being as a union of soul and body, and, from this conviction, found it easy to accept both the immortality of the soul and the idea of the resurrection of the body. He also accepted Aristotle's recognition of the importance of *sense data* (information from which knowledge of the material world is inferred) as a source of knowledge and of the way in which further knowledge can be derived by the mental formation of concepts derived from these data. He was persuaded that the existence of God could be proved by such reasoning from sense data, and he believed human knowledge and Christian revelation to be different facets of a single truth. Thomas was canonized in 1323, and his system of theology has, in general, become the official doctrine of the Roman Catholic Church.

Cartesian dualism

The French philosopher René Descartes (1596–1650; ➪ p. 162) was particularly concerned with the relationship of mind and body, and his writings remain influential to this day. Descartes was determined to believe only those things about which he could be entirely certain. To this end he used various hypotheses to try to test his beliefs. One of these was the hypothesis that an evil genius existed whose whole purpose was to deceive him, so he could not, therefore, logically trust the data from his senses. This method seemed powerful, but immediately raised the difficulty that the evil genius might be deceiving him into believing that he, the philosopher, existed, when, in fact, he did not. Descartes' answer to this was the axiom '*Cogito ergo sum*' – 'I think, therefore I am' – the theory that the proof of existence was the very consciousness that could question it.

Descartes held that human beings are composed of two kinds of substance – mind and body. The mind was a dimensionless and indivisible but conscious being, capable of volition, understanding, imagination, perception and will. The body was an entity located in space and was infinitely divisible. The mind, being dimensionless, was capable, in theory, of surviving the death of the body. These two parts, he believed, were capable of interacting, and the mind was able to make the body move by moving a small part of the brain. This it did by way of the pineal gland. Such movements could also cause sensation and emotion.

Some of Descartes' ideas on the mind–body problem have had to be set aside in the light of more recent physiological knowledge, but the general notion of the mind and the body as distinct, but intrinsically related, entities has held sway for over three centuries. It is today, however, being increasingly questioned.

Leibnitz and Berkeley

The German philosopher, mathematician and polymath Gottfried Wilhelm von Leibnitz (1646–1716) could not accept Descartes' views on the mind. He also rejected the idea of the British philosopher John Locke (1632–1704) that, at birth, the mind is a 'blank page' (*tabula rasa*), to be written on only by what we learn through the senses. Instead, Leibnitz held that the mind and the body were not causally related but were two perfectly synchronized entities, set running by God like two clocks that kept perfect time, thus remaining together for ever.

The Irish philosopher Bishop George Berkeley (1685–1753) was a leading exponent of philosophical idealism – the principle that only minds have real existence and that the material world only seems to exist. Berkeley analysed the processes of perception and pointed out, for instance, that the colour of objects could not be said to have any real existence because perception of colour necessarily involved the action of light, eyes and minds. Berkeley dealt similarly with the other properties of matter, showing

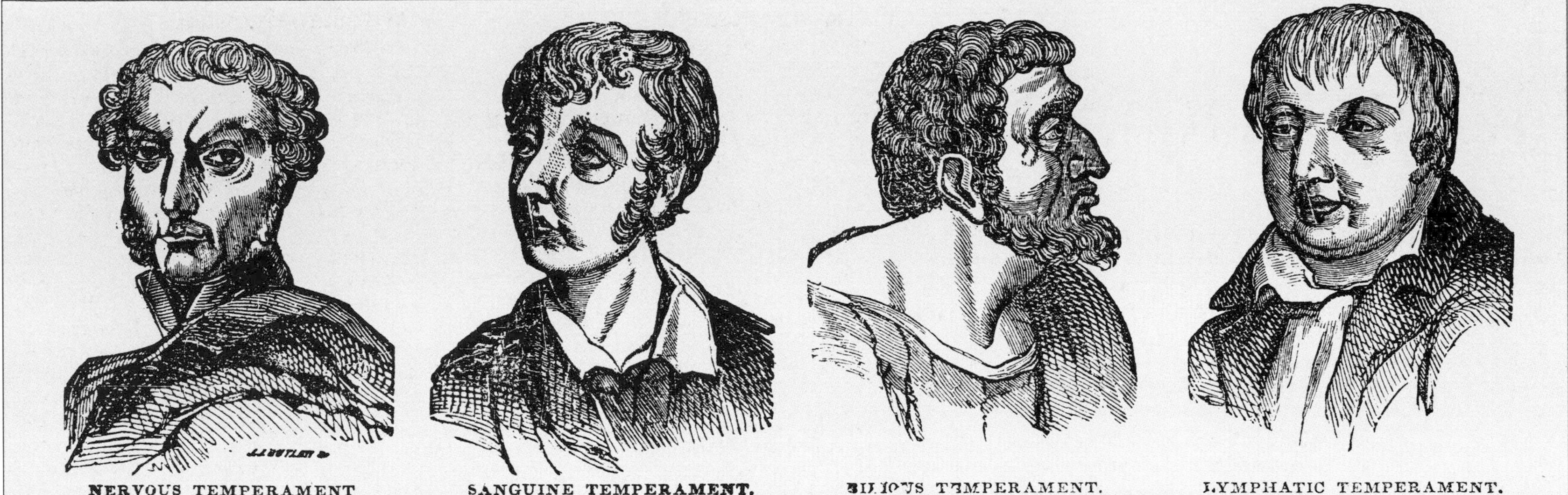

Since Greek times medical and philosophical theories have linked the state of the mind with the health of the body. This 19th-century print depicts the prevailing notion of the four temperaments – nervous, sanguine, bilious and lymphatic – and their perceived physical manifestations. From R. Trall, *Hydropathic Encyclopaedia*, (1874). (AR)

that our perception or conception of them depended wholly on the existence of the mind. These views still carry much weight, and it is now clear, from the convergence of physics and physiology, that whatever may be the true nature of the physical world, it is not as it seems.

Berkeley took an extreme position and held that *all* the qualities of matter depended on the mind. The criticism – that common-sense experience tells us that objects have continuing existence independent of our perception of them – he met triumphantly by affirming the existence of a greater mind, by simply stating that the whole universe exists by being perceived in the mind of God. This argument has been found by some to be an adequate proof of the existence of God.

Gilbert Ryle and the 'ghost in the machine'

The British philosopher Gilbert Ryle (1900–76), Waynflete Professor of Metaphysics and Philosophy at Oxford University and editor of the periodical *Mind*, made a notable modern contribution to the debate. In his most important book *The Concept of Mind* (1949) he categorically rejects Descartes' distinction between the body and the mind (which he ironically describes, 'with deliberate abusiveness', as the 'ghost in the machine'), claiming that this concept is what he calls a 'category mistake'. Such mental concepts as the idea of the mind–body duality work well in everyday life but do not, he claims, stand up to philosophic criticism. It is a category mistake to consider the mind as being something that can be regarded in the same way as one regards the body. As an example of a category mistake, Ryle describes a foreigner visiting Oxford, who, after being shown all the colleges, libraries, museums, laboratories, playing fields, and so on, asks: 'But where is the University?' The concept of the University is, of course, in a different, and noncorresponding, category from that of the physical structures; and concepts in different categories cannot be treated in the same way.

Transmigration of Souls

The idea of the transmigration of souls (also known as reincarnation or metempsychosis) – the notion that a soul may leave one body at the time of death and take up residence in another body – is widespread and very persistent. It was familiar to the ancient Greeks and was accepted by Plato and the Pythagorean school of thought. It attracted adherents to some of the gnostic and occult forms of Christianity and Judaism and has survived in many religions. It appears, however, in its most fully articulated form in the religions originating in India – Sikhism, Jainism, Buddhism and, especially, Hinduism, where it was developed, around 600 BC, in the *Upanishads*. The Hindu doctrine holds that human beings die and are reborn many times. The soul, or *atman*, is indestructible. It is identified with the divine spirit, *Brahman*, the ultimate source of everything, from which it comes and to which it will ultimately return. The soul passes from one body to another in a continuous process. At each rebirth, the condition of existence is determined by the standard of virtue, or behaviour, reached in the previous incarnation.

The idea is intimately linked to the concept of *karma*, in which the happiness or distress of the individual is a just reward or punishment for conduct in earlier lives. Transmigration and the cycle of karma extends through a sufficient number of lives to allow realization of the truth and the Absolute so that, ultimately, the soul may be reabsorbed into the ocean of divinity whence it came. Although Buddhism also incorporates the idea of karma, it rejects the Hindu notion of transmigration of the soul. Buddhists believe that the soul ceases to exist at the time of death but that one aspect of it, the vijñana, or germ of consciousness, is implanted in the womb of a new mother.

The concept of transmigration has proved attractive to many and has been adopted by many latter-day religious movements, including Theosophy and many orientally inspired cults.

Cremation of the body of Rajiv Gandhi, ex-prime minister of India, 1991. (Gamma)

SEE ALSO

● ELEMENTS AND HUMOURS p. 34
● TOWARDS A SCIENCE OF THE MIND p. 162

The Human Computer?

The modern debate on the mind–body problem is wide-ranging and is never far from the surface in the more fundamental discussions of neurobiologists and psychologists. Underlying all such debate is the unquestioned premise that the brain is a complex functioning machine subject to the same laws of physics as any other machine. The most immediate obvious analogy is with the computer, but although this analogy can be helpful it must be used with caution.

Almost all current computers are serial devices in which simple functions are performed consecutively on a single stream of electrical pulses. We know that similar functions are performed in the brain, but that the brain is not a serial device. A more accurate analogy would be to a massive parallel computer consisting of many thousands of serial computers acting simultaneously and sometimes linked laterally. Such devices are now beginning to be developed.

Can we, by using our brains, understand the brain? And can the brain make sense of the mind? These are difficult questions. So far, however, scientists do not appear to have been restricted by such philosophic arguments in their comprehension of brain function. The nature of the mind–brain relationship is a constant challenge to science, and, in recent years, major advances have been made in revealing it. Some neurobiologists take the radical view that the concept of mind should be abandoned altogether. These scientists claim that certain physiological or physicochemical states of the brain *are* mental states and that all intellectual activity can – in principle – be explained in terms of processes analogous to those occurring in a computer. Humans, they believe, are machines, and differ only quantitatively from other machines.

In contrast, many scientists take a strongly dualistic view. They regard brain and mind as two equally irreducible aspects of a person, requiring quite different methods of study and separate vocabularies. Those who take this view do not necessarily deny that the mind is a product of neurological (nerve-impulse) activity. They are, however, inclined to be sceptical of the brain/computer analogy, believing that there is a qualitative difference between brain and machine. The gulf between the obviously mechanical and the apparently metaphysical aspects of the human being has led to the polarization of brain-function studies into neurology on the one hand, and psychology and psychiatry on the other. Neurologists are knowledgeable about brain structure and about the consequences of damage to that structure. They know a great deal about the broader aspects of brain function and about localization of function in the brain. A neurologist, after clinical examination, can often identify with precision the site of organic disorder or injury in the brain. Psychologists and psychiatrists, on the other hand, tend to concentrate their attention on the mental functions and on descriptions of behaviour.

Recent research has, to some extent, reduced the separation between these two approaches. Many organic syndromes are now recognized in which known brain damage from disease or injury affects the state of the mind. In most cases these disorders involve widespread loss of brain tissue, so they tell us little about the location of specific mental functions. However, techniques such as positron-emission scanning, magnetic-resonance imaging, the electroencephalogram, brain electrical activity mapping, and the study of the effects of drugs that modify brain function are gradually beginning to show how some, at least, of the mental or behavioural disorders are caused by localized organic brain damage.

The brain/computer analogy

The brain is composed of some 100 000 000 000 nerve cells (*neurons*) (⇨ p. 74–5), each consisting of a cell body, a long, often branched, fibre called the *axon*, and anything from one to several thousand short fibres called *dendrites*.

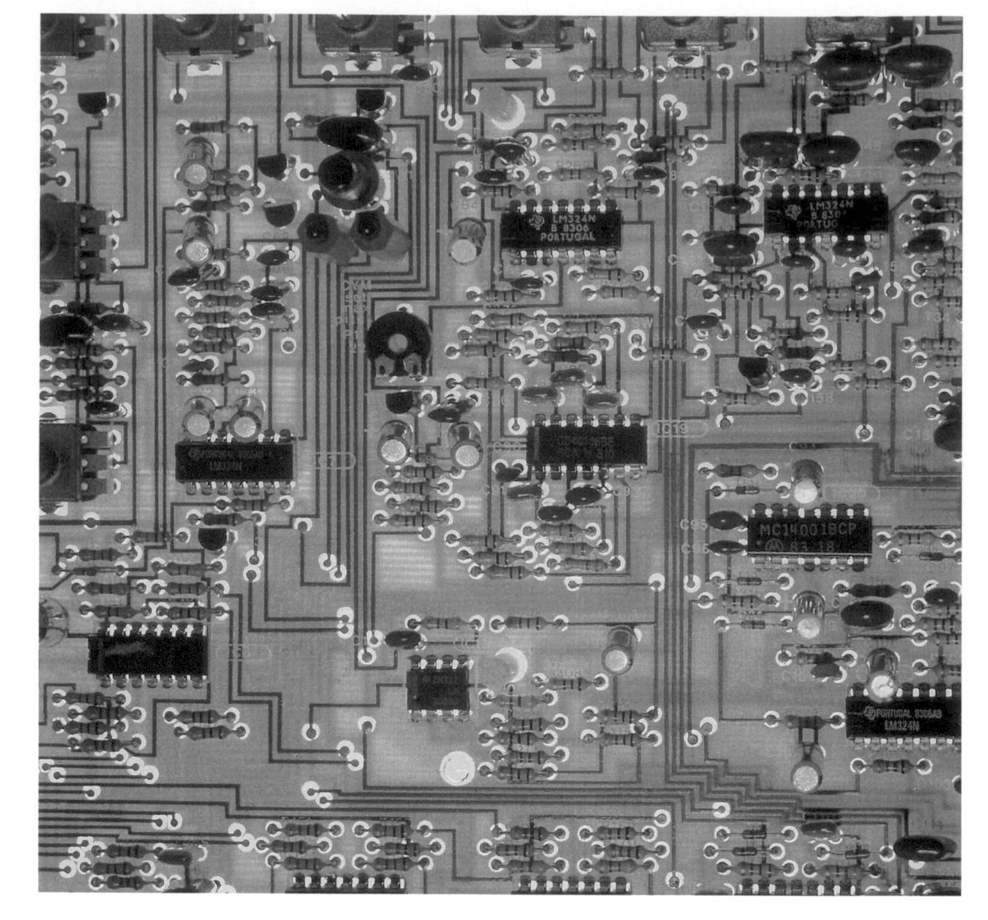

Two complex, efficient communication networks – an industrial computer circuit board (left), contrasted with human neurons (nerve cells) and their axons shown in scanning electron micrograph. (Telegraph/RB)

Connections between adjacent or widely separated neurons are numerous and complex, and single neurons commonly receive several hundred to several thousand inputs from other nerves. In general, the cell body and the dendrites receive incoming signals, which are integrated in the cell body, and the output signal leaves by way of the axon.

This almost incredible complexity of interconnection could hardly be more different from the inherent simplicity of the structure of the ordinary digital computer. Even so, the arrangement of stimulatory and inhibitory connections in the brain closely mimics the logical 'gates' (⇨ pp. 90–1) that are the basis of digital computing. The biological system would allow for all of the logical elements found in the central processing unit of a digital computer, these elements being simple electronic circuits by which logical processes can be realised. Logical gates would be readily implemented by such an arrangement of neurons, but in view of the number of modified connections it seems that some kind of modified (statistical) gating system is normally involved. In such a system, if the number of stimulatory impulses exceeds the number of inhibitory impulses they, the former, would prevail; otherwise inhibition would succeed. Nerve impulses,

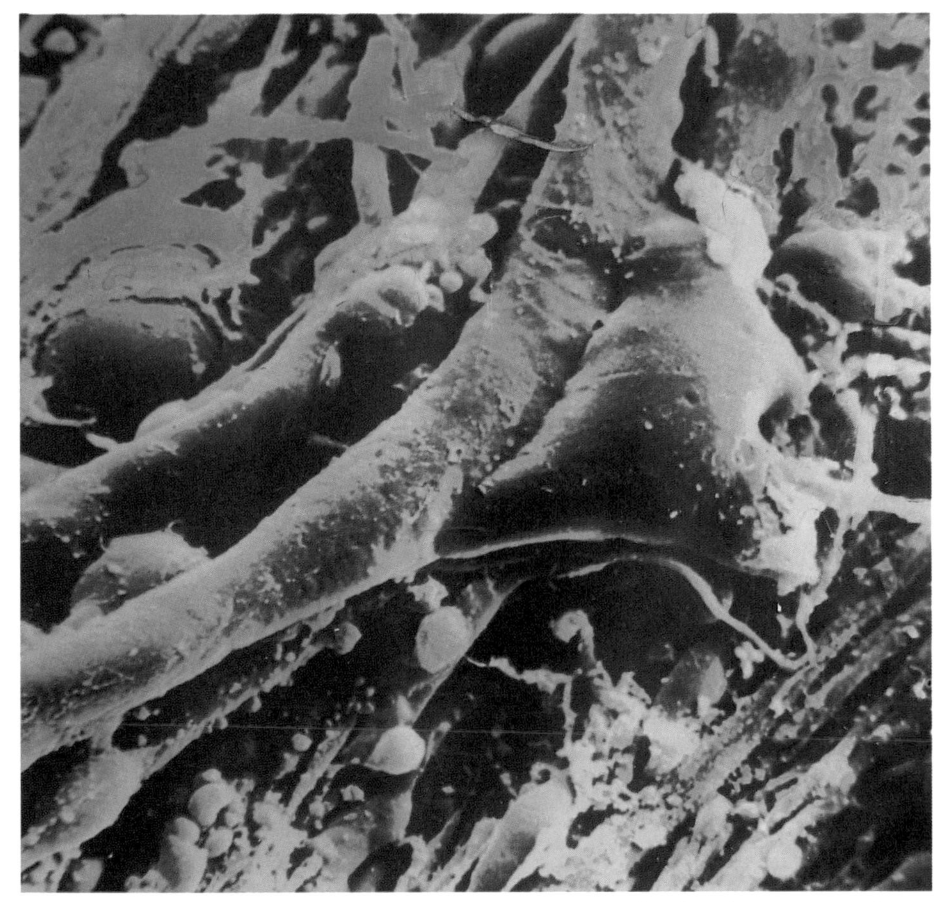

although electrical in nature, travel at very low speeds compared to the speed of propagation of electricity in computers; and the timing of impulses in the brain is probably much less important than in digital computers, where everything is organized by a clock (oscillator or pulse generator) running at many millions of cycles per second (MHz).

Tracing the brain connections

Another major difference between the brain and present-day computers is that whereas computers are general-purpose machines performing different functions in accordance with instructions provided in the software, the brain appears to be 'hard-wired' to carry out its functions. Major subdivisions of brain function – movement, sensation, speech, vision, hearing, and so on – are subserved by particular known parts of the brain, and these are connected by bundles of nerve fibres to the effector or sense organ concerned, as well as to other parts of the brain. The tracing of these connections was a vital step in furthering the comparison between the brain and the computer, and much of the early work was done by the Spanish anatomist Santiago Ramòn Y Cajal (⇨ box). More recently, newer and better techniques have carried this work further. In the early 1950s it was found that a killed nerve cell, while degenerating, can be stained so that its entire length is made distinguishable. In this way it is possible to show how particular cells in certain parts of the brain are connected to quite remote parts. The method has been used to work out the neuron map of the brain.

It is not enough merely to demonstrate the structure of the brain. Neurons are functional units that may or may not be firing at any particular time. Much attention has been paid recently to this functional aspect. Glucose is the brain's 'fuel', and this is consumed, in any particular area of the brain, at a rate that depends on the level of function in that part. Various methods have been devised to measure the rate of glucose consumption in different parts of the brain. Researchers have made glucose molecules containing radioactive atoms, and the degree of radioactivity from local accumulation of radioactive substances has been measured. The most sensitive technique, to date, uses glucose labelled in this way with positron-emitting isotopes. These can be detected in the living brain by a positron-emission tomography (PET) scanner outside the skull. If this is done in the course of any activity, such as hearing or seeing, the part of the brain especially involved can be determined.

Golgi, Cajal and the Nerve Connections

Probably the greatest single advance in the technique of determining the nerve connections in the brain was the discovery by the Italian microscopic anatomist (*histologist*) Camillo Golgi (1843–1926) of the use of silver salts to stain nerve tissue. This method brings out detail in the nerve cells never previously seen, and allows detail of the connections of nerve cells to be worked out. Golgi's method allows the investigator to pick out a few neurons from among the masses and to demonstrate all its branches and connections. Once the method had been demonstrated, in the 1880s, it was taken up by the obscure Spanish histologist Santiago Ramòn Y Cajal (1852–1934), who improved it and set to work to use it to discover how the brain and nervous system was organized. Cajal soon became a master microscopist, in whose hands the technique proved wonderfully fruitful. In 1889 he lectured at the Berlin conference of the German Anatomical Society where he astonished the leading neurologists of the day with his demonstrations of brain structure. Cajal, encouraged, devoted the rest of his life to the study.

Golgi was none too pleased with some of Cajal's ideas, especially his belief that the brain consisted entirely of nerve cells. Golgi was convinced, with other influential scientists, that nerve cells were connected by a network of non-nervous fibres – the reticularist theory. Eventually Cajal was able to show that this theory was wrong, that the nervous system consisted only of nerve cells and their supporting tissues, and that nerve cells contacted each other by their own fibres. In 1904 Cajal published an enormous work, *Histologie du Système Nerveux de L'homme et des Vertébrés*, which established his reputation as the world's leading neuroanatomist, and which is still regarded as the major single work on the subject. In this, he showed that there was nothing random about the nerve connections in the brain but that these were elaborately structured. He demonstrated how the different regions of the brain differed in architecture and, to a limited extent, how they were interconnected.

In 1906 Cajal and Golgi shared the Nobel Prize in Physiology or Medicine.

Electrical differences between the brain and the computer

Digital computers use the binary system of representing information in which the presence of a voltage level represents a 1 and its absence (or a much lower level) represents a 0. All information can be represented in terms of 1s and 0s by arranging these in codes. In the commonly used ASCII (American standard code for information interchange) code, A is 1000100, B is 0100100, C is 1100100, D is 0010100, and so on. Information is conveyed as a very rapid succession of electrical pulses. In the nervous system, on the other hand, no such system of coding is used. Information travels in nerves in the form of zones of reversal of the normal positive outside/negative inside state (⇨ pp. 74–5). Such reversal is called depolarization and causes a pulse that travels along the nerve fibre much more slowly than electricity travels along a wire. Depolarization pulses follow each other at a repetition rate that varies. This is called frequency modulation (FM), a form of electrical signal much used in electronic communication but not as the basis of computing.

SEE ALSO

- WHAT THE BODY IS MADE OF p. 36
- CELLS, THE STUFF OF LIFE p. 38
- THE CONTROL SYSTEM pp. 74–81
- MENTAL DISORDERS pp. 184–95

The Control System 1: the Nervous System

The only way in which we can derive any information about the outside world is by way of our sense organs – eyes, ears, nose, tongue, skin, and muscle and tendon nerve endings – all of which send nerve impulses (⇨ box) to the brain. And the only way in which we can respond to the outside world and change it is by brain action initiating the passage of nerve impulses to our muscles and glands. Thus the brain is a responsive entity with an input and an output. Without input, it is questionable whether anything much would happen in the brain. Total sensory deprivation in a mature adult would probably lead to total immobility. We know that, in a child, sensory deprivation is seriously damaging and results in failure of brain development. It is clear that information input is essential if the brain is to function.

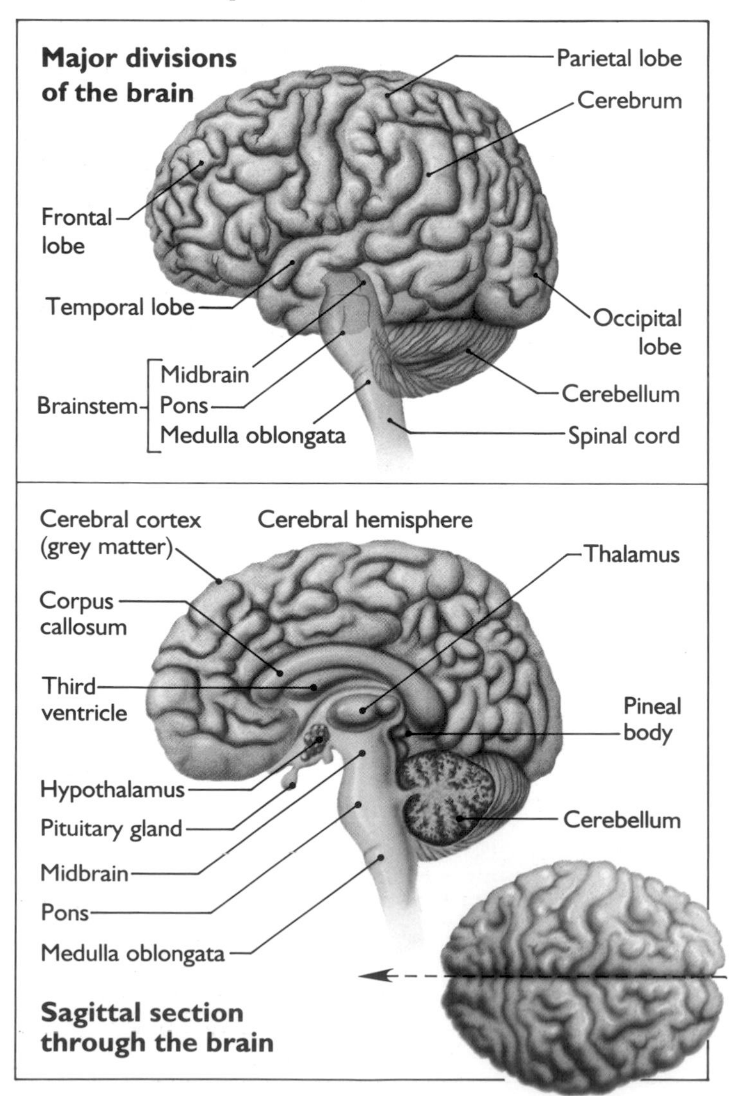

The brain is the central, and by far the most important, organ of the body. The rest of the body is essentially a means of supporting, protecting, transporting, and carrying out the instructions of the brain. The brain is the seat of consciousness, pleasure and emotion, and the information centre of the body. It is the storage site of everything we have learned since we were born, and contains a unique database that underlies our whole personality and capability. The brain also stores much information inherited from our ancestors that is manifested as instinct, patterns of response, and so on. The brain is a pleasure-seeking organ, intent upon its own gratification and directing us, most of the time, to act in such a way as to stimulate pleasure areas within it. For some people, this process seems an adequate explanation of the purpose of existence. But there are other aspects of brain operation about which we know little or nothing. Unlike many of the more easily understood brain functions, these higher aspects – consciousness, perception, thought, imagination – cannot be localized to any particular part of the brain.

The general structure of the brain

The most obvious neurological difference between humans and other animals is the great development in the human brain of two large, almost mirror-image masses known as the *cerebral hemispheres* or, together, as the *cerebrum*. The two hemispheres are almost isolated from each other except for a massive multi-cable junction, called the *corpus callosum*, that connects them. The cerebral hemispheres are conspicuous for the complex and repeated folding (infolding) of the outer surface, the *cortex* (⇨ also p. 76). The effect of this infolding is to allow a very large area of cortex to be accommodated in a small space. The cortex is the most advanced part of the brain and it is here that the best-understood functions of the brain are located (⇨ pp. 76–7).

Running down to connect the middle of the underside of the cerebrum to the spinal cord is the *brainstem*. This is a thick stalk of nervous tissue containing the great longitudinal tracts of nerve-fibre bundles running into and out of the spinal cord. The brainstem also contains the collections of cell bodies (nuclei) of most of the 12 pairs of nerves, the *cranial nerves*, which emerge directly from the brain. These nuclei are interconnected and are linked to other parts of the brain. The brain and spinal cord constitute the *central nervous system* and the great complex of nerves outside both the brain and spinal cord form the *peripheral nervous system*.

Lying beneath the undersurface of the cerebrum, at the back, is the *cerebellum*, a separate brain, distinguishable by its narrower surface corrugations. The cerebellum is largely concerned with unconscious, automatic functions such as balance and the control and coordination of voluntary movements. It receives numerous connections from the motor parts of the brain (i.e. those parts that are concerned with movement), from the balancing mechanisms inside the inner ears (⇨ p. 87) and from the *basal ganglia* (⇨ below).

For good reason, the brain is better protected than any other organ. It is cushioned in water, wrapped in three layers of membranes called the *meninges*, and enclosed in a strong bony case, the skull. The brain has an exceptionally large requirement for fuel – a need that varies with brain activity – and can function properly only if provided with an unceasing supply of glucose, oxygen and

The Nervous System: Anatomy

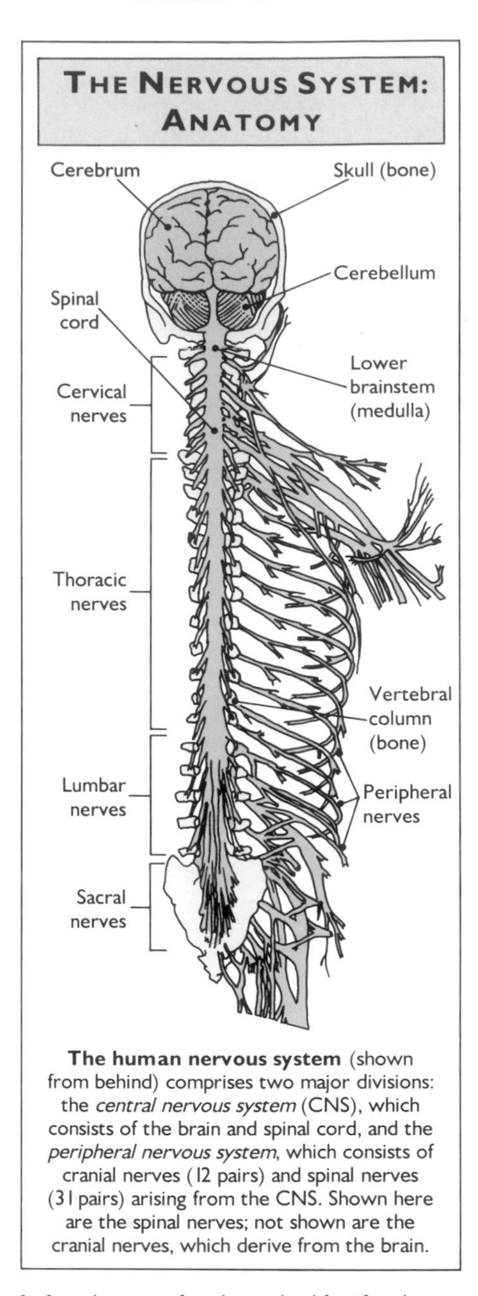

The human nervous system (shown from behind) comprises two major divisions: the *central nervous system* (CNS), which consists of the brain and spinal cord, and the *peripheral nervous system*, which consists of cranial nerves (12 pairs) and spinal nerves (31 pairs) arising from the CNS. Shown here are the spinal nerves; not shown are the cranial nerves, which derive from the brain.

Motor neuron in the grey matter of the spinal cord. (RB)

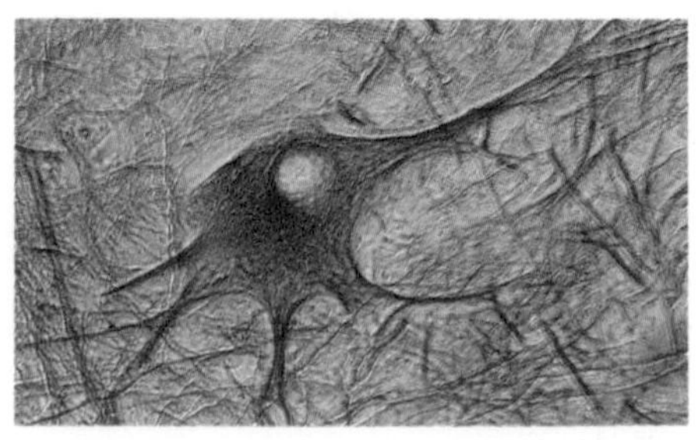

other nutrients, by way of the bloodstream. The blood supply to the brain is profuse, carried by four major arteries (the two *carotids* and the two *vertebrals*), which provide branches to all parts. Blockage of these arteries or their branches, or bleeding from any of the branches in or around the brain, results in destruction of brain tissue and is always serious. This is the cause of a stroke, and is a common cause of death. Any interruption to the blood supply to the brain, even for short periods, is dangerous. Heart stoppage for only three or four minutes will cause permanent damage, and death is inevitable if, at normal temperatures, the blood supply ceases for more than six to eight minutes.

Sensory input

The receptor organs of the four main types of information input – vision, hearing, smell and taste – connect directly to the brain by short nerve tracts. By way of these nerves, a mass of data is supplied to the conscious brain, and this is analysed, correlated with existing stored data, stored and, if necessary, acted upon. At the same time, a mass of sensory information enters the brain from receptor nerve endings in the skin, muscles, tendons, joints and internal organs. This sensory input informs the brain about the state of the environment and about the relative position of the limbs. If the internal organs are suffering some disorder, information from them may be supplied to the brain; otherwise we are not generally conscious of them. Much of the incoming sensory information passes, first, to one or other of a pair of large collections of nerve cells, the basal ganglia, lying in the brain near its underside. The main sensory ganglion is called the *thalamus*. Incoming sensory information may result in unconscious, automatic compensatory or adjusting action, often by way of the hypothalamus (➪ pp. 78–9) lying just under the thalamus. Other sensory input results in conscious awareness of some bodily function or state and prompts voluntary action.

The effector (action) system

Voluntary movement of any part of the body is mediated by the brain. It is initiated in a particular part of the cortex (➪ above and p. 76) and is brought about by a centrally placed part of the brain, the *pyramidal system*. This consists of a massive pair of inverted pyramids of nerve fibres running down through the whole vertical extent of the brain. In the upper part of the brainstem these *motor fibres* cross from one side to the other, and link to connecting neurons that then pass down through the brainstem into and down the spinal cord – the *lower motor neurons*. There, the long motor fibres trigger off the nerve cells whose axons form the peripheral nerves to the muscles. Most of the peripheral nerves also carry sensory fibres bringing in information from the skin and other parts. The connections of these run up the spinal cord to the thalamus and the sensory cortex (➪ pp. 76–7). All nerves outside the spinal cord and brain are part of the peripheral nervous system.

Brain output is not limited to causing the contraction of skeletal muscle (➪ pp. 42–3). The other output function is so important that it is commonly regarded as a system in its own right – the *autonomic nervous system*. This is concerned with the operation of the smooth (involuntary) muscle in the walls of arteries and hollow organs, with the control of the heart muscle, and with causing glands to secrete. The autonomic nervous system can be divided into the *sympathetic* and *parasympathetic* systems, which are involved in preparing the body for action and calming it down respectively. The autonomic system is not, however, a separate nervous system as the name might imply, but simply a peripheral extension of the central nervous system.

The limbic system

Surrounding the basal ganglia on each side is a massive sickle-shaped collection of linked nerve-cell masses known as the *limbic system*. This system, which includes the hypothalamus (➪ pp. 78–9), can be thought of as a primitive brain with connections from the organs of sight, smell and hearing. The limbic system is that part of the brain concerned with the processing of information so that it can be stored in memory. It is also the seat of the emotional states (➪ p. 204). Stimulation of various parts of the limbic system causes a variety of emotions – joy, fear, rage, sexual excitement and so on – and behaviour associated with emotion. If the hypothalamus is isolated from the limbic system the emotional components of the behaviour are absent.

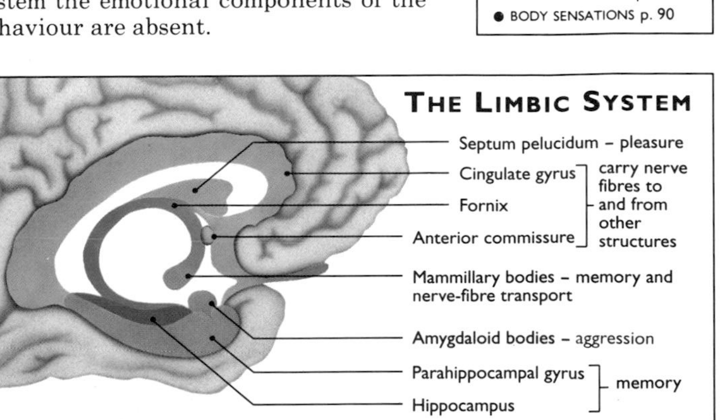

SEE ALSO
- THE HUMAN COMPUTER? p. 72
- THE CONTROL SYSTEM 2: THE AREAS OF THE BRAIN p. 76
- SPEECH p. 82
- VISION p. 84
- HEARING AND BALANCE p. 86
- TASTE AND SMELL p. 88
- BODY SENSATIONS p. 90

NEURONS, NERVE IMPULSES AND SYNAPSES

THE NERVE CELL

Cell body
Dendrites
Region of initial segment
Nucleus
Axon collateral
Myelin sheath
Axon
Synapse
Target cell
Axon terminal
Receptor
Axon terminals

The nerve cell or *neuron* has much in common with other body cells (➪ p. 38). It has a nucleus and cytoplasm containing the same general cell organs (*organelles*) as other cells. It differs, however, in two main particulars – shape and excitability. Although the body of the nerve cell is of roughly the same general size as that of other cells – much too small to be seen without a microscope – it possesses a fine extension, in the form of a fibre called the *axon*, that may be many hundreds or thousands of times longer than any other type of body cell. This fibre is insulated by a fatty covering, the *myelin sheath*, and may be 60–100 cm (24–40 in) long. Running into the cell body are numerous shorter branching fibres, called *dendrites*. The cell body seems to be concerned largely with the nutrition and maintenance of the axon, and has little to do with the origination and propagation of the nerve impulse.

Excitability, the most important property of nerve tissue, depends on the presence of potassium ions (electrically charged potassium atoms) within the axon and sodium ions outside it. Normally, the exterior of the axon carries a positive charge and the inside is negative. This is because unlike charges attract one another, while like charges repel each other. A nerve impulse is the movement along the axon of a zone in which positively charged sodium ions have passed into the axon, making the inside locally positive. The outside, at this point, immediately becomes negative and this zone of reversed polarity automatically moves along the axon in both directions because of the attraction of opposite charges.

Nerve fibres do not contact one another directly but by way of special links, called *synapses*. The number of synapses in the brain almost defies belief and can be only estimated. So far as can be judged, the number is of the order of 100 000 000 000 000 (one hundred trillion). The main feature of the synapse is the actual gap between the two nerves. This gap is very narrow and lies between a slight widening or bulb at the end of the axon (the *axon terminal*), and the surface of another nerve cell or a muscle cell. The gap, of course, holds tissue fluid, containing plenty of sodium ions, but the nerve impulse cannot pass across it. When the impulse reaches the end of the axon it causes one of a range of particular chemical substances called *neurotransmitters* to be released by the axon. The substance quickly diffuses across the gap and its molecules fit into receptors on the other surface. This changes the local permeability of the target-cell membrane to sodium, and sodium ions pass in causing the surrounding area to reverse its electrical charge. As a result, a nerve impulse is propagated or a muscle fibre contracts. Many drugs that act on the brain either act like neurotransmitters or block their action.

Some synapses cause an impulse to pass to the next neuron, while others are inhibitory and can prevent impulses starting in the second neuron. Integration of the excitatory and inhibitory effects will determine whether the second neuron will fire or not.

The Control System 2: the Areas of the Brain

Much of what happens in the brain between the receipt of incoming stimuli and the resulting response is still unexplained, especially in the case of perception, thought, emotion, memory and creativity. The 'wiring' and neurophysiology underlying these 'higher' activities is far less well understood than the brain areas and connections concerned with more primitive functions such as movement and sensation. For the higher functions it is often necessary to think of the brain as a 'black box', a self-contained system whose circuitry need not be known for its function to be understood.

SEE ALSO

- THE HUMAN COMPUTER? p. 72
- THE CONTROL SYSTEM 1: THE NERVOUS SYSTEM p. 74
- SPEECH p. 82
- VISION p. 84
- HEARING AND BALANCE p. 86
- TASTE AND SMELL p. 88
- BODY SENSATIONS p. 90
- PSYCHOLOGY pp. 162–81
- MENTAL DISORDERS pp. 184–95

Every part of the brain has a specific purpose and, although we know little of the neurological basis of consciousness, there remain hardly any parts of the brain whose function is not at least partly understood. A possible conclusion from this is that the higher functions are different in nature from functions that can be localized – that they are, in fact, products of the simultaneous functioning of many parts of the brain.

Some of the simpler connections between input to the nervous system and output from it are particularly well understood. When the number of synapses (➪ p. 75) between input and output is small, the function involved is usually described as a *reflex*. The sudden withdrawal of the hand from contact with a red-hot iron is a reflex involving only three or four sets of synapses and occurs by way of the spinal cord without the brain being necessarily involved. The nervous system is interested in changes, especially sudden changes, and all the sense-input systems are so designed that they send in strong signals in response to change, but very weak signals when nothing much is happening. This is an efficient arrangement. We do not need to be constantly reminded that we are sitting in a chair, but the brain needs to know when we get up.

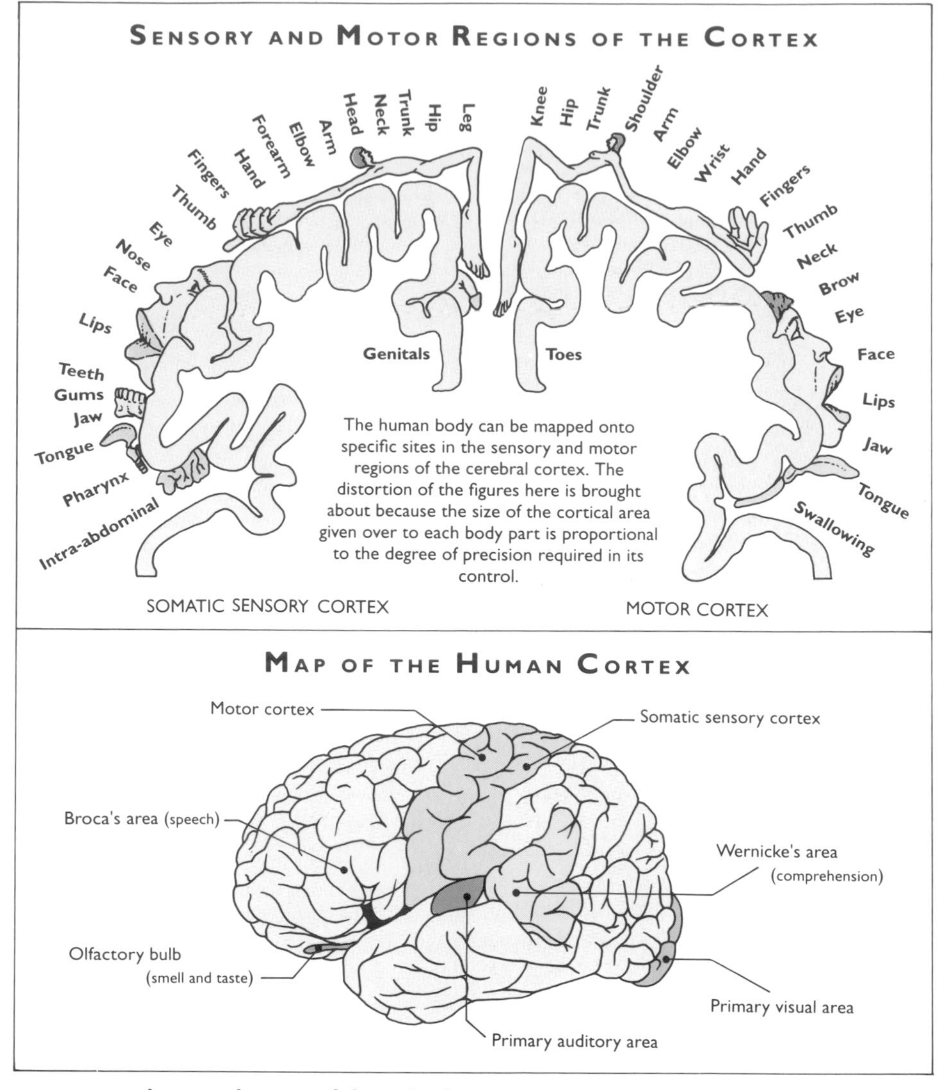

The cerebral cortex

We know much about what goes on in the brain, especially in connection with sensation and movement. The location in the brain of the areas for movement and bodily sensation (motor and sensory functions), vision, hearing, speech, taste and smell has long been well known (➪ diagram). These areas are located in specific zones in the outer layer of the brain – the *cerebral cortex*. The cortex is remarkable for its sheer size, which is concealed by the extraordinary degree of infolding. If it were fully spread out it would have about the same area as a full double sheet of a broadsheet newspaper. It contains some 50 billion nerve-cell bodies (➪ p. 75) of different types arranged in six layers.

Because most of the cortex consists of nerve-cell bodies, giving it a grey colour, it is known as 'grey matter'. From these cell bodies, a mass of nerve fibres (axons) runs under the cortex to interconnect the various areas, to connect them to other collected masses (nuclei) of nerve-cell bodies deep in the brain, and to connect to the spinal cord and hence to the rest of the body by way of peripheral nerves (➪ pp. 74–5). Nerve fibres, en masse, give a white appearance and are called 'white matter'.

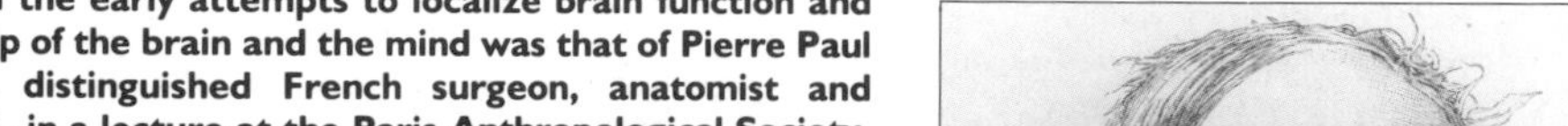

BROCA AND THE LOCALIZATION OF FUNCTION IN THE BRAIN

The most important of the early attempts to localize brain function and to show the relationship of the brain and the mind was that of Pierre Paul Broca. Broca was a distinguished French surgeon, anatomist and anthropologist. In 1861, in a lecture at the Paris Anthropological Society, Broca demonstrated the brain of a former patient. This man had lost the ability to speak or write, although his intelligence and comprehension were unaffected. Prior to his death he had been able to communicate only by gestures, nods and facial expressions. Broca had conducted a postmortem examination and had found an area of damage the size of a golf ball in the left hemisphere of the brain. Broca's conviction that this was the cause of the language defect (*aphasia*) was disputed by many. But within a decade other neurologists began to correlate brain damage in the same general area with various forms of aphasia. The area of the brain (area 44 in the diagram) affected is still called 'Broca's area' and the type of aphasia first described by him is called 'Broca's aphasia'.

The acceptance of Broca's findings prompted others to search for further correlations between recorded loss of function and subsequent observable brain damage. Later, in the course of brain surgery, it became possible to discover the effects on the body of mild electrical stimulation of the cortex of the brain. By the end of the 19th century much about the function of the cortex of the brain was known.

Pierre Paul Broca (1824–80). (Sygma)

THE FUNCTIONAL AREAS OF THE CORTEX

The cortical functional areas are of basic importance. Much of what we know about the function of each area has been established by examining what bodily functions have been lost in individuals who have suffered damage – either through stroke or injury – in a particular area. We know that mental experience of the various types of sensation, as well as volition concerned with movement, are associated with neurological activity in the appropriate areas. Adjacent to the main areas for the sensory functions are what are called 'association areas'. Perception of events in the world usually involves activity in the association areas as well as in the primary functional areas. All these areas have been allocated numbers, although only a few of the most important numbered areas are discussed here.

PERSONALITY

The prefrontal areas (9, 10, 11 and 12 in the diagram) relate to personality and to some of the higher functions. These areas are connected to the thalamus (⇨ below) and the hypothalamus (⇨ pp. 78–9). The function of the prefrontal areas was revealed to the world in 1868 in a paper describing the case of Phineas Gage, a 'capable, God-fearing foreman' who, after a crowbar had been driven through the front of his brain by a gunpowder explosion, became 'irreverent, dissipated, irresponsible and vacillating'. Prefrontal lobotomy – removal of the prefrontal lobes – has a calming and normalizing effect on people with severe psychiatric disorders, but damages initiative, spontaneity and the inclination to make use of intelligence. Problem-solving capacity is affected by a defect of conceptualization, and there is general loss of resoluteness of character ('ego strength'), sensitivity and compassion. Prefrontal lobotomy, as a treatment of psychiatric disorder, has now been largely abandoned.

MOVEMENT

The primary motor area (area 4 in the diagram) is concerned with voluntary movement, and different parts serve different parts of the body (⇨ diagram). Damage to this area on one side causes paralysis of voluntary movement on the other side of the body. This is because the nerve fibres from this area cross to the other side before running down in the spinal cord. Irritative disturbances in this area cause major epileptic seizures. Area 6 is concerned with automatic functions such as the control of muscles in standing and the maintenance of posture. Area 8 is concerned with the relationship of head movement to eye movement. Area 44 and the nearby part of area 4 are concerned with the movements of the mouth, tongue, throat and larynx concerned in speech and swallowing. Damage here causes severe speech defects of a purely motor kind.

SKIN, TENDON AND MUSCLE SENSATION

Immediately behind the motor area 4 lie the sensory areas 1, 2 and 3. Areas 5 and 7 are the sensory association areas. The sensory areas receive massive bundles of nerve fibres carrying sensory information from every part of the body, especially the skin, the tendons and the muscles. Like the motor cortex, the sensory cortex is mapped out in terms of body areas (⇨ diagram). These fibres reach the cortex via the large sensory nucleus, the thalamus, near the base of the brain, which is also connected to the hypothalamus. Destruction of parts of the sensory cortex causes loss of sensation in various parts of the opposite side of the body; irritative damage causes 'pins and needles' or a crawling sensation (*formication*) in corresponding parts of the skin. The association areas 5 and 7 are necessary for the correlation of sensation with other data, such as names and functions. Damage to these areas affects the ability to identify an object by feel alone.

LANGUAGE

Area 44 (Broca's area) is the primary speech area, concerned with the motor aspects of speech. But the whole function of language involves much more than speech and has a much wider representation in the cortex. The language areas, in addition to 44, are 37 and 39, on the inside surface, and 41 and 42 on the outside. Area 37 is concerned with the relationship of speech to vision and hearing, and area 39 with the perception of written language. Areas 41 and 42 are concerned with the perception of spoken language. Damage localized to these areas causes corresponding defects. There is, in consequence, a wide range of disorders of the language function, depending on the part of the cortex affected. Loss of production or comprehension of language is called *aphasia* and this may be expressive (motor), sensory or both. There may be word blindness, word deafness or the inability to communicate by writing (*agraphia*). Language is also concerned with cerebral dominance and is, in most people, represented on the left side. In some left-handed people it is represented on the right side.

HEARING

Area 41 is known as the primary auditory cortex and is concerned with hearing. Destruction on both sides causes deafness, but loss of this area of the cortex on one side has little effect. Area 42, and the adjacent areas 21 and 22, are the auditory association areas. Damage in areas 21 and 22 does not affect the perception of pure tones and sounds but seriously impairs the ability to recognize sounds or to identify music. To a person affected in this way, all sounds, however diverse, seem to be alike. The creaking of a door is indistinguishable from the tinkle of a bell.

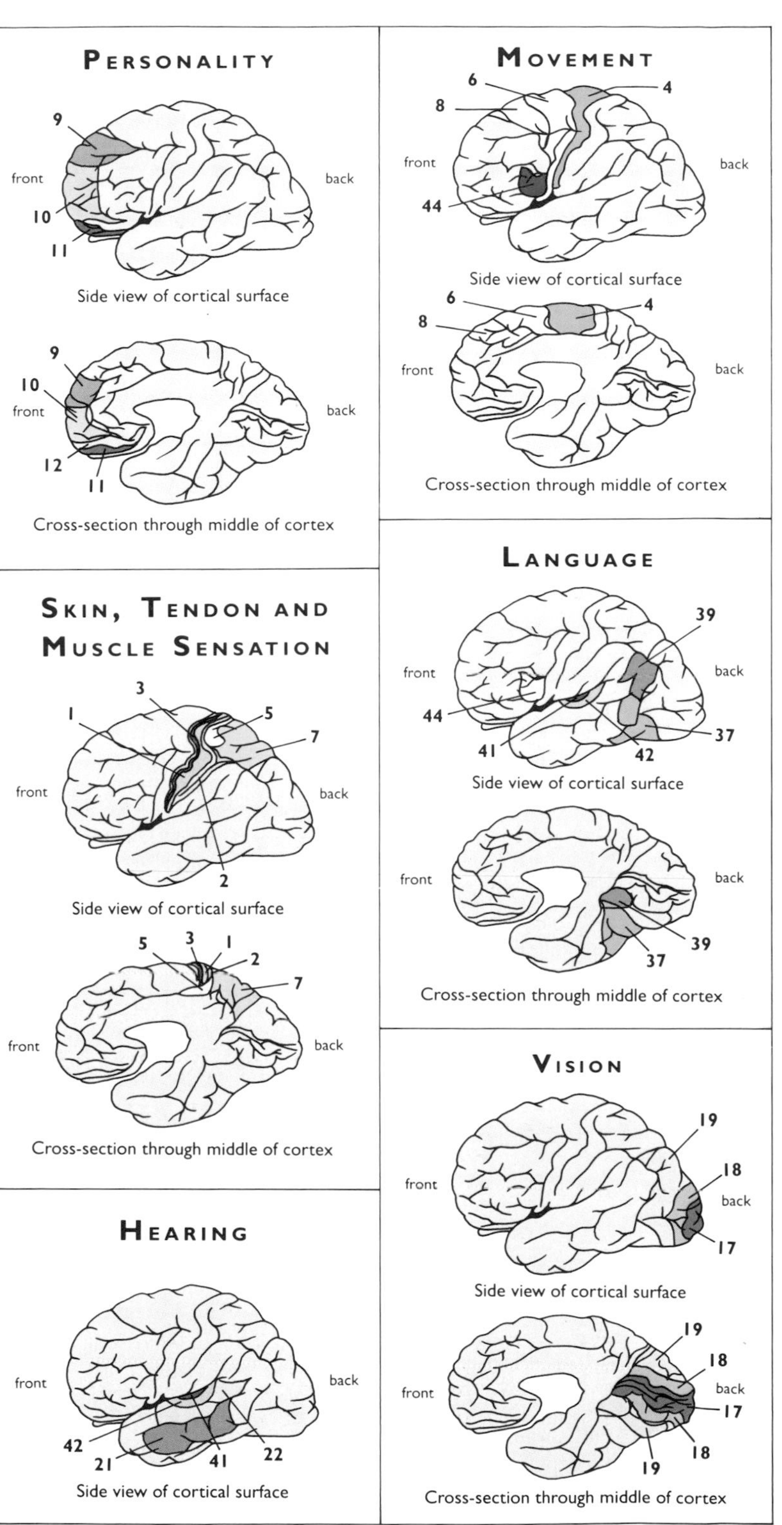

VISION

If the visual area 17, right at the back of the brain, is destroyed by disease or injury, the person concerned will be completely blind although the eyes are entirely undamaged. If the visual association areas, 18 and 19, are destroyed the affected person will be blind but will deny it, and will rationalize the tendency to bump into things by saying, 'It's too dark,' or, 'I've lost my glasses.' Damage to the visual association area can cause a variety of effects. Objects may seem distorted in size, form or colour; there may be hallucinations of light flashes, stars, geometric forms, or even of persons or animals. The latter may appear of normal size, or very tiny, or very large. There may be total inability to name, or to state the function of, objects seen, although such objects can at once be identified by feel, taste or smell. This is called *visual agnosia*.

The Control System 3: the Endocrines

The endocrine system is as important as the central nervous system as a communication and control system for the body. The *endocrine glands* that make up the system secrete their products, the *hormones*, directly into the bloodstream. The hormones, or 'chemical messengers', are thus carried to all parts of the body where they may influence the action of millions of different cells – the target cells for the particular hormone. Hormones are powerful substances that act in very low concentration. They are taken up by receptors on the cell membranes – proteins to which they bind specifically – thereby starting a sequence of events in the cell that alter its function.

The endocrine system automatically controls and integrates many important body functions. These functions include body growth, the rate of energy production and build-up and breakdown of tissues (metabolism), the action of the heart, the tension of the blood vessels and hence the blood pressure, body temperature, the water content of the body, digestion and absorption of food, the development of the secondary sexual characteristics, reproduction, and the body's response to stress. Although the elements of the system – the endocrine glands – are scattered throughout the body, the action of most of them is tightly controlled by a part of the brain called the *hypothalamus*, which works through the *pituitary gland*. These two structures are the coordinating elements for nearly all of the other endocrine glands – the thyroid, parathyroid and adrenal glands (⇨ pp. 80–1), and the sex glands. The hypothalamus has often been described as the 'conductor of the endocrine orchestra', and the pituitary as the 'leader'. The endocrine system does not work in isolation, but is intimately linked to the central nervous system, and each profoundly influences the other.

Oestrogen crystals shown here in a polarized-light microscope image. Along with the other sex hormones – androgens and progestogens – oestrogen is a steroid. (RB)

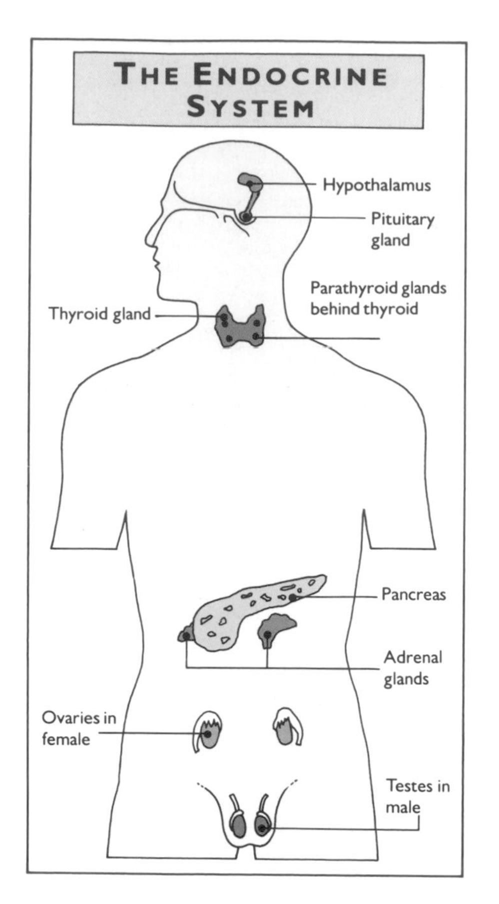

What are hormones?

There are three classes of hormone. The great majority of them are proteins or peptides (⇨ pp. 36–7), and they vary greatly in size, ranging from small peptides containing only three amino acids to small proteins containing hundreds of amino acids. In most cases they are formed initially as large peptides, or *pro-hormones*, from which the active hormones are split off by enzymes. The second class, the *amine hormones*, include adrenaline and the hormones from the thyroid gland. Adrenaline is called a *catecholamine*. The third class, the *steroid hormones*, are formed from cholesterol and have chemical structures made up of four interconnected rings of carbon atoms. They are produced by the adrenal glands, the sex glands and the afterbirth (*placenta*) in pregnancy.

The hypothalamus

Near the surface at the centre of the underside of the brain is the region known as the hypothalamus. This contains several groups of aggregated nerve cells (brain nuclei), most of which are connected directly by nerve fibres to the pituitary gland immediately below, although some are connected to other parts of the brain.

The hypothalamus, although very much a part of the nervous system, is also an endocrine gland. It produces various hormones: two of these (*oxytocin* and *vasopressin*; ⇨ below) act remotely, while the others act only on the pituitary gland, controlling the output of its hormones. The hypothalamus also acts on the pituitary by way of interconnecting nerve fibres (⇨ below). The hypothalamus is itself part of a feedback hormone-control system. Hormones in the blood affect any cells carrying hormone receptors. As the cells of the hypothalamus carry such receptors, they are affected by these blood hormones, which exert a controlling negative-feedback influence, damping down oversecretion from any of the endocrine glands.

The hypothalamus is especially important as it is the point at which the neural and hormonal systems of the body interact. Many parts of the brain connect by nerve

PITUITARY DISORDERS

If, for any reason, the pituitary fails early in life, there is severe retardation of body growth. This results in a type of dwarfism in which the individual fails to reach sexual maturity and is sterile. This is accompanied by atrophy of the adrenal gland, with failure of production of adrenal hormones, and atrophy of the thyroid gland, resulting in slowing of all body processes. Pituitary failure later in life causes marked loss of weight, severe weakness, and the effects of undersecretion of the thyroid and adrenal glands (⇨ pp. 80–1).

Pituitary growth hormone affects not only the growth of the long bones (⇨ pp. 40–1), but also the growth of many tissues and organs of the body. Overproduction of growth hormone, as a result of a tumour arising from the cells that produce the hormone, has different effects, depending on when this occurs. Excess growth hormone before puberty causes *gigantism* – excessive growth of the entire body. If the excess occurs after the epiphyses have fused, the result is the condition known as *acromegaly*, in which the bones of the jaw, hands and feet are abnormally enlarged.

Pituitary tumours can also result in oversecretion of prolactin. This results in milk production from the breasts of either sex. Milk secretion is normally associated with underfunction of the sex glands (as was noted by Hippocrates over 2000 years ago), and a prolactin-producing tumour results not only in absence of menstruation and relative sterility in women, but also in loss of libido or impotence in men. Because of the close relationship of the pituitary gland to the crossing of the optic nerves, a tumour of this kind often causes blindness unless detected early and treated surgically.

Loss of the hormone vasopressin from the rear lobe of the gland results in the condition of *diabetes insipidus*, in which large volumes of water are lost in the urine and there is great thirst. The condition can be controlled by giving the hormone.

The Discovery of Hormones

Ernest Henry Starling (1866–1927) and his brother-in-law William Maddock Bayliss (1860–1924) were two English physiologists who formed a lifelong intellectual alliance. In 1902 they were investigating the puzzling question of how it was that the pancreas invariably produced its digestive juice as soon as food passed from the stomach into the duodenum. Pavlov (➪ pp. 172–3) had claimed that this was the result of messages passing along nerves, but Starling and Bayliss proved that the pancreas continued to secrete on time even if all the nerves to it were cut. Careful research showed that when stomach acid was in contact with the lining of the small intestine, a substance, which they named 'secretin', was produced. This substance passed to the pancreas by way of the bloodstream and prompted it to start secreting. Later, as it became clear that there were other chemical messengers of this kind, Starling suggested the general term 'hormones' for these substances, derived from the Greek *hormao*, meaning 'I set in motion.'

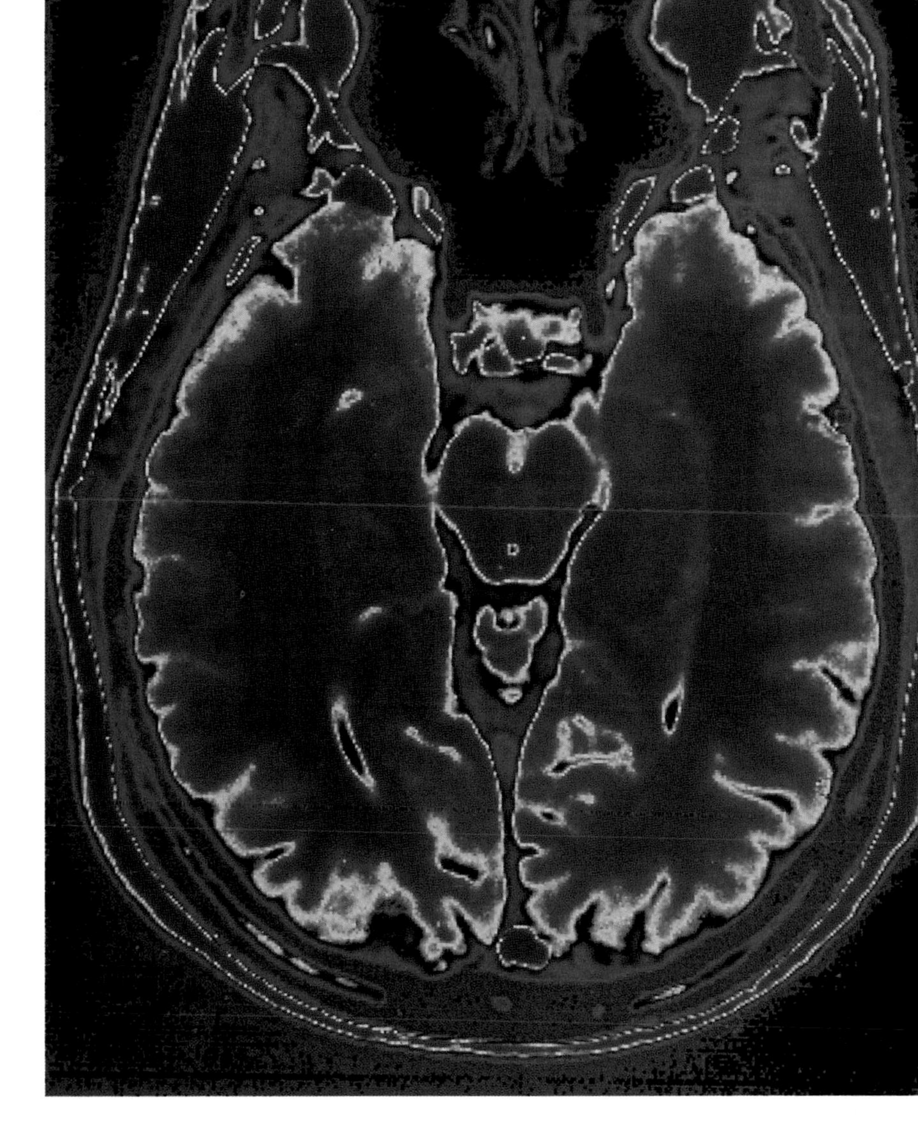

Normal brain shown in this colour-digitized magnetic resonance imaging scan. The pituitary gland is clearly visible as the brown/yellow structure just above the central, heart-shaped pons. The pons links the brainstem with the brain. Situated on either side of it are the cerebral hemispheres. (J. Croyle, Custom Medical Stock, SPL)

fibres to the hypothalamus, which constantly receives information relating to such matters as the current state of bodily and mental stress, perception of danger, and the need for physical activity. Such electrical brain action causes changes in the hypothalamus, and these changes are coordinated with other hormonal information, resulting in messages being sent to the pituitary gland. This, in turn, sends hormones to any or all of the other endocrine organs to prompt them into activity. Depending on the current general body situation, the result may vary from an outpouring of adrenaline and cortisol in a situation of high stress, to a calming reduction in the production of thyroid or sex hormones.

Whether emotion causes adrenaline and steroid hormone production or the hormones cause the emotion is a question not yet resolved. Some scientists believe that all emotions are the necessary concomitants of hypothalamic and endocrine action prompted by mentally significant information. Drugs that block the action of certain hormones can largely eliminate the emotional response without in any way altering the intellectual awareness of the stimulating information.

The pituitary gland

A conspicuous feature of the centre of the underside of the brain is the pea-sized pituitary gland, hanging down from the hypothalamus by a narrow stalk of blood vessels and nerves, and accommodated in a bony hollow, the *sella turcica*, or 'turkish saddle', on the base of the skull, just behind the back wall of the nose. The pituitary is divided into two lobes. The rear lobe is an outgrowth of the hypothalamus and consists mainly of nerve tissue, but together with the hypothalamus, it secretes the two hormones vasopressin and oxytocin (➪ below). All the other pituitary gland hormones are secreted by the front lobe. Some of these hormones act directly to produce an effect, but most of them act to stimulate the other endocrine organs into producing their own hormones.

The front-lobe hormones are:

- *Growth hormone*, which directly controls body growth up to early adult life (➪ pp. 106–9).
- *Thyroid-stimulating hormone*, which controls the output of thyroid hormones from the thyroid gland (➪ p. 80).
- *Adrenocorticotrophic hormone (ACTH)*, which controls the output of cortisol from the adrenal glands (➪ pp. 80–1).
- *Follicle-stimulating hormone*, which controls the production of eggs (ova) from the ovary (➪ p. 96).
- *Luteinizing hormone*, which is necessary to maintain pregnancy after fertilization (➪ p. 102).
- *Prolactin*, which promotes milk production at the end of pregnancy (➪ p. 105).
- *Melanocyte-stimulating hormone*, which stimulates the growth of pigment cells in the skin (➪ pp. 56–7).

The rear-lobe hormones are:

- *Vasopressin*, also called the *antidiuretic hormone*, which increases the reabsorption of water in the kidneys and controls water loss (➪ p. 64).
- *Oxytocin*, which releases milk from the breast tissue and causes the womb muscle to contract at the end of pregnancy (➪ p. 105).

The action of pituitary hormones can most easily be understood by considering the effects of pituitary under- or over-action (➪ box on pituitary disorders).

SEE ALSO
- THE HUMAN COMPUTER? p. 72
- THE CONTROL SYSTEM 1: THE NERVOUS SYSTEM p. 74
- THE CONTROL SYSTEM 4: GLANDS AND HORMONES p. 80
- BODY SENSATIONS p. 90
- THE EMOTIONS p. 204

The Control System 4: Glands and Hormones

The operation of hormones exemplifies the interrelationship of the functions of mind and body. Hormones control many of the body's workings, their release coinciding either with physical needs or the emotions.

The Discovery of Thyroid Hormone and the Steroids

Inspired by the concept of hormones introduced by Starling and Bayliss eight years earlier (➪ pp. 78–9), a young American biochemist, Edward Calvin Kendall (1886–1972), began, in 1910, to look into the workings of the thyroid gland. It was known that overactivity of the thyroid was associated with increased metabolic activity throughout the whole body, and it was also known that loss of the thyroid led to a severe slowing of all body functions. So, since it had such a widespread effect, it seemed likely that the thyroid acted by way of hormones. After four years of work on the iodine-containing protein thyroglobulin found in the gland, Kendall eventually isolated the active principle, thyroxine. This he found to be a fairly simple substance, related to the common amino acid tyrosine, but containing four iodine atoms in its molecule. Presently thyroxine was being widely used in the treatment of thyroid-deficiency disorders.

By now established at the famous Mayo Clinic in Minnesota, Kendall was not content to leave hormones alone. He and his colleagues had noted that the condition of a woman with rheumatoid arthritis was much improved while she was pregnant. Convinced that this was due to a hormonal effect, Kendall worked on the problem until, in 1934, he succeeded in isolating the corticosteroid hormones from the adrenal gland. This was a major breakthrough. With his associates he then developed a way of partially synthesizing cortisone and hydrocortisone, thus making these invaluable substances available for therapeutic use in arthritis and many other conditions. In 1950 Kendall was awarded the Nobel Prize for Physiology or Medicine.

Once the hypothalamus (➪ p. 78) has determined the body's immediate needs and has relayed appropriate information to the pituitary gland (➪ pp. 78–9), there is an immediate outpouring of control hormones from the gland into the bloodstream. Most of these hormones are secreted by cells in the front lobe of the pituitary gland, but some of them are released from nerve endings in the hypothalamus and in the rear lobe of the pituitary. Hormones released from nerve endings are called *neurohormones*, and this is one of the ways in which the central nervous system and the endocrine system interact. The nervous system can also act directly on several of the endocrine glands, and the hormones produced by endocrine glands can alter the function of the nervous system and affect behaviour. Pituitary and hypothalamic hormones control the endocrine glands, but most of the hormones secreted by the various endocrine glands into the bloodstream are monitored by the hypothalamus, which then adjusts its hormone output according to the levels found. High levels result in reduced output and low levels in increased output. This kind of automatic control is called *negative feedback*, and it creates a very stable system. The output of all the endocrine organs is controlled in this way for as long as the body's needs are met.

Three females showing the characteristic signs of myxoedema (left) and cretinism. Unless treated with thyroid hormone, myxoedema may result in death from failure to maintain body temperature. Cretinism arises when the thyroid becomes underactive because of iodine deficiency early in life. From George Gould and Walter Pyle, *Anomalies and Curiosities of Medicine* (1900). (AR)

The thyroid gland

The thyroid lies in the neck, like a bow tie on the front of the windpipe, just below the 'Adam's apple' (the larynx). It produces two iodine-containing hormones: *thyroxine* (T4), with four iodine atoms in the molecule; and *tri-iodothyronine* (T3), with three iodine atoms in the molecule. It also produces a third hormone, *calcitonin*. T4 and T3 act directly on almost all the cells in the body, controlling the rate at which they burn up fuel (➪ pp. 52–3) and hence the rate of breakdown and build-up of chemical substances within the cells (metabolism).

Excess production of thyroid hormone is the cause of the condition known as *hyperthyroidism*, in which there is an abnormally fast tissue breakdown (*catabolism*) and increased heat production. Glucose and fat stores are rapidly depleted and muscles waste. The affected person is hyperactive, has a rapid pulse (often with palpitations), is jumpy, anxious and emotionally changeable, dislikes warm weather, and has sweaty hands. The appetite is often good but weight is lost rapidly. The eyes often protrude (*exophthalmos*), causing a characteristic staring appearance, and the upper eyelids lag behind on looking down. Hyperthyroidism affects women far more often than men, and is usually caused by disease of the thyroid gland, but may sometimes result from excess production of the thyroid-stimulating hormone of the pituitary.

Inadequate output of thyroid hormone in adults is called *hypothyroidism*. This may be caused by thyroid gland disease, by various drugs, or by overdosage of thyroid hormone (usually taken in an attempt to lose weight). The excess hormone causes the pituitary to shut down its production of thyroid-stimulating hormone. Hypothyroidism causes a slowing down of physical and mental processes, weight gain, undue sensitivity to cold, a hoarse voice, dry and flaky skin, loss of hair and of the outer parts of the eyebrows, and puffiness of the tissues (*myxoedema*). There may be deafness, depression and even psychosis (➪ pp. 192–3). Lack of thyroid hormone from birth causes *cretinism*, with irreversible brain damage unless the condition is treated very early. In older children inadequate thyroid hormone leads to failure of growth and development. A *goitre* is an abnormal enlargement of the thyroid gland from any cause.

The third hormone, calcitonin, although secreted by cells in the thyroid gland, has nothing to do with metabolism. It acts on bone, limiting the release of calcium into the blood. The cells in the thyroid gland that produce calcitonin also monitor blood-calcium levels continuously and secrete calcitonin accordingly. If the level falls they produce less calcitonin, so the bones release more calcium; if the blood-calcium level rises these cells produce more calcitonin. In this way the level of

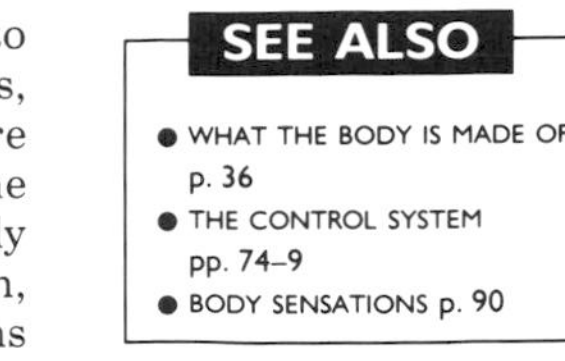

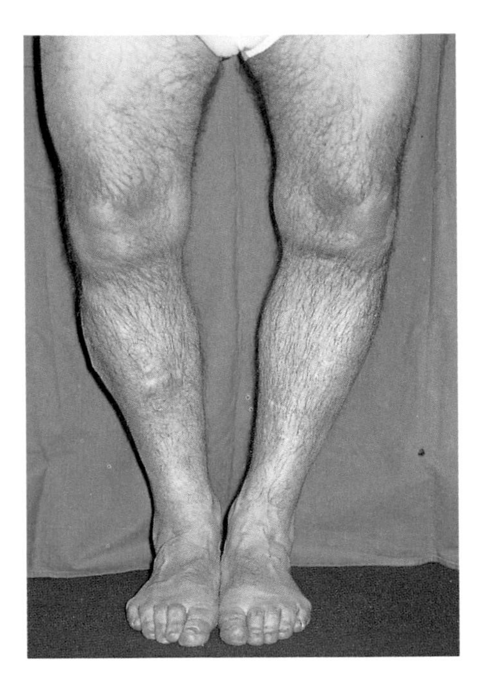

Hyperparathyroidism, the result of overactive parathyroid glands, causes the bones to soften as too much calcium is removed from them and deposited in the body's soft tissues. In extreme cases, the consequent deformation of the bones strongly resembles rickets.
(Dr P. Marazzi, SPL)

blood calcium, the constancy of which is critically important for nerve and muscle function, is kept within narrow limits. Strict maintenance of blood calcium is so important that another hormone mechanism is also involved. This is the function of the parathyroid glands.

The parathyroid glands

The parathyroid glands are four inconspicuous, bean-shaped structures, each about 0.5 cm ($^1/_5$ in) long, partly buried in the substance of the thyroid gland. Their secretion, the hormone *parathormone*, also regulates calcium levels in the blood. If the level of calcium in the blood drops, the parathyroids secrete more parathormone, which acts on the bones to increase the rate of release of calcium so that the level of blood calcium rises again. The hormone also acts on the tubules of the kidneys (⇨ pp. 64–5) to reduce calcium loss and on the lining of the intestine to increase calcium absorption. Abnormally high levels of parathormone, as may sometimes occur from a tumour of one of the parathyroid glands, leads to excessive loss of calcium from bones, with serious softening and distortion. Insufficient parathormone results in a dangerous drop in blood calcium, resulting in abnormal muscle excitability and uncontrollable spasms of contraction. This condition is called *tetany*.

The adrenal glands

The two adrenals sit, like triangular caps, one on top of each kidney. Each gland has two parts with different functions. The inner part, or *medulla*, produces *adrenaline*, and the outer part, or *cortex*, produces three kinds of steroid hormones – cortisol, aldosterone and male sex hormones. *Cortisol* prompts cells to increase the synthesis of proteins – mainly enzymes and structural proteins – by acting on messenger RNA (⇨ pp. 26–7). It also stimulates the conversion of amino acids (⇨ pp. 36–7) into the fuel glucose. These actions help the body to react to stress. *Aldosterone* controls water balance by determining the rate of reabsorption of water in the kidneys (⇨ pp. 64–5). The *male sex hormones* (*androgens*), which are produced by women as well as by men, strongly stimulate protein synthesis (*anabolism*) in many parts of the body, and are important in determining body growth and development. This effect is greater in men than in women because of the additional powerful androgens (⇨ below) produced by the testicles.

Adrenaline is produced when unusual efforts are required of the body, especially in emergency situations. It increases the heart rate, raises the blood pressure, speeds the rate and depth of breathing, widens the arteries to the muscles and narrows those to the intestines, mobilizes glucose for fuel, and increases alertness and excitement. It has been called the hormone of 'fright, fight or flight'.

The pancreas

The pancreas has two functions; it produces digestive enzymes (its *exocrine function*) and it synthesizes the hormones insulin and glucagon. These two hormones respectively lower and raise the amounts of sugar (glucose; ⇨ pp. 36–7) in the blood (its *endocrine function*). *Insulin* promotes the movement of the vital fuel glucose through cell membranes into cells. If deficient, glucose accumulates in the blood. Muscle protein is converted to glucose and fats are burnt in excess, producing toxic acidic by-products. There is severe wasting of muscles, and the urine is loaded with glucose that the body is trying to dispose of. This condition, *diabetes*, can be controlled by injections of insulin.

Glucagon acts in a manner opposite to that of insulin. It causes liver glycogen (⇨ pp. 48–9) to break down to glucose, so increasing the amount of sugar in the bloodstream. Glucagon also mobilizes fatty acids for energy purposes.

The sex glands

Sexual differentiation at puberty (⇨ pp. 108–9) is initiated, in both females and males, when the pituitary begins to produce hormones called *gonadotrophins*. These cause the ovaries and the testicles to increase production of their own hormones, respectively oestrogens and androgens, principally testosterone. In girls, *oestrogen* causes breast development, growth of pubic hair, widening of the pelvis, deposition of fat under the skin in certain areas to produce female contours, and the onset of ovulation and menstruation. In boys, *testosterone* causes the penis, testicles and scrotum to enlarge, the beard and pubic hair to appear, the prostate gland and seminal vesicles to mature, and the larynx to increase in size so that the voice deepens.

In both sexes there is a considerable anabolic effect at puberty with a spurt in body growth and weight gain. Body weight may almost double during this period. The increase in boys is mainly due to increase in muscle bulk. In girls, weight gain is also caused by the large increase in fat deposition.

Adrenaline in action. With increased levels of the hormone coursing through their bodies these US troops are alert and prepared to run for cover during an air attack.
(AKG)

Speech

Speech is a remarkably complicated activity involving the operation of many parts of the brain. The use of language involves comprehension, the formulation of thoughts, the acquisition, storage and recall of words and their relationships, the selection of words, their arrangement in meaningful sequences, and their articulation.

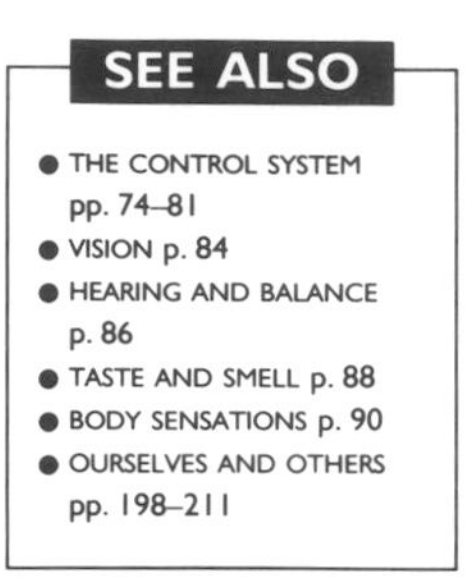
SEE ALSO
- THE CONTROL SYSTEM pp. 74–81
- VISION p. 84
- HEARING AND BALANCE p. 86
- TASTE AND SMELL p. 88
- BODY SENSATIONS p. 90
- OURSELVES AND OTHERS pp. 198–211

Verbal thinking and the articulation of speech, although closely connected, involve quite different neurological activities. Human speech differs qualitatively from the communication sounds made by other animals, and involves considerable activity in the cortex of the brain (⇨ pp. 76–7). In the other primates, vocalization appears to be initiated and controlled by the limbic system (⇨ pp. 74–5) rather than by the cortex. So it is probably correct to suggest that speech is one of the distinguishing attributes of the human being.

However, there is some controversy about the extent to which the speech function is innate in humans. While no one disputes that the brain is organized and programmed for speech, some authorities take the view that all verbal behaviour is learned. Others, notably Noam Chomsky (⇨ pp. 174–5), believe that the brain is 'hard wired' to subserve important aspects of language and is capable of translating experience into a kind of basic grammar common to all languages. There is considerable linguistic evidence for this view, but little or nothing is known about the anatomy or physiology that would support such a function.

Speech centres in the brain

There are five readily distinguishable brain areas concerned with language (⇨ pp. 76–7 for further details and diagrams). The area in the frontal lobe, near the motor cortex (*Broca's area* – area 44), is concerned with the control of the many muscles in the face, tongue, throat and jaw involved in the articulation of speech. The other areas, further back along the temporal lobe of the brain (conveniently lumped together as *Wernicke's area*), are concerned with all the many sensory aspects of speech. Wernicke's area is connected to Broca's area by a thick tract of nerve fibres. Speech is believed to be formulated in Wernicke's area and the information passed to Broca's area where the programme for the necessary complex, muscular activity is put together. This is then passed to the motor cortex so that the appropriate muscles can be activated. Wernicke's area receives input from the visual cortex (areas 17, 18 and 19) for written or printed language and from the auditory cortex (areas 41, 42, 21 and 22) for heard speech.

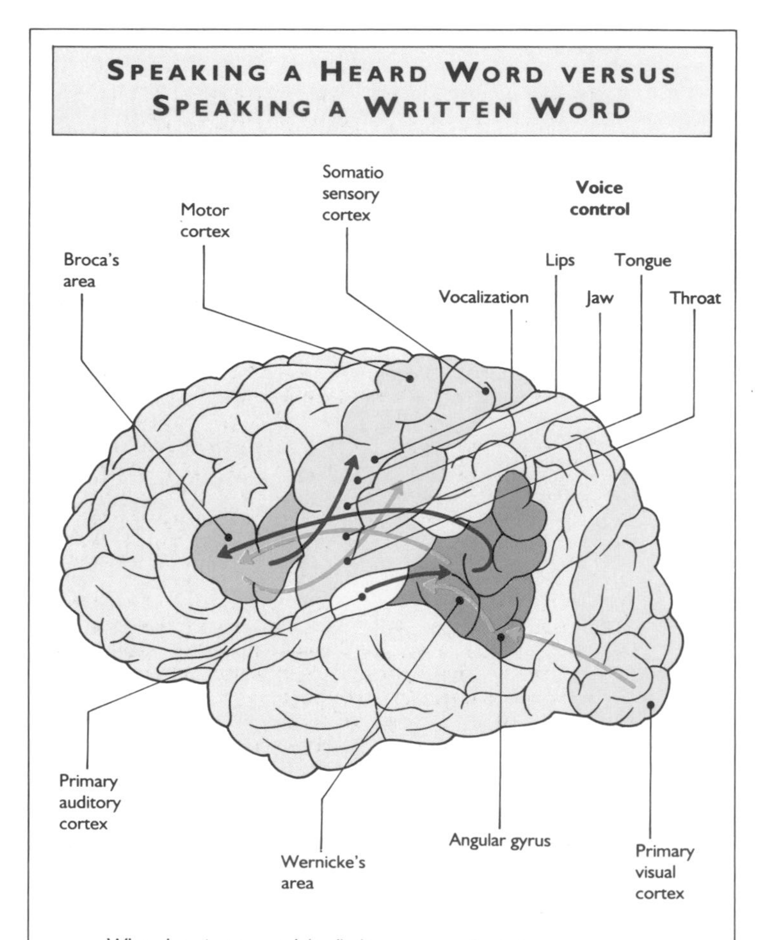

When hearing a word (red) the ears transfer information to the primary auditory cortex. It cannot be understood, however, until the signal has been processed in Wernicke's area. If the word is to be spoken, Broca's area arranges a program for its articulation, which is transferred to the motor cortex.

When a word is read (blue), the primary visual area registers a signal, which it relays to the angular gyrus. This structure associates the visual form with the corresponding auditory arrangement in Wernicke's area.

Phonation

The vocal cords are folds of mucous membrane inside the voice box (larynx). They have associated muscles that change their tension and bring them together, edge to edge. During normal breathing the cords are held apart; the production of voice sounds begins with the moving together of the cords during expiration. As they come together the pressure in the lungs rises, the degree of the rise varying with the loudness of the sounds we wish to make. Air passing between the tightly pressed cords forces them suddenly apart. This leads to a sudden drop of pressure so that the cords are able to come together again and the pressure under them rises once more. The effect is a rapidly repeated series of separations and closures so that a succession of compressions and easings is imposed on the column of air in the throat, mouth and nose. The vibrations thus set up in the column of air are known as sound waves.

The pitch of the sounds produced by the vocal cords depends on the frequency of vibration: the higher the frequency, the higher the pitch. This, in turn, depends on the tension on the vocal cords and on the length of cord allowed to vibrate. A short length produces a high pitch, and the full length and lowest tension produces the deepest pitch possible. However, the operation of the cords is not a simple matter as they vibrate in a complex manner and different parts can vibrate simultaneously at different frequencies. In general, women and children have shorter vocal cords than men, so their voices are pitched higher. The range of the lowest pitch of the voice extends from about 80 Hz (hertz; cycles or vibrations per second) in men (nearly two octaves below middle C) to about 400 Hz in women

Scientist performing a computer analysis on the word 'baby' as part of continuing research on and development of high-level systems able to recognize the human voice. (Hank Morgan, SPL)

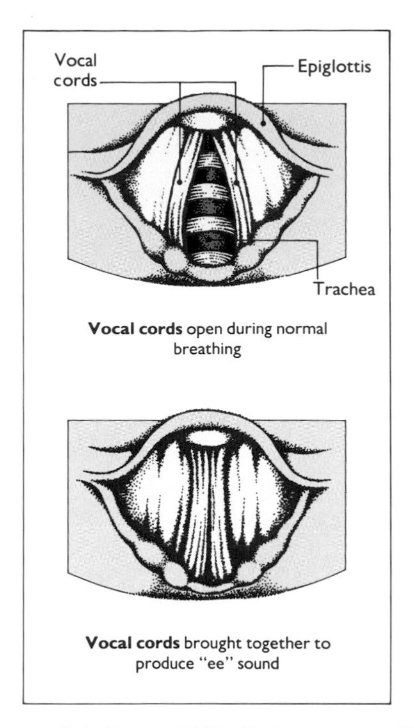

Vocal cords open during normal breathing

Vocal cords brought together to produce "ee" sound

(around A above middle C). A soprano will easily produce a singing note of 1000 Hz (about two octaves above middle C, which has a frequency of 261 Hz).

Voice quality

The basic sound produced by the vocal cords has a waveform described as a sawtooth. This is the kind of wave produced by any double-reed musical instrument such as an oboe or a bassoon. The sawtooth waveform is quite different from the sine (bell-shaped) waveform produced by a tuning fork or a flute played softly. Pure sine waves have a fundamental tone of a particular frequency (the recognizable pitch of the tone) and nothing else. While sawtooth waves have a fundamental tone they also have a rich collection of harmonics. These are additional tones, or overtones, having frequencies that are simple multiples of the fundamental frequency. The harmonics are usually at a lower volume (or amplitude) than the fundamental tone and so are not heard separately but blend with the tone to alter its quality (or timbre). Different musical instruments produce different combinations of overtones, which is why they have their own distinctive characters.

However, the quality of the voice has little to do with the richness and relative amplitude of the various harmonics produced by vibration of the vocal cords. Its final quality depends instead on the principle of *sympathetic resonance*. Like musical instruments many objects can vibrate at a characteristic frequency within the audible range so as to produce a particular note, the pitch of the note being determined by the size of the object. Such objects will sound if exposed to a range of pitches that includes those of the natural resonant frequency of the object. The effect of this 'sympathetic vibration' is to amplify that particular tone in the same way as small, well-timed (in-phase) pushes on a swing quickly build up the amplitude of movement. Hollow cavities also have a natural frequency of vibration, as can be shown by blowing across the top of a bottle. There are many such cavities in the human head and neck – the throat cavity, nasal cavity, mouth cavity and the bony sinuses. Each of these reinforces, to a greater or lesser degree, the particular harmonics in the basic tone produced by the larynx that correspond to their natural resonant frequency.

The effect of these cavities on voice quality varies with their volume, with the force of the expiration of the air, and with the degree of communication between them. Speech quality is, for instance, greatly affected by alterations in the degree of swelling of the lining of the nose, as during a cold. Laryngeal tone, by itself, is thin, weak and lacking character, as was demonstrated by a postmortem report on the larynx of the great tenor Francesco Tamagno, Verdi's original Otello, which concluded: 'The organ differs from that of a normal person only in that it exhibits an unusually large number of scars on the wall . . . caused by catarrh.'

Formants

Resonances have an even more important function – to produce vowel sounds. Each vowel sound involves a combination of two frequencies to produce what is called a *formant*. These pairs of frequencies are common to all vowels, whoever produces them. Formant frequency has a bearing on the ease with which vowel sounds can be produced when singing high notes. Opera composers have long known to avoid writing high notes for those vowels with a low upper formant frequency, such as the 'oo' in 'pool', while formants with a high upper frequency, such as the 'ee' in 'team', are easy to sing on a high note. Any ignorance of formants would be quickly pointed out to composers in rehearsal by the sopranos.

Articulation

Modulation of the basic pitch-varying tones to produce the vowels and consonants of speech is primarily the function of the mouth cavity. This is done by ever-changing variations in its shape and volume and in the area and shape of its three openings – to the throat, to the nose and to the outside via the mouth opening (i.e. between the lips). Quick changes in the mouth opening, in particular, allows a wide range of different sounds to be produced. The relationships of the tip of the tongue to the teeth, lips and the roof of the mouth, and of the lips to each other, are especially important in forming the outlet shape necessary to produce these sounds from the harmonic-rich vocal cord tones. Thus articulation of sound involves precise and accurately timed contractions and relaxations of the muscles of the tongue, the lips, the soft palate and the face. Vowel sounds are produced by the shape of the lips, whereas the production of consonant sounds often involves sudden separation of the lips, or movements of the tip of the tongue away from the back of the teeth, or from the palate.

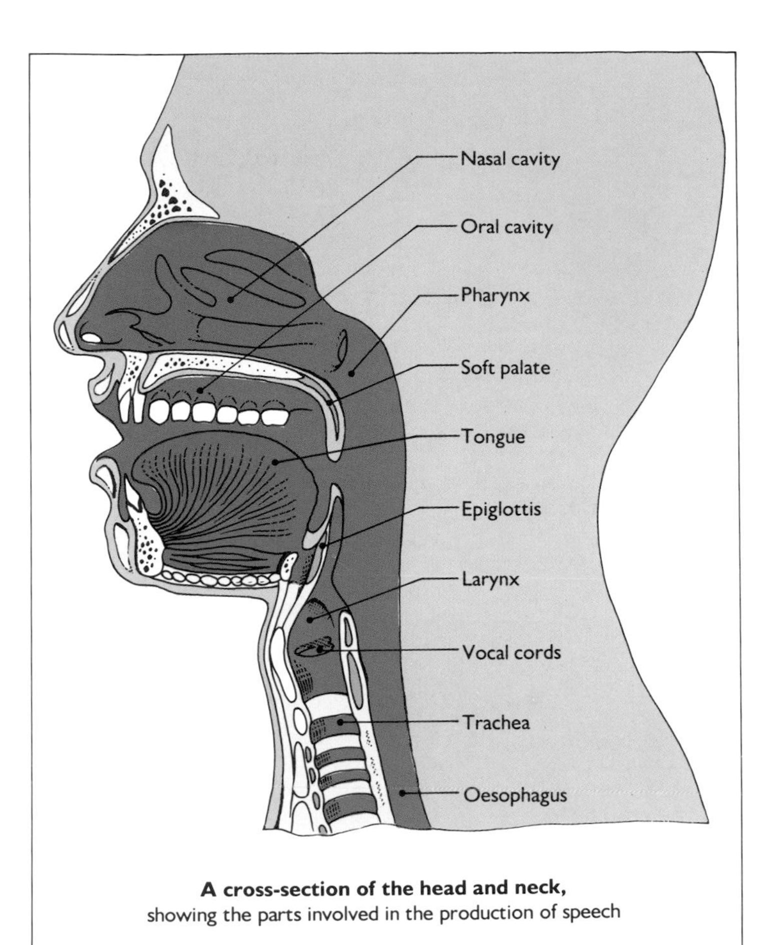

A cross-section of the head and neck, showing the parts involved in the production of speech

— Carl Wernicke and Aphasia —

Aphasia is an acquired speech disorder, resulting from brain damage, which affects the understanding and production of language rather than the articulation of speech. It may take various forms (➪ below). Carl Wernicke (1848–1905) was a young professor of anatomy at Breslau (now Wroclaw in Poland). Stimulated by Paul Broca's demonstration of a case of speech defect arising from temporal-lobe brain damage (➪ pp. 76–7), Wernicke decided to investigate the temporal lobe of the left side of the brain by careful postmortem study of the brains of people with a known history of acquired speech defects. During the 1870s he developed a theory of speech function that is generally accepted today and that has had an important influence on ideas of brain function generally.

Broca had already shown that there was an area (area 44) near the front of the temporal lobe that controlled speech movements. Wernicke now showed that there were areas further back, connected to Broca's area, that were concerned with the more mental aspects of speech. His published work includes a full description of the effects of brain damage in these areas – effects known as '*receptive*' or '*sensory*' *aphasia*. Such damage affected the understanding of spoken language or the ability to recognize or to understand the meaning of printed or written words (*alexia*). Alternatively, it might have destroyed the ability to produce meaningful writing (*agraphia*). In some cases, while speech remained fluent, the effect of the damage deprived it of meaning. All this was in marked contrast, however, to Broca's aphasia, in which comprehension was normal but the affected person was unable to speak.

Wernicke came to the conclusion that the mental functions involved in speech, and other higher functions, were not performed in discrete areas of the brain cortex as in the case of vision, hearing, smell and tactile sensation. These higher functions depended on the complex links between all these areas, and between them and various parts of the temporal lobe. Wernicke's ideas have been supported by subsequent research, and the areas he described and the aphasia resulting from their damage now both bear his name.

Vision

We know a great deal about how the eye works. We also know a lot about the neurological connections between the eye and the brain – the optical pathways – and about the mapping of the fields of vision in the cortex (➪ pp. 76–7) of the brain. But as to the physiological events that underlie the actual experience of vision, we know almost nothing at all.

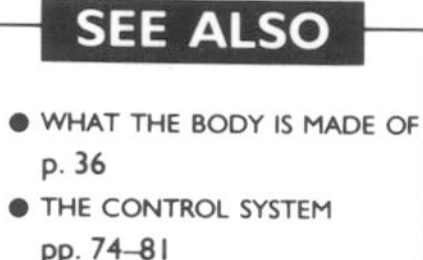

SEE ALSO

- WHAT THE BODY IS MADE OF p. 36
- THE CONTROL SYSTEM pp. 74–81
- SPEECH p. 82
- HEARING AND BALANCE p. 86
- TASTE AND SMELL p. 88
- BODY SENSATIONS p. 90

This illustrates one of the great paradoxes of research into neurophysiology: the closer we come to understanding the functioning of the brain, the further we seem to be from any explanation of one of its most striking manifestations – the consciousness of sensory experience, our perception of the external world. Human beings rely more heavily on vision than on any other sense. This central importance of vision is reflected in the intimacy and size of the link between the eye and the brain (➪ below).

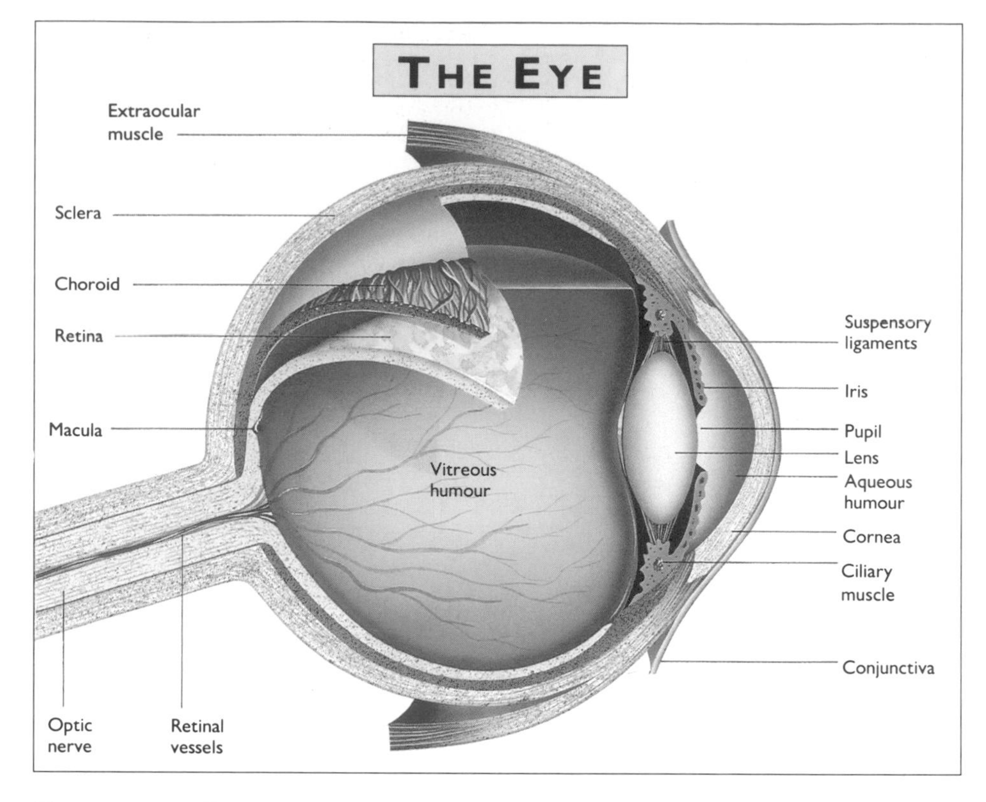

How the human eye works

The outer coating of the eye is a tough, white layer called the *sclera* (➪ diagram) partly covered by a transparent membrane, the *conjunctiva*. Under the sclera is a thin layer of blood vessels and black pigment, the *choroid*, and under this is the *retina*. The main chamber of the eye is filled with a delicate, watery gel, the *vitreous humour*, while the narrow chamber at the front is filled with water (*aqueous humour*).

Early spectacles in use. Detail from *The Outpouring of the Holy Spirit* by Konrad von Soest (c.1370–c.1422). (AKG)

The eye comprises two main parts – an optical, image-forming lens system consisting of the *cornea*, *iris diaphragm* and internal *crystalline lens* – and a complex neurological device, the retina, which converts the images falling on it into patterns of electrical signals. These pass back to the rear part of the brain by way of the *optic nerves* and their nerve tract continuations. The eye has a maximum aperture (focal length divided by maximum pupil diameter, the f number) of about f2. The iris readily constricts, however, to stop down the optical system and reduce the aperture to around f22. It does this through constriction of the muscle fibres surrounding the *pupil* (central hole). Stopping down occurs automatically on exposure to bright light and on near viewing. The main lens, the cornea, is wide-angled – we can see out to 180° – but vision is not distorted because the image plane (the retina) is spherically concave. The focal length of the eye (the distance from the cornea to the retina) is around 22 mm and the depth of field (depth of the zone of sharp, straight-ahead vision) is correspondingly great.

Focusing is effected by an automatic process known as *accommodation* – an adjustment of the optical power of the lens, which lies immediately behind the pupil. This lens is elastic in younger people and, if allowed to relax, assumes an almost spherical shape. It is stretched into a flatter shape, however, by the delicate protein strands of the *suspensory ligaments* around its equator that are pulled tight by a muscle ring, the *ciliary body*, surrounding the lens. When this muscle ring contracts, the tension on the supporting strands is released and the lens surface becomes more curved by virtue of its own elasticity. This allows the lens to converge rays from a near object, these being more divergent than rays from a distant object. The stimulus for accommodation arises from the detection by the retina of the degree of coming together or separation (vergence) of the rays passing through it.

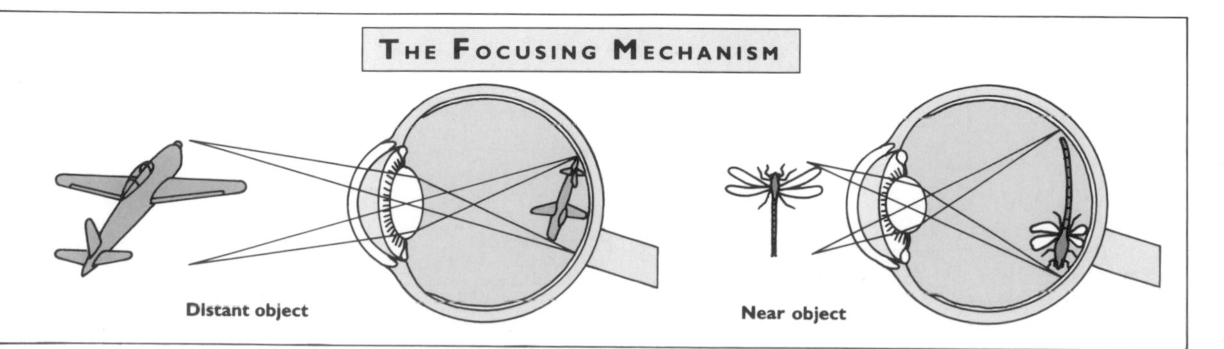

The retina

The images formed on the retina by the optical system are inverted and are bit-mapped (constructed of tiny, discrete points of determinable location). There is an almost one-to-one correspondence between points on the retina and points on the visual cortex of the brain, and the image can be thought of as represented there, also in bit-mapped form. The retina is not simply a passive light-to-nerve impulse converter. Its light-sensitive cells (*photoreceptor cells*) are interconnected in such a way as to form a computer that increases the range of contrast sensitivity and codes the output signals. There are two types of photoreceptor cell – *rods* and *cones*. The rods are the more sensitive to light, but are colour-blind, and are concentrated most at the edge of the retina. The cones are more numerous at the centre of the retina, and are colour-sensitive. They are of three types, each type giving its maximum nerve-impulse output when one of the three primary colours – red, green or blue-violet – falls upon it. In this way, colour information is coded in the patterns of nerve impulses passing along the optic nerve fibres. The procedure is analogous, in reverse, to the way a colour image is produced in a TV tube. The packing of photoreceptor cells is tightest at the centre of the retina and only here, at the *macula lutea*, is high resolution vision possible. But automatic 'fixation' reflexes ensure that the eyes are always aligned accurately on objects the individual wishes to see, and in this way

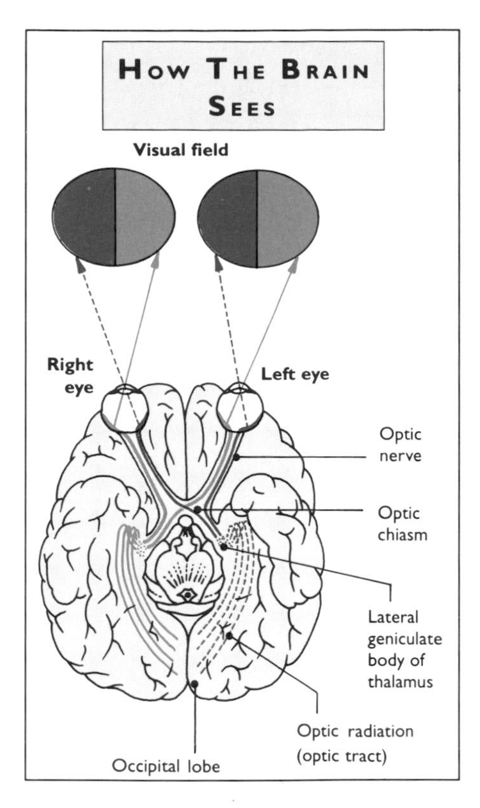

the image always falls on the macula. It is impossible to read normal print if the gaze is steadily fixed even a few degrees to one side of it.

The eye and the brain

Between the object of perception and the experience of vision lies a remarkable pathway in which information alters its form and nature several times. We see by virtue of rays of light of varying intensity and direction, reflected from the surface of, or transmitted through, external objects. Some of this light enters the eye and is focused on the retina where it stimulates millions of photoreceptor cells to produce nerve impulses. These are coordinated and pass out of the back of the eye along the million or so separate fibres of each optic nerve. Some of these fibres remain on the same side, while some cross over to the other. All of them end in one of half a dozen layers in a kind of junction box (known as the *geniculate body*) situated on the underside of the brain, one on each side.

Some of the layers in each geniculate body are connected to one eye, the rest to the other. These layers consist of synapses (⇨ pp. 74–5) between the nerve fibres (axons) arising in the retina and a new set of nerve cells whose axons run in great sweeping curves through the substance of the brain, right to the back. There, in the outer layer (the *occipital cortex*, or area 17; ⇨ pp. 76–7), lie further nerve cells that are, in some unexplained way, concerned with the processes of visual experience. The link between the synapses in the geniculate body layers are made as a result of visual experience in childhood. If, for any reason, one eye is not used for the first few months of life, the geniculate body layers on each side corresponding to that eye do not develop properly and the eye remains permanently blind.

Binocular vision requires that the two eyes should work together, aligning themselves with high precision on the object of interest so that the image of the object formed on each retina should fall upon corresponding areas of the two retinas. This is a demanding requirement calling for a precise feedback-control mechanism. It is achieved under brain control by means of which the contractions of the six tiny muscles that move each eye (the *extraocular muscles*) are coordinated by a computing nerve network in the brainstem (⇨ pp. 74–5). The eyeballs can rotate freely as they are embedded in pads of fat within the *orbits*; these are bony caverns in the skull that provide protection from injury.

Perception

The only way we can obtain information about the external world is through the operation of our sense organs. Perception is the awareness or consciousness of the external world gained through the senses – sight (⇨ main text), hearing (⇨ pp. 86–7), taste and smell (⇨ pp. 88–9), and touch (⇨ pp. 90–1). It is also the process of obtaining the information needed to experience this awareness. Perception is a complex process involving the integration of past experience with current data input, and an elaborate process of interpretation. In making perceptions we are constantly forming hypotheses about reality. Often, the hypotheses are correct, but sometimes they are wrong and we are misled.

The study of human perception is important because, so far, we know very little of the true nature of awareness. Perception is entirely a personal (subjective) matter and we can never know what any other person's awareness is like. This makes the objective or scientific study of the subject difficult. For this and other reasons, psychologists and psychiatrists are especially interested in two classes of what appear to be defects of perception – the many puzzling optical and other illusions any of us can experience, and those visual, auditory, olfactory (smell) or gustatory (taste) hallucinations that can arise in the course of physical or mental disease. A study of such defects has thrown much light on the way in which the brain responds to its sense data, and on the way in which brain function can become disordered.

Perception is also of great interest to philosophers trying to arrive at some knowledge of the true nature of the external world, and some have made a convincing case for the proposition that the world is, in reality, very different from how it appears to us via the senses. This proposition is supported by physical science. For instance, when we see a red box, its colour appears fixed along with other of its physical properties, such as dimension and mass. The box's colour, however, is a result of the capacity of the box's surface molecules to absorb certain wavelengths and reflect others. Red surfaces absorb all but the wavelengths of red light, which they reflect. White surfaces, by contrast, reflect all wavelengths.

Eye Defects

Optical defects arise in various ways. The commonest result from a disparity between the focal length of the lens system (the distance from the lenses to the point of focus) and the front-to-back (axial) length of the eye. If the lens is relatively too powerful the image of distant objects will tend to form in front of the retina. Only the light rays from near objects are sufficiently divergent to focus on the retina. This condition is called short-sightedness, or *myopia*, and is usually due to excessive curvature of the cornea or to excessive axial length of the eyeball. If, on the other hand, the cornea is relatively too flat or the eyeball too short, the eye is *hypermetropic* and near objects may be hard to see unless the individual can focus (accommodate) the eyes sufficiently. This is the condition commonly called long-sightedness. Hypermetropia often becomes manifest only when the focusing power deteriorates (⇨ below). In the normal cornea, the degree of curvature in any direction is uniform. If, however, there is a greater degree of curvature in any one direction (meridian) than in another, the result is *astigmatism*. The effect of this is that the eye cannot sharply focus image lines lying in all orientations. If, for instance, vertical lines are sharp, horizontal lines will be blurred. The corneal meridia of maximal and minimal curvature need not be vertical and horizontal but may lie, like a rotated cross shape, in any two perpendicular planes.

The internal crystalline lens continues to secrete fibres throughout life and, in consequence, becomes 'tighter' and less elastic with increasing years. This results in a progressive loss of focusing power, usually manifesting itself by an inability to read comfortably around the mid-40s. This is called *presbyopia* (from Greek, *presbos*, 'old man'). Abnormal opacification of the crystalline lens is called *cataract* and should not be confused with corneal opacification caused by scarring from disease or injury. Damage to the retinal nerve fibres at the head of the optic nerve from increased pressure within the eye is called *glaucoma*. Damaging disorders of the retina are called *retinopathies*.

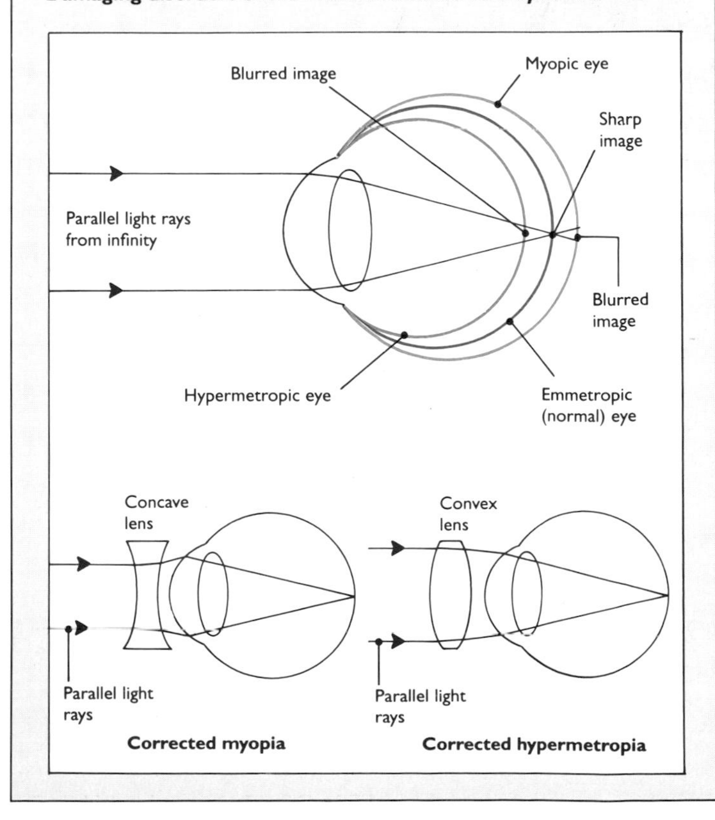

Hearing and Balance

In many of the simpler animals the functions of hearing and balance are served by separate mechanisms, often in different parts of the body. Hearing in vertebrates, however, evolved out of balancing mechanisms and in all vertebrate species, including ourselves, the ear is the organ both of hearing and of equilibrium. In humans and most other mammals, the two functions are so intimately linked that they share a common nerve pathway to the brain.

Man using an ear trumpet (right) from a satirical British cartoon of the 1830s. (AR)

Disorders of balance are commonly associated with disorders of hearing. Hearing is an essential medium of social intercourse, so important that its severe loss is often more damaging to relationships than loss of vision. The atmosphere in which we live is seldom free from the periodic vibration that we interpret as sound. We hear it all but our brains filter out most of it, and present only what is important to us. For example, we will not pay attention to background conversation until we hear someone mention our name. Hearing has the advantage over sight in that it can attract our attention without our active participation. Hearing music is also a source of pleasure. Modern living, though, offers threats to our hearing apparatus, and a knowledge of the structure of the ear can help us understand how these threats arise and how they may be avoided.

The structure and function of the ear

It is convenient to divide the ear into three parts – the outer, middle and inner ears. The *outer ear* consists of the *pinna*, the visible part, and the *external auditory canal*, or meatus, a tube about 2.5 cm (1 in) long that ends in the eardrum. The pinna serves little more than a decorative function in humans and no longer has much sound-gathering power – in contrast to the proportionately larger external ears of many other mammals. The *eardrum*, or *tympanic membrane*, is a delicate circular surface of stretched skin and fibrous tissue capable of free vibration in response to the most subtle changes in the pressure of the air in contact with it. The sensitivity of the drum to force is remarkable; the footstep of a mosquito upon it is clearly audible.

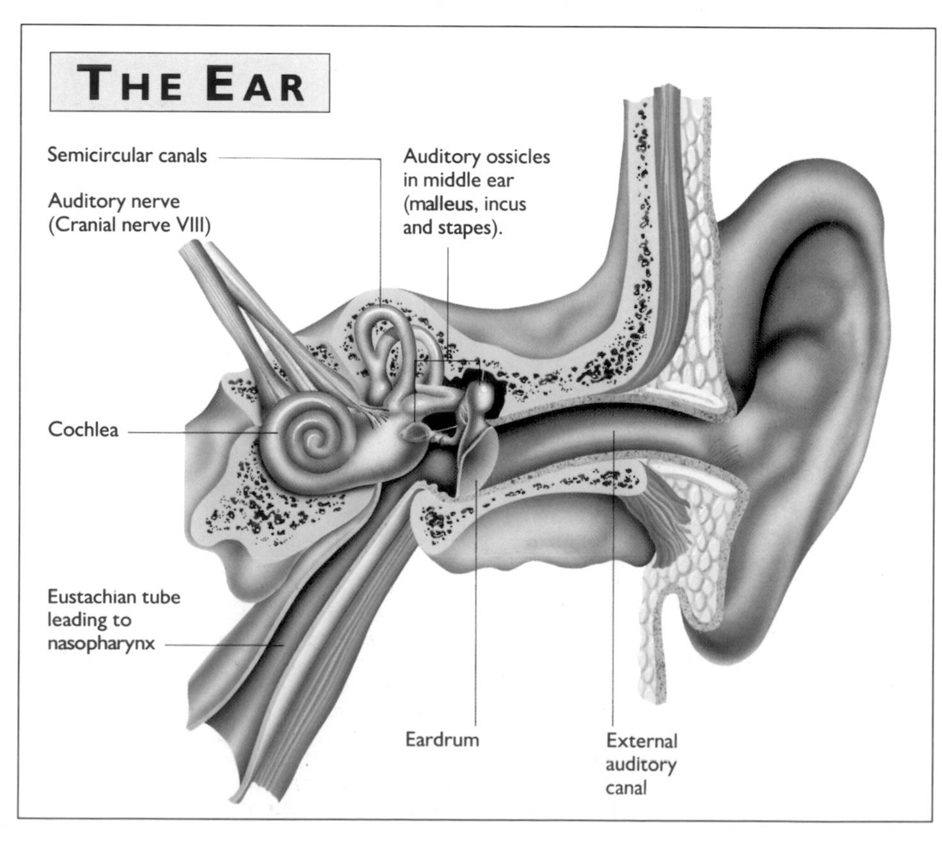

The *middle ear* lies between the eardrum and the outer bony wall of the inner ear. It is a narrow cleft crossed by a chain of three tiny articulated bones, the *auditory ossicles*, that link the eardrum to a window in the inner ear. The eardrum must be very free to vibrate, but the force of the vibrations must be transferred to fluid in the inner ear. This fluid offers considerable resistance to movement and vibrations could not be effectively conveyed to it were the drum connected directly. The purpose of the ossicles is to act as a system of levers, converting the relatively large-amplitude vibrations at the drum to small-amplitude but more powerful vibrations at the window. The middle ear is lined with mucous membrane containing blood vessels, and oxygen in the air in the middle ear is constantly being absorbed into the blood in these vessels. The resulting tendency for the structure to develop a partial vacuum and for the drum to be forced inwards is prevented by the presence of a tube joining the middle ear to the back of the nose. This is the *Eustachian tube*, through which air can pass into the middle ear to equalize the pressure on both sides of the drum. The front end of the tube opens each time we swallow.

The *inner ear* is the most complicated of the three parts. The hearing sensory apparatus is contained in the *cochlea*, a structure shaped like a snail's shell. The balancing mechanism (⇨ below) consists of a widened extension to the cochlea carrying the three *semicircular canals* (⇨ diagram). The whole combined apparatus, called, for obvious reasons, the *labyrinth*, is filled with fluid. The inner of the three auditory ossicles, the *stapes* or stirrup, has a footplate that fits neatly into the oval window of the outer wall of the labyrinth. Vibration of the stapes conveys the sound vibrations to the fluid in the labyrinth. The cochlea contains a spiral membrane, the *basilar membrane*. This can be likened to a kind of elongated harp with strings (fibres), of different length and tension, free to vibrate in sympathy with vibrations in the cochlear fluid. Under the influence of different frequencies, different parts of the basilar membrane vibrate. High frequencies cause fibres near one end to vibrate; low frequencies cause vibration near the other end. Resting on the basilar membrane are many rows of *hair cells*. These are the sensitive elements that convert vibrations into nerve impulses. Movement of these cells causes their protruding hairs to wipe against an overlying membrane, and this causes the hair cells to fire, sending nerve-impulses along the auditory nerve to the brain. Loudness (amplitude) is conveyed by a higher frequency of nerve-impulse transmission, while pitch (frequency) is communicated by the location of the particular nerve fibres stimulated. In this way the brain is informed of the amplitude and frequency of the sounds impinging on the eardrums. The quality of the sounds is inherent in the waveform (⇨ box).

The hair cells of the inner ear are delicate and easily destroyed, especially by sustained loud noise or by high-intensity impulsive noise, such as explosions.

Types of deafness and their causes

There are basically two distinct types of deafness – *conductive deafness* and *sensorineural deafness* – and the causes are quite different. Conductive deafness is caused by disorders occurring between the exterior and the oval window in the outer wall of the inner ear; sensorineural deafness results from disorders of the inner ear.

Conductive deafness may arise from any cause that interferes with the transmission of vibrations from the air to the inner ear. Common causes include wax in the auditory canal, perforation of the eardrum, middle-ear inflammation (known as *otitis media*), excessive sticky secretion in the middle ear so that movement of the

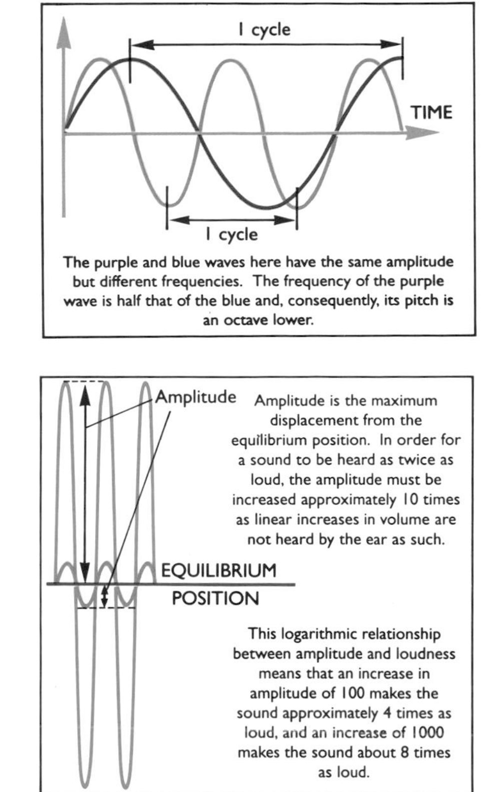

The purple and blue waves here have the same amplitude but different frequencies. The frequency of the purple wave is half that of the blue and, consequently, its pitch is an octave lower.

Amplitude is the maximum displacement from the equilibrium position. In order for a sound to be heard as twice as loud, the amplitude must be increased approximately 10 times as linear increases in volume are not heard by the ear as such.

This logarithmic relationship between amplitude and loudness means that an increase in amplitude of 100 makes the sound approximately 4 times as loud, and an increase of 1000 makes the sound about 8 times as loud.

ossicles is impeded (*glue ear*), damage to the ossicles from injury, and the condition of *otosclerosis*, in which the footplate of the inner bone, the stapes, becomes fused by new bone formation to the edges of the oval window. Most cases of conductive deafness are susceptible to treatment and many can be cured. Otosclerosis will often respond to delicate microsurgery on the stapes.

Sensorineural deafness is, in general, more serious, as it usually implies destruction of some or most of the cochlear hair cells. The commonest cause of this is *acoustic trauma* – the progressive damage to hair cells (⇨ above) from excessive noise. Damage can result either from noises of high intensity for long periods, or from sudden impulsive noises of very high intensity such as explosions or those caused by a blow on the ear. Very loud, sudden noises can literally shake the hearing mechanism to pieces, leaving the destroyed hair cells floating in the cochlear fluid. Whatever the cause, damage to the hair cells is permanent and irremediable. Hearing aids are of little or no value in the management of severe sensorineural deafness; they are useful only to amplify sounds that can still be heard. Even the latest multichannel electronic cochlear implants offer only crude sound perception and cannot, for instance, allow speech to be understood. Other causes of damage to hair cells include those rare cases of overdosage with various drugs, such as the aminoglycoside antibiotics, aspirin, quinine and certain diuretics; Ménière's disease; and the effects of ageing (*presbyacusis*).

The balancing mechanism

The part of the labyrinth concerned with balance (equilibrium) is known as the *vestibular system*. As in the cochlea, the principal feature of the balancing mechanism is the action of hair cells. In this case, however, the hair cells detect not fluid vibrations but fluid movement. Any change in the motion of the head, especially angular motion, causes the fluid in the vestibular system to move. And because of the position of the vestibular hair cells, movement can be detected in three different dimensions. The cells can all detect bodily acceleration or deceleration in any direction. The detecting hair cells are situated in each of the three semicircular canals and in the adjacent swelling of the labyrinth. Any movement of the fluid in which they are immersed will cause the corresponding hair cells to fire off impulses and these pass along nerve fibres of the same cranial nerve that carries impulses from the cochlea. The brain is thus informed of the plane in which movement of the head is occurring. However, only a *change* in the rate of motion causes fluid movement, as movement at constant velocity does not cause stimulation of the hair cells.

BASIC ACOUSTICS

• Sound can be propagated through air and other gases, through liquids and even through solids, but not through a vacuum. In air, sound radiates from a source as a succession of rapidly alternating compressions and decompressions, spreading outwards at a speed of about 300 m per second (1080 km/h, or 670 mph). For audible sounds, the frequency of these alternations ranges from about 20 Hz (hertz; cycles per second) to about 20 000 Hz. Almost everyone can hear low-frequency sound, but the upper limit of audibility almost always reduces with increasing age.

• The air vibrations associated with musical notes can be represented by comparatively simple repetitive (periodic) waveforms that can be analysed into a small number of different frequencies. Noise waveforms are usually more complex, however, and cover a wide range of superimposed frequencies. Noise is often impulsive rather than sustained.

• Audible sounds are characterized by pitch, volume and tone quality. *Pitch* is determined by the frequency of the vibrations. A high frequency causes a high-pitched note; a low frequency a low-pitched note. *Volume* is determined by the size (amplitude) of the vibrations. The greater the amplitude, the louder the sound. *Tone quality*, or *timbre*, depends on the presence of *harmonics* – frequencies that are simple multiples of the basic frequency and are in phase with it.

• However complex the pattern of sounds imposed simultaneously on the air, the air vibrations form a single periodic waveform, and the movement of the eardrums corresponds to this waveform. The brain, though, is able to analyse the resulting pattern of nerve impulses and reproduce, subjectively, the effect of separate sound sources.

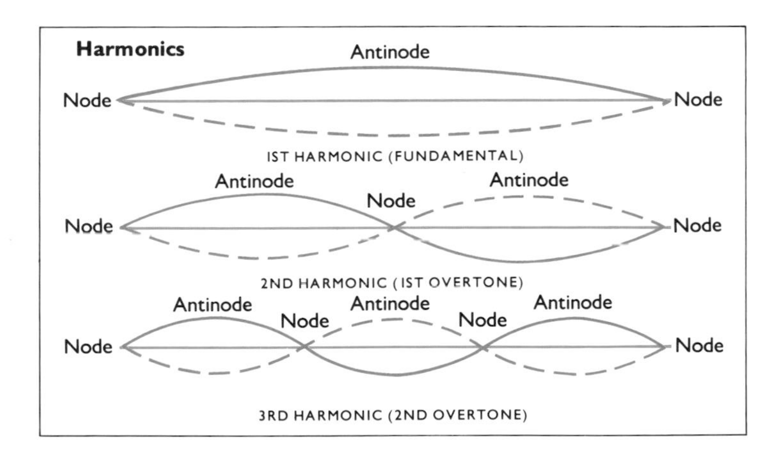

SEE ALSO

- WHAT THE BODY IS MADE OF p. 36
- MOVEMENT pp. 40–3
- THE CONTROL SYSTEM pp. 74–81
- SPEECH p. 82
- VISION p. 84
- TASTE AND SMELL p. 88
- BODY SENSATIONS p. 90

Taste and Smell

By itself the faculty of taste (*gustation*) is very limited, having some utilitarian value but few or no aesthetic possibilities. However, when accompanied by a normal sense of smell (*olfaction*), taste allows us to enjoy the whole range and subtlety of gastronomic experience.

The reason for this seeming anomaly is that when we use the word 'taste' we are really referring to the simultaneous employment of both faculties. Both functions emerged relatively early in the course of evolutionary history but it seems that these senses are now of comparatively lesser importance to humans than to many other animals. This is especially true of the sense of smell, which is no longer consciously regarded by most people as an important source of information. We should not, however, underestimate the effect of olfaction as a means of conveying subtle and perhaps barely recognized information to and fro between individuals. An ability to appreciate flavour, as a means of identifying and enjoying substances, is important.

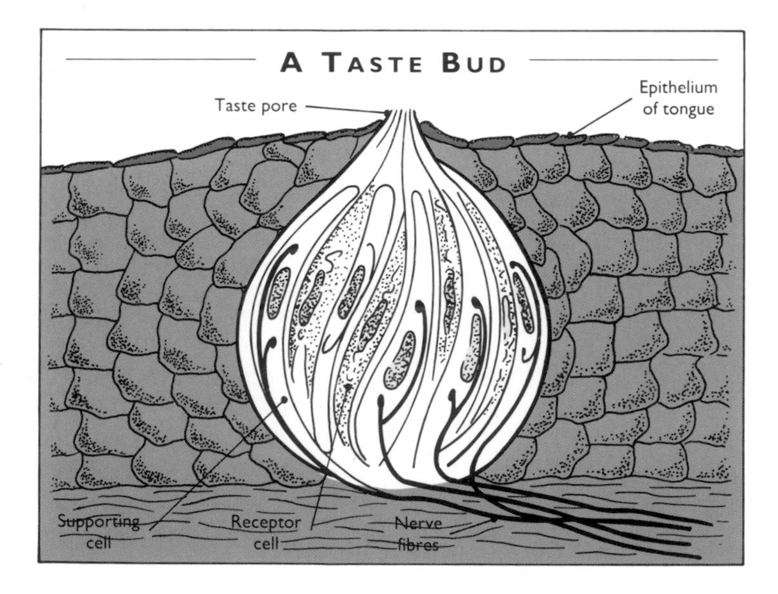

SEE ALSO
- THE CONTROL SYSTEM pp. 74–81
- SPEECH p. 82
- VISION p. 84
- HEARING AND BALANCE p. 86
- BODY SENSATIONS p. 90

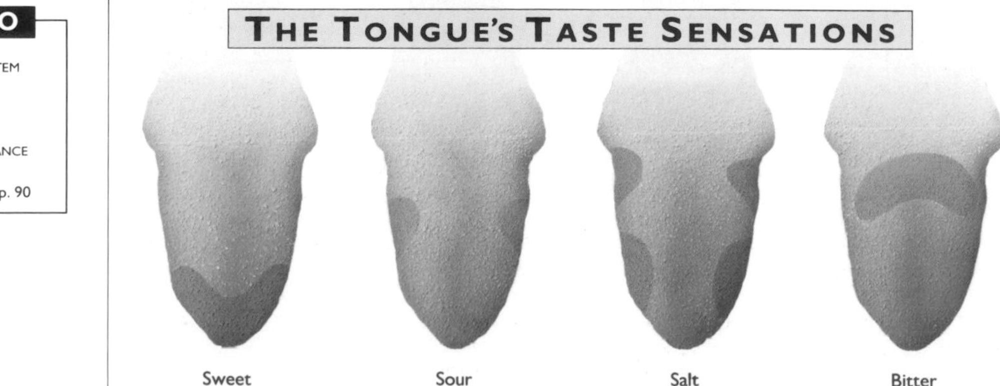

The taste buds

Taste buds are specialized nerve endings situated mostly on tiny mushroom-shaped protrusions on the tongue known as *fungiform papillae*, but also occurring on the roof of the mouth, the throat and the upper third of the gullet (oesophagus). There are some 10 000 taste buds altogether and each has a life of seven to ten days, being replaced by new structures that develop from the surrounding epithelium (⇨ pp. 62–3). Although they cannot be distinguished from each other anatomically, it is commonly stated that taste buds are of four kinds, responding, respectively, to the four types (modalities) of taste sensation – sweet, sour, salt and bitter. These are the only sensations that the sense of taste alone can appreciate.

Some authorities deny that there are four kinds of bud and point out that single taste buds can respond to substances in more than one of the four categories, but whatever the case, different parts of the tongue are more sensitive to different tastes. The taste buds that respond to salt are distributed on all parts of the tongue, but especially on the edges; those sensitive to sweet flavours are concentrated mainly on and around the tip of the tongue. Sour tastes are detected mainly by buds situated on the sides of the tongue, and bitter flavours are experienced at the back of the tongue.

Taste buds are constructed like miniature oranges or bunches of bananas (⇨ diagram). Each contains some 50–7 slender curved cells with outer protrusions ending in small open pores that are bathed in the salivary fluid layer on the lining of the mouth and throat. The presence of this fluid is essential for their function, and the chemical substances that give rise to the sensation of taste cannot operate on the taste buds until they have been dissolved in the fluid. Once these dissolved substances enter the pores they are able to come into contact with the special taste receptors on the outer membrane of the receptor cells in the taste bud. It seems to be likely that the membranes of single receptor cells carry different receptors for different substances. When the receptor cells are stimulated in this way they release neurotransmitters, which spread across synapses (⇨ pp. 74–5) and give rise to nerve impulses in nerve fibres running from the taste buds to the brain. These fibres run in the glossopharyngeal, lingual and vagus nerves (the vagus is another of the cranial nerves arising directly from the brain and extending through the chest, into the abdomen).

These nerve connections for taste pass up the brainstem to the sensory cell nucleus, the thalamus (⇨ p. 77), and end eventually in the sensory cortex of the brain in the region mapped for the mouth (⇨ pp. 76–7). As in other parts of the nervous system, intensity of sensation is coded as an increase in the frequency of repetitive nerve impulses (frequency modulation) in the fibres. For a particular nerve fibre, differences in the frequency of nerve impulses also appear to code for the different modalities of taste. Some fibres fire at a high rate in the presence of a sweet substance while others fire slowly. Some respond only to one modality but most respond to two. Yet others respond to three or four. It is probable then that perception of taste depends on the pattern of response of a group of nerve fibres.

The olfactory receptors

Odour is conveyed to us by molecular particles of the substance concerned being carried in the air. These particles must then reach the olfactory sense receptors in the nose and be dissolved in the layer of fluid mucus on the nose lining before they can be appreciated. The smell receptors lie in a small patch (2.5 cm^2 / 0.4 sq in) of mucous membrane lying immediately under the thin, bony plates that form the roof of the nose. These plates are perforated with small holes through which pass the nerve fibres of the olfactory nerves that run from the mucous membrane directly to the brain. The olfactory nerve bodies are long and narrow and lie in the mucous membrane. Their lower ends swell into knob-like structures from which 6–12 fine, hair-like fingers called *cilia* pass to the surface of the membrane so that their tips lie in the mucus layer. Mucus readily dissolves fine particles in the inspired air, especially if it is sniffed in through the nose. The dissolved odorous particles interact with receptors on the cell membranes, causing the cells to fire and despatch nerve impulses to the brain. As in the case of taste, the intensity of the stimulus and the differentiation of olfactory quality are coded by differences in the frequency of these impulses in a given time. No structural differences are apparent between the different olfactory nerve cells.

The olfactory nerves enter the olfactory bulbs – elongated, stem-like projections from the underside of the brain that lie on top of the perforated bone plates. From the bulbs, the nerve fibres pass to a part of the brain cortex lying on the underside of the frontal lobes, known as the limbic system (⇨ pp. 74–5). This system is a kind of primitive brain concerned with some of the more basic functions, such as the satisfaction of the various appetites. It is also concerned with the hypothalamus (⇨ p. 78) and its regulatory functions.

Smell and Behaviour

A *pheromone* is a chemical signal or 'messenger' that is released by an animal and has a specific effect (behavioural or physiological) on another member of the same species. Bees entering a strange hive are identified by smell and killed. The antennae of the male silk moth carry receptors that respond only to the pheromones given off by the female silk moth, making them fly in the direction of maximum pheromone concentration.

The behaviour of mammals, too, can be strongly influenced by olfactory stimuli. The 'marking' activities of dogs and cats are well known. In many species attraction of males to sexually receptive females is mediated by smell – even by the smell imparted to objects that have been touched by the female. Moreover, the secretion of reproductive hormones can be powerfully affected by olfactory stimuli, even to the extent of terminating a pregnancy when a recently mated female baboon smells the odour of a strange male.

Some vestige of this kind of mechanism persists in humans. Because pheromones are so important in the sexual attraction of most mammals it seems probable that they are involved, at least to a limited extent, in human sexual responses. Certain fatty acids have been identified in human vaginal secretions that are identical to those known to act as pheromones in other mammals. It has also been shown that female human sensitivity to musklike smells is greater at the time of ovulation than at other times. Musk, one of the central ingredients in many perfumes, is derived from the sex glands of the musk deer, and is chemically related to the human sex hormones. These facts must, however, be interpreted in the light of the greatly reduced significance of the sense of smell in humans. It is true that some men are attracted by certain kinds of body odour in women, and vice versa, but perfumes and aftershaves may have taken the place of the natural body odours that are now removed by washing, and many other factors are involved in human sexual attraction. Stories of people being sexually irresistible because of their pheromones are probably apocryphal.

Chemistry and the sense of smell

The olfactory system is very sensitive and is capable of detecting odours from material present in a concentration of only a few parts per million. Sensitivity to smell is said to be 100 000 times greater than sensitivity to taste. Because of this and because of the rapid diffusion of molecular particles in air we are able to detect odours that arise a considerable distance away. Although we can identify many thousands of different odours they probably arise from combinations of comparatively few different stimuli.

Odorous substances must be volatile enough to give off particles into the air, and they must also be sufficiently soluble to go into solution in the nasal mucus; indeed, many fat- and water-soluble substances are strongly odorous. But volatility and solubility in themselves are not enough to explain olfaction fully. Most odorous substances are organic, and it is possible to relate many chemical groups to particular kinds of smells. For example, a wide range of molecules that are derived from the benzene ring have a pleasant aroma, which has led chemists to describe the whole class of benzene-ring substances as the aromatic compounds. Many of the smells of plants derive from aromatic compounds such as the essential oils.

The relationship between chemistry and smell is not, however, an obvious one, and very different molecules may have similar smells. On the other hand, different spatial arrangements of identical molecular structures (stereoisomers) can produce very different smells, and clear distinctions can result from sometimes quite subtle changes in the molecular structure of odorous material. Many organic compounds can exist in two forms that are identical except that one molecule is a mirror image of the other; such forms are known as enantiomers. Even such a minor difference as this can affect the way in which the olfactory apparatus reacts: for example, one enantiomer of the compound limonene smells strongly of lemons, while the other smells of oranges.

Factors affecting the sense of smell

Volatility is influenced by temperature, which is one reason why perfumes are more effective on the body than in the bottle, and why refrigerators often smell stale when switched off. Hunting dogs perform better in warm weather than in the cold.

Any factor that interferes with ready access of air to the olfactory nerve endings will affect the sense of smell. When the nasal mucous membrane is swollen as a result of a cold or other inflammatory disorder (*rhinitis*) the sense of smell may be severely inhibited. An injury to the base of the skull that causes a fracture of the roof of the nose will commonly lead to permanent and total loss of the sense of smell (*anosmia*) from tearing of the olfactory nerve filaments. Increase in nasal sensitivity occurs when we are very hungry, and women are said to experience an increased sensitivity to substances related to the sex hormones at certain phases of the menstrual cycle.

Adaptation to persistent smells occurs within a matter of minutes. This can be useful, allowing people to work comfortably in conditions that may seem uncomfortable to somebody new. Adaptation is not wholly a function of the olfactory nerve endings but is also the result of some form of inhibition occurring in the brain, and it is one of the characteristics of the nervous system as a whole that it tends to ignore persistent stimuli and shows the greatest sensitivity to change. To some extent, cross adaptation also occurs between different strongly smelling substances: for example, a nose adapted to one essential oil, such as oil of wintergreen or eucalyptus, may have difficulty in detecting another.

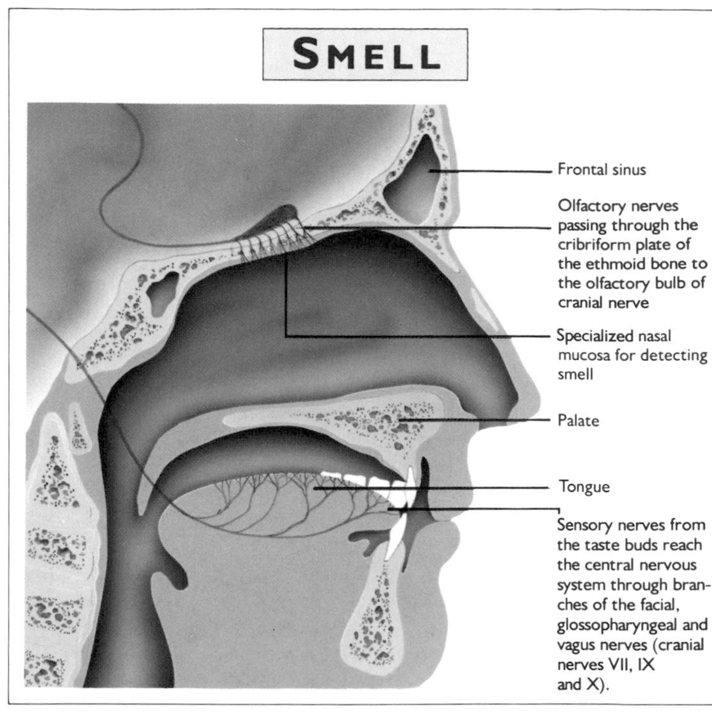

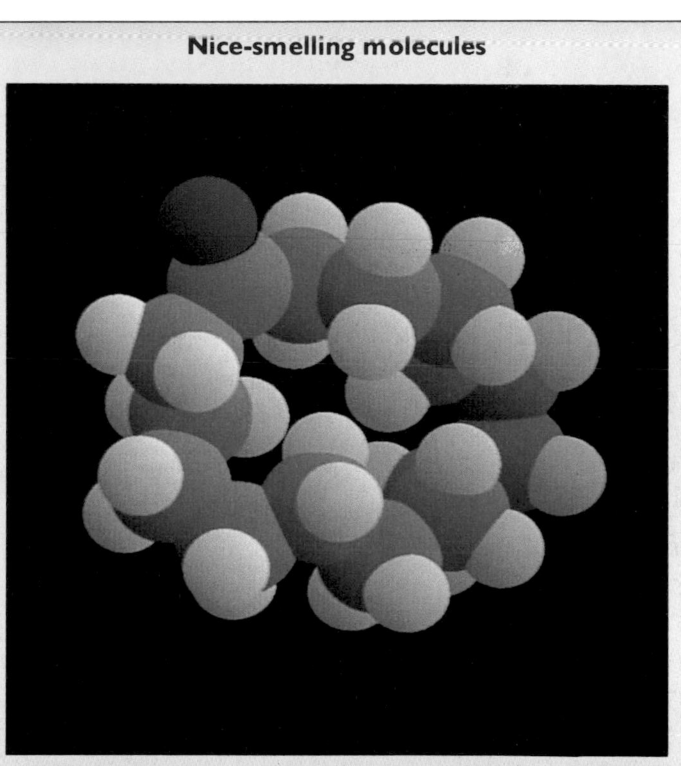

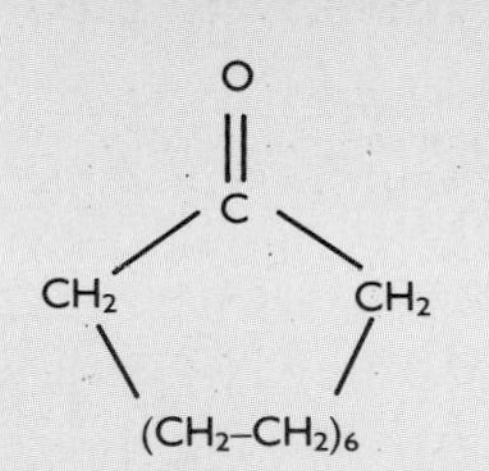

Exaltone (above) was the first synthetic chemical to be used in the manufacture of perfumes. Now a major industry is based on the production and use of chemicals with a wide range of smells. These chemicals obviate the need to extract tiny amounts of chemicals from animals, such as musk from deer and civettone from civets.

Body Sensations

The five senses are vision, hearing, taste, smell and touch. Touch is a composite sense since, through it, we can perceive temperature, vibration, pressure, pinprick and pain.

This list of the sources and routes of information input to the brain is not exhaustive and the sensory receptors also include those sensitive to head position in the inner ears (➪ p. 87), the stretch receptors in the muscles and tendons, those in the joints, and the pressure sensors in the neck arteries, aorta and heart that inform the brain of the blood pressure and volume. All the information passing to the brain is coded in the form of repetitive nerve impulses, and all these nerve impulses are the same. In this, sensory information differs radically from, say, the electrical information passing from an amplifier to a loudspeaker. Individual nerve impulses follow the 'all or none' law. This means that they occur either fully or not at all. There is no question of amplitude modulation as in the case of audio signals in a microphone cable. The only thing that can vary, when information is conveyed along a single nerve fibre, is the repetition frequency. The maximum repetition frequency of nerve impulses is not high – only a few hundred per second (Hz). Fibres in the acoustic nerve seldom fire at a frequency of more than 200 Hz. But in spite of this seemingly severe limitation, a considerable range of intensities (amplitude) can be represented to the brain by differences in nerve-impulse frequency. Sound frequency (pitch) variations are conveyed by differences in the location of the fibres stimulated.

In general, information in the nervous system is conveyed in terms of the numbers and location of fibres firing. Electronic information systems operate with single channels that are modulated in various ways to convey data. Biological information systems, on the other hand, use bundles of thousands or millions of channels, each one representing to the brain a position or a particular type (modality) of sensation. Each optic nerve contains about one million separate fibres, every one coming from a particular point on the retina and informing the brain that that point has, or has not, been stimulated by a photon of light. The point-to-point correspondence persists all the way from the retina to the visual cortex. The same kind of point-to-point correspondence occurs in all the senses. Medical illustrators have often drawn grotesque human figures (➪ pp. 76–7) to represent the mapping on the brain's sensory cortex of the skin sensation. Such figures are distorted because different weighting has to be given to the higher concentration of sensory nerve endings in different parts of the body, such as the tongue and the sex organs. Simultaneous pinpricks between the shoulder-blades must be separated by several millimetres to be appreciated as separate; on the fingertip they are felt as distinct even if very close together.

In electronic systems the coded information is converted back, at the receiving end, into a familiar form – sound, pictures, text on a screen. In the brain, no such conversion occurs. The coded information itself is perceived as sound, vision, touch, taste, smell, and so on. Thus the means by which information is conveyed to the brain differs fundamentally from the thing it represents. The ultimate philosophical question is whether our perception of the 'outer world' bears any meaningful resemblance to what is really out there.

Much of the information received by the brain is not perceived in the conscious mind. This is fortunate, as we have quite enough to attend to without being constantly informed of such parameters as the levels of oxygen and carbon dioxide in our blood or the degree of tension in the arteries supplying our intestines. Even the sensory modalities of which we are often fully aware do not intrude at all times on our consciousness. Touch and pressure sensation tend to get through to us only when a significant change occurs or when they become excessive. The stimulation of the nerve endings subserving pain may or may not even give rise to awareness of pain; much depends on the circumstances.

The nature of pain

Pain differs from the other kinds of sensation in that not only is it unpleasant but it usually has a major psychological component. Pain is a localized sensation caused by stimulation strong enough to damage tissue or to threaten damage to tissue. Unless chronic (long lasting), it commonly serves as a warning of danger and prompts action tending to end it. People who, because of disease, such as leprosy, tabes dorsalis (degeneration of part of the spinal cord, caused by syphilis), or other conditions affecting the sensory nerves, are unable to experience pain, invariably suffer serious and cumulative bodily damage.

The response to pain may be reflex, involuntary and rapid, or conscious, deliberate and purposeful. Persistent pain is usually associated with distress and anxiety – and often with fear – and there may be physiological changes similar to those experienced during anger and aggression. The heart beats faster, the blood pressure and the rate of respiration rise, the pupils dilate and the skin sweats. There is an increased secretion of adrenaline from the adrenal glands (➪ pp. 80–1) and increased mobilization of glucose from the glycogen stores that are contained in the liver (➪ pp. 48–9).

Our perception of the significance of pain is often more related to these secondary effects than to the intensity of the pain itself. If pain is separated from its mental component, as is possible by the use of drugs such as morphine, it may still be felt but may no longer be unpleasant. The distress caused by pain depends also, to a large extent, on our awareness of the cause and is modified by past experience. Even minor pain inflicted by a torturer may seem more severe than the same physical hurt resulting from an innocent cause such as an accident.

The origins of pain

The nerve endings subserving pain are called *nociceptors*. These do not appear to differ physically from one another, but

SEE ALSO

- WHAT THE BODY IS MADE OF p. 36
- CELLS, THE STUFF OF LIFE p. 38
- MOVEMENT pp. 40–3
- EXERCISE AND BODY UNITY p. 54
- THE BODY COVERING p. 56
- THE HUMAN COMPUTER? p. 72
- THE CONTROL SYSTEM pp. 74–81
- SPEECH p. 82
- VISION p. 84
- HEARING AND BALANCE p. 86
- TASTE AND SMELL p. 88

Spinal Column in Cross-section

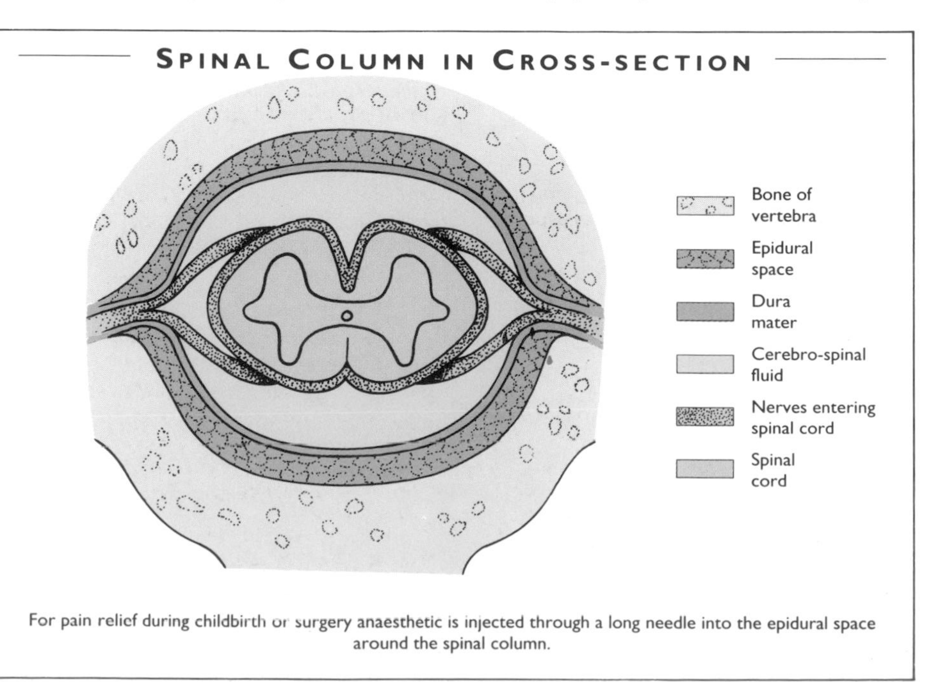

For pain relief during childbirth or surgery anaesthetic is injected through a long needle into the epidural space around the spinal column.

ENDORPHINS

Remarkable pain tolerance brought about by mental transcendence of bodily sensation – an Indian fakir, asleep on a bed of nails. (Popperfoto)

Drugs such as morphine (derived from opium) act on specific receptor sites in the brain that appear to have no other function. This seemingly remarkable coincidence led scientists to expect, and some to predict, that natural morphine-like substances must be produced by the body. In 1975 two such substances were isolated from the brain and called *enkephalins* (*enkephalon* is Greek for 'brain'). Later, several more of these active substances were found, all with the same opioid (opium-like) core of five amino acids. Because of their morphine-like chemical structure and properties, and because they come from within the body, they have been named *endorphins* – a contraction of 'endogenous morphine-like substances'.

Endorphins are neurotransmitters with a wide range of functions. They control the perception of pain during highly stressful events, such as sudden severe injury; in such cases pain may begin to be felt only some time after the injury. They also help to regulate the action of the heart, exert a controlling influence on hormones, and help to reduce dangerous surgical shock from blood loss. They seem to be involved, in some way, in controlling mood, emotion and motivation. They act on the centres of the brain concerned with the heartbeat and the control of blood pressure, and on the pituitary gland. One extraordinary research finding was that the levels of circulating endorphins were found to be higher after subjects took a tablet that they *believed* to contain a pain-killing drug, but that was in fact an inactive placebo.

They also appear to be released during very hard, prolonged, repetitive physical exercise. Long-distance runners often report a sense of elation after a certain period, and it is suggested by some that they can become addicted to their own endorphins. A similar effect may account for the ecstatic feelings or trance-like states reported by participants in religious ceremonies involving prolonged dancing and chanting, and may explain how, in such circumstances, people can, apparently painlessly, pierce their bodies with sharp implements.

different nociceptors seem to respond to different kinds of painful stimuli – mechanical, thermal or chemical. Tissue damage results in the release of various strongly stimulating substances such as prostaglandins, and these are the principal stimulators of nociceptors, causing the nerve fibres to fire and conduct impulses to the brain. Drugs such as aspirin inhibit the enzymes that cause the release of prostaglandins from damaged cells and this is how they act as painkillers. Aspirin has no effect on pain caused by pinprick, however, in which nerve endings are directly stimulated without the intermediate stage of tissue damage.

Different nociceptors show different sensitivities. Some are stimulated by low-grade 'warning' events of insufficient force to cause actual pain. Others respond only to strong stimuli such as pricking, cutting or burning. The stronger the stimulus, the higher the frequency of the nerve-impulse sequence sent to the brain. If a nerve fibre subserving pain is stimulated, pain will be perceived in the area of the nerve ending, irrespective of how the nerve impulse originated. Such a fibre can be stimulated at a point much nearer the brain than the remote nerve ending. In the condition of *post-herpetic pain* following shingles, for instance, sensory nerves are stimulated near the spinal cord by an inflammatory reaction caused by herpes zoster viruses. This causes pain that is perceived as coming from the skin.

Although the nerves carrying pain impulses terminate in the brain, and give rise to neurological activity there, the pain is usually felt in the region in which the nerve endings are situated. If the conduction of pain impulses is prevented, by, for example, injecting a local anaesthetic around the trunk of the sensory nerve or near the spinal cord, no pain will be felt, although the damaging events at the nerve ending are continuing unabated. Referred pain is pain experienced in an area other than the site of the cause because the same sensory nerve supplies both areas. Gall bladder inflammation, for instance, causes pain in the tip of the right shoulder because both the diaphragm and the skin of the shoulder are supplied by the same nerve. After amputation, phantom-limb pain, experienced exactly as if the limb were still present, can occur from irritation to the cut nerve ends in the stump. The brain can interpret the resulting nerve endings only as coming from the lost limb.

Passage of pain nerve impulses may also be blocked by the arrival of impulses caused by the stimulation of other sensory nerves. These second impulses may be stimulated by rubbing, scratching or stroking the skin, by electrical stimulation applied through the skin, or by acupuncture.

The 'gate' theory

Computers operate by an elaboration of logical 'gates' through which a stream of electrical pulses (1s or 0s) passes. These impulses may be blocked by a secondary controlling electrical signal of the same kind or may be compared with other similar pulses. Most physiologists now accept that the nervous system contains analogous arrangements of neurons operating as gates, and that pain impulses travelling up the spinal cord pass through such gates and can be blocked, or allowed to pass, by controlling signals. It seems probable that all the neuronal signals that are concerned with pain must pass through such gates.

Nerve fibres carrying pain impulses may be large or small. Both, however, affect the state of the gates. The tendency is for small fibres to open the gates and large fibres to close them. Large-fibre stimulation also sends messages to a higher level, which in addition also act to close the gate. As far as we know gates are also probably under the control of the brain.

THE GATE CONTROL SYSTEM

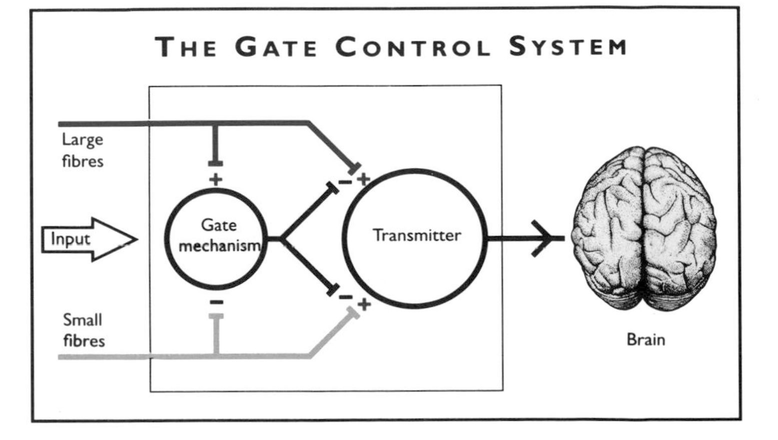

THE CYCLE OF LIFE

Childbirth Throughout History

That conception involves sexual intercourse between men and women appears to have been known for thousands of years. But what happened biologically remained a source of speculation until the 19th century.

The similarity between the seeds of plants and male semen must also have been understood for thousands of years; the term 'seed' for seminal fluid is frequently found in biblical, Talmudic and much other literature. Although little in the way of formal medical literature on the subject of reproduction has survived from ancient times, nonmedical literature commonly deals with the subject. There are many references to obstetrical matters in Homer's *Iliad* and *Odyssey*. The ancient Hebrews had a rich store of obstetrical and gynaecological knowledge. The books of the Pentateuch (the first five books of the Bible) contain numerous references to virginity, contraception, midwifery, abortion and sexually transmitted diseases.

Sixteenth-century engraving of a woman giving birth sitting up and using a birthing stool (a chair that has the front portion of the seat cut away, here allowing the midwife access). In the West birthing stools have recently come back into use as women have expressed their desire for flexibility of position when in labour. (LSI)

Theories of conception

However, early notions of what happened in reproduction were confused and diverse. Many people believed that new individuals were produced as a result of the mingling of menstrual and seminal fluids. Others held that new individuals arose from large eggs hidden in the body. Hippocrates of Cos (460–377 BC), in his book *The Seed and the Nature of the Child,* said that conception resulted from a mixing together of male and female seed.

William Harvey (1578–1657), in his book *De Generatione Animalium* (1651), claimed that all animals were derived from an 'ovum' or 'primordium', contained within the female body, which had the capacity to develop into a new individual when activated by male semen. The idea of the universality of the female ovum gained wide support when Regnier de Graaf (1641–73) and Jan Swammerdam (1637–80) demonstrated that there were follicles in the ovaries of mammals (Graafian follicles), which they took to be eggs. However, they were mistaken; the mammalian ovum was not in fact discovered until the 19th century. But although Harvey believed that the sperm exerted some influence on the egg, beginning a process of development from which a new individual was formed (the doctrine of *epigenesis*), this view was widely rejected. Most scientists believed that the seminal fluid merely acted as a stimulus to start the growth of the *preformed* individual in the egg. Even when spermatozoa were discovered by Anton van Leeuwenhoek (1632–1723) in 1677, few people considered that they contributed materially to the new individual. Most took them to be parasitic organisms with no part to play in the promotion of reproduction.

For a long time, the idea of preformation was paramount. There were two schools of thought, both of whom held that the new individual developed from a perfect, preformed, miniature person, complete in everything but size. The first school, the 'ovists', held that this little individual was encapsulated in the egg. After Leeuwenhoek discovered spermatozoa, the second school, the 'animalculists', sprang up. The animalculists believed that every

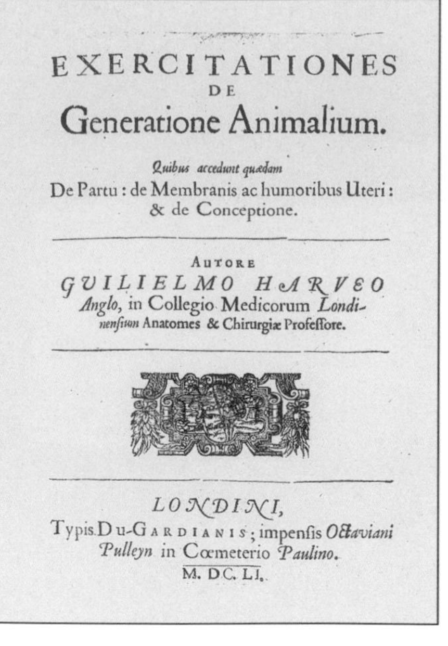

EXERCITATIONES
DE
Generatione Animalium.

Quibus accedunt quædam
De Partu: de Membranis ac humoribus Uteri: & de Conceptione.

AUTORE
GUILIELMO HARVEO
Anglo, in Collegio Medicorum Londinensium Anatomes & Chirurgiæ Professore.

LONDINI,
Typis Du-Gardianis; impensis Octaviani Pulleyn in Cœmeterio Paulino.
M. DC. LI.

Title page of Harvey's *De Generatione Animalium* (1651). (Wellcome)

sperm contained a tiny homunculus (miniature person) able to grow within the female body. Argument raged between the two schools throughout the whole of the 18th century, and it was not until the beginning of the 19th century that Harvey's correct ideas of epigenesis were once again entertained.

Even as late as the 18th century the actual role of the sperm remained unknown. The proof of the correctness of epigenesis owed much to the work of the German-Russian embryologist Karl Ernst von Baer (1792–1876). In 1827 he published an account of his researches showing that the ovarian follicles described by Graaf were not in fact eggs but rather that each mature follicle contained an egg. In 1828 and 1837 von Baer published the two volumes of a major textbook of mammalian development (embryology). This showed how a mammal's egg developed through numerous stages as an embryo and a fetus before becoming a recognizable individual.

Midwifery

From ancient times, assistance in childbirth seems to have been largely the province of women. There are numerous references to midwives in ancient texts from India, China, Egypt, Mesopotamia and Greece.

In 1513, shortly after the invention of printing, a short handbook for midwives was published by Eucharius Röslin, a physician from Worms, in what is now southern Germany. This seems to be the earliest surviving formal treatise on the subject of obstetrics. This book was extensively plagiarized and imitations of it subsequently appeared in many languages. The English version, *The Byrthe of Mankynde*, was first published in 1540 by Thomas Raynalde during the reign of Henry VIII. Several subsequent editions appeared, the last in 1676. The book describes the female genitalia, the causes and functions of menstruation, the fetus and its coverings, the causes of vaginal discharge, the manner of birth, the delivery of the baby, the processes of milk production (lactation), infant care and feeding, and a collection of household remedies. Although much of the content of this book is valid and useful, it contains many inaccuracies, misunderstandings and incorrect assumptions. Other books followed, notably Caspar Wolff's *Gynaeciorum* in 1566. The first reliable account of the female reproductive organs appeared in 1543 in Vesalius's classic work *De Humani Corporis Fabrica*. This was enlarged upon by subsequent anatomists such as Bartolomeo Eustachio, Gabriello Fallopio and Ambroise Paré.

Most midwives of the time were illiterate and uneducated and were therefore unable to take advantage of such published information. As the only group with knowledge of the subject, however, they were indispensable. They were also, unfortunately, very vulnerable, often being held accountable for such misfortunes as stillbirths, deformities and maternal deaths from haemorrhage or childbed fever. Those found guilty of procuring abortion were often excommunicated and usually executed. Many, for no particular reason save the fact that someone had simply denounced them, were accused of witchcraft and were tortured and burned. Much of the rigour of, and authority for, such persecution of midwives came from statements in the notorious *Malleus Maleficarum* ('The Hammer of the Witches'), written around 1486 by two German members of the Inquisition, Heinrich Kraemer and Johan Sprenger. This popular book went through 28 editions.

Early incubators in use. From *The Illustrated London News*, 8 March 1884. (AR)

The male *accoucheur*

Harvey's book, *De Generatione Animalium* (⇨ above), also contained a chapter on labour and the delivery of the child. This seems to have been the first original account of obstetrics by an English writer. In 1671 Jane Sharp published her *Complete Midwife's Companion* – a useful and systematic study of the subject. Around this period the physicians, who were always male, began to invade the province of the midwives and started to deliver babies themselves. This was done entirely by feel, the whole of the woman's genital area being covered by a cloth extending from her waist to the physician's neck. The invention of obstetrical forceps, which the doctors did not allow midwives to use, further increased male dominance in this area.

Among the wealthy it became conventional to have a male surgeon-*accoucheur* (obstetrician) to conduct deliveries, and many medical men built up lucrative obstetrical practices. Highly successful 18th-century medical men like William Smellie, William Hunter and Charles White devoted themselves largely to the art of obstetrics. Smellie published a major textbook, *Midwifery*, in 1752 – an excellent and practical work that was the first to describe in detail the use of forceps to aid delivery. At the turn of the century, Charles White of Manchester was teaching the importance of cleanliness as a means of avoiding childbed fever, fifty years before Semmelweiss's celebrated quarrel with his Viennese colleagues (⇨ box).

By the middle of the 19th century obstetrics was a recognized medical discipline and midwives had become wholly subservient to doctors, both in practice and by law. At the same time, however, the status and professionalism of midwives began to improve and, by the end of the century, they were beginning to be regarded as a body of properly trained and educated specialists. Today, in the West, midwives are highly trained individuals with established professional standards and defined relationships with doctors, and are able to conduct the majority of births without medical intervention. In some countries though, notably the USA, most deliveries are still managed by doctors.

The Semmelweiss Scandal

Ignaz Philipp Semmelweiss (1818–65), a Hungarian physician, took up an appointment as an assistant physician in the General Hospital in Vienna in 1846. There were two maternity wards. In Ward 1, where the babies were delivered by medical students, the death rate of the mothers from 'childbed fever' was horrifying – sometimes almost 30%. In Ward 2, where deliveries were conducted by midwives, the rate was about 3%. Pregnant women begged and prayed not to be admitted to the notorious Ward 1. Semmelweiss noticed that women admitted just after the baby had been born rarely suffered from the deadly disease. He was also deeply affected by the case of his friend and colleague Jakob Kolletschka, a professor of forensic medicine, who died after cutting his finger during a post-mortem examination. The changes found in Kolletschka's body were identical to those in the bodies of women who had died of childbed fever. Putting two and two together, Semmelweiss decided that the medical students must be carrying some poison from the corpses in the autopsy room to the labour ward. So he insisted that anyone delivering a baby must first wash his or her hands in chlorinated water. Within a year the mortality rate had dropped to 1%.

Semmelweiss's chief, Dr Klein, and others, resented this interference and the suggestion that they might have been responsible for the women's deaths. Like other doctors of the time, they were proud of the 'hospital odour' they carried on their hands. Semmelweiss was forced out of his job, his work was forgotten, and the death rate among the women rose again to record heights. Semmelweiss ended up in a mental hospital where he died of blood poisoning from a cut finger. Fourteen years after his death, a notable French gynaecologist was about to deliver a public lecture in Paris condemning the idea of contagion in childbed fever when he was interrupted and silenced by the great Louis Pasteur, who proceeded to announce his discovery of the organism responsible for childbed fever – the streptococcus bacterium.

- PASSING ON LIFE p. 96
- HOW WE ARE BORN p. 104
- DISEASE IN HISTORY p. 118

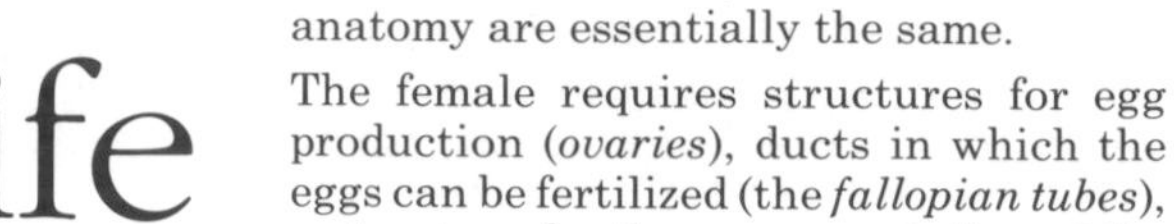

Passing on Life

Reproduction is not necessarily sexual; indeed the great majority of living things, such as the cells of our bodies and many microorganisms, reproduce asexually. But nearly all highly organized animals and plants reproduce sexually and possess specialized organs for the purpose.

Asexual reproduction means that the offspring are identical to the parent because no new genetic information is involved. Sexual reproduction, on the other hand, is the basis for the remarkable diversity of individuals within a species. In humans, sexual reproduction also involves major emotional and social elements relating to sexual attraction, mating, family formation and the protection of the young.

The essential feature of human reproduction is that each new individual begins life as a single cell called a *zygote*, which results from the penetration and fertilization of the mother's egg (or *ovum*) by the father's sperm (or *spermatozoon*). Both the unfertilized ovum and the sperm are single cells, but differ from normal body cells in that they contain only half the normal number of chromosomes. Half the genetic characteristics of the person-to-be come from the mother and half from the father (⇨ pp. 20–1). Fertilization and the subsequent nurturing of the growing embryo and fetus occur within the body of the mother; sperm production and emission are all that are required of the father. These facts are associated with the considerable differences between those parts of the male and female anatomy concerned with reproduction. In some other respects, although there are many quantitative differences, male and female anatomy are essentially the same.

SEE ALSO

- THE PATTERN OF INHERITANCE p. 20
- THE TRANSMISSION OF GENETIC CHARACTERISTICS p. 28
- LIFE IN THE WOMB p. 102
- HOW WE ARE BORN p. 104

The female requires structures for egg production (*ovaries*), ducts in which the eggs can be fertilized (the *fallopian tubes*), a structure for the accommodation of the growing embryo and fetus (the womb or *uterus*) and a suitable duct, accessible from the outside, that allows sperm to be deposited near the site of the egg (the *vagina*). Because the new individual must be nourished within the body of the mother until it is mature enough to exist independently, the womb must be capable of extending considerably, of providing large quantities of all the substances needed for growth, and of expelling the mature fetus at the appropriate time.

The female reproductive organs

The externally visible parts of the female genitalia (the *vulva*) consist of the large and small lips (the *labia majora* and *minora*) and the *clitoris*. Although they play their part in intercourse, they are not essential for reproduction. The large lips are usually in contact so that the entrance to the vagina (the *introitus*) is covered. Just within them are the thin skin folds constituting the labia minora. These join at the front to form a kind of hood for the clitoris. Both the clitoris and the labia minora contain many blood vessels, which, during sexual arousal, become widened so that the parts become engorged, swollen and hot. The same stimulus causes secretion of clear mucus from the two *Bartholin's glands*, which lie in the rear parts of the labia majora. This mucus lubricates the entrance to the vagina and facilitates sexual intercourse. Between the clitoris, at the front, and the vaginal opening near the back of the vulva, lies the small, pouting opening of the *urethra* – the short tube that carries urine from the bladder to the exterior.

The clitoris, like the penis, contains spongy tissue and is capable, like its male counterpart, of becoming engorged with blood and erect. During sexual intercourse (⇨ below), movement from the vagina causes the hood of the labia minora to massage gently the sensitive bulb of the clitoris. This stimulation may result in one or more orgasms of varying intensity and character. These are accompanied by retraction of the clitoris, thickening and deep red colouring of the labia minora, repeated contractions of the outer third of the vagina and rhythmical contractions of the womb. The face and neck may also develop a mottled flush.

The vagina is a remarkably extensible tube of muscle and fibrous tissue, about 10 cm (4 in) long, but capable of extending considerably under conditions of sexual arousal. It contains no glands, but fluid can pass through its lining from the underlying tissue. It runs upwards and slightly backwards, and is normally passively closed. Tight closure may occur voluntarily or from fear by contraction of the pelvic-floor muscles. Large numbers of bacteria, known as *lactobacilli*, inhabit the healthy vagina. These act on a carbohydrate stored in the surface cells to produce lactic acid, which is important in discouraging growth of undesirable organisms such as the thrush fungus.

The uterus is a small, hollow, pear-shaped, thick-walled, muscular organ, lying inclined forwards almost at right angles to the vagina. The neck of the uterus, or *cervix*, protrudes into the upper part of the vagina. The cavity of the uterus is about 8 cm (3 in) long, but expands greatly during pregnancy. The lining of the uterus varies in thickness, from 1 to 5 mm ($^1/_{25}$ to $^1/_5$ in) depending on the stage of the menstrual cycle. From the right and left sides of the upper part of the body of the uterus, narrow muscular tubes, the fallopian or uterine tubes, run outwards to the ovaries. The outer end of each of these tubes is open and bears a large number of finger-like structures (*fimbriae*) that almost cover the surface of the corresponding ovary. These fimbriae ensure that eggs released from the ovaries are carried into the fallopian tubes, where fertilization occurs. Each ovary is an almond-shaped body, 3 to 4 cm (1 to $1\frac{1}{2}$ in) long, hanging from the back of a membrane that descends from the fallopian tube on each side. The eggs develop in cell collections, known as *Graafian follicles*, on the surface of the ovaries.

The male reproductive system

The pair of *testes*, or testicles, are the site of sperm production. They are accommodated in the hanging skin bag, the *scrotum*, where the temperature is somewhat lower than the internal body temperature. This is necessary for sperm production. Each testis is divided into 300–400 tiny lobes, each containing one to three coiled tubes in the walls of which the sperms develop. The total length of these *seminiferous tubules* is about 500 m (1640

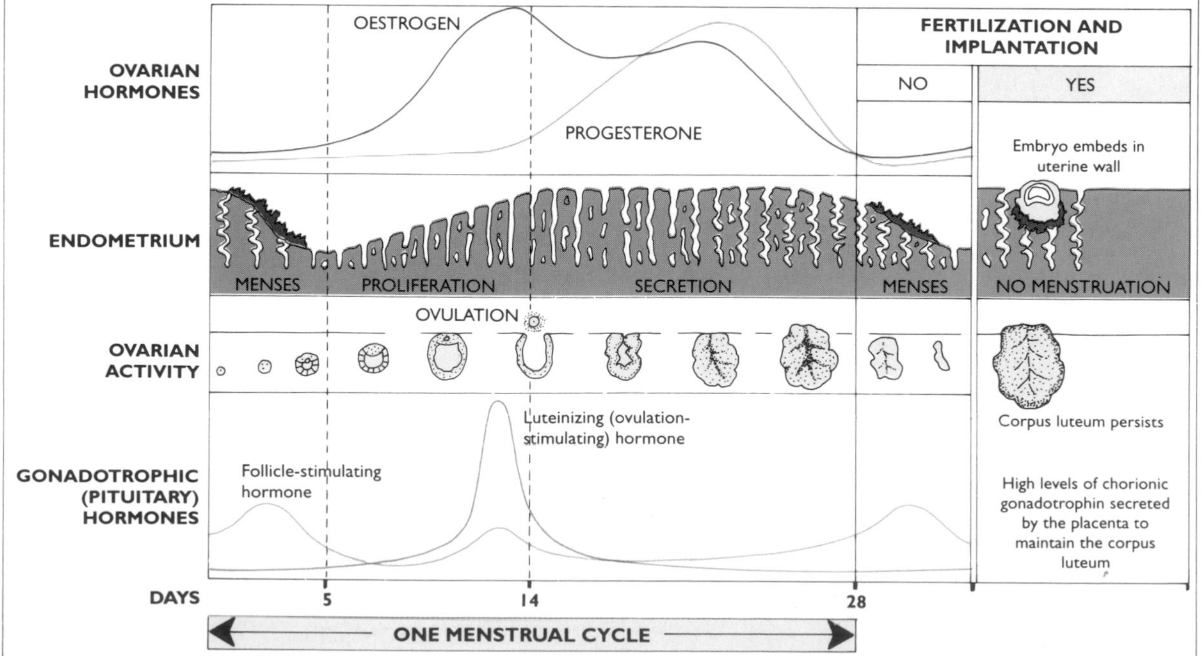

The Menstrual Cycle

The menstrual cycle is the result of the production of a sequence of sex hormones from the ovaries (▷ pp. 78–9). These act on the lining of the uterus (the *endometrium*), causing it to thicken and become suitable for the reception of a fertilized egg. If pregnancy occurs, the cycle proceeds no further, but if it does not, the thickened lining is cast off over the course of three or four days. This is called *menstruation*. The whole cycle takes about 28 days and is timed around the event of ovulation. In the absence of pregnancy, menstruation normally occurs about 14 days after ovulation.

The thickening of the womb lining is caused by oestrogen hormones, the output of which begins to rise at the end of menstruation. During this time a follicle is maturing in an ovary as a result of follicle-stimulating hormone from the pituitary gland (▷ pp. 78–9). The follicle cells produce the oestrogen. After the egg is released a mass of large, yellowish cells (the *corpus luteum*) forms in the empty follicle. These cells secrete increasing quantities of another hormone, *progesterone*. This causes the womb lining to thicken further by means of fluid retention. If pregnancy occurs, the corpus luteum persists and continues to maintain the womb lining. If it does not, the corpus luteum shrinks and the output of both oestrogen and progesterone drops. As a result, the womb lining is shed as a flow of blood, mucus and tissue – known as the menstrual period.

ft), and they join up into a single, highly coiled tube that forms a slug-shaped mass on the outside of the testis, the *epididymis*. This is several metres long and it takes a number of days for sperms to traverse it. This time is necessary for their maturation. Sperms do not 'swim' up these tubes; they are pushed up, as a dense sludge, by the pressure of fluid from below.

Each epididymis ends in a *vas deferens* – the tube that carries the sperms up to be stored in the *seminal vesicles*. These are small, elongated sacs lying on either side of the *prostate gland*, just under the urinary bladder. The prostate secretes about one third of the fluid content of the seminal fluid, and ducts pass through it from the seminal vesicles to enter the urine tube (the *urethra*) that runs along the inside of the penis. Most of the fluid content comes from the seminal vesicles.

The main bulk of the penis consists of three longitudinal columns of spongy erectile tissue – the two *corpora cavernosa*, which lie side by side along the upper part of the organ, and the *corpus spongiosum*, lying centrally behind and enclosing the urethra. The front end of the corpus spongiosum expands into an acorn-shaped swelling called the *glans penis*, and this is enclosed in a sheath of skin, the foreskin or *prepuce*. Sexual interest causes the arteries supplying the erectile tissue to widen so that a quantity of blood flows under pressure into the three columns. As a result they expand considerably and the penis becomes stiffened, enlarged and erect. This causes compression of the veins that drain blood from the columns, and the erection is maintained until sexual interest gradually diminishes.

Penile erection is readily induced, especially in the young, by anticipation of sexual activity, sexual thoughts and fantasizing, erotic literature or pictures, physical contact of a suggestive nature or genital stimulation. The same stimuli cause secretion of clear mucus by glands near the base of the penis, and this appears at the urethral opening at the tip of the penis. This lubrication of the rigid organ aids relatively easy insertion into the vagina of a cooperative partner.

Sexual intercourse and fertilization

Rhythmical massage of the sensitive glans penis by the walls of the vagina produces a strongly pleasurable sensation. This, together with the psychic stimulus of close human intimacy, soon produces the male orgasm. In most cases this occurs within four or five minutes of penetration – often much sooner. First the contents of the seminal vesicles are squeezed into the urethra by contraction of the muscles in the vesicle walls and the prostate. Then the muscles on the floor of the pelvis and around the base of the penis contract rhythmically and repeatedly, driving the seminal fluid out of the penis (*ejaculation*). The male orgasm is associated with a pleasurable release of tension.

Within a few minutes of ejaculation, some sperms will have passed into the canal of the cervix. The state of the cervical mucus greatly influences the ability of sperms to pass. Although sperms have a long tail that gives them *motility* (the ability to swim), how they complete their journey remains uncertain. Because of their very small size relative to the length of the journey, it is unlikely that they are entirely unaided. It seems probable that they are also carried by fluid currents set up in the uterus by the action of 'hair' cells (*cilia*) in the uterine lining. Only one sperm in a hundred reaches the inside of the womb, and for 30 to 45 hours they remain capable of penetrating an egg. For pregnancy to occur, eggs must be fertilized within the fallopian tube, and this must occur between 6 and 24 hours after release from the ovary.

The relatively few sperms that reach the ovum surround it, butting with their heads and releasing an enzyme, *hyaluronidase*, that softens and thins the outer membrane of the egg. Normally only one sperm penetrates and, as soon as this has happened, a physical barrier to further sperm penetration forms as a result of a change in the electrical charge on the egg. If more than one sperm does gain admission, the number of chromosomes of the zygote exceeds the normal, which leads to gross abnormality in the embryo, and usually results in an early miscarriage.

Reproductive Organs

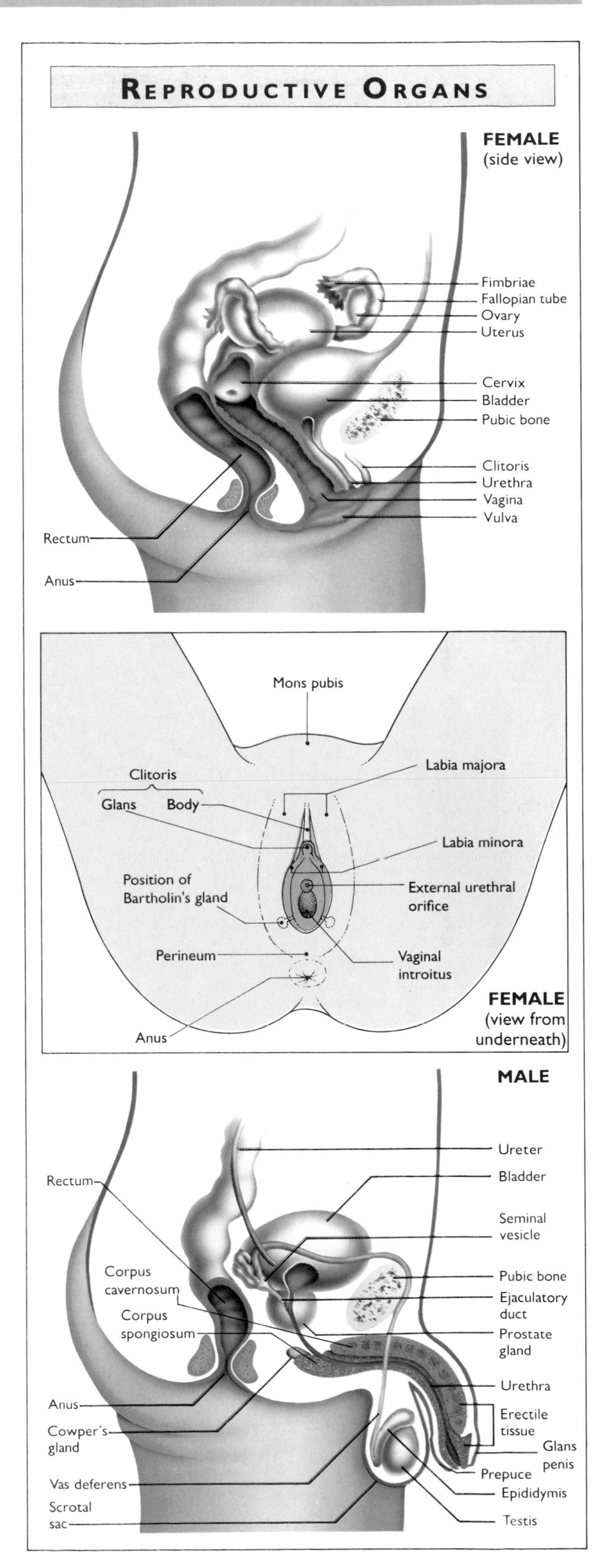

Infertility: the Causes and Options

Most healthy couples who want a baby, and who engage in sexual intercourse at least twice a week, can expect to achieve a pregnancy within a few months. Failure to do so after a year is usually regarded as an indication of infertility. This can be doubly distressing: firstly because of the frustration of the often powerful parental instinct; and secondly because of the implication that one or other partner is in some way abnormal. Modern medicine, though, can do much to resolve this unhappy state of affairs.

In the days before reproductive physiology and pathology were well understood, infertility was invariably attributed to the woman. The Bible and other early literature is full of references to 'barren women'. We now know that in one fifth to one quarter of cases, the failure to conceive is due to sperm problems in the male. But because the female reproductive system is considerably more complicated than that of the male, it is true that, in most cases, infertility can be attributed to a gynaecological disorder.

Causes of failure to conceive

After menstruation starts, the first few cycles may occur without egg production (ovulation) taking place, and so may be infertile. This should not be relied upon as a contraceptive. Once ovulation is well under way, fertility soon rises to its highest level and remains so throughout a woman's late teens and twenties. After the age of 35, fertility declines slowly. In males, active sperm production starts at puberty and continues throughout life. Sperm numbers and motility may reduce with age, but many men remain fully fertile well into their eighties or nineties.

Human sperm shown here in a light micrograph. The sperm on the left has two heads; if more than about a third of the sperm are abnormal like this, fertilization is unlikely to occur. (Magnification x 400 at 35 mm across.)
(John Walsh, SPL)

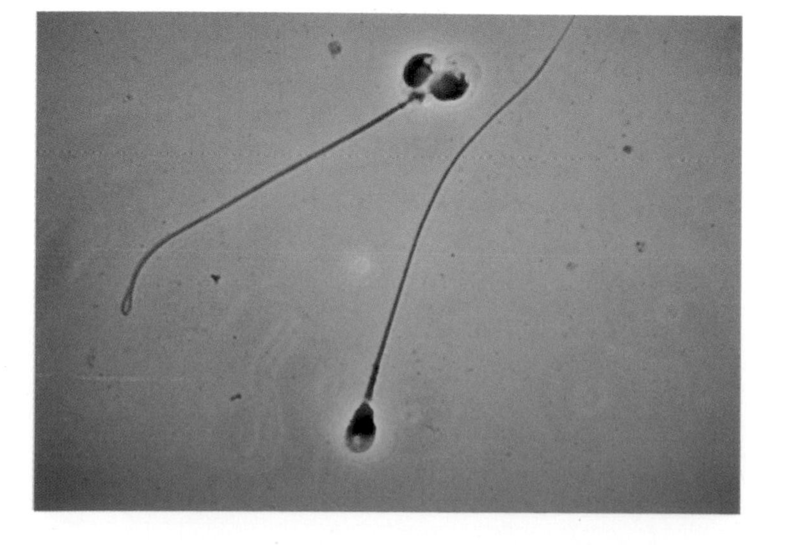

Even in these enlightened days, failure to conceive is occasionally due to total ignorance of the elementary facts of life, accompanied by an unusual degree of shyness about experimenting. Conversely, infertility may also occur in those who are only too well informed but who have an unspoken fear of pregnancy or a strong but unacknowledged disinclination for children. Even in people desperate for children, tension and anxiety can take their toll and may even affect ovulation by interfering with its hormonal control (⇨ pp. 96–7). Anxiety is also a very common cause of impotence in men. Prolonged psychosomatic impotence is uncommon but can, of course, lead to failure to conceive. Organic impotence is also rare but, if complete, will lead to failure of conception by normal means. Some men have a penile abnormality called *hypospadias* in which the urethra opens on the underside of the penis just in front of the scrotum. This means that the semen is not deposited high enough in the vagina. Recent developments in plastic surgery have, however, made it possible to correct this problem. Various other penile abnormalities can also cause failure to conceive.

Fertility is little affected by a person's general state of health or nutrition unless this involves the reproductive system. But women who diet too strictly, especially those who suffer from anorexia nervosa, often cease to menstruate and ovulate, and thus become sterile. A high level of athletic activity may also prevent ovulation. Underactivity of the thyroid gland is a cause of infertility and is associated with obesity. Badly controlled diabetes may also lead to infertility.

Infrequent ovulation is commoner than absence of ovulation, but may still prevent conception. Irregular, unpredictable and infrequent egg production greatly reduce the chances of the egg and sperm meeting, which is, of course, necessary for fertilization. Many complex hormonal problems – involving the *hypothalamus* (⇨ pp. 78–9), the *pituitary gland*, the *ovaries* and the *thyroid gland* – can lead to infrequent ovulation or even total absence of ovulation. Various tests can be carried out to establish the presence or absence of ovulation, and ovulation failure can be treated with hormones or other drugs that stimulate ovulation or prevent its inhibition.

Another important cause of infertility is blockage of the fallopian tubes (⇨ pp. 96–7), which is almost always the result of infection and inflammation (*salpingitis*). Sexually transmitted gonorrhoea or chlamydial infection commonly lead to blockage, but infection may also follow the normal delivery of a baby or occur after an abortion. Infection can also be the result of peritonitis from a burst inflamed appendix or of tuberculosis. Blockage of the fallopian tubes prevents the ova from passing along, or even entering, the tube, so that fertilization cannot occur.

Monitoring Ovulation

The presence or absence of ovulation can be checked in a number of ways:

- **The Graafian follicle (⇨ p. 96), which releases the egg, can be seen on ultrasound scanning.**
- **A blood test can often show the rise in the level of progesterone (⇨ pp. 96–7) that follows ovulation.**
- **A sample of the womb lining may show the characteristic changes that follow ovulation.**
- **Immediately after ovulation the mucus in the cervix, normally thick and impenetrable to sperm, becomes more watery and can be drawn out into a fine thread that dries with a fern-like crystalline pattern.**
- **The body temperature rises about 0.5° C (1° F) after ovulation occurs, and remains up for the second half of the menstrual cycle.**

The state of the fallopian tubes can be investigated by injecting a harmless dye, methylene blue, through a tube that fits tightly into the cervix. The passage of the dye through the outer open ends of the tubes can be observed through a *laparoscope* (a type of endoscope; ⇨ pp. 128–9), if the tubes are not blocked. Laparoscopy also allows the tubes to be inspected for visible abnormalities, and the ovaries for the presence of a Graafian follicle or a corpus luteum (⇨ p. 96). Alternatively, a solution opaque to x-rays may be injected and an x-ray taken. This will readily show up any obstruction. Tubal blockage can sometimes be treated effectively by microsurgery. Some women have their tubes deliberately blocked as a method of sterilization (laparoscopic sterilization). Should a woman then change her mind and decide to try to become pregnant this operation can be reversed with about a 70% success rate.

Male infertility

Apart from impotence, male infertility is almost entirely determined by the sperm count and whether the sperms are physically normal and sufficiently motile (actively mobile). For infertile couples, seminal fluid analysis, which is easy and quick, is always done before subjecting the female partner to infertility investigation. If there are no sperms in the semen (*azoospermia*) there is no possibility of fertilization, and the most likely cause of the infertility has been found. In seminal fluid analysis a sample is obtained by masturbation and is examined within two hours. The sperms present are all at least two months old. Seminal analysis gives varying results, but, for fertility, the ejaculate should have a volume of 2–6 ml (0.07–0.2 fl oz) and should contain 20 000 000 spermatozoa per ml. More than 40% of these should be energetically motile and fewer than 30% should have an abnormal appearance. Chronic alcohol excess causes a low sperm count but this is usually reversed if drinking is then moderated. In some cases of low sperm count, treatment with synthetic male sex hormones is effective.

Artificial insemination

When the male partner is impotent or suffers from hypospadias or any other penile abnormality that prevents normal coitus, fertilization can usually be achieved by artificial insemination using the partner's seminal fluid, which is obtained by masturbation (still known as artificial insemination by husband, or AIH, even though the couple may not be married). This is a simple technique that can readily be performed by the woman herself, using a sterile syringe to inject the semen high into the vagina, or into a plastic cap that is then fitted over the cervix. The fertilization rate is much the same as that obtained with normal sexual intercourse. If the male partner does not have any spermatozoa in his semen, artificial insemination with donor semen (artificial insemination by donor, or AID) may be considered, but this raises new and emotive issues that should be fully understood and discussed before treatment is decided upon. Donors are anonymous and are screened for AIDS and syphilis. Frozen semen is available from males who are considered desirable, including Nobel Prize winners. Donors are matched to resemble the male partner's physical characteristics – skin, hair and eye colour, height, and build – as closely as possible.

In vitro fertilization research taking place. The needle (right) is being used to inject sperm directly into the ovum (centre), which is held steady by a glass pipette.
(Hank Morgan, SPL)

In vitro fertilization (IVF)

The term 'in vitro' means 'in glass' and refers to a procedure in which living eggs are taken from a woman's ovary, fertilized by sperm in a sterile glass dish, and replaced in the womb. The procedure is fraught with difficulty and, at present, is successful in only about 10% of attempts. Egg production is stimulated with drugs or hormones and the growth of the ovarian follicles checked by ultrasound. When a follicle reaches a size about 1.5 mm ($^1/_{16}$ in), a dose of hormone is given to prompt the release of eggs. These are collected using a fine needle guided by the ultrasound image. The eggs are incubated at body temperature in a culture medium for 4–6 hours and then the sperm are added. The fertilized egg is kept in the culture medium for about two days, and is then placed in the woman's womb through a fine tube.

The practice of in vitro fertilization is currently the least successful way of managing infertility and is generally regarded as a last resort.

The Ethical Issues

One of the consequences of the development of IVF treatment has been to make embryo research possible, as couples undergoing IVF techniques can elect to donate 'spare' embryos to medical science. This research will unquestionably lead to major life-saving and life-enhancing advances, especially in the field of genetics, but one of the many important questions that it raises is whether we should value the sentient life of a disease-stricken child or adult more or less than the potential life of an embryo. Any activity that seems to run counter to the natural processes and methods of human conception has always raised ethical, moral or even theological questions.

Contraception, for instance, is still officially condemned by the Roman Catholic Church. From a strictly materialistic point of view, an embryo consisting of a few unorganized cells, with no possibility of awareness, cannot be considered as being in any real sense a human being – although, of course, having the potentiality to become one. Millions of such embryos are discarded naturally every day.

So much public concern was voiced in the 1970s over such matters as surrogacy, in vitro fertilization and the research work necessary to establish such methods, that in 1982 the British government set up a committee, chaired by Dame Mary Warnock, then Mistress of Girton College, Cambridge, and now Baroness Warnock, to consider the ethical and legal implications. The Warnock Committee members were divided in their views but the consensus was that legitimate research was justified on the early embryo – up to 14 days after fertilization. The committee also said that it should be an offence to operate a surrogacy agency, whether or not for profit. Children born after AID or IVF, it maintained, should be legitimate in law, and the practice of storing frozen sperm for AID and of freezing spare embryos should continue.

The findings of the Warnock Committee were widely felt to be sensible and valuable but settled no arguments. People's attitudes continue to be determined not so much by logic as by the nature of their fundamental beliefs. In 1986 the American Fertility Society produced a set of ethical guidelines for these matters. There was no objection, in principle, to artificial insemination. In vitro fertilization was considered ethically acceptable so long as couples were made aware of the low success rate. Surrogate motherhood was deemed ethically questionable and should, they suggested, be undertaken only in institutions with ethical committees.

Surrogacy

A surrogate mother is one who agrees to become pregnant and then hand over the baby to an infertile couple for adoption. She may be fertilized by artificial insemination using the semen of the adopting male parent, or by another partner. Some 1000 children have been brought into the world by this method in the United States, and the American experience has often been far from happy. Surrogacy often means desperate heartache for the biological mother, and may lead the participants into a legal minefield. The judgement of the New Jersey Supreme Court in one notorious case was, 'Whatever idealism may have motivated any of the participants, the profit motive predominates, permeates, and ultimately governs the transaction.' Commercial surrogacy is now illegal in the UK.

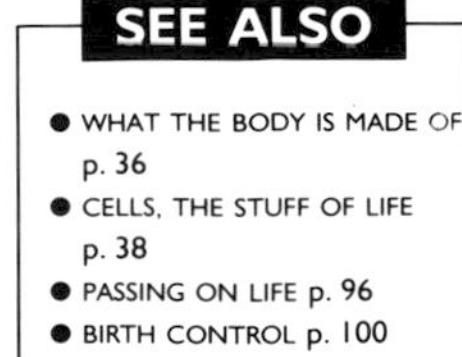

SEE ALSO
- WHAT THE BODY IS MADE OF p. 36
- CELLS, THE STUFF OF LIFE p. 38
- PASSING ON LIFE p. 96
- BIRTH CONTROL p. 100

Birth Control

World overpopulation is probably the greatest threat facing humanity today. Populations do not grow linearly – by addition; they increase exponentially – by multiplication. This means that it is not only populations that get larger; the rate of growth of populations also increases – an alarming fact.

Thirty years ago world population was rising by about one million every eight days. Today, a million more babies are born every four days. We can take little comfort in the knowledge that people are now having fewer babies; the real trouble is that every year there are far more people to have them. The present rate of growth of about 1.7% on today's world population of 5.5 billion means an annual increase of 93.5 million. This is a far greater increase than that which occurred a few decades ago, when there was a smaller total population. Most of the growth takes place in the less well-developed countries, and these have an average rate of increase of about 2.4%.

SEE ALSO

- CHILDBIRTH THROUGHOUT HISTORY p. 94
- PASSING ON LIFE p. 96
- INFERTILITY: THE OPTIONS p. 98

Abortion

In many countries, the deliberate termination of a pregnancy is legal so long as the continuation of the pregnancy is deemed, on medical grounds, to involve risk to the life or physical or mental health of the pregnant woman. It is also legal if there is a substantial risk that the child, if born, would suffer such physical or mental abnormalities as to be seriously disabled. In Britain, abortion may also be legally performed if continuation of the pregnancy would involve risk to the physical or mental health of existing children of the woman. In 1990, the British Parliament supported these grounds for abortion but stipulated that they should be allowed only up to a limit of 24 weeks – the point at which the fetus is deemed to be *viable* (capable of surviving outside the womb). In some countries, the regulations are stricter. In Ireland, for instance, where abortion was previously illegal, a change in the law was made to allow a teenage girl to have an abortion after she became pregnant as a result of rape. In others, such as China, where the population is increasing at an alarming rate, women who have already had one child and become pregnant again may find themselves liable for compulsory abortion. In Russia, where contraceptives are of limited availability, abortion is currently very common.

The methods used to procure abortion differ according to how far the pregnancy has advanced. Widening of the cervical canal and scraping of the lining of the womb (dilatation and curettage or 'D and C'), once the standard method in early pregnancy, has now largely given way to vacuum suction through a tube. In very early pregnancy a very fine tube can be used, under local anaesthesia. The procedure takes about five minutes. Later, the cervical canal must be dilated and a wider tube used. Suction methods can be used up to 12 weeks. Still later in pregnancy, drugs known as prostaglandins are given, either into the bloodstream by injection or directly into the womb. These cause strong contractions of the womb. If necessary, an abdominal operation similar to a Caesarian section can be done.

Not surprisingly, abortion is a subject about which there is fierce debate. 'Pro-life' campaigners abhor abortion on the grounds that it violates the sanctity of human life, and amounts to murder. 'Pro-choice' supporters, on the other hand, argue that a woman has the right to decide what should happen to her body, and thus to her life. These positions, which stem from differences in fundamental belief, share little or no common ground, and the issue is therefore unlikely ever to be resolved by rational argument.

A general increase in life expectancy compounds the problem.

Birth control is an emotive issue. Prohibition of contraception and abortion by the Roman Catholic Church and disapproval by the growing number of fundamentalist religious groups have had major political and economic implications. For example, well-meaning international agencies with programmes that include birth control are likely to have funding withdrawn.

Gross inequalities between nations in food production and distribution conceal the problem of a growing population from the more prosperous nations, but the ordinary people in Ethiopia and Bangladesh know very well what it means for a rising population to try to subsist on fixed or dwindling amounts of food. Recent years have seen repeated disastrous famines – tragedies that have shamed the developed countries into mounting massive aid programmes. Unfortunately, there is no reason to suppose that the problem will not get steadily worse.

Birth control in the past

Attempts at birth control date back to the earliest times. It is likely that, as soon as the relationship between seminal deposition in the vagina and subsequent pregnancy became apparent, some couples will have practised *coitus interruptus* (withdrawal of the penis before ejaculation). Genesis, chapter 38, records how Onan repeatedly 'spilled his seed on the ground', thereby incurring the wrath of God, who killed him. This passage, along with its interpretation as a prohibition of contraception and masturbation, has been responsible for immeasurable distress to humanity ever since.

The ancient Egyptian Ebers papyrus, which dates from 1550 BC, includes accounts of contraceptive methods; the subject is also dealt with by the early authors Pedanius Dioscorides (c. AD 77) and Soranus of Ephesus (c. AD 100). Recommended methods included vaginal wiping after intercourse, the use of vaginal sponges or cloths soaked in alum, lactic acid or honey, douching, and jumping backwards seven times by the female partner, presumably in the hope of dislodging the semen.

The earliest condom seems to have been the idea of Gabriello Fallopio (1523–63), who had a special interest in reproductive anatomy and physiology (he gave his name to the fallopian tubes), and whose account was published a year after he died. The term 'condom' probably derives from the Latin *condus* ('receptacle') rather than from the mysterious Dr Condom of folklore fame. The early condoms were made from animal intestines, but the development, in 1839, of vulcanized rubber by Charles Goodyear (1800–60) allowed a much superior article to be manufactured, and effective condoms have been available since the middle of the 19th century. But even before this, vaginal barrier methods, such as caps and

Main Contraceptive Methods

THE CAP

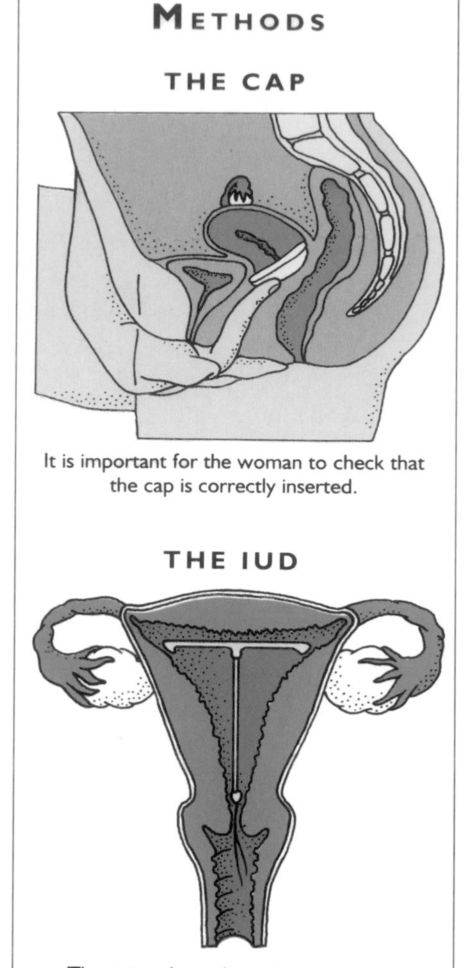

It is important for the woman to check that the cap is correctly inserted.

THE IUD

The strings hang down into the vagina allowing the woman to check that the IUD is in place, and facilitating removal.

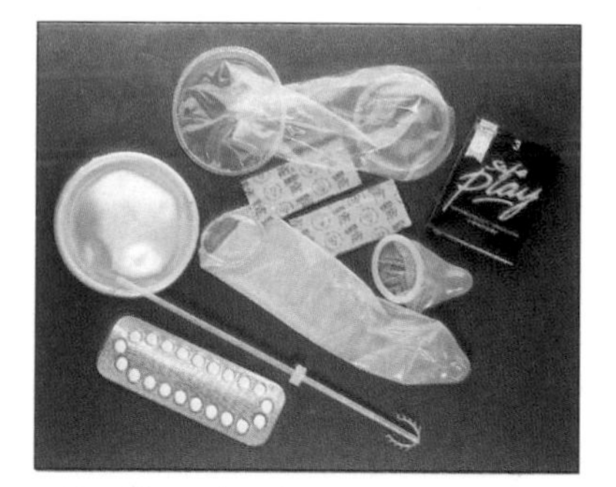

Female and male contraceptives – (clockwise from top) the Femidom, condoms, the IUD still in its insertion tube, the pill, and the cap. (Image Select)

diaphragms (rubber devices, placed over the woman's cervix) used with spermicides (sperm-killing creams or gels) had been medically described and employed, as had vasectomy, the cutting and tying of the man's sperm-carrying tubes. Crude intrauterine devices (IUDs) were introduced in 1868. An early form, the Gräfenberg ring, caused chronic inflammation and disease, and for a time the IUD was widely condemned.

By 1900 all present methods of birth control, except the hormonal contraceptive pill, were available. Advocacy, in several books, of the widespread application of these methods – birth control for all – by the courageous botanist Marie Stopes (1880–1958) earned her the gratitude of the ordinary mother oppressed by endless childbearing. But it also brought her persecution and vilification, not only by the Establishment, especially the Roman Catholic Church, but also by the more conservative members of the medical profession. Marie Stopes' Mothers' Clinic for Birth Control, established in Holloway, London, in 1921, bravely set a precedent for other similar establishments all over the world.

The effect of the female sex hormones on egg production (ovulation) was known as early as the 1920s, but it was not until 1956 that the American physiologist and endocrinologist Gregory Pincus (1903–67) reported that ovulation could be temporarily stopped by means of progesterone and allied hormones. Working in collaboration with gynaecologists, he developed an oestrogen–progestogen combination that could be taken by mouth and that was virtually 100% effective in preventing pregnancy.

This was the contraceptive pill, approved by the American Food and Drug Administration in 1960, and first put on the market in Britain in 1962. Since then it has been used by many millions of women throughout the world. Side effects of the pill include breast tenderness, nausea, headache and weight gain through fluid retention. Women taking it are at a slightly higher risk of suffering from the condition known as thrombosis, a blockage in the arteries or veins that can lead to heart attacks and strokes. This risk is increased if the woman smokes.

New developments

The birth-control methods most commonly used today are detailed in the box. However, various new methods are being developed. Long-acting hormonal contraceptives can be given by injection every 8–12 weeks, or by skin patch, and are very effective. They cause irregular vaginal bleeding and absence of normal menstruation, however.

Contraceptive capsule implants, placed under the skin, provide effective cover for five years with a pregnancy rate of less than 1% (i.e., only 1 woman out of a 100, using this method for a year, would become pregnant), although, again, irregular bleeding may be a problem. Fertility returns soon after the capsule is removed. A silicone rubber ring, containing progesterone and oestrogen, placed high in the vagina for three weeks at a time, allows regular menstruation and affords contraceptive protection as effectively as the pill.

The 'abortion pill', mifepristone or RU486, which acts by blocking the action of progesterone, is not, of course, a method of contraception. It can be used up to the ninth week of pregnancy, and is being widely used in the West. Hundreds of thousands of women have had their pregnancies terminated by this method.

Extensive research has been done on male oral contraceptives, all with negative results. Substances tried proved to be too toxic or reduced sperm count and/or libido. The exception, however, cottonseed gossypol, has undergone extensive testing, especially in China, but there are doubts about its safety.

General practice nurse giving advice on contraception at a well woman clinic. It is important that people are aware of all the methods available, and of any possible side effects, as not all forms of contraception suit all individuals. (Simon Fraser, SPL)

Birth Control Methods Today

Most women who want to avoid pregnancy use some form of contraception, but about one third rely on male or female sterilization. Of all women using contraception in developed countries, one third are on the pill.

The available methods of contraception, roughly in order of increasing reliability, are:

THE RHYTHM OR CALENDAR METHOD

This method is the only birth-control method (apart from total abstinence) that is sanctioned by the Roman Catholic Church. It is notoriously unreliable, mainly because it depends on predicting the length of the woman's next menstrual cycle, which can never be precisely done. The idea behind this method is that the couple should restrict intercourse to those days in the woman's cycle when an egg will not be present. Temperature checks (0.5° C / 1° F rise at ovulation) and checks of the viscosity of the cervical mucus (much reduced at ovulation) may help to improve the results.

CONTINUING BREAST FEEDING

Breast feeding has a contraceptive effect because of the hormonal changes associated with it. So long as a woman continues with it without menstruating, her chances of becoming pregnant are about one in fifty. But, as ovulation occurs before menstruation, long-term reliance on this method would be risky.

COITUS INTERRUPTUS AND COITUS RESERVATUS

Coitus interruptus (withdrawal of the penis from the vagina before ejaculation) and coitus reservatus (avoidance of ejaculation) are also unreliable as contraceptive methods because, during intercourse, semen may leak from the penis, even if ejaculation has not occurred.

SPERMICIDES ALONE

Creams, gels or films that kill sperm may be placed high in the vagina, over the cervix, prior to intercourse. They are, however, much more reliable when used in conjunction with a barrier device such as a condom or diaphragm.

VAGINAL SPONGE

This disposable sponge fits over the cervix. It is impregnated with spermicide and is effective for 24 hours. It must be left in for at least 6 hours after intercourse.

CONDOM (MALE AND FEMALE)

The male condom (sheath) is fitted over the erect penis just prior to intercourse. The female condom consists of a loose rubber bag with a flexible metal ring at the closed end that fits over the cervix so that the sides of the bag line the inside of the vagina, protruding slightly. Many condoms are made with a thin coating of spermicide. Condoms also considerably reduce the risk of sexually transmitted disease, including AIDS.

OTHER BARRIER METHODS

The vaginal diaphragm is a large rubber dome with a covered springy edge that fits high in the vagina, covering the cervix. The cervical cap is smaller and fits snugly over the cervix. Neither is particularly reliable unless used in conjunction with a spermicide.

INTRAUTERINE DEVICES

The intrauterine device (IUD or coil) is produced in a variety of shapes and is made of plastic or metal wire. It is inserted into the womb using a special applicator and has a fine cord left hanging from the cervix to allow easy removal. The IUD works partly by preventing the implantation of the fertilized egg in the uterine wall, and partly by interfering with the access of the sperm to the ovum. The IUD is capable of preventing pregnancy provided that it is fitted at least five days after unprotected intercourse. Its disadvantages include the possibility of heavy, painful periods, and the risk of infection or infertility. However, the IUD does allow a woman to forget about contraception for several years at a time.

ORAL CONTRACEPTIVES

The pill is by far the most reliable contraceptive after abstinence and sterilization. It contains the hormone progesterone or a combination of oestrogen and progesterone, and works by preventing ovulation. The 'morning-after' pill, which contains a higher concentration of hormones, may be used to prevent pregnancy within about 72 hours of intercourse.

STERILIZATION

Women can be sterilized by an operation to tie or clip the fallopian tubes (tubal ligation). Male sterilization involves cutting the *vas deferens* on both sides (vasectomy). The vas is the channel through which sperms pass up from the testis.

Life in the Womb

Although all human life originates in only a single fertilized egg (zygote), this single cell – only about 0.1 mm in diameter – contains all the planning data necessary for the production of a unique individual. Subsequent life in the womb (uterus) features more rapid and more radical physical changes than occur at any other period of life.

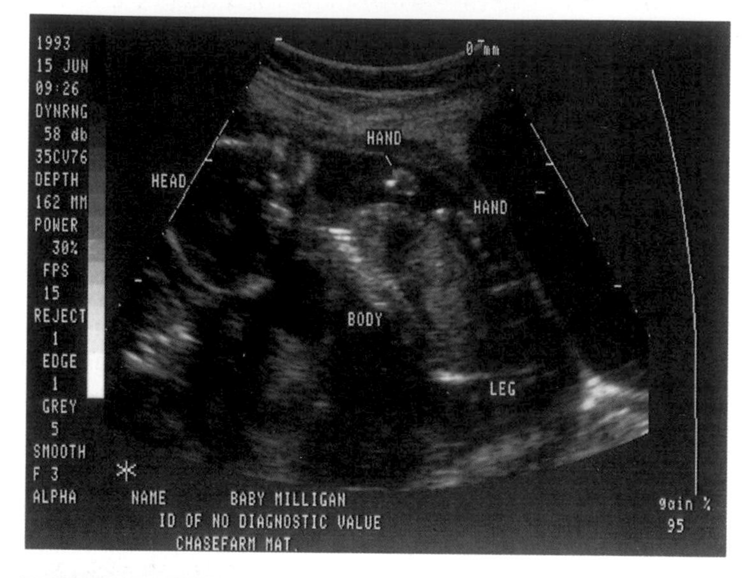

Fetal scan. The fetus is approximately 18 weeks old. (Courtesy of Tracey Milligan)

At the zygote stage all the major bodily characteristics have been determined, including the sex, and only heredity operates on its development. Later, and progressively, as the embryo becomes a fetus and then a fully formed baby, environmental influences become increasingly important. The many millionfold increase in weight from a fraction of a milligram to several kilograms is brought about entirely by material entering the embryo and fetus from the mother's body. The first five or six days of the new life are spent in the fallopian tube (⇨ pp. 96–7), where the initial cell divisions occur. Implantation into the womb lining then occurs. Prior to implantation the human organism is called a *pre-embryo*. Between the time of implantation and the end of the seventh or eighth week of life it is called an *embryo*. Thereafter, the developing individual is known as a *fetus*.

The embryo's development

In the first three days after fertilization and the fusion of the sperm and egg nuclei (⇨ p. 96), the egg cell splits in half five times within its membrane to form a 32-cell mass called a *morula*. At this stage none of the cells has become specialized and it is impossible to tell which of them will develop into the fetus and which will become the placenta, the organ that later nourishes the fetus. These cells are called *totipotential*. Soon, however, the first signs of differentiation appear as the cell mass organizes itself into a hollow, fluid-filled ball called a *blastocyst* (from the Greek *blastos*, 'bud', and *kustos*, 'pouch'). Inside this, at one side, is a small collection of cells that will become the fetus. Alongside these cells is the yolk sac derived from the egg. This provides nourishment for the growing embryo until the blood circulation is established in the placenta and nutrition is provided through it by the mother.

The cells on the outer side of the blastocyst develop tiny projections called *microvilli*, and these become the placenta. These outer cells have a powerful tendency to burrow into things and the six-day embryo is now at the stage when it is capable of implanting itself into the lining of the womb. If it does not reach the womb it will try to burrow into any nearby tissue and this is how pregnancies sometimes occur in the fallopian tubes or even in the abdominal cavity (*ectopic pregnancies*; ⇨ box on miscarriage).

At these early stages embryos are remarkably tough. At the blastocyst stage they can be chopped in two to produce identical twins. In fact, each cell of an eight-cell embryo is capable of developing into a complete individual if the cells are separated from each other at this stage. Such an embryo can be transferred to another person without incurring rejection problems. In theory, it could even be transferred to the body of a man, but because men do not have wombs its burrowing tendency would almost certainly cause dangerous bleeding.

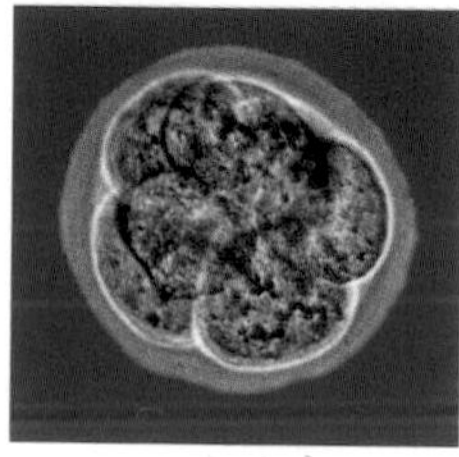

Human zygote (early embryo) at around the eight-cell stage of division. (Zefa)

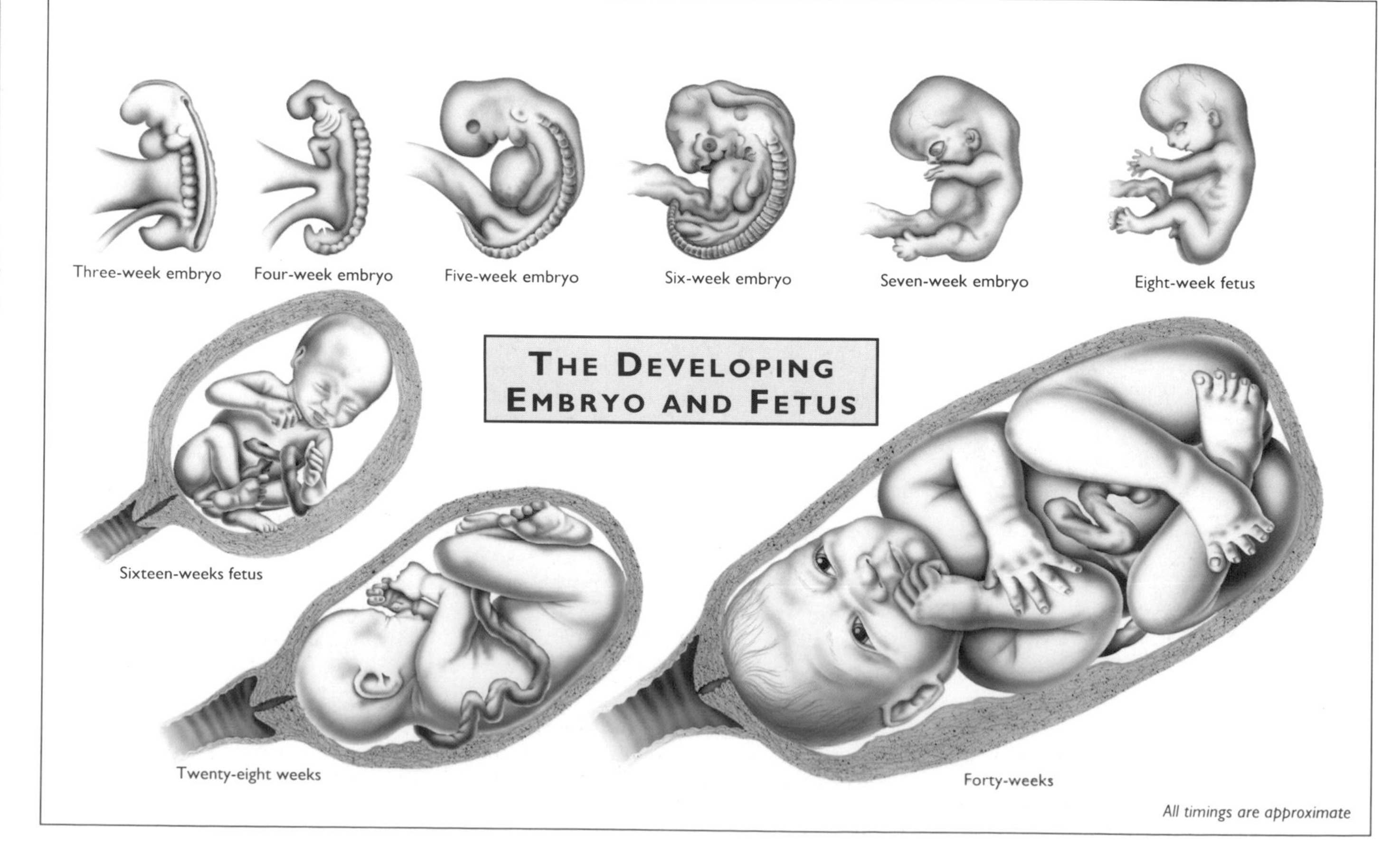

SEE ALSO

- CHILDBIRTH THROUGHOUT HISTORY p. 94
- PASSING ON LIFE p. 96
- INFERTILITY: THE OPTIONS p. 98
- BIRTH CONTROL p. 100
- HOW WE ARE BORN p. 104

About two weeks after fertilization, the embryo is fully implanted and is buried in the womb lining. The outer layer of the blastocyst turns into a membranous sac enclosing the growing embryo and the fluid around it. The small projections on the outside (the villi) branch and acquire tiny blood vessels. Now known as the *trophoblast* (from the Greek *trophe*, 'food'), this layer produces the *chorionic villi*, which then develop into the placenta. The cells of the chorionic villi are genetically identical to all the others in the embryo and can be sampled in order to provide genetic information (for instance, about possible abnormalities), without harming the embryo.

The next stage is the separation of the inner cell mass into three distinct layers. Each of these will eventually develop into a specific group of adult tissues. The outer layer is the *ectoderm*, the middle the *mesoderm* and the inner the *endoderm*. Definite head and tail ends appear and a groove forms along the line of the back into which cells from the ectoderm settle to form a sunken tube in the mesoderm. This tube will form the spine, the central nervous system and the eyes. The ectoderm becomes the skin, hair, nails and breast tissue. From the mesoderm develop the skeleton, muscles, heart and blood vessels, kidneys and sex organs. The endoderm becomes the digestive system, the liver and the glands that produce the digestive enzymes.

In the process of differentiation, cells gradually lose their universal potentiality and collect together to form distinct and specific tissues with particular structures and functions. Some acquire the ability to contract and become muscle cells, while others develop a secretory function and become glands. Still others become able to respond to stimuli by producing electrical impulses and these develop into nerve cells. This process of development of different cell types and their incorporation into tissues and organs is called *morphogenesis*. While this rapid differentiation is going on, the embryo becomes suspended in the surrounding fluid by a stalk of cells that connects it to the centre of the region of chorionic villi forming the placenta. This stalk encloses the remainder of the yolk sac and eventually becomes the *umbilical cord*. Soon blood vessels develop within this stalk and connect the developing circulatory system of the embryo to the placenta, which is intimately related to the blood vessels in the wall of the womb, and thus to the circulation of the mother.

The size of the fetus

By the fifth week of pregnancy the sac surrounding the embryo is about 8 mm (1/3 in) in diameter, and this grows at a rate of about 1 mm (1/25 in) a day. At 10 weeks the body of the embryo itself is about 2.5 cm (1 in) long, measured from the crown of the head to the rump. By now it is recognizably human and its genitals can be distinguished as either male or female. The face is formed but the eyelids are fused together. The brain is in a very primitive state, with a smooth outline and little development of the outer layer – the cerebral cortex – which is concerned with thought, awareness and consciousness. By three months, the fetus is about 5 cm (2 in) long (crown to rump), and by four months it has grown to about 10 cm (4 in) in length. In the sixth month, the fetus is up to 20 cm (8 in) long (head to toe) and weighs up to 800 g (1¾ lb). A fetus born at this stage is unlikely to survive, but as the length of time in the womb increases so does the probability of survival. Nowadays, because of advances in fetal care and incubation, most fetuses over 2000 g (4½ lb) can survive outside the womb. Interestingly, although most lay people equate early birth with premature birth, these are not necessarily the same thing. Obstetricians cite only one criterion for fetal maturity, and this is size. As in later life, some babies mature earlier than others. Thus, two babies may be conceived at the same time and born simultaneously, but if one is significantly underweight, it will be 'premature'.

Throughout pregnancy the fetus floats freely in the growing quantity of fluid that surrounds it, anchored by the umbilical cord. This fluid is called the *amniotic fluid* and at full term (i.e. just prior to birth) its volume is about 1 litre (2 pt). It cushions the fetus against physical shock and against the pressures of the contracting womb during labour, maintains a constant temperature and allows some fetal movement. Amniotic fluid is constantly swallowed by the fetus and excreted as dilute urine, so it contains material that can give information about the health, especially with regard to abnormalities, of the fetus; it is obtained by sampling the fluid through a needle (amniocentesis). As the fetus grows and the volume of amniotic fluid increases, the womb must grow to accommodate them. Just prior to birth, the uterus has risen to the top of the woman's abdominal cavity, displacing the bowels, and she suffers considerable discomfort.

THE FETAL CIRCULATION

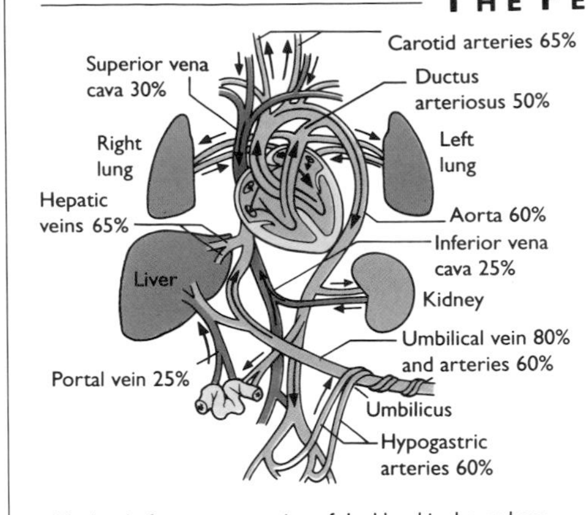

The level of oxygen saturation of the blood is shown here by the percentage figures. The ductus arteriosus allows the fetus's blood to bypass the lungs; it closes soon after birth.

Because the fetus is surrounded by fluid, its lungs do not expand until after birth, and an alternative source of oxygen is required. All the oxygen and nutrients that the fetus requires are provided by way of the placenta, an organ in which the mother's blood comes into close proximity to, while not actually mixing with, that of the fetus. The fetus's heart is not required to pump all its blood to its own lungs (which are largely bypassed during pregnancy). Instead, both sides of the fetal heart cooperate to pump blood to all parts of its body and along the umbilical cord to the placenta. Blood, rich in oxygen and nutrients, returning from the placenta, reaches the heart partly by way of the liver. After birth, the lung bypass channel and the vessels to the umbilical cord close.

By the end of pregnancy the placenta has become a thick, disc-shaped object about 15–20 cm (6–8 in) in diameter, firmly attached to the inside of the uterus. The mother's blood enters the placenta through the wall of the womb and the umbilical cord comes off from the other side of the placenta. Oxygen, carbon dioxide, sugars, amino acids, fats, vitamins, minerals, as well as many drugs, can pass freely across the placental barrier.

MISCARRIAGE

In lay terminology, 'miscarriage' means the loss of the embryo or fetus at any stage. However, doctors usually restrict the term to loss of the fetus in the later stages of pregnancy. Loss in the earlier stages they call 'abortion', without any implication that the termination is deliberate. A high proportion of pregnancies end in spontaneous abortion without pregnancy having ever been suspected. Even among known pregnancies, some 10–15% end in spontaneous abortion within the first 12 weeks. For reasons that are not clear, this is commoner in the first pregnancy than in later pregnancies.

Although there are many causes of spontaneous abortion, the reason, in any particular case, is seldom apparent. Possible causes include:

- **Chromosomal abnormalities or mutations that cause severe fetal malformations such that it is impossible for the fetus to survive.**
- **Genetic disorders compatible with survival (such as Down's syndrome) but that, in a high proportion of cases, cause abortion.**
- **Failure of the embryo to develop into a normal fetus. This may result in a fleshy growth called a *mole* (from the Latin *mola*, 'millstone').**
- **Physical abnormalities in the uterus. These may be congenital (present from birth) – such as a double uterus – or may be acquired – such as a tear in the cervix from a previous delivery.**
- **Fetal damage from x-rays or other forms of radiation. This is why pregnant women are not, if possible, exposed to x-rays.**
- **Severe accidental injury to the uterus. (Falls and other minor injuries are unlikely to cause abortion.)**
- **Various drugs – especially dangerous are the prostaglandins, ergot preparations and quinine.**
- **Foreign bodies inserted into the uterus.**
- **Surgical intervention in the pelvic region.**
- **Various diseases (⇨ pp. 126–7) suffered by the mother during early pregnancy. The most important of these are German measles (rubella), poorly controlled diabetes, listeriosis, toxoplasmosis, malaria and brucellosis.**
- **Failure of the mother's immune system to produce the antibodies needed to block the normal rejection process. If this happens, the fetus is treated as a foreign body, and is attacked and destroyed.**
- **Ectopic pregnancy – a pregnancy occurring outside the uterus.**

The major warning sign of possible spontaneous abortion is vaginal bleeding. Although this does not indicate that abortion is inevitable, it is always a cause for concern. Bleeding is usually painless, but if it is associated with lower abdominal pain, there may be an ectopic pregnancy. This may occur in a fallopian tube or in the abdominal cavity, often as a result of partly blocked tubes. The ability of the early embryo to burrow makes an ectopic pregnancy very dangerous; large arteries may be eroded, with a severe risk of internal bleeding.

How We are Born

The process by which a baby is expelled from the womb after its 40-week growth period is called *labour*. The term is appropriate, for delivering a baby is very hard work, but it says nothing about the pain that is almost inseparable from giving birth. In spite of this, however, to most women childbirth is one of the highlights of their lives and an intensely rewarding experience.

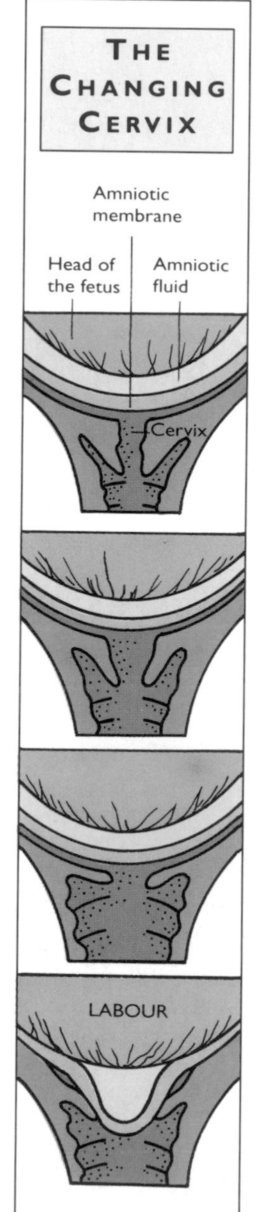

The cervix is partially taken up during pregnancy. It is then fully taken up and dilated during labour.

For various reasons, childbirth was long subject to prudery and mystery and, in many cultures, has, as a result, been cloaked in superstition and myth. Today women and their partners are far better informed about the processes of child-bearing and childbirth than they ever were in the past. However, some of this ignorance persists and women are sometimes taken by surprise by the reality of giving birth, despite the fact that experience has shown that women need reliable and frank information so that they can be prepared and can participate fully in this remarkable process. For this reason, classes in preparation for childbirth are now available in many maternity hospitals, clinics and other venues, and these may also be attended by the male partner, who is now usually encouraged to be present at the birth. In many cases, mothers-to-be prefer to give birth at home in relaxed surroundings and with minimal medical intervention, although doctors do not generally encourage this practice for first-time births or in cases where complications seem likely.

The onset of labour

The natural tendency of the stretched and stimulated womb (uterus) is to contract strongly and expel its contents. This sometimes happens before the fetus has reached full term, giving rise to a miscarriage or to premature labour. Because the uterus grows with the growth of the fetus rather than merely increasing in size by passive stretching, however, it does not normally reach the state of tension that causes it to contract strongly enough to expel the fetus before the right time has come. Minor contractions occur from time to time during pregnancy but these are not strong enough to overcome the resistance of the narrow outlet (*cervix*).

As far as we know, it is likely that strong contractions at full term are triggered by natural body substances known as *prostaglandins* (➪ pp. 90–1), which are released by rupture of the membranes surrounding the fetus or by stretching the canal of the cervix (➪ below). Prostaglandins are also used artificially to induce labour, when, for instance, the baby is more than two weeks overdue.

In the majority of cases the baby is lying head down at the end of pregnancy and is delivered head first. Often, about four weeks before delivery, the head sinks down into the brim of the bony pelvis so that the womb takes up less room in the abdomen. This may reduce the mother's discomfort and make breathing easier. The phenomenon, when observed by the mother, is known as 'lightening'.

Labour is divided into three stages. The *first stage* is the period from the onset of labour pains until the opening of the cervix is fully widened. This stretching of the cervix is called *dilatation* (or dilation). The *second stage* is the much shorter period from this time until the baby has been born, and is the period during which the mother actively pushes the baby down through the cervix and out through the vagina. The *third stage* lasts from the time of the birth of the baby until the afterbirth (placenta; ➪ pp. 96–7) and its membranes have been delivered, and the uterus has contracted down firmly enough to compress and close the blood vessels that connected to the placenta.

First stage

In a woman having her first baby (a *primigravida*), the first stage should not last longer than 12 hours; in subsequent pregnancies the first stage of labour is often much shorter. Labour starts when contractions begin to occur at regular intervals and become painful enough to attract the attention of the mother.

Labour Pains

Many women who have never had babies are familiar with the experience of labour pains, which are of the same character as severe period pains (*spasmodic dysmenorrhoea*), but are, of course, much worse than ordinary period pains. The cause of the pain is the same as that in a heart attack – loss of blood supply to the muscle. Because the contractions are brief, however, the fetus does not suffer harm from this temporary deprivation. Between contractions the blood supply to the placenta is restored.

Nitrous oxide combined with oxygen ('gas and air') is a commonly used analgesic and has the advantage of allowing the woman to avoid full anaesthesia. Pethidine, administered by injection, is another effective painkiller but can induce an unpleasant effect of giddiness and nausea. An *epidural* is an injection into the base of the spine that numbs labour pains completely by temporarily paralysing the woman from the waist down. Many women, however, prefer to try to do without painkillers and rely on special breathing techniques instead.

During each contraction the uterus can be felt to harden. Each contraction lasts for 40–50 seconds, during which time the pain rises to a maximum and then dies down again. To begin with, the interval between pains is usually as long as 20 minutes, but quite often pains occur at about 5-minute intervals from the outset. As labour progresses the pains come more frequently, become more severe, and last a

Caesarian Section

Contrary to popular belief, Julius Caesar was not born by Caesarian section. The term is of uncertain origin but may come from the Latin word *caedere* ('to cut') or from the name of the Roman law *Lex Caesarea*, which required that any woman who died in late pregnancy should have her abdomen cut open so that the baby might be saved. Caesarian section has become very common in recent years; some 25% of babies born in the USA are delivered in this way. A Caesarian section may have to be performed when there are the following indications:

- **Signs of fetal distress during labour.**
- **Failure of normal uterine contractions (*uterine inertia*).**
- **Severe bleeding before delivery (*antepartum haemorrhage*).**
- **Severe high blood pressure in the mother. This is a feature of *pre-eclamptic toxaemia* and is very dangerous as there is a real risk of the mother having a stroke.**
- **Severe rhesus incompatibility disease. The longer the pregnancy continues the greater the danger to the baby as the attack on its red blood cells by antibodies increases progressively.**
- **A placenta abnormally placed over the outlet of the womb (*placenta praevia*).**
- **Gross disproportion in size between the baby's head and the mother's pelvis.**
- **An unduly large baby, as in maternal diabetes.**
- **Severe prematurity. Very small babies are at risk of brain haemorrhage during normal delivery.**
- **The appearance of the umbilical cord before the baby's head.**
- **Presentation of the fetal shoulder or arm, instead of the head. Normal delivery is impossible and turning the baby in the womb can endanger it.**
- **Breech (buttocks first) presentation.**
- **Twins that have become locked together.**
- **Serious heart or other disease in the mother so that labour would be dangerous for her.**
- **Presence of a genital infection in the mother. If a normal birth were to occur during an outbreak of, for example, herpes, the virus could be transmitted to the baby's eyes, possibly causing blindness.**

The modern Caesarian section is performed through a horizontal incision low in the abdomen and uterus. This low incision avoids unnecessary weakening of the womb muscles and allows the operation to be performed on the same woman again in subsequent pregnancies, if necessary.

little longer. It is rare, however, for a pain to last longer than a minute. By the end of the first stage pains may occur every 2 or 3 minutes.

As the canal of the cervix is widened and pulled up by the contracting wall of the uterus (⇨ illustration), the lower part of the membranes are separated from the wall of the uterus. This causes a slight oozing of blood. At the same time the mucus in the cervix is squeezed out. The mixture of blood and mucus is called 'the show', and is an indication that labour is getting under way. Towards the end of the first stage the fluid-filled membranes surrounding the fetus are forced down into the cervix and eventually rupture, releasing a gush of amniotic fluid (⇨ pp. 102–3). This is known as the 'breaking of the waters'. Sometimes the membranes spontaneously rupture at an earlier stage; when they do, labour is likely to follow almost immediately. Quite often, the membranes are ruptured deliberately by the obstetrician during labour, as this allows the baby's head to descend and press on the cervix – widening it more efficiently.

Towards the end of the first stage, as the cervix is reaching full dilatation, the pains may be very severe, frequent and distressing. Because she does not feel the cervix widening, the woman in labour has no indication, other than what she is told, that any progress is being made. She may therefore think that her pains are ineffective. However, in normal labour, severe pains almost always indicate that all is going as it should. There are many ways in which doctors and midwives can help to control pain at this stage so that mothers are not required to bear too much in labour.

Second stage

After the cervix is fully dilated the second stage begins. Now the resistance of the cervix has been overcome so that the baby's head can descend to the muscular floor of the pelvis. For the mother to be able to expel the baby she needs to overcome the resistance caused by the pelvic floor. She does this by pushing downwards, using much the same method employed to empty the bowel when constipated. This is called 'bearing down' and is partly voluntary, partly automatic (*reflex*). The second stage should not last longer than an hour in a primigravida and may be much shorter in a women who has already had babies (*multipara*). (Sometimes forceps are used to help get the baby out quickly and avoid further fetal distress but in a small number of cases they can damage the baby.)

As the baby's head is forced through the pelvis, the mother may experience cramping pain in her legs from pressure on the nerves emerging from the lower part of the spine. Many women these days choose to adopt a sitting or squatting position during the second stage of labour. These positions are often more comfortable than the flat-on-your-back one, which is increasingly regarded as unnatural, and allow the woman to contribute more actively and take advantage of the effects of gravity. The vagina itself is remarkably distensible (able to expand), but the vaginal outlet can often become painfully stretched, so to allow it to stretch gradually, the midwife will often hold the advancing baby's head back for a time by external pressure. Sometimes, in order to prevent the outlet tearing back into the anus, a doctor will make a sideways cut, called an *episiotomy*. Once the head is born the body follows quickly, during either the same or the next contraction. One end of the umbilical cord remains in the mother until the placenta is delivered.

Third stage

As the baby leaves the uterus the walls close down behind it, and, as the area of the inner wall shrinks, the placenta is separated by the resulting shearing force. A few further contractions force it downwards into the lower part of the uterus or the upper part of the vagina. This causes the mother to feel the need to bear down once again and the placenta is delivered along with a small quantity of blood. Soon the uterus has come down so that it lies just below the level of the navel, where it can be felt as a hard muscular ball.

During pregnancy the levels of the pituitary hormone prolactin (⇨ pp. 78–9) are rising. At the end of pregnancy the levels are high enough to promote milk production, but this is prevented by the hormones progesterone and oestrogen until after the baby has been born. Soon after birth the stimulus of the baby's sucking at the nipple promotes a copious flow of milk. Indeed, this stimulus can maintain milk production almost indefinitely.

The Moment of Birth

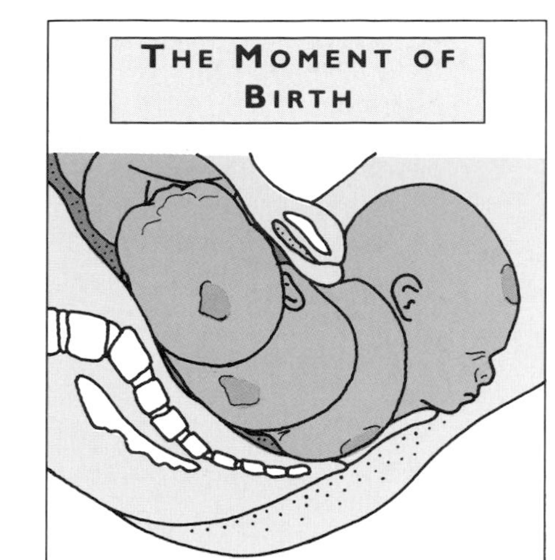

During birth the head moves down the birth canal, rotating as it goes to face forward at birth.

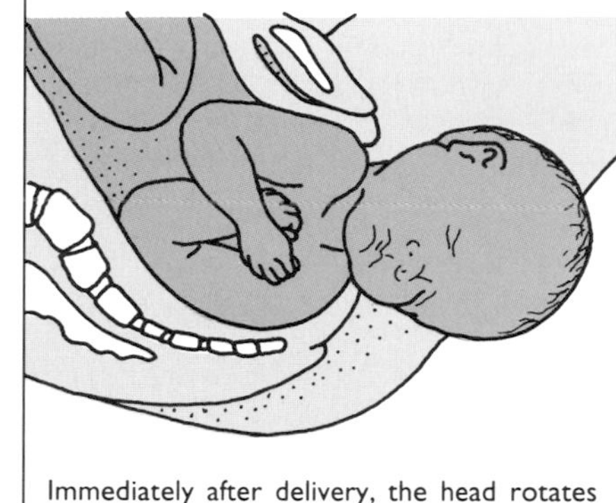

Immediately after delivery, the head rotates again and the anterior shoulder turns forward to pass under the subpubic arch.

How the Head Engages

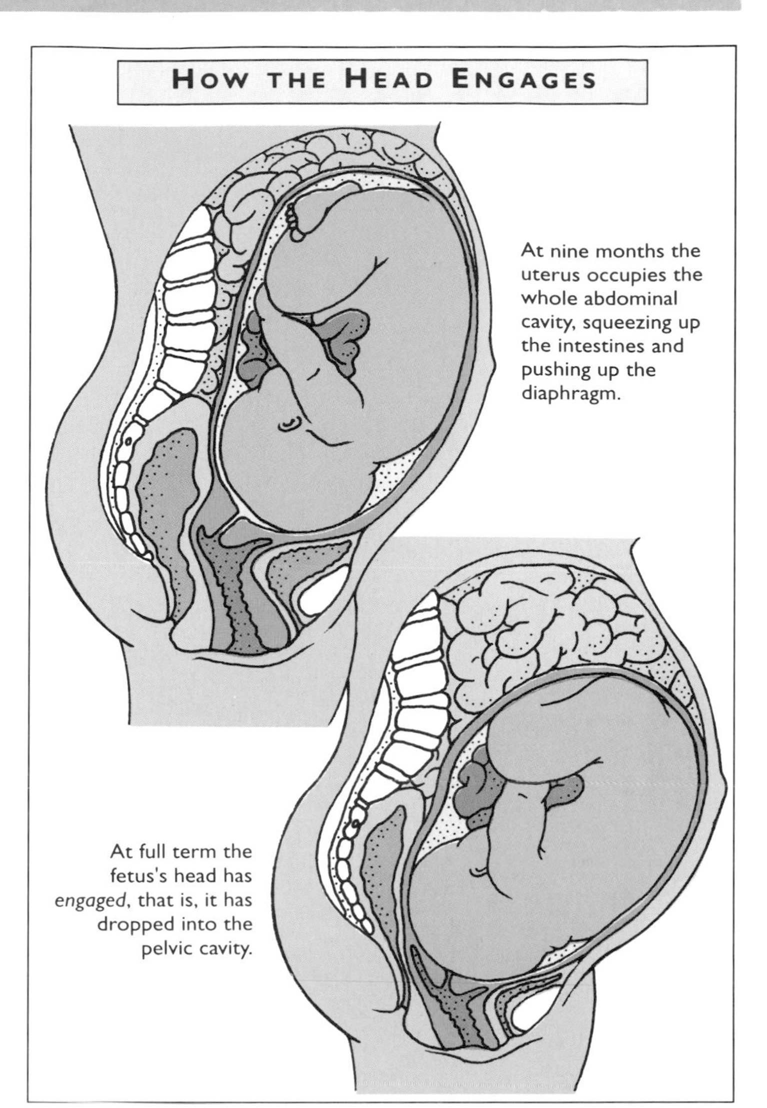

At nine months the uterus occupies the whole abdominal cavity, squeezing up the intestines and pushing up the diaphragm.

At full term the fetus's head has *engaged*, that is, it has dropped into the pelvic cavity.

SEE ALSO

- CHILDBIRTH THROUGHOUT HISTORY p. 94
- PASSING ON LIFE p. 96
- LIFE IN THE WOMB p. 102

The moment of mother–child bonding. This photograph was taken almost immediately after the birth. (Nicky Blakeney)

From Baby to Child

The first few years of life are critically important. The things that happen then to the developing infant and child colour and condition the rest of its life.

Mother and baby (right) having fun. Parental stimulation is vital for a child's mental and social development. (Images)

Early impressions go deep and, especially if traumatic, may have a life-long effect. Indeed, some psychologists assert that early environmental influences play the most important part in determining the whole personality. The consensus view among biologists is that the outcome is the result of the interaction of environmental forces on the genetically programmed organism. Growth and development are as much a matter of interior change as of a simple increase in the size of the body. The most important internal changes are in the brain and nervous system – changes that make it possible for the young person to register and store information, to acquire social skills, and to mature emotionally. Above all, the maturing of the nervous system allows the individual a greater degree of interaction with the environment and the means of obtaining more information.

Twins, just a few days old, demonstrating the innate sucking response. (Popperfoto)

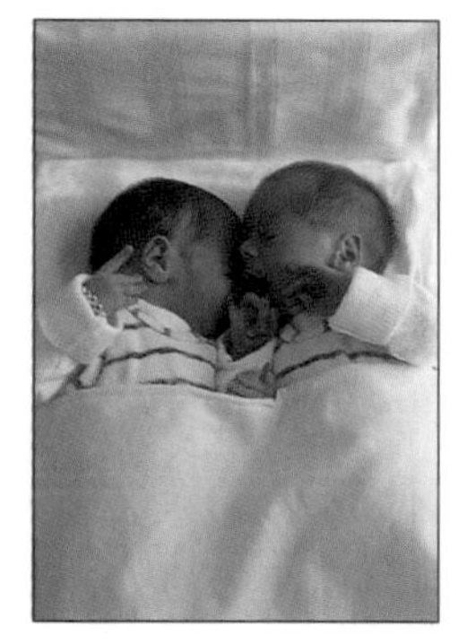

Information input is essential for the brain's development; unless the brain receives normal input from all the sense organs, it will be retarded. Sensory deprivation is as damaging as deprivation of food. Well-fed babies who are deprived of human relationships and contacts soon fall behind their contemporaries both in mind and in body.

At birth a baby's head is about three quarters of its eventual adult size and a quarter of its total body length. An adult's head is only about one eighth of the body length. The relatively large size of the baby's head is because most of the growth of the brain has already occurred before birth. Even though brain growth and development continue at a rapid rate, by the end of the first year of life half of the life-time growth of the brain has been completed. Thereafter, most of the growth is in new nerve fibres and complex connections, formed largely as a result of the effect of sensory input.

The causes of growth

As the nervous system develops, so the bodily organs gradually mature. All of these are present at birth, but many are structurally and functionally immature. Their growth and development is brought about largely by the action of *growth hormone* (➪ pp. 78–9), produced by the pituitary gland, and given out in pulses, mainly during the night. This hormone has many specific effects, one of which is to stimulate the production of proteins (➪ pp. 36–7), the substances out of which much of the body is made. It does this in two ways: firstly by increasing the supply of amino acids, the building blocks for proteins, to the cells; and secondly by increasing the rate at which DNA (the blueprint for proteins) is consulted for details of the correct order in which these amino acids are put together (➪ pp. 24–5). Growth hormone also increases the rate at which cells divide and thus reproduce. Finally, it acts on the growing zones in the long bones (the *epiphyses*), causing them progressively to extend. Body growth is also greatly influenced by other hormones, namely thyroid hormone, insulin and the sex hormones (➪ pp. 80–1). All these, acting together, bring about a steady increase in the bulk and dimensions of tissues and organs.

The newborn baby

Because every individual has a unique genetic make-up and experiences a unique set of influences while it is in the womb, each newborn baby is different from all others. Babies differ in temperament and responsiveness and in the way they react to their environment. All possess a set of built-in reflexes that help them to survive. They are, for instance, able to react to stimuli, often in a self-protective way, and know, without instruction, how to feed from the breast or from a bottle equipped with a teat. Hearing is well established and different sounds produce different reactions. Sudden loud sounds are startling and are disliked, while the sound of the female human voice tends to cause alert attention, and the lower tones of the male voice tend to be soothing. However, babies are distressed by high-pitched crying sounds produced by other infants. Within a few weeks of birth they are able to distinguish the sound of the mother's voice from that of another woman.

The sense of smell is also well developed at birth and strong prejudices in favour of pleasant smells and against unpleasant smells seem to be innate. Within a week or two of birth, babies can usually identify their mothers and fathers by smell. They also show definite taste preferences at birth, enjoying sweet flavours and rejecting sour or bitter substances. Vision is somewhat blurred at birth, but the ability to fix the eyes on an object of interest and to follow it is well established by two months. Babies have definite visual preferences, favouring images with high contrast, bright colours and curved lines. Their favourite object of gaze is the human face.

Early mental and social development

The first few weeks of life are spent mainly sleeping (initially up to 20 hours per day), crying and feeding. Crying increases to a maximum around 12 weeks because this is the only way the baby can respond to stimuli such as hunger, discomfort, pain, fear or overstimulation. Later the baby develops other ways of responding and begins to cry less. During the first year of life, the baby's perception of reality is limited to what can be seen, heard, touched or sucked. Objects outside the field of vision no longer exist as far as the infant is concerned. But around 9–12

Mimicry (right) is part of the process of learning through play. (JW)

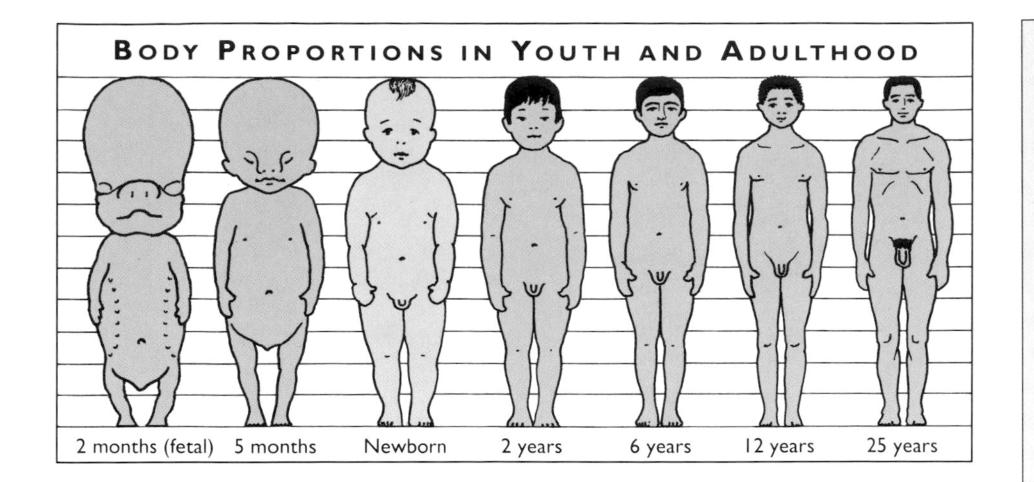

months infants begin to develop the idea that objects may have permanent existence, even when not seen. This idea is first related to the mother (or to whomever is the main carer), who at this point acquires great emotional importance in the child's consciousness. 'Peek-a-boo' games delight the child because of this. The parents' response to the baby's attempts at communication is also vital for normal development. If they are depressed or emotionally cold and cannot respond to the baby's expressions, future development and the ability to form close human attachments are likely to be seriously damaged.

By about 15 months, the ability to recognize that a stranger is not the mother or father reaches a peak, and, at this stage, the child's resulting anxiety may be severe. By two years the development of the nervous system allows the infant to recognize that there are many different people in the world and 'stranger anxiety' largely disappears. The child's perception of its parents may be so strong that even a photograph is recognized.

Early physical development

Many factors determine a baby's size and weight at birth, but the most important of these are the size, nutritional state and general health of the mother before and during pregnancy. A baby usually loses some weight in the first few days after birth but this is made up again in about 10 days. Thereafter weight is usually gained at a rate of about 30 g (1 oz) per day. If the baby is genetically destined to be large but is born small as a result of environmental effects (such as, for instance, its mother having smoked during the pregnancy), it will generally grow faster than average so as to catch up its deficit. In most cases, such babies will have reached their full, genetically ordained size for age by about 18 months. Similarly, babies genetically coded for physical smallness, but who are large at birth as a result of such conditions as maternal diabetes, may seem to be failing to grow at an expected rate. In fact, they are simply reverting to type.

The average rate of growth is such that most babies are 50% longer at the end of their first year and have grown to treble their birth weight. This rapid rate of growth soon declines, however, and by about 2 years of age it will have reached the fairly constant rate characteristic of most of childhood. On average, infants and young children gain weight at a rate of 2–3 kg (4–6 lb) per year, and increase in height by 5–7.5 cm (2–3 in) per year. In the early months of life almost half the calories consumed in food are expended on growth. This proportion also declines rapidly, so that by about 2 years of age only 3% is devoted to growth.

Walking

Most babies can sit up at 6 months and may start to crawl shortly afterwards, but the age at which they begin to walk varies widely within the normal range of 9–17 months. We should not be surprised or concerned by this as different parts of the nervous system may mature at different rates, and walking is, in any case, a complex accomplishment, calling for the effective action of many different parts – the brain, the nerves, the balancing mechanism, the muscles, and so on.

Walking is a vital stage in development as it allows the child to explore and extend its environment, thus greatly increasing the possibilities of acquiring information. At the same time, the sense of relative independence that the child experiences is an important developmental factor. Conflicts commonly arise at this stage between parental concern for the safety of the child and the child's quest for knowledge of the limits of what it can control. Exploration and new experiences are important for development and a balance is therefore necessary between over-restriction and safety.

Speech

By about two months a baby is capable of producing cooing sounds in response to sounds from adults. By 6 months or so the baby can make spontaneous, repetitive babbling sounds, and by 1 year the child is beginning to understand that there is some relationship between particular sounds and objects. Once it becomes apparent that uttering a particular sound can result in the acquisition of a desired object this connection is quickly reinforced. By around 18 months some children have a vocabulary of as many as 50 words. At this stage parents can help children to enlarge their vocabularies by using simple, normal language to refer to objects, rather than by making baby noises. The vocabulary of understood words is usually larger than the list of words that can be articulated, and during and after the second year there is usually a sudden and remarkable increase in the number of words that can be understood and spoken. Children brought up in a stimulating environment among articulate people now rapidly develop their powers of expression and are soon prompted to enlarge their environment even more by learning to read.

PIAGET AND CHILD DEVELOPMENT

The Swiss psychologist Jean Piaget (1896–1980), Professor of Child Psychology at the University of Geneva, has been one of the most influential figures in the study of the intellectual, logical and perceptual development of children. Piaget's original intention was to try to find the biological basis of logic. To this end he began to study the development of thought in children so as to throw light on human thought processes generally. But he became so fascinated by what he found in his studies of children that he devoted the rest of his life to the subject. He concluded that mental growth was the result of the interaction between environmental influences and an innate and developing mental structure. Input of new information from the environment, he claimed, challenges the child's concept of the world (*paradigm*) so that this needs to be modified or expanded, and this process occurs repeatedly. The process, he held, was one of *assimilation* (taking in new experiences) and *accommodation* (changing thought processes in order to adapt to new influences). Such a concept implies the existence of an innate form of logic, and proposes that intelligence is developed by a process of progressive refinement of the logical system as a result of experience.

Piaget defined four stages in development, for which he set specific age limits, and taught that each one had to be fully developed before the child could pass on to the next. After the first (*sensorimotor*) stage the child enters the *preoperational* stage (2–7 years). Here the child learns about the difference between fantasy and reality and, in particular, of the relation between cause and effect. As this happens the child may mistakenly construe links between its own behaviour and major events in its life, such as a death or the divorce of parents. Thinking that events may have been brought about by magic is common as the child is too inexperienced about the world to identify the real causes. From the ages of about 7–12 the child gradually acquires the capacity for logical thought about objects (the *concrete operations* stage), but it is not until the period from 12 years onwards that logical manipulation of complex, abstract concepts (*formal* operations) becomes possible. Indeed, many adults never reach this stage of cognitive development.

SEE ALSO
- HOW WE ARE BORN p. 104
- PUBERTY AND ADOLESCENCE p. 108
- LEARNING IN PRACTICE p. 174

This little girl, aged one, is totally immersed in play. As well as enjoying herself she is making important discoveries about the relative sizes and capacities of these plastic beakers. (Popperfoto)

Puberty and Adolescence

Puberty is the period of physical development during which certain physical changes occur in the body leading to the individual becoming sexually mature and capable of sexual activity and reproduction. Adolescence, on the other hand, is a social concept, describing the sometimes prolonged period of transition from childhood to adulthood.

It is during adolescence that young people first experiment with the idea of pairing up. These bonds may be fairly loose, however, since membership of the group is also considered very important. (JW)

Puberty occurs during adolescence, but adolescence involves mental and emotional changes in addition to those concerned with the reproductive function. Prior to puberty, children grow in body and mind but the changes that occur are only quantitative. At puberty, qualitative changes begin that fully differentiate relatively asexual girls and boys into women and men.

On average at around the age of 10 in girls and 12 in boys there is a striking *growth spurt*. The rate of growth increases to some 9 cm (3½ in) per year in girls and over 10 cm (4 in) per year in boys. This prepubertal growth spurt affects different parts of the body at different times so that, for a while, the body may appear to be disproportioned. Growth acceleration affects first the feet, then the legs, followed by the trunk, and finally the face, especially the lower jaw. Because puberty occurs about two years later in boys than in girls, boys have a longer period of preadolescent growth spurt and tend to gain a significant advantage in height over girls.

Puberty in girls

Since the middle of the 19th century, puberty in Western European and American girls has been taking place progressively earlier. The average age of the onset has been getting lower at a rate of four to six months every decade; and this phenomenon is thought to be due to improvements in nutrition. Puberty in girls occurs at any age between 10 and 13 years and takes 3 or 4 years in all. Because of this variability, some girls may have reached physical sexual maturity while others of the same age may still have a childlike physique. A similar phenomenon occurs with boys and, in both sexes, can be a cause of distress. The effect is, however, temporary, and by the age of 16 or 17 almost all the late starters will have caught up. Late onset of puberty has no significance for future sexuality.

The period of puberty features considerable body growth, changes in body proportions and major changes in the sexual and reproductive organs. The first sign is usually breast budding, although pubic and underarm hair may sometimes appear first. Breast buds may appear as early as the age of 8 but, on average, they appear at about 11. From this point it is usual for about a year to pass before menstruation (⇨ pp. 96–7) starts. By the time the girl menstruates the other changes of puberty are well established. Breast growth is often rapid, and, for a time, one side may grow more rapidly than the other. It is rare, however, for the breasts to remain of significantly different size. Changes in the bones of the pelvis cause it to widen relative to the rest of the skeleton, and the visual effect of this is emphasized by new deposits of fat laid down around the hips. The general contours of the female body are considerably influenced by these specific fat deposits, which also occur under the skin of the breasts and

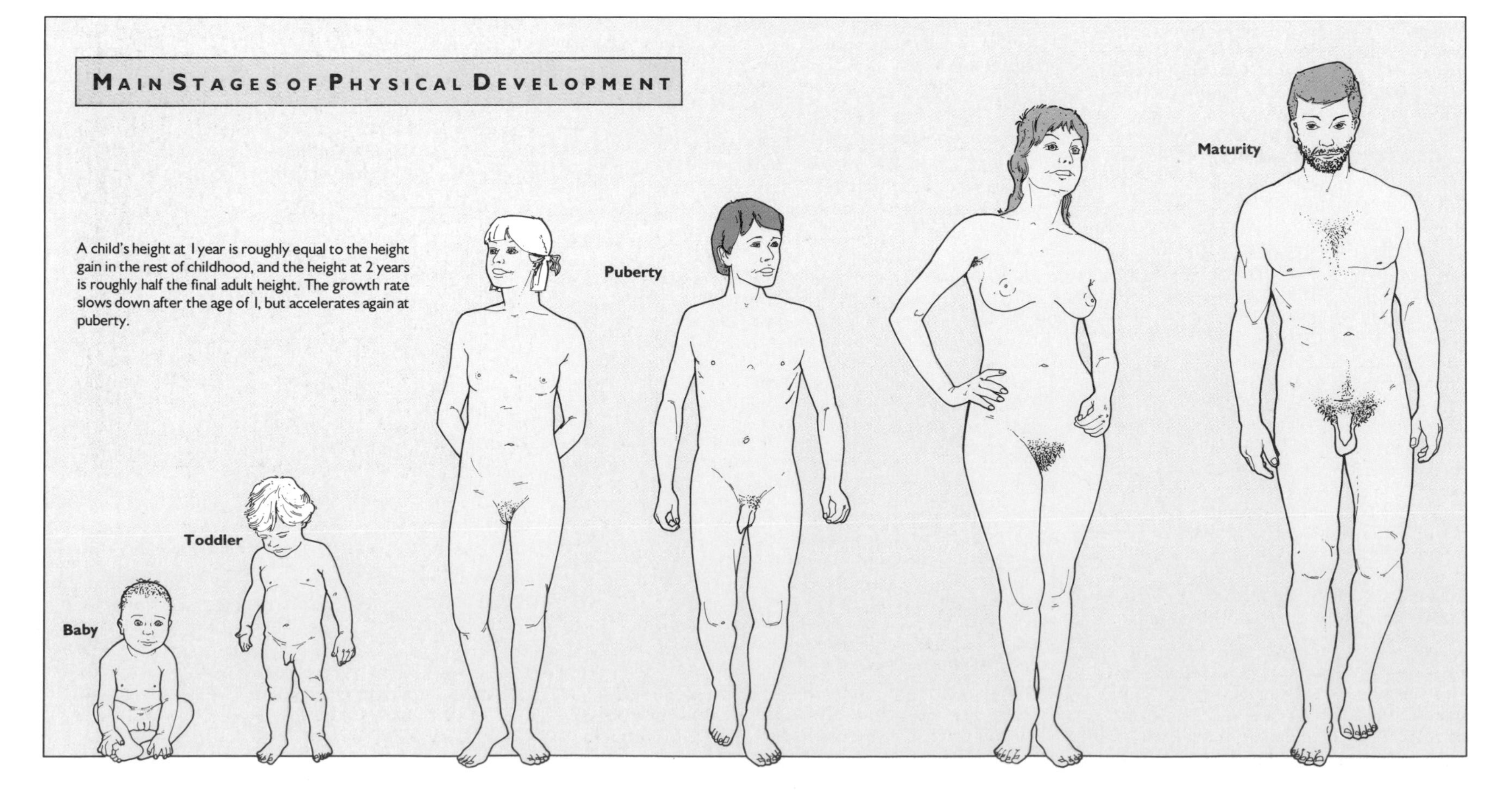

Abnormal Puberty

The whole process of puberty is initiated by the production of gonadotrophin-releasing hormone by the hypothalamus of the brain (➪ pp. 78–9). Why this occurs is not clear, but we do know that this hormone acts on the pituitary gland causing it to begin to secrete a new group of hormones, the gonadotrophins, at around, 10–14 years. The term means 'sex gland stimulators'. Gonadotrophins cause the ovaries to secrete oestrogens, and the testicles to produce testosterone. Very rarely, abnormal changes in the hypothalamus, such as a brain tumour, may start this sequence going at a much earlier age than normal. As a result, puberty may occur at almost any earlier age, even in infancy. Because the adrenal glands (➪ pp. 80–1) also secrete male sex hormones, a tumour of an adrenal can have a similar effect in a small boy. There have been cases of extreme precocious sexual development, in which full sexual maturity and reproductive capacity have been reached at the age of 5 years in both boys and girls. The youngest-known mother gave birth to a healthy baby four months before her sixth birthday.

buttocks. In girls, puberty is deemed complete when menstruation is occurring regularly. This means that ovulation is taking place and conception is possible. The onset of regular menstruation can, however, be affected by various factors such as obesity, which tends to bring on the periods earlier, and malnutrition and excessive athletic activity, which tend to delay onset.

Puberty in boys

All the physical changes that occur during male puberty are caused by the male sex hormone testosterone (➪ pp. 78–9). This is an *anabolic steroid* (growth-promoting hormone), produced in the testicles by cells lying between the sperm-producing tubules (the *interstitial cells*). Until shortly before puberty the testicles contain only numerous solid cords of pale cells, and there is no sign of sperm production. At puberty, however, the cells at the centre of these cords atrophy so that the cords become tubes. These are the seminiferous tubules (➪ pp. 96–7) in which the sperms develop. This process is completed early in puberty, and thereafter sperm production is active and rapid. Very large numbers of sperms are produced – about 300–600 per gram (10 000–20 000 per oz) of testicle every second.

Testosterone is a powerful hormone and has many effects. It causes the testicles, scrotum and penis to enlarge, and enables the penis to erect fully. It also causes the seminiferous tubules to begin to produce sperms, and makes the sperm-carrying ducts and the semen-storage sacs (the *seminal vesicles*) increase in size and mature. It enlarges the prostate gland and causes it to begin to secrete fluid that makes up part of the seminal fluid. It makes the voice box (larynx; ➪ pp. 82–3) bigger and the voice to drop in pitch as a result. It promotes the growth of pubic and underarm hair and of the beard, and may cause hair to grow on the chest and abdomen. Finally, it accelerates general body growth and muscular development.

These changes usually occur roughly in the order recounted above. Puberty may start at any age from about 10–15 and usually takes about two and a half years, so it may not be complete until after the age of 17. Most boys have reached about 80% of their adult height before puberty, but there is nearly always a large growth spurt during puberty too.

Psychological aspects of puberty

The main psychological consequence of puberty is the initiation and growth of sexual interest and sexual drive. Both boys and girls quickly become aware of the physical differences between the sexes and this awareness often arouses intense interest in the opposite sex. In many cases this becomes the chief preoccupation. There is often a new and uncharacteristic concern over personal appearance and clothes, and adolescents may become increasingly anxious to conform to current fashions in dress.

Adolescence is a difficult time for many. Growing knowledge and intellectual powers often bring with them a kind of arrogance and a contempt for the opinions of older and more experienced people. Adult ways and views are often rejected by adolescents and they may be extremely impatient with imposed rules and regulations. There is a growing recognition of the importance of success, and if they fail to achieve this, as it is defined by adults (good school reports, for instance), adolescents may then make attempts to succeed in the eyes of their peer group. This commonly leads to delinquency, but this is usually only a temporary phase.

Many adolescents do not reach Piaget's fourth stage of development until adulthood, if at all. This stage is the fully adult (formal) way of thinking (➪ pp. 172–3) in which abstract reasoning predominates. In many adolescents, thinking continues to be concerned with the particular rather than with the general, the concrete rather than the conceptual. Thinking often remains egocentric (unable to see things from another person's point of view) and some adolescents have difficulty in appreciating the relationship between present action and future consequence.

SEE ALSO
- PASSING ON LIFE p. 96
- HOW WE PRESENT OURSELVES TO OTHERS p. 198
- SEXUALITY p. 202

The Revolt of Youth

The adolescent years are a time of important self-discovery, when a child's emergent sense of identity is further shaped and solidified. Adolescents often form informal peer groups, adhering to certain patterns of behaviour, attitudes and appearance. Membership of such a group serves to reinforce the individual's own inclinations and to steer his or her thoughts, opinions, likes and dislikes. In the developed world, where families do not depend on the income that their children can earn, adolescents have the leisure time and the money to take part in group activities, making it possible for them to display attitudes and resentments that may previously have been submerged.

Folk heroes have always had a strong appeal as role models for adolescents, and, with the proliferation of films, television and popular music in the 1950s and 1960s, cultural icons such as Elvis Presley, James Dean, the Rolling Stones and the Beatles had an enormous influence. One of the earliest post-war indications of youth revolt among boys was the growing of long hair as a silent protest against the prevailing fashion for short hair and its militaristic connotations. Deliberate eccentricity in dress followed, epitomized by the 1950s 'Teddy (Edwardian) boys' in their drainpipe trousers, long, draped jackets and thick, crêpe-soled shoes. The elegant, Italianized 'mods' on their scooters and their rivals the leather-clad 'rockers' or 'greasers' on their motorcycles were a feature of the early 1960s. These gave way, in the late 1960s, to the menacing and aggressively leather-clad, tattooed and uniformed 'Hell's Angels'. The late 1960s also saw the growth of the 'hippy' movement, whose membership came from a more middle-class population and were devoted to 'flower power', communal living and experiments in mind expansion by means of hallucinogenic drugs.

The 'skinheads', who first appeared in the late 1960s and are often associated with extreme, right-wing political movements, and the 'punks' of the 1970s and 1980s, with their spiky, colourful hairstyles and their anti-Establishment ideas, tried to carry visual protest against orthodoxy to the limit.

Mean and moody – the characteristic attitude of the 1950s Teddy boy. Activities such as smoking and gambling (dice are just visible in this boy's hand) were guaranteed to provoke the disapproval of older generations. (Popperfoto)

Generally little attempt is made by society to interfere with the group activities of young people unless these have consequences that impinge on the freedoms of other groups or individuals.

Ageing

A progressively increasing life expectancy, coupled with developments in medical science, means that elderly people, particularly in the West, can now look forward to the prospect of a longer, healthier and more independent old age.

Longer life expectancy together with a declining birth rate have changed the structure of Western populations, the number of elderly people having grown in both relative and absolute terms. If this trend continues, the growing financial dependency of a larger and larger elderly population on a shrinking younger one is likely to have serious consequences for the world economy.

The natural life span (that is, unimpeded by disease) of the human being appears to be between 100 and 110 years, and there is no reason to suppose that this is changing. The increasing expectation of life is largely due to advances in medical science that allow more and more people to approach or reach the natural age limit. Some experts dispute this figure, however, and hold that the human body is programmed to begin to fall apart at about 85 (⇨ box); they point out that, at that age, tiny insults – injuries that would hardly be noticed at 20 – are often enough to cause death.

An artificial hip joint, shown in a false-colour x-ray. Nearly all people of 60 and over are affected to some degree by the disease of the joints known as osteoarthritis, although in some cases the symptoms are mild or absent. The lives of countless people virtually crippled by arthritis in the hips have been transformed by routine hip-replacement operations. (SPL)

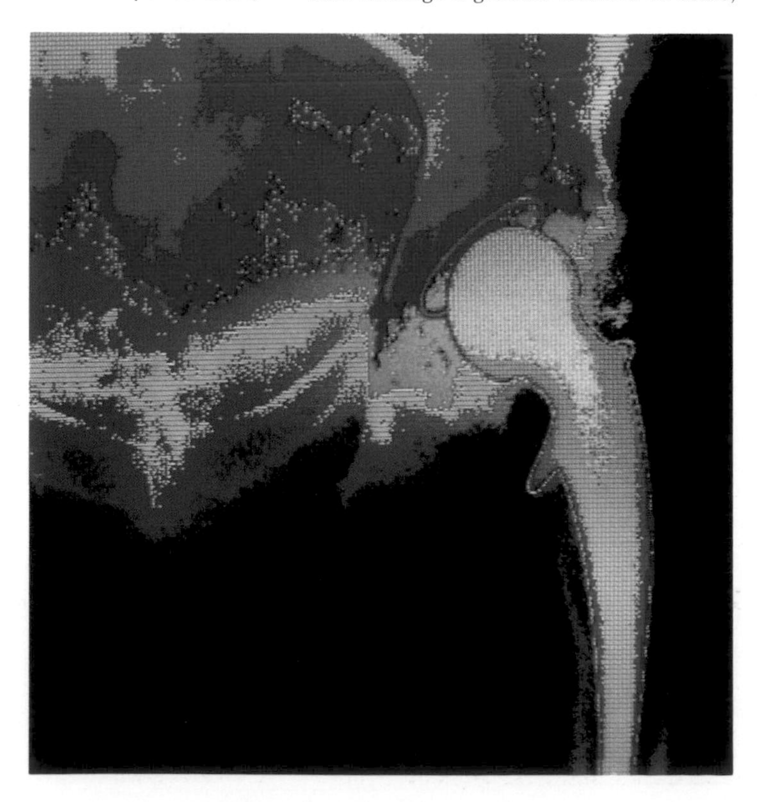

Environmental influences

In most people, physical and mental development continue up to the mid-20s and then undergo a gradual decline. In some, physical and mental capacities continue to increase until well after that age. Environmental influences and life style clearly have a major part to play in determining the outcome of a person's development, and the signs of ageing appear earlier and are more severe in some people than in others. The body is strongly affected by outside factors that are known to accelerate the process of ageing. Among the most obvious of these are lack of interest in external affairs, inattention to one's own health, lack of exercise of body and mind, unsuitable diet, smoking, excessive alcohol intake, use of drugs, and undue exposure to sunlight.

The progressive reduction in physical and mental capacities deemed to be characteristic of the latter part of life is often merely a reflection of cultural expectations. For the old as well as the young, the strengths and skills of the body and mind increase the more they are tested. Researchers are now beginning to understand the central importance of remaining mentally and physically occupied, and the part such activity plays in allaying the signs of ageing.

The effects of ageing

Some of the commonest effects of ageing include:

- Loss of elasticity in the arteries leading to increasing rigidity.
- Thickening of the walls of the arteries.
- Even deposition of cholesterol in the arterial walls.
- Widening, elongation and *tortuosity* (twisting) of arteries.
- Local ballooning of arteries (*aneurysm* formation).
- A gradual rise in the blood pressure.
- Loss of elasticity of the skin leading to wrinkling and sagging.
- Accumulation of skin pigment in patches.
- Loss of bone bulk from reduction in amounts of protein and mineral content (*osteoporosis*).
- Increase in bone brittleness.
- Loss of muscle bulk and power.
- Development of a white ring around the edge of the cornea (*arcus senilis*).
- Greying and then progressive whitening of the head hair.

The deposition of cholesterol in the walls of arteries that occurs with age is a normal and largely harmless process, and is to be distinguished from the formation of plaques of cholesterol and other materials that is a feature of the disease known as atherosclerosis.

The emphasis given in the above list to changes in the arteries is not arbitrary; as the French physician Pierre Jean Georges Cabanis (1757–1808) rightly said: 'A man is as old as his arteries.' This is now known to be all too obviously true. In the Western world, arterial disease, especially atherosclerosis, is by far the commonest cause not only of death but of serious diseases in later life. These include strokes, senile dementia, heart attacks and leg gangrene.

In spite of every effort, ageing (or *senescence*, in contrast to *senility*, with its connotations of mental deterioration), has certain inescapable features or associations. Loss of muscle bulk causes a decrease in power. The lungs lose elasticity and become more prone to local areas of collapse (*atelectasis*) and to pneumonia, especially after prolonged bodily immobilization. Progressive loss of nerve tissue may lead to such symptoms as unsteadiness, vertigo and reduced powers of memory. Loss of elasticity of the internal lenses of the eyes leads to difficulty in focusing on close objects (*presbyopia*), while gradual denaturing of the protein in the lenses leads to opacification (*cataract*). Degenerative changes in the retinas, especially in the most sensitive central zones, may lead to blindness, as may a gradual failure of the internal fluid drainage mechanisms of the eyes (*glaucoma*). The immune system declines in efficiency with age, so susceptibility to infection increases. Various endocrine changes occur with age (⇨ pp. 78–9). The tissues become relatively resistant to insulin, so higher blood sugar levels occur, bringing with them an increased tendency to arterial disease. The output of sex hormones decreases, gradually in men but acutely in menopausal women (⇨ box), and this commonly has serious consequences for the strength of the bones.

Theories of ageing

Throughout life, DNA undergoes constant changes (mutations) as a result of external agencies and mistakes in replication (⇨ pp. 24–5). Because of its double-strand structure and the existence of certain repair enzymes, mutations are readily corrected. According to one theory of ageing, there is an increased susceptibility to such external influences and/or a decrease in the efficiency of the repair mechanisms. So far, however, no evidence has been found of defective repair and the rate of spontaneous mutations from external causes is insufficient to account for the number of

Threescore Years and... Twenty-Five?

The Stanford University rheumatologist James Fries has analysed how life expectancy has changed since the beginning of this century and has shown that the curve of life expectancy from birth and the curve from age 65, if extrapolated, intersect at about age 85 around the year 2010. This, he suggests, indicates the natural limit. Others question this reasoning and point out that many of the conditions that have traditionally carried away the elderly – conditions such as cancer, pneumonia, osteoporosis and atherosclerosis – are increasingly susceptible to medical intervention. Studies of mortality rates for people of 85 and over confirm that they have dropped substantially in the last 50 years. Even for people of 100, the proportion dying in 1 year has dropped.

LIFE AFTER THE MENOPAUSE

For women, the most striking and important effect of ageing on the endocrine system is the sudden drop in the output of the hormone oestrogen (➪ p. 96–7) at the time of the menopause. There is no corresponding drop in male sex hormones and the idea of a male menopause is fictitious. Sex hormones are powerfully anabolic – capable of building up and maintaining body tissue; the sudden loss of oestrogen at the menopause can lead to rapid loss of bone bulk, as a result of a reduction in the protein (collagen) from which bone is made and in its mineral hardener. This is called *osteoporosis.*

Oestrogen also gives protection against arterial disease, especially atherosclerosis. This is why heart attacks and strokes are very much more common in men than in women under menopausal age. This protection is lost after the menopause, so that within a few years the incidence of these diseases is as high in women as in men.

Osteoporosis is a very real threat to postmenopausal women. The bones become fragile and may fracture under quite minor stress. Some 25% of women over the age of 70 suffer serious fractures, some of which occur almost spontaneously. Fracture of the neck of the hip bone is especially common in this group. The spine may soften and can quickly become seriously bowed (the so-called 'dowager's hump'). Severe backache and fractures of the forearm are also common.

All post-menopausal woman, especially small women with delicate bones, should seriously consider the advantages of hormone-replacement therapy (HRT) in order to minimize osteoporosis. Once established, little can be done to help. Regular physical exercise, a good calcium intake, and the avoidance of smoking also help.

The spongy bone tissue within a spinal vertebra affected by osteoporosis. (Magnification x 30 at 6 x 7 cm) (Prof. P. Motta, Dept. of Anatomy, University 'La Sapienza', Rome, SPL)

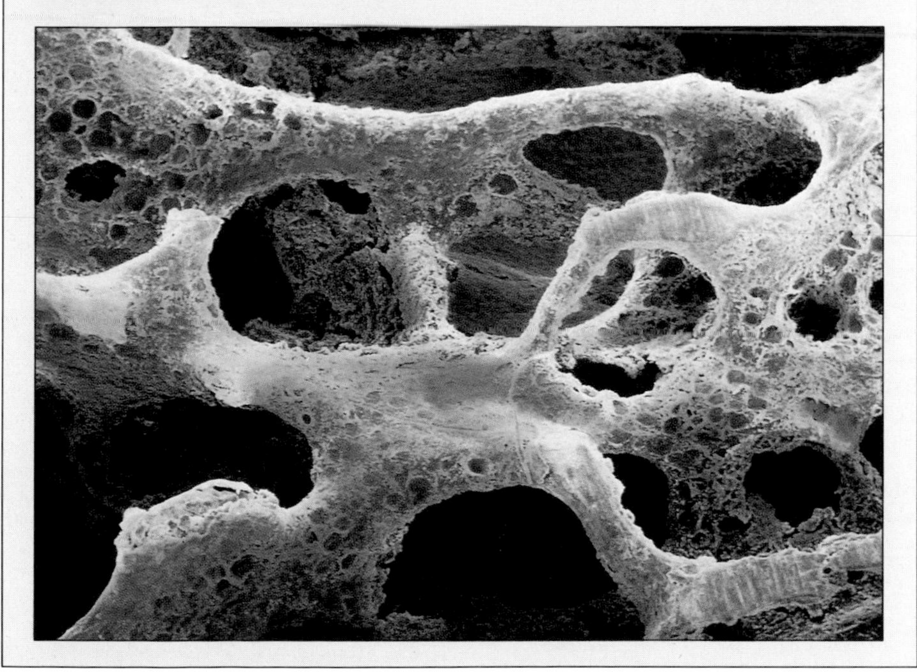

changes that occur with ageing. An alternative hypothesis, known as the 'error catastrophe theory', proposes that errors occurring in the replication of DNA and RNA and in protein synthesis (➪ pp. 26–7) gradually and increasingly augment each other until they reach the stage of catastrophic failure at which cells can no longer replicate.

The most generally accepted theory at present is that ageing is regulated by particular genes. Cells in the adult body fall into three groups: those that reproduce continuously, those that reproduce when needed, and those that never reproduce. In the first category are the cells of the skin, intestinal lining and the blood-forming tissues; in the second are cells of the liver; and in the third are the cells of the brain and nervous system and the heart and voluntary muscles. Cells in the first group will readily reproduce in artificial culture.

In the early 1960s Leonard Hayflick of the Wistar Institute in Philadelphia demonstrated that artificial cultures of fibroblasts – the main constructional cells of the body (➪ pp. 38–9) – would not continue to divide (replicate) indefinitely, even if provided with ideal conditions and nutrition. The maximum number of population doublings of the cells appeared to be about 50. The most interesting finding in this work, however, was that the number of doublings was apparently determined by the age of the person from whom the cells were taken. Cells from a baby would undergo the full number of divisions; those from an old person only one or two. Likewise, cells from a person with premature senility (➪ below) would also divide only a few times. This finding has had a strong influence on current thinking and has been taken to indicate that there is a finite limit to human life. Not all experts agree, however, that one can extrapolate from the culture flask to the human organism.

When old fibroblasts that are no longer replicating are fused with those from young donors, the synthesis of DNA in the young nuclei is stopped. The same effect occurs even if only the material surrounding the nucleus (the cytoplasm) of the old cells is inserted into the young cells. This suggests that the inhibiting factor (thought to be a protein) is present in the cytoplasm. This factor is believed by some to be the principal cause of ageing.

Premature ageing

There is a rare disorder called *progeria* in which a young person begins to age prematurely, gradually acquiring over the course of about 10 years all the characteristics of old age – wrinkled skin, white hair, arterial degeneration, high blood pressure, cataracts, increasing susceptibility to disease, and so on. Death occurs in early adult life, usually from coronary artery disease, by which stage the sufferer is in a state of apparent senility.

This disease may take two forms, distinguished by the age of onset. In infantile progeria, which starts in early childhood, the whole process of ageing may be complete by the age of 30, while adult progeria, or *Werner's syndrome*, usually starts in the 30s or 40s, often with premature baldness or greying of the hair. The skin quickly becomes wrinkled, arterial disease rapidly develops, the sex glands waste away, and diabetes is common. Death occurs often within 10 years of onset. The cause of progeria is unknown and there is no effective treatment. Although few people are affected by it, the condition is of intense interest to scientists as it sheds light on the various causes of ageing.

SEE ALSO

- CELLS, THE STUFF OF LIFE p. 38
- EXERCISE AND BODY UNITY p. 54
- HEALING AND REPAIR p. 62
- BIORHYTHMS p. 66
- HOW WE DIE p. 112

The Spanish artist Pablo Picasso, whose creative activity remained undiminished into his 80s and 90s. The achievements of the old often occasion surprise, but there is in fact an impressive record of their accomplishments in every field of human creativity – art, literature, music and science. (Popperfoto)

How We Die

There are few certainties in life, but we can be confidently sure of at least one thing – that we will die. Death is the cessation of all biological functions. It occurs because cells are deprived of the substances needed for continuing operation, or because they are unable to use them because of poisoning or degeneration.

Mask wearers form part of a funeral procession in modern-day Cameroon. Death rituals are invested with a traditional importance as the final rite of passage. (KP)

Different functions cease at different times and some, for example, hair and nail growth, persist for hours after the heart beat and respiratory movements have stopped. Even after the heart has stopped beating, it remains alive and can be transplanted into another person to sustain life for years. Many cells continue to survive for a time after somatic (bodily) death but, because all cells need oxygen and fuel and these are provided by a functioning circulation and respiratory system, the failure of the heart and lungs to maintain the supply is soon followed by cell death. Immediately after death the skin becomes pale, also as a result of the cessation of the blood supply. The digestive tract is the first part of the body to decay since its enzymes are no longer prevented from attacking the body itself. This process is aided by the effect of ever-present bacteria in the gut, which are no longer attacked and killed by the immune system. Rigor mortis, the stiffening of the muscles, caused by the solidification of muscle protein, sets in within two or three hours. The speed with which it takes place varies according to the temperature and the level of exertion undergone by the deceased prior to death – it is quicker to appear in somebody who was running, for example. As the muscle proteins degenerate, rigor mortis passes off, usually after a day or two. Bacteria in the environment carry enzymes (biological catalysts) that break down tissue molecules to simpler substances, and it is these that are mainly responsible for the final biological consequence of death – the return of the body to its chemical constituents (decomposition). Nowadays it is common for us not to wait for bacterial action to take place after burial but to accelerate the process by burning the body (cremation).

Demons and angels gathered around the bed of a dying man in preparation for the struggle for his soul (small gold figure). From a German woodcut c. 1480. (Images)

The fear of death

While children often display acute fear of death, young people do not really believe that they are going to die, and it would be unhealthy if they were obsessed with their own mortality. A study of attitudes to death has shown that 90% of university students hardly ever consider the matter in relation to themselves while, in the case of older people, 70% often do. Fear of death is, however, usually much more acute in a young person when the imminent prospect must be faced, or in an older person suddenly struck by a heart attack or facing extreme danger. But for people who consider they have had a reasonable life span, death is natural and normal and is usually accepted as such. The elderly, having many more 'intimations of mortality', usually take a more realistic view, and recognize that they are not going to be spared the fate that has already overtaken so many of their contemporaries. A man of 65 can reasonably expect only 17 or 18 more years of life, but few older people find this fact unduly distressing. For most, age and maturity bring an acceptance of the inevitable. Characteristically, the very old seldom fear death; many of them, even those in good health and free from pain, welcome it, feeling they have lived long enough. Such people die without struggle or resentment and, if their affairs are in order, they are usually carried off with an easy and accepting mind.

The process of death after illness or severe injury, even in the young, is usually well ordered by a beneficent nature. Whichever process is causing the death, its effect on the brain brings its own panacea so that distress, horror and anxiety all appear to be abolished. This has been reported by doctors and other observers of the dying through the ages. Some have even reported their own experience. William Hunter (1718–83), brother of the great anatomist John Hunter, and himself an experienced surgeon, said on his deathbed: 'If I had strength enough to hold a pen, I would write how easy and pleasant a thing it is to die.'

The French essayist Michel de Montaigne (1533–92) also put the matter well: 'It is not without reason we are taught to take notice of our sleep for the resemblance it hath with death. How easily we pass from waking to sleeping; with how little interest we lose the knowledge of light and of ourselves. For, touching the instant or moment of the passage, it is not to be feared that it should bring any travail or displeasure with it, forasmuch as we can have neither sense nor feeling.'

Near-death experiences

Reports of what they experienced by those who have come near to death, or who believe that they did, have a remarkable similarity and have aroused a great deal

EUTHANASIA

People who are mortally ill and are in severe bodily and mental pain often beg that their sufferings be ended by death. This raises the question of the responsibility of relatives and medical attendants in the matter. Some doctors take it to be their first responsibility to preserve life, whatever its quality and regardless of the sufferings of the dying. Most, however, see it as their duty, in the case of terminally ill patients, to take all necessary steps to relieve suffering *even if these steps might shorten life.* Drugs that are highly effective in relieving pain and distress of mind inevitably interfere to some extent with brain function, including that of the respiratory centres in the brainstem. In the case of many severely ill patients, respiratory lack might shorten life, so some doctors avoid the use of such drugs, but most feel that their use is nevertheless justified.

There is another sense in which many doctors practise euthanasia – the term is a combination of two Greek words, meaning 'pleasant death'. Terminally ill patients are commonly carried off not by the condition from which they are dying but by a simultaneous infection such as pneumonia (the 'old person's friend'). While some doctors feel that all such infections must be energetically treated, there are many who, in the presence of severe pain and the wish for death, do not believe that they are necessarily required to do so.

Some doctors and relatives, out of a deep sense of pity, have overtly performed euthanasia in the sense of deliberate mercy killing. In such cases the act is often openly reported to the police. In the UK prosecution for murder or manslaughter may follow, but in the special, extenuating circumstances the sentence is usually the lowest prescribed by law. Most European and many other countries have legislation allowing the distinct category of compassionate murder, for which sentences also tend to be lenient. In Switzerland a doctor who, for proper reasons, provides a patient with the means to commit suicide is not liable to prosecution. Euthanasia is now accepted in the Netherlands where I death in 250 is by nonvoluntary euthanasia. This is where doctors and relatives decide to terminate a patient's life support, on the basis that that person's condition is total and irreversible, as, for instance, when brain death has occurred.

Although there is widespread support for the general idea of euthanasia, there is also considerable opposition to it, often on religious grounds and for reasons concerned with the sanctity of life, similar to those put forward by people who oppose abortion (⇨ pp. 100–1). Even among those who support the general idea of euthanasia, the matter is still very contentious with many questions still to be settled. Who should decide whether euthanasia should be performed? How can it be established that the person making the request is of sound mind? At what stage in an incurable illness can it be considered permissible to end a human life? Might not the supplicant change his or her mind? How can suffering be quantified? How much influence should the relatives have in the decision? Is there not a danger that gradually the motives may change so that such sanction might be used as a means of getting rid of old and 'inconvenient' people? Such questions are unlikely to be decided by logical argument, and it seems improbable that the many different views of ethical philosophers, theologians, medical practitioners and legislators will ever be reconciled.

SEE ALSO

- BODY AND SOUL p. 70
- AGEING p. 110
- RITES OF PASSAGE p. 216

of interest among doctors, philosophers and religious thinkers. Such reports are taken by some to provide proof of, and insight into the nature of, an afterlife. Some have even suggested that the nature and similarity of reports of near-death experience may have been the origins of similarities among religious beliefs from widely separated cultures.

When totally deprived of oxygen and glucose, the brain dies in about 5–10 minutes at normal temperatures. Brain death may take much longer at low temperatures because cell metabolism is greatly slowed and fuel reserves last longer. Near-death experiences occur during the period between oxygen deprivation – usually from cardiac or respiratory arrest – and the point of actual brain death. Although this period is only a matter of minutes, it may seem much longer to the person concerned. Such hallucinations involving time distortion are well-known consequences of brain malfunction, as from the effects of various drugs or, as in this case, from lack of oxygen (*hypoxia*). Other effects of such brain dysfunction, such as a feeling of omniscience and a heightened sense of the spiritual dimension in life, have also been documented.

According to one authoritative study, 60% of the subjects of near-death experiences reported a feeling of peace and contentment; nearly 40% had a sense of detachment from their own bodies; 23% believed that they were entering darkness; 16% reported seeing a great light; and 10% believed that they had entered the light. These experiences have plausible physiological explanations. The release of natural morphine-like substances (*endorphins*), which act on the emotional centre of the brain (the limbic system; ⇨ pp. 74–5), produces a sense of well-being and allays anxiety, even in the most stressful circumstances. It seems likely that endorphins are released in increasing quantities as the effects of hypoxia increase. This would cause more marked effects on the limbic system and could account for the out-of-body hallucination. Progressive loss of oxygen to the cells of the part of the brain concerned with vision (the visual cortex at the back of the brain) could cause first a blacking out of vision and then perception of bright light. At the same time, abnormal stimulation of the motor cortex of the brain (the part concerned with voluntary movement), or of the cerebellum (the part that coordinates movement and equilibrium), could cause a sensation as if one were walking.

In the great majority of people who reach this stage, all consciousness is now lost as the functioning of the brain cells ceases from oxygen lack. A few, however, are resuscitated or recover spontaneously, and are able to report their experiences. There is no reason to suppose that such reports provide us with any information about what happens after death.

In the West, women, on average, outlive men. The main reason why this should be so is that, during their reproductive years, women secrete the hormone oestrogen, which has a protective effect against atherosclerosis. This is a degenerative disease of the arteries that can lead to coronary heart disease and stroke. In the developing countries, environmental factors such as famine, disease and war generally mean that members of both sexes die before this statistical difference can emerge.

The funeral procession of John F. Kennedy, 35th President of the United States, who was assassinated in Dallas in 1963. Kennedy's importance as head of state is reflected in the grandeur of his burial. (Popperfoto)

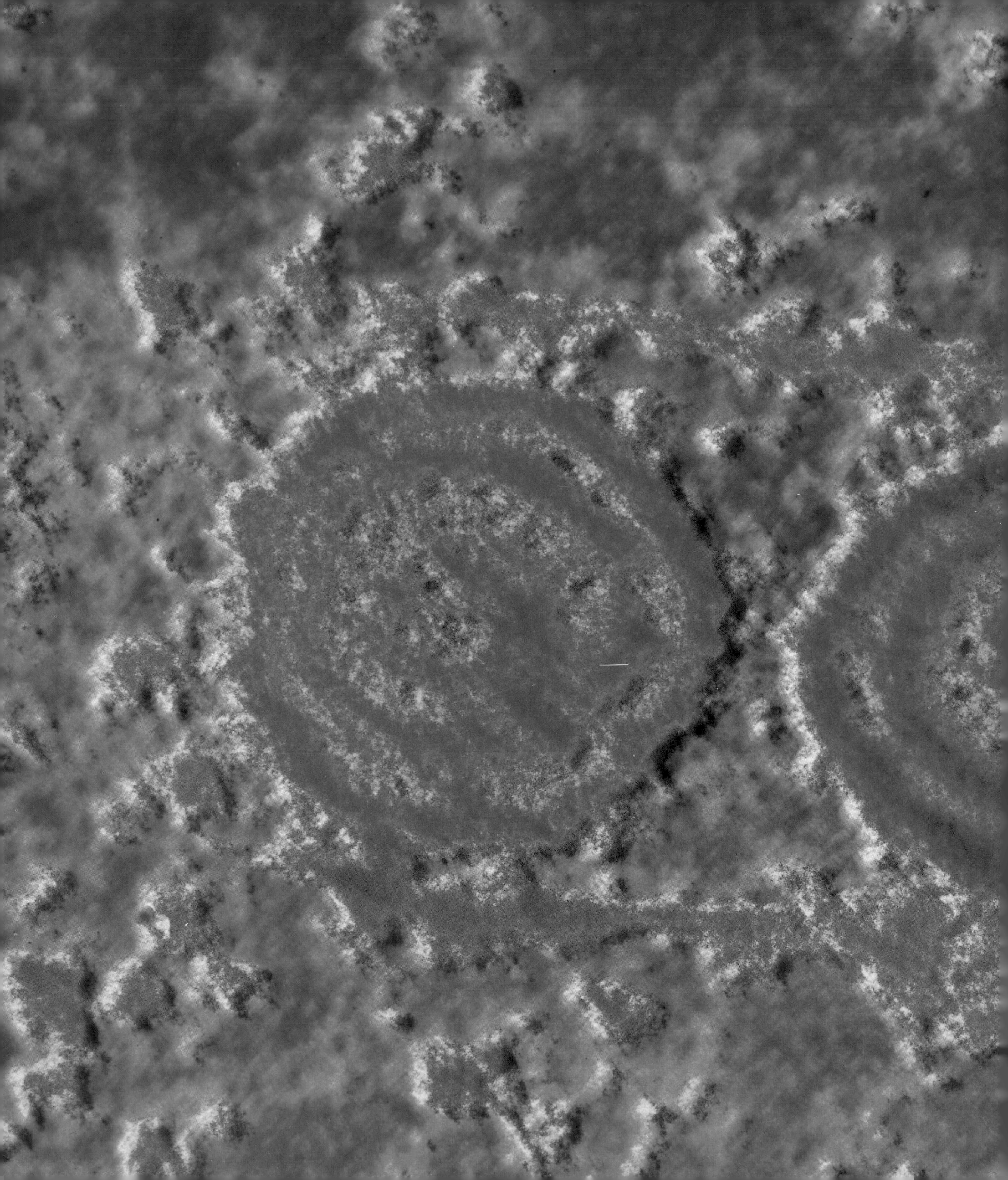

HEALTH AND DISEASE

Concepts of Disease

The study of disease is called *pathology*, and, as a serious scientific discipline, it is fairly young. Although interested laypeople have observed and recorded the effects of disease processes on the body from the times of the earliest written records – especially those causing visible changes – these observations for a long time aroused little general interest.

Early ideas about the causes of disease were based on superstition or mere conjecture. The first real discoveries were not made until around the 17th century; and it was not until the 19th century that any significant advances were made in understanding the causes of disease.

Bodily disease always used to be linked to spiritual matters. In ancient societies, the sick commonly visited temples in the hope of receiving relief or cure. Small sculptures or clay models representing disease processes were often left in these temples as offerings or symbols of thanksgiving. From these we can gain some idea of the conditions suffered by their donors – breast cancer, hernias, fluid collection in the abdomen, skin ulcers, varicose veins, and so on. Because few effective treatments existed, the signs of disease (also called pathology) were often more markedly advanced than would be allowed to happen now, and the ancients were thoroughly familiar with the appearances of a great number of disorders.

One of the most important sources of information on ancient perceptions of disease is an Egyptian papyrus dating from 1550 BC and translated by the German Egyptologist Georg Ebers. The Ebers papyrus records details of major cancers, worm infestation and other intestinal parasites, dysentery, spontaneous fractures of bone – probably osteoporosis (▷ pp. 110–11) or bone cancer, head injuries and fatty tumours. Severe bodily decline (*cachexia*) from long-term infection was also well known. The process of embalming dead bodies for mummification often involved the removal of the bodily organs, and the embalmers must have been thoroughly familiar with the appearance of those that were normal and of the changes brought about by disease. This huge source of pathological material does not, however, seem to have aroused much interest at the time and there is no recorded evidence of any attempt on the part of the ancient Egyptians to correlate these unusual appearances in death with illness in life.

Recent examination of Egyptian mummies has shown that the ancient Egyptians suffered all the diseases with which we are familiar today. Pathologists who have examined them have found clear evidence of such diseases as bone tuberculosis and bone cancer, various kinds of arthritis, bone infection (*osteomyelitis*), the now very common arterial disease of atherosclerosis, and various lung disorders such as pneumonia, anthracosis and pleurisy.

Galen and Hippocrates, depicted in a 13th-century fresco in the domed church in Anagni, a small city near Rome. (AKG)

Early theories of disease causation

Prior to the work of the Greek physician Hippocrates of Cos (c. 460–377 BC) most people seem simply to have assumed that disease was caused by evil spirits or the malice of the gods. Hippocrates rejected this view and sought to find a mechanistic explanation. To this end he adopted and extended an ancient Greek theory and proposed that all disease was due to disorder of one or more of four humours – blood, phlegm, yellow bile and black bile. This idea was derived from an older philosophic notion that all matter was composed of earth, air, fire and water, and possessed the qualities cold, dry, hot and wet. Blood was warm and moist; phlegm cold and moist; yellow bile warm and dry; and black bile was cold and dry. Blood came from the heart; phlegm from the brain; yellow bile from the liver; and black bile from the spleen. The humoral theory was adopted by Galen (c. AD 130–201) and his ideas dominated medicine, not always to its advantage, for fifteen hundred years (▷ p. 34).

During this period it was believed that when the humours were well blended the body was healthy; when there was an excess or deficiency of any one or more, the result was disease. Too much phlegm in the lungs caused consumption (tuberculosis); too much in the bowels caused dysentery. An excess of blood caused apoplexy (stroke); too much yellow bile caused jaundice, while excess black bile brought out melancholy (depression). A gross and uncontrollable imbalance in the humours, it was thought, caused cancer. Depending on which humour predominated, a person's personality would be shaped accordingly. The individual could be sanguine (blood), phlegmatic (phlegm), choleric (yellow bile) or melancholic (black bile) – terms that are still in use today, in literary, if not in medical, contexts.

The dawn of pathology

Even before the publication, in 1628, of the English physician William Harvey's germinal work *Movement of the Heart and Blood* (▷ p. 35) there had been some scientific investigation into the nature of disease. The first textbook of pathological appearances (morbid anatomy), by the Florentine surgeon Antonio Benivieni (1443–1502), was posthumously published in 1507. This was called *On the Hidden Causes of Disease* and included the findings at 20 postmortem examinations and an account of 100 clinical cases. The French physician Jean Fernel (1497–1558) published an important medical textbook in 1554, a large part of which was devoted to pathology. Nevertheless Fernel believed implicitly in the humoral theory and these efforts had little real influence on knowledge of the causes of disease.

The seeming authority with which Galen had propounded the humoral theory, and the subsequent influence of his writings, held back any real advance in pathological understanding until Harvey (1578–1657) showed that Galen had been wrong in some very fundamental particulars. This broke the hold Galen had had over medical thinking and led to the abandonment of the humoral theory.

The most important figure after Harvey was the Italian anatomist and founder of modern pathology Giovanni Battista Morgagni (1682–1771). At the age of 30 Morgagni became Professor of Anatomy at the University of Padua and held the appointment almost until he died at the age of 90. His most important work, the monumental *De Sedibus et Causis Morborum* ('On the Sites and Causes of Disease'), published in 1761, contained the results of his examination of about 700 cases. In all of these he correlated his postmortem findings with what was known of the patients' illnesses during life – a procedure now called *clinicopathological study*. In 1793 Matthew Baillie (1761–1823), a Scotsman practising in London, published a well-illustrated book *Morbid Anatomy*. This included an account of emphysema in the lungs of the lexicographer, Dr Samuel Johnson.

The microscope was first used to investigate pathological changes in tissues by Marie François Xavier Bichat (1771–1802). His book *Traité des Membranes (Treatise on Membranes)*, published in 1800, opened up a new and important field in the understanding of the nature of disease processes. Bichat was well before his time in his perception that disease was a matter of changes in the body's cells rather than in the shape, size and functional capacity of its organs. Today scientists would qualify his view by asserting that disease is essentially a cellular and biochemical phenomenon.

Louis Pasteur and the germ theory of disease

The French chemist and founder of the science of microbiology Louis Pasteur (1822–95) must be considered one of the greatest figures in the development of pathology and of modern medicine. He was appointed Professor of Chemistry at the Lille Faculty of Science in 1854, and soon afterwards was asked by the French government to investigate the reason why, under certain conditions, wine goes sour – a matter of huge economic importance. In a series of experiments he was able to show that if a substance in a flask was made sterile by boiling, and if air was prevented from reaching it, no fermentation or putrefaction would occur. If contact with the outside was allowed, though, these changes did occur. Clearly, something was entering the substance from outside the flask. Later he was able to show that the fungus *Mycoderma aceti* was responsible for turning the wine to vinegar. But it was found that heating the wine to 55° C (131° F) prevented the problem. This process, which became known as *pasteurization*, was then applied to beer and milk, thereby saving humanity much suffering from bovine tuberculosis. Needless to say, the brewers also gained financially from the discovery.

On the basis of this work came the recognition that a great many diseases were caused by invisible 'germs' (microorganisms). The English surgeon Joseph Lister (1827–1912) read about Pasteur's discovery and, following his principles, dictated practices that were to change the face of surgery. Lister adopted *antisepsis* (the destruction of microorganisms) using scrupulous cleanliness and a spray of carbolic acid to kill germs, but this was soon followed by a method in which organisms were avoided altogether by the creation of a germ-free environment (*asepsis*).

Pasteur went on to demonstrate that all organisms were derived from other organisms and that the idea that they arose by a process of spontaneous generation was wrong. He went on to identify the bacillus responsible for the serious disease anthrax. Then, inspired by the work, eighty years earlier, of Edward Jenner (⇨ pp. 58–9), he showed that immunization against the disease was possible with a strain of anthrax whose virulence had been weakened by repeated passage through a series of animals. His researches provided a logical basis for hygiene and sanitation, and he showed how sterilization by heat or other means could prevent the spread of disease. In 1882 Pasteur established the cause of rabies and showed that it was transmitted by an organism too small to be seen with a microscope – a virus. Three years later he then made an effective vaccine against the disease. In 1888 Pasteur was appointed the first director of the new Pasteur Institute in Paris.

Robert Koch and modern bacteriology

Robert Koch (1843–1910) was a German country doctor practising near Breslau, Silesia (now Wroclaw in Poland), when an epidemic of anthrax broke out in the local cattle. Koch, who was familiar with Pasteur's recent research on the subject, decided to investigate, and discovered that he could culture the bacillus responsible, in warm blood serum (the clear liquid left when blood is allowed to clot), outside the animal body. In this way Koch was able to follow the life cycle of the *Bacillus anthracis* and show how it formed spores that enabled it to survive in adverse conditions. Now intensely interested, Koch contacted a bacteriologist at Breslau University who encouraged him to proceed and to give what was a very influential lecture on the life cycle of the bacillus at a seminar in Breslau in 1876.

Koch moved to Berlin where he soon showed that bacteria could be much more easily seen under the microscope if they were first stained with special dyes. He then went on to develop an improved method of cultivation of organisms using a solid medium of agar-agar (a gelatinous substance obtained from seaweed) mixed with a nutrient broth. Growing in these circumstances, bacteria formed colonies readily visible to the naked eye. Pure subcultures could then be made. In 1882 Koch identified the dreaded tubercle bacillus – the cause of tuberculosis – and then in 1883 he isolated the organism responsible for cholera. In 1885 he was appointed Professor of Hygiene at the University of Berlin.

Louis Pasteur, here seen holding two of the rabbits used in his research into rabies. Cartoon from *Vanity Fair*, London (1887). (Image Select)

Koch laid down a set of rules for establishing that a particular disease was caused by a certain organism, rules now known as 'Koch's postulates'. To fulfil them, four criteria must be satisfied. The organism must be present in every case; it must be possible to make a pure culture of the organism; an animal inoculated with the cultured organism must develop the disease; and the same organism must be recoverable from the infected animal and again grown in pure culture.

Following Koch's example and methods other scientists quickly established the organisms responsible for a wide range of diseases such as boils, diphtheria, gonorrhoea, leprosy, meningitis, plague, pneumonia, syphilis, tetanus, tonsillitis, typhoid fever and *undulant fever* (which is characterized by lethargy, and aches and pains).

SEE ALSO

- PROTECTION AGAINST ATTACK pp. 58–61
- HEALING AND REPAIR p. 62
- DISEASE IN HISTORY p. 118
- WHAT IS DISEASE? p. 120
- WHAT IS HEALTH? p. 122
- PREVENTING DISEASE p. 124
- HEALTH CARE IN THE PAST p. 150

Disease in History

From earliest times disease has been a major influence on the history of humankind, and it is only in recent years that disease has ceased to be the major determinant of life expectancy. In particular, population size has been greatly affected by disease – far more than by war or famine. The epidemics that have ravaged humanity have also had immense social, political and economic effects.

During periods of peace and freedom from the plague human populations have grown rapidly; but in times of plague they have decreased at a rate that has sometimes threatened their extinction. However, this is only one of the ways in which history has been affected by disease. The actions of tyrants and dictators who have determined the fate of nations have sometimes been influenced by diseases they suffered. Wars have been lost by disease, and conquered territories have had to be abandoned for the same reason. The very evolution of humans has been to a large extent influenced by disease.

Small populations are less susceptible to severe epidemic disease than large and overcrowded ones. Any increase in population predisposes to infectious disease because proximity increases the chance of infection and thus larger populations are exposed to greater concentrations of the infective agent. Therefore plagues are a very efficient means of preventing population growth. They are also very influential in limiting social and economic development. In times of pestilence life is short. Social advances are not made by children or adolescents, so if life expectancy is only 20 or 25 years, the average period of individual productivity is very brief.

Epidemics – outbreaks of a particular disease in a large number of people over a short period of time – have ravaged humanity from the dawn of history. Thucydides (471–400 BC) in his *History of the Peloponnesian War* provides a graphic description of one – probably plague – in Athens in 430 BC. 'Men in perfect health were seized in a moment with violent fever, inflammation of the eyes, sore throat and tongue. The breath was foetid, there was coughing and vomiting, violent convulsions, livid skin breaking out in pustules and ulcers . . . Either they died on the seventh or ninth day or, if they lived, the disease descended into the bowels and there produced violent ulceration, severe diarrhoea and later exhaustion that carried them off.'

The Black Death

Bubonic plague is a disease of rats that is spread to humans by rat fleas when their hosts die. In people who survive the bubonic stage it can become septicaemic and pneumonic. Pneumonic plague is highly infectious, being spread by coughing. There were epidemics of plague in Rome in the years AD 68, 79, 125 and 164, the last of which persisted for 16 years and threatened to destroy the army. Tacitus records that, 'the houses were filled with corpses and the streets with funerals'. Another devastating outbreak occurred in south-east Europe in the 6th century AD. Up to one third of the population of the Byzantine Empire may have died. The Black Death – mainly bubonic plague – appeared again in Europe in 1347. After such an interval it found a highly susceptible population and for four years the plague took a greater toll of life than any previously recorded disaster. Spread to Europe is said to have been caused deliberately by Tartar soldiers who, besieging a trading post in the Crimea, catapulted plague-infected corpses into the town. The disease appeared the same year in Sicily and within two years it was appearing all over southern and western Europe and soon spread to Scandinavia and the Baltic states, ravaging populations wherever it arose. Further outbreaks were to occur at regular intervals up to the beginning of the 17th century in England and the 18th century in continental Europe.

The virulence varied. Estimates based on local archives suggest that between one eighth and two thirds of the affected populations died from the disease. The contemporary French historian Jean Froissart (1333–1419) believed that about one third of the population of Europe – some 25 million people – succumbed in the first epidemic, an estimate that is now generally accepted. Local depopulation was ubiquitous: a thousand English villages disappeared. There were too few labourers to cultivate the land, much farmland was abandoned, and landowners ruined. Wages and prices were forced up, and serfdom had to be dispensed with. In spite of measures introduced by governments to maintain their economic power, the surviving peasantry enjoyed greatly improved conditions. Later outbreaks were not as devastating overall, but still hit individual communities equally catastrophically.

Plague and superstition

To many people, the visitation of the Black Death was God's punishment for sin. Religious teaching often reinforced this view and people were taught to accept its horrors with resignation and to await the end of the world, which the plague presaged. Many people tried to deflect divine wrath by prayer, fasting and self-flagellation. Many others gave all their possessions to the Church. But not everybody was so amenable. Faced with the imminent threat of death some people decided to get what pleasure they could out of life and adopted a philosophy of 'Let us eat, drink and be merry for tomorrow we die.'

Inevitably, the Jews, who were hated by non-Jews for their usury, were accused of being the cause of the plague or of being responsible for spreading it. It was

THE SPREAD OF THE BLACK DEATH 1346–53

INITIAL FOCUS 1346
CRIMEA
Kaffa
Constantinople
Genoa
Marseilles
Messina
Alexandria
INFECTED AREAS
1348
1349
1350
1351
1352
1353
SPREAD
1347

THE GREAT PLAGUE OF LONDON

In late 1664 the plague made one of its periodic returns to England. Writing 50 years later Daniel Defoe recorded in his *Journal of the Plague Year* how in December of that year, 'Two men, said to be Frenchmen, died of the plague in Long Acre, or rather the upper end of Drury Lane . . . in the last week in December 1664 another man died in the same house, and of the same distemper.' Over the ensuing weeks a steady rise in the number of deaths in the city alarmed the inhabitants. By mid-July 1665 the weekly mortality bill showed 71 deaths in the area now known as Tower Hamlets, but this was nothing to what was to come. In the week 8–15 August 3880 deaths from plague were recorded. The figures for ensuing weeks were: 4237, 6102, 6988, 6544, peaking in the week 12–19 September at 7165. Between 8 August and 10 October 49 705 people died of the plague.

Again, Pepys: 'December 31st. – Many of such as I knew very well, are dead; yet to our great joy the town fills again, and shops begin to open again. Pray God continue the Plague's decrease . . .'

Pepys's prayer was answered and the number of deaths fell as rapidly as it had risen. Nine months later London suffered the worst fire in its history, but a great purification. Between 2 and 5 September 1666 much of the city, including 13 000 houses, was destroyed.

Fleeing the plague during an outbreak in Europe in 1630. The better-off often left cities during the plague for the relative safety of the less densely populated countryside. (ME)

claimed, on absurd evidence, that they were poisoning wells or practising black magic. Pogroms, torture and execution abounded and thousands were burnt in supposed retaliation.

Other diseases

Although, by virtue of its high infectivity rate and extraordinary virulence, the plague has been by far the greatest killer of humankind, it has not, of course, been the only major disease in history. Smallpox has been endemic in almost all populations, and, until it was eradicated in 1977, caused millions of deaths and much disfigurement from scarring. Where infectious diseases, such as measles, have been introduced into populations that had never known the disease and that were, in consequence, highly susceptible, the results have been devastating. The Carib indians of the Caribbean, for instance, were wiped out in this way in the 16th century, and the Mexican Indian populations were devastated by smallpox introduced by the Spaniards. Similarly, the introduction of syphilis into populations with no immunological protection has led to innumerable deaths and considerable suffering. Cholera, spread in water contaminated with human excrement, has also been endemic in many cities and has caused major epidemics. More than 500 people died in 10 days from cholera acquired from the Broad Street pump in Golden Square, London, in 1824; and there were nearly 4500 deaths from cholera in New York in 1849.

Attempts to move people into areas where disease organisms and their *vectors* (transmission agents) are common have sometimes been disastrous. The first attempt to cut the Panama canal, for instance, had to be abandoned in 1889 after thousands of workers had died from yellow fever. It was not until stringent public health measures were taken to control the *Aedes aegypti* mosquito, that carried the virus, that the work could safely be resumed and the canal finished in 1914. Changes in the virulence of existing infective organisms has also had horrifying effects on populations with no natural resistance. This was the cause of an appalling epidemic of influenza that swept through Europe in 1919 killing even more people than had died as a result of the 1914–18 War.

AIDS – the modern plague

In June 1981, in Los Angeles, a strange outbreak occurred of a very uncommon disease – a form of pneumonia caused by an organism called *Pneumocystis carinii.* All five of the patients involved were homosexual men. Within a short time further unusual outbreaks were discovered, this time of another, quite different disease – Kaposi's sarcoma. Initially 26 cases were reported, occurring in New York and in California. Again, all these patients were homosexual men. In addition to Kaposi's sarcoma, four of these men also had *Pneumocystis carinii* pneumonia. These events presaged a new disease that would tax the ingenuity and demand the dedicated labour of the best medical minds in the world.

Soon the numbers of such cases were doubling every six months. In addition to pneumocystis pneumonia and Kaposi's sarcoma, these men were developing widespread herpes and thrush infections, tuberculosis and a variety of infections with organisms normally hardly able to attack humans. It was not long before it became clear that the men were suffering from a rare but recognizable clinical syndrome – immune deficiency. They were developing these rare 'opportunistic' infections and cancers because they had lost their resistance to them. The term *acquired immune deficiency syndrome,* soon abbreviated to AIDS, was coined.

In 1983, scientists in Paris and Maryland were able to establish that AIDS was caused by a virus of a type related to other known viruses. It is called the human immunodeficiency virus (HIV), and is spread mainly by sexual contact. HIV is an RNA virus that carries an enzyme, reverse transcriptase, that enables it to make DNA copies of itself in the helper-T cells (➪ pp. 58–61) it invades.

The initial infection is usually symptomless, and the interval between infection and the first appearance of symptoms is 1–10 years. The effects vary. Some 50% develop the full-blown disorder with serious opportunistic infections and sometimes Kaposi's sarcoma. About 30% develop the AIDS-related complex (ARC) featuring weight loss, thrush, diarrhoea and fever. This commonly proceeds to the full syndrome. In over 50% of cases the brain is affected, either directly by the virus to cause dementia, or by opportunistic organisms such as toxoplasmosis, cytomegalovirus, herpes, fungi or tuberculosis.

We now know that AIDS appeared first in the homosexual community simply because, at the time, that was the only group engaged in a sufficiently high level of sexual promiscuity to ensure adequate transmission of a relatively small number of a new virus. As the virus has become commoner and more people have been infected, heterosexual transmission has become more frequent. In highly promiscuous and heavily infected groups AIDS is now spread as readily by heterosexual as by homosexual activity.

Once a person becomes infected that person is infectious for life. Eighty per cent of people with AIDS die within two years of diagnosis. Although all the various manifestations of AIDS can be treated and life prolonged, no specific treatment for HIV infection has appeared, and no effective vaccine has yet been developed. In the United States, AIDS is now second to heart disease as a leading cause of death, and in some parts of Africa AIDS has become the leading cause of death in adults. The World Health Organization estimated in 1993 that 14 million people were infected with HIV. Nearly all of these will develop the syndrome.

SEE ALSO

- PROTECTION AGAINST ATTACK pp. 58–61
- HEALING AND REPAIR p. 62
- WHAT IS DISEASE? p. 120
- TRANSMISSION OF INFECTION p. 130

Queen Elizabeth I of England painted by Nicholas Hilliard (1537–1619) A common theme in contemporary writings on plague and disease was that they were no respecters of rank. Queen Elizabeth I bore the marks of smallpox on her face but covered them up with a thick layer of powder. (BAL)

What is Disease?

Literally construed, the term 'disease' (from the Latin *dis* 'not', and the Old French *aise* 'comfort') may be misleading, as the early stages of many serious diseases do not produce any symptoms at all. Even so, disease is generally thought of as any condition involving an impairment of function and tending to produce observable effects.

Organic disease involves physical changes in the body's tissues. These changes may be temporary and reversible – changes such as those of mild inflammation (redness and swelling from widening of nearby blood vessels) – or structural and permanent. Permanent changes in tissues and organs are not necessarily serious as the body possesses remarkable powers of repair and, in some cases (as in that of the liver; ⇨ p. 48), of regeneration, so that recovery of full function and appearance may follow even a serious injury. But in many cases, structural damage from disease is of grave and life-long consequence. The implications of such structural change depend largely on the importance of the part affected. Severe skin wounds may heal well and have a permanent effect that is cosmetic only. But repeated small insults, for instance, to the brain or heart muscle because of an inadequate blood supply, will lead, on the one hand, to dementia (progressive loss of brain function), and, on the other, to heart failure.

The human body is a complex of interrelated systems with a physiology that is dedicated to maintaining the healthy status quo. This is achieved by an elaborate and effective set of control mechanisms that operate on a constant supply of information and use feed-back loops (servo-mechanisms). These are both chemical and neurological (⇨ pp. 36 and 74–81). The normal physiological balance so achieved is called *homeostasis* (literally, 'the same state'), and the effect of serious disease is to upset this state. A major breakdown of homeostasis – as, for instance, when the acidity of the blood moves outside narrow limits, or when the levels of sugar or oxygen in the blood drop too low – is often fatal.

The causes of disease

The causes of disease are legion but can, nevertheless, be usefully classified. Diseases may arise as a result of causes originating in the body of the individual (also called endogenous, primary or essential causes), or as a result of external causes. Endogenous disorders may be communicable to offspring (since many of them are of genetic origin), but are not communicable to contemporaries. Exogenous disorders, such as infectious diseases, however, are often communicable to others. Many exogenous disorders are the result of environmental factors that can operate in common on large sections of the population, while others are due to special localized environmental factors. Industrial diseases are of the latter kind in the majority of cases.

Congenital abnormalities

Causes of disease may operate on the individual before or after birth. Congenital abnormalities – structural defects present at birth – may affect almost any part of the body and their results are often permanent. They range from the trivial to the devastating. Most of us are born with some minor abnormality that is likely to go unnoticed, but some people have the misfortune to be born with seriously defective hearts, incomplete spinal columns, cataracts, abnormal kidneys, and so on. The causes of these abnormalities may be genetic, environmental, or a combination of both.

Some congenital abnormalities are the result of external influences, operating during the pregnancy, that lead to failure of normal development. These influences include infections (especially viral), the toxic effects of poisons such as alcohol and the absorbable products of cigarette smoking, and the side effects of certain drugs. If the mother is infected with the German measles (*rubella*) virus early in her pregnancy her child may suffer from a range of physical defects, including severe congenital heart disease, opacities of the corneas or crystalline lenses (*cataract*), an abnormally small brain (*microcephaly*), or deafness. Excess alcohol intake during pregnancy causes a general retardation of the growth of the baby's brain and body. And if the mother smokes during pregnancy fetal growth may be adversely affected, possibly with long-term consequences for the child's intellectual development. One of the most striking examples of drug side effects was the tragedy of the 1960s in which an allegedly safe sedative drug (thalidomide or Distaval) interfered so severely with the early development of the fetal body that thousands of babies were born with gross abnormalities, especially small 'flippers' in place of limbs (*phocomelia*).

Genetic causes of disease

Some diseases have genetic origins because of defects in the DNA (⇨ p. 24) of cell nuclei or mitochondria (⇨ p. 38), or

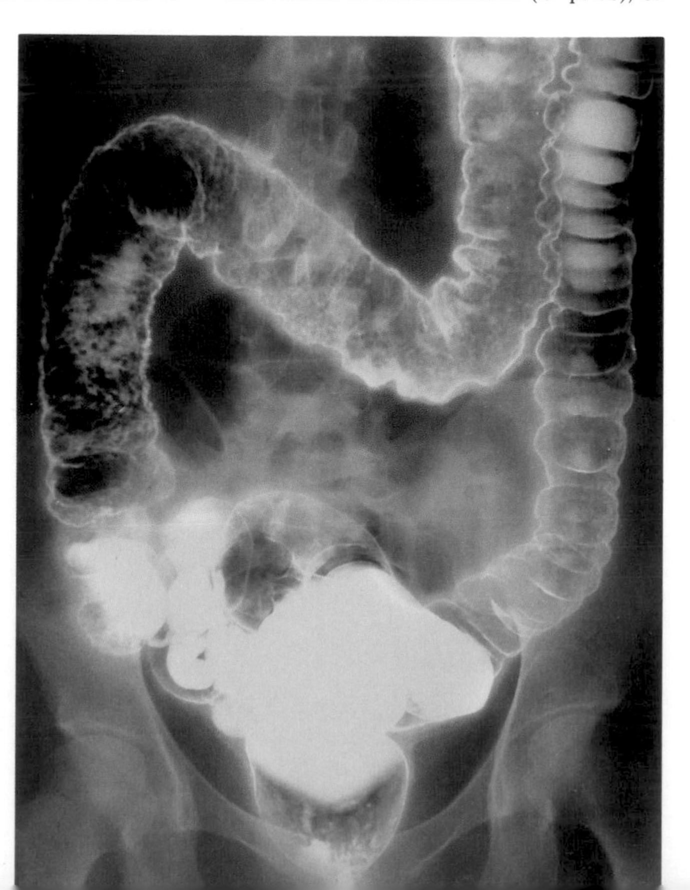

The term 'disease' covers all disorders of the human organism, from the minor and temporary, such as inflammation caused by a bee sting, through to the serious and possibly life-threatening, one example being ulcerative colitis (right). This condition, characterized by painful inflammation of the colon and recurrent diarrhoea, predisposes to cancer. (RB/SPL)

from abnormalities of entire chromosomes (⇨ p. 22). As a result of these defects, genes are either ineffective or missing. Mutations in the DNA may have occurred in the parents of the afflicted person, or may have been transmitted through several generations (⇨ p. 28). They may also be due to radiation, to replication errors or to other, unknown, causes. It is possible, for instance, that viruses or toxic substances may be involved in causing mutations. Many single- or multiple-gene defects cause enzymes to be absent or defective, giving rise to metabolic disorders. Several thousand such diseases are known, and these, too, may range from the minor to the disastrous. Their effects may be present at birth or may not appear until later in life. Whole chromosome abnormalities tend to cause more marked and widespread effects, as exemplified by Down's syndrome, which is the result of an additional chromosome number 21 (trisomy 21).

Physical injury

Physical injury implies much more than the simple application of mechanical force as it includes the effects of heat, cold, high-voltage electricity, sudden air compression (blast), high-velocity missiles, the whole spectrum of electromagnetic radiation (including light, microwaves, x-rays and gamma rays), and cosmic rays. Physical injury causes damage to tissues in a number of ways: by fracture, contusion (bruising), laceration (tearing), incision, burning, freezing with ice crystal formation, coagulation of protein and blood, tissue disruption, and direct breakdown of biochemical molecules by means of ionization. So long as infection is avoided, the body's healing response to mechanical injury is usually good. The immediate effect of ionizing radiation depends on the dose. Long-term effects, however, include cancer and genetic mutation.

Chemical injury

The most important chemical causes of disease are poisoning and the effect of corrosive substances on the tissues. Poisoning is brought about either by the chemical action of certain substances that damage cells when present in very small quantities, or by certain other substances, almost unlimited in number, such as water, salt, vitamins A and D and many minerals and metals, that are normally harmless, or even essential, but are damaging when present in excess. Most useful drugs are poisons and must be taken in a strictly limited amount – this is known as the *therapeutic dose*. The ratio between the useful and the dangerous dose is known as the *therapeutic index*. For most drugs this is large and offers a fair margin of safety, but for some, the therapeutic index is small and great care is needed when prescribing and measuring the dose. Many poisons are specific in their function and act on only certain tissues. Many are limited in their effect because the body can eliminate them rapidly, while others become dangerous because the body does not excrete them and they accumulate gradually. Several sorts of bacteria cause their damaging effect on the body by releasing powerful poisons known as *toxins* (⇨ below).

The action of poisons is often to inactivate cell enzymes so that vital biochemical processes are blocked. Very poisonous substances include arsenic, lead salts, salts of mercury, cyanide (potassium cyanide or hydrocyanic acid gas), methyl alcohol and various plant and bacterial toxins. The study of the action of poisons is called *toxicology*.

Infectious agents

Infectious agents can be divided into two broad categories – the microscopic and the visible. Of all the vast range of microorganisms known to exist, only a tiny proportion normally cause disease. Even so, these disease-causing (*pathogenic*) organisms cover a wide range, including thousands of different strains of viruses, rickettsia (microorganisms intermediate in size between viruses and bacteria), bacteria, fungi, single-celled parasites (protozoa), amoebae and microscopic worms (microfilariae). Visible infectious organisms include a large range of worm parasites that inhabit the intestine, bloodstream, liver, lungs, muscles, brain and other parts.

Viruses live and reproduce within cells, causing their death by interfering with their normal functions. Huge aggregations of viruses in cells may be visible microscopically as *inclusion bodies*. Viruses that have replicated within cells then break out and invade further cells. In the case of pathogenic bacteria, however, three key characteristics apply: they function best at body temperature; they are able to invade the body; and they carry toxins that destroy cells or interfere with their functions. *Exotoxins* are those poisons that are released by bacteria and can act remotely. *Endotoxins* act at the site of bacterial concentration. Unlike viruses bacteria are found inside cells only when they are being destroyed – by *phagocytes* (cells that form part of the body's immune system; ⇨ p. 58).

Endotoxins are lipopolysaccharides (molecules composed of lipids and polysaccharides), while exotoxins are made of proteins. The diphtheria toxin, for instance, is an exotoxin with two polypeptide chains (⇨ p. 36). One of these punches a hole in the cell membrane allowing the other to enter and to attach itself to the factor that moves ribosomes (⇨ p. 26) along messenger RNA and inactivates it, thus preventing the cell from manufacturing any protein. Botulinum toxin is probably the most powerful poison known to humanity – a quarter of a gram (0.009 oz) could kill about 60 million people, approximately equal to the population of, for example, the UK. This toxin acts by binding to the *synapses* between nerves and muscles and preventing the release of the neurotransmitter acetylcholine. The result is complete paralysis and total loss of motor function.

DEGENERATIVE DISEASE

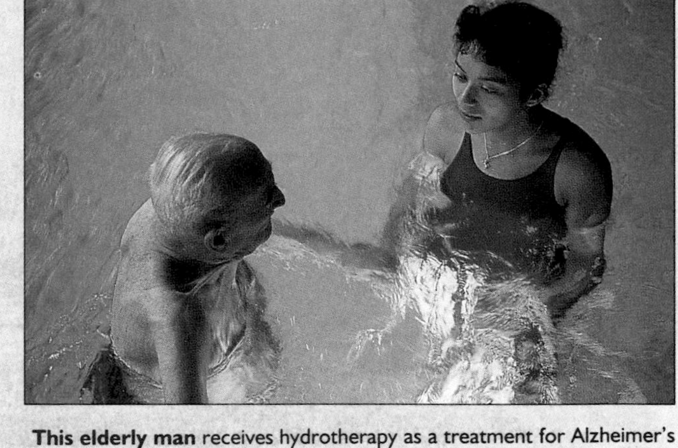

This elderly man receives hydrotherapy as a treatment for Alzheimer's disease, the most common type of dementia in old age, which is characterized by memory disorders, personality changes and confusion. Although Alzheimer's disease is incurable it is thought that its progress may be prevented by physical and mental exercise. (Catherine Pouedras, SPL)

Many changes occur in the body with increasing age, and it is a matter of debate which of them can be thought of as 'ageing' and which 'degenerative disease'. Perhaps the most obvious change is in the skin, which loses its elasticity and becomes thin and wrinkled, bruises easily and becomes unevenly pigmented. The joints – especially those subjected to weight, such as the hip and knee – wear out due to loss of cartilage (causing the condition osteoarthritis). In addition, the arteries lose their elasticity, veins become tortuous (twisted) and dilated, and the blood pressure tends to rise. A variety of 'diseases' result from arterial changes; some of these can be prevented (stroke, for instance, by the treatment of high blood pressure) but others cannot.

Fungal infections of the skin (*epidermophytoses*) are very common but internal fungal disease is serious usually only when the immune system has become defective (a condition called immunodeficiency). Diseases caused by protozoa, such as malaria, leishmaniasis (a group of infections affecting the skin and internal organs), toxoplasmosis (an infection often acquired before birth that may affect the nervous system or other organs) and trypanosomiasis (sleeping sickness) are very common in the developing countries and are the cause of an enormous number of deaths. Amoebic dysentery and worm infestations also take a huge toll in human health.

Neoplasm

Neoplasm is the general term given to an abnormal proliferation of cells that serves no useful purpose. Neoplasms – commonly called 'growths' or 'tumours' – may be benign or malignant, depending on whether they remain localized and encapsulated or invade tissue and spread. The latter group are known as cancers (⇨ p. 140). There are hundreds of different kinds of neoplasm, and the causes of most of them are still unknown. Some, such as the benign skin tumour, the wart, are known to be caused by viruses. And there is a strong suspicion that a number of cancers, especially Burkitt's lymphoma, are also caused in this way. Other cancers are known to be caused by exposure to radiation and by a wide range of chemical substances (known as carcinogens).

SEE ALSO

- HEREDITY pp. 20–31
- WHAT THE BODY IS MADE OF p. 36
- CELLS, THE STUFF OF LIFE p. 38
- PROTECTION AGAINST ATTACK pp. 58–61
- HEALING AND REPAIR p. 62
- HEALTH AND DISEASE pp. 116–19 and 122–59
- MENTAL DISORDERS pp. 184–95

What is Health?

Health is generally considered to be the state of excellent functioning of body and mind, unhindered by disease. The term itself means 'wholeness', and derives from the Old English word *haelth*, meaning 'whole'. And it is related to the verb 'to heal' – that is, 'to make whole'.

The Greeks did not consider health to be a single entity and used two separate words – *heuexia*, meaning 'a good habit of body', and *hygieia*, meaning 'a good way of living'. They were also alive to the importance of the distinction between mind and body in this context. In the Platonic dialogue *Charmides*, Socrates argued that a physician who tried to heal the body without also healing the mind was a fool. One should, he said, attend first to the state of the mind, for health was dependent on virtue. Being well in body required that we should establish good habits of living.

The World Health Organization's (WHO) official definition of health reads: 'Health is a state of complete physical, mental, and social wellbeing and not merely the absence of disease and infirmity.' It is not at all easy to quantify health in this positive sense, however, and, for practical purposes, health professionals, especially community physicians and statisticians, must rely on negative indices – such as mortality rates and illness and disability figures – in trying to assess the health of a community.

From the individual point of view, however, some rough measure of quantification is possible. Health implies 'fitness' – the ability to perform and to live a full life. Large numbers of people are free from organic disease but are so unfit that only the most sedentary of lifestyles is possible for them, and such people cannot be considered healthy. Positive health is usually associated with a feeling of wellbeing and is almost always the result of high motivation for living. There is an important relationship between lack of fitness and ill health; ultimately the kind of lifestyle that results in unfitness is apt also to be the kind that leads to actual disease. Factors that adversely affect a person's fitness levels include physical idleness, an unsuitable diet, overeating, drinking too much alcohol, and smoking.

The meaning of fitness

Fitness is essentially a matter of the amount of exercise an individual can tolerate. The concept of exercise tolerance is important as this allows useful comparisons of the state of fitness to be made. It is also important as a way of assessing heart and respiratory-system disease. The exercise tolerance scale is a kind of fitness hierarchy with long-distance athletes at the top and bed-ridden cardiac patients at the bottom. The scale may be calibrated in terms of the distance that can be travelled, at a brisk pace, before the person has to stop because of severe breathlessness, or other effects. Most people are somewhere near the bottom of the scale.

People who are unfit but free from disease can, whatever their age, readily increase their tolerance by taking regular exercise. Even people with a heart or respiratory condition can often also do so, but close medical supervision is necessary. In the investigation of such cases doctors will always try to assess the exercise tolerance. A patient with *angina pectoris* (a condition in which the coronary arteries are unable adequately to supply the heart with blood) may be able to walk 50 m before being stopped by pain. Later, if the condition gets worse, he or she may be able to go only 20 m before the symptom occurs. Fit people can easily run up several flights of stairs, two or three steps at a time, while those who are less fit can only walk, one step at a time. Some instinctively avoid stairs, finding that even talking is difficult while climbing them, because of breathlessness.

Breathlessness and a fast pulse rate are the normal response to exertion. Their degree, however, for a given amount of exertion, varies widely between people of different standards of fitness. This is because of the many physiological and even structural changes that occur in the heart, lungs and muscles as a result of training (⇨ p. 54). Because of their higher cardio-respiratory efficiency, athletes usually have slow resting pulse rates, and

Health is a relative term. Even though this young man is HIV-positive (is infected with the virus that leads to AIDS) he can consider himself to be healthy until (or unless) he begins to develop AIDS-related symptoms. He is undergoing computer-monitored exercise testing at the Institute for Immunological Disorders, Houston, Texas. (Hank Morgan, SPL)

a long-distance runner may have a resting pulse rate of as low as 40 beats per minute. Above a certain heart rate, pumping efficiency drops, so if a heart can work efficiently up to 160 beats per minute (which is normal) an unfit person will only be able to exert it to the point where it is working at twice its rate when resting, without being stressed, while an athlete can quadruple its rate. Circulatory efficiency is not, however, simply a matter of heart rate. The important parameter is cardiac output – the amount of blood that can be pumped in a given time. This is the product of heart rate and stroke volume (the volume pumped per contraction). Stroke volume varies with the power and efficiency of the contraction of the heart muscle with each beat. In a trained athlete, the heart output may rise from the resting value of 5 litres (approximately 9 pt) per minute, to a level as high as 35 (62).

How to assess fitness

The degree of breathlessness on effort is a sensitive index of the degree of fitness. Men or women well into middle age, who are not particularly concerned with athletic pursuits, should still be able to do the following:

- Walk ten miles.
- Walk at a brisk pace for a mile without embarrassing breathlessness.
- Continue after this, at a more normal pace, without difficulty.
- Run up a flight of 20 steps in seven bounds with no overt sign of distress.
- In an emergency, run 30 m (about 33 yd) in under seven seconds.

The significance of symptoms

The list of possible symptoms is very long, but there is a small group of commonly occurring ones that may cause particular concern. Some idea of the possible significance of these factors may have a bearing on the assessment of the state of an individual's health.

Breathlessness. Apart from simple unfitness, breathlessness has many possible causes. A sudden change in the degree of breathlessness for a given amount of exertion, or breathlessness at rest, are always significant and require investigation. Other causes include: smoking, an over-full stomach, well-advanced pregnancy, obesity, anaemia, general debility, heart disorder, pneumonia or other respiratory infections, emphysema (break-down of the lung sacs so that a smaller area is available for oxygen transfer), asthma, spontaneous pneumothorax (local rupture of the lung covering so that air escapes from the lung and collapses it), and obstruction of the larynx or trachea (windpipe).

Chest pain. The chief concern with this symptom is that it may signify heart disease (⇨ p. 138). This is essentially a male problem because women are protected by their sex hormones until after the menopause and enjoy a high degree of immunity to coronary artery disease. Most chest pains have an innocent explanation, however, and pain that is constantly, or frequently present, unrelated to exertion, emotion or stress, and that in no way interferes with activity, is unlikely to be due to a heart problem.

Common causes of chest pain include 'heartburn' (*reflux oesophagitis* in which acid wells up into the gullet from the stomach), spasm of the oesophagus, 'wind' from air swallowing in attempts to belch, hunger, part of the stomach pushing up through the diaphragm (*hiatus hernia*), inflammation of the lung coverings (*pleurisy*), inflammation of the heart's outer capsule (*pericarditis*), a virus infection causing pain in the chest muscles (*Bornholm's epidemic myalgia*), pain in the nerves supplying the respiratory muscles between the ribs (*intercostal neuritis*), disease of the joints between the ribs and the breastbone, bronchitis or tracheitis, bulging (*aneurysm*) of the aorta, and cancer of the lung. This list is by no means exhaustive.

In the case of angina pectoris, the pain begins beyond a certain degree of exertion, and on resting it is quickly relieved. It is usually felt just behind the breastbone and may be very severe or no more than a vague ache. A pain of this kind that is not quickly relieved by rest may be due to a heart attack.

Headache. Headache is so common and brain tumours so rare that the chances of a particular headache being caused by a tumour are very small, and, in fact, three quarters of all headaches are caused by muscle tension. Of the remainder most are migraines (headaches due to changes in the arteries supplying the scalp and brain) and these usually involve visual disturbances. The third, very minor, group contains many different organic conditions that can cause headache. Headaches in this group are nearly always associated with other, often severe, symptoms. For instance, brain tumours commonly cause loss of peripheral vision, double vision, sudden projectile vomiting, weakness, loss of sensation, speech difficulties, and so on.

Other symptoms and signs. New, persistent or frequently recurring pain in any part of the body should be medically investigated. Other areas of concern include: an unexplained and persistent cough, coughing up or vomiting of blood, blood in the urine or in the stools, tarry black or stringy stools, any unusual or unexplained change in the bowel habit, urinary incontinence, burning pain on urination, undue frequency of urination with small output, or excessive urinary output with constant thirst. Any of these might indicate a serious problem and should be examined medically.

SCREENING TESTS FOR DISEASE

Routine screening has been the subject of controversy but most authorities would agree that if we could ignore economic considerations it would be highly desirable for everyone, however free from symptoms, to have a medical history taken and a routine physical examination performed at regular intervals by an experienced practitioner. In addition, many would recommend the following tests:

For women

- **Breast examination to detect lumps or other changes in the tissue.**
- **Mammography (special breast x-ray that can pick up tumours that cannot be detected by hand).**
- **Pelvic examination to pick up irregularities that might indicate infection.**
- **Pap (or cervical) smear test to detect precancerous changes in the cervix.**

For men

- **Rectal examination to find signs of an enlarged prostate gland.**

For all

- **Sigmoidoscopy (direct visual examination, through a tube or endoscope, of the inside of the lower colon to look for cancerous tissue).**
- **Stool test for *occult* (not visibly apparent) blood.**
- **Blood-pressure checks.**
- **Blood-cholesterol tests.**

If these tests pick up serious diseases, such as cancer, at an early enough stage, its eradication or control may be possible and relatively easy.

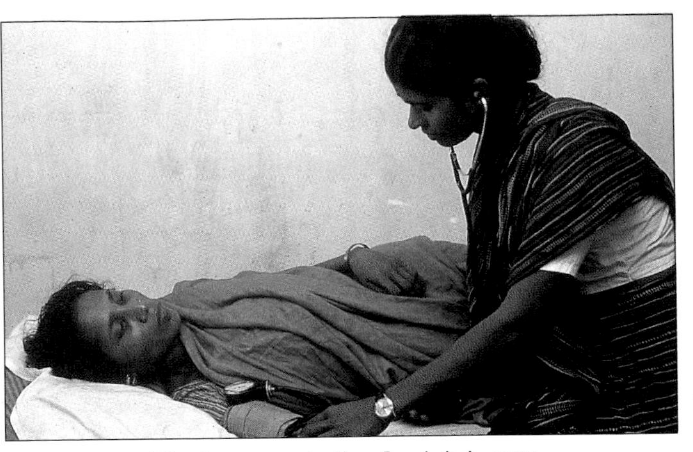

Blood-pressure testing, Bangladesh. (WHO)

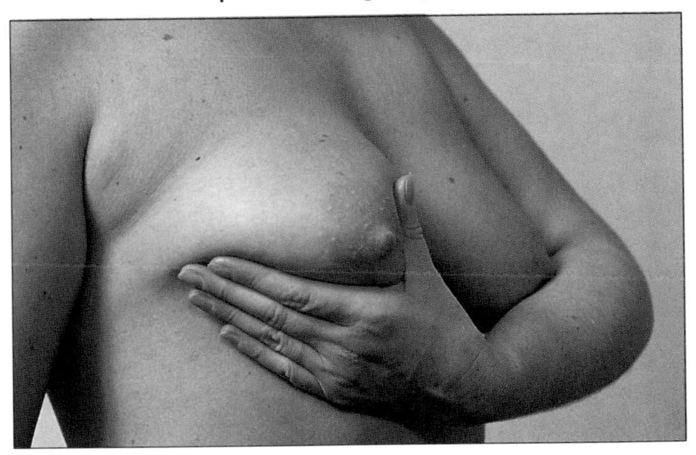

A woman should regularly check her breasts for unusual lumps, or for changes to the nipple (such as puckering of the skin or the appearance of a discharge). The best time for such a check, known as *autopalpation*, is about four days after the end of her period. (Gamma)

- FUEL AND MAINTENANCE 2: DIET p. 46
- EXERCISE AND BODY UNITY p. 54
- PROTECTION AGAINST ATTACK pp. 58–61
- HEALTH AND DISEASE pp. 116–21 and 124–59

Preventing Disease

Given present-day knowledge, many diseases are preventable. Indeed, the high level of health currently enjoyed in the West owes much more to preventive medicine and improved living conditions than to advances in therapy or the availability of high-quality medical care.

Preventive medicine, a complex and wide-ranging science, has had notable success. But, for various reasons, there are often delays in the application of knowledge and much of what is now known has not yet been exploited. In addition, the prevention of disease is unglamorous compared with successful treatment, and it is always the latter that hits the headlines.

Preventive medicine has a longer history than is commonly appreciated. When the work of Hippocrates (⇨ p. 116) demonstrated that diseases had natural or environmental causes, the way was open for the application of the principle that the promotion of public health depends to a great degree on standards of public administration. In his book *Airs, Water and Places*, and in the first and third book of *Epidemics*, one reads that it is necessary, in studying the diseases of an area, to consider the winds, water, soil, siting, local diet and other features. Hippocrates did not, however, go so far as to suggest that one could prevent disease by modifying causal factors, but the implication remained.

The most advanced civilizations of the ancient world were, in fact, able to do much to improve public health. The Romans constructed elaborate aqueducts to bring clean water to their cities, and built effective sewers, such as the Cloaca Maxima, to carry away human excreta. The most affluent were able to heat their houses in winter by means of underfloor fires that circulated hot air between double walls (the *hypocaust system*). After the decline of the Roman Empire (from about the 4th century AD) it was to be many centuries before such standards were to be achieved again.

Smallpox, a highly infectious viral disease, has now been eradicated thanks to a process of worldwide immunization. (WHO)

Control of infectious disease

Infectious disease in a community cannot be controlled by antibiotics alone but by the application of knowledge about the causes of disease, mode of spread, means of prevention of infection and means of active combat within the body against those infecting organisms that have gained access. The real weapons of preventive medicine are: personal cleanliness, socially responsible behaviour, restriction of access of exotic diseases to a community, the destruction of insect carriers of disease, an uncontaminated water and food supply, effective and safe sewage disposal, and widespread acceptance of immunization against disease. These have freed us from the worst of such diseases as plague, smallpox, cholera, tuberculosis, typhoid, typhus and poliomyelitis.

Immunization

All microorganisms that produce disease carry on their surfaces thousands of specific molecules, known as antigens (⇨ pp. 58–61). These differ from similar molecules carried on the surfaces of human cells, providing the immune system with a way of recognizing such organisms as 'foreign'. If an individual has previously been invaded by the same organism, or has been immunized against that organism, large quantities of specific proteins known as antibodies, produced by B lymphocyte cells, will be present in the body. These bind to the antigens of the invader and make the organisms susceptible to destruction by other cells of the immune system – the phagocytes (macrophages).

In the case of immunization against viruses, however, the process is more complicated. Invading viruses enter body cells and cause viral protein to appear on the outside of those cells. This alerts the immune system to the fact that that cell is now undesirable, activates recognition T cells and prompts cytotoxic T lymphocytes to attack and kill the cells. Some of these killer cells remain as memory cells, so that if the same virus reinvades, they are immediately available to attack it.

Vaccines are of three general types. They may contain dead bacteria or viruses, live but safely damaged (attenuated) organisms, or bacterial toxins that have been rendered harmless. The latter group are known as *toxoids*. All three types prompt the immune system to behave as if an actual invasion of virulent organisms had occurred. Appropriate B lymphocytes are

Occupational Medicine

The control of infectious disease is by no means the only concern of preventive medicine. Another major activity is occupational medicine, which concentrates on the hazards to their health that people are exposed to at work. Throughout the centuries, workers have suffered innumerable disorders, many of them fatal, that could be directly attributed to their working conditions, and millions of workers have died or have suffered serious damage to health as a result of exposure to industrial by-products. A few of the many hazardous substances to which workers have been exposed are: silica dust (which causes the lung disease silicosis), coal dust (causing another lung disorder, anthracosis), asbestos dust (asbestosis), shale lubricating oil sprays and chimney soot (scrotal cancer), lead and other metallic salts (brain damage), and contaminated wool (anthrax) and phosphorus (necrosis, or structural damage of the jawbone). The English doctor Charles Thackrah (1795–1833), an early commentator, wrote: '... the miners rarely work for more than six hours a day, yet they seldom attain the age of forty ... the fork-grinders who use a dry grindstone die at the age of 28 or 32, while the table-knife grinders who work on wet stones survive to between 40 and 50.'

More recently, workers have been endangered by exposure to radioactive paints (causing mouth cancer), vinyl chloride (liver cancer and fibrosis), nickel compounds (nasal cancer), asbestos (lung cancer), naphthylamine (kidney and bladder cancer) and a wide range of dangerous and toxic industrial solvents, liquids, fumes and gases, such as carbon bisulphide, benzene and its derivatives, tetrachloroethane, dioxin, methyl bromide, chlorinated naphthalene, nickel carbonyl and nitrogen oxides. A much reduced incidence of industrial disease and injury can be achieved by the use of protective clothing and masks, increased ventilation, dust suppression, segregation of hazardous processes, automatic guards against physical injury, safety helmets, education of the workers, and, whenever possible, the substitution of safe for dangerous substances.

selected from the body's enormous variety of types, are cloned, and produce antibodies (immunoglobulins) and memory cells, from which renewed quantities of antibodies can be produced later at short notice.

There now exist effective vaccines for most infections, and these have virtually eliminated many of the most serious infectious diseases. A notable exception is a vaccine to the AIDS virus, HIV. This virus certainly results in the body producing antibodies, but these do not neutralize new HIV invaders. Moreover, the virus specifically destroys cells of the immune system responsible for protection against itself. HIV also inserts a perfect copy of its own genetic material (RNA; ➪ p. 26) into the DNA of the human cell. This means that when such cells reproduce, they automatically generate more HIV. So far these facts have made the production of an effective vaccine very difficult.

Taking regular exercise is important for a sense of overall well-being, but one of its specific benefits is to increase the individual's ability to resist infection. (Popperfoto)

Public awareness

The effectiveness of preventive measures against disease depends in large part on public awareness of the hazards and their consequences. In the West, public health authorities and government health administrators now see it as an important part of their function to keep the public informed. Subjects such as the dangers of cigarette smoking, overindulgence in alcohol, sexually transmitted diseases, and the importance of a healthy diet along with regular health screening have now become familiar through messages sent out by these bodies on television and radio and in newspapers.

Countries in the developing world suffer greatly from diseases that could otherwise be wholly preventable but for the lack of resources available for health care. What medical attempts there have been to implement preventive medicine have often been hampered by factors such as poor infrastructure, unsafe water supplies, war and famine.

Because of economic deprivation, many of the inhabitants of such countries suffer from nutritional deprivation, parasitic diseases, frequent gastrointestinal infections, tuberculosis, and insect-borne infections of many kinds.

An Afghan refugee woman receiving health education in Pakistan. (UNHCR)

SEE ALSO
- THE GENETIC CODE p. 26
- PROTECTION AGAINST ATTACK pp. 58–61
- HEALTH AND DISEASE pp. 116–23 and 126–59

Symptoms and Signs

Medical diagnosis is the process of making a choice between a wide range of possible explanations for a patient's condition. Each condition carries qualifiers that make it more or less likely in a given context. Making a decision, although often an easy task, is sometimes very difficult because of the complexity of choices or the paucity of information available. The doctor's first concern is, therefore, to have as much information about the case as possible.

Doctor using a spatula, to hold down the tongue while he examines this girl's throat. (Blair Seitz, SPL)

SEE ALSO

- EXERCISE AND BODY UNITY p. 54
- PROTECTION AGAINST ATTACK pp. 58–61
- HEALING AND REPAIR p. 62
- HEALTH AND DISEASE pp. 116–25 and 128–59

In general, good doctors can be distinguished by the skill they display, and the time they spend, in history taking. Doctors use the word 'history' to mean a record of everything relevant to the patient's health. It includes the details of the present complaint, previous illnesses, and the patient's social, familial and occupational background. Factors such as diet, sleep pattern, amount of exercise taken, sexual behaviour, smoking, alcohol intake and use of drugs are all of possible relevance. Even in this era of high-tech medicine the medical history is the most important part of the whole process of diagnosis, and doctors will often spend at least as much time in taking the medical history as in carrying out the examination. Medical diagnosis is very like detective work; the answer to a single question immediately brings up a range of possibilities in the mind of the doctor, who then has to work on the balance of probabilities. A few carefully selected questions will indicate the most promising avenue, and the doctor will follow this route as long as the responses are consistent with this provisional idea. He or she will try to avoid leading questions – questions that suggest a particular answer is required – but will, at the same time, tactfully keep the patient from straying into irrelevance.

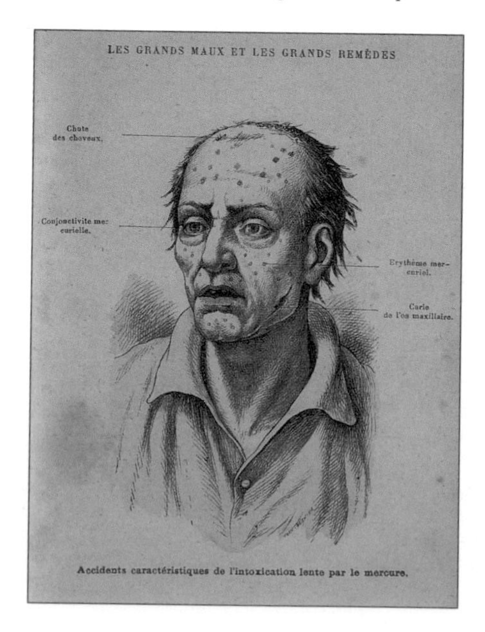

Signs of mercury poisoning, taken from a late 19th-century French medical text. The hat trade, at this time, saw numerous cases of such poisoning, as mercury was used in the manufacturing process. Constant inhalation of the toxin caused workers to suffer the disorders depicted here, and to twitch violently, which some interpreted as madness. Thus was coined the phrase 'as mad as a hatter'. (AR)

Doctors know that if a probable diagnosis has not been reached by the time the patient's history is taken, the problem, in the case of serious illness, is likely to require lengthy investigation. When a tentative, but probable, diagnosis has been made, this may often be confirmed by x-ray, CT (computer-assisted tomography – a method for taking x-rays of internal organs) or MRI (magnetic-resonance imaging – another method for producing images of the body's organs), laboratory testing or a 'therapeutic trial'. Good history-taking narrows down the range of special investigations required.

The medical examination

At the heart of the medical examination is the distinction between *symptoms* and *signs*. Symptoms are things felt by the patient and are not directly perceptible by anyone else. Signs, on the other hand, are changes or features that can be observed in the affected individual by another person. Because symptoms are hearsay, they are less reliable than signs. Nevertheless, in making a diagnosis, the doctor is heavily dependent on an accurate description of them.

Signs are elicited by careful, informed inspection and by carrying out certain simple clinical tests in the course of the examination. The routine for a general check-up includes:

- A check of the state of nutrition, skin colour (jaundice, *cyanosis* – blueness of the skin caused by insufficient oxygen in the blood – anaemia, pigmentation).
- A check for deformities.
- Measurement of temperature, blood pressure, pulse rate, respiration rate and depth.
- A check of the eyes for movement, squint, protrusion, jerkiness (*nystagmus*), lid droop, pupil size and reaction to light.
- A check of the facial appearance.
- Examination of the mouth and throat.
- Neck examination for vein congestion, enlarged lymph nodes, visible pulsation of the veins or arteries and enlargement of the thyroid gland.
- A check of the armpits for enlarged lymph nodes.
- A check of the fingernails for clubbing or hollowing.
- Testing of muscle power and tone.
- Testing of reflex jerks at forearm, knee and ankle.
- A check of joint freedom.
- A note of the chest expansion and equality of movement on the two sides.
- A note of the position of the apex beat (the furthest to the left and down that the heartbeat can be felt).
- Percussion of the chest to note quality of sound.
- Use of the stethoscope to hear breath and heart sounds, and to observe voice-resonance conduction.
- A check of the movements of the spine.
- A check of the abdomen for size, symmetry and distention.
- A note of visible pulsation or bowel activity.
- A check for *hernias* (abnormal protrusion of an organ or tissue through a wall or the cavity it is normally situated in).
- Careful feeling of the abdomen for tenderness, rigidity, splashing, enlargement of liver, spleen, kidneys, gall bladder or urinary bladder.
- A check whether lymph nodes in the groin can be felt; if they can this implies swelling.
- Rectal examination for enlarged prostate or tenderness, or for cancer.
- A check of the legs and feet for *oedema* (swelling due to accumulation of fluid), varicose veins, muscle power, coordination, pulses and foot temperature.

AN A-TO-Z OF SYMPTOMS AND SIGNS

No list of symptoms and their possible implications, however complete, can replace skilled investigation by an experienced and knowledgeable doctor. Even so, such lists are useful, and, by indicating unsuspected possibilities, will often prompt people to seek medical attention. Here is a list of some of the most common symptoms and signs with a brief account of their possible significance.

Abdominal pain
Acute gastritis, amoebic dysentery, appendicitis, bacillary dysentery, colon cancer, *diverticulitis* (inflammation of the colon), food poisoning, gall-bladder inflammation, gastroenteritis, haemolytic anaemia, irritable-bowel syndrome, pancreatitis, peritonitis, rectal cancer, sickle-cell anaemia, stomach cancer, stomach or duodenal ulcer, strangulated hernia, ulcerative colitis.

Appetite loss
Anorexia nervosa, anxiety, cancer, *coeliac disease* (gluten intolerance), emotional disturbances, *endocarditis* (inflammation of the inner lining of the heart), gastritis, glandular fever, hepatitis, menopause, oesophagitis, *sprue* (malabsorption of nutrients), tuberculosis, vitamin B deficiency.

Black stools
Bleeding high in the intestinal tract, especially from the stomach or duodenum; iron medication.

Blood in the stools
Bacillary dysentery, cirrhosis of the liver, colon cancer, haemorrhoids, intestinal obstruction, *purpura* (a general bleeding disorder) or other bleeding disorders, rectal cancer.

Bowel-habit change (persistent)
Colon cancer, ovarian cancer, rectal cancer.

Breathlessness
Asthma, bronchial foreign body, *bronchiectasis* (permanent widening of the bronchi), bronchitis, congenital heart disease, *croup* (inflammation of the air tubes in children), cystic fibrosis, emphysema, heart failure, high altitude, lung abscess, lung cancer, lung collapse, neurosis, obesity, *pericarditis* (inflammation of the membrane surrounding the heart), pleurisy, *pneumothorax* (the presence of air in the pleural cavity), pulmonary tuberculosis, severe anaemia, shock, spontaneous pneumonia.

Chest pain
Achalasia of the oesophagus (inability of the oesophagus to relax), angina pectoris, aortic *aneurysm* (widening), *Bornholm's disease* (a viral infection), cancer, coronary thrombosis, endocarditis, indigestion, *intercostal neuritis* (inflammation of the nerves between the ribs), lung cancer, lung collapse, neurosis, oesophageal cancer, oesophagitis, perforated peptic ulcer, pericarditis, pleurisy, pneumonia, pneumothorax, *pulmonary embolism* (blockage by a blood clot of a main lung artery), *reflux oesophagitis* (heartburn), rib-joint arthritis, shoulder-joint bursitis.

Constipation
Anal fissures, anxiety, dehydration, deliberate fasting, insufficient roughage intake, lead poisoning, neglect of call of nature, painful piles, severe debility, starvation.

Cough
Allergic alveolitis (inflammation of air sacs in the lungs), aortic aneurysm, asthma, bronchiectasis, bronchitis, croup, cystic fibrosis, emphysema, *empyema* (pus in the pleural cavity), heart failure, lung abscess, lung cancer, lung oedema, measles, nasal catarrh with postnasal drip, pneumonia, pulmonary embolism, pulmonary tuberculosis, smoking, spontaneous pneumothorax, whooping cough.

Coughing blood
Allergic alveolitis, chronic or acute bronchitis, excessive coughing, foreign body in the lung, lung abscess, lung cancer, pneumonia, pulmonary embolism, pulmonary tuberculosis.

Diarrhoea
Bad eating habits, coeliac disease, food allergy, food poisoning, irritable-bowel syndrome, neurosis, postgastrectomy syndrome, sprue, stress, thyroid-gland overactivity, travel, undue excitement.

Dizziness: ➪ vertigo.

Energy loss
Alcohol excess, boredom, *cardiomyopathy* (heart disease), neurosis, poor motivation, unfitness; ➪ weakness.

Flatulence
Air swallowing, alcohol abuse, bad eating habits, gobbling food.

Frequent urination
Chronic nephritis (inflammation of the kidneys), prostate cancer, prostate-gland enlargement, untreated diabetes, urinary infection (such as cystitis).

Headache
Acute glaucoma, any feverish illness, benign intracranial hypertension, brain tumour, cluster headache (occurring several times daily for weeks at a time), contraceptive pill, encephalitis, heat exhaustion, hypertension, *malignant hypertension* (very high blood pressure), meningitis, migraine, muscle tension, *phaeochromocytoma* (a rare adrenaline-secreting tumour of the adrenal gland), *polycythaemia* (excess haemoglobin in the blood), premenstrual tension, sexual intercourse, sinusitis, stroke, *temporal arteritis* (inflammation of arteries in the head), *trigeminal neuralgia* (disorder of one of the facial nerves), *uraemia* (accumulation of waste products in the blood as a result of kidney failure), vitamin A overdose.

Heartburn
Chronic gastritis, hiatus hernia, oesophageal ulcer, stomach cancer.

Hiccups
Acute gastritis, acute nephritis, alcohol excess, common diaphragmatic spasm (of unknown cause), hepatitis, pancreatitis, pleurisy, pneumonia, uraemia, vitamin B deficiency, whooping cough.

Itching
Anaphylactic shock (severe allergic reation), cancer, chickenpox, chilblains, contact dermatitis, diabetes, fungal skin infections, gout, hepatitis, hives, jaundice, leukaemia, lice, *lichen planus* (an uncommon skin disease), *lupus erythematosus* (inflammation of the connective tissue), measles, *pityriasis rosea* (a common skin complaint), pregnancy, scabies, thrush.

Jaundice
Acute yellow atrophy of the liver, *biliary cirrhosis* (inflammation of the bile duct, leading to liver failure), cancer of the head of the pancreas, cirrhosis of the liver, drug reactions, gall-bladder cancer, gallstone obstruction, *Gilbert's disease* (a form of mild jaundice brought on by fasting), hepatitis, liver cancer, yellow fever.

Memory loss
Alzheimer's disease, anxiety, brain tumour, *cerebral atherosclerosis* (build-up of fat on the inside of the arteries), deafness, dementia, encephalitis, *fugue* (rare psychological reaction to a stressful situation, in which the affected person wanders away, suffering amnesia), head injury, inattention, meningitis, poor concentration, stroke, syphilis, vitamin B deficiency.

Pain on urination
Bladder cancer, bladder infection, bladder stones, gonorrhoea, kidney infection, thrush, *trichomoniasis* (infection of the urethra or prostate gland).

Palpitations
Atrial fibrillation (irregular pulse), atrial flutter, *cardiac neurosis* (unjustified belief that one is suffering from a heart condition), certain drugs, emotional stress, excess alcohol, excess coffee, *extrasystoles* (premature contractions of the heart), *paroxysmal tachycardia* (sudden onset of rapid pulse).

Paralysis
Brain abscess, brain tumour, cerebral palsy, cervical osteoarthritis, diabetes, dystrophy, *Friedreich's ataxia* (inherited disorder of the cerebellum), *Guillain-Barré syndrome* (an immunological disorder), head injury, hysteria, intervertebral-disc prolapse, motor neurone disease, multiple sclerosis, muscular dystrophy, spinal-cord injury, *myasthenia gravis* (causes muscle weakness), Parkinson's disease, poliomyelitis, spinal-cord tumour, stroke, transverse myelitis (inflammation of the spinal cord), vitamin B12 deficiency.

Seizures
Alcohol excess, brain tumour, epilepsy, head injury, high body temperatures, stroke.

Skin rashes
Acne, chickenpox, contact dermatitis, dermatitis herpetiformis, drug eruptions, eczema, fungal infections, herpes simplex, impetigo, lichen planus, measles, *molluscum contagiosum* (a viral infection of the skin), nappy rash, pityriasis rosea, psoriasis, purpura, *rosacea* (a persistent skin disorder), *roseola infantum* (or sixth disease), *rubella* (or German measles), scabies, scarlet fever, *seborrhoeic dermatitis* (severe dandruff), shingles, urticaria, *vitiligo* (a disorder of pigmentation).

Speech difficulty
Brain tumour causing *aphasia* or *dysphasia* (language disorders), cleft palate, congenital deafness, developmental disorders, *dysarthria* (articulation problems), multiple sclerosis, paralysis of the larynx, Parkinson's disease, stroke, stuttering, uncorrected lisps and errors of articulation.

Staggering
Alcoholic intoxication, barbiturate poisoning, cerebellar degeneration, cerebellar tumour, chronic alcoholism, hysteria, *labyrinthitis* (inflammation of the inner ear), *Ménière's disease* (a middle-ear condition), multiple sclerosis, Parkinson's disease, *tabes dorsalis* (degeneration of the sensory nerve columns in the spinal cord).

String-like stools
Rectal cancer.

Swallowing problems
Aortic aneurysm, *atresia* (closure) of the oesophagus, cancer of the oesophagus, foreign body in the oesophagus, myasthenia gravis, oesophageal pouch, *oesophagitis* (inflammation of the oesophagus), *pharyngeal pouch* (sac of mucous membrane bulging from the junction of the pharynx and oesophagus), spasm of the oesophagus, *stricture* (narrowing) of the oesophagus, stroke, thyroid-gland enlargement.

Tics and muscle jerks
Akathisia (inability to sit still because of drug side-effects), *blepharospasm* (uncontrollable winking), Creutzfeldt–Jakob syndrome, *Gilles de la Tourette syndrome* (a condition featuring involuntary body movements and compulsive utterances), habit spasms, myoclonic epilepsy, *myoclonus* (brief involuntary muscle contraction), *spasmodic torticollis* (involuntary twisting of head or neck to one side), viral encephalitis, Wilson's disease, writer's cramp.

Vertigo
Anaemia, brain injury, brain tumour, drug sensitivity, ear wax, faintness from temporarily low blood pressure, *heart block* (dissociation between contractions of the upper and lower chambers), hypoglycaemia, labyrinthitis, Ménière's disease, menopause, *otitis media* (inflammation of the middle-ear cavity), perforated eardrum, severely high blood pressure, stroke, unexplained causes.

Vomiting
Addison's disease (disorder of the adrenal glands), appendicitis, brain tumour, bulimia, *cholecystitis* (inflammation of the gall bladder), diabetes, encephalitis, excessive intake of food or alcohol, food poisoning, gastritis, gastroenteritis, head injury, hepatitis, *intussusception* (serious bowel disorder), labyrinthitis, Ménière's disease, meningitis, migraine, pancreatitis, pregnancy, *pyloric stenosis* (narrowing of the stomach outlet), stomach or duodenal ulcer.

Vomiting blood
Duodenal ulcer, *Mallory-Weiss syndrome* (tear in lower end of oesophagus), severe gastritis, stomach cancer, stomach ulcer.

Weakness
Advanced cancer, anaemia, debility, diabetes, lack of exercise, muscular dystrophy, neurotic fatigue, obesity, old age, poor motivation, postviral fatigue syndrome, severe liver or kidney disease, severe loss of weight from anorexia, thyroid underactivity, vitamin deficiency.

Diagnostic Aids

The growth of technology in medicine has made accurate diagnosis possible in a higher proportion of cases than ever before. But, while invaluable though technology may be in numerous medical areas, this is not to say that it will necessarily replace all existing practices, and good medicine looks on technology as a tool to be employed judiciously and sparingly.

People with potentially serious conditions are usually afraid and in need of authoritative opinion, advice and, if possible, reassurance and comfort. But reassurance cannot be honestly provided until the cause of the patient's problem is known and the outlook (*prognosis*) established. Advances in technology have, in many instances, helped to make the skilled physician's task easier.

The stethoscope

The stethoscope, the very symbol of the medical profession, was invented in 1816 by the French physician René Théophile Hyacinthe Laënnec (1781–1826). It consists simply of a pair of hollow plastic or rubber tubes connecting a chestpiece to a pair of earpieces. The chestpiece is usually double-sided, with a 'bell' on one side and a 'diaphragm' on the other. Low-pitched sounds and murmurs are best heard with the bell and high-pitched sounds with the diaphragm. In Western medicine the stethoscope is now much less important than it has been, but listening to the chest with it (*auscultation*) can still provide the experienced clinician with valuable information about the state of the lungs, the lung coverings (*pleurae*) and the heart. Lung collapse, solidification, cavities, and sputum in the air tubes can all be detected. Auscultation of the heart is a skilled procedure calling for long experience. There are four normal heart sounds and an analysis of the time relationship of these to any abnormal sounds (or murmurs) allows the diagnosis of heart-valve defects and other disorders.

The sphygmomanometer

The sphygmomanometer is used to measure a person's blood pressure. An inflatable cuff connected to a pressure gauge is wrapped round the upper arm, and a stethoscope bell is applied to the skin over the main artery of the arm, just below the cuff. This cuff is then inflated with air until the artery is compressed and the blood flow through it stopped, and the pressure in the cuff is then gradually released. When the blood pressure is just able to force some blood through the narrowed artery, a loud knocking sound is heard by the doctor in the stethoscope with each beat. The pressure in the cuff at this point, which is registered on the gauge, is the same as the peak (*systolic*) pressure of the blood. The cuff pressure is allowed to drop again until no further sounds are heard, and this is taken to be the running, between-beat (*diastolic*) blood pressure. A typical reading is 120/80; 120 being the systolic pressure. Generally, readings up to 150 and 90, for systolic and diastolic pressures respectively, are considered healthy.

CT scanning control room. A CT scan provides computerized x-rays of the internal organs. Images can be constructed so as to appear in any desired plane or orientation. (RB)

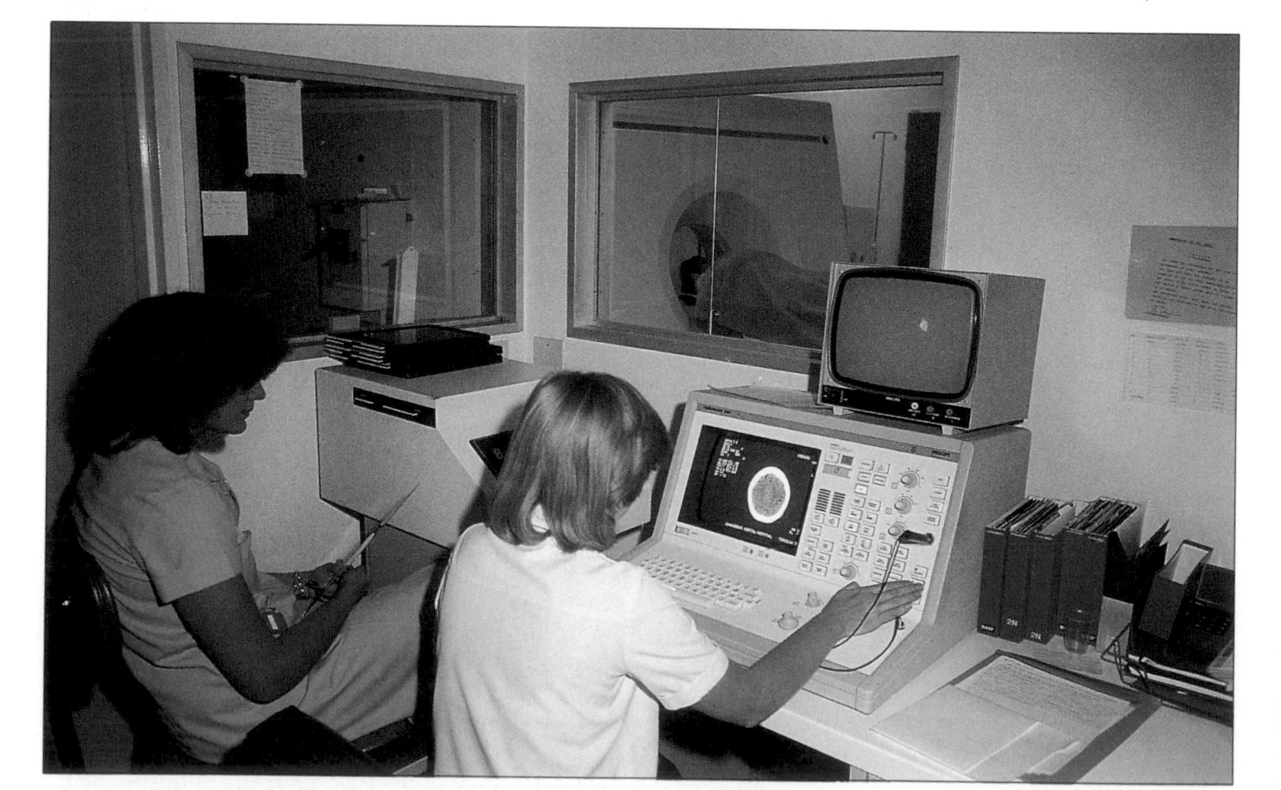

The electrocardiograph

Electrical changes in the heart muscle can indicate heart disease. They can be measured using the *electrocardiograph*, which produces a tracing known as an *electrocardiogram* (ECG). The currents involved are small and amplification is needed. They are picked up by leads attached to the patient's wrists and ankles and to various points across the front of the chest. Contact is made using conductive jelly and metal electrodes held in place by straps or suction cups. The amplifier feeds a normal milliammeter (device for measuring an electric current), the needle of which carries a heated wire pen that draws a tracing on a strip of moving calibrated paper. The test may be conducted with the patient either resting or exercising strenuously on a treadmill. The latter method can bring out abnormalities not seen on the resting record.

The pattern drawn on the paper is highly standardized in health but undergoes a number of characteristic changes if there is any heart disease. The number of possibilities is limited and most modern ECG machines are able to read the tracings automatically, compare them to stored computer data, and print out a diagnosis.

The electroencephalograph

The electroencephalograph is a multichannel device that records the electrical activity of the brain, producing a multiple tracing known as an *electroencephalogram* (EEG). Several pairs of leads are attached to the patient's scalp and the currents detected taken to separate amplifiers, each of which actuates a separate pen-bearing electric meter. Brain electrical activity involves the action of many billions of nerve cells, and all the resulting electrical signals are superimposed to give only a statistical result. Because of this, the EEG, unlike the ECG, does not produce specific tracings that represent specific brain disorders. Rather, it produces a recognizable pattern of brain-wave normality that may be disturbed in characteristic ways. Some conditions, such as epilepsy, disturb this pattern in a striking and obvious manner; others do so in a more subtle way and require skilled interpretation. Much use is made of comparisons between the signals received from one part of the scalp and another. The EEG can detect the massive electrical discharges characteristic of epilepsy, and the changes occurring around areas of brain-tissue loss in stroke and brain tumours. It is also helpful in diagnosing brain death.

Laboratory tests

The modern pathology laboratory conducts a very wide battery of tests. These are so numerous that laboratories have to be divided into quite separate departments to administer them, and must employ automation to get through the work. Routine tests include:

- Blood tests to check the red cells, the levels of haemoglobin within them, the white-cell counts and counts of the numbers of the different types, and the sedimentation rate of blood (*haematology*).

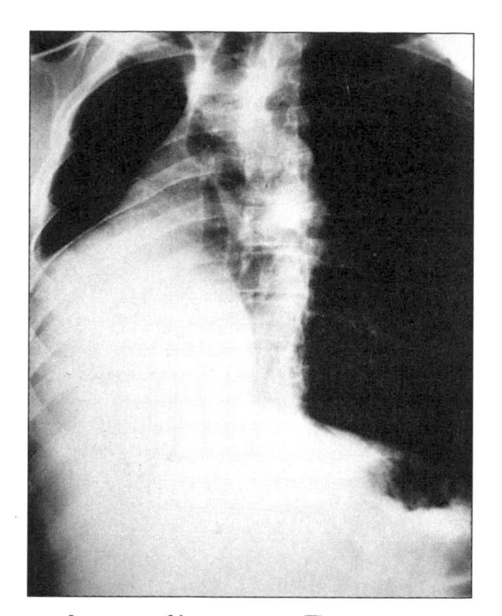

An x-ray of lung cancer. The cancerous tumour is visible as a white 'shadow' in the bottom left-hand side of the image. (RB)

- Examination of blood smears for organisms such as malarial parasites, trypanosomes, Leishman–Donovan bodies, the spirochetes causing relapsing fever, and microfilarial worms (*tropical pathology*).
- Tests to determine the levels of the wide range of biochemical substances in the blood (*serology*).
- Tests of other body fluids such as urine, pus, gastric juice, pleural fluid and cerebrospinal fluid (*clinical chemistry*).
- Tests on faeces for abnormal constituents and worm eggs.
- Tests to identify infecting organisms (*microbiology*).
- Immunological tests to determine the presence of antibodies across an enormous range of conditions.
- Examination of cells derived from scrapings such as cervical (or Pap) smears and bronchoscopy examinations (*exfoliative cytology*).
- Microscopic examination of tissue samples (biopsies) to identify specific disease processes (*histopathology*) and genetic tests conducted on DNA samples derived from mouth scrapings, or amniocentesis or chorionic villus samples (*molecular biology*).

X-rays and scanners

X-rays were discovered by Wilhelm Konrad Röntgen (1845–1923) while experimenting with a vacuum tube through which a stream of electrons was made to pass by high-voltage attraction. When the stream struck a metal anode, invisible rays were produced that could pass through many opaque substances and cause fluorescent materials to glow or a photographic plate to be exposed. This discovery was quickly exploited for medical purposes, and, without any further development in principle, soon became used worldwide for the diagnosis of a wide range of conditions. The only major advance in x-ray application was the development of the CT (computerized tomography) scanner in 1971. CT scanners allow a remarkable improvement in resolution over plain x-rays and do not suffer from the disadvantage of low-density tissue details being obscured by high-density tissue, such as bone. Many organs, including the kidneys, heart and blood vessels, can be outlined by injecting iodine-containing compounds, which are opaque to x-rays. Similarly, barium highlights the intestinal tract. CT scanners, however, involve a larger dose of radiation than most plain x-rays do; a scan of the abdomen delivers a dose of 8.8 mSv (millisieverts) – eight times the dose of a plain x-ray.

Magnetic-resonance imaging (MRI) scanners do not involve x-rays but work on the principle of atomic nuclear-spin resonance. They use massive magnetic fields – so far thought to be harmless to the body – and radio signals (which are at the other end of the electromagnetic spectrum from x-rays) to deflect atoms and cause them to emit tiny signals that can be localized. These produce computer-constructed images of amazing detail. MRI scanners can resolve detail in the brain and spinal cord so fine that the individual plaques in multiple sclerosis, for instance, can be seen. Changes in the heart muscle following a coronary thrombosis, and images of the abnormality in congenital heart disease, can be made out. As in CT scanning, these images can be reconstituted in any plane, and MRI can readily distinguish local areas of tissue death from the surrounding living tissue.

Radionuclide scanning

Short-lived radioactive isotopes of various elements can safely be injected into the body and their destination tracked by a sensitive instrument called a gamma camera. Isotopes are selected for their ability to concentrate in particular organs or tissues such as the thyroid gland, the bones, the heart, the kidneys or in tumours of various kinds. Radionuclide scanning can then reveal abnormal organ activity or unsuspected tumours. Positron-emission tomography (PET) is even more sophisticated and uses isotopes containing positive electrons that react with normal electrons to convert matter to radiation. This method can give invaluable information on the state of a person's vitality and metabolic activity. Areas that have been deprived of their blood supply, or areas no longer living, can readily be detected using this method. The short half-life of these isotopes limits the use of PET scanning to hospitals equipped with the large and expensive cyclotrons (a type of particle accelerator) in which the isotopes are made.

Ultrasound imaging

High-frequency sound reflects off interfaces and the echoes can be detected and represented as images. Used in scanning beams, it can build up pictures of internal structures. While image resolution is improving steadily, it is used very much less than other scanning methods. Ultrasound is, however, believed to be completely safe, and this is why it is so widely used in examining pregnant women to confirm that all is well with the baby and the placenta.

Ultrasound waves pass easily through fluids and soft tissues, and so are especially suitable for examining fluid-filled organs such as the pregnant uterus and the gall bladder and soft solid structures such as the liver. Ultrasound scanning can detect the sex of the fetus, multiple pregnancies, malplacement of the placenta, and congenital abnormalities of the fetus's limbs and organs. It can also detect failure of fetal growth and is used to allow doctors to perform safe amniocentesis and even life-saving operations on the fetus in the womb. Outside obstetrics, it is used for imaging the beating heart, the kidneys, pancreas and gall bladder, and for measuring the parameters of the eye so that the appropriate power of lens implants can be determined.

Endoscopes

Endoscopes are surgical instruments that use fibre optics for illuminating and viewing the various organs from inside the body, and they may be either rigid or flexible and steerable. Increasingly, surgery and biopsies are being performed through endoscopes introduced into the body via natural orifices or through small ('button-hole') incisions. It has now become commonplace, for instance, for gall-bladder removal operations to be done through an endoscope, and an increasing range of operations is being performed in this way; much gynaecological surgery, for instance, is now endoscopic. The primary purpose of endoscopy, however, is still diagnostic – the direct examination of internal parts for the identification of disease. In particular, both ends of the intestinal tract, the major air passages of the lungs and much of the abdominal cavity, can be closely examined.

SEE ALSO

- HEALING AND REPAIR p. 62
- HEALTH AND DISEASE pp. 116–27 and 130–59

A transverse section of the abdomen, shown by CT scan, revealing the kidney (small, crescent-shaped object) and liver (large green–purple structure). (RB)

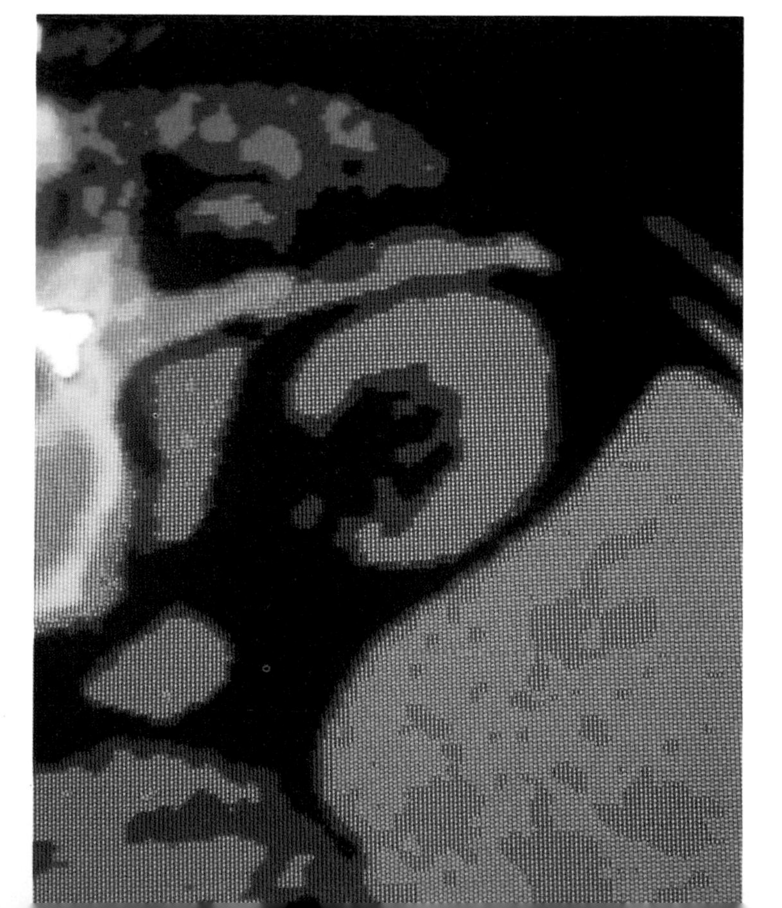

Transmission of Infection

The world is heavily contaminated with viruses, bacteria and other microorganisms, and although only a small proportion of these can cause disease, our bodies are never free from disease-producing organisms. Fortunately, the body is equipped with a highly efficient defence system against infection (⊳ pp. 58–61), and it is only when this system is breached by injury, assaulted by an overwhelming dose of virulent organisms or is rendered ineffective by inherent disorder (*immune deficiency*) that infection occurs.

Sixteenth-century physician (bottom, centre) visits a patient suffering from the plague. He is holding a perfumed sponge to his nose while his assistant burns incense, as it was believed that these measures could prevent the spread of disease. (RB)

While most organisms do not cause disease, some are inherently very dangerous. In theory, it takes only a single bacterium to cause a disease, but in practice one organism would never be able to establish an infection because it would be destroyed by the body before it had a chance to reproduce and establish a colony. The dose of organisms needed to cause an infection varies with the characteristics of the organism. Some need to be present in very large numbers, some in a much smaller dose. Whether or not infection occurs is really a matter of the balance of forces – on the one hand, the dose, invasiveness and virulence of the organism, and, on the other, the resistance of the potential host.

The external defences

Many organisms on or in the body remain innocuous so long as they stay in certain protected places, such as the surface of the intact skin, the nose, mouth, vagina or the interior of the intestinal tract. Indeed, such areas are permanently colonized by organisms that coexist with the body quite harmlessly. If, however, breach of the surfaces of these areas, as by injury or disease, allows the organisms access to the normally sterile interior tissues, infection is inevitable. The large intestine (the *colon* and *rectum*) contains countless millions of organisms – the great bulk of the dried weight of faeces consists of bacteria – but these cause no harm. If, however, a perforating ulcer were to allow bowel contents to pass into the abdominal cavity around the intestine, the result would be a severe and dangerous infection known as *peritonitis*. This is one of the reasons why the inside of the bowel is often considered, topologically and practically, to be as much part of the outside of the body as the skin.

Even before the immune system comes into action, the body has other defences against infection. The nose contains sticky mucus and hairs to trap contaminated material, and employs a sneeze mechanism to rid itself of irritants. The saliva and tears have antibiotic properties, and the throat is protected by an automatic cough reflex. The stomach contains a strong acid that kills most of the bacteria it encounters.

Local and systemic infection

Most infections are confined to a relatively small area of the body. These are called *local* infections and are very common. Examples include pimples, boils, styes, tooth decay, herpes attacks, impetigo and conjunctivitis. Such infections seldom amount to much, and hardly ever cause real danger. In some, however, the organisms, or the poisons (toxins) they produce, get into the bloodstream and quickly involve the whole body. These are called *systemic infections*. Again, many of these systemic infections are, in most cases, minor – diseases such as measles or chickenpox – but there are also many systemic infections that can be very serious or even fatal. Diphtheria, tetanus, staphylococcal septicaemia, tuberculosis, plague, typhoid and Legionnaire's disease are among them.

Inflammation

Local infections imply a breach of a protective barrier and multiplication of organisms within body tissues. The response to this is *inflammation* – an important bodily reaction that usually keeps the infection localized and brings it under control. A few million organisms are needed to produce obvious inflammation, and as a result of the action of chemical substances released by them, and by cells that have been damaged by them, local blood vessels widen and the flow of blood through them increases. This causes the typical redness and warmth of an inflamed area. At the same time, watery fluid leaks out of the vessels, causing swelling. The increased permeability of small vessels makes it easier for white phagocyte cells of the immune system to reach the area and concentrate their attack on the organisms present. In most cases this attack is successful in preventing the infection going any further, and the whole process dies down. But if enough organisms are present – many millions – the body also responds by forming a wall of fibrous tissue around the area to localize the trouble. Inside the wall, affected tissue may be replaced by a collection of dead white cells, dead body cells, dead bacteria and cell debris. This is called *pus* and the whole formation is called an *abscess*.

Serious systemic infections always involve fever. A fever starts when immune-system cells are damaged and release into the bloodstream a substance called interleukin-1. This acts on the body's temperature-control thermostat in the brain, adjusting it, effectively, to a higher setting. The normal body temperature is then interpreted as being too low and

Sexually Transmitted Disease

The mucous membranes of the genitalia of both sexes offer less resistance to the passage of organisms than skin. They also come into close contact during sexual activity, and provide a fluid interface that encourages movement of organisms in either direction. As a result, disease transmission can occur very readily in the course of vaginal, anal, and oral sexual practices, and the number of diseases spread in this way is larger than is generally supposed. The conditions spread in this way include:

- **Gonorrhoea.**
- **Syphilis.**
- **AIDS.**
- **Genital herpes.**
- **Chlamydial infections (non-specific urethritis).**
- ***Candidiasis* (thrush).**
- ***Gardnerella vaginalis* infection (vaginal inflammation).**
- ***Trichomoniasis* (causes irritation, burning and itching in the vagina in women, and in the urethra or prostate in men).**
- ***Molluscum contagiosum* (virus infection of the skin).**
- ***Lymphogranuloma venereum* (causes genital ulcers, enlargement of the lymph nodes, fever, and enlargement of the liver and spleen).**
- ***Chancroid* (causes painful genital ulcers and swelling of the lymph nodes in the groin).**
- **Crab lice.**
- **Scabies.**

extra heat is automatically produced by the action of shivering. As a result, the body temperature rises to a level beyond which many disease-causing organisms can survive. However, doctors often act to bring a person's temperature down for fear that it could otherwise reach a dangerously high level. Many systemic infections also feature rashes, which consist of multiple local areas of skin damage produced by the circulating organisms or their toxins. Although the rashes are the only visible signs of this effect, similar damage may be occurring widely within the body.

The incubation period

After the invasion and before the first sign of the disease it causes, an organism needs time to incubate. This is called the *incubation period*. During this time foreign organisms multiply until there are enough of them present to cause discernible effects in the patient. Incubation periods vary greatly in length, from a few hours to several years. Cholera, for instance, can appear within a few hours of drinking contaminated water; AIDS does not appear until 5–10 years after the HIV virus has been acquired. The incubation period of rabies varies from about 14 days to several months, the length of the period being proportional to the distance of the site of the bite by the rabid animal from the infected person's brain. This is because the viruses travel up inside the sensory nerves from the point of the bite to the brain, where they have their effect. Most diseases, however, have a fixed incubation period, and this can be a help in the process of diagnosis.

The spread of infection

Most infections in human beings are acquired, directly or indirectly, from other human beings. A few are acquired from animals that are suffering from a certain disease. These are called *zoonoses* and tend to affect people occupationally involved with animals. Some diseases are transmitted by animals, including insects, who are not themselves suffering from the disease, acting instead as a *vector* for that disease. This is commonly the case with insects. Diseases caused by animal sources and vectors are numerous in the less well-developed parts of the world, but play only a small part in the causation of disease in the West. Insect vectors are the sole effective mode of spread of malaria, yellow fever, *dengue* (which leaves the sufferer feeling weak for several weeks), sleeping sickness (*trypanosomiasis*), *onchocerciasis* (river blindness), sandfly fever, *leishmaniasis* (a disorder of the skin or internal organs), typhus and *Rocky Mountain spotted fever* (also known as tick typhus).

Infection can be transmitted from person to person in many ways, one being direct skin-to-skin contact. The intact skin offers considerable resistance to infection, but if the dose of organisms is large enough to breach the defences or the virulence high, disease may be acquired in this way. Some organisms, especially staphylococci, readily penetrate the sweat glands and hair follicles of the skin, causing pustules and boils. Some streptococci can cause impetigo and a spreading skin infection called *cellulitis*. Several common fungi can become established in the outer layer of the skin (the *epidermis* ⊳ p. 56) to set up infection, and are called the *epidermophytoses*. These are readily acquired through contact with contaminated objects such as swimming pool duck-boards and changing-room floors.

Other common routes of transmission are those of finger-to-eye and finger-to-nose. The large range of rhinoviruses that cause the common cold are mainly transmitted in this way – picked up on the fingers that then rub the eyes or touch the insides of the nostrils. These and other organisms can also be transmitted on articles contaminated with human secretions. Colds are more readily transmitted by hand-shaking than by sneezing; that is, they are genuinely 'contagious'. Diseases like chickenpox, on the other hand, are spread mainly in droplets ejected from the nose and throat by coughing and sneezing. Viral particles can be transmitted through air for several metres in this way and many diseases, including measles, meningitis, pneumonia, tonsillitis and tuberculosis, are spread by the inhalation of sneezed or coughed aerosol droplets. Much less commonly, some infective agents are acquired by inhaling the dried, but surviving, spores in dust. This is a mode of spread of the diseases *Q fever*, *coccidioidomycosis* (both influenza-type infections) and *histoplasmosis* (a lung infection), rare in the West.

Although stomach acid provides excellent protection against most organisms, some survive it and pass through to cause infection either of the intestinal tract or of the body generally. Organisms capable of causing infection in this way are acquired in contaminated food and include the viruses that cause hepatitis A and poliomyelitis, the bacteria that cause gastroenteritis, food poisoning and typhoid, the vibrios and spirochaetes that cause cholera and brucellosis, and the entamoebae that cause amoebic dysentery. Food poisoning can also result from food contaminated with discharge from septic spots on the hands of food handlers. This may contain the virulent toxin from staphylococci and result in an acute attack within a few hours.

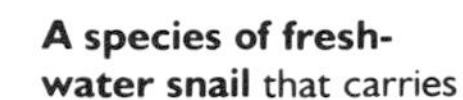

A species of freshwater snail that carries the disease schistosomiasis (also known as bilharzia), which is caused by the parasitic flatworm *Schistosoma*. Eggs excreted by infected people undergo part of their larval development inside the bodies of these snails. The larvae released by the snails then penetrate the skin of anyone bathing in infected water and colonize the blood vessels of the intestine. The features of this disease, which can be fatal, include diarrhoea, kidney failure, cirrhosis of the liver, epilepsy and coma. (RB)

Faecal contamination is a very common cause of infection as it contains millions of coliform (intestinal) organisms that can cause diarrhoea. But faecal matter often also contains pathogenic organisms such as salmonellae, and food can become contaminated with these organisms if prepared by somebody who has failed to wash properly after defecating. Enteroviruses, such as those that cause poliomyelitis and hepatitis A, are spread by means of the fecal–oral route.

Infection can also be spread directly from an infected mother to her baby by way of the placenta during pregnancy or, less commonly, by way of the breast milk afterwards. Probably the most serious disease to be spread via the placenta is German measles (*rubella*), since, if this occurs early in the pregnancy, the result is likely to be devastating congenital abnormality in the baby. Other diseases passed to the fetus during pregnancy are *cytomegalovirus disease* (causing liver enlargement and blood disorders) and *toxoplasmosis* (which damages the nervous system). During birth, babies may acquire infections such as gonorrhoea, syphilis or herpes from the vagina.

Other modes of spread of infection include the sharing of unsterilized needles; tattooing and ear piercing with unsterile equipment; and the sharing of unsterilized shaving equipment and of toothbrushes. The main risks from these modes are hepatitis B and AIDS. Syphilis is also sometimes transmitted in this way.

SEE ALSO

- THE BODY COVERING p. 56
- PROTECTION AGAINST ATTACK pp 58–61
- HEALING AND REPAIR p. 62
- HEALTH AND DISEASE pp. 116–29 and 132–59

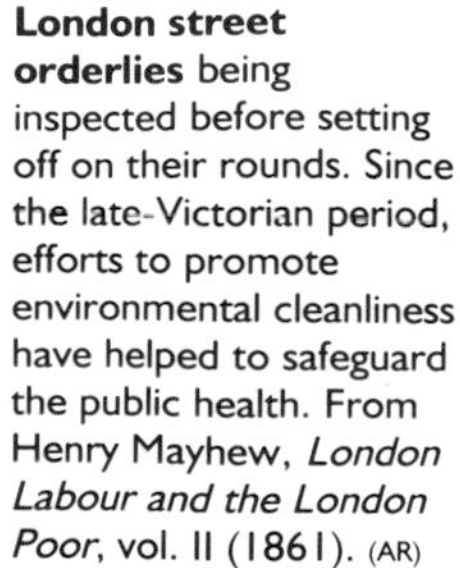

London street orderlies being inspected before setting off on their rounds. Since the late-Victorian period, efforts to promote environmental cleanliness have helped to safeguard the public health. From Henry Mayhew, *London Labour and the London Poor*, vol. II (1861). (AR)

Infectious Diseases 1: Viruses

Viruses are the smallest disease-causing organisms that are known to carry the genes for their own replication. They are, in fact, no more than little packets of DNA or RNA (⇨ pp. 24–7), and they are everywhere. All other known organisms, even bacteria, are parasitized by them. The body cells of all mammals, birds, reptiles, fish, amphibia, insects, plants, algae and fungi all play host to them.

Romanian orphan suffering from AIDS. (Popperfoto)

Viruses are not cells and cannot, in fact, perform any metabolic processes on their own. Except under laboratory conditions they cannot reproduce outside living cells, so it is arguable whether they themselves should be described as 'living'. Even so, they are capable of wreaking destruction on living organisms, including, of course, the human being.

The term 'virus' has had an interesting career. It is a Latin word, meaning a slimy, poisonous liquid or venom. In the 19th century it was used to describe a morbid or poisonous principle produced in the body by disease, especially one that could be introduced into the body of another person to cause disease. During the period of this usage, no notion of an organism was yet implied. In 1892, Dimitri Ivanovsky, a Russian botanist, discovered that tobacco mosaic disease could be transmitted from plant to plant by sap from affected tobacco plants that had been passed through a filter of such fine pore size that it held back all known bacteria. The cause of the disease was unknown, so the fluid was said to contain a 'filterable virus'. This usage became widespread and, in the manner of such phrases, was soon reduced to the single word 'virus'. When it later became known that there were organisms so small that they could pass through these filters, they became known as 'viruses'.

The structure of viruses

Viruses have a central core of DNA or RNA surrounded by a protective protein coat called a *capsid*. They vary in size from about half the diameter of the smallest known bacterium down to a size not much more than that of a large molecule. They also vary considerably in shape because of different arrangements of the protein subunits that make up the capsid. These may be arranged in a single layer, as a helical tube, or as multi-layered shells. In some, the capsids are pyramidal or twenty-sided (*icosahedral*).

DNA viruses carry their genetic material (*genome*) in the familiar double-helical form; RNA viruses have only a single strand, but this is often present in two identical copies. RNA viruses cannot replicate their genetic material without first producing a complementary strand to the RNA. This is done by means of an enzyme called *reverse transcriptase* that the virus itself carries. Once this complementary strand is formed, the replication can occur normally, as with DNA.

As they enter cells, some viruses discard the capsid, exposing their genome. Others pass in whole and then uncoat their nucleic acid. This then acts in much the same way as the cell's own DNA, making use of that cell's energy-providing and metabolic organelles (⇨ p. 38). Some of the viral genes are transcribed into messenger RNA, which synthesizes the transcription enzymes needed for the replication of copies of the viral genome. Viral nucleic acid then competes with the cell's own genetic material for control of the cell activities and biochemical processes. Some even synthesize an enzyme that destroys the DNA of the cell so that the virus can proceed unhindered in the synthesis of more viral particles. Sometimes the cell DNA is not destroyed, but the viral nucleic acid is inserted into it by means of enzymes that break the cell DNA so that viral DNA can be spliced in. The first new virus is produced within 2–24 hours and, within a few days, thousands are generated. Invaded cells can become packed with viral particles and even mechanically disrupted and killed by the sheer profusion of viral replication. As these particles leave the cell, each acquires a new capsid.

Virus Reproduction

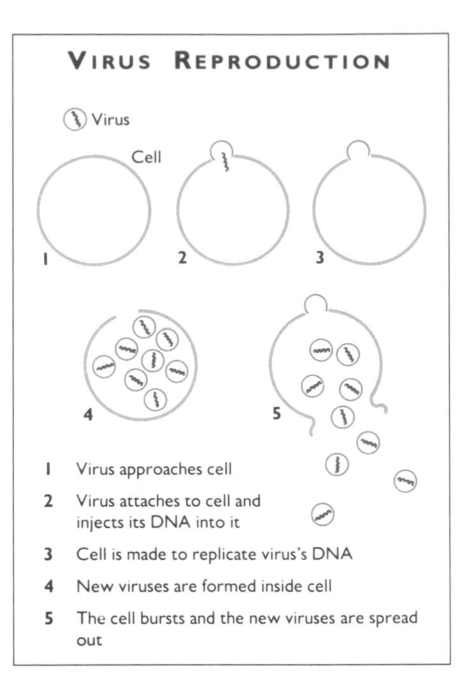

1 Virus approaches cell
2 Virus attaches to cell and injects its DNA into it
3 Cell is made to replicate virus's DNA
4 New viruses are formed inside cell
5 The cell bursts and the new viruses are spread out

How viruses enter the body

Viruses are so small that they can enter the body by several different routes. They may be swallowed in food and fluids or inhaled in airborne droplets. They may pass through minute abrasions in the skin, and through the mucous membranes of the mouth, eye (conjunctiva), genital tract and rectum, or they may pass through the skin in the saliva of biting insects, on the teeth of rabid dogs, or on the shared needles of drug addicts. As soon as they have gained entry they make for and invade local cells, spreading

Cancer-Causing Viruses

Viruses that are known to cause, or are strongly suspected of causing, cancer in humans include:

- **Hepatitis B virus – liver cancer.**
- **Wart (papovavirus) – skin cancer, cancer of the cervix.**
- **Herpes viruses – Burkitt's lymphoma and possibly cancer of the cervix and Kaposi's sarcoma.**
- **Human T-cell leukaemia virus (HTLV-I and II) – causing leukaemia and lymphoma.**

RETROVIRUSES

The medical world has been much preoccupied in recent years with the family of *retroviruses*, which includes the HIV (human immunodeficiency virus) that causes AIDS (acquired immune deficiency syndrome; ⇨ pp. 58–61). Retroviruses are RNA viruses whose genomes consist of two copies of a single strand of RNA (⇨ pp. 24–7). The enzyme reverse transcriptase allows them to make a complementary copy of the RNA and produce, in effect, a strand of DNA, thus allowing replication to occur. The discovery that DNA could be formed from RNA ran counter to the previously accepted dogma that the direction of transcription was always from DNA to RNA, and much attention has since been given to reverse transcriptase in the hope that, by finding a way of attacking it, an effective treatment for AIDS might be found.

in exponentially increasing numbers from cell to cell. Many are taken up and killed by phagocyte cells, which are equipped with destructive enzymes especially for this purpose. Some, however, are able to resist these enzymes and even to replicate within the phagocytes. Other viruses, meanwhile, are carried to the lymph nodes and killed there. Those that reach the bloodstream, however, are quickly spread to all parts of the body.

How viruses cause disease

Viruses affect cells in different ways, but their net effect is either to close down the normal biochemical processes of cells so that they die or to so damage the cell's DNA that its reproduction becomes impossible. No harm is done if only a few cells are affected in this way, but it is in the nature of virus infections that the number of cells affected rapidly multiplies. The outcome depends on the shift of the balance of forces between the effectiveness of the viral invasion (its virulence) and the effectiveness of the body's immune system. This, in turn, determines the number of cells destroyed or inactivated. Some parts of the body, such as the liver, can tolerate considerable cell loss and can even replace them by regeneration (⇨ p. 48), while others, such as the brain, have no regenerative power at all.

When a virus genome is incorporated into host-cell DNA, without in fact killing the cell, it can cause a fundamental transformation in that cell, since this new DNA will now be passed on with each generation of the cell line, and the transformed cells acquire new properties. One important change may be the loss of what is called 'contact inhibition' – the control of cell growth and reproduction that results from cells being in contact with other cells – producing tumours. Viruses that transform cells in this way are called *oncogenic* (tumour-forming), and are capable of causing a number of different cancers in animals and a few in human beings. Certain other human cancers are strongly suspected of being caused by different viruses; both DNA viruses and RNA retroviruses are being investigated.

All normal human cells contain sequences of some 20 genes called *oncogenes* or *proto-oncogenes* that are essential for the growth and differentiation of cells. Their action in causing cell reproduction is normally held strictly in check, but may be adversely affected when certain retroviruses, carrying almost identical oncogenes in their RNA, invade body cells. These viruses incorporate their nucleic acid into the host genome for the purposes of reproduction, and may pick up the cell oncogenes instead of their own. All future generations of these viruses (clones) will now contain cellular oncogenes and these would require only a small mutation to become capable of causing cancer in any cells they invade (⇨ box).

The now most notorious way in which viruses can cause disease is by their interference with cell-mediated immunity. This is how the human immunodeficiency retrovirus (HIV) acts. This virus invades, and interferes with the function of, helper-T lymphocytes (⇨ pp. 58–61). As a result, the normal immune response to all the infections normally combated by the T-cell mechanism is progressively lost. This leads to many infections, including some that do not occur in people with a normal immune system. Certain rare cancers, notably Kaposi's sarcoma, are also combated by the T-cell mechanism and these are a feature of AIDS.

Virus interference

When viruses invade host cells, these cells produce and release substances known as *interferons*. These pass to other normal cells and act to render them less susceptible to attack by the same and other viruses. Interferons do not necessarily protect only against the same virus but often have a general protective effect. They confer this protective effect either by changing the surface receptors on the cells so as to deter the viruses from entering the cells or by preventing viral replication within the cells. Clearly, such substances are of great therapeutic potential and much research has been directed to producing them in quantity. Interferons have been made commercially by genetic engineering, and are increasingly being used in clinical trials.

THE MAIN VIRUS FAMILIES

Viruses that cause human disease are classified into some 20 or so large families. The most important of these are:

FAMILY	CONDITION CAUSED
Adenoviruses	Respiratory and eye infections
Bunyaviruses	California encephalitis, haemorrhagic fever, Rift Valley fever
Coronaviruses	Common cold
Herpesviruses	Cold sores, genital herpes, chickenpox, shingles, glandular fever, *cytomegalovirus* infections (liver enlargement)
Orthomyxoviruses	Influenza
Papovaviruses	Warts
Paramyxoviruses	Measles, rubella, mumps and croup
Picornaviruses	Poliomyelitis, hepatitis A, chest infections, encephalitis, myocarditis, haemorrhagic conjunctivitis
Poxviruses	*Molluscum contagiosum* (skin infection), smallpox
Retroviruses	AIDS, brain degeneration, possibly cancer
Rhabdoviruses	Rabies
Togaviruses	Yellow fever, *dengue* (tropical disease causing fever, headache, aches and enlargement of the lymph nodes), Japanese encephalitis.

The prions

Prions are disease-causing organisms that are smaller even than viruses. They were discovered in 1983, and found to consist exclusively of protein. Although they contain no DNA or RNA, prions can nevertheless replicate to cause brain-degenerating diseases such as kuru and Creutzfeldt–Jakob disease in humans, and scrapie in sheep, and there is speculation as to whether Alzheimer's disease (⇨ p. 186) may also be caused by prions.

SEE ALSO

- CELLS, THE STUFF OF LIFE p. 38
- THE BODY COVERING p. 56
- PROTECTION AGAINST ATTACK pp. 58–61
- HEALING AND REPAIR p. 62
- WHAT IS DISEASE? p. 120
- PREVENTING DISEASE p. 124
- SYMPTOMS AND SIGNS p. 126
- TRANSMISSION OF INFECTION p. 130
- INFECTIOUS DISEASES pp. 134–7

SHAPES OF VARIOUS VIRUSES

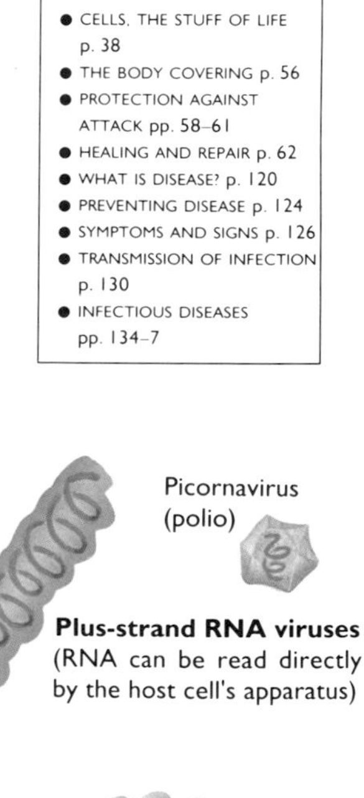
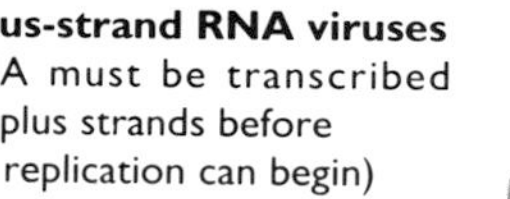
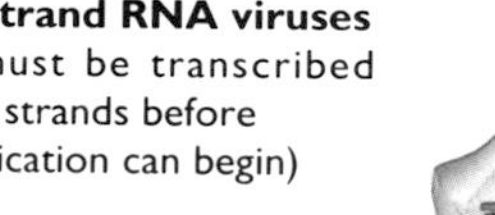
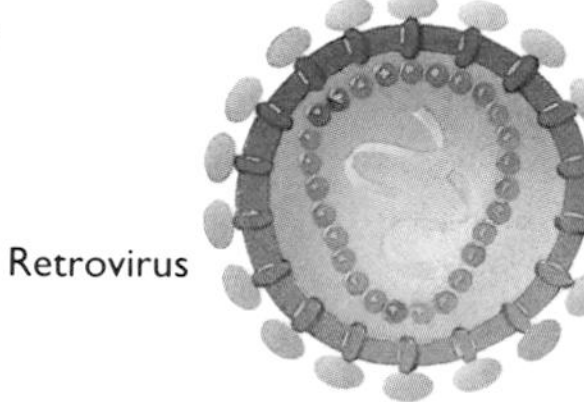

Infectious Diseases 2: Bacteria

The bacteria constitute an enormous group of single-celled organisms, capable, unlike viruses (⇨ p. 132), of surviving and reproducing independently, but, in many cases, living as parasites on other organisms.

Bacteria exist everywhere; soil, for example, contains countless billions. Most are wholly or relatively harmless to human beings. Those that cause human disease form only a small proportion of all bacteria and are specially adapted to life inside the human organism. They survive and reproduce best at 37° C (about 98.6° F) – human body temperature.

It is believed that bacteria probably closely resemble the earliest forms of cell life ever to evolve on this planet, since fossilized bacteria have been identified in rocks estimated to be at least three billion years old. Research findings such as this form part of the discipline of bacteriology, a branch of the wider subject of microbiology, which includes the study of viruses, bacteria, fungi, protozoa and algae – all of which may have medical importance. Medical bacteriology is concerned with the study of those types of bacteria that can cause infectious disease.

SEE ALSO

- CELLS, THE STUFF OF LIFE p. 38
- THE BODY COVERING p. 56
- PROTECTION AGAINST ATTACK pp. 58–61
- HEALING AND REPAIR p. 62
- HEALTH AND DISEASE pp. 116–33 and 136–59

Gram and Bacterial Staining

Hans Christian Gram (1853–1938) was a Danish bacteriologist who was troubled by the difficulty of identifying bacteria under the microscope. Robert Koch (⇨ p. 116) and Paul Ehrlich had already shown that organisms could be stained with aniline and gentian violet dyes. Gram followed up the Ehrlich method by soaking the slides on which he had placed samples of cocci, from the lungs of people who had died of pneumonia, in a solution of iodine and potassium iodide. This, he found, turned the previously dark blue sections a dark purple–red colour. When he then tried the effects of alcohol in decolourizing the stained bacteria he was interested to find that some organisms retained the stain while others did not.

What he had discovered was to become a technique of fundamental importance in medical bacteriology, and his paper was the first report of the bacteriological staining method universally used today to divide all bacteria into two groups. Gram staining separates all bacteria into a 'Gram-positive' or a 'Gram-negative' group, and the members of each group have been well documented. Gram-positive bacteria appear a deep violet colour while Gram-negative bacteria appear red. There is also a useful correlation between the Gram reaction and antibiotic sensitivity. In general, Gram-positive bacteria are sensitive to the penicillins, while other antibiotics are needed to kill Gram-negative bacteria.

Prokaryote Cell Structure

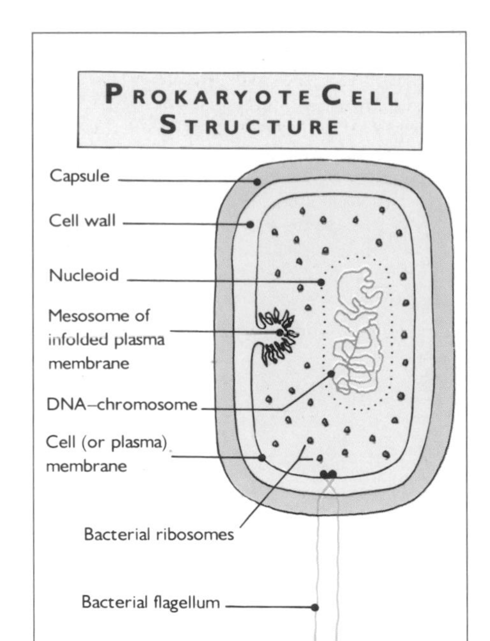

Diagram of the main components of a bacterial cell. It is surrounded by a plasma (or cell) membrane, which is infolded at the region of the mesosome to create extra membrane surface for various biochemical processes, and abuts the cell wall. Outside the cell wall there is a capsule. Inside the plasma membrane is the protoplasm, which contains ions, enzymes and other dissolved molecules, together with free ribosomes for the synthesis of proteins. Within the protoplasm, the nucleoid (the clear space with no limiting membrane) houses the normally circular pieces of DNA. Some bacteria are mobile and swim with the aid of flagella made from a special protein, flagellin.

Historical concepts

The obvious facts of the spread of disease from one person to another must, throughout history, have prompted people to suspect that some agent of infection was passing from one to the other. In 1546, the Italian physician, poet, classicist and researcher Girolamo Fracastoro (1483–1553) published a remarkable book on infectious diseases entitled *De Contagione, Contagiosis Morbis et eorum Curatione* ('On Contagion, Contagious Diseases and their Treatment') in which he suggested that each infectious disease was spread by 'seminaria' (seeds) specific to that particular disease. He also suggested that these seminaria could be transmitted through the air, by direct contact or by articles ('fomites'), and that they multiplied in the body. This extraordinarily accurate perception anticipated the conclusions of Louis Pasteur and Robert Koch (⇨ p. 116) by several centuries. In *De Contagione* he describes how syphilis is transmitted during sexual intercourse and gives a detailed and accurate account of the effects of the disease.

Bacteria and other microscopic organisms were first seen by the Dutch linen draper Anton van Leeuwenhoek (1632–1723) by means of his single-lens microscopes. In a letter written in 1683 to the Royal Society of London and entitled 'Containing some Microscopical observations about Animals in the scurf of the teeth . . .' he stated: 'Tho my teeth are kept usually very clean, nevertheless when I view them in a Magnifying Glass, I find growing between them a little white matter as thick as wetted flower (flour) . . . I therefore took some of this flower and mixt it with pure rain water wherein were no Animals . . . and then to my great surprize perceived that the aforesaid matter contained very many small living animals, which moved themselves very extravagantly.' This is thought likely to be the first account of the direct observation of bacteria.

What are bacteria?

Attempts to classify bacteria have brought about controversy. Are they animals or plants? This question immediately demonstrates the inadequacy of a system that insists that all living things must be one or the other. Bacteria, like plants, have rigid cell walls, but the vast majority do not use photosynthesis (the process whereby plants use light energy to connect carbon dioxide and water to carbohydrates). Like animals, bacteria move about freely and assimilate organic food. The earlier classifiers, however, plumped for plants, and in the two-kingdom (plant–animal) system, bacteria are still regarded as plants. In another system that proposes five kingdoms, the bacteria and algae (*prokaryotes* – single-celled organisms that do not have a defined cell nucleus) have one kingdom to themselves and the fungi and protozoa (*eukaryotes* – organisms whose cell or cells do possess a well-defined nucleus) occupy another.

The characteristics and uses of bacteria

Bacteria are too small to be seen by the naked eye, although colonies are readily visible. Bacteria are measured in microns (μ), each micron being one millionth of a metre (or one thousandth of a millimetre) long. Most bacteria are in the range of 0.1–4.0 μ wide and 0.2–50 μ long. The common cocci, however, such as streptococci and staphylococci, are spherical and about 1 μ in diameter.

Bacteria are remarkably resistant to extremes in their physical surroundings; many of the 2000 or so species that have been identified survive conditions that would kill most other organisms. Some of them live in hot springs at temperatures close to boiling point; others can survive at an altitude so high as to be almost airless; yet others survive long freezing; and others still flourish at oceanic depths of 10 km (about 6 mi), seemingly indifferent to the massive pressures that exist at such depths. To resist extreme conditions, many bacteria can enter a sporing stage in which each is surrounded by a dense protective capsule.

Although bacteria are often thought of as being invariably harmful, very many of them are of incalculable biological, ecological and economic importance. Indeed, a large number are essential to plant and animal life. For instance, most carry enzymes that can break down dead organic matter (decomposition), reducing it to elements that can then be assimilated by other organisms. If this did not happen, the world would soon become uninhabitable. The same process enriches the soil by the breakdown of compost. Nitrogen-fixing bacteria, especially those found in nodules in the roots of leguminous plants, convert atmospheric nitrogen to nitrates that are then taken up by other plants. Bacterial enzymes have had widespread industrial uses for centuries and their applications are constantly being extended. Bacterial fermentation is essential in the production of leather, tobacco, cheese, vinegar, buttermilk and many other products. Sewage disposal plants make use of bacterial fermentation to reduce dangerous organic waste to a harmless form.

Details on the way in which bacteria are spread and cause disease may be found elsewhere (⇨ pp. 116–21 and 124–5).

How are bacteria destroyed?

All bacteria can be killed by sufficiently high temperatures, and most can be destroyed by temperatures that do little harm to the materials they are contaminating. Most disease-producing bacteria (*pathogens*) are killed at the temperature of boiling water (100° C, 212° F). Those that occur in milk can be destroyed by keeping the milk at 62° C (143.6° F) for half an hour, or at 71° C (159.8° F) for 15 seconds. Spore-bearing organisms (⇨ above) are much harder to kill, however, and require the higher temperatures of compressed steam.

Antiseptics and disinfectants either chemically alter the structure of organisms or prevent them from reproducing. Many substances, such as carbolic acid (phenol), alcohol, chlorine gas, and various mercurial poisons, also have these effects. Most antiseptics are as dangerous to the human being as to bacteria, however, so, in the treatment of bacterial infection in the human body, antibiotics that kill bacteria or prevent their reproduction, while being relatively harmless to humans, are used. Most antibiotics are capable of surviving passage through the stomach, and so can be taken by mouth rather than injected. From the stomach they are transported into the bloodstream, where they attach themselves to the cell membrane of the bacteria they are targeted to find, and inhibit the action of the bacteria by interfering with their metabolic processes.

THE CLASSIFICATION OF BACTERIA

Bacteria are generally classified according to their different shapes. The most common of these are:

- **Bacilli – rod-shaped, often with many small, whiplike structures (*flagella*) by which they are propelled. They include the huge *Salmonella* genus that causes typhoid and food poisoning, and the organisms that cause dysentery, tetanus, diphtheria and anthrax.**
- **Cocci – spherical and can aggregate in bunches (staphylococci) or in strings like beads (streptococci). They include many organisms that cause a wide range of infections such as boils, styes, tonsillitis, pneumonia, food poisoning and tooth decay.**
- **Diplococci – small, kidney-shaped cocci that occur in pairs. Gonorrhoea and meningitis are caused by different species of diplococci.**
- **Spirochaetes – delicate helical organisms such as the *Treponema pallidum* that causes syphilis or the *Leptospira* or *icterohaemorrhagiae* or *L. canicola* that cause Weil's disease (or *leptospirosis*, an infection transmitted in the urine of rats or dogs, causing headache, muscle aches, loss of appetite and a skin rash).**
- **Mycoplasmas – these are very small organisms, more like amoebae than the more rigidly conformed bacilli, cocci and spirochaetes, and they may take up any shape.**

Bacteria may also be classified by whether or not they need free atmospheric oxygen. Those that do are called *aerobic* organisms; those that do not, *anaerobic*. Anaerobes derive oxygen from the products of fermentation, which they bring about by their own intrinsic enzymes (⇨ main text).

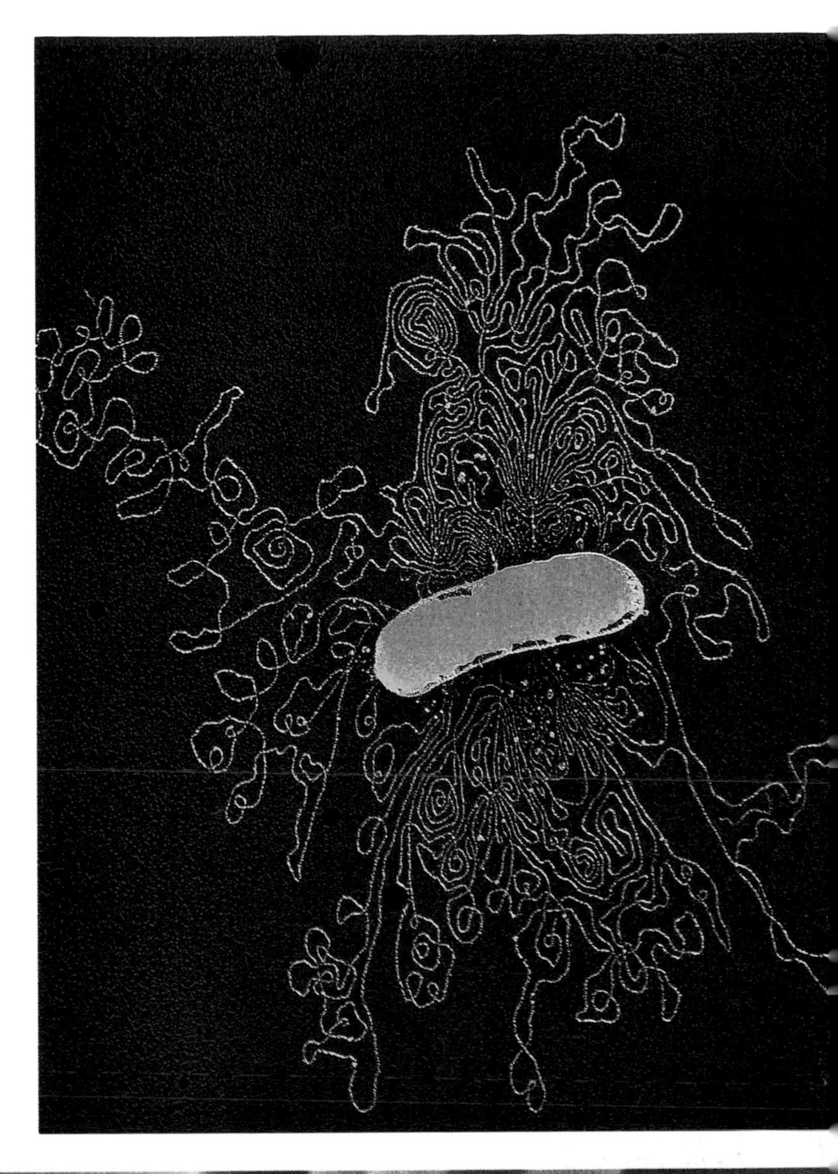

The good and the bad. The bacterium *Escherichia coli* (right) is a usually harmless inhabitant of the human intestine, present in its millions. It is shown here, surrounded by its DNA, after being treated with an enzyme to weaken the cell wall. *Salmonella* bacteria (below) can cause *salmonellosis* (food poisoning) in humans when ingested in large numbers through contaminated food. (Dr Gopal Murti and London School of Hygiene and Tropical Medicine, SPL)

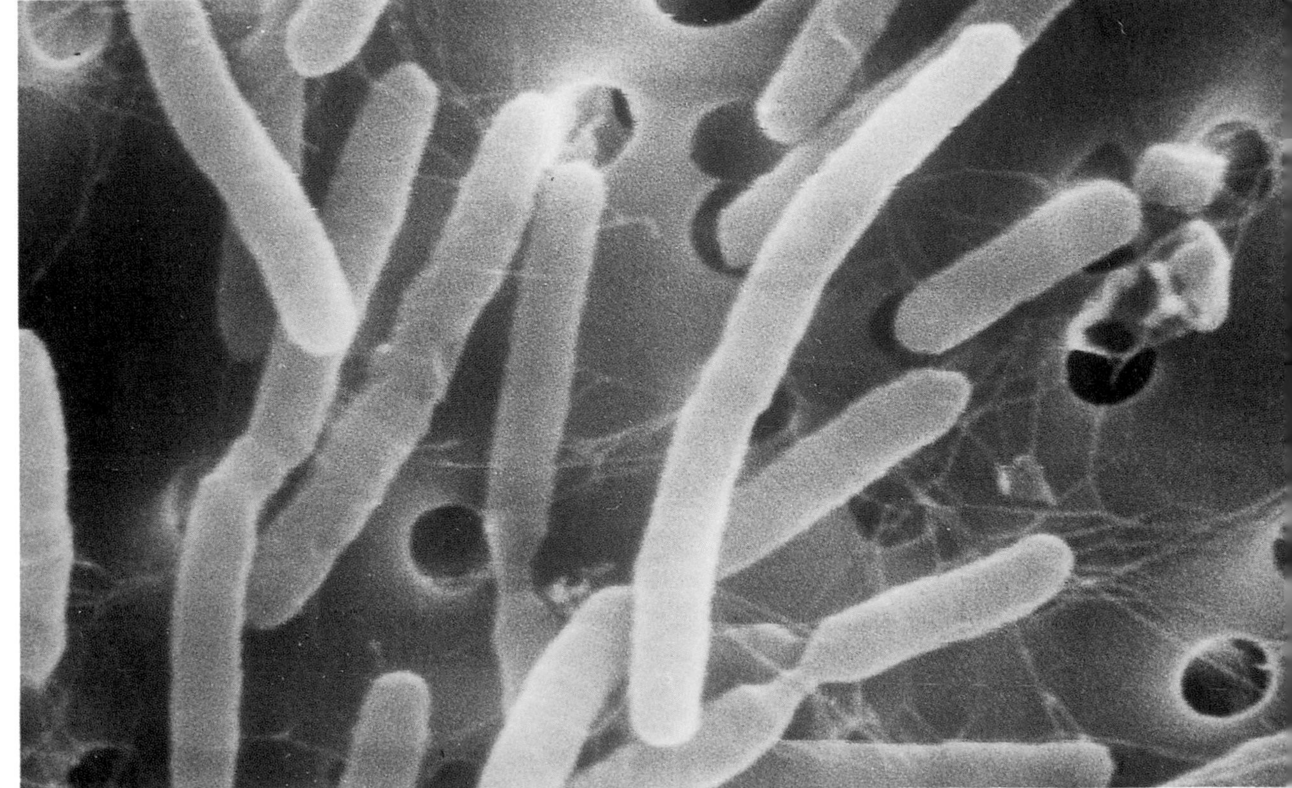

Infectious Diseases 3: Parasites

Every human being, however scrupulous about personal hygiene, is host to a large number of parasites. We provide sustenance for millions of viruses and bacteria, fungi, protozoa, and the occasional louse, mite or threadworm. A large section of the world's population regularly harbours, in addition to these, a range of other worms, various flukes, fleas, and a range of leeches.

SEE ALSO
- PROTECTION AGAINST ATTACK pp. 58–61
- HEALING AND REPAIR p. 62
- HEALTH AND DISEASE pp. 116–35 and 138–59

A parasite is any organism that lives on another so as to derive some advantage, usually nutritional, and may be classified in two types – internal (*endoparasites*) and external (*ectoparasites*). Some are long-term residents, spending their whole lives on or in the body of the host; others spend only a part of their life cycle in association with humans. Parasitism is a highly successful process and is so common that the number of parasitic organisms considerably exceeds the number of those parasitized. The human intestinal worm population of the world, for instance, is very much larger than the human population. Parasites never offer any advantage to their hosts, and, in some cases, their presence may be dangerous. A few cause serious diseases, and some are commonly responsible for the death of their host.

'Malaria' Ross

For centuries, malaria, as the name implies, was believed to be caused by 'bad air' – the miasma from swamps. In 1880, the French surgeon Charles Louis Alphonse Laveran (1845–1922) discovered single-celled protozoan organisms in the blood of malaria patients, and was able to demonstrate that these were the cause of the disease. Surgeon Major Ronald Ross (1857–1932) of the Indian Army Medical Service heard of Laveran's discovery and tried, without success, to find these parasites in his own patients. On a visit to England in 1894, Ross called on the tropical-medicine researcher Patrick Manson, and was shown how to stain blood smears on microscope slides so as to demonstrate the protozoa. Manson told Ross he believed that mosquitoes carried malaria.

Ross was deeply impressed by this idea and resolved to follow it up on his return to India the next year. He took a new microscope with him and for two years worked unremittingly, travelling up one blind alley after another. He tried to show that, as Manson had suggested, the water in which mosquitoes had died was infectious, and followed up the possibility that the disease was transmitted via the insect's excreta. He also investigated transmission by biting, and conducted numerous unsuccessful experiments using the wrong kinds of mosquitoes – the *Culex* and *Aedes* species instead of *Anopheles*. At last, however, in 1897, he found malarial parasite oocysts (the structure that develops from the fertilized malarial parasite) on the stomach walls of *Anopheles stephensi* mosquitoes that had fed on the blood of a malarial patient. Encouraged, he proceeded to work out the entire life cycle of the parasite in the mosquito, and was finally able to prove transmission by bringing infected mosquitoes to London and allowing them to bite two healthy volunteers, one being his only son. Both developed malaria.

Ross was awarded the Nobel Prize for Physiology or Medicine 1902 and Laveran was honoured in the same way in 1907. Ross was then knighted in 1911.

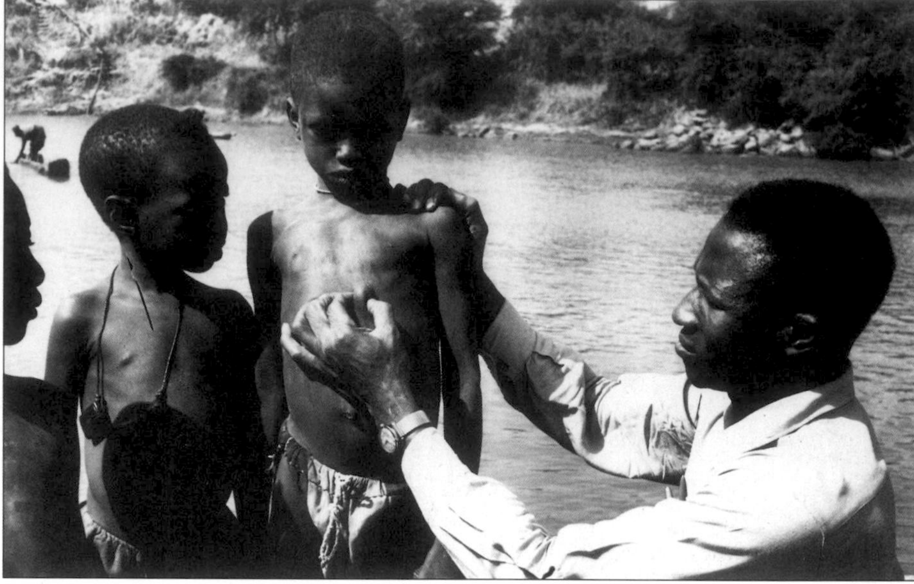

'River blindness', or *onchocerciasis*, can be detected by examining the skin for lumps betraying the presence of the parasitic worm that causes the disease, as here in Chad, Central Africa. This relatively little-known disease frequently forces people away from fertile agricultural valleys in order to avoid the disease and the flies that carry it. (WHO)

In the living world, parasitism tends to be hierarchical. Many ecologists would agree that humans parasitize the globe. Human beings and other complex animals are, in turn parasitized by arthropods, such as lice, which carry many bacteria (➪ p. 134). These, in turn, carry many viruses (➪ p. 132).

The major parasites in humans can be divided into the single-celled *protozoa*, the worms (*helminths*) and the large group of jointed-leg creatures, the *arthropods*.

Protozoan parasites

The protozoa are so numerous as to require a classificatory subkingdom, or *phylum*, of their own – a subdivision of the *Protista* kingdom. This phylum contains over 65 000 species, but taxonomists are still arguing about which organisms should be included and how the subkingdom should be subdivided. It includes a large number of amoebic organisms that move by pushing out protrusions ('false feet' or *pseudopodia*) and flowing into them. The commonest disease-causing amoebic human parasite is *Entamoeba histolytica*, the organism that causes amoebic dysentery and amoebic liver abscesses, and which infects millions of people every year. The flagellate protozoa (the *Mastigophora*) move by the lashing action of a whip-like organ, the flagellum. This phylum contains several parasites: *Trypanosoma gambiense*, which causes African sleeping sickness; *Trypanosoma cruzi*, which causes Chagas' disease in South America and elsewhere; *Giardia lamblia*, which causes severe diarrhoea; and *Leishmania donovani*, which causes Leishmaniasis. Ciliated protozoa (*Ciliophora*) move by means of numerous tiny, waving hairs, or *cilia*. Although these are of little medical importance, the same cannot, however, be said of the *Sporozoa*, a group of protozoan parasites that includes all the *Plasmodium* species of malarial parasites.

Worm parasites

The helminths include the roundworms or *nematodes*, the flukes (*trematodes*) and the tapeworms (*cestodes*). Roundworm eggs can enter the human body through the medium of food or water, and the larvae of some species of roundworm can enter through the intact skin, causing irritating 'creeping eruption' or *larva migrans*. As these larvae develop in the body, they pass through the lungs, where they can cause severe inflammatory reactions or organic damage. Some species invade the muscles and other tissue, causing severe or even fatal reactions.

Head (scolex) of a tapeworm. (Polarized-light image, magnification x 12 at 35 mm depth.) (RB)

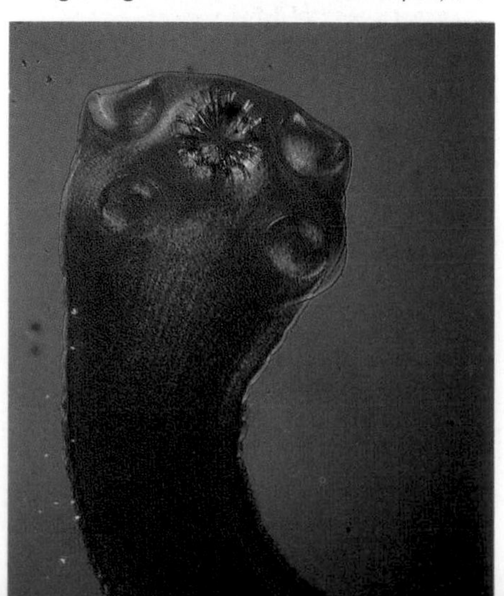

Tapeworm larvae also invade muscles, and sometimes the brain, to form permanent and often harmful cysts. Other worm parasites invade the skin to form substantial lumps containing coiled-up worms; some invade the eye, causing blindness (*onchocerciasis* or 'river blindness'); some cause the disease filariasis, blocking the lymphatic channels so that tissue fluid cannot drain away, and *elephantiasis* (swelling) of limbs or other parts results; several flukes concentrate in the liver, where they can form abscesses and destroy liver tissue; many others cause intestinal damage that may even lead to cancer; and some concentrate on the bladder, causing bleeding into the urine and occasionally bringing about malignant changes.

The arthropods

Arthropods (mites, ticks, bugs, lice, fleas, houseflies, sandflies, mosquitoes) are more commonly transmitters (vectors) of infectious disease than parasites.

- Forage mites cause dermatitis in people handling stored food.
- Straw itch mites cause severe dermatitis in workers handling grain or cotton cargoes.
- Red poultry mites cause dermatitis in poultry workers.
- The *Sarcoptes scaboi* mite causes scabies.
- The blood-sucking rat mite causes severe itching and transmits typhus.
- Harvest mites cause intense itching and also transmit typhus.
- Soft ticks transmit the tick-borne relapsing fever.
- Hard ticks cause painful bites and paralysis, and transmit typhus, *Q fever* (an influenza-like infection) and *tularaemia* (which features enlargement of the lymph nodes and small areas of tissue death all over the body).
- Bed bugs cause painful bites.
- Human body and head lice cause itching and dermatitis, encouraging secondary infections such as impetigo and boils.

Parasitism in the Tropics

Contrary to popular belief, the high incidence of parasitism in the tropics has little to do with the environmental temperatures. The incidence of parasitism and parasitic disease depends far more on general living standards, standards of hygiene and public health, water and food quality, effectiveness of sewage disposal, and general standards of education and public information. Less well-developed countries have fewer resources with which to provide these things, and they suffer a high incidence of human parasitization. People who have no shoes, for instance, and who walk on bare ground heavily contaminated with hookworm larvae cannot avoid becoming parasitized with hookworm. Similarly, those who spend part of each day washing clothes in rivers into which people with schistosomiasis (⇨ p. 130) have urinated will inevitably contract the disease.

They also transmit epidemic typhus and relapsing fever.

- Cone-nosed bugs transmit Chagas' disease.
- Crab lice cause dermatitis.
- The common flea transmits the bubonic plague.
- The pregnant female 'jigger' flea burrows into the skin of the feet and there grows large, causing a painful inflammation.
- Sandflies transmit *sandfly fever* (an influenza-like illness), *Oroya fever* (which features high temperature, anaemia and enlargement of the spleen lymph nodes), and *leishmaniasis* (infection of the skin or internal organs).
- Mosquitoes transmit malaria, yellow fever, filariasis and dengue fever.
- Midges cause painful, disabling and occasionally fatal bites; they also transmit filariasis.
- Biting black flies cause painful bites with marked swelling and ulceration; they also transmit 'river blindness'.
- Gadflies cause painful bites and transmit tularaemia and filariasis.

The human flea, *Pulex irritans*, has remarkable athletic powers; it is capable of a horizontal leap of 13 in (33 cm). (Polarized-light image, magnification x 12 at 35 mm across.) (RB)

- Some flies lay their eggs in parts of the body (*myiasis*) and allow the maggots to develop there to the adult form. Usually, this occurs only in open wounds or sores, but a few species can parasitize the nasal cavities or penetrate the intact skin.
- Tsetse flies transmit African sleeping sickness.
- House flies transmit typhoid, bacillary and amoebic dysentery.

The Malaria Cycle

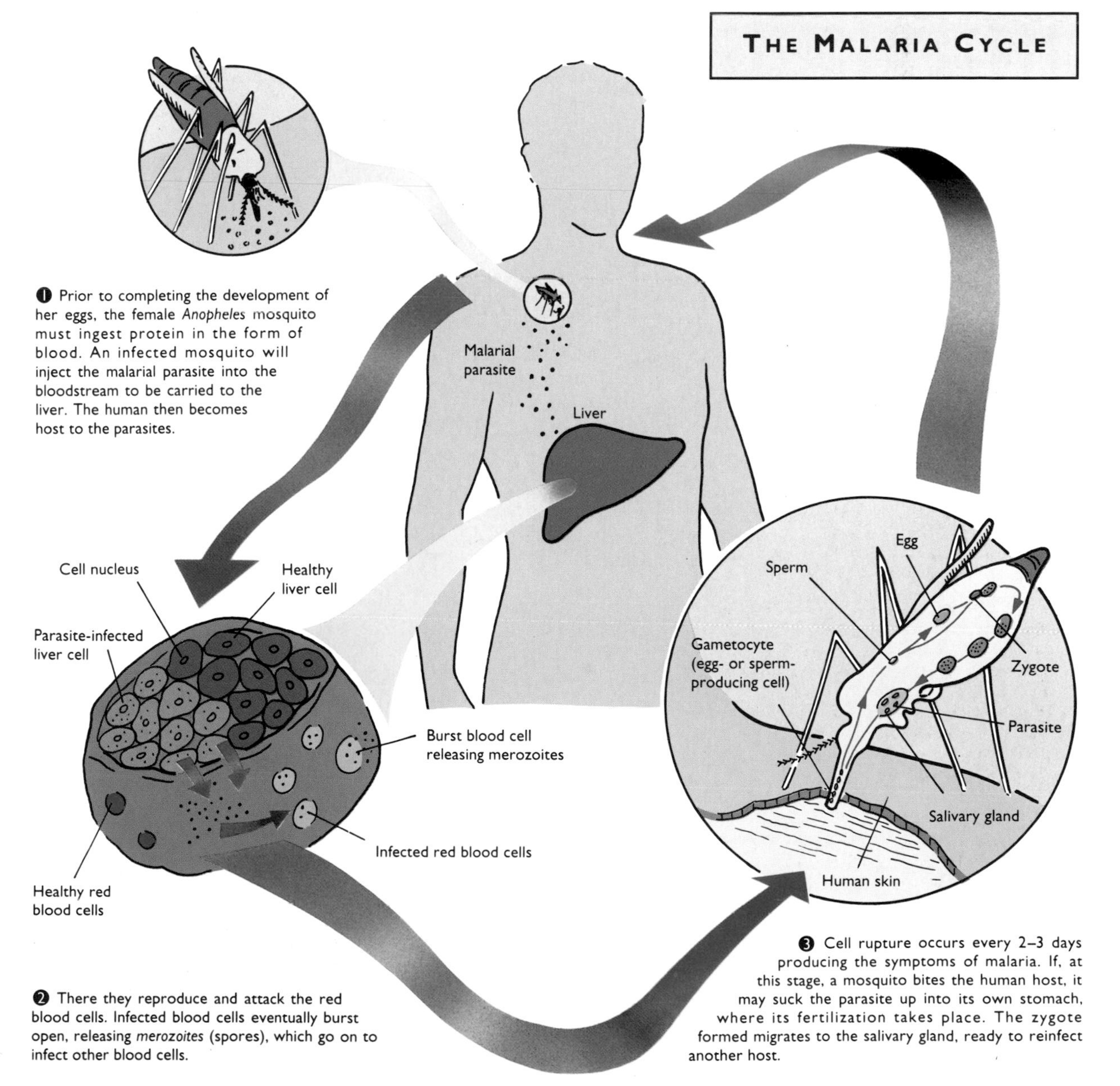

❶ Prior to completing the development of her eggs, the female *Anopheles* mosquito must ingest protein in the form of blood. An infected mosquito will inject the malarial parasite into the bloodstream to be carried to the liver. The human then becomes host to the parasites.

❷ There they reproduce and attack the red blood cells. Infected blood cells eventually burst open, releasing *merozoites* (spores), which go on to infect other blood cells.

❸ Cell rupture occurs every 2–3 days producing the symptoms of malaria. If, at this stage, a mosquito bites the human host, it may suck the parasite up into its own stomach, where its fertilization takes place. The zygote formed migrates to the salivary gland, ready to reinfect another host.

Heart and Lung Diseases

Oxygen is acquired via the lungs and distributed in the blood by the pumping action of the heart (⇨ pp. 50–3). In addition, the heart must maintain the circulation and supply of all the body fuel, especially glucose and fatty acids, and the many other cell nutrients required to maintain body function. These secondary functions become academic, however, if the heart or lungs fail to keep up the supply of oxygen.

In the West, diseases of the heart and lungs together cause more deaths and serious loss of quality of life than all other diseases put together. Heart and lung diseases cause over half the deaths in developed countries and, ironically, much of this disease can be attributed directly to the indulgence that high standards of living allow.

Girl using an inhaler to control an asthma attack, a disorder of the bronchial tubes that leads to breathing problems. The inhaler administers a dose of bronchodilator, an agent that widens the air passages, bringing about rapid relief. (John Durham, SPL)

The significance of atherosclerosis

The most important disorders of the heart – heart attack (*myocardial infarction*) and *angina pectoris* (chest pain caused by a failure of the coronary arteries to supply enough blood) – are secondary to one particular disease of the arteries. This is atherosclerosis, a disorder from which few of us, however young, are entirely free. Essentially, atherosclerosis is a disease of large and medium-sized arteries throughout the body, on the inner surfaces of which raised plaques of cholesterol, protein and degenerate cells appear. These plaques are composed of atheroma, which first makes its appearance in early childhood as fatty streaks in the artery linings. In the teens these become perceptibly raised, and, by the age of thirty, most people have obvious plaques that reduce the bore of their arteries. The effects seldom become manifest, however, until middle age or later, and the event that usually demonstrates the presence of atheroma is the formation of a blood clot, on top of a plaque, large enough completely to close off the artery. Blood will not clot inside a healthy artery because, in that circumstance, some of the substances necessary for coagulation are absent. But damage to the internal lining by atheroma allows the blood to have access to these substances, which are present in the tissues. Atherosclerosis also weakens arteries, allowing them to balloon out under pressure to form *aneurysms* (widened sections).

Atherosclerosis affects all arteries, but obstruction is particularly liable to occur in arteries of a certain size, because of the proportions of the elastic and muscular elements in their walls. For this reason sudden closure occurs more commonly in the coronary arteries of the heart, in branches of the four arteries that supply the brain, and in branches of the arteries that supply the legs. Obstruction of these causes, respectively, heart attacks, strokes (*cerebral thrombosis*) and the death of tissues in the legs (*gangrene*).

Angina and coronary thrombosis

The heart is a muscular pump that works continuously and very hard. To maintain this workload, a constant supply of oxygen and fuel is needed by the heart muscle. The coronary arteries, which run over the surface of the heart, can provide these essential supplies so long as they are not unduly narrowed. If the heart is called on to perform work in excess of the available fuel supply, it inevitably protests. When muscle contracts without an adequate blood flow through it, the acidic waste products of the process of metabolism accumulate and these then cause the pain known as angina. Angina, however, is not a disease but a symptom of coronary narrowing, and is usually brought on by a certain amount of exercise, such as walking a particular distance. The pain varies greatly in intensity and character but is often of a gripping or crushing nature, with a burning quality. It starts in the centre of the chest and may extend through to the back, up into the jaw and down the left arm or both arms. As may be expected, it is rapidly relieved if whatever exertion has caused it is then stopped. It may also be relieved by taking a medication, such as nitroglycerine, that temporarily widens arteries. Angina may be brought on or exacerbated by intense emotion and is often worse after large meals or when the affected person is out of doors in cold, windy weather.

Severe pain that is not relieved by rest or nitrates probably signals a heart attack. The affected person often suffers an acute sense of anxiety and impending death. Severe breathlessness and vomiting are common, and the affected person's pallor and expression immediately convey the gravity of the situation. This condition is due to obstruction of a coronary artery branch from the formation of a clot on top of an atheromatous plaque – a coronary thrombosis – with the effect that part of the heart muscle is immediately deprived of its blood supply, and dies (myocardial infarction). Death of a small part of the heart muscle may be followed by healing and scar formation, with a variable degree of weakening of the pumping action. If the blocked artery branch is large, however, so much heart tissue dies that recovery may be impossible. Cardiac arrest occurs, and, unless the individual is resuscitated by means of cardiac massage, death occurs.

Other heart disorders

- Heart failure – the condition in which the heart is unable to pump blood efficiently enough to prevent stagnation and the accumulation of fluid in the tissues (*oedema*). Left heart failure causes oedema of the lungs (blood from the lungs returns to the left side of the heart), leading to a damming back of blood in the lungs, with leakage of water from the blood into the lung tissues. Right heart failure causes oedema in the general body tissues, because, in this case, the blood from the body returns to the right side of the heart.
- Heart block – the result of damage to the conduction tissues of the heart, usually from myocardial infarction. Heart block interferes with the regulation of the heart rate and often requires pacemaker implantation.
- Heart-valve disorders – inflammation of the heart lining, usually from rheumatic fever or congenital defects. Valves may be narrowed (*stenosis*) or incompetent and many severe secondary effects follow.
- Acute myocarditis – inflammation of the heart muscle, which may be caused by various infections and toxins, such as diphtheria.
- Pericarditis – inflammation of the two-layered fibrous bag that surrounds the heart.
- Congenital heart disease – a variety of structural heart defects present at birth.
- Cardiomyopathy – a heart-muscle disorder that may be caused by, among other things, anabolic steroids or even excessive athletic activity.

Chronic obstructive airway disease (COAD)

COAD is a general term for a group of conditions in which, for one reason or another, the passage of oxygen from the atmosphere to the blood is hindered. This may occur because of structural damage to the air tubes, as in chronic bronchitis and *bronchiectasis* (⇨ below), or because the immediate interface between the air in the lung and the haemoglobin in the blood has become diminished.

Chronic bronchitis involves inflammation and thickening of the air-tube linings, excess mucus production, diminished mucus transport and infection so that the mucus turns to pus. There is constant coughing of purulent (pus-containing) sputum. Because the obstruction is greater during expiration, air becomes trapped in the air sacs and these become overdistended with the result that they rupture, causing the disease of *emphysema* (⇨ below).

Bronchiectasis is a persistent inflammation of the smaller air tubes, leading to softening and local widening of areas in which infected sputum becomes trapped.

SMOKING – THE CONSEQUENCES

Smoking cigarettes is now known to be the most important single, readily preventable factor in the causation of diseases of the heart and lungs. Although atherosclerosis (⇨ main text) is clearly also related to other risk factors such as diet, heredity and cholesterol metabolism, there is an unequivocal correlation with smoking. Cigarette smoke contains over 1200 different chemical substances, some of which act directly on the lung tissues, and others of which are readily absorbable into the blood and from there are able to act on any of the body tissues. Mortality studies have shown that in men who smoke one or more packets of cigarettes a day, the death rate from coronary heart disease is from 70–200% higher than in non-smokers. Recorded increases in coronary heart disease in women are attributed to the same cause.

Comparisons of postmortem examination between the arteries of smokers and nonsmokers also show an obvious correlation between the degree of atherosclerosis and smoking. People who have heart attacks are more likely to die if they are smokers than if they are nonsmokers. The obstructive arterial disease *thromboangiitis obliterans*, which commonly leads to amputation of both legs because of gangrene, occurs almost exclusively in smokers.

The correlation between smoking and lung cancer, chronic bronchitis and emphysema is equally well established. The evidence is three-fold – statistical, clinical and experimental. Heavy smokers produce 20 times as many lung cancers as nonsmokers. Those who stop smoking, however, return to control base-line in about 10 years. Examination of the lungs of smokers shows typical precancerous cell changes in 10% of cases. Of those who die of lung cancer, 96% show precancerous changes in other parts of the lungs also. In laboratory experiments, many of the substances found in cigarette smoke have been found to be capable of causing cancer in animals.

Although smoking is on the decrease in the West (above), the take-up rate among young women is going up. Among smokers as a whole, women are also more reluctant than men to give up. (Adam Hart-Davis, SPL)

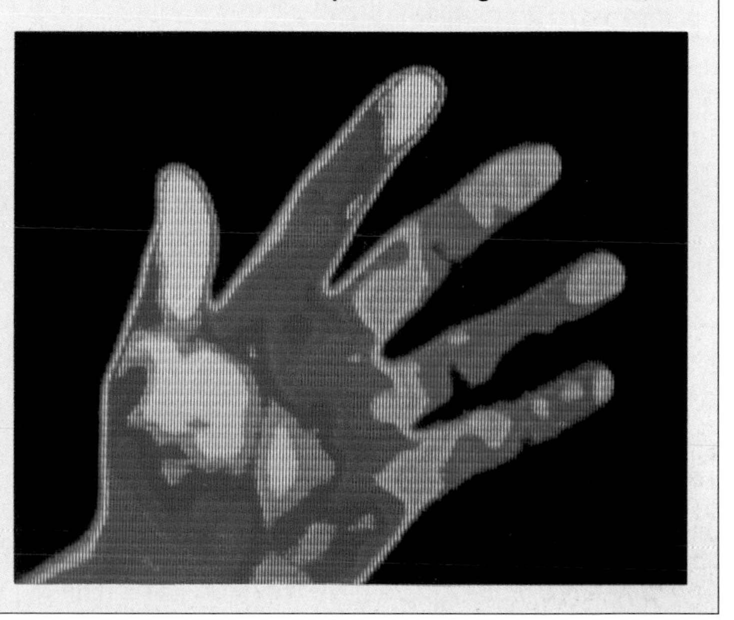

Smoking damages the blood circulation, as this thermogram (heat-distribution image) of a smoker's hand demonstrates. The three fingers on the right, coloured blue, are cooler than they should be, because of a reduced blood supply. (Alfred Pasieka, SPL)

The infected person suffers a constant cough and produces copious sputum, which is occasionally blood-stained. The condition features weight and appetite loss, lassitude, night sweats and, in children, failure to thrive. Bronchiectasis is common in people with *cystic fibrosis* (an inherited disease affecting the glandular tissues) and following pneumonia and other lung diseases.

In emphysema, the tiny air sacs, the *alveoli*, break down to form larger cavities whose internal surface area for gas exchange is much less than the former total surface area. The result of COAD, from this or any other cause, is a reduction in the oxygenation of the blood, leading, in some cases, to severe disability, with blueness of the skin (*cyanosis*), and severe breathlessness, even at rest.

Other lung disorders

- Bronchial asthma – spasm of the circular muscles of the air tubes (*bronchi*), often as a result of allergy, causing obstruction to the free flow of air into the lungs.
- Lung cancer – cancer of the lining of the air tubes (*bronchial carcinoma*), which is nearly always fatal.
- Respiratory distress syndrome – in newborn premature babies, this is due to lack of a wetting agent (*surfactant*) that allows the lungs to expand. In adults it is due to any circumstance that causes the lungs' air sacs to fill with fluid so that oxygen cannot pass. Causes include pneumonia, inhalation of toxic gases, drowning, surgical shock, and overdosage with certain drugs.
- Lobar pneumonia – lung inflammation in which the lung becomes solid with fluid.
- Bronchopneumonia – patchy lung inflammation involving the smaller air tubes.
- Tuberculosis – areas of local lung damage with breakdown to form masses of cheesy material (*caseation*) and cavity formation. Now rare in the West except in people whose immune systems have been compromised (as with AIDS).
- Silicosis – a form of lung scarring caused by inhalation of silica dust.
- Asbestosis – a chronic disease induced by the inhalation of asbestos fibres, causing widespread lung scarring, predisposing to lung cancer.
- Fungus infections of the lungs – rare except, again, in people whose immune systems have been damaged.

SEE ALSO

- THE BODY MACHINE pp. 50–3
- HEALING AND REPAIR p. 62
- HEALTH AND DISEASE pp. 120–7
- CANCER p. 140
- ENVIRONMENT, OCCUPATION AND HEALTH p. 144

Cancer

Cancer is much more readily treatable now than it was even a few decades ago as our understanding of its nature has increased remarkably in recent years.

As a cause of death in the West, cancer comes well behind diseases of the heart and blood vessels. These are responsible for about half of all deaths, while only about 20% of deaths are due to cancer. In less well-developed countries, however, the proportion of deaths caused by cancer varies somewhat since specific types of cancer are much commoner in some areas than in others.

Benign and malignant tumours

Cancer is characterized by the presence of malignant tumours in the body. Unlike a benign tumour, which remains localized and is usually surrounded by a dense capsule, a malignant tumour grows into adjoining tissue in an irregular manner, invading and destroying it. Malignant tumours also commonly seed themselves off to other parts of the body by making their way into blood and lymph vessels. Spread to areas remote from the site of origin is called *metastasis*, and new colonies of tumours that have arisen in this way are known as *metastases*.

Because there is a diversity of tumour types and they have different locations and effects, a rigid distinction between benign and malignant tumours is not always realistic. Some 'malignant' tumours are hardly dangerous at all; some 'benign' tumours may, because of their location, be highly dangerous. A *meningioma* growing inside the skull is a 'benign' tumour, but, because it pushes into and compresses the brain, it may cause severe damage and even death. Since they are encapsulated, many benign tumours can, however, be 'shelled out' and completely removed by surgery, effecting a complete cure. The possibility of complete surgical removal of a malignant tumour, such as a *glioma* (tumour of the binding or 'glial' tissues) in the brain, is much lower, first, because the limits of the tumour may be impossible to define, and, second, because many malignant tumours have already metastasized before they can be treated.

All tumours consist of both a cell mass that divides and reproduces to an abnormal degree, and the supporting mass of connective tissue and blood vessels. In common with other body cells, tumours cannot survive without a good blood supply. This they provide for themselves, but not always very efficiently. If a tumour outgrows its blood supply, part of it will inevitably die. Tumours occur only in cells that are capable of reproduction. They are found most commonly in organs' surface cells (*epithelia*) that reproduce frequently. And they occur often in the lining of the large intestine and of the air tubes (*bronchi*) of the lungs.

The nature of cancer

In a sense, cancer is a single disease – an uncontrolled and often dangerous growth of a particular tissue. But this concept is less useful, medically, than the recognition that there are hundreds of different cancers, each arising from a particular type of cell or tissue in a particular way, and each having its own characteristics. Some cancers, for instance, are so malignant that a diagnosis is tantamount to a death sentence. Others have such low malignancy that doctors will usually decide that no treatment is required. Some cancers remain localized to the area of origin for years and spread only, if at all, at a very late stage in their development. Others, as we have seen, seed themselves throughout the body at such an early stage that this will almost always have happened by the time they are diagnosed. Some cancers are very sensitive to radiotherapy and to chemotherapy while others hardly respond at all. Some cancers that, a few years ago, were invariably fatal, are now usually cured; others continue to resist any form of treatment other than early surgery.

Normal cells grown in a tissue culture in a laboratory continue to divide until they form a complete single layer over the surface of the culture medium. At this point they stop dividing because of the phenomenon of *contact inhibition*. Normal cells that are in firm contact with other cells or some barrier, stop reproducing. If cancer cells are allowed to grow in culture – which they do readily – they do not show contact inhibition but reproduce to form piled-up disorderly masses. Normal

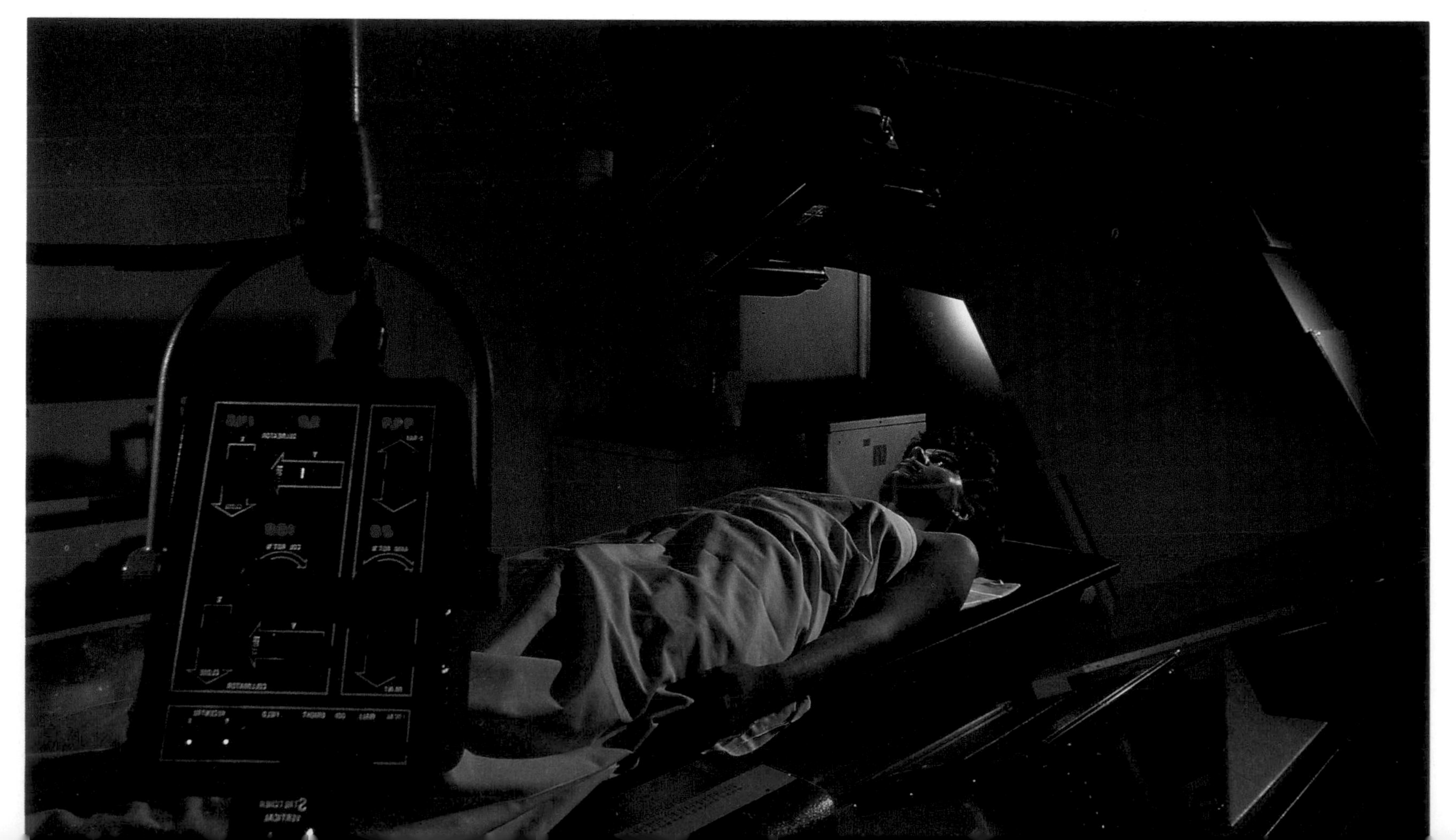

Massive x-ray machines are used in the treatment of cancer. The heavy doses of radiation can be focused so as to have the maximum effect. (Images)

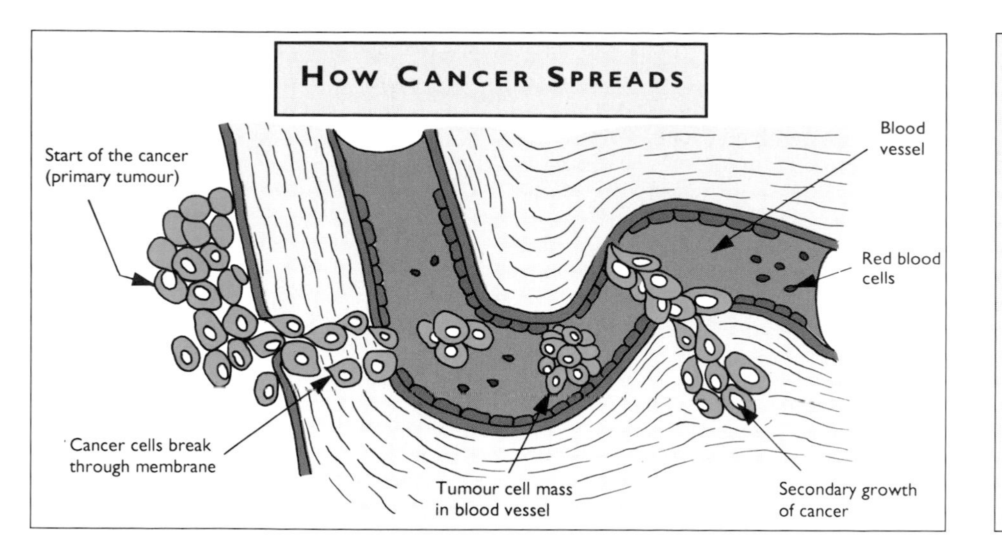

Cancer Terminology

The names for tumours always end in 'oma', which is Greek for 'root' or 'tuber'. In the case of benign tumours the ending is simply added to the name of the tissue. Thus, a benign tumour of fat is a *lipoma*; of muscle, a *myoma*; of bone, an *osteoma*; of glandular tissue, an *adenoma*; of fibrous tissue, a *fibroma*, and so on. Malignant tumours can be divided into two large groups: those that arise from surface tissues and those that arise from deeper connective tissues such as bone, cartilage, muscle, fibrous tissue and glial tissue. Tumours of surface tissues are called *carcinomas*, and tumours of connective tissues are called *sarcomas*. So an *adenocarcinoma* is a malignant tumour of glandular epithelium, while an *osteosarcoma* is a malignant tumour of bone. Tumours are also often named from the tissue or organ from which they arise. Thus the common lung cancer is usually a *bronchial carcinoma* because it actually occurs in the lining of the air tube (*bronchus*). A *melanoma* is derived from the pigment cells, the melanocytes (➪ p. 56). Carcinomas are much commoner than sarcomas.

cells will usually grow only when firmly anchored to a solid surface; cancer cells will grow regardless. Cultures of normal cells have a definite life span and eventually die, while cultures of cancer cells appear to be immortal and can be subcultured indefinitely. Indeed, some tumour cell lines that were derived from a single cancer have been maintained and used in laboratories all over the world for decades – long after the death of the human beings in whom they originated.

Most human cancer cells show chromosomal abnormalities with some cells having several hundred chromosomes rather than the usual 23 pairs (➪ p. 22). These abnormalities are common to all cancers of a particular type and are probably the reason for the change in the behaviour of cancer cells. Specific gene abnormalities have been identified for most *leukaemias* (blood cancers) and for a number of other cancers.

The causes of cancer

Numerous agents have been shown, unequivocally, to cause cancer. These can be divided into three groups – chemical agents, radiation and viruses. Cancer researchers now believe that at least 80% of cancers are caused by environmental factors, including a person's lifestyle. Causal agents may act separately or co-operatively and there is evidence that the transformation of a cell from a normal to a malignant form does not occur suddenly, but is progressive, involving many stages and repeated attacks by the causal agent.

Research also suggests that the attack on cells that renders them malignant operates on two sets of genes that are present in every cell – the *proto-oncogenes* (the precursors of cancer genes, the *oncogenes*) and the cancer suppressor genes, the *anti-oncogenes*. Proto-oncogenes play an essential part in regulating normal growth and differentiation of cells, but mutations in these genes, caused by chemical carcinogens, viruses or radiation, alter their function so that the cell concerned begins to grow in an abnormal way.

The number of chemical substances known to cause these mutations is very large. Some of them have been known to be carcinogens for well over 200 years. Scrotal skin cancer was a well known hazard for chimney sweeps until the English surgeon Percivall Pott (1714–88) identified soot as the cause and eliminated the problem by recommending daily washing. Shale lubricating oil thrown out by high-speed spindles caused the same disease in Lancashire thread spinners. Since then hundreds of carcinogenic substances have been identified that act on DNA to cause the chemical changes in the bases (➪ p. 24) that we call mutations.

Radiation acts on DNA in a similar way to cause alterations in the bases and to change the genetic code. Research has demonstrated that the ultraviolet component in sunlight is implicated in causing the three common kinds of skin cancer – basal-cell carcinoma, *squamous* (flat cell) *carcinoma* and malignant melanoma (➪ p. 56). Ionizing radiation, such as x-rays, gamma rays, alpha and beta particles, protons and neutrons, are all carcinogenic and are known to cause leukaemia and other cancers.

Many viruses are known to cause cancer in animals, and a few, such as the papilloma virus, the Epstein–Barr virus, the hepatitis B virus and the human T-cell leukaemia virus, have been shown to cause cancer in humans. Viruses that invade human cells commonly incorporate a part of their own DNA – sometimes the oncogene section – into that of the host cell (➪ p. 132) and at least 20 viral oncogenes have been identified. Inserted into the host DNA, they cause the cell to become malignant. It is also known that certain viruses act on normal body cells to convert their proto-oncogenes to oncogenes.

How does cancer affect the body?

Cancers are parasitic and thrive at the expense of the host. Primary and secondary tumours (metastases; ➪ above) function in essentially the same way, causing local tissue destruction, ulceration, haemorrhaging, obstruction of hollow organs and endocrine effects by the secretion of large quantities of hormones or hormone-like substances. In addition, advanced cancers that have spread widely nearly always produce, in the terminal stages, severe loss of body fat and muscle, so that the affected person becomes greatly wasted and weakened. This is called *cachexia* and its cause remains obscure. In such cases it is usual for an otherwise inconsequential infection to prove fatal.

Treatment

The most important factors determining the success of cancer treatment are the nature of the tumour and the stage at which the diagnosis is made. All cancers are curable if detected early enough, and this is why regular medical checkups, screening programmes and awareness of the early signs of possible cancer are so important (➪ p. 124). Mammography, breast self-examination, cervical (or Pap) smear tests, stool tests for hidden (*occult*) blood, chest x-rays, rectal examinations and *sigmoidoscopy* (direct inspection of the lower colon through a tube); and early reporting of symptoms (such as unexplained cough, changes in the bowel habits, blood in the stools or urine, suspicious lumps and unhealing ulcers) all help to minimize the tragedy of cancers that are detected too late for anything useful to be done. If medical suspicion is aroused, current methods of body imaging, endoscopy, and the use of tumour markers can detect tumours at a stage at which modern surgery (➪ p. 154), chemotherapy and radiotherapy (➪ p. 156) can often be curative.

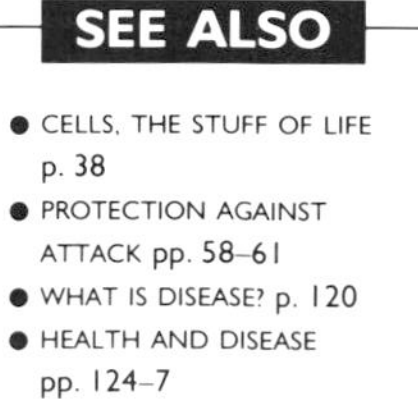
SEE ALSO
- CELLS, THE STUFF OF LIFE p. 38
- PROTECTION AGAINST ATTACK pp. 58–61
- WHAT IS DISEASE? p. 120
- HEALTH AND DISEASE pp. 124–7
- HEART AND LUNG DISEASES p. 138

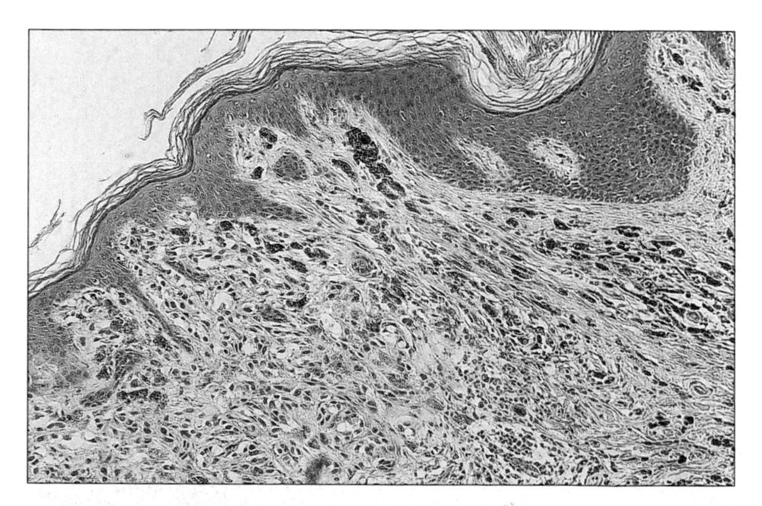

Skin cancer – a malignant melanoma. About half of all malignant melanomas arise out of moles on the skin, especially where the skin has been exposed, unprotected, to sunlight. Only about one mole in a million ever becomes cancerous, but changes to watch out for include darkening or irregularity of colour, a change in shape, itching or pain, or softening and crumbling of the mole's surface. Treatment, if given early enough, is highly successful. (RB)

Allergies

Allergy is a hypersensitivity of the body to any one of a wide variety of substances, so that when the sufferer comes in contact with the substance to which he or she is allergic, an unpleasant bodily response occurs. Allergic reactions take different forms, depending on the part of the body affected, and may concern the upper respiratory system, the lungs, the skin or the intestines. While some of these effects are comparatively mild, others can be severe or even fatal.

The term 'allergy' is often used loosely and incorrectly by the layperson. An allergy is, in fact, a disorder of the immune system (⇨ pp. 58–61), and many disturbances of the body, attributed to allergy, have nothing to do with that system. A number of substances can affect the body adversely and do so in a variety of ways, but these effects are not necessarily allergic. The idea of allergy has, however, caught the public imagination, and much misunderstanding has resulted. The 'total allergy syndrome' is a case in point. This is a disorder in which affected people are said to be allergic to almost anything, and readily suffer symptoms such as breathlessness and dizziness on exposure to certain stimuli. But the great majority, if not all, of such cases are of psychological rather than immunological origin. Food allergies, too, are less common than is generally supposed. Although about 25% of parents believe that their children are allergic to some food material, tests have shown that only about 1 child in 50 has a genuine allergy to a food constituent. Food *intolerance*, which causes a particular and local irritation of the gastrointestinal tract (without involving the whole of the immune system), is relatively common, but this has nothing to do with allergy.

- CELLS, THE STUFF OF LIFE p. 38
- OXYGEN: THE VITAL ELEMENT p. 52
- THE BODY MACHINE pp. 56–61
- HEALTH AND DISEASE pp. 120–3
- SYMPTOMS AND SIGNS p. 126
- ENVIRONMENT, OCCUPATION AND HEALTH p. 144

The nature of allergy

A substance that causes an allergic reaction is called an *allergen*. A particular allergen produces a particular response in the allergic individual, while a similar substance will not. This is called *specificity*. Every time the allergic person comes into contact with the allergen he or she will suffer the known effect, a phenomenon called *memory*. And this effect is usually greater with each subsequent contact (*amplification*).

Although many different substances can act as allergens, the majority fall into a comparatively small number of groups. These include tree and grass pollens, proteins from fungi, house dust-mite proteins, animal skin and scales, insect-sting venoms, drugs, chemicals, some metals, some food proteins and a number of vaccines.

The type of allergic reaction depends on which part of the body the allergen contacts. When an allergic reaction occurs in the nose or eye membranes, as in hay fever, the affected person begins to sneeze and suffers excessive watering, swelling of the membranes, along with itching in the nose and throat. These effects can be caused either by pollens or by animal dander (small particles of hair or feathers). When allergy occurs in the skin, there is local inflammation, either in the form of raised, pinkish areas (known as *urticaria*), or in the form of a much more persistent eczema or dermatitis. Allergy affecting the linings of the air tubes (*bronchi*) produces *asthma* – a tight spasm of the circular muscles in the wall of the tubes. When an allergen reaches the lining of the intestine, the resulting inflammation causes upsets such as diarrhoea and vomiting.

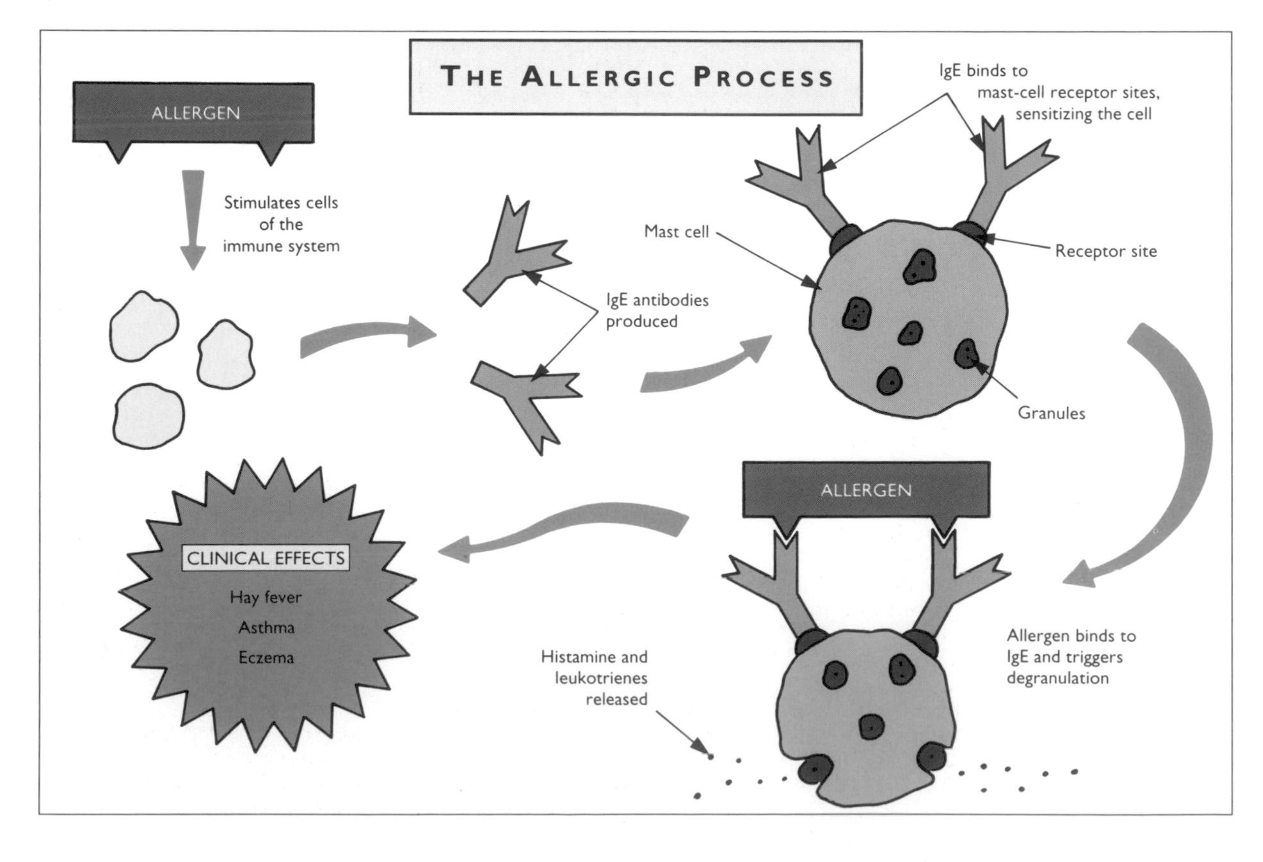

The causes of allergy

Most cases of allergy are caused by a reaction between the allergen and a particular kind of antibody, one of the five types, known as *immunoglobulin type E* (IgE). Throughout the body, especially near the body surfaces such as the *mucous membranes* (the linings of the mouth, nose, eyelids, intestine and vagina), are millions of fixed individual cells known as *mast cells*. These attract IgE molecules, which are commonly found firmly attached to particular receptor sites on the mast-cell outer membranes. If these cells are stained, they are seen to be filled with many large granules. These contain a number of powerful chemicals, including histamine and the leukotrienes, both of which have important roles to play in helping immune-system cells escape from small blood vessels so that they can attack disease-causing organisms. The release of these substances from the mast cells is called *degranulation*, and is controlled by the action of IgE.

Unfortunately, allergens that become attached to the IgE molecules on the mast cells can act as 'triggers' to this action, and can cause the simultaneous degranulation of large numbers of mast cells and an outpouring of histamine and leukotrienes into the surrounding tissues, causing inflammation. The pattern and severity of the resulting reaction depends on the number of mast cells in the area and the amounts of IgE they carry on their surfaces.

Skin allergies

Urticaria, commonly called 'nettle rash' or 'hives', features raised, intensely itchy, pinkish areas, surrounded by paler areas of skin, which last for half an hour to several days and then disappear completely. Urticaria can be caused by insect bites, jellyfish stings, contact with plants, drug allergy or sensitivity to food materials or food additives. Food dyes such as tartrazine, food preservatives, yeast, aspirin, penicillin and other drugs have also been found to have an allergic effect. Not all cases of urticaria are allergic in nature as it may also be caused by heat, cold or physical injury. In addition, emotional factors are often cited as being a cause, but whether they really are has not been proven. Sometimes the lining of the voice box (*larynx*) is affected by allergy and may swell to the point of cutting off the air supply. In such a case only urgent action, such as performing a *trachaeostomy* (operation to insert a breathing tube into the windpipe), can save life.

The skin complaint eczema is not a single specific condition, like impetigo or acne, but is the response of the skin to a wide variety of damaging influences, and is a

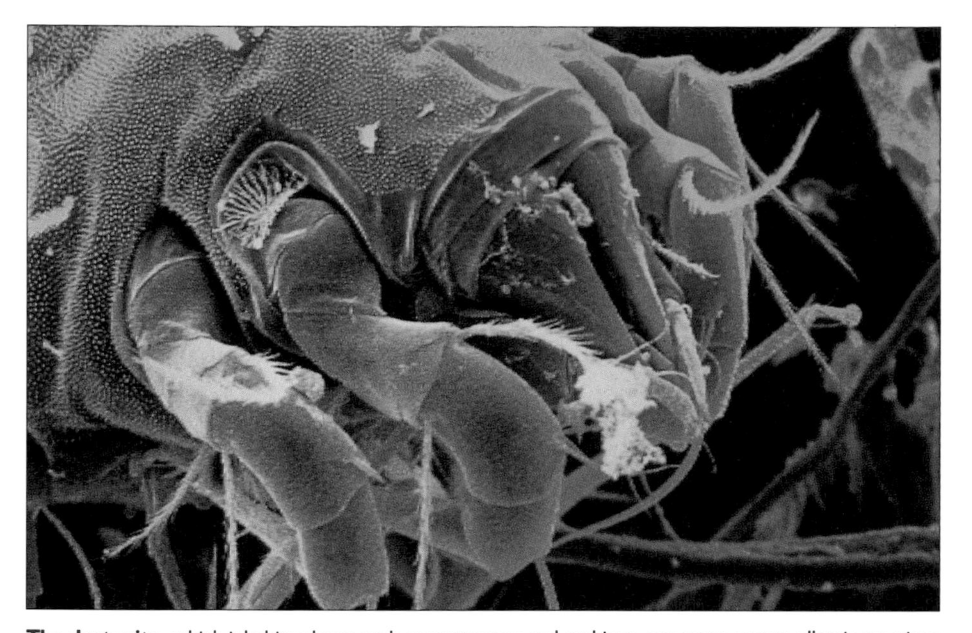

The dust mite, which inhabits places such as mattresses and cushions, can cause severe allergic reactions in the unlucky few, who must regularly vacuum their homes, keeping dust traps to a minimum. (Microscopix)

feature of many different skin disorders. 'Eczema' is a general dermatological term meaning inflammation of the skin, without implying any particular cause, and for practical purposes the terms 'eczema' and 'dermatitis' mean the same thing. Allergy is, however, a very common cause. Eczema features scaly, red, itchy patches, which when scratched lead to thickening and shiny areas with deepened skin markings. This thickening is called *lichenification*. Often, there are small fluid-filled blisters that leak serum, so that the skin becomes 'weepy'. Exuded serum quickly clots to form crusts. The damage to the skin surface may allow germs access to the deeper layers, where they can flourish, causing more severe inflammation and pus formation. This is called *secondary infection* and is a complication of the eczema, not a basic feature.

Eczema may affect any part of the skin, although certain parts are more commonly affected. In adults it normally occurs on the hands as a result of allergy to substances such as washing-up liquid; on the groin, from contact with biological washing powders; on the wrists, back or ears from contact with nickel watchstraps, bracelets, bra fasteners or spectacle frames; or anywhere else where there is contact with materials to which allergy has developed. Emotional upset and stress may also precipitate the eczema. Contact dermatitis also occurs in babies and children, but for them, the common type of eczema is called *atopic*. This word comes from the Greek *a-*, meaning 'not', and *topos*, 'a place', and simply means that, unlike contact dermatitis, the effect – the eczema – does not occur at the same place as the cause. In atopic eczema the allergic tendency is hereditary and the condition often appears in the first year of life. Affected children, or other members of the family, often have other allergies, such as asthma or hay fever.

Atopic eczema usually affects babies between the ages of 2 and 18 months. There are many causes, a common one being allergy to protein in wheat, milk and eggs. A rash appears behind the knees, in the creases of the elbows, behind the ears and on the face. It is usually mild but very itchy, rather scaly and with small red pimples. It is almost impossible to prevent the baby from scratching, and soon the pimples begin to ooze serum and the affected parts may join to form large weeping areas. Secondary infection may then occur, particularly in the nappy area. Atopic eczema often clears up, even without treatment, as the child grows, and although it may come and go over the years, it usually settles completely by early adolescence.

Respiratory allergy (allergic asthma)

Asthma is usually atopic. During an asthma attack the circular smooth muscles of the bronchi go into a state of sustained spasm so that they become narrowed and the passage of air impeded. Often inspiration is easier than expiration, causing the lungs to become inflated and making the person wheeze when breathing out. The commonest cause of bronchospasm is sensitivity to an allergen, but it can also be induced by infection, emotion, and, in many asthmatics, exertion. Allergens responsible for causing asthma include pollens, house-dust mites, animal dander, feathers, drugs, foods and industrial chemicals.

Asthma is not a trivial condition and has a distressing, and rising, annual mortality rate. Self-help in avoiding danger is important, but this is possible only if the sufferer has knowledge of the signs of worsening of the condition and of the steps to be taken to overcome them. *Status asthmaticus*, a prolonged attack of severe asthma, is very dangerous and calls for urgent medical attention. The same applies to cases of progressive worsening, with reduced response to the normal remedies.

— How Allergies Are Treated —

Prevention of allergy is better than cure, and allergens should be avoided if at all possible. The second line of attack, however, is drug treatment.

Antihistamine drugs prevent the histamine released by mast-cell granules from acting on other tissues. These drugs are chemically similar to histamine and occupy the same tissue receptors, though without stimulating them. In this way they block the action of the histamine. These drugs are usually given by mouth but may be administered by intravenous injection if necessary. Unfortunately, they have little effect in allergic asthma. Antihistamines should be taken with care as they often cause sleepiness.

Sodium cromoglycate is a drug that stabilizes the cell membranes of the mast cells and so reduces their tendency to release histamine and leukotrienes. This drug is useful in any IgE-mediated allergic condition, including asthma.

The hormone adrenaline and similarly acting drugs such as isoprenaline, aminophylline and salbutamol have effects that immediately oppose those of histamine. In severe cases of allergy such drugs, or adrenaline, given by injection, can be life-saving. Corticosteroid drugs have powerful actions in suppressing inflammation and these, too, can be life-saving. Even in less severe cases the use of steroids can be helpful, especially if they can be directed to the affected area, as in their use by means of inhalation in the treatment of asthma.

Because all allergic reactions are dependent on the immune system it may sometimes be necessary, in severe cases, to interfere with the functioning of this system, as by deliberately attacking the cells that produce antibodies – the B lymphocytes. Drugs that prevent such cells from dividing are called *immunosuppressive* drugs.

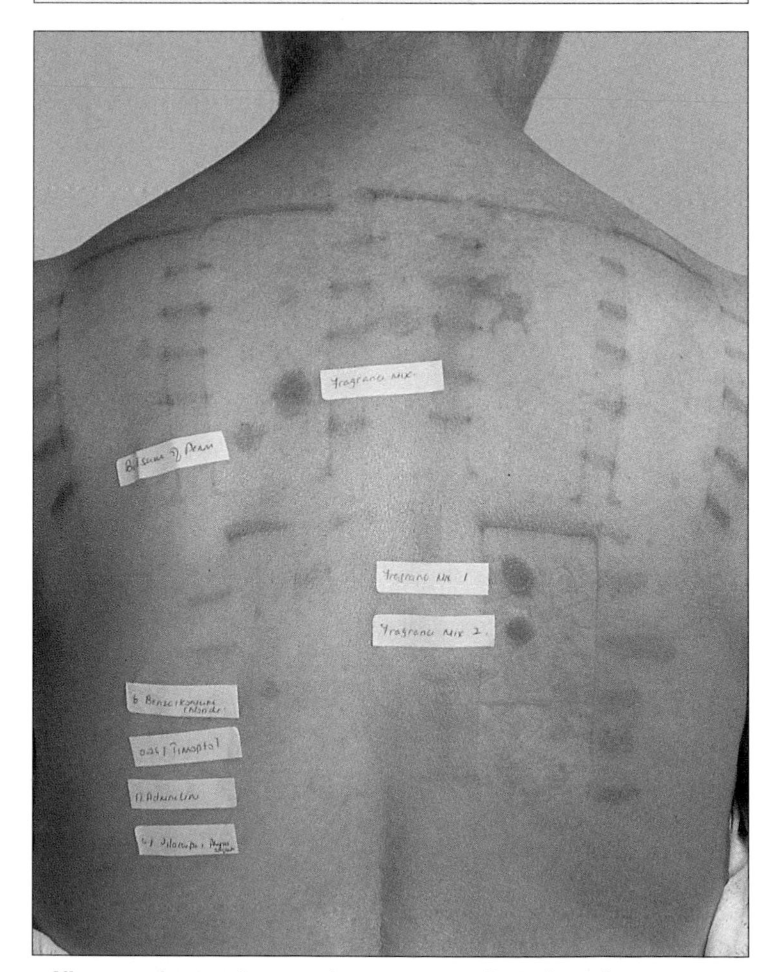

Allergy testing: patch test results on a patient suffering from allergic contact dermatitis. Small quantities of various known allergens are applied to the skin and covered with adhesive tape. Often an allergic reaction will produce a flaring of the skin within about 15 minutes, but a delayed reaction could take up to 72 hours to develop. (John Radcliffe Hospital, SPL)

Environment, Occupation and Health

Once our basic constitution has been determined by genetics, our health depends on the balance between the many potentially damaging environmental factors that operate on us and on our ability to resist them. The general environment may be hostile; we may be exposed to excessively low or high temperatures, damaging solar radiation, wind storms, atmospheric pollution, a multitude of infective agents, polluted food and water, and other hazards, such as the motor car. The workplace, too, may contain many dangerous environmental factors.

SEE ALSO

- HEALTH AND DISEASE pp. 120–7 and 132–43

The first defence against damage to our health by the environment is information. Millions of human beings have, in the past, been adversely affected by environmental factors of which they had no knowledge. This was especially true of chemical or cancer-producing (*carcinogenic*) industrial hazards. Miners died of anthracosis, silicosis and asbestosis; hatters went mad and died of mercury poisoning; chimney sweeps and cotton spinners died of scrotal cancer from soot and shale oil; workers in the match industry died of jawbone erosion from phosphorus poisoning ('phossy jaw'); and pottery workers died of lead poisoning from contact with the glazes. Future generations may well look back with horror at the hazards to which we, today, are blindly exposing ourselves.

Bad housing and insanitary conditions, as here amongst the poor in Nepal, are important contributory factors to ill health. The difficulty in keeping food, for instance, uncontaminated by bacteria can lead to a host of diseases and disorders. (WHO)

Climatic factors in disease

Understanding climatic hazards can enable one to circumvent most of them. The ultraviolet component of sunlight, especially the short wavelength ultraviolet light (UVL) of 290–320 nanometres, is very damaging to the skin, particularly of white people. Regular sunbathers usually succeed in avoiding sunburn, but suffer damage to the elastic collagen tissue of the skin (⇨ p. 56) and experience premature skin ageing with atrophy, laxity, wrinkling, pigment patches and local overgrowth of the epidermis (*hyperkeratosis*). They also suffer a much increased incidence of the skin cancers – basal-cell carcinoma, squamous-cell carcinoma and malignant melanoma. Australia has the highest incidence of skin cancer in the world. In some especially sunny areas people over the age of 65 are particularly prone to developing basal-cell carcinomas, principally because of the fact that, being older, their cumulative exposure to damaging sunlight is greater.

Heat exhaustion in hot climates is caused by inadequate intake of fluid and salt to replace that lost in sweat. This condition causes headache, nausea, giddiness, loss of appetite and muscle cramps. The condition of *heat stroke*, however, is more serious and is due to a failure of the body's heat-regulating mechanism, as a result of the brain overheating. The affected person finds that he or she stops sweating, and the body-core temperature rises to dangerous levels. Unless the body is rapidly cooled, this condition can be fatal.

Hypothermia, on the other hand, is due to the body having inadequate insulation in very cold conditions, so that the core temperature of the body drops – sometimes as low as 25 °C (or 77 °F; normal body temperature is about 37 °C/98 °F). The person in question becomes pale, stiff and loses consciousness, and will die unless gradually rewarmed, preferably in a space (aluminium-insulated) blanket. The person may also require medical treatment to safeguard the heart and prevent pneumonia.

Mountain sickness is a serious disorder affecting people who climb to high altitudes too quickly. The condition occurs above altitudes of 2500 m (8200 ft), although some people can go to 5500 m (18 000 ft) with impunity. The low oxygen tension (⇨ p. 52) that these altitudes induce causes headaches, lassitude, nausea, vomiting, dizziness, muscle weakness and breathlessness. The main hazard is the risk of suddenly developing fluid accumulation (*oedema*) of the lungs or the brain. Both conditions are rapidly fatal and can be prevented only by recognizing the early symptoms and bringing the victim quickly down to lower altitudes. Giving oxygen as soon as the first symptoms are observed is also beneficial.

Atmospheric pollution

The exhaust emissions from internal combustion engines produce nitrogen oxides and other substances that are acted on by sunlight to produce high levels of oxidants, especially ozone, and these are very hazardous to health. For example, they increase the risk of respiratory diseases such as chronic bronchitis and emphysema (⇨ p. 138). While general atmospheric pollution cannot begin to rival the local air pollution implicit in cigarette smoking, it is, nevertheless, also a contributory factor to the development of chronic bronchitis. Statistics have shown that people exposed to severe urban and industrial air pollution suffer more from these conditions and have a higher mortality than those who are not. Atmospheric pollution is especially hazardous for elderly people and for those already suffering from lung or heart diseases, and there have been numerous occasions when heavy smog has resulted in a sharp rise in the mortality rate in such groups. This happened in London in 1952 and 1956 during the domestic coal-burning era, and in Los Angeles in 1954 and 1955. From time to time atmospheric pollution reaches disaster levels, as in the methyl isocyanate release in Bhopal, India, in 1984 when 2000 people died from acute poisoning and lung oedema, and many others suffered persistent corneal ulceration and other severe effects.

Lead poisoning is another important element in atmospheric pollution. Most of it comes from vehicle exhausts, the rest originating in lead smelters or the burning of old car batteries. Most of such atmospheric lead is deposited on the ground, but some may be unwittingly inhaled. It may also contaminate growing fruit and vegetables. The main source of serious lead poisoning in children, however, comes from lead-based paint. If they ingest such paint, the lead becomes bound to their red blood cells and can lead to brain poisoning. Normally, children around 10 years of age will have about 10 micrograms (μg) of lead per 100 ml of blood. But clear evidence of learning difficulties is found in children with about 50 μg per 100 ml.

Industrial hazards

The main industrial hazards fall into seven categories. These are: mechanical, chemical, biological, radiational, acoustic, ergonomic and psychological.

Mechanical hazards. Much attention has been paid to the risks from dangerous machinery and a great deal of legislation has been drafted to ensure the safety of workers from this cause. If a potentially dangerous piece of machinery is well designed, it will automatically shut itself off if any part of the worker's body is interposed between dangerous machine parts. Even so, serious industrial accidents from machinery are still common, usually because the machinery has been used carelessly or because the safety precautions have been flouted. Workers' limbs and scalps continue to be torn off; hands are crushed; and eyes are still lost because of penetrating high-speed particles. Agricultural machinery, in particular, causes many serious accidents.

Chemical hazards. Among the more common disease-causing chemicals are:

- Alcohols, especially methyl alcohol, which can cause blindness.
- Ammonia vapour. Inhalation can cause

death from lung or from laryngeal oedema.
• Arsenic, which causes vomiting, diarrhoea, skin rash, nerve paralysis and death.
• Carbon disulphide. This grain fumigant and solvent causes fatigue, insomnia, loss of appetite and weight, arterial disease and coronary thrombosis.
• Carbon monoxide, which causes loss of memory, mental disability, irritability, personality changes and violence.
• Chlorine, which causes severe respiratory irritation, coughing, chest pain, dizziness, oedema of the larynx and cardiac arrest.
• Corrosive poisons. Accidental ingestion causes damage to the mouth, throat and gullet and leads to severe narrowing or even obstruction of the latter.
• Cyanide, which is one of the most poisonous substances known. Enormous quantities are used in industry. Acute poisoning is rapidly fatal.
• Glycols. Ethylene glycol (antifreeze) produces drunkenness and breaks down to toxic products. A dose of 100 ml can be fatal.
• Hydrocarbons. There are thousands of these products, some of them highly toxic or fatal on inhalation. Many deaths occur from deliberate solvent abuse, of which these are constituents.
• Hydrogen sulphide. This gas, which smells of rotten eggs, can cause eye irritation, mental confusion, lung oedema, coma and death.
• Metal salts. The most dangerous of these are salts of lead, mercury, cadmium, thallium and aluminium.
• Pesticides. The organophosphorus group of pesticides can produce paralysis and death.
• Phosphorus. Red phosphorus is fairly safe, but yellow phosphorus, used as a rat poison and in fireworks, is very toxic, causing nausea, vomiting, diarrhoea, collapse and liver failure.
• Polychlorinated biphenyls. These liquids are used as electrical insulators, heat-transfer agents, coolants and in hydraulic systems. They are persistent and are widely distributed. Poisoning causes skin rashes, watering of the eyes, headache and sometimes severe liver damage. Pregnant mothers exposed to them may miscarry or produce small babies with congenital abnormalities.
• Dioxin. This organo-halogen compound is a highly poisonous by-product of the manufacture of chlorinated phenols. It was a contaminant of the herbicide 2, 4, 5-T (Agent Orange), widely used in the Vietnam war. It is believed to have brought about the deaths of thousands of animals and to have caused many human abortions and congenital abnormalities.
• Solvents. These are most dangerous if abused for their intoxicant effects. They include acetone, amyl acetate, benzene, bromochlorodifluoromethane, butane, carbon tetrachloride, toluene, trichloroethylene and Xylene.

Biological hazards. Livestock handlers and veterinary workers often contract diseases from animals (*zoonoses*; ⇨ p. 130). These include anthrax, tuberculosis, Q fever, salmonella infections, sleeping sickness, rabies, toxoplasmosis, trichinosis and very many other diseases. Exposure to plants and plant products may often induce allergic reactions in the skin as well as asthma and hay fever (⇨ p. 142). It goes without saying that health-care workers are exposed to infectious disease, and special concern has arisen about the risks of HIV infection to surgeons and others exposed to contact with patients' blood. The risk of this seems smaller, however, than had at first been feared. Of 5425 cases of AIDS in American health-care workers only three had well-documented exposure to patients' HIV-infected blood. It seems probable that, at the most, about 5% were infected as a result of their work. Concerns have also been voiced as to whether patients may be at risk while being treated by HIV-positive medical personnel, although the risks seem similarly small.

Radiational hazards. There is clear evidence that radiation causes cancer. Early x-ray workers, unaware of the risk, often suffered cancers of the hand. And people employed to paint the dials of watches and clocks with radioactive luminous paint, accustomed as they were to licking their brushes to give them a fine point, developed bone cancer. In the modern era, evidence has accumulated about the raised incidence of leukaemia and other cancers in those exposed to radiation from fallout of atomic weapons. There seems to be no safe minimum dose, but the risk is clearly in proportion to the total cumulative dose. Even in the absence of nuclear fallout, though, all of us are exposed to ionizing radiation at all times, from radioactive materials in the earth and the atmosphere, from manufactured radiation, and from cosmic radiation arriving from outer space.

Acoustic hazards. Long-term exposure to loud noise is a more serious health hazard than is generally appreciated. It causes damage to the delicate hair cells in the cochlea of the inner ear with irremediable loss of hearing (⇨ p. 86). The increasing use of high-power audio amplifiers by rock groups and others will almost certainly lead to considerable high-tone deafness later in life for them and members of their audiences. Unnecessarily loud personal headphones, high-powered stereo amplifiers in cars and the use of rifles and shotguns without ear defenders are also likely to do permanent damage. Noise can also affect health by causing stress and sleep deprivation.

Ergonomic hazards. Ergonomics is the study of people in relation to their working conditions, and of how these conditions can be adjusted in the interests of comfort and health. A great deal of illness results from ergonomic failure. Many thousands of cases of back pain and resulting disability, for instance, occur every year simply because of unsatisfactory working conditions. Poor seating conditions and faulty lifting methods are often involved, and people working in offices are as liable to this problem as are manual workers. Other conditions attributable to ergonomic shortcomings include a range of musculo-skeletal disorders (including, perhaps, repetitive strain injury), postural defects, headache, eye-strain and general loss of skilled function.

Much of modern industry necessitates the use of dangerous corrosive or flammable chemicals, as in this paint-manufacturing plant. Protective clothing, face masks, and even, in some cases, a portable oxygen supply make it possible for human beings to work safely in otherwise hostile conditions. (Images)

Psychological hazards. Many people are forced by economic considerations to continue in working conditions that have become unpleasant or even abhorrent. There are many possible reasons for this, but not all such cases are obvious. Some problems that seem, at first sight, to have a purely physical basis are shown, on analysis, to originate mainly because of psychological stresses. Many cases of repetitive strain injury (⇨ above), for instance, have a strong psychological element.

People forced to work in unsatisfactory conditions may suffer depression, irritability, loss of appetite, insomnia and even loss of self-esteem. The problem can lead to severe psychological distress with family or marital upset, and alcohol or drug abuse.

Some people still face tremendous hazards in their working day. These gold miners operate in noisy, dusty, cramped conditions, with little light and not much air, all the time running the risk that the roof above them might collapse. (RB)

Drug Abuse

Drug abuse is the deliberate use of any drug for recreational purposes. Sustained abuse can lead to physical and emotional dependence, causing changes to the state of the affected person's physical and mental wellbeing.

People have always used drugs for pleasure, the most commonly used being alcohol, which, although dangerous in excess, is largely sanctioned by Western society. Until recently, nicotine (in the form of tobacco) was also widely sanctioned, but attitudes to this are changing because of the health risks that smoking presents to smokers and passive smokers alike. In many societies the use of other drugs, such as coca (cocaine) and hashish (cannabis) has also been widely accepted.

The medical definition of a drug is a very wide one and encompasses any substance outside the range of normal foodstuffs that has an effect on the body. Drugs may take any form – solid, liquid or gas – and may enter the body by any route.

Drugs of abuse

Drugs of abuse have conventionally been regarded as drugs that are prohibited by law. This is an unsatisfactory definition because legal drugs are far more often abused than those that are outlawed. Since the beginning of modern pharmacology, every drug with any noticeable effect on the state of the mind has been widely abused. In the 19th century *laudanum* (a solution or 'tincture' of opium in alcohol) was taken as commonly as aspirin or paracetamol is today. From the beginning of the 20th century the early sedatives, such as potassium bromide, were abused to the point of inducing toxic 'bromism'; then most of the *barbiturates* such as phenobarbitone, sodium amytal and sodium seconal were overused to the extent that millions of people became addicts. Today, the *benzodiazepines* (such as Valium and Librium) are building up their own toll of addicts. In each case the new group of drugs was initially hailed as a safe and nonaddictive sedative, and for each in turn medical attitudes were gradually amended in the same way. Much the same thing happened with drugs such as cocaine and the amphetamines. Both were regarded as medically respectable until they became widely abused.

Opium smoking reached the height of its popularity in Europe at the end of the 19th century when artistic, moral and social values were in a state of flux. It usually took place in opium dens, as depicted here in Paris c. 1900. Western society, however, now disapproves of the recreational use of opium and its derivatives, and regards the behaviour of those in the last century to be part of a general air of *fin-de-siècle* decadence. (AR)

Legitimate Use of Drugs

Any drug with a powerful action on the body, even if this action can be harmful, is likely to have some medical value. If drugs are abused, however, the disadvantages of continued use commonly outweigh the medical benefits, and these drugs may largely fall out of legitimate use. There remain, however, a number of drugs of abuse that retain substantial medical value.

The amphetamines, once widely prescribed, are now little used by doctors except to treat certain syndromes in children characterized by lack of attention coupled with hyperactivity, and to reduce the number of epileptic 'absence' attacks (*petit mal*). These drugs are strong appetite suppressants but are no longer generally approved for the treatment of obesity.

Nearly all Western countries have proscribed diamorphine (heroin) as too dangerous for medical use. British doctors, however, have retained the right to prescribe this powerful pain-killing and mood-enhancing drug to people suffering from incurable, painful and terminal diseases. Morphine continues to be widely used for pain relief and is considered by many to be indispensable.

The cannabinoids – the active principals in marijuana – have very limited medical use. The severe nausea and vomiting caused by anticancer treatment with *cytotoxic* (cell-damaging) drugs or radiation, however, can be relieved by the drug nabilone (trade name Cesamet), which is a synthetic cannabinoid substance.

The psychedelic drug phencyclidine ('angel dust') was used, under the trade name Sernyl, as a potent pain-relieving drug, but its hallucinogenic effects led to its withdrawal. A chemically related drug, ketamine, was, however developed, and has since been in wide use as an anaesthetic agent. Ketamine allows limited surgery, such as the setting of fractures, on conscious patients, although hallucinations may still sometimes occur.

Abuse and addiction

One of the main objections to drug abuse is that it is so often associated with *addiction* (dependence). This objection arises partly out of the fact that abuse and addiction are sometimes confused. Most of the drugs that are abused for pleasure are addictive if used repeatedly. The length of time for which a drug must be used before addiction is established varies with the personality and motivation of the user, and with the characteristics of the drug. In general, the more powerful

Freud and Cocaine

Sigmund Freud's (1856–1939; ⇨ p. 164) interest in cocaine was aroused by reports on its use to invigorate German Army troops during manoeuvres. Freud read an account of these successful experiments and decided to try the drug on patients with heart disease and on those with 'nervous exhaustion' following withdrawal from morphine. But first he decided to try it on himself when fatigued. The results so pleased him that he began to administer cocaine to friends, colleagues and patients.

In his paper 'Über Coca', published in 1884, he gave an account of the use of the leaves of the coca plant by South American Indians and described its effects on animals and humans. His conclusion was that the greatest value of the drug was in temporarily strengthening the body. He also recommended it as a means of relieving indigestion, and in the treatment of anaemia, asthma and addiction to alcohol and morphine. He was also convinced that the drug was an aphrodisiac, and referred briefly to the anaesthetic properties of the drug.

In Freud's absence, the idea that cocaine might aid ophthalmic surgery was tried out with great success by his colleague Carl Koller. Koller announced how easily the cornea could be completely anaesthetized for surgery. Within weeks the news had spread around Europe, the drug came into routine use, and Koller was famous. Thus, ironically, Freud was denied credit for the one really valuable application of the drug – its use as a local anaesthetic – while the other uses he suggested came, in the end, to be seen, by himself as well as others, to be as undesirable as they were misguided. After about 1886, Freud no longer took cocaine, or advocated its use as a general stimulant.

Drug Trafficking

The illegal manufacture and sale of drugs has become a major industry. The sums of money involved are enormous. In the USA, for instance, the annual turnover on drugs amounts to at least $80 billion – mostly for cocaine and marijuana (the dried leaves of the cannabis plant), with only about 10% of the value of the traffic being in heroin. Most of the cannabis is grown in Egypt, Morocco and Mexico, while cocaine is produced almost entirely in the South American Andes, especially in Bolivia and Peru, where the coca plant grows freely. The poppies from which opium is derived, and heroin synthesized, are grown mainly in the 'golden triangle' region of Thailand, Burma and Laos, and also in India, Afghanistan and Mexico.

Tens of thousands of tons of these drugs are illegally imported into the West, often by circuitous routes and under the aegis of criminal organizations that employ methods such as smuggling, bribery, intimidation, and even murder to maintain their lucrative business.

Heroin is usually smuggled in an almost pure form by couriers operating under various criminal organizations such as the Mafia and the Chinese triads, and then is mixed or 'cut' with inert powder to a purity of only about 5%, with proportionate increase in the profits. A vast organization of paid administrators and 'pushers' then distributes the drugs in the streets. One kilogram of base morphine can be purchased from the producers for about £10 000. After cutting, it has a street value of about £1.5 million.

Other drugs of the stimulant amphetamine class, and hallucinogens like LSD (lysergic acid diethylamide), mescaline, and psilocybin, are mainly produced synthetically in illicit laboratories throughout the world. Many of these laboratories concentrate on the production of what have become known as 'designer drugs'. These are established drugs that are chemically modified so as to be classifiable as new substances not covered by prohibitive or restrictive legislation. Some designer drugs, such as those derived from the drug fentanyl, have proved highly potent and very dangerous.

Different societies have had different ideas about what drugs were acceptable. Alcohol was banned in Prohibition America (1920–33), as it is today in Islamic countries. But controlling the individual use of banned drugs is very difficult. During Prohibition, bootleggers (suppliers of illicit alcohol) found a huge market for their wares, for which they often found ingenious hiding places.
(Image Select)

SEE ALSO

- WHAT IS DISEASE? p. 120
- WHAT IS HEALTH? p. 122
- PILLS AND POTIONS: THE HISTORY OF DRUG THERAPY p. 152

the effect and the more the effect is desired, the more readily addiction will occur. Similarly, if the person concerned finds that the drug seems to supply a deeply felt want or seems to compensate for a defect in the personality, then addiction is likely.

Medical professionals recognize an established pattern of drug-taking. Illicit drug takers usually escalate from 'soft' drugs, such as cannabis, cocaine and amphetamine, to the 'hard' drug heroin. Many de-escalate as they get older, however. A basic characteristic of addiction is that the affected person finds great difficulty in discontinuing the use of the drug. Many, indeed, make no effort to give up even after it becomes apparent that the habit is seriously affecting their quality of life. Indeed, maintaining the continuing supply of the drug seems to become almost the most important thing in the addict's life. Personal ethics or morality frequently become less important than keeping up the supply, and addicts who cannot obtain their drug otherwise, will, if necessary, lie, steal, or resort to prostitution. Addiction is also associated with self-destructive behaviour, in that addicts will often drop out of school, leave home or give up good jobs. The symptoms experienced when a person is forcibly deprived of a drug vary in severity. In general, however, the more easily addiction occurs, the more unpleasant tend to be the effects of physical and mental withdrawal. Most of these changes in behaviour are not caused by the effect of the drug on the brain. They are due either to the need to obtain supplies at all costs, or they occur as a response to society's attitude to drug-taking.

Dangers to health

The use of heroin is considerably hazardous, especially if the drug is used intravenously in order to obtain a 'rush' or 'flash'. Many deaths result from poisoning with morphine, of which heroin is a derivative, due to unexpected alterations in the degree of 'cutting' (mixing with other substances, such as talc or scouring powder) of the drug. It is said that when addicts hear of a death from overdose they seek out the vendor, who is clearly selling 'good stuff'. Intravenous drug users have a mortality rate about forty times above normal, and, if needles are shared, are at substantial risk from blood-borne infections such as AIDS and hepatitis B.

Drug addict being injected with a heroin cocktail at a 'drug party' in Budapest, Hungary.
(Gamma)

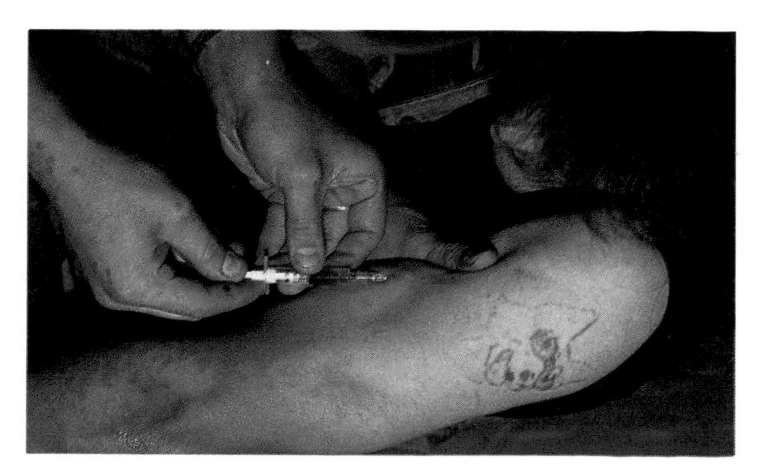

Hereditary Diseases

Since the middle of the 20th century, when the chemical basis of inheritance began to be understood (⇨ pp. 20–31), there has been a veritable explosion of interest in hereditary disease, with the result that many hitherto incurable diseases can be treated successfully.

SEE ALSO

- HEREDITY pp. 20–3 and 26–9
- WHAT THE BODY IS MADE OF p. 36
- SYMPTOMS AND SIGNS p. 126

Most diseases, even those with an apparently obvious cause, occur for complex reasons. The term *multifactorial* is applied to such diseases. Increased susceptibility to disease may be as important as causal factors such as bacteria or physical injury, and genetics is one of the many causes of individual variation in susceptibility to disease. An increasing number of multifactorial major disorders – conditions such as diabetes, atherosclerosis, heart disease, high blood pressure and peptic ulceration – are now known to have an important genetic component in their causation. Although such conditions often 'run in the family', they are not, however, of a kind normally thought of as being hereditary. The hereditary diseases are those caused by single defective genes or by abnormalities in complete chromosomes.

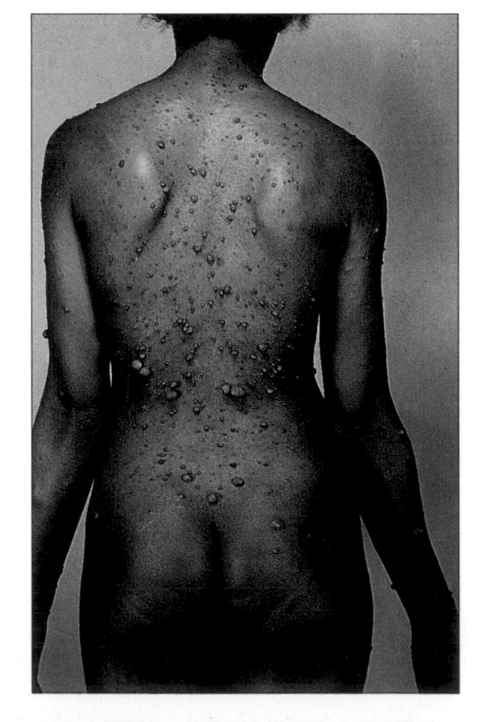

Neurofibromatosis, also known as von Recklinghausen's disease, is passed on genetically by means of dominant inheritance. It features multiple soft tumours of the fibrous sheaths of nerves in the skin, the spinal cord, brain and elsewhere. (NMSB)

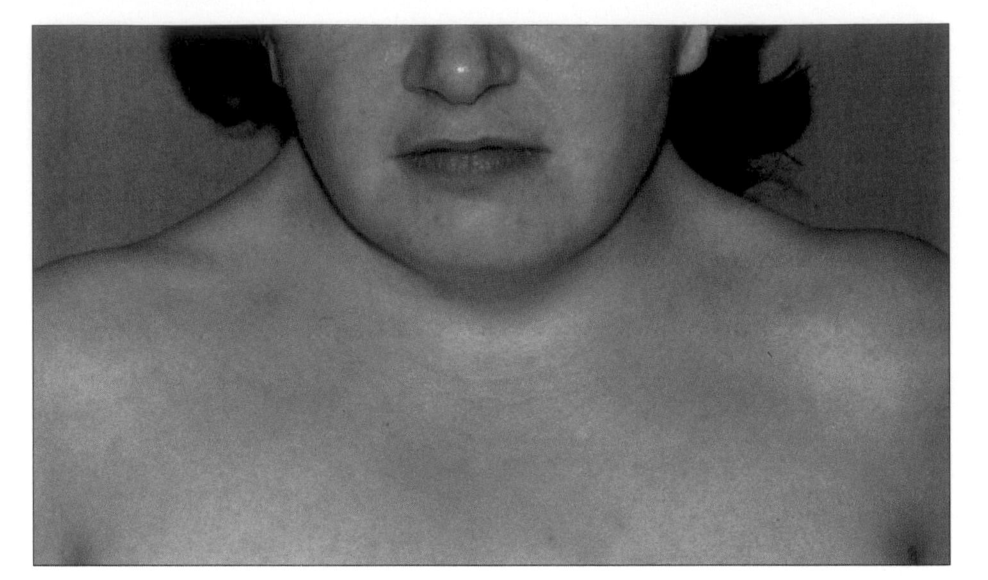

Turner's syndrome is a genetic disorder affecting females, and is caused by the absence of one of the two X chromosomes. Women with the syndrome are short in stature, have webbed skin on each side of the neck, and misshapen ears, and suffer abnormalities of the eyes and bones. In addition, these women have malformed vaginas, wombs and breasts, and do not menstruate. (NMSB)

Single-gene disorders

More than 4000 different single-gene disorders are now known to medical science and these are commoner than is generally supposed. About one person in 100 has a single-gene disorder. They may be dominant or recessive (⇨ p. 28) and the gene causing the trouble may be on the X chromosome, causing a sex-linked disorder, or on one of the other chromosomes, causing an autosomal disorder. In dominant disorders, only one of the pair of corresponding genes need be abnormal to produce the disorder; in recessive disorders, both genes must be affected.

Genes are the biological codes giving the order and selection of amino acids for proteins (⇨ p. 26), generally taking the form of enzymes essential for some chemical reaction. Defective genes will usually produce a protein of some kind, but this may have little or no enzyme action. In recessive disorders, when one gene is normal this gene will produce enough enzyme to allow the body to function almost normally. A *trait* – a tendency to the disorder – may, however, be present. But in dominant disorders, or in recessive disorders when both genes are affected, the enzyme action is absent and a biochemical reaction necessary for normal bodily functioning or development cannot proceed. The resulting effect may be local or widespread, depending on the importance of the reaction to the body as a whole.

GENETIC COUNSELLING

People who have a child with a hereditary disease, or have a family history of a hereditary disease, naturally want to know the risks that the disease will appear in subsequent children. They may also be in doubt as to whether a disease that seems to run in the family is in fact hereditary, or may be concerned about the risks of having babies after about the age of 40. The purpose of genetic counselling is to resolve matters such as these.

Genetic counsellors are doctors or scientists skilled in the complexities of genetics and well informed on the features of hereditary diseases. Some hospitals have teams that include, in addition to geneticists, a paediatrician, an obstetrician, a social worker and a nurse. These teams have ready access to special consultants who are experts in particular diseases.

A counselling session starts with the taking of a careful family history. The counsellor then performs a full physical examination of anyone known to be suffering from, or suspected of having, the disease in question. In some cases, the condition is obvious and can immediately be recognized. In others, laboratory or other tests may be necessary. If the disease is thought to be genetic, the mode of inheritance must be investigated. The counsellor will try to draw a family tree for the affected person, through the generations, showing all cases in which the condition occurred. This is called a *pedigree*, and, if sufficiently complete, will show the mode of inheritance.

Laboratory tests can unveil such things as, for instance, the presence of an extra chromosome in Down's syndrome, abnormal haemoglobin in sickle-cell anaemia, or the absence of a particular blood-clotting factor in haemophilia. Once the disease is reliably identified, the risk of it happening again can usually be stated accurately. For certain chromosome disorders the risk may be 100%. In single-gene disorders the risk may be 25% or even 50%. If the condition is sex-linked, the mode of inheritance and the risks can be clearly explained (⇨ p. 28).

The Most Common Single-gene Disorders

• *Agammaglobulinaemia* – a sex-linked or autosomal recessive immune deficiency disorder, affecting about one person in 10 000. The antibodies of the gamma globulin group are absent so that the affected person is greatly susceptible to bacterial infection. Treatment is by gamma globulin injections and antibiotics. Diagnosis is possible before birth.

• *Alpha1-antitrypsin deficiency* – an autosomal dominant disorder that causes a change in a protein produced in the liver. This leads to liver disease that progresses to cirrhosis in about 10% of affected children. It also causes emphysema in adult life, especially in cigarette smokers. The condition can be diagnosed before birth.

• *Alport's syndrome* – like all X-linked recessive disorders, this affects, in general, only males and is carried by females. The gene causes a defect in a protein needed to form a membrane in the kidneys and in the inner ear. Affected males suffer kidney disease and deafness. Female carriers can be identified.

• *Apert's syndrome* – an autosomal dominant disorder usually caused by a new mutation. There is abnormally early closure of the bones of the vault of the skull and healing together of the bones of the hand and fingers, leaving the thumb free. Other features sometimes found are cleft palate and mild brain damage.

• *Congenital adrenal hyperplasia* – an autosomal recessive disorder that affects about one child in 20 000. It is caused by a defective or absent gene that normally produces an enzyme needed for the synthesis of certain steroid hormones in the adrenal glands. Because the pituitary gland detects the absence of these hormones in the blood, it stimulates the adrenals to increase their output. The result is a great increase in other steroids, especially male sex hormones. Females become virilized and males sexually precocious. Diagnosis is possible before birth, and full recovery is possible with treatment.

• *Cystic fibrosis* – an autosomal recessive disorder affecting about one person in 1500. All the glands in the body secrete excessively sticky mucus. The lungs become severely congested and suffer *bronchiectasis* (permanent widening of the bronchi) and *fibrosis* (scarring and thickening) and put a strain on the heart. The digestive tract and other systems of the body are also affected. Carriers of the gene can be detected by a simple mouthwash test.

• *Cystinuria* – an autosomal recessive disorder, affecting about one person in 10 000. There is abnormal excretion of certain amino acids, especially cystine, in the urine, which leads to stone formation in the kidneys and bladder. Carriers can be detected and the condition can be diagnosed before birth. Treatment is effective.

• *Ehlers–Danlos syndrome* – a rare condition affecting about one person in 150 000 that can be inherited as an autosomal dominant, as an autosomal recessive, or as an X-linked disorder. A defect in the production of the protein collagen leads to extraordinary elasticity of the skin and laxity of the joints. The skin may tear easily and the lenses of the eyes dislocate. Wounds heal badly and bruising occurs on minor trauma. There is no effective treatment.

• *Galactosaemia* – an autosomal recessive disorder, affecting about one person in 80 000, and caused by the absence of an enzyme that breaks down the milk sugar galactose. Galactose, which is toxic, accumulates in the body causing vomiting, diarrhoea, liver damage, cataracts and brain damage. Early diagnosis and a galactose-free diet allows normal development and health. The gene disorder can be detected before birth by amniocentesis (➪ p. 102).

• *Gilles de la Tourette's syndrome* – an autosomal dominant disorder causing repetitive twitching (tics) that progress to sniffs, snorts and involuntary vocalization and then a compulsive need to touch other people and often to utter obscenities. Various drugs have proved useful.

• *Haemochromatosis* – a common autosomal recessive disorder, affecting about one person in 500. Iron is absorbed from the diet to an abnormal degree, and is deposited in the tissues, where it causes scar tissue. The result is widespread damage, especially to the liver (cirrhosis), pancreas (diabetes), heart (causing heart failure), testicles (loss of libido) and joints (arthritis). Treatment is by repeated bleeding to rid the body of iron.

• *Haemophilia* – an X-linked recessive blood-clotting disorder causing a life-long tendency to excessive bleeding. The condition is treated by means of repeated injections of blood-clotting factors, obtained from donated blood.

• *Lesch–Nyhan syndrome* – an X-linked recessive disorder causing progressive spasticity, writhing movements, mental dysfunction and a tendency to self-mutilation. The brain damage that causes these effects results from very high levels of uric acid. Detection is possible before birth and female carriers can be identified.

• *Malignant hyperthermia* – an autosomal dominant disorder in which certain anaesthetic drugs cause dangerously, sometimes fatally, high temperature from muscle overactivity. About one person in 40 000 is affected. The condition can be diagnosed and the drugs avoided.

• *Myotonic dystrophy* – an autosomal dominant disorder caused by a gene on chromosome number 19. There is progressive weakness of the face, neck, forearm and hand muscles, with inability to relax the grasp. Cataracts also occur. About one person in 20 000 is affected.

• *Osteogenesis imperfecta* – 'brittle-bone' disease, which may be inherited as an autosomal dominant and affects about one person in 20 000. Defective collagen leads to bone fragility and multiple fractures. Tooth defects and deafness also occur.

• *Otosclerosis* – this autosomal dominant condition affects the inner of the three tiny bones that conduct sound vibrations across the middle ear so that it becomes fused and fixed. Surgery is effective.

• *Retinitis pigmentosa* – this may be autosomal dominant or recessive, or sex-linked and affects about one person in 4000. There is progressive degeneration of the retinas leading to poor night vision and sometimes eventual blindness. Female carriers of the X-linked variety can be detected.

• *Sickle-cell disease* – an autosomal recessive disease, caused by a mutation in a single base in the DNA of chromosome number 11. This causes the amino acid valine to be substituted for glutamic acid in the haemoglobin of the red blood cells. In homozygous people (➪ p. 28) the haemoglobin molecules can readily change in such a way as to distort the red cells and cause them to form a sludge that can block small blood vessels. In this state, the red cells are also more readily destroyed by the immune system, causing serious illness. People who carry only one defective gene of the pair (heterozygous people) are said to have the sickle-cell trait, and, in most cases, this causes them no trouble. The diagnosis can be made before birth, and the carrier state can be detected by a simple blood test.

• *Steroid sulfatase deficiency* – this sex-linked recessive disorder affects about one male in 6000, causing severe scaliness of the skin (*ichthyosis* or 'fish-skin'). It also affects vision by causing opacities in the corneas. It is due to partial or total deletion of the gene that codes for the enzyme steroid sulfatase, the absence of which can be detected by hair root analysis.

• *Testicular feminization syndrome* – this X-linked recessive disorder results in people who appear to be normal females, with breasts and female external genitalia, but who are, in fact, genetically male (that is, they have XY chromosomes). The gene defect blocks the action of the male sex hormone testosterone so that male characteristics cannot develop and the affected babies are brought up as females. The disorder occurs in only one in 60 000 males. Carriers can be detected.

Other single gene disorders include *muscular dystrophy*, in which sufferers undergo progressive muscle weakness; *pyloric stenosis*, in which the outlet from the stomach is narrowed; *spina bifida*, which is a defect in the bones of the spine; *phenylketonuria*, which can result in brain damage; and *Huntington's chorea*, which causes ceaseless, uncontrollable jerky movements of the body and limbs and serious brain damage.

In addition, a number of major conditions are caused by defects in the number or structure of whole chromosomes (➪ p. 28).

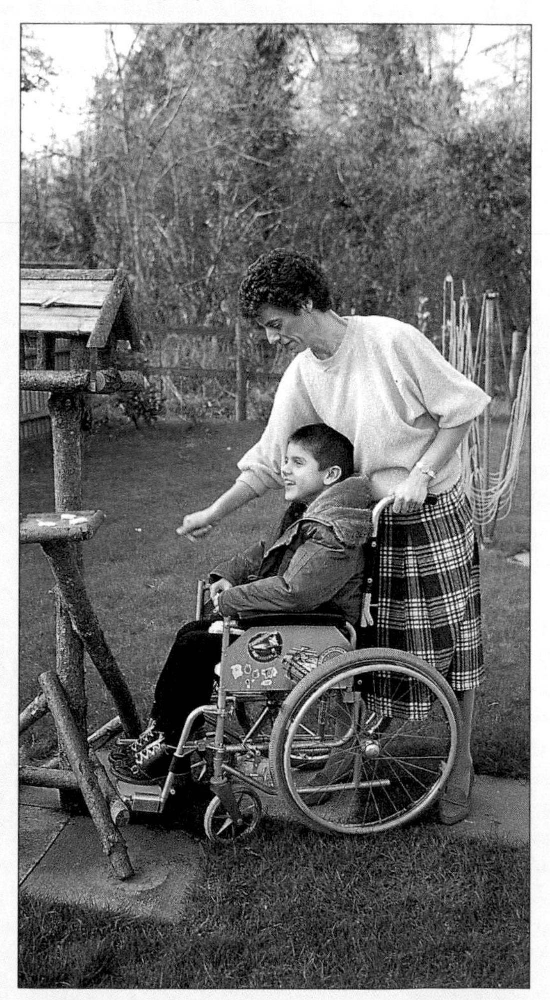

Boy with muscular dystrophy, a hereditary disorder that features gradual and progressive muscle degeneration. All forms of muscular dystrophy are genetically determined, but the process by which the muscles waste is still unknown. The muscle protein is normal, but there is probably a defect in the muscle-cell membrane that allows enzymes to enter and digest the contents of the fibre. (MD)

Health Care in the Past

Early humans, ignorant of the causes of disease, ascribed them to malevolent spirits and turned for help in time of illness to those who claimed to be in touch with supernatural powers. Only with the introduction of medicine as a scientific discipline in the universities did knowledge about the real nature of illness grow.

SEE ALSO

- THE EMERGENCE OF MEDICAL SCIENCE p. 34
- BODY AND SOUL p. 70
- CHILDBIRTH THROUGHOUT HISTORY p. 94
- HEALTH AND DISEASE pp. 116–19
- PILLS AND POTIONS: THE HISTORY OF DRUG THERAPY p. 152
- A HISTORY OF MENTAL ILLNESS p. 184

Magic, religion and healing were, at first, inextricably associated. Eventually, however, healers began to claim powers of their own and medicine became separated from religion. In the medieval period this separation was formalized by the foundation of distinct university faculties of theology and medicine. Gradually, with the growth of knowledge, specialization and expertise increased.

Even in the earliest times healers must have been capable of a reasonable range of useful health-care procedures. The management of fractures by the use of splints, the stitching (*suturing*) of wounds and the opening and draining of abscesses were understood and practised early on. We know that soothing lotions and various plant-derived drugs, such as opium, were used, and that a range of surgical operations was performed – caesarian section, for instance, was known in India several thousand years BC. Such knowledge must have been valued and must have conferred status. It was not necessarily infallible, though, as not all treatments were beneficial. Boring a hole in the living skull (*trephination* or *trepanning*), presumably to release evil spirits in 'possession' of the patient, was a widespread practice, which, not surprisingly, often led to death.

Wundartzney:
Zů allen Gebrechen des gantzen Leibs/
Vnd zů iedem Glid besonder/ Mit was zůfällen die entstehn/ Vnnd eim Wundartzt zůkommen mögen. Vil edler/bewärter Artzneien/Rath vnnd Meysterstuck. Des vil erfarnen Chirurgen
IOANNIS CHARETANI.
Rechte Kunst vnd Bericht der Aderläß.
Für die Aderlässer vnd Scherer.
Zu Straßburg. Chr. Eg.

Title page of a 16th-century German medical text, depicting the contemporary physician's range of instruments. (Images)

Blood letting, or 'breathing a vein', was a commonly used technique, right up until the early 19th century. It was employed in the belief that it could cure a variety of conditions, including melancholy, fever, and, in this case, gluttony. (AR)

Mesopotamian and Egyptian medicine

In Babylon, around 1780 BC, King Hammurabi produced a Code of Laws that included rates of payment for successful medical treatment and penalties for failure. An important medical papyrus was found in 1875 in a tomb in Thebes, near Luxor, by the German Egyptologist and novelist Georg Ebers (1837–98). The Ebers papyrus, accurately dated to 1550 BC, gives a notable insight into the state of medicine at the time. There is not much to admire, since the emphasis is largely on spells and spurious medicines, and the physician is exhorted to pronounce an incantation when administering the remedies. Little emphasis was placed on diagnosis, and the identification of disease was crude and very general. It is notable that the most enthusiastic commendations are applied to the least plausible of the remedies.

Greek medicine

The civilizing and rational tradition of the Greeks did much to dispel the magical elements that previously prevailed in medicine. Although the Greeks were generally more interested in philosophic speculation than in empiricism and the amassing of factual knowledge, their preoccupation with logic and abilities in detecting errors in argument were useful checks on superstition, so that their care of the sick at least contained a good deal of common sense. Thinkers such as the physician Hippocrates of Cos (c. 460–377 BC; ➪ p. 34), the philosopher Socrates (c. 470–399 BC) and the physician, scientist and philosopher Aristotle (384–324 BC) brought a new rigour into medical thought and fresh standards into medical practice. Doctors began to observe closely and to keep good records, and diagnosis became more precise. The recording of observations preserved experience so that the prediction of a likely outcome (*prognosis*) became possible. The great collection of Hippocratic writings formed a textual basis for a practice of medicine that, while still full of fanciful theory, came very close to being scientific. The Greek-born physician Galen (AD c. 130–200; ➪ p. 34), who worked in Rome, left many medical treatises, and, for 1500 years, was considered the only real authority on medical matters.

Arabian medicine

The Islamic influence, which arose and spread widely after the birth of the prophet Muhammad in AD 570, helped to propagate the important developments that were taking place in Arab medicine. The physician Rhazes (860–932) wrote extensively in Arabic, commenting on the older writers and correcting errors by his own observations. His descriptions of smallpox and measles are the first reliable accounts in medical literature.

The Persian philosopher-scientist Avicenna (980–1037) was the chief physician to the hospital at Baghdad and court physician to a succession of Arab Caliphs (rulers). He was a prolific writer on medicine: his *Canon of Medicine* attempted to codify all medical knowledge and contains many accounts of the subject that still stand today. He recommended the *cautery* (an instrument for cauterizing or sealing wounds by means of heat) rather than the surgeon's knife, for instance, and his account of the clinical manifestations of pleurisy could hardly be bettered by modern texts. Avicenna was the leading medical authority of the Middle Ages with influence second only to that of Galen.

In two of the stories in *The Arabian Nights Entertainment*, which originated in the 10th century, a slave girl, Tawadudd, gives a surprisingly detailed account of the state of Arabian medicine at the time, revealing a remarkable knowledge of anatomy, the significance of symptoms and physical signs, some ideas on the causation of disease (including sexually transmitted disease) and advice on diet and the avoidance of alcohol. The date of this tale is uncertain.

The Arabs built many fine hospitals, especially in the West, with flower gardens and fountains, in which were succoured people who were mentally unstable, or who were destitute or sick. The Spanish city of Córdoba alone had 50 hospitals. The Moorish surgeon Albucasis (936–1013) of Córdoba was remarkable in the standard of his health care. He used a red-hot iron cautery, described the lithotomy position, in which the patient lies on his or her back with the knees bent and the thighs widely apart (used for gynaecological operations or surgical procedures on the perineum), practised tracheostomy, tied off bleeding arteries, conducted eye surgery, and was able to distinguish goitres from cancers of the thyroid gland.

Medicine in the Renaissance

The 15th and 16th centuries saw the development of printing and a widespread proliferation of medical knowledge. Leonardo da Vinci (1452–1519) dissected thirty corpses and made hundreds of brilliant anatomical drawings. Paracelsus (1493–1541) denounced Galen's theory of the four humours and burned his books. Andreas Vesalius (1514–64; ⇨ p. 34) and Gabriello Fallopio (1523–62; ⇨ p. 94) put anatomy on a sound scientific basis, and the physician Geronimo Cardano (1501–76) infuriated his colleagues by publishing *The Bad Practice of Healing Among Modern Doctors*. Meanwhile, Ambroise Paré (1510–90) notably advanced the art of surgery, and Dr William Chamberlen (1540–96) that of the conduct of childbirth.

The Düsseldorf physician Johann Weyer (1516–88) wrote *De Praestigiis Daemonum et Incantationibus ac Veneficiis* (1563), a plea against the folly and cruelty of the witchcraft trials, and a book that aroused the fury of the clergy and was included in the *Index of Prohibited Books*. Weyer's book, something of an antidote to the notorious *Malleus Maleficarum* (⇨ p. 95), was one of the first attempts at a textbook of psychiatry, and must have saved many unfortunate eccentrics and lunatics from the terrible fate of burning. These and many other thinkers began the important work of overthrowing some of the erroneous dogmas of Galen so as to bring a new and more scientific dimension into medical care.

Medicine in the 17th and 18th centuries

Standards of health care steadily improved as medical knowledge grew and science and enlightenment advanced under the influence of intellectuals such as the mathematician and physicist Isaac Newton (1642–1727), the philosopher Francis Bacon (1561–1626), the pioneer of magnetism William Gilbert (1540–1603), the astronomer Galileo Galilei (1564–1642), the thinker René Descartes (1596–1650; ⇨ pp. 70 and 162) and the chemist Robert Boyle (1627–91).

These advances were slow but, nevertheless, inexorable. Sanctorius of Padua (1561–1636) invented the clinical thermometer and a method of counting the pulse rate, as well as making important studies in physiology. William Harvey (1578–1657; ⇨ p. 34) demonstrated the circulation of the blood and pioneered the science of embryology. Anton van Leeuwenhoek (1632–1723) developed the first practical microscope, and Thomas Willis (1621–75) described the detailed structure of the brain and its blood supply.

Outstanding in the 17th century was the Oxford physician Thomas Sydenham (1624–89), who revolutionized 'bedside' clinical medicine by his determination to concentrate on what could be done to relieve the suffering of the sick. He wrote voluminously and accurately on such conditions as gout, scarlet fever, measles, bronchopneumonia, pleurisy, dysentery, malaria, and hysteria. He was also the first to describe a form of *chorea* (a disorder featuring uncontrollable, jerky movements) that still bears his name (Sydenham's chorea, also known as Saint Vitus's dance). And he rationalized medical treatment by concentrating on the few drugs used at the time that had real medical value, also recommending fresh air and exercise. His influence was so important that he has often been described as the English Hippocrates.

The 18th century was dominated in medicine by Hermann Boerhaave of Leyden (1668–1738), and in surgery and obstetrics respectively by John Hunter of London (1728–93) and his brother William (1718–83). Boerhaave, the most sought-after physician of the day, made a fortune of 2 million florins from medicine but found time also to teach constantly and to write two very influential works, *Institutiones Medicae* (1708) and *Aphorismi de Cognoscendis et Curandis Morbis* (1709), which were translated into several languages including Arabic. Important advances were also made by the surgeon Percivall Pott (1714–88; ⇨ p. 140) and the ophthalmologist Jacques Daviel (1696–1762).

The Birmingham physician William Withering (1741–99), on a hint from a country woman, investigated the properties of the foxglove flower for the treatment of fluid retention (or 'dropsy' – now known as *oedema*). He found that oedema of cardiac origin was indeed relieved by the foxglove and established the use of the drug extracted from it, that we now call digitalis. In a similar way, Edward Jenner (1749–1832; ⇨ p. 58) found a way to prevent smallpox.

The naval surgeon James Lind (1716–94) interested himself in the serious bleeding disease of scurvy, and in 1753 he published *A Treatise on the Scurvy* containing an inquiry into the nature, causes and cure of the disease. This showed how scurvy could be prevented and cured by taking fresh fruit and vegetables, which we now know to contain vitamin C, a deficiency of which causes the condition.

HIGHLIGHTS IN 19th-CENTURY MEDICINE

- **1812 Charles Bell distinguishes the sensory from the motor element in spinal nerves and describes facial palsy.**
- **1819 René Laënnec describes *auscultation* (attentive listening) with the stethoscope.**
- **1823 The medical journal, *The Lancet*, is inaugurated.**
- **1827 Carl Ernst von Baer first describes an ovum in the ovary, and Richard Bright describes kidney disease.**
- **1841 Jacob Henle writes on microscopic anatomy.**
- **1844 Horace Wells uses nitrous oxide gas ('laughing gas') in dentistry.**
- **1846 William Thomas Morton uses ether as an anaesthetic.**
- **1847 James Simpson experiments with the liquid chloroform as an anaesthetic.**
- **1849 Thomas Addison describes the function of the adrenal glands.**
- **1851 Hermann von Helmholtz introduces the ophthalmoscope.**
- **1853 Queen Victoria accepts chloroform in childbirth, so paving the way for pain relief during labour for millions of women.**
- **1858 Rudolf Virchow publishes *Cellular Pathology*, and Louis Pasteur attacks the theory of spontaneous generation of life.**
- **1859 Charles Darwin publishes *The Origin of Species*.**
- **1861 Ignaz Philipp Semmelweiss demonstrates the origin of puerperal fever.**
- **1863 von Helmholtz writes on the physiology of hearing.**
- **1864 Joseph Lister initiates antiseptic surgery and operates under a spray of carbolic acid.**
- **1876 Robert Koch publishes *The Aetiology of Anthrax*.**
- **1877 Patrick Manson describes microfilaria.**
- **1879 Pasteur grows streptococci, and Manson suggests that mosquitoes might transmit malaria.**
- **1881 Pasteur develops a vaccine against anthrax.**
- **1882 Koch discovers the tubercle bacillus.**
- **1885 James Leonard Corning introduces lumbar puncture.**
- **1890 William Stewart Halstead introduces surgical gloves.**
- **1893 Wilhelm His describes the conducting system of the heart (bundle of His).**
- **1894 Willem Einthoven develops the electrocardiograph.**
- **1895 Wilhelm Conrad Röntgen discovers x-rays.**
- **1896 Scipione Riva-Rocci invents the sphygmomanometer, and Joseph François Babinski describes the plantar (foot) reflex.**
- **1897 Ivan Pavlov describes the conditioned reflex.**
- **1898 Marie Curie isolates radium.**

The 19th century

The enormous medical advances of the 19th century (⇨ box) were counterbalanced by the insanitary conditions suffered by many in the West. For these, the general move to the towns in the wake of advancing industrialization meant abject poverty, slum housing, dreadful working conditions, long hours, poor water supplies, high infant mortality and a short life expectancy. Thus the benefits of the many medical discoveries made during this time were not generally to be felt until the next century.

Florence Nightingale (1820–1910) founded the profession of nursing, and, in 1854, went with a group of nurses to Scutari, where conditions of squalor prevailed, to attend those wounded in the Crimean War. Her efforts reduced the hospital death rate from 42% to 2%. From Edwin Hodder, *Heroes of Britain* (c. 1880). (AR)

Pills and Potions: the History of Drug Therapy

The development of drug therapy can be traced through a number of stages. First, empirical observation of the bodily effects of various plants led to the growth of folk medicine, which was later systematized and recorded in *herbals* and *pharmacopoeias* (lists of drugs with their uses and dosages). This gave way to a more critical and selective approach, with many alleged remedies being discarded. Later, the active elements in natural remedies were identified, followed by the elucidation of drug action and the formulation of many new and powerful drugs.

The effects of certain natural substances on the human body have been known since earliest times. In Europe it was discovered that the berries of the deadly nightshade (*Atropa belladonna*) caused blurred vision, a dry mouth, flushing of the skin, and even convulsions and death. In South America it was found that chewing the leaves of the coca plant, although numbing the mouth, relieved hunger and fatigue, and induced a pleasant state of mind. Also in South America it was discovered that chewing the bark of the cinchona tree, which contains quinine, caused excitement, confusion and singing in the ears, but could also cure a common fever (malaria). Similarly, in North America the native Indians found that the dried bark of the cascara buckthorn (*Cascara sagrada*) had a purgative effect and could be used to relieve constipation.

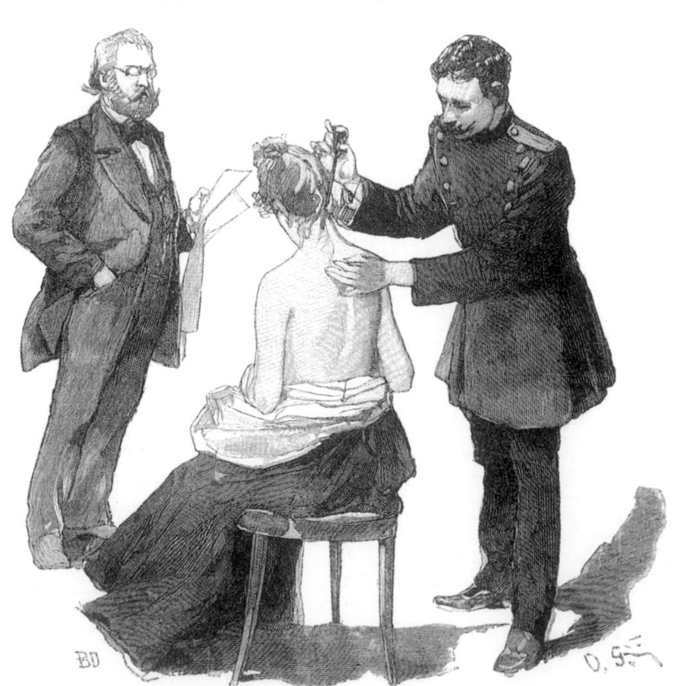

Robert Koch directing an experiment in which tuberculin (a substance he derived from tubercle bacilli) was injected into a patient with tuberculosis in an attempt to cure her. Unfortunately tuberculin's curative powers were disappointing, and its value as a diagnostic aid was, for a time, overlooked. From *La Science Illustrée* (1891). (AR)

The dawn of pharmacology

A good deal of the content of the early herbals was medically questionable but some of the empirical fact was remarkably sound. The foxglove, *Digitalis purpurea*, was widely recommended and used for the dropsy (oedema), long before the English physician William Withering (1741–99) published an account in 1785 of the action of the dried leaves on the heart (⇨ p. 150). Today it is a medical commonplace that the oedema of heart failure can be cleared by the action of digitalis in improving the heart's efficiency.

Early scientific trends

One of the first to bring a scientific approach to the study of the effects of drugs was the 16th-century Swiss physician and alchemist Theophrastus Bombastus von Hohenheim, more commonly known as Paracelsus (1493–1541; ⇨ pp. 34 and 150). Although he still practised magic and dabbled in the occult, Paracelsus taught strict observation of the effects of drugs, and abandoned the old ideas of the 'humours' as the cause of disease. He also introduced several important drugs, including *laudanum* (a solution, or tincture, of opium in alcohol), mercury, sulphur, iron and arsenic. Remarkably for his time, he taught that surgery should be performed only under the cleanest possible conditions – a lesson forgotten by many who followed him.

The scientific influence of Paracelsus did nothing to abolish belief in the magic he had practised, however, and, as late as the beginning of the 18th century, apothecaries were still selling the *elixir universale* – a 'cure all' said to contain dried human brains, powdered lion's heart, earthworms, gold, witch hazel and Egyptian onions.

Scientific empiricism was, however, beginning to be put into action. In 1753 the Scottish naval surgeon James Lind (1716–94) had discovered that scurvy in sailors could be completely prevented by a small regular dose of citrus fruit juice (⇨ p. 150). The reason for this remained quite unknown, and vitamins were not recognized until the 20th century, but the method worked and was eventually, many years later, adopted by the conservative British Admiralty. In 1796 the English physician Edward Jenner (1749–1823) introduced vaccination against smallpox (⇨ p. 58). Again, the mechanism for this was unknown, but the treatment was effective and the idea eventually led to the now enormous subject of immunology.

19th-century advances

The entire pharmaceutical industry was initially founded on natural remedies, and even by the end of the 19th century, much of what doctors prescribed was still of little or no medical value. The pharmacopoeia, although impressively large, actually contained only a small number of genuinely effective drugs. The most important of these were: opium to control pain, the anaesthetics ether and chloroform, digitalis to strengthen the action of

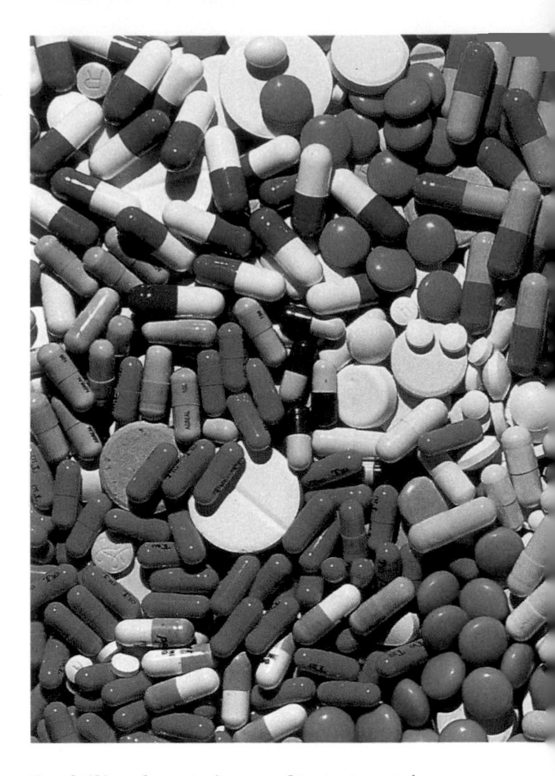

the failing heart, iron salts to treat iron-deficiency anaemia, bicarbonate of soda to neutralize stomach acid and relieve dyspepsia, atropine to relax the focusing muscle of the eye and promote the resolution of internal inflammation, iodine to prevent goitre, quinine to suppress or prevent malaria, salicylates (the natural precursors of aspirin) to control the pain of rheumatism, and colchicine to treat the excruciatingly painful condition of gout.

Laudanum, while obviously a powerful analgesic and tranquillizer, was, until the end of the 19th century, extremely popular in Europe and America for the 'treatment' of everything from toothache to tuberculosis. Hundreds of other plant and mineral derivatives – mostly of no benefit – were compounded by the pharmacist into pills, infusions, tinctures and mixtures in accordance with often elaborate prescriptions written in Latin to conceal the contents.

In the course of the 19th century, however, analytical chemists had increasingly been determining the chemical nature of the active elements in useful plants. Once identified, the useful principle could often be made synthetically. For instance, the isolation of salicylic acid from willow trees led, in 1853, to the synthesis, by Charles Gerhardt (1816–56), of the drug acetylsalicylic acid (aspirin), whose many properties are still being discovered. Indeed, one recent addition to its list of uses is as a preventative of coronary heart disease.

It is hard to exaggerate the effect on medicine when, in 1863, the French chemist Louis Pasteur (1822–95) proved the existence of bacteria that could cause disease (⇨ pp. 116 and 134). Following this, in 1876, the German bacteriologist Robert Koch (1843–1910) grew cultures of the anthrax bacillus, first in blood serum

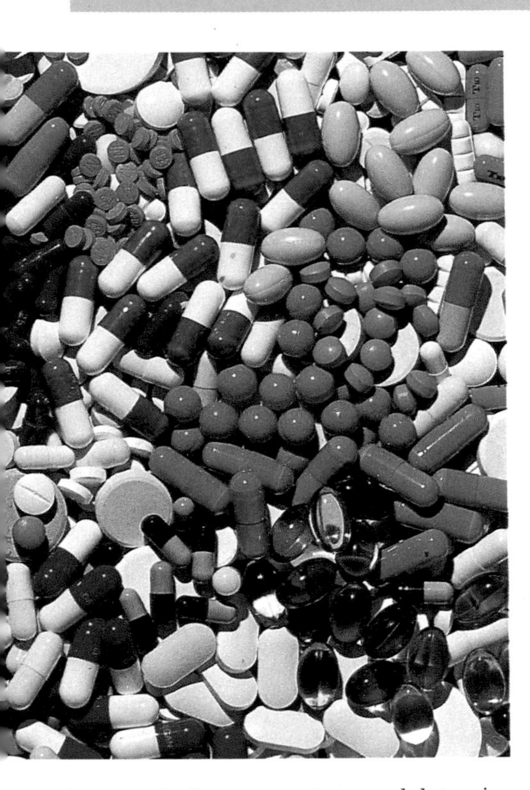

Pills, capsules and tablets galore: modern medical discoveries have meant an end to pain and distress to millions. (Images)

kept at body temperature and later in solid media such as gelatin and agar-agar. Pasteur followed up this work and, by heating the anthrax organisms so as to destroy their virulence, was then able to make the first artificially prepared vaccine. He then went on to develop a vaccine against rabies. In 1890 Emil von Behring (1854–1917) developed a natural antidote (an *antitoxin*) to the powerful poison produced by diphtheria organisms – a development that saved countless lives.

Drug therapy today

In 1909, the German medical scientist Paul Ehrlich (1854–1915) was able to show that the organic arsenical compound salvarsan – the 'magic bullet' – was effective against syphilis. This notable advance promoted an ever-widening search for drugs that could destroy disease organisms in the human body without harming the patient.

The German bacteriologist Gerhard Domagk (1895–1964), aware that certain artificial (or 'azo') dyes were effective in staining bacteria, wondered whether they might be damaging to these organisms in the living body. After trying many of these dyes he hit upon an azo compound, prontosil, which had been synthesized in 1932. This he found to be remarkably effective against streptococci bacteria in mice, while apparently causing the animals no harm. At this stage in his work Domagk's daughter pricked her finger and developed a severe streptococcal infection with blood poisoning (*septicaemia*). She became critically ill and was expected to die. In desperation, Domagk gave her a large dose of prontosil, and she made a complete recovery. The effectiveness of prontosil was due to its conversion in the body to sulphanilamide. This discovery started an investigation into many hundreds of related compounds for possible antibacterial activity. The drug company May and Baker struck lucky with the 693rd trial, and, in 1938, marketed M&B 693-sulphapyridine. Sulphathiazole appeared in 1940, sulphadimidine in 1941, and these were followed by scores of other effective sulpha drugs, which saved millions of lives in World War II.

Modern developments

The discovery of penicillin (⇨ box) led to a world-wide hunt for other moulds with antibiotic properties, with the result that many hundreds of antibiotic substances were discovered. Streptomycin was isolated from a strain of *Streptomyces griseus* in 1943, neomycin was discovered in 1949, kanamycin in 1957, and gentamicin in 1964.

The elucidation of the chemical structure of the penicillins led to a major industry concerned with the production of semi-synthetic penicillins. A whole range of new penicillins, with different spectrums of antibacterial action, was produced – penicillin G, penicillin V, methicillin, oxacillin, cloxacillin, ampicillin, amoxycillin, and many others.

The first cephalosporin antibiotic was isolated in 1948 from sea water near a sewage outlet off the Sardinian coast. This led to an explosive expansion of new antibiotics – cephalothin, cephalexin, cefaclor, cefadroxil, cefuroxime, and scores of others. The first tetracycline antibiotic, chlortetracycline, was isolated from the fungus *Streptomyces aureofaciens* in 1948. Other similar drugs, including semi-synthetic tetracyclines, soon followed – oxytetracycline, demeclocycline, methacycline, doxycycline, and minocycline.

By the middle of the 20th century an acceleration of intensive research and development in biochemistry and pharmacology had led to a huge increase in the availability of other powerful and useful drugs. By the 1950s doctors had at their disposal, in addition to the antibiotics, the full range of vitamins, the hormone insulin, the barbiturates, new and better anaesthetics, anti-inflammatory and other corticosteroid drugs, synthetic hormones, oral contraceptives, tranquillizers, anticoagulants, antiepilepsy drugs, anticancer drugs, and antiviral drugs.

A massive research programme into drug action, especially in the second half of the century, had shown how drugs interact with the body. Many had been found to operate at cell-membrane receptors or nerve junctions (*synapses*; ⇨ p. 74), either in a manner similar to that of the natural stimulating substances, or by blocking them so as to inhibit natural action. Many were found to simulate natural physiological substances, such as neurotransmitters in the brain and nervous system, so as to potentiate their action. Several were found to interfere with the action or reproduction of bacteria, viruses, fungi and other human parasites, and others were able to interfere with the growth of cancers. Drugs were developed for their beneficial action on the heart and the arteries, on the kidneys, on the blood-forming tissues, on the digestive system, on the lungs and on virtually every other part of the body.

FLEMING AND PENICILLIN

In 1928 Alexander Fleming (1881–1955), a Scottish bacteriologist working in London, noticed that one of a number of bacterial culture plates that had accumulated in the laboratory sink had grown a colony of mould. This was a common event on old, exposed plates, but Fleming also observed that around the mould there was a zone almost completely free from bacterial growth. Intrigued, Fleming carried out many careful experiments using the mould on the plate and reported his findings in a paper published in 1929 in the *British Journal of Experimental Pathology*. This very full paper describes how he subcultured the mould and found that the broth in which he grew it at room temperature had acquired the property of preventing the reproduction of many common bacteria. For convenience he gave the name 'penicillin' to the filtered broth in which the mould – a penicillium species – had been grown. He proved it nontoxic by injecting it into a mouse, and confirmed that it was not irritant when applied to the human conjunctiva every hour for a day. He also showed that it did not interfere with the action of white blood cells.

Fleming thought that the new substance might be useful as a surface dressing for septic wounds but did not anticipate its use as an antibiotic for internal use. In 1940, referring to its possible use as a local antiseptic, he said that the trouble of making it did not seem worth while. The subsequent development of the drug had to wait until 1938, when Ernst Chain (1906–79) and Howard Florey (1898–1968), after reading Fleming's paper, succeeded in purifying enough penicillin to conduct human trials and demonstrate the range of properties of this new drug. Penicillin was found to be present in several moulds of the *Penicillium* and *Aspergillus* genera. It was soon shown to be almost completely nontoxic to the body, and was the first fully successful and safe antibiotic for acute internal bacterial infections in humans. In 1945 Fleming, Chain and Florey shared the Nobel Prize for Physiology or Medicine.

Sir Alexander Fleming (1881–1955), Scottish bacteriologist and discoverer of penicillin. (Popperfoto)

SEE ALSO

- THE EMERGENCE OF MEDICAL SCIENCE p. 34
- HEALTH AND DISEASE pp. 116–19
- PREVENTING DISEASE p. 124
- HEALTH CARE IN THE PAST p. 150

Surgery

We usually think of surgery as a modern medical procedure, but archaeological evidence of crude surgery, especially drilling (*trepanning* or *trephination*) of the skull, or amputation of limbs, is widespread.

With the dawn of civilization in Ancient Mesopotamia (modern-day Iraq), written records begin to appear, containing many references to surgical treatment. Surgery also featured importantly in the early Egyptian and Indian civilizations and in the Babylonian, Chinese, Greek, Roman, Arabic and early European cultures.

The seal of a Babylonian surgeon called Urlugaledin dating back to about 2300 BC has survived and can be seen in the Louvre museum in Paris. Much information on surgical practice has been obtained from the Code of King Hammurabi of Babylon (⇨ p. 150), which was engraved on a stone column and is also preserved in the Louvre.

SEE ALSO

- THE EMERGENCE OF MEDICAL SCIENCE p. 34
- DIAGNOSTIC AIDS p. 128
- HEALTH CARE IN THE PAST p. 150

The Modern Operating Theatre

Operating theatres are designed to ensure the maximum safety for the patient, especially by avoiding infection of the open wound. Fixed or very large features such as floors, walls and tables cannot be sterilized but they are washed down every day and bacterial counts are done regularly. Forced ventilation with filtered air keeps out air contamination. X-ray and CT scan (⇨ p. 128) viewing boxes are often built into the walls, and, hanging conspicuously in the middle of the ceiling, there is a large operating lamp capable of being moved into almost any position to produce shadowless illumination. Sometimes an operating microscope is also mounted on the ceiling.

The operating table is of heavy construction and immobile while in use, but can be moved freely to clear the floor for cleaning. It has a wide range of adjustments and can be raised or lowered, tilted about its long or central axes, or tilted independently up or down at either end. It is covered with a thick slab of rubber for the comfort and protection of the patient. Near the table is the anaesthetic trolley fitted with cylinders of oxygen and anaesthetic gases and with valves and gauges for controlling the rate of delivery of the anaesthetic agents. It also contains a ventilator to maintain respiration in the anaesthetized and paralysed patient, and an electrocardiogram monitor, often with an audible bleep.

Several stainless steel wheeled tables surround the operating table proper. These are covered with sterile towels before use and the operating instruments are laid out on them. Various special devices may also be present, as required. These may include special frames for the support of the patient, various imaging devices such as endoscopes, monitors for closed-circuit TV, an electric diathermy machine (a device that uses high-frequency alternating current to heat or burn tissue), an electromagnet for the removal of metallic foreign bodies and a heart–lung bypass machine.

Operating theatres have annexes with foot or elbow-operated taps where the surgeons, the assistants and the participating nurses scrub their hands and arms before donning sterile gowns and rubber gloves. Adjoining the central operating room are various ancillary rooms, a sterilizing room for instruments, an instrument store, an anaesthetic room where patients are anaesthetized, and a staff changing room.

The first surgical texts

The Edwin Smith papyrus, found in Egypt at Luxor in 1862, is dated about 1600 BC. This fifteen-foot-long roll refers to surgical practices probably dating back some thousands of years before that. It is, in fact, incomplete as a surgical textbook, but it covers a variety of topics, including the management of various injuries, wounds, fractures, dislocations and tumours. Unfortunately, it breaks off after dealing with only the upper part of the body. Much of it is sound and effective, and the descriptions of the management of fractures of the arm and nose are particularly impressive. The account is written in the form of actual case histories and distinguishes between those disorders that should be treated and those that are best left alone.

The surgical writings of Hippocrates (⇨ p. 34) are extensive and medically valuable but cannot compare with the contribution of the 1st-century Roman writer Aulus Cornelius Celsus, whose major textbook, *De Medicina*, written about AD 30, served as a source of useful information for surgeons for hundreds of years.

Surgery in the Renaissance

During the Middle Ages there was little advance in surgical technique. Celsus reigned supreme until challenged in the 16th century by Theophrastus Bombastus von Hohenheim (1493–1541), who, with some justification, considered himself so far ahead of Celsus that he proclaimed this publicly by adopting the name of Paracelsus (para meaning 'beyond'; ⇨ pp. 34 and 150).

Ambroise Paré (1510–90) developed many important surgical and obstetrical techniques – treating hernias, tying off arteries, managing gunshot wounds, turning babies in the womb – always trying to keep the patient's pain and distress to the minimum. Dr William Chamberlen (1540–96) invented the obstetrical forceps and used them extensively. There were many other anatomists and surgeons also advancing the craft during this period and in the following centuries (⇨ p. 150).

18th- and 19th-century surgery

The Scottish Hunter brothers were immensely successful, both commercially and educationally, in the field of medicine. William (1718–83) was an eminent obstetrician and surgeon, but it was John (1728–93) whose experimental approach to surgery put it firmly on the map as a scientific discipline. Another important surgeon of the time was Percivall Pott (1714–88), who advanced the management of fractures, head injuries, tuberculosis of the spine, hernias and *hydrocoele* (fluid accumulation in the scrotum), and who established that cancer of the scrotum in chimney sweeps could be prevented by regular washing off of soot from their skin (⇨ p. 144). Jacques Daviel (1696–1762) developed methods of removing opaque cataractous lenses from the eye, thus restoring sight. He performed this operation successfully on hundreds of patients.

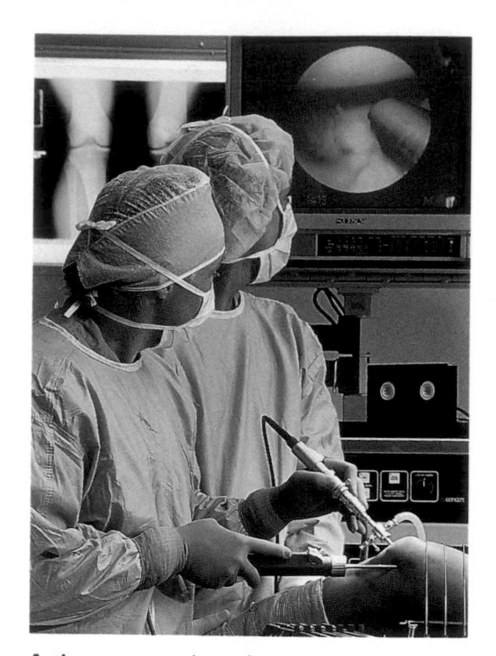

Arthroscopy, as shown here, is the examination of the inside of a joint (in this case, the knee), usually by means of a fine-bore fibreoptic endoscope. (Zefa)

Surgery today

The accelerated growth of scientific understanding, begun in the 19th century, continues apace in the 20th. Contemporary surgery is dominated by technological advances, which have saved countless lives, avoided immeasurable distress and greatly improved the outlook in many conditions and the quality of life of sufferers.

Microsurgery. The operating microscope is a binocular device that the surgeon can use and control hands free. A zoom magnification of up to about 40 times is used, and focus, zoom and X and Y shift (respectively, left and right, and up and down) are controlled by means of pedals. The microscope is arranged so that both the surgeon and the assistant see the same field and do so while sitting comfortably upright and looking horizontally into the microscope. Microsurgeons and those using laparoscopy (⇨ below) must learn to move their fingers so as to control instruments in one place while looking at the results of this manipulation in another. This seems strange at first but soon becomes second nature.

Microsurgery was initiated during the 1960s by eye surgeons dissatisfied with the limited degree of precision possible using only the naked eye or magnifying spectacles. It soon became apparent that not only could far better results be obtained by using a microscope, but also that operations impossible by any other means could become routine. An entirely new range of instruments, of a previously unheard-of delicacy, had to be designed and manufactured. Ophthalmic microsurgery led to such advances as the routine use of lens implants following cataract surgery, new drainage operations for *glaucoma* (an increase in the pressure of the fluids in the eye), operations on the vitreous gel of the eye and even direct surgery on the delicate retina (⇨ p. 84).

Ear, nose and throat surgeons were quick to see the possibilities of this new idea and soon microsurgery became routine, especially for operations on the middle and inner ear. Procedures such as the stapedectomy operation for the treatment of the deafening disease of otosclerosis, or reconstruction operations on the tiny bones of the middle ear (☞ p. 86), became possible. Soon, the opportunities presented by microsurgery became apparent to surgeons in other disciplines too. Vascular surgeons, anxious to be able to join up small arteries and veins, and neurosurgeons, aware of the importance of precise placement of severed peripheral nerves, also adopted the method. This made it possible, for the first time, to move large blocks of tissue, together with their blood supply, from one part of the body to another and even successfully to rejoin complete limbs that had been traumatically amputated. Gynaecologists, too, found microsurgery essential for the more delicate procedures on the fallopian tubes (☞ p. 96). Even some orthopaedic surgeons are now using microsurgery to allow bone grafts to be done complete with a new blood supply (*vascularized grafts*) in cases in which fractures have failed to heal.

Endoscopic diagnosis. In the past, much surgical diagnosis was made by inference from history, symptoms and the observation or eliciting of physical signs. An important recent advance in surgery is the ability to make such direct observations using optical devices, known as *endoscopes* (☞ p. 128), that can be introduced into the body either through natural orifices or through very small surgical openings.

Endoscopes, which may be rigid or flexible, use fibreoptic technology. Flexible endoscopes are usually steerable, with rotating knobs that can turn the tip of the instrument in any direction. An inevitable development from endoscopy was the idea of passing instruments through a narrow channel in the endoscope to allow small samples of tissue to be taken or small areas destroyed with a hot-wire cautery. Soon, a range of special instruments of this kind was produced and these now include forceps of various kinds, scissors, tying instruments, cup-shaped grabbers and lasers. An endoscope used in this way is called a *laparoscope*. Another important advance was the incorporation of a closed-circuit TV channel in the endoscope so that the operation site could be viewed on a monitor by both the surgeon and theatre assistant, and instrument manipulation made easier.

Laparoscopic surgery. The use of the laparoscope to perform a range of abdominal and even chest operations is likely to be a major feature of surgery in the future. The so-called *keyhole surgery* was originated by gynaecologists, who found that by blowing the harmless gas carbon dioxide into the abdomen, the bowels could be pushed up out of the way of the pelvic organs so as to allow endoscopic access for a range of gynaecological operations, especially procedures on the fallopian tubes such as sterilization. Surgeons interested in the gall bladder next took up the procedure, and the laparoscopic removal of the gall bladder (*laparoscopic cholecystectomy*) has now become almost a standard method.

The minimally invasive experience of laparoscopic gall-bladder removal has proved so much better for the patient than previous open surgery methods – patients are up the next day and back to work in a week – that surgeons have naturally been tempted to extend the idea to other operations. Obvious limitations in removing organs are imposed by the size of the opening, but procedures such as the division of *adhesions* (tissues abnormally healed together) and *appendicectomy* (removal of the appendix) are now commonplace. Various ingenious ideas have been evolved to extend the scope of the method. Separated organs can be put into small plastic bags within the abdomen and then pulled out through the small incisions after the laparoscope has been removed.

Leg amputation in 16th-century Europe. Needless to say, the patient has not had the benefit of an anaesthetic; indeed, it is the job of the man on the right, with the padded fist, to knock him out again should he show signs of regaining consciousness before the operation is over. From Hans van Gersdorff, *Veldt Boek van den Chirugia Scheel-Hans* (1593). (AR and E.P. Goldschmidt & Co., Ltd)

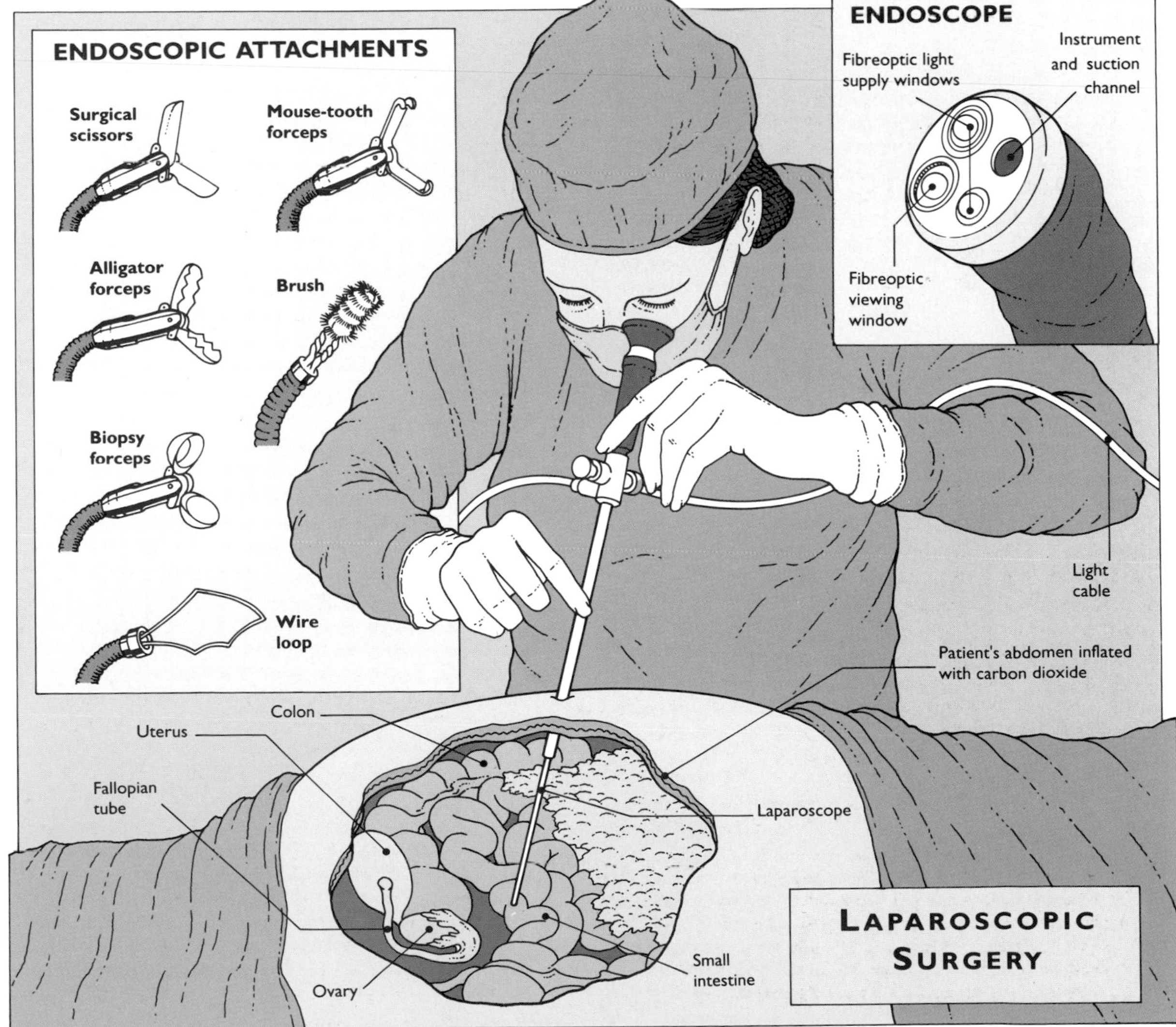

Special Forms of Treatment

Although the great majority of diseases are treated either medically (by drugs) or surgically, the range of treatments available for the sick extends far beyond these measures.

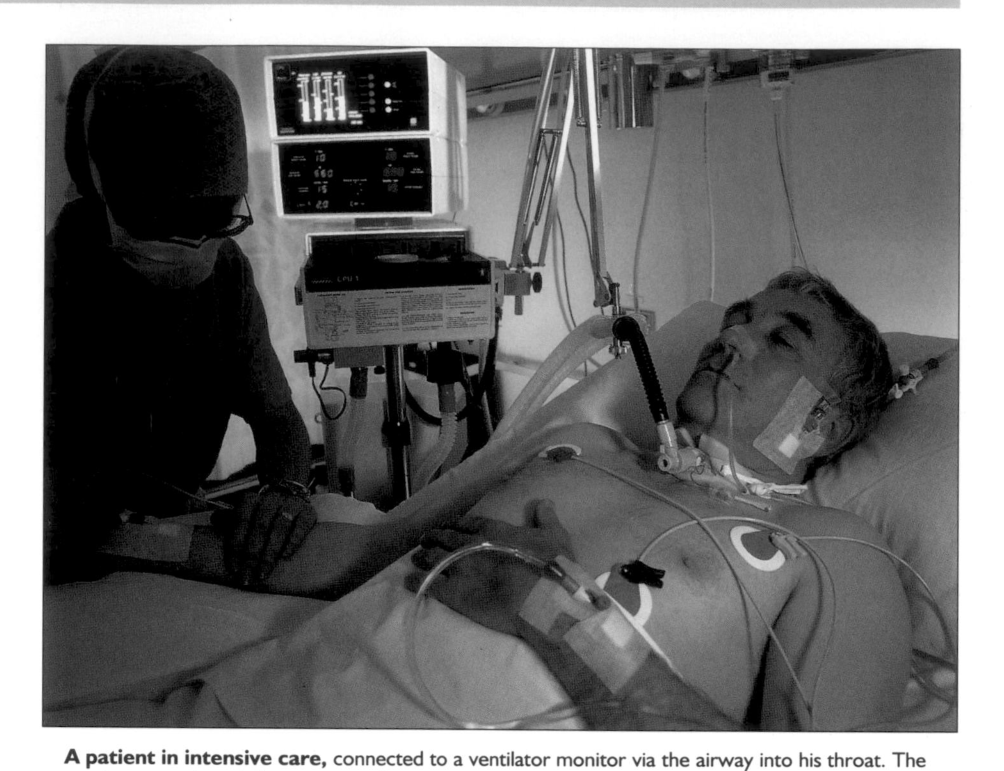

A patient in intensive care, connected to a ventilator monitor via the airway into his throat. The ventilator monitors information on the oxygen and carbon dioxide volume of his exhaled air, and is equipped with an automatic alarm in the event of emergency. (Malcolm Fielding, the BOC Group plc, SPL)

Nonmedical and nonsurgical treatments are used for some of the most life-threatening diseases. Cancer, probably the most dreaded of all diseases, is commonly and successfully treated by radiotherapy. Physical medicine has an important part to play in the rehabilitation of the sick and injured. Supportive treatment by machine is available to sustain people who have suffered heart and kidney failure until more definitive measures, including transplantation, can be taken. People with heart block and cardiac-rhythm disorders can be maintained in health by means of implanted artificial pacemakers. And when certain body parts finally succumb they can be replaced by effective prosthetic devices.

Of the various forms of radiant energy used in medical treatment – heat, ultrasound, light, radio waves, x-rays and gamma rays – the last two are probably the most useful. They lie at the high frequency end of the electromagnetic spectrum, and the difference between them is a matter of wavelength. X-rays and gamma rays are both forms of *ionizing radiation*, which has more profound biological effects than nonionizing radiation (such as visible light and heat) because it is capable of breaking chemical bonds, thus damaging molecules. Large molecules, such as DNA (➪ p. 24), are especially susceptible to such damage and are not always able to repair themselves. The energy of such ionizing radiation, produced by machines, is measured in terms of the voltage existing between the electrodes of the tube that produces them. For high-energy radiation this is of the order of millions of volts (megavolts or MeV). The higher the voltage, the greater the acceleration of the electrons striking the metal *anode* (positive electrode) and the more penetrating the radiation given off.

Radiotherapy

Radiotherapy or radiation therapy (➪ p. 140) has, in the past, been used for a wide range of conditions. We now know that, except for the treatment of cancer, most of such attempts were misguided and destructive. Ionizing radiation is useful precisely because it is damaging, and its use in cancer is based on the fact that rapidly reproducing cells (such as cancer cells) are significantly more likely to be killed by radiation than cells dividing slowly (such as normal body cells).

Different kinds of normal body cells, however, reproduce at different rates, according to the body's needs. Brain cells do not reproduce at all and are relatively resistant to radiation. The cells of the testes and ovaries, those of the lining of the intestinal tract, those of the bone marrow and those of the hair follicles are among the most rapidly reproducing in the body. A given dose of radiation will thus affect these parts of the body more than it will affect the brain or other tissues. Thus, people accidentally exposed to high doses of radiation suffer diarrhoea and vomiting, anaemia, loss of hair and temporary or permanent sterility (the condition known as *radiation sickness*).

Radiotherapy aims to expose the area of the cancer to ionizing radiation in doses designed to kill the cancer cells without doing too much damage to the rest of the body. The site of the tumour must be known with precision and account must be taken of its depth below the skin's surface and the type of tissue surrounding it. The response to radiation depends on the intensity and duration of the exposure, the distance of the tumour from the radiation source and the absorption of radiation by intervening tissue. It also

Prosthetic Devices

The most successful internal prosthetic devices to date have been the ***artificial hip*** and the ***intraocular lens implant.*** The artificial hip prosthesis is used as a total replacement for a hip joint that has become severely damaged by arthritis or by the loss of its blood supply following fracture of the neck of the thigh bone (*femur*; ➪ pp. 40 and 110). The prosthesis is in two parts: a socket, usually made of a tough plastic, and a ball-and-shaft part, made of corrosion-resistant metal. The existing socket in the side of the pelvis (the *acetabulum*) is reamed out and the prosthetic socket cemented into place with acrylic adhesive. The whole of the top of the femur, including the neck, is sawn off and the shaft of the metal part of the prosthesis is pushed down the canal of the bone and cemented into place. The two parts are then articulated. The results are excellent: painless mobility is restored to the patient almost at once.

Lens implants have now become almost universal following removal of the opaque natural lens (*cataract*; ➪ pp. 84 and 110). Formerly, the resulting deficit in focusing had to be compensated for by very strong spectacles that were heavy and produced greatly magnified and peripherally distorted images. The optically perfect polymethylmethacrylate lens implant, however, produces far better results. It weighs only a fraction of a gram and is inserted into the capsule of the original natural lens after all the opaque lens matter has been removed. It is self-centring and is held in place by fine, springy plastic supports. Although of fixed focus, the power can be selected to give sharp distance vision, and glasses can then be prescribed for reading.

Contrary to popular belief, although a number of *artificial hearts* have been made and used experimentally in animals, in no case has an artificial heart yet been able to provide even a reasonable quality of life for a human patient. So far, these devices have been used on a small number of people, and a few have been kept alive until a human heart has become available for transplantation. Some cardiologists and heart surgeons believe that artificial hearts have a place in the support of people dying from heart disease, but few at present believe that they can be used on a permanent basis. Early experience has been very disappointing. Three of the first five patients given permanent artificial hearts died within a short time, and the other two suffered severely incapacitating strokes. The main problem seems to be the difficulty of avoiding the repeated passage to the brain of blood clots that form in the heart. Other difficulties include massive internal haemorrhaging, infection and kidney failure.

depends on the radiosensitivity of the particular tumour, which can vary widely. The calculations needed to assess the dosage, angle of incidence of radiation and type of radiation are complex and are performed by computer. Heavy metal shielding is used to protect normal tissue, so far as possible, from radiation, and treatments are usually given for periods of a few minutes, several times a week for a number of weeks.

Radiotherapy may be provided by high-energy x-ray machines, by linear accelerators, by *cyclotrons* (another type of particle accelerator), or by radioactive sources such as cobalt-60 bombs. In all cases the beam of radiation is focused and directed by heavy *collimating* (making parallel) lead shielding. Small, readily accessible tumours may be irradiated by means of small sources that are placed within the body close to the tumour (*intra-cavitary therapy*) or even, sometimes, within the tumour (*interstitial therapy*). Such treatment may last for periods of hours to days.

Physiotherapy

Physiotherapy is treatment by exercise, active and passive movement, massage and the application of physical agents such as heat, radio waves, ultrasound, ultraviolet and visible light, and water buoyancy. It is a subdivision of the larger field of physical medicine, which also incorporates diagnosis and the planning of rehabilitation. Physiotherapy aims to promote physical and psychological recovery from injury and disease, to improve muscle power and stamina, to increase the range of movements in joint and hence bodily mobility, to relieve pain and to assist in the restoration of specific local functions such as grasping, use of utensils and tools, and the fine skills involving the use of the fingers.

As a separate discipline physiotherapy arose from the massive need for rehabilitation of the millions of men injured during World War I. Today, every major general hospital has its physiotherapy department, usually centred on a large gymnasium, with side rooms for individual treatments. Many have a special swimming pool dedicated to the use of people with muscle weakness too severe to overcome the effects of gravity. Such people can often, however, move and exercise while immersed in water and thus gradually gain the strength to exercise out of the pool.

Physiotherapists use various forms of massage, intense beams of radio-wave energy (*short-wave diathermy*), heat lamps and ultrasound to promote an increased flow of blood through damaged tissue to improve its oxygenation and immune competence. Their chief function, however, is to supervise judicious exercising, to restore confidence, build up muscle power and re-establish coordination. They are also concerned to teach the proper use of crutches, walking sticks and artificial limbs. Some physiotherapists specialize in the management of people with physical disabilities.

Dietary treatment

Medical dietetics is an established speciality based on the known principles of nutritional science. These include a recognition of the energy value of the range of foodstuffs in terms of the amount of heat produced when a known quantity is burnt (*calorific value*), the daily allowances recommended by experts of each of the three classes of foodstuff (carbohydrates, proteins and fats), the minimum requirements of the range of vitamins and minerals, and the requirements of water and trace elements.

All major hospitals have one or more dieticians on the staff, and the treatment that they provide is concerned with supplying appropriate menus for the generality of patients whose nutritional needs are normal, for people with special dietary requirements resulting from either under- or over-nutrition or for those with malabsorption disorders, and with providing formulations for the few patients who need to be fed either by a tube that bypasses the stomach or even directly into the bloodstream by way of a vein (*parenteral feeding*).

There are a number of specific disorders for which a scrupulously careful examination of the affected person's diet is essential. One example is *galactosaemia*, a genetic disorder due to the absence of an enzyme necessary for the breakdown of milk sugar (*galactose*) to glucose. The build-up of galactose causes cataracts of the eye and severe brain damage. But, under a dietitian's supervision, an infant fed on a galactose-free diet can develop entirely normally.

Physiological support

One of the greatest successes of the technological revolution in medicine has been the use of machines to support life in people who would otherwise die or suffer severe disablement. Machines called *ventilators* are used to maintain respiration in people recovering from serious illness, while others are used to perform the function of the kidneys in abstracting the waste products of metabolism from the blood (*dialysis*), which would otherwise accumulate dangerously. Some are used temporarily to maintain both the circulation of the blood, and its oxygenation and conditioning (*heart–lung bypass machines*). And yet others (*artificial pacemakers*) provide a repetitive electrical shock to maintain the regularity of the heartbeat (*ventricular contraction*) at a rate appropriate to the body's needs.

Certain devices are used temporarily to provide assistance in pumping blood to a heart that has failed and cannot on its own maintain sufficient blood pressure to supply its own muscle and the brain. The most successful of the current heart-assist devices consists of a plastic balloon (the *intra-aortic balloon pump*) that is inserted into the main artery of the body (the *aorta*; ➪ p. 50) and is inflated and deflated synchronously with the heartbeat so as to force the blood more energetically through the arteries and to raise the blood pressure at the openings of the coronary arteries. Intra-aortic balloon pumps are inserted via an artery in the groin and have been used for periods of up to nearly 11 months following heart operations.

- DIAGNOSTIC AIDS p. 128
- HEALTH AND DISEASE pp. 152–5

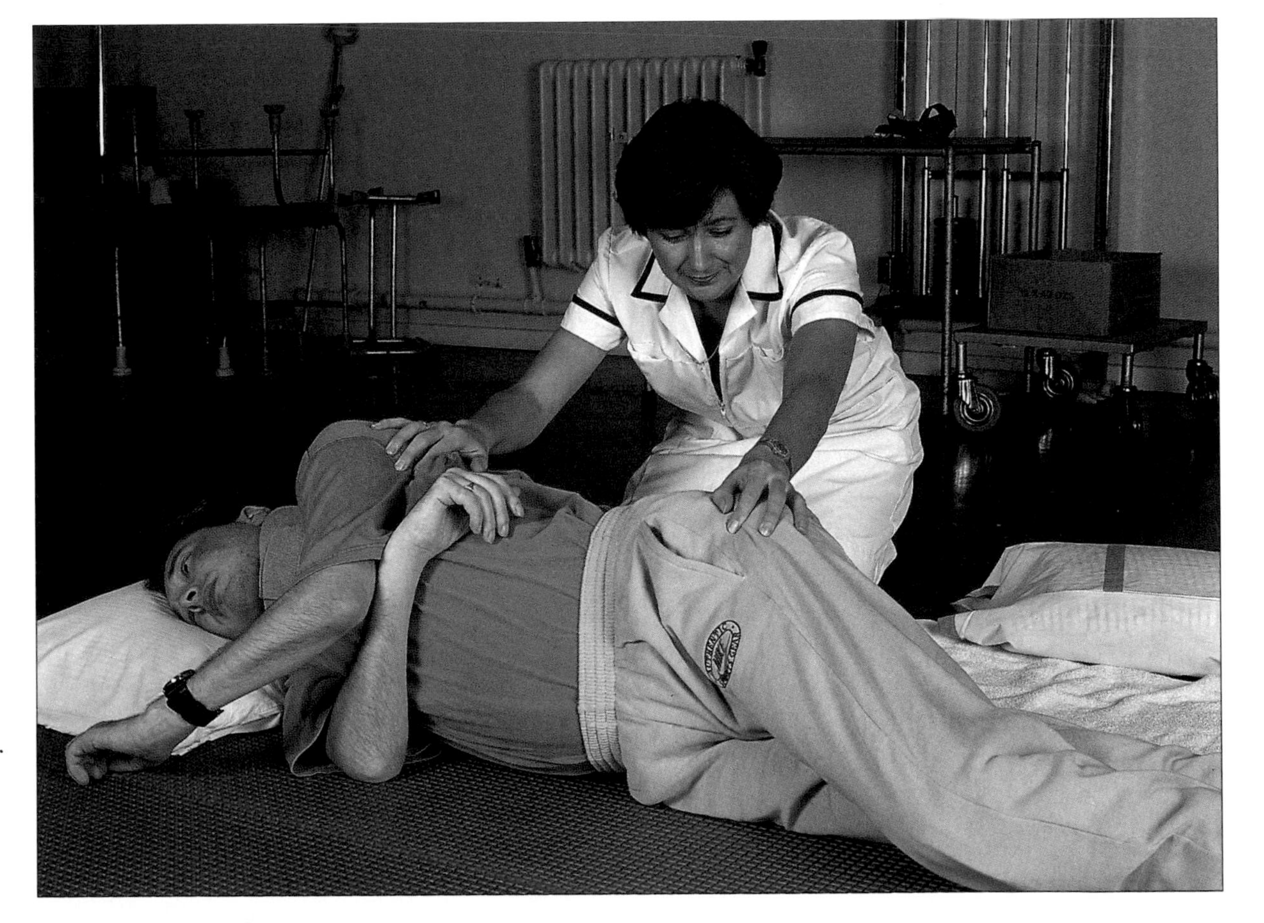

Physiotherapist treating a young man who is paralysed in all four limbs (*tetraplegic*), as the result of a broken neck. Although he has limited control of his shoulders and arms, and has problems balancing, physiotherapy is useful, not least as a means of exercising immobilized muscles and joints. (Simon Fraser, Hexham General Hospital, SPL)

Complementary Medicine

In spite of the manifest blessings that orthodox medicine has conferred on the human being, some people seek an alternative regimen. They do so often because they find unacceptable conventional medicine's sometimes overenthusiastic reliance on drugs that may have unpleasant or even dangerous side effects, and technology that seems, at times, to displace human concerns.

The term 'alternative medicine' prevailed until recently to imply methods of therapy that can safely and reliably replace scientific medicine; this is not a safe inference. Even the most partisan adherent would hesitate to trust the management of, say, a compound fracture or a breast cancer to an 'alternative' therapist. Therefore, the term 'complementary' has become an acceptable substitute. The common feature that seems to run through every one of the complementary treatments is the importance placed on the whole person, not just on specific symptoms – this is known as the *holistic* approach.

Samuel Hahnemann (1755–1843), founder of the practice of homeopathy. (ME)

Herbalism

Herbalism is an ancient form of medicine. From the dawn of humanity, people have been using plants to cure their illnesses. From the Middle Ages, *herbals* – manuals listing the names of plants and what they could be used for – were widely employed. In the 17th century Nicholas Culpeper (1616–54) combined herbalism with astrology in his *Herbal*. Herbalists today use the roots, leaves, stems, flowers and seeds of plants to produce medicines. A large number of orthodox modern medicines are also derived from plants – the painkillers known as *opiates*, for instance, are derived from the opium poppy.

Once a diagnosis has been made, the herbalist will dilute a concentrated extract of a certain herb in water or mix it into a paste to form a cream or ointment. Conditions such as arthritis, colds and coughs, skin problems, digestive disorders and minor injuries are regarded as the most likely to benefit from herbalism.

Homeopathy

The German doctor Christian Friedrich Samuel Hahnemann (1755–1843) was a sensitive physician who became disenchanted with the medical treatments of his day – largely blood-letting and the use of dangerous drugs – and became increasingly aware of the damage such treatments were causing to patients. He gave up orthodox practice for a time and in his *Organon der Rationellen Heilkunde* of 1810 proposed a new method of treatment based solely on symptoms.

Hahnemann had noticed that a number of drugs produced effects similar to the symptoms of various diseases, and decided that diseases could be cured by giving such drugs – the process of like curing like. At first he used large doses and caused serious symptoms to appear in his patients. As he reduced the doses, these effects, predictably, became less. He then became convinced that the effect of the drug would be increased in proportion to the amount it was diluted, and proposed a system of successive dilutions, each by one hundred times. His favourite dilution was the 30th – a reduction by an astronomical figure. These repetitive dilutions, called *potentiations*, were to be accompanied by strong shaking or pounding of the container (*succussion*). Hahnemann's system was called *homeopathy* – treatment by like substances – to distinguish it from conventional *allopathy* – treatment by opposing substances.

Homeopathy continues to be practised to this day and commands wide support, especially in Europe. Homeopathic practitioners take detailed medical histories of their patients, and select their remedies with care from a large official pharmacopoeia. The system has, however, aroused strong controversy among orthodox medical practitioners.

Aromatherapy

Aromatherapy is principally a massage technique in which essential oils derived from herbs, flowers and spices are rubbed into the skin and inhaled by the person being massaged. The fragrances these oils produce are said to be particularly effective for psychological complaints such as anxiety or depression, but are also said to help a range of conditions, including skin disorders and burns.

Other Complementary Therapies

- ***Anthroposophical medicine*** **– essentially an attitude to health rather than a therapy. Using homeopathic remedies, it stresses the need for a doctor to attain a spiritual understanding of both the patient and the plants used in 'treatment'.**
- ***Applied kinesiology*** **– often used in conjunction with chiropractic. Practitioners use a system of 'muscle testing' to identify weaknesses in organs and muscles. The aim is to balance these weak elements by touch.**
- ***Bach flower remedies*** **– used by practitioners for conditions such as anxiety and depression.**
- ***Biochemics*** **– invented by the German chemist Dr W. H. Schnessler, who recognized 12 inorganic salts and oxides in the body. Using homeopathic methods, practitioners seek to redress any imbalances in these – and in any of an additional 30 trace elements since identified.**
- ***Colour therapy*** **– different coloured light is used for a variety of conditions, including stress.**
- ***Hydrotherapy*** **– water is used in a number of therapies, for example to stimulate the circulation, or in colonic irrigation.**
- ***Iridology*** **– the study of the iris as a tool for 'diagnosing' illness.**
- ***Naturopathy*** **– avoidance of any contaminants and drugs.**
- ***Rolfing*** **– a system of deep massage.**
- ***Shiatsu*** **– Japanese deep massage used to stimulate acupuncture points. Claimed to be a preventative as well as a cure for a range of disorders, including stress.**

Acupuncture

Acupuncture originated in China over 5000 years ago. The technique uses fine needles inserted at specific points on the body in order, it is said, to restore the balance of an inner 'life force' known as *chi* energy and believed to flow along a number of *meridians* or channels in the body. Each of the 12 main meridians is believed to have its own pulse – six in each wrist – and the acupuncturist checks these carefully in order to decide which points to stimulate.

The technique has been shown to be remarkably successful at stopping pain, and in China major operations have been carried out using only acupuncture for pain relief. Scientists have discovered that the needles appear to make the body produce its own natural painkillers, *endorphins* (⇨ p. 90). Acupuncture is also claimed to be effective in treating a range of diseases, including respiratory, digestive, bone and muscle disorders.

Reflexology

Like acupuncture, reflexology is based on the idea that the body contains channels of 'life force'. Reflexologists believe that this force exists in 10 'zones' of energy that each begins in the toes and ends in the fingers.

By touching and feeling the toes and feet, reflexologists claim to be able to feel blocks in these channels of energy (they say these feel like crystals below the skin surface), and by manipulating and massaging the foot in a specific way they try to move the blockage, thus, it is believed, curing the illness. Like acupuncture, reflexology is used for most conditions.

Osteopathy

Osteopathy is a manipulative technique founded by the American doctor Andrew Taylor Still (1828–1917). Joints are pushed and occasionally pulled so as to restore them to their normal positions, thus relieving tensions on surrounding muscles, tendons and ligaments.

Osteopaths tend to concentrate their work on the spine since this contains the spinal cord and all the nerves that control the body. Back pain is the disorder most commonly treated by an osteopath.

The Alexander Technique

The Alexander Technique is a method of producing postural changes, which are claimed to relieve a number of physical disorders. The technique was developed in the 19th century by an Australian actor, Matthias Alexander (1869–1955). He realized that the position of his head and neck were the cause of his frequent loss of voice during performances, and found that by altering the use of his body he could cure himself.

During a series of lessons – 12 or more – the person 'relearns' how to use the body, breaking harmful postural habits. The technique is claimed to be beneficial for everyone, but in particular for those who have suffered long spells of general ill health – lethargy or poor sleeping, for example.

Biofeedback

Biofeedback is a technique used in the hope of helping people learn to control physical phenomena governed by the autonomic nervous system (➪ p. 74), such as blood pressure, heartbeat and temperature. Electrodes placed on the body pick up electrical impulses produced by physical changes. The impulses are transformed by the biofeedback machine into an electronic sound, or shown by the rise and fall of a needle on a dial. The person concentrates on changing the tone of the sound or on causing the needle to move, and in doing so learns, for instance, to lower the blood pressure or slow down the heartbeat.

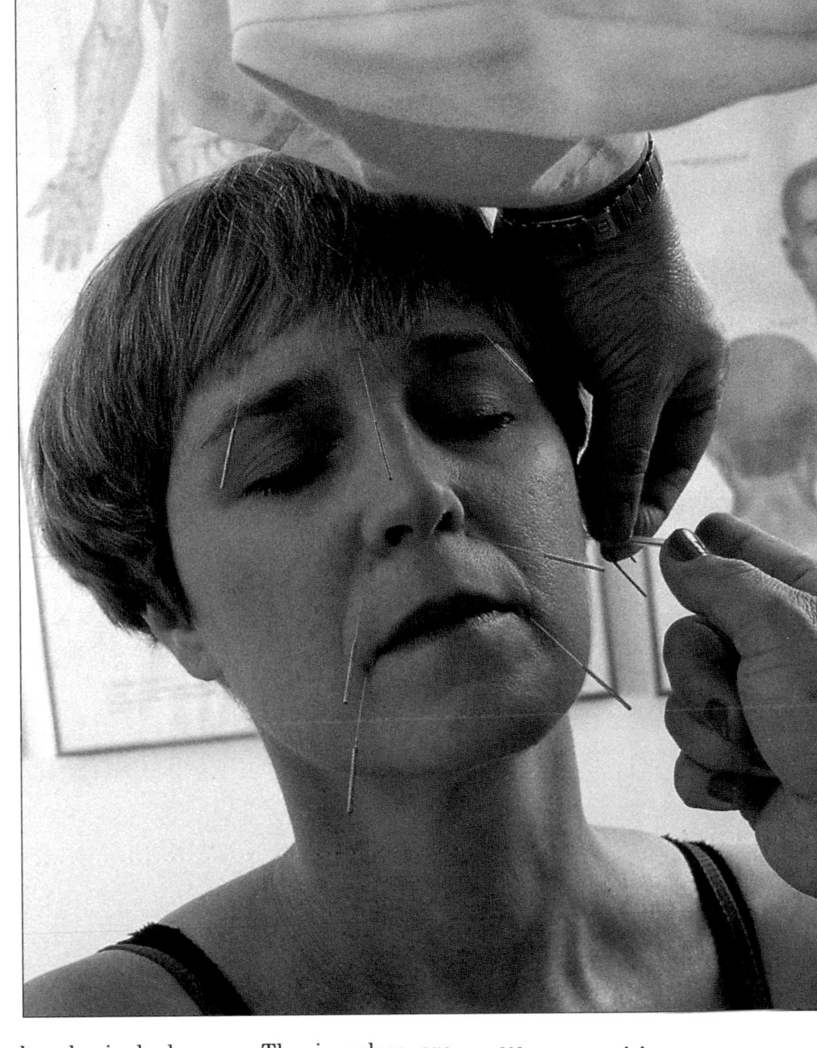

Woman receiving acupuncture therapy. This branch of Chinese medicine is now also used by some Western doctors as a method of producing anaesthesia. (Zefa)

Chiropractic

The central philosophy of chiropractic is that malalignments of the bones in the spine cause disturbances of the nervous and vascular systems leading to disease not only in the bones and muscles themselves, but in any organ of the body.

Chiropractitioners work with the help of x-rays to discover where the malalignments are, and to identify 'intersegmental dysrelationships'. They then manipulate the bones using short, but very forceful thrusts to the joint, thus relieving the root cause of the problem. However, chiropractic should not be used in any case of bone malignancy (cancer) or where the spinal cord is compressed, as the rapid, forceful thrusts can lead to fracture and paralysis in these cases.

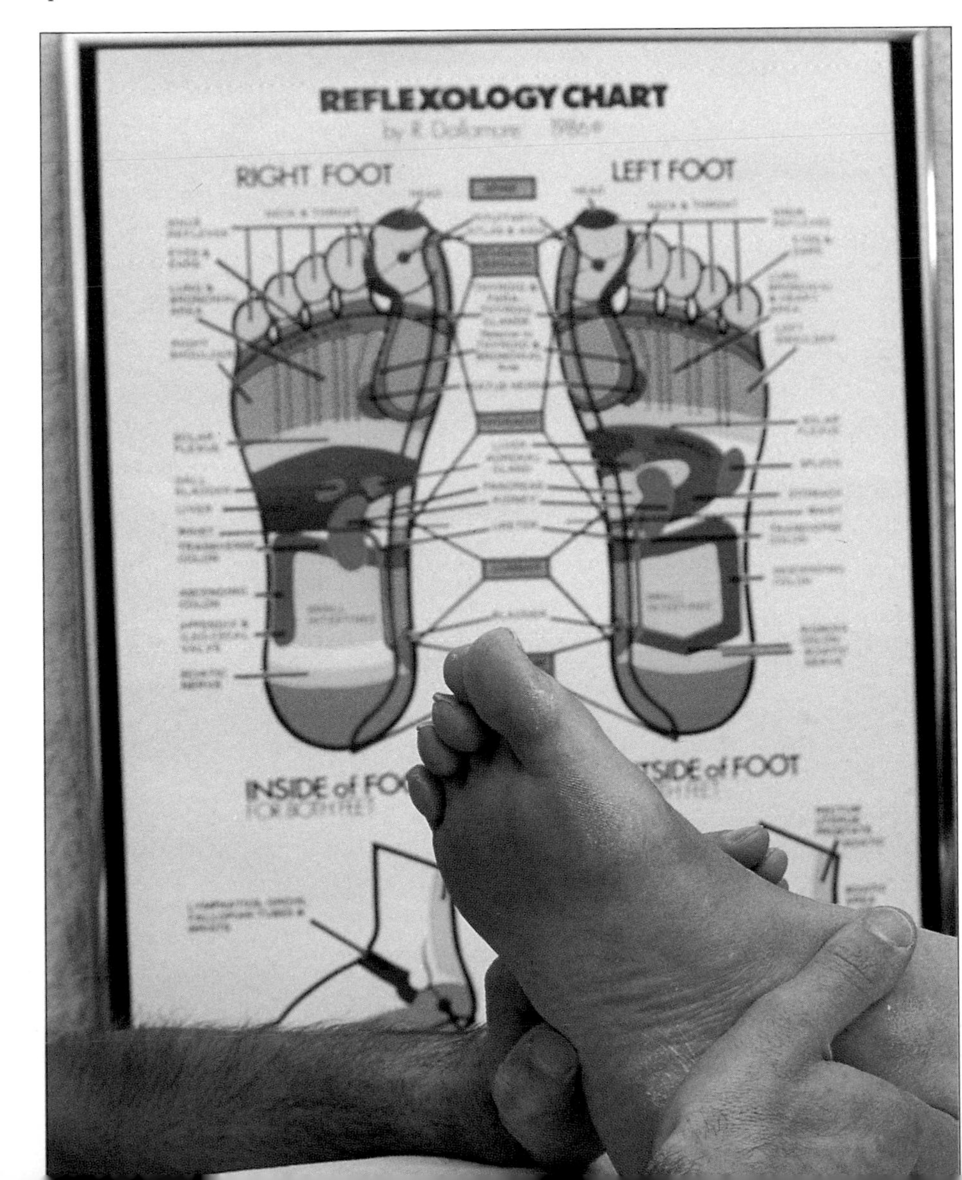

Reflexology is based on the belief that certain areas of the feet correspond to various organs of the body, as shown in this therapists' chart. (JW)

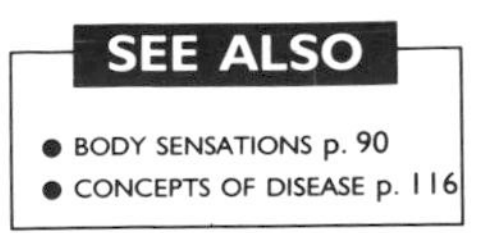
SEE ALSO
• BODY SENSATIONS p. 90
• CONCEPTS OF DISEASE p. 116

PSYCHOLOGY

Towards a Science of the Mind

Throughout the ages philosophers and scientists have been deeply interested in the functioning of the mind – how people think and feel, form opinions and beliefs, how they develop and mature mentally, how they change and, above all, how and why they act the way they do. Today, these questions are the province of psychology.

SEE ALSO

- BODY AND SOUL p. 70
- THE HUMAN COMPUTER? p. 72
- THE PSYCHOANALYTIC MOVEMENT p. 164
- PSYCHOLOGY TODAY p. 166

Although modern psychology is essentially concerned with mental function, given the present state of science, this is primarily assessed by the observation of behaviour. In this context, the term 'behaviour' is used in a very wide sense, and includes all detectable responses, although factors such as language and representation are also studied. One preliminary difficulty is the need to decide exactly what we mean by the word 'mind'. The history of psychology is inextricably linked with, and often inseparable from, that of other reaches of human thought and speculation – philosophy, theology, demonology and magic. While some of the early ideas now seem ridiculous, they were often plausible in the light of what was then known. The study of the history of human ideas is, in itself, a source of information about the mind.

Historical ideas of mind and personality

For many centuries, a clear verbal distinction was made between the mind and the soul (⇨ p. 70), the latter being an entity distinct from the body and often regarded as the seat of the emotions or sentiments. The Greek philosopher Aristotle (384–322 BC), however, believed that the soul had the power to think and called the thinking soul the mind (*nous*).

Early Western thinkers, prompted by the Greek physician Galen (c. 130–201), conceived the world as compounded of four elements – earth, air, fire and water. Corresponding to these, in the human body, were the four humours – black bile, yellow bile, blood and phlegm (⇨ p. 34). Our emotional and temperamental natures were, it was believed, determined by the relative proportions of these humours. If there was any preponderance of one humour, the resulting characteristic state of mind might be, respectively, melancholic, irascible, sanguine or phlegmatic. The English clergyman Robert Burton (1577–1640), in his *Anatomy of Melancholy* (1621), described the melancholic man as 'of a deep reach, excellent apprehension, judicious, wise and witty' (i.e. of quick intelligence).

The concept of the humours persisted throughout the Middle Ages and the Renaissance, being gradually modified with time. In the 16th century all forms of psychological upset were described as melancholy. During that period, virtually all scholarship and philosophy was concerned with the support of religious faith. While the doctors ascribed melancholy to the humours, the churchmen, in accordance with theological tenets, were more inclined to attribute severe emotional disturbances and madness to demonic possession.

Descartes and dualism

The central figure in the renaissance of rationalism in the 17th century was the philosopher René Descartes (1596–1650). He was well aware of the mind–soul difficulty and stated in his definition that mind was a substance distinct from matter. He used the term 'mind' rather than 'soul' since the latter was ambiguous. Subsequent philosophers tended to ignore this ambiguity.

In his *Discourse on Method* (1637) Descartes wrote, 'I am a substance whose whole nature or essence is to think, and which for its existence needs no place, nor depends on any material thing.' This idea was incorporated into the concept that came to be described, after his time, as *Cartesian dualism*. There were, he held, two fundamentally different kinds of substance – material substance (*res extensa*), which could be seen and touched, and thinking substance (*res cogitans*), which had no dimensional qualities. The human body was of the former substance, while the mind – which included thoughts, emotions, desires and volitions – was of the latter.

Cartesian dualism – the mind–body concept – has had a dominant effect on humanity's idea of mind, and was accepted by most philosophers until well into the 20th century. It is, unthinkingly, held by many people to this day, and is only now – with the growth of knowledge of the relationship between body function and its psychological concomitants – beginning to be challenged seriously. The concept of dualism raises a number of grave difficulties, some of which were recognized by Descartes himself. One of these is how to explain the close causal link between mind action and body action – how an immaterial substance can cause changes in a material substance.

Early developments in modern psychology

The first steps towards establishing psychology as a scientific and experimental discipline were taken towards the end of the 19th century when Professor Wilhelm Max Wundt (1832–1920) set up the first psychological laboratory in Leipzig, Germany, in 1879. Wundt, who is acknowledged to be the 'father' of experimental psychology, also started the first journal on the subject in 1881. His most important book, *Principles of Physiological Psychology* (1873–4), established that behavioural psychology was essentially related to the structure and function of the nervous system. The only way to study the mind

Bust of Aristotle – Greek philosopher whose writings ranged over subjects as diverse as logic, ethics, metaphysics, poetics, politics, biology and zoology. (AKG)

directly was by *introspection*, and he trained observers to apply introspection rigorously under strictly specified conditions. One criticism of this approach is, however, that introspectors are liable to find almost anything they are looking for, and there was much disagreement among them. Wundt's most valuable work was concerned with sensation and perception, especially vision and hearing. His laboratory quickly became the centre of world interest of those seriously concerned with the subject, and was the model for all other psychological laboratories set up at the time.

William James and the *Principles of Psychology*

One of the major influences in the development of modern psychology was the American philosopher and scholar William James (1842–1910). As much a philosopher as a psychologist, James wrote an immensely influential work, *Principles of Psychology* (1890), which took him twelve years to write. He adopted no fixed position and, although he founded an experimental psychology laboratory at Harvard, he was more interested in general observations on human nature, behaviour and experience. One of his most often quoted remarks is, 'Sow an action, and you reap a habit; sow a habit and you reap a character; sow a character and you reap a destiny.'

Functionalism and behaviourism

Under the influence of Darwin's theory of evolution (⇨ p. 8) many psychologists came to believe that the purpose of mental processes was to enable the organism to adapt to its environment. The main proponent of this movement – known as *functionalism* – was the American philosopher and educationalist John Dewey (1859–1952), who worked at the University of Chicago. Dewey was a pragmatic philosopher, and he believed that the experimental approach should be applied to all social and educational problems. Human beings, he said, were organisms trying to adapt, and when difficulties occurred they had to stop and think. Such 'operational' thinking was necessary to judge the value of anything, including moral questions. Dewey had a considerable influence on American educational thought and policy.

Another important figure who reacted against the introspective school of psychology was the American John B. Watson (1878–1958), the founder of the school known as *behaviourism*. Influenced by Pavlov's work on conditioned reflexes (⇨ p. 172), Watson came to believe that all human behaviour, however complex, could be accounted for as the sum of many conditioned reflexes, and that all behaviour was learned. He concluded that objectively observed behaviour was the only legitimate subject-matter for psychology. Watson's ideas had a great and lasting influence on American psychology. Even today, many still think of psychology as the science of behaviour rather than of the mind.

René Descartes (left) provided the first mechanistic basis for the theory of mind–body dualism and because of this is widely regarded as the founder of modern philosophy. (ME)

Today one of the best-known behaviourists is the American B.F. Skinner (1904–90), who believed that 'most behaviour is selected by its consequences'. In other words, we continue to do things we have done successfully in the past, and behaviour that we think we choose deliberately and freely is merely behaviour selected because of the consequences that similar behaviour has had in the past. Much of our mental life – even seemingly abstract thought and the formation of concepts – is, he held, modelled on our experience of the physical environment. Reasoning is essentially the formation of rules for our own use based on what has been profitable for us in the past. Skinner's work has also been strongly influential on contemporary psychological thought.

In terms of the influence exerted on 20th-century psychological theory and practice, behaviourism is rivalled only by the discipline of psychoanalysis (⇨ p. 164). At the same time, behaviourism has attracted criticism for its alleged reductionism of the human being, and has been shown to have disturbing applications, for instance to the control of the individual by society or the state. Today, almost all psychologists and psychiatrists acknowledge the merits of at least some of the ideas of Watson and Skinner. Some, such as H.J. Eysenck, have applied these ideas, in the techniques of behaviour therapy, to achieve the first successful treatment of a number of severe psychoneurotic conditions such as the range of phobic disorders (⇨ p. 190). In general, however, extreme behaviourism, with its prohibition of 'introspection' as a source of knowledge of the mind, has now been superseded. Contemporary research into the nature and workings of the mind is largely the province of neurophysiologists and cognitive psychologists (⇨ p. 166).

Modern ideas of the mind

The Cartesian notion that the mind is a separate, non-physical entity is gradually being replaced by ideas more in keeping with modern physiological knowledge. An emotion such as fear, for instance, whether occasioned by perception of danger or by the memory of such danger, is invariably associated with the release into the blood of hormones, such as adrenaline, that cause the heart to beat rapidly, the breathing to deepen, the palms to sweat and the hair to rise. A drug that blocks the action of adrenaline can abolish the emotion, although the purely intellectual awareness of the danger persists. Conversely adrenaline can induce a like emotion even in the absence of external causes of fear. Many drugs are known whose pharmacological properties are similar to those of the natural neurotransmitters of the brain and nervous system. (Neurotransmitters are chemical substances that propagate the action of nerve cells, mainly on other nerve cells.) People who take these drugs experience mental events of various kinds recognizably akin to those occurring during the normal action of the neurotransmitters.

Such examples suggest that at least some mental experience is inextricably linked with bodily function so as to be inconceivable apart from it. Many physiologists now believe that *all* mental experience will eventually be shown to be a product of neurological activity (⇨ p. 72) – although there is still much that is unknown about the nature and processes of this activity.

During the medieval period it was believed that a person's thoughts and behaviour could be influenced by demons. This 15th-century woodcut depicts a woman possessed by the devil. (AKG)

The Psycho-analytic Movement

In the 19th and early 20th centuries, psychological research moved in two main directions. On the one hand was the objective, rather dull – but important – scientific work of the behaviourists on stimuli and their responses (⇨ pp. 163 and 166); on the other was a sustained attempt by people such as Freud, Jung and Adler to try to find out directly, by questioning and the use of 'free association', what was going on deep in people's minds. It was this approach – *psychoanalysis* – that caught the popular imagination.

Sigmund Freud (1856–1939), father of psychoanalysis. (Sygma)

The results obtained from psychoanalytical methods, and their varied interpretations, have been widely criticized as lacking in scientific rigour. The kinds of 'experiments' involved are not possible to repeat precisely, and the theories based on them – being subjective interpretations rather than deductive inferences – cannot, by their nature, be either proved or disproved. It has also been remarked that psychoanalytical theories tend to reflect the personalities of their proponents, and their social and cultural backgrounds, in a way that established scientific laws – such as Newton's laws of gravity and motion – do not.

However, the psychoanalysts have given us many insights into the human psyche, and their work involved heroic – if flawed – attempts to provide complete and integrated theories of why we feel, think and behave as we do. These theories have formed the basis of various – often successful – techniques of psychotherapy (⇨ p. 195), although the benefit to the patient is more likely to result from the wisdom, maturity, experience and sympathy of the analyst rather than from the application of any of the theories.

The Freudian school

Sigmund Freud was born in Freiberg, Moravia (now in the Czech Republic), in 1856 but spent most of his life in Vienna. He was a child prodigy who read Shakespeare and Goethe at the age of 8, and was versed in the Greek, Latin, French and German classics. In spite of a taste for philosophy, he studied medicine and qualified in 1881. After some excellent early work on aphasia (⇨ p. 82), infant cerebral palsy, neurological connections in the nervous system (⇨ p. 74) and the psychological effects of cocaine, he worked for some time in Paris with the French neurologist Jean Martin Charcot (1825–93). Fascinated by Charcot's work on hysteria, Freud turned his attention to the psychological basis of mental disorder, and at first used hypnotism to elicit from his patients painful memories that seemed to be at the root of many psychological problems.

This work led him to the idea of repressed memories and impulses and to a profound consideration of the nature of the unconscious mind. He used the term *sublimation* (⇨ p. 188) to describe the diversion of psychic energy derived from sexual impulses into nonsexual activity, especially of a creative nature. Later he abandoned hypnotism and adopted the method of free association. This 'talking cure' appeared to relieve symptoms such as hysteria. Freud also discovered that his female patients tended to fall in love with him and to become emotionally dependent on him. This phenomenon he called *transference*, and, after some time, he concluded that transference was essential for the success of the method.

Freud's interest in the significance of dreams as a source of insight into the unconscious mind led to the publication in 1900 of *The Interpretation of Dreams*, which was largely ignored. In 1903 he published *Three Essays on the Theory of Sexuality*, which at first caused a storm of abuse, but gradually his ideas gained acceptance. His subsequent publications included *Totem and Taboo* (1913), *Beyond the Pleasure Principle* (1920), and *The Ego and the Id* (1923).

On the German invasion of Austria in 1938 Freud, who was of Jewish origin, was immediately in great danger. But, under pressure from the Americans, the Nazis let him leave unmolested. He settled in London, where he died in 1939 from cancer of the upper jaw, from which he had suffered stoically for years. His daughter Anna Freud became, in her own right, a leading figure in the psychoanalytic movement.

Throughout his life, Freud continued to develop, revise and amend his ideas. In the manner of a religion, he attracted many followers and disciples. And, like many religions, the Freudian school has had its full share of schisms, and is now greatly fragmented. However, Freud's ideas (⇨ box) have had an immense and pervasive cultural influence on the 20th century, mainly because of their emphasis on sex, and the licence they provide for a general discussion of the subject in an age when such talk has been largely taboo. In addition, although Freud did not 'discover' the unconscious mind, as is sometimes asserted, the light he cast upon

- TOWARDS A SCIENCE OF THE MIND p. 162
- PSYCHOLOGY TODAY p. 166
- PSYCHOLOGY pp. 170–3 and 176–9

it changed the whole aspect of human thought.

The Jungian school

A deeply religious man, the Swiss psychiatrist Carl Gustav Jung (1875–1961) was steeped in mythology and had an exceptionally rich dream life. These dreams had a large religious and mythological content, and were a major factor in forming his ideas about the mind. At first, Jung was greatly influenced by Freud and became a close associate and colleague. In 1911 he was appointed the first president of the International Psychoanalytic Association, but he soon became disenchanted with Freud's emphasis on the erotic, which he considered reductionist and unbalanced, and broke off the relationship.

Jung came to believe that human beings were motivated by a kind of creative energy, for which he used the term *libido*, that could act in different ways. People whose chief interests were external to themselves he called *extraverts*; those whose main preoccupation was with the things of the mind and the spirit he called *introverts*. He identified what he called the *anima* and *animus*, respectively the feminine principle as present in the male unconscious, and the masculine principle in the female unconscious. The animus/anima was the 'inner being', the part of the psyche in intimate contact with the unconscious. And he adopted the term *persona*, taken directly from the Latin word for an actor's mask, using it to refer to the role that one is expected to play in society – a role often at variance with the animus or anima. Thus the persona operates as a mechanism for concealing true thoughts and feelings.

Jung was deeply preoccupied by symbols, and detected a common pattern of symbolic images underlying the mythology, magic and religion of many different cultures throughout history. This led him to believe in a *collective unconscious* shared by all humanity that contained what he called the *archetypes* – patterns of thought and behaviour common to all human beings and realized in all the major religious and mythological systems. Jungian therapy is much concerned with dreams and symbols, and is ostensibly calculated to put the patient in touch with his or her unconscious mind and with the archetypal significance of its main symbols.

Like Freud, Jung has had a large cultural influence. His ideas of personality types and of the animus/anima are attractive but may easily be applied too rigidly. Personality traits tend to fall in a middle range, not at opposite ends of a spectrum. As for the collective unconscious, there are much simpler and more plausible explanations – mainly to be found in the common elements of human neurophysiology (⇨ p. 74) – for the universality of symbols.

The Adlerian school

The Austrian psychologist Alfred Adler (1870–1937) was also originally a Freudian who defected and formed his own school. He came to reject Freud's ideas of infant sexuality and unconscious motivation. Sex, according to Adler, was not a significant factor until the development of mature sexual interest – by which time the patterns of personality are already formed. He decided that the child, conscious of its weakness and inferiority, tries to improve its status by self-assertion and the aspiration to grow up and achieve superiority. Compensation for this sense of inferiority can be achieved either by direct effort to overcome it or by efforts directed to achieve success in another field. A major part of motivation, he believed, was the tendency to avoid situations that emphasized one's sense of inferiority and to seek those that diminished it.

Adler used the term *life style* to refer to the totality of one's perception of oneself in relation to the world – especially one's goals and how one tries to achieve them. A healthy life style incorporated a balance between selfish and social impulses, and the use of means to overcome inferiority that were not damaging to oneself or others. Courage, cooperation and a healthy and balanced attitude to sex were required. Neuroses, he believed, were caused by a lack of balance between selfish and social drives.

Adler was a pioneer in child guidance and set up many child-guidance clinics in Vienna. He was a successful public lecturer who popularized the phrase 'inferiority complex' and his ideas on how children should be brought up were widely influential, especially in America, where he lectured extensively. His theory and practice were, however, condemned by most psychoanalysts as naive, superficial and hasty. Laypeople, however, continue to give unconscious and unacknowledged support to much that he taught and believed.

Freudian Theory

The primary assertions of Freudian psychoanalytic theory are that psychosexual development passes, from birth to sexual maturity, through various stages in each of which pleasure is centred on a part of the body. The *oral* stage lasts for the first year of life when the child's pleasure is focused on feeding. The *anal* stage lasts from the ages of 1 to 3 when the child is learning to control its bowels. The *phallic* stage, with the penis or clitoris as the centre of attention, follows at the age of 3 or 4. There is then a *latency* period during which sexual impulses are 'sublimated' into other pursuits – social, intellectual, athletic. Finally, the mature *genital* stage is entered at puberty. 'Fixation' can occur at any of the earlier stages, thereby determining the personality.

Freud asserted that boys of age 3 to 5 are sexually jealous of their fathers, whom they wish to kill, and are in love with their mothers (the *Oedipus complex*). With girls the situation is reversed (the *Electra complex*). Furthermore, the psychological health of the adult is determined mainly by the success or otherwise with which the child copes with these perfectly normal sexual and aggressive impulses. Freud believed that everything of importance in mental life takes place in the unconscious mind, and much of what goes on there is concerned with sex and aggression.

Every verbal slip or lapse of memory has a cause – usually a concealed expression of sexual or aggressive inclination (the Freudian slip). Dreams, Freud stated, are the most obvious manifestations of the unconscious mind, and are full of highly significant symbolism, much of it of a sexual nature. Unpleasant experiences or guilt-provoking desires are repressed, but give rise to such symptoms as anxiety, depression, phobias and 'hysterical' paralysis. Patients, he said, could be cured only by successful psychoanalytic interpretation of their dreams, slips of the tongue and neurotic behaviour.

Freudian theory stipulates that the mind is divided into three parts. The *id* is concerned with basic instincts. The *ego* deals with the tasks of reality and the sense of self. And the *superego* acts as a conscience that represents ideals and values, and controls the impulses of both the id and the ego.

In addition, Freud posited two other separate instincts. The presence of the first, *narcissism*, or self-love, is thought to be healthy in the early stages of development, but unhealthy later. It involves a withdrawing of the libido (⇨ main text) from concern with external persons and objects, and a redirecting of it to the self. The second, *thanatos*, the death instinct, is the drive leading to self-destruction. It implies a denial of the libido, and is, therefore, outside the realm of the id.

Carl Gustav Jung (1875–1961). Following his disagreement with Freud, Jung went on to found analytical psychology. (ME)

Psychology Today

The rapid growth in the knowledge of brain structure (neuroanatomy) and brain function (neurophysiology) – together with the new sciences of information theory, digital computing and artificial intelligence – has placed psychology in a new and ever-widening perspective calling for more sophisticated techniques of research. At the same time the practical applications of psychology have expanded in many directions, and applied psychology now has a part to play in almost every area of life.

In the 1950s and 1960s psychology was dominated by behaviourism (⇨ p. 163) and, even today, many psychologists call themselves behaviourists, insisting that they are not interested in any data purporting to be a report on consciousness. This position can be epitomized in the concept of stimulus and response. This, claim the behaviourists, is all that can be observed. The body is a 'black box', a system whose workings we do not need to understand in order to know its function, and it is operated on by environmental stimuli to which it responds in various observable ways. Psychology based on this view is often called S–R (stimulus–response) psychology. To some extent, this position is a reaction against the claims of the psychoanalytic movement (⇨ p. 164). There are many psychologists, however, to whom this is an unnecessarily extreme attitude, and among these are the proponents of cognitive psychology.

Applied Psychology

CLINICAL PSYCHOLOGY is not to be confused with psychiatry (⇨ pp. 184–95), which is a medical speciality practised by doctors concerned with the treatment of mental disorders of all kinds. Clinical psychology is a discipline allied to medicine and is concerned, broadly, with the diagnosis and non-medical treatment of emotional disorders and behavioural problems. It may also be concerned with the treatment of interpersonal problems, marital difficulties, sexual problems and drug abuse. Clinical psychologists may work in hospitals, in the community (perhaps running clinics in a GP's surgery), in prisons, or they may work privately or in clinics often providing psychotherapy (⇨ p. 164). Clinical psychologists in hospitals are often concerned with the assessment of learning difficulties, whether present from birth or the result of later brain damage, and with the progress of functional recovery from brain injury.

OCCUPATIONAL PSYCHOLOGY is the application of psychology to the problems of industry, management, business and the workplace. It is a large subject, encompassing such matters as personnel selection and placement, training, career planning and working conditions. Inevitably, industrial psychologists experience conflicting aims, even divided loyalties. Employers may be interested primarily in productivity and profit, while employees may be more concerned with working conditions and individual rights. Most operate on the assumption that these aims are not inherently antagonistic.

EDUCATIONAL PSYCHOLOGY is concerned with the practical problems of teaching and learning, and is directed both to the production of improved teaching methods and to the assistance of people with learning difficulties. Educational psychologists work in association with teachers and children, and are also concerned with counselling older students about courses and personal educational problems. Although the mechanisms of learning are a central concern of research, educational psychologists are often more concerned with the effects on learning of such factors as ability, intelligence, aptitude, personality traits, the classroom environment, social status and emotional development. Educational psychologists may subscribe to any of the schools of psychology. Behaviourists (⇨ p. 163) believe in progressing by small stages with the application of strong positive reinforcement. Developmental psychologists follow the theories of the Swiss psychologist Jean Piaget (⇨ p. 175), applying his ideas of the way in which children develop in stages by constructing their own concepts of reality through the solution of problems encountered in real life. Freudians concentrate on the role of the emotions and the exploration of feelings, and attempt to ensure the satisfaction of basic emotional needs. Others take a humanistic standpoint and base their work on the assumption that children should be allowed to direct their learning themselves within a loving and accepting environment.

Cognitive psychology

Cognition is the mental act or process by which knowledge is acquired, and is also used to mean the knowledge that results from such a process. There are various methods by which we gain knowledge: by perception; by the correlation of past-remembered data with recent data acquisitions (what might be called acts of creativity); by solving problems; and possibly by a process described as 'intuition'. Cognitive psychology is concerned not only with these, but also with how knowledge is stored, correlated and retrieved. Thus it is concerned with attention, the processes of concept formation, information processing, memory, and with the mental processes underlying speech (*psycholinguistics*).

To some extent, cognitive psychology is a reaction against the inflexibility of behaviourism and the austerities of stimulus–response psychology, both of which it holds to be seriously incomplete as theories of mental functioning. Unlike the behaviourist, the cognitive psychologist asks questions about mental events and hopes for explanations of them.

Cognitive psychologists see the brain as an information-processing system that operates on, and stores, the data acquired by the senses. Much of this activity is performed without conscious awareness. Although cognitive psychology is very much concerned with the actual processes going on in the mind, it does not investigate these by conscious introspection. What the cognitive psychologists do is to carry out experiments to measure and analyse human performance in carrying out mental tasks. They might, for instance, measure and compare the time taken for different mental activities, such

Patient (female) and therapist during a counselling session. (Will & Deni McIntyre, SPL)

Psychologist (left) uses the Token and Reporter test to evaluate a woman's ability to understand language and to express herself. This test is particularly helpful in screening for dysphasia, a disorder of language affecting comprehension rather than articulation. (Will & Deni McIntyre, SPL)

SEE ALSO

- TOWARDS A SCIENCE OF THE MIND p. 162
- THE PSYCHOANALYTIC MOVEMENT p. 164
- MENTAL DISORDERS pp. 184–95

as pattern recognition, the perception of logical relationships or the selection of words used to express particular concepts. They may conduct experiments to assess how much data can be stored in short-term memory (➪ p. 170) and for how long. And they might perform tests to measure reaction times or to assess the accuracy with which mental tasks are performed. Even noting and recording characteristic errors have proved valuable. Possible models of the underlying mental processes can be constructed from analysis of this data.

Although most cognitive psychologists are peripherally interested in neurophysiology, they do not pretend that these models represent the actual processes occurring in the brain. Nevertheless, as these models are refined, by testing and criticism, it is hoped that they will approach ever closer to reality and will gradually lead to a clearer understanding of how the brain functions.

The brain/computer analogy

Today it is difficult for cognitive psychologists to describe their models of mental processes without using the language of digital computing. Whether this implies that the brain is a digital computer and the mind a computer program is, however, a different matter. Most digital computers process a single stream of numerical data in binary form, i.e. the data is encoded as a series of 0s and 1s, which are represented physically as on and off electrical pulses. In contrast, the brain operates in parallel mode, with millions of different streams being processed simultaneously, and it does not appear to handle data in binary form. Even so, with every advance in knowledge of mental functioning, the analogy between the brain and the computer seems closer (➪ p. 72). Many papers written by cognitive psychologists are full of computer terms such as 'buffer', 'coding', 'executive routine', 'algorithms', 'data store', 'central processing unit' and 'information management'.

The techniques of artificial intelligence (AI) have also become important tools in cognitive psychology, and a succession of computer programs have been written that seem to be approximating to the way in which the mind works. Artificial intelligence means more than simply the ability to solve problems, though – computers have been doing that for decades. AI involves a dynamic learning process whereby the program alters itself so as to change its responses as a result of past experience.

Practical applications of psychology

A central concern of psychology is the systematic study of human behaviour. Psychologists try to understand, describe, predict, and sometimes change human behaviour. They are concerned with behaviour at all stages in life, and with the behaviour of people in all kinds of contexts, occupations and groupings. We can distinguish between the schools of psychology and its branches. In general, the schools are sub-divisions based on the different theoretical systems to which their adherents subscribe. The branches, on the other hand, are defined by the subject matter they deal with, such as abnormal or developmental psychology. Practical psychologists often profess not to belong to any school, and may select ideas from more than one.

PSYCHOMETRICS

Psychometrics literally means 'measuring the mind', but in practice it is concerned with the somewhat more limited goal of trying to find ways of quantifying the various forms of mental ability. It is concerned with assessing intelligence, aptitudes, talents, skills, potential in various fields, personality traits and predisposition towards psychological breakdown.

Psychometricians are constantly aware of the need to ensure that tests really do measure what they purport to measure, that their results are reliable and valid over the long term, and that they can be properly related to statistical norms in large groups. The methods used include interviews, direct observation, formal written tests, manual-skill tests, matrix tests, vocational tests, personality inventories, situational testing and such 'projective' methods as the Rorschach (ink blot) test (➪ p. 200).

Computer test designed to assess non-verbal planning abilities and visual organization skills. (Will & Deni McIntyre, SPL)

What is Intelligence?

Intelligence manifests itself in various abilities, such as a general effectiveness in putting two and two together and coming up with four, or in grasping the essence of difficult ideas, or accumulating mental data. But to describe the manifestations of something is not to define it, and there is still no general consensus among psychologists over a satisfactory definition of intelligence.

When people first began seriously to consider the nature of intelligence it seemed obvious that it was a matter of inherent brain power. Different people inherited brains of different quality and that was that. Intelligence obviously ran in families. Every now and then an exceptionally powerful brain turned up and the possessor became a genius. Sometimes people were born with very low-grade brains and these unfortunates were mentally 'deficient'.

Further consideration, however, showed that this scheme was inadequate. The idea of innate brain power, independent of educational and environmental factors, did not fit the facts. A mass of evidence showed that the development of intelligence depended largely on the input to the brain after birth. Children of highly intelligent parents became intelligent only if their environment was conducive to the development of intelligence. And the kind of intelligence that developed was very much determined by their environment, in particular by the way they were educated. It seems clear that two factors are necessary for the development of intelligence – the inheritance of good 'hardware', and the subsequent provision of good 'operating software' and data.

Does inherent brain power exist?

The early idea that intelligence was a simple power of the mind independent of its separate abilities has also had to yield to scrutiny. Numerous attempts to find this inherent power have failed. It seems that intelligence, on close inspection, resolves into a large complex of different abilities or skills present to different degrees. None of these skills – not even reasoning power (which, on examination resolves itself into the ability to perform one or other special skill) – can be unequivocally selected as a central, innate entity we could call intelligence.

The list of mental skills is a long one and includes such things as verbal comprehension, word fluency, numerical ability, the ability to detect significant associations, the power of spatial visualization, speed of perception, the ability to acquire and memorize information, the power of recall, the ability to adjust to change, planning ability, the ability to choose between alternative courses of action, and so on. The disparity in the degree to which these different skills may be present in any one person is remarkable, and two people can be generally acknowledged to be intelligent yet may have few abilities in common. In fact, it has been claimed that human intelligence comprises 120 distinguishable elementary abilities, each one involving an operation on something to produce a product.

Psychologists have been deeply divided on the question of how many distinguishable mental skills constitute intelligence, and whether there is anything more than the totality of these skills. Some claim that there is no general factor in human intelligence. Most, however, subscribe to the view that there is such a factor but are a little vague as to its nature. Psychologists are also at variance in their definitions of intelligence. Fourteen well-known experts, when asked, produced different answers. These included such ideas as the ability to learn by experience, to adapt to changing environments, to think in abstractions, to perceive truth and to acquire skills. One seemingly disingenuous expert defined intelligence as the ability to do well in intelligence tests.

A definition still fairly widely accepted is that of Charles Spearman (1863–1945), who was Grote Professor of Mind and Logic at University College, London. Spearman taught that there *is* a general ability, which he called 'g', and that this general ability is necessary for the performance of all mental tasks. Surrounding this general ability are a number of separate specific abilities, present to different degrees and capable of being separately measured. Spearman's concept has been widely discussed and many modifications suggested, but has not been seriously challenged, even by modern cognitive psychologists (⇨ p. 166), who try to understand intelligence in terms of information processing. The nature of the basic quality 'g' has been a source of much argument, although a strong case has been made that it is the capacity to detect new and non-chance associations.

Intelligence tests

The *intelligence quotient* (*IQ*) is the ratio of the mental age to the chronological age. When these are equal, the IQ is given as 100. This is the average (though unofficial) score across the whole population of a certain age. The problem in assessing intelligence is to find fair, realistic and reliable ways of measuring the mental

Cyril Burt and the Great Intelligence Fraud

Cyril Burt (1883–1971) was one of Britain's most famous and respected psychologists. As a young man he met and was deeply influenced by Francis Galton (⇨ pp. 30–1), whose ideas on the importance of heredity in determining individual ability he quickly assimilated and never abandoned. Burt's central interest was in intelligence, and from the beginning of his career he was involved in the collection of data on intelligence and in formulating tests. In 1932 he was appointed Professor of Psychology at University College, London, in succession to Charles Spearman (⇨ main text). There he upheld and reinforced Spearman's ideas of a 'g' factor in intelligence.

After his appointment he produced a number of papers and books inspired by Galton's ideas on the central role of heredity. An increasing number of scientists, however, were becoming critical of Burt's ideas. In response to these criticisms, Burt published a succession of papers, especially in 1956 and 1966, purporting to show, on the alleged evidence of comparative studies of a large number of twins reared together and apart, that intelligence was overwhelmingly innate and not the result of environmental factors. These papers had great influence on psychological and educational thought. It was not until after his death in 1971 that it was argued that much of the 'evidence' quoted in these papers had been fabricated, and that his alleged collaborators, Miss Howard and Miss Conway, did not exist.

Burt was the effective founder of educational psychology and was knighted for his services to British psychology. But his almost religious adherence to the Galtonian tradition in the face of contrary evidence, and his inability to accept adverse criticism, has cast serious doubt on his scientific judgement.

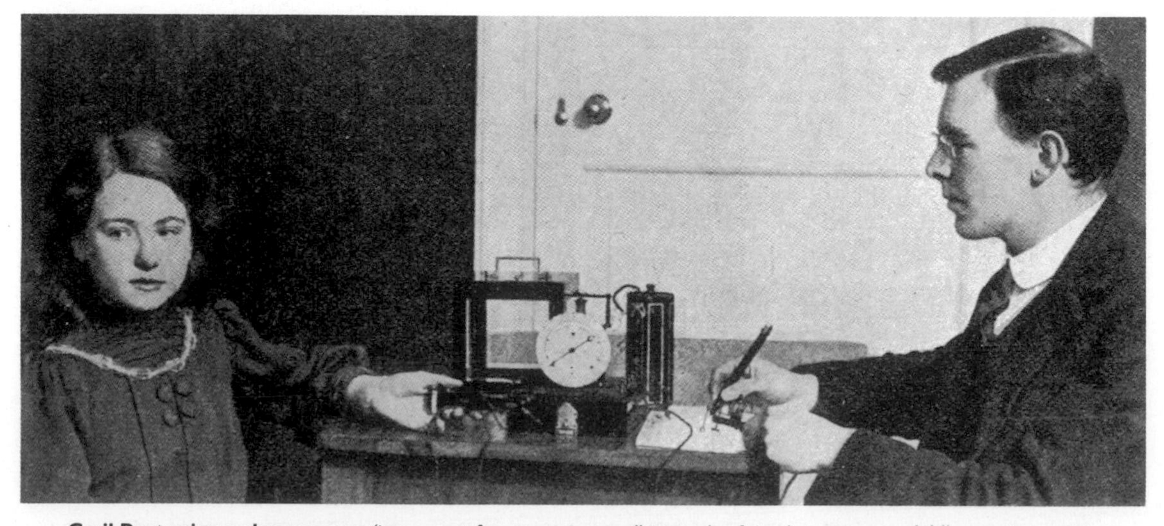

Cyril Burt using a chronoscope (instrument for measuring small intervals of time), to assess a child's reaction speed. From J. Arthur Thomson, *Outline of Science* (c. 1925). (AR)

The highest childhood IQ score was achieved by Marilyn vos Savant, born in 1947 in Missouri, USA. At the age of 10 her IQ measured 228.
(Courtesy of Marilyn vos Savant)

age. This difficulty relates closely to the difficulty of defining intelligence (⇨ above) and of trying to find abilities to test that are independent of educational and cultural influences. Intelligence tests compiled without due regard to these factors have, rightly, been condemned as being unfair to those candidates who do not share the educational and cultural background of the testers. Such criticisms have tended to bring all intelligence testing into disrepute.

The first formal tests of intelligence were devised, at the request of the French government, by the French psychologist Alfred Binet (1857–1911), working with Théodore Simon. The purpose was to determine which children were worthy enough to receive education. Binet originated the idea of IQ in response. These tests were subsequently repeatedly modified at Stanford University in California by Louis Terman and others, and the original Binet-Simon test (1908) became the Stanford-Binet tests. The current Stanford-Binet tests provide tasks for individuals aged two to adulthood. Very young children are asked to draw copies of objects, to string beads, build with blocks and answer questions on familiar activities. Tests for older children involve such things as detecting absurdities, finding what various pairs of words have in common, completing sentences with omitted words, explaining proverbs, and so on. Such tests are, of course, strongly educationally oriented and test scholastic ability.

The most widely used intelligence tests today are the *Wechsler tests*, compiled by the New York psychologist David Wechsler. These are of two basic kinds – the Wechsler Adult Intelligence Scale (WAIS) and the Wechsler Intelligence Scale for Children (WISC). The tests involve progressively increasing difficulty. Each test has verbal and performance parts, and these can be applied independently for those with language difficulties, or combined to give an overall score. The verbal parts test vocabulary, verbal reasoning, verbal memory, arithmetical skill, and general knowledge. The performance sections involve completing pictures, arranging pictures in a logical order, reproducing designs with coloured blocks, assembling puzzles, tracing mazes, and so on. Again, these tests cannot be said to assess much more than the general educational level, as the level of definitive intelligence can be assessed in only the most general way in infancy. There are, however, many preschool tests for slightly older children, based on the work done by educational and clinical psychologists with preschoolers.

Intelligence tests are not, except in the most general way, accurate predictors of achievement. Scholastic achievement depends on a number of factors other than intelligence, especially the quality of instruction, parental expectations, and a rich early educational environment. Achievement later in life is even less accurately predictable on the basis of intelligence tests, and is determined by many other factors, including personality, physical appearance, gender, class, opportunity, luck, and the possession of special skills. Good motivation can compensate for restricted intelligence, while low motivation can result in little effective use being made of high intelligence. However, all things being equal, it is the case that there is a general positive correlation between intelligence, as measured by tests, and professional achievement, although class is a greater predictor in the West.

Although intelligence tests purport to measure innate mental ability, they actually tell us little or nothing about the relative importance of heredity and environment in determining intelligence.

SEE ALSO

- NATURE VERSUS NURTURE p. 30
- THE HUMAN COMPUTER? p. 72
- THEORIES OF LEARNING p. 172
- LEARNING IN PRACTICE p. 174

CREATIVITY IN UNEXPECTED PLACES

Stephen Wiltshire – shown here in 1987, aged 13, is autistic and possesses remarkable artistic powers. His architectural drawings display a sophisticated understanding of proportion and perspective, and show the correct numbers of doors and windows. (Popperfoto)

Individuals who have an IQ of less than 70 are described as having learning difficulties. A very small proportion of people in this category have an extraordinary talent of some kind – often for music or for the ability to memorize certain categories of fact or to perform certain kinds of mental arithmetic. Some are able to play chess to a high standard or to produce drawings or sculpture with great skill. The 'idiot savant', however, has little appreciation of his or her capabilities.

Sometimes these individuals also have a physical disability and manifest clear signs of brain damage or autism – a serious childhood disorder in which the child is withdrawn, self-absorbed, interested in objects rather than people, and often unable to communicate clearly through speech. Sometimes, when their talents are encouraged and promoted, however, they may show some improvement in their general abilities.

The observation of the creative ability of some 'idiot savants' has prompted the idea that the phenomenon might be explained on the basis of left-sided, but not right-sided, brain damage. The left half of the brain is concerned with verbal and general intellectual activity; the right half with artistic and creative activity (⇨ p. 176). The suggestion is that following loss in left-side brain function, the right brain becomes over-developed in compensation.

What is Memory?

When we try to imagine the nature and physical basis of human memory we are apt to think of mechanical analogies such as digital computers, tape recorders, photographs, and so on. These analogies can be helpful but they are often misleading.

Marcel Proust (1871–1922) (right) whose 13-volume work *A la Recherche du Temps Perdu* ('Remembrance of Things Past') explored the nature of memory, often relying on long, evocative sentences to do so. (Museum of Illiers Cambray)

The processes involved in human memory are more complex than they seem, certainly much more complex than those involved in the processing and storage of data on the floppy or hard disk of a computer. Nevertheless, a great deal has been discovered in recent years, and the precise nature of human memory is gradually being uncovered. Research into this area produces conflicting evidence, and there are some differences of opinion among the experts.

Aristotle (384–322 BC) came to the conclusion that memory was situated in the heart, not the brain, and his ideas held sway for about 500 years until overturned by Galen (c. AD 130–201), whose anatomical dissections indicated that the brain was a much more likely candidate. Medieval ideas of memory were, as today, based on analogies with the technology of the time. Then, the most advanced technology was hydraulic, and memory was conceived as a flow of fluid through pipes, controlled by valves. René Descartes (1596–1650) also subscribed to a hydraulic analogy, the fluid being 'animal spirits'. Descartes believed that when we wanted to remember something, the pineal gland drove these spirits to different parts of the brain where they encountered physical traces left by the object we wished to remember.

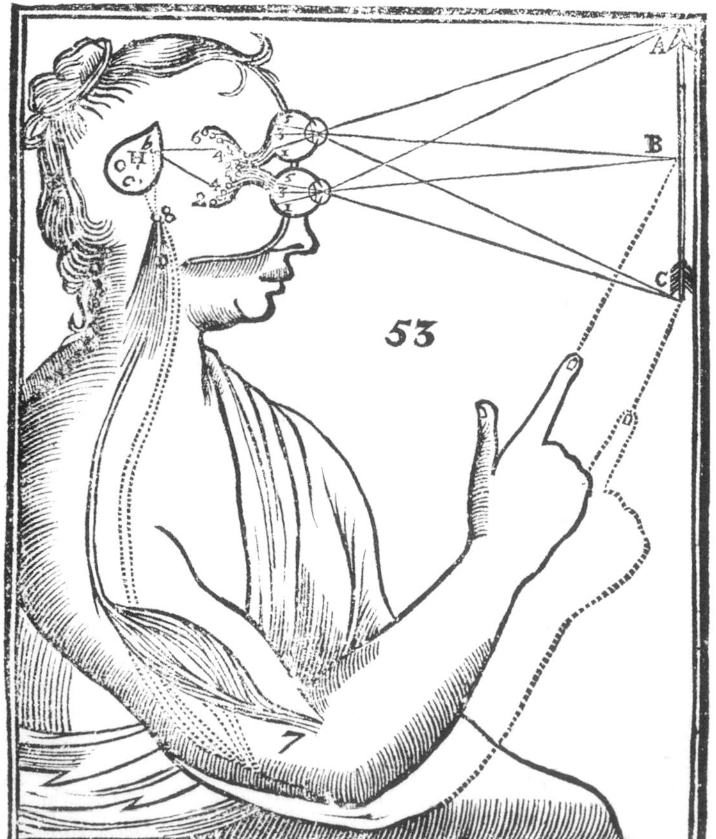

Illustration from Descartes' *Opera Philosophica* (1692) showing the pineal gland that he thought essential for memory interacting with the eyes and muscles. (AR)

Today, memory is regarded as a function of the brain, and defined as the ability to store and retrieve information. The term 'memory' is also used for the actual information store. Unlike the arrangements in a computer, however, human memory is not a discrete and recognizable part of the brain where data can be stored. While it is true that certain known parts of the brain are involved in the temporary storage, registration, processing and recall of information that is to be permanently stored, there is no single part corresponding to a computer's disk drive for long-term store.

The nature of the input

The sensory experience that precedes the storage of data in memory is very complex, suggesting that there may be many different kinds of memory – visual, auditory, olfactory (smell), gustatory (taste) and tactile (touch). Moreover, data is never presented to us in simple form, but always as part of a complex context in which it is embedded. This context is likely to form part of the memory; indeed daily experience shows us how important context and associations are for effective memorizing. For example, a single item of information conveyed to us by way of speech will be set in a context of other data – the appearance of the speaker's face, its spatial relationship to other things, the quality of the voice, and the displays of emotion. Research has shown that there is good reason to believe that this information is distributed to those parts of the brain known to be concerned with the different sensory functions (⇨ pp. 76–7).

Short-term memory

Complex perceptions cannot all be stored separately at the very moment that they are perceived, so some kind of analysis and selection is necessary to determine what should be registered and kept. What happens is that only significant changes in familiar contexts are likely to be registered by the brain. Such analysis and selection cannot be performed, however, unless the data is stored temporarily so that it can then be operated upon. This means that there must be two levels of memory store – iconic (short-term) and working (long-term) memory.

Iconic memory has to be constantly refreshed. Most of us can look up and remember a new telephone number if we repeat it to ourselves a few times before dialling. But if we are interrupted the number will be lost as new incoming data displaces the current contents of the short-term memory. There is also a strict limit to the length of any item of new data if it is to be held in this short-term store. Again, most of us can readily hold a seven- or eight-digit number in our heads, but a twelve-digit one is too long. Iconic memory can also be emptied by a blow or an electric shock to the head, suggesting that short-term memory operates via some kind of dynamic neuronal circuit, possibly of circulating nerve impulses. A useful analogy is that of the volatile random access memory (RAM) of a computer, the contents of which are lost when the power is turned off.

Mnemonics

Each of us carries in the brain an almost immeasurable amount of data, recorded, somehow, via short-term memory and preserved, often for a lifetime, in permanent storage. Some evidence suggests, however, that stored information need not always pass through the short-term memory, but may go straight into long-term store. This mass of data is highly organized in terms of meaning and association; the better the organization, the more accessible it is. Clues, mnemonics and, in particular, cues, all speed up the efficacy of information retrieval. For instance, one might forget the beginning of Hamlet's celebrated soliloquy, but the cue, 'To be . . .' is likely to evoke the continuation, '. . . or not to be'. Similarly, efficient registration for long-term storage demands good organization and strong association. The scholar with a profound grasp of his or her subject assimilates new information on that subject with the greatest of ease, so long as it can be related to existing stored data. Entirely new matter, unrelated to any previous experience, is much more difficult to memorize.

The interface circuits

We now know the exact sites in the brain through which sensory data must pass to be stored in memory. These interface circuits – through which long-term memory is recorded and recalled – are contained in two large structures on the inner surfaces of the temporal lobes of each cerebral hemisphere – two massive collections of nerve cells known as the amygdala and the hippocampus. Together these make up the limbic system (➪ pp. 74–5). The amygdala is connected to all the sensory areas of the cortex by two-way pathways. The hippocampus also has extensive connections to these areas. Destruction of these two structures leads to profound loss of memory.

The basis of long-term memory

The exact physical basis of long-term memory remains unknown; several hypotheses have, however, been put forward. The sheer size of the database in relation to the size of the brain implies that the unit of information – whether it be a binary digit (bit) or some other code – must be very small. This has led some scientists to suggest that a protein molecule provides the basis. It would be naive, however, to think of long-term memory storage as based on two kinds of protein molecule corresponding to the presence or absence of a spot of magnetization on a computer disk. For one thing we have no reason to suppose that the nervous system operates as a computer does on the binary system (i.e. coding 1 for 'on' and 0 for 'off'). For another, the life-time of protein molecules is very much shorter than the length, as far as we know, of human memory.

There is evidence that the sites of memory are the same as the areas of the brain where the corresponding sensory impressions are processed – in various parts of the outer layer (the cortex). It is now almost certain that, in memory recall, the amygdala and the hippocampus engage in a kind of feedback dialogue with the appropriate part of the cerebral cortex – playing back the kind of neurological activity that occurs during sensory experience, rather in the manner envisaged by Descartes.

This being so, it seems likely that long-term memory store takes the form of the interconnection of nerve cells in a particular way. Evidence suggests that particular connections – activation of the nerve-to-nerve junctions known as synapses (➪ pp. 74–5) – occur as a result of repeated mental stimulation of the kind that occurs during sensory experience. Scientists are beginning to understand the way in which repeated stimulation leads to permanent link-up. The addition of a phosphate group (the process of phosphorylation) to a brain protein, called F1, as a result of the action of an enzyme, protein kinase C, is now known to be capable of causing the necessary changes in the synapses. Unfortunately for laboratory studies, though, protein modified in this way has a limited life span, ranging from minutes to weeks. Various ingenious suggestions for ways round this difficulty have been proposed, including processes that automatically regenerate protein molecules and a special form of gene expression that gives rise to long-life protein.

DÉJÀ VU

Almost everyone is familiar with the phenomenon of *déjà vu* ('already seen') in which there is a brief but powerful conviction that what is currently taking place has happened to us before. Usually there is a strong sense of familiarity accompanied by a compelling, but mistaken, conviction that one already knows what is around the next corner. There is as yet no consensus of opinion on the explanation of this interesting phenomenon.

***Déjà vu* has been claimed as evidence of reincarnation or of the possession of 'second sight' or telepathic powers. Freudian psychologists (➪ p. 164) claim that the experience of *déjà vu* has actually happened before but has been repressed.**

Another suggestion is that *déjà vu* is the result of data relating to the current perception reaching the memory store a fraction of a second before it reaches consciousness, so that the effect is as if it is being remembered. In support of this idea is the undoubted strength of the experience of *déjà vu* – which is just what one would expect from such a recently registered memory. In this context it is significant that *déjà vu* is a very common manifestation of certain types of brain damage, such as those that cause temporal-lobe epilepsy.

Another possible explanation is based on the hypothesis that memory recall is a process of synthesis, or reconstruction, from the stored items of the different sensation components that made up the experience, each successive episode of recall being the recall, not of the originally stored data, but of the last such synthesis. If the synthesis involved mistakes, omissions or false elaborations we would have a sense of familiarity but would quickly see that the event, as now apparently recalled, never really happened.

SEE ALSO

- BODY AND SOUL p. 70
- THE HUMAN COMPUTER? p. 72
- THE CONTROL SYSTEM 1: THE NERVOUS SYSTEM p. 74
- THE CONTROL SYSTEM 2: THE AREAS OF THE BRAIN p. 76
- TOWARDS A SCIENCE OF THE MIND p. 162

Dustin Hoffman starred in the film *Rain Man* as an autistic man who had an extraordinary capacity for remembering series of numbers, so much so that he could memorize whole pages from telephone directories. (Kobal)

Theories of Learning

One of the fundamental differences between humans and other animals is the extent to which our behaviour is determined not by instinct but by learned patterns.

SEE ALSO

- NATURE VERSUS NURTURE p. 30
- THE HUMAN COMPUTER? p. 72
- PSYCHOLOGY TODAY p. 166
- WHAT IS INTELLIGENCE? p. 168
- LEARNING IN PRACTICE p. 174

The way we behave is largely, but not wholly, a product of learning, but some of our more basic patterns of behaviour are determined by our inherited body structure – in particular, the fine structure of the nervous system. Behaviour that is the result of 'built-in' neurological circuits is often called instinctual or instinctive and is fairly stereotyped; a particular stimulus will nearly always result in a predictable response. *Homo sapiens* is a learning animal, equipped by evolution to be the most efficient learning machine in existence. Learning involves the acquisition and storage of data, by any means and through any of our senses. It also involves the integration of these data with information that is already stored so that the behaviour is subsequently modified. Data acquisition causes permanent internal changes (➪ pp. 170–1), but these are not discernible. The only way to discover that learning has occurred is to observe changes in behaviour.

Unfortunately, as in most fundamental matters, psychologists and philosophers have long differed widely in their ideas about how we learn. For most of the 20th century, however, their arguments have been concerned with the interpretation of known scientific fact, and with the question of how far one can extrapolate the findings on animal experiments to human beings. These are sometimes emotive issues, and the arguments have not always been detached. One influential theory says that human beings learn through a process of conditioning, but this idea is inherently distasteful to many people, as it seems to deprive us of our freedom of will. But this is only one of a broader class of philosophical problems that arise from the view that the more physiological knowledge advances the more closely we appear to resemble – in many respects – a machine (➪ p. 72).

Pavlov and classical conditioning

Modern thinking about learning started with the then startling findings of Ivan Petrovich Pavlov (1849–1936), a distinguished Russian physiologist who made many important contributions to the field of physiology, especially on the subject of digestion. His great discovery, however – on which much subsequent learning theory was based – was made almost by accident.

Pavlov was studying the stimulus to the secretion of saliva in dogs, using meat powder that caused salivation when put in dogs' mouths, even when given to puppies who had never before tasted meat. This was an inherent, unlearned, response, unaffected by experience. Pavlov noticed that dogs who had already tasted the powder also salivated in anticipation of it. This objective indication of a 'psychic' stimulation to salivation fascinated Pavlov. He decided to study the effects of associated stimuli that would not normally occur in a dog's life – such as the clicking of a metronome or the ringing of a bell – and soon found that dogs that salivated when allowed to see food while a metronome clicked or a bell rang soon salivated to the sound of the metronome or the bell alone. This he called a *conditional response*, and he named the process *a conditional reflex* (later altered by his translators to 'conditioned').

Pavlov found that the time relationship of the two stimuli was critical. For a conditioned reflex to be established the stimuli must occur together or the conditioning stimulus (the metronome) must be applied very soon before the unconditioned stimulus (the food). If the food was presented before the bell, conditioning did not occur. If the bell was repeatedly rung without food being given, the reflex was gradually extinguished. Pavlov found that his dogs tended to give the same conditioned response to a number of similar stimuli, such as bells of different pitch, but could also be trained to discriminate between bells of only slightly different pitch. Other kinds of closely similar stimuli could also be distinguished.

In studying the limits to which closely similar stimuli could be told apart, Pavlov found that animals straining at these limits became agitated and aggressive, and tended to salivate strongly in response to all stimuli. Such animals lost the ability to develop normal conditioned reflexes. The implications of these extraordinary findings were not lost on Pavlov, who spent the rest of his life exploring the possibility that much of human behaviour could be accounted for by such conditioned reflexes, and he was awarded the Nobel Prize for Physiology or Medicine in 1904. He went on to develop theories of language, mental illness, neuroses, and human temperament – all based on the principle of conditioned reflexes. His principles, as applied to the process of learning, are now described as the *classical conditioning theory*.

Thorndike and the 'trial-and-error' theory

Edward Lee Thorndike (1874–1949) was a formidable scholar and writer on psychological subjects, whose views, for a time, dominated American psychological thought. Thorndike conducted very large

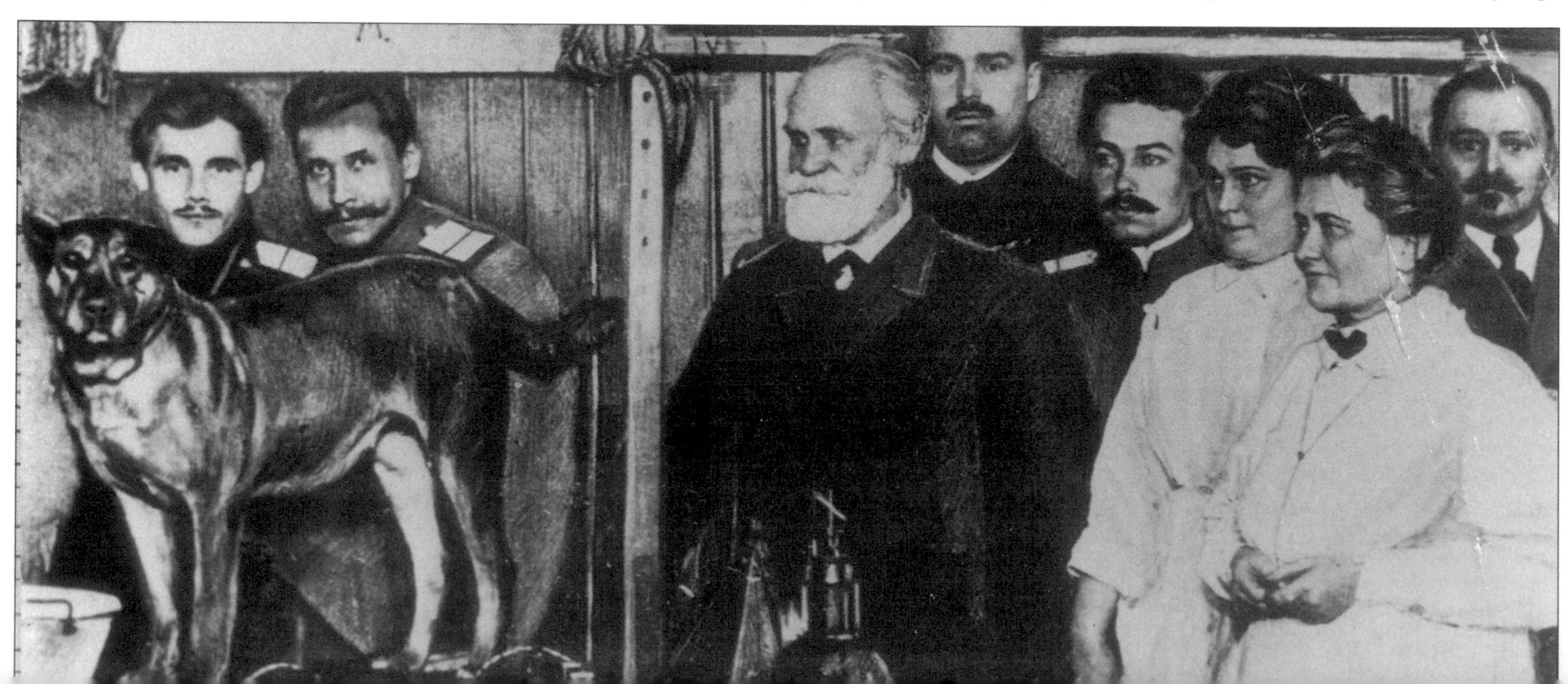

Ivan Petrovich Pavlov (centre) surrounded by his laboratory assistants in Petrograd, 1914. (Keystone)

B.F. Skinner (1904–90), American behaviourial psychologist, pictured here as a young Harvard fellow. (UPI/Bettmann)

numbers of experiments in which he observed how animals learned to solve problems with increasing ease on repetition, and his data comprised the first 'learning curves'.

Thorndike's conclusions were that learning occurs by random trials in which accidental successes lead to rewards of some kind. Actions that lead to satisfying results are 'stamped in' by the subject, wrong attempts lead to dissatisfaction or annoyance and are 'stamped out'. Eventually, only the actions that lead to success remain, and something has been learned. This analysis allowed the researcher to ignore questions of consciousness, and in this respect Thorndike's ideas were an important precursor of the behaviourist school of psychology (⇨ p. 163).

Skinner and operant conditioning

Operant or *instrumental conditioning* differs from classical conditioning in that a certain desired consequence will occur only if the subject makes the required response. The response, or action, is instrumental in obtaining the desired reward. Instrumental conditioning is universally employed in the training of animals of all kinds and, according to many psychologists, in particular the behaviourist B.F. Skinner (⇨ p. 163), the idea can be extended to a theory that can account for most human learning.

According to Skinner, behaviour results in *reinforcement* or *punishment*. Behaviour that leads to reinforcement is more likely to be repeated than behaviour that leads to punishment. Reinforcement can be positive or negative. A positive reinforcer is any factor that increases the strength of the response that preceded it. A hungry cat trying, for instance, to discover how to get out of a cage will try harder if some food is placed outside. A negative reinforcer is any factor whose removal will increase the strength of the preceding response – as, for instance, if an aggressive dog outside the same cat's cage is taken away. Primary reinforcers – such as the provision of food – produce their effect without previous association; conditioned reinforcers are those that act as a result of association with primary reinforcers. Money, for instance, is a conditioned reinforcer because we associate it with needs such as food, shelter and clothing.

Skinner applied his ideas of operant conditioning to the development of crude teaching machines. These employed 'programmed learning' in which reinforcement was applied as needed to achieve the desired result. This involved stages of gradually increasing difficulty with continuous reinforcement of correct responses and extinguishing of incorrect responses. The machines allowed the subject to continue only if the correct response was made. While such machines have been fairly successful, as in language laboratories, the method has not been widely employed, even since the development of personal computers. This may be because of the restrictions in the flexibility and content of the machines' programs. For certain limited purposes such methods are probably ideal, but for education in the broader sense, only the human teaching machine seems adequate.

Cognitive views on learning

Cognitive psychology (⇨ p. 166) takes a rather broader view of learning than that of the behaviourists, and has provided us with some new perspectives on the subject. Cognitive studies on animals indicate that their behaviour is often based on some kind of internal representation of past experience. Rats and mice will quickly learn to run a complex maze correctly if food is always provided at the centre. If no food is provided, they merely explore casually. If, however, rats, for example, are allowed to explore an unbaited maze for 10 days or so, and are then made aware that food has been placed at the centre, they will run the maze as efficiently as the rats who were rewarded with food each time. This suggests that rats can form an accurate mental picture of a maze. It may seem naïve to announce this as proof that animals can remember, but it is objective evidence of a particular kind of mental process – the application of consciousness and memory – something that the behaviourists have insisted does not exist.

Observational learning

A great deal of learning occurs in a vicarious way by our observation of the behaviour of other people and of its consequences (social learning). Such learning is especially efficient if the consequences of others' behaviour is painful to them. A child will learn to respect road traffic by seeing a dog run over and killed. Children with a natural, healthy fear of large domestic animals may lose their fear by observing other children petting them. Even the observation of fear in others can teach fear.

The concept of observational learning has wide implications. Seeing reward or punishment being imposed on other people has been shown to affect the observer as if the reward or punishment had been directly applied to him or her. One trial demonstrated that children who observed a subject being rewarded for aggressiveness later behaved more aggressively than another group that had observed the subject being punished for aggressiveness. Clearly, the general nature of their environment has important implications for the mental development of children and adolescents.

White mouse in a maze. Experiments such as this allow cognitive scientists to observe the behaviour of laboratory animals, and then to make inferences about the mental processes that prompt such behaviour. (IB)

Learning in Practice

The content of the mind is not limited to information that can be expressed in words but also includes the data underlying visual imagery, physical skills and creativity. All stored information is of interest to researchers looking into how we learn.

Verbally expressible data include those derived from educational and social experience, and will have a content relating to ideas of morality, social responsibility and so on. If we are to have any criterion for the assessment of the individual's ability to contribute to society, it must be based largely on the nature of this mental content. So the ways in which such data are acquired, and their acquisition promoted, are of major concern. The acquisition of knowledge and skills is largely a matter of learning. The influence of hereditary factors (⇨ p. 30) has been endlessly debated and has, in the past, sometimes been overrated. There can be no denying that some people inherit better 'hardware', for a particular purpose, than others. But good hardware is of little value unless it is put to good use, and this requires the input of a large number of well-selected data from the individual's environment. The problem is to find how these can be facilitated or improved. Early in life, an essential requirement is the means to communicate and to formulate ideas – language.

Language acquisition

A baby a month old can begin to resolve and discriminate various single syllables, such as 'ma', 'pa' and 'ba'. Soon the child can resolve, and derive some meaning from, about one hundred combinations of pitch, intonation and expression. It seems likely that some comprehension of speech precedes speech production, and that when words are first uttered, they are meaningful. Most children begin to put two or three words together, to form simple sentences, at around the age of 18 months. These groups of words are, at first, mainly nouns and verbs without definite and indefinite articles or prepositions and inflected forms.

As might be expected, the initial vocabulary relates exclusively to environmental matters of everyday experience and importance to him or her. Terms are quickly acquired for body parts, for members of the family, for food, toys, and so on. Comments relate to very fundamental concepts such as presence or absence, location, likes and dislikes, and activities relating to familiar objects and people. Sentences gradually increase in average length at a rate that varies considerably from child to child, and the speed with which complexity and sentence length are acquired does not appear to be predictive of adult intelligence.

It is now that the richness of the environment begins to have a notable effect. Observational and imitative learning will ensure that the mental achievements of the child closely reflect, in quality if not in quantity, the nature of the cultural environment. A child brought up in an articulate, educated family will have an enormous advantage, in terms of acquiring verbal facility and associated forms of education. This child will soon acquire a substantial vocabulary with a good grasp of meanings – mainly derived from context, a grasp of grammatical usage that later seems almost 'instinctive' and is effortlessly applied, and even a naïve and almost unconscious command of etymology so that, later, sense can be made of many new and unfamiliar words. By comparison, the child exposed to a less rich verbal environment will have a less varied context from which to learn. Should such a child later wish to acquire these characteristics, a much greater effort will be required.

A child starting school may speak highly grammatically but be without any awareness of the formal rules of grammar. To

This child's meal time is also an important learning experience about eating rituals such as sitting at the table and using cutlery. (JW)

Jean Piaget's idea that a child's mental development occurs in a series of identifiable stages introduced a whole new approach to the study of cognitive functioning. He began his research by observing his own children. (Popperfoto/UPI)

some extent, early schooling involves the elucidation, or exteriorization, of rules for existing practice. The child has already unconsciously assimilated many such rules, along with their exceptions, and the use of such rules allows him or her to formulate and articulate an endless variety of sentences without having to remember them all. The theory that this process is purely imitative is widely challenged (➪ box). It would, indeed, be a remarkable feat of analysis if the child were able unconsciously to formulate the rules of 'correct' grammar merely from observing the language of others.

Acquiring skills

Normal life involves the acquisition of a wide range of skills. Some are so basic to normal living that life would be severely restricted without them. Some of the more obvious include motor skills, such as articulating words, walking, running, and making small, controlled movements as in writing; social skills, such as learning to relate successfully to others; arithmetical and other problem-solving skills; and musical and artistic skills. The acquisition of the more basic skills – such as standing, walking and speaking – appears to occur without any conscious volition, but the small child is, in fact, strongly motivated by the environment to practise these skills, and will do so until a high degree of efficiency is reached. Everyday observation of the behaviour of toddlers shows that this motivation operates even before the nervous system is sufficiently mature to allow them to be fully exercised.

Learning new skills is a slow process requiring repeated attempts. Improvement is not a smooth, gradual process but a series of quantum leaps. Often these are small, but progress usually occurs in substantial jumps. These will not occur spontaneously but only if many attempts are made – this is why practice is so important in the acquisition of skills. A person trained in a skill has achieved some general advantage, which, to a varying extent, can be transferred from one activity to another. Manual precision and control, acquired in learning to write, for instance, immediately confer improved efficiency in other delicate manual skills. Such skills can be analysed and the components common to different activities recognized.

Ethical learning

Traditionally, questions of personal morality have been very much the province of philosophers and theologians, whose approach to these important matters has tended to be either speculative or dogmatic. But from the mid-20th century such questions have been widely investigated as behavioural phenomena, and have been the subject of intense scientific study. There have been several different approaches.

The Swiss psychologist Jean Piaget (1896–1980), who based his ideas on a close study of behaviour patterns in developing children, held that the sense of morality is based on nothing more than respect for an imposed system of rules. At first the child has no inherent morality, but ideas of right and wrong are soon imposed by parents and other environmental influences and these are, at first, accepted without question or even thought. Some time between the ages of about 6 and 12 the child gradually rationalizes this into a seemingly logical system of moral principles. Other psychologists have described various stages in this process, and suggest that the progress through these stages depends on the child's capacity for abstract reasoning. Most adults do not, they suggest, proceed beyond an intermediate stage of sophistication in their concepts of morality.

Some psychologists working in this field concentrate their attention on the significance of values and attitudes. Values are described as persistently held beliefs or standards that relate to a person's goal in life and that have an important role in determining behaviour. Attitudes are less profound and are manifested by whether our responses to various ideas, concepts, people and objects are favourable or unfavourable. When, in practice, attitudes are found to conflict with values, they tend to change. However, no convincing proof exists that values are particularly more influential on behaviour than are attitudes.

The 'social–behavioural' approach studies how far the social environment determines moral awareness and the way social contexts affect moral behaviour. There is, unfortunately, no clear evidence of any direct causal relationships. It seems evident that, for most of us, behaviour does not quite come up to the standards of how we like to think we should behave. More commonly, we behave in a manner that varies with different circumstances, and this may either reinforce or weaken conformity with our private concept of morality. Ethical behaviour would thus seem to be at least partly dependent on social learning.

The truth about an individual's ethical learning probably lies in a combination of all these theories. Moral character is an unstable general attribute, importantly determined by early influences, but thereafter modulated by the effects of different environments on our conduct. The concept of good behaviour is arbitrary and context-related, probably requiring sound emotional, intellectual and informational integration. It is a social construct that differs from society to society and between groups in the same society.

SEE ALSO

- PSYCHOLOGY TODAY p. 166
- WHAT IS INTELLIGENCE? p. 168
- WHAT IS MEMORY? p. 170
- THEORIES OF LEARNING p. 172

Chomsky and Language

The American linguist Noam Chomsky (1928–) believes that our ability to produce an endless variety of sentences cannot be explained by any theory involving learning by experience and observation. He holds that such an ability relies on there being an inborn knowledge of the linguistic rules for sentence formation.

Chomsky points out that sentences have 'surface' and 'deep' structures. Surface structures vary widely – and can, for instance, be expressed in the active or passive voice – without any change in the essential meaning (the deep structure). This deep structure captures the functional relationships, such as that between the subject and the object, and specifies all the information needed to allow the correct interpretation of the sentence. Alternatively, a sentence can have two or more deep structures – and can, therefore, mean two or more entirely different things. The sentence, 'They are washing cloths,' can mean, 'These people are engaged in washing cloths,' or, 'These cloths are used for washing something or somebody.'

Chomsky believes that the understanding of deep structures is an inherent human capacity, and asserts that certain very general principles governing all human languages (a *universal grammar*) are built into the human brain, regardless of nationality or intelligence. These ideas are the subject of fierce debate among linguistic experts.

The linguist Noam Chomsky observed that the complexity of children's early language, and the speed with which this complexity increases, were beyond anything that could be explained by their experience of the language use of adults around them. This led him to postulate the 'language-acquisition device', a structure in the brain (as yet undiscovered) that has an inborn and universal understanding of grammar. (Gamma)

Imagination and Creativity

Creativity presents a challenge to scientific thinking as it is not the product of logical thought. The study of creativity involves exploration of the nature of the brain.

Normal scientific thinking tends to be *convergent* – a process of synthesis in which causally related phenomena are put together in a conventional way. Creative thought is *divergent*, and often appears irrational – a leap in the dark, a process of 'lateral thinking', of dreaming, fantasizing or engaging in free associations and analogies. The processes adopted by creative artists and scientists vary greatly. People of comparable status appear to create with varying degrees of difficulty. Mozart composed masterpieces at top speed, with no hesitation and few amendments; Brahms would weep and groan in the agony of his composition. Mozart could see every note in the score in his mind before picking up his pen; other composers can do nothing without an instrument.

Albert Einstein (1879–1955), physicist and mathematician, formulated his special theory of relativity after daydreaming whilst lying under a tree. Seeing the sunlight through the leaves, he began to imagine what it might be like to travel along a light beam – and the theory that was to change the nature of physics forever was born. (ME)

The zip fastener is a well-known example of how imaginative thinking can create an effective yet surprisingly simple device. (Image Select)

Experience shows that we often solve important problems, not at the height of concentrated thought, but in the period of relaxation afterwards. For example, solutions may appear to us on waking from a night's sleep especially if we have dwelt on the matter beforehand. It is clear from this that much of creativity is a function of the unconscious mind, which, once supplied with the necessary data and given time to work, may come up with a surprising answer. Einstein, noted for the fundamental novelty of some of his most important ideas, pithily remarked that the really creative scientists are those with access to their dreams. To produce something completely new, it is often necessary to forget conventional wisdom.

The nature of the creative process

It is said that the difference between a craftsman and an artist is that the craftsman knows exactly what the outcome of all the effort will be before the work is even begun, while an artist must wait until the work is complete before discovering what has been achieved. The latter process is probably most typical of the creative artist, but there are those who know in advance exactly what they want to achieve and appear merely to be concerned with the means of doing so. Many novelists, on being asked about the creative process, agree that they have no idea in advance how a book is going to evolve. They need an idea to get them started and then, if all goes well, the characters take over – though some writers, such as Vladimir Nabokov (1899–1977), have scorned the idea.

Some artists find that the production of a work of art involves distinct stages such as preparation, incubation, inspiration and elaboration. Others experience all these stages repeatedly in the same act of creation, or describe other stages. Yet others begin in a state of confusion with fragmentary ideas competing for attention in the mind, and find that some kind of definitive entity solidifies from the mist as ideas are rejected or put together.

The theories of Freudian and Jungian psychologists (➪ pp. 164–5) have been at the forefront of attempts to explain the creative process. Freud initially saw creativity as the working out of unconscious desires (wish fulfilment). Later, as his ideas changed, he came to see the creative act as a process of defence by the ego against indictments by the superego. Jung, in his life-long preoccupation with symbols saw creativity as an unconscious symbol-making process.

It would be unreasonable to expect to be able to understand the creative process at a mechanistic level. The most that one can say is that nothing can come out of the brain that has not previously gone in, but that the possibilities of synthesis, by the interaction of new with stored data, are infinite. The components, before synthesis, may be familiar; once incorporated into a new creation they may no longer be identifiable, so that it may seem that something completely new has been made.

Creativity and intelligence

Intelligence is regarded by psychologists as a complex of a large number of separate definable abilities, among which are usually included several that are creative in nature. The idea that creativity is a component of intelligence is not, however, universally accepted. This may reflect the difficulty in adequately defining intelligence (➪ p. 168), but laypeople often nevertheless make an intuitive

A Scientist's Creative Daydream

The German chemist Friedrich August Kekulé von Stradonitz (1829–96) was one of the first important theorists of organic chemistry. The benzene molecule, containing six carbon atoms and six hydrogen atoms, was of considerable importance in organic chemistry because of the range of new synthetic dyes that were being developed from it. Unfortunately, benzene appeared to defy the normal rule that carbon combined with four hydrogen atoms (i.e. was tetravalent). Its structure was unknown, and this was holding up progress. One day in 1865 Kekulé was daydreaming on a bus, thinking casually of atoms. It seemed to him that he could see chains of atoms whirling in a dance. As he watched, he saw the tail of one chain attach itself to its own head, forming a spinning ring of atoms. Suddenly wide awake, Kekulé realized that this was the answer to the problem of the structure of benzene. The six carbon atoms were arranged in a closed ring with alternate double and single bonds and a hydrogen atom attached to each carbon. This intuition also solved another problem, that of valency, and was hence to be one of the major advances in all chemistry.

distinction between the two qualities. When studies have been made to try to correlate intelligence and creativity they have produced conflicting results, which suggests that the originators of different trials may have had different ideas of the nature of creativity. The consensus of opinion, however, is that the two correlate well at low and average levels, but that, when exceptionally gifted people are studied, intelligence and creativity are often mutually incompatible.

We know that many adults of great originality – people like Albert Einstein and Isaac Newton – were unremarkable as children. We also know that the highest achievers in science – the people who make the major advances – do not always have exceptionally high IQs. They can be people of high average ability whose interest and imagination are caught by a particular subject, and who then concentrate very hard on it.

Wolfgang Amadeus Mozart (1756–91), Austrian composer, displayed an extraordinary musical ability very early on. He had begun to play the harpsicord by the age of three, and gave his first public performance at five, by which time he had already begun composing. His talents were so impressive that he was summoned, aged six, to play at the court of the Empress Maria Theresa. (AKG)

Creativity and the right side of the brain

In the great majority of people the left hemisphere of the brain contains the nerve centres for speech, language and language-related functions such as rational thought. It also controls movement of the right side of the body. Because speech, language and writing are central to our higher activities, the left hemisphere is called the dominant hemisphere. In a small proportion of people the right hemisphere is dominant in this way.

It has been argued that the right side of the brain is concerned with a wide range of non-verbal activities – things like spatial relationships, patterns, styles, design, data synthesis, metaphors, new combinations of ideas, and so on. The right brain is intuitive rather than logical, holistic rather than specific, and relational rather than factual. This suggests that all the important functions concerned with creativity are centred in the right side of the brain. However, this raises the question of literary art as verbal ability – command of the language – must, of course, substantially involve the left side, but literary *creativity* is mainly a function of the right side. We can make sense of this apparent paradox by dissecting the practice of writing. A much-attested method when composing is to jot down ideas as they occur (right brain). This quickly appears not to be random, as ideas connect and one leads to another. When this has run its course, the writer begins to organize these joltings into coherent argument (left brain).

In summary, since the relative use of the different parts of the brain determines our abilities, we can infer that highly creative people have outstanding access to certain parts of the right side of their brains, relative to other parts. But high levels of access to certain parts of the brain do not happen by chance: innate abilities, or the structural basis for the development of such abilities, are often inherited. In such cases there are often also powerful early environmental influences operating to encourage the use and development of these faculties – a process associated with the development of the part of the brain concerned.

'An infinite capacity for taking pains'

Children showing outstanding early ability – child prodigies – are of special interest to those studying creativity. Most of them seem to have been very one-sided – geniuses in one area but otherwise very ordinary. A clear distinction should be made between prodigies who are otherwise normal and those – the so-called 'idiot savants' (⇨ pp. 168–9) – who have learning difficulties in other respects.

The general view that child prodigies achieve their remarkable success without great labour is almost certainly wrong. Studies of many composers have shown that most of them worked intensively for at least ten years before producing anything of merit. Even Mozart was drilled ruthlessly in composition by his father before, at the age of 12, he showed the first signs of his supreme grasp.

It seems probable that in many child prodigies, unusual achievement is the result of an exceptional quality of mind that allows single-minded concentration on the recording and organization of experience so that great achievement is possible without help or even against opposition. When the 17th-century French mathematician Blaise Pascal (1623–62) was a child, his father, anxious that he should study the classics, deprived him of the mathematical textbooks in which he was showing interest. So young Pascal secretly worked out geometry for himself.

SEE ALSO

- THE HUMAN COMPUTER? p. 72
- WHAT IS INTELLIGENCE? p. 168
- WHAT IS MEMORY? p. 170
- THEORIES OF LEARNING p. 172
- LEARNING IN PRACTICE p. 174
- DREAMS p. 178

Can Creativity be Measured?

Whether creativity can be measured is a matter still open to question. Various attempts to assess creativity have been made. One has been to assess what is sometimes called *ideational fluency* – the number of different ideas a person can generate in a particular context. A psychologist might, for instance, ask a test subject to suggest as many uses as possible of an empty champagne bottle. Other tests might involve interpretation of Rorschach ink blots (⇨ p. 188), made by dropping ink on a sheet of paper and then folding it in half; or writing an account of the story behind various posed photographs. A major difficulty in such testing is the highly subjective nature of the examiner's response to the subject's answers; some answers might seem highly original to one examiner but banal to another.

Dreams

Extraordinary things have been believed about dreams. In some cultures it is believed that the soul leaves the body during dreams and that it is dangerous to wake a dreaming person. In other cultures a dream of a person's adultery is taken, by some people, to justify a partner's repudiation.

SEE ALSO
- BODY SENSATIONS p. 90
- IMAGINATION AND CREATIVITY p. 176

Each of us has dreams every night, but we remember only a small proportion of their content. Moreover, researchers can obtain information on the content of dreams only from reports of them, and so it is not particularly reliable. But different reports have so much in common that some of the general characteristics of dreams can be determined. They commonly feature a familiar location or ordinary everyday surroundings, rarely a bizarre or exotic setting. Visual imagery dominates, but most dreams have some auditory features. The majority seem to reflect, or relate to, events, emotions and thoughts that have been experienced in the recent past – often within a day or two – and their content can often be recognized as a mixture of past experience, personal interests, wishes, and inclinations. Stimuli currently operating on the body are also important. A full bladder, for instance, will frequently make its tensions known in a dream, and if a man's seminal vesicles (➪ p. 96) are full, he may have an erotic ('wet') dream involving orgasm. For reasons that are not clear, the emotional content of dreams – if there is any emotional content – tends more often to be unpleasant than pleasant, with the commonest emotions experienced being anxiety or fear, followed by anger.

Freud on dreams

To Freud (➪ p. 164), dreams were the 'royal road' to the unconscious mind, revealing to the psychoanalyst the buried secrets of the inner life. It seemed to him that thinking during sleep was less repressed than conscious thoughts so that the unconscious preoccupation with sex and aggression had freer rein. He suggested that every dream has a *manifest content* – the remembered details – and a *latent content* – the repressed infantile, sexual and aggressive wishes of the dreamer. He believed that the dreaming psyche disguises the real nature of the inner life, distorting the details or representing them symbolically. Without this disguise the true nature of the inner life would be so shocking as to wake the dreamer.

Sometimes a number of latent elements were, he claimed, represented by a single manifest element – a process Freud called *condensation*. Similarly, emotions felt towards one person or object were transferred, in the dream, to another person or object. This he called *displacement*. Freud also claimed that people unwittingly altered the accounts or recollection of their dreams so as to make more sense of them. This complex of processes, by which the latent content is converted into the manifest content, Freud called the *dreamwork*. His theories of dreams have had a considerable influence on the practice of psychotherapy, but are not widely used in academic psychology today, as they cannot be empirically tested.

REM sleep and dreams

In the early 1950s, studies of human behaviour during sleep revealed that there are periods during which the eyes move rapidly, and that this is associated with other changes. It is now well known that about an hour after we fall asleep, our voluntary muscles, which have been normally tense, suddenly relax, our eyes move rapidly from side to side under our lids, our breathing deepens, our heart rate becomes irregular, and the electrical pattern of the brain waves, the electroencephalogram or EEG, comes to resemble that of an alert, awake person. In 95% of men, the penis becomes erect.

The whole phenomenon is known as *rapid eye movement* (*REM*) sleep, and each period of REM sleep lasts for 5–30 minutes. When wakened from REM sleep, 80–90% of people report that they have been having vivid dreams. The dreaming appears to be continuous during REM sleep, and people commonly report long dreams following extensive periods of uninterrupted REM sleep. These episodes of REM sleep are followed by more quiescent periods of deeper sleep during which the EEG pattern becomes much slower with higher amplitude waves, and the muscles become somewhat more tense. People woken up during this phase of sleep seldom admit to having real dreams but only to having been in a state of calm thoughtfulness. These two phases alternate on a 30- to 90-minute cycle, REM sleep constituting, on average, about 20% of the total sleeping time. The time spent in REM sleep increases towards the end of an undisturbed night.

When attempts are made to deprive people of REM sleep by waking them as soon as the REM stage starts, a remarkable phenomenon can be observed. On each

Wood engraving of a nightmare demon from Collin de Plancy, *Dictionnaire Infernal* (1863), but derived from a painting by Salvator Rosa (1615–73). (Images)

successive night in which this is done there is an increase in the number of times the subjects try to enter REM sleep. To begin with, they may have to be wakened 7 or 8 times. By the fifth night they may have to be wakened 30 times. If they are then left undisturbed they spend a far higher than normal proportion of the time, during the next few consecutive nights, in REM sleep. Deprivation of REM sleep has striking effects during waking hours, causing irritability, anxiety, loss of concentration, suspiciousness, apathy and even a tendency to hallucinations.

In babies and infants REM sleep persists for half the sleep cycle – more than twice as long as in adults. This probably implies that infants dream much more than adults do. Some scientists believe that REM sleep in babies is necessary for the normal growth and functional connection of nerves. Old people, too, sometimes show an increase in the proportion of time spent in REM sleep. REM sleep is not limited to humans; many vertebrates, including all mammals, exhibit the symptoms, and this, taken with other evidence, suggests that many animals have a rich dream life.

The undeniable need for REM sleep does not necessarily imply that it is dreaming that is essential to us. So far, research has not been able to throw light on whether the dreaming is a central part of the phenomenon or merely a kind of by-product.

A Dream of Latmos, by Sir Joseph Noel Paton (1821–1901), which depicts a dream as the expression of unconscious desire. (FAPL)

Modern theories of dreams

Ideas on the function of dreaming are still entirely speculative. We know that the brain needs constant stimulation to continue to function normally, and one suggestion is that dreaming is a way of helping to maintain brain function during sleep, when the amount of external stimulation is much reduced. Some neurophysiologists reject the notion that dreams have any psychological meaning or importance. In their view, dreams are simply the result of random stimulation of areas of the cerebral cortex (➪ p. 76) that subserve conscious experience. When this happens, an attempt has to be made to make sense of the resulting impressions and these, they say, are our dreams. It has also been suggested that dreams are the result of the brain's actions in trying to erase false associations by disconnecting erroneous neural links between brain cells. This process might serve the essential function of allowing correct memory processing (➪ p. 170). The bizarre content of dreams, then, is in fact the record of material that is to be purged from the individual's memory.

On the basis of experimental evidence, other neurophysiologists have concluded that formerly acquired information vital to an animal's survival is accessed during REM sleep and integrated with immediate past experience so as to modify future behaviour and increase the chances of survival. According to this theory, our own dreams may have a similar mechanism and function. It has been suggested that this mechanism may have been inherited from our prehuman ancestors, and that the largely visual content of our dreams reflects the absence of speech in other animals. Supporting evidence has been derived from studies of the reported REM dreams of people suffering from marital problems, where the contents of the dreams are strongly related to the ways in which these people are coping with their real-life crises.

DREAMS AND DIVINATION

Historically, the most prevalent view about dreams was that they predict the future. Written records of dream interpretation date back at least 4000 years and are found in the archives of the ancient Egyptians, the Babylonians, the ancient Indians, the Sumerians, the ancient Greeks and Romans, and others. In the Greek and Roman world, many enterprising soothsayers and priests made a profitable business of dream divination, and the Greeks also incorporated a kind of dream therapy into their medical practice.

The Bible, especially the Old Testament, is rich in dream interpretation and prophecy. The story of Joseph's dream in the 37th chapter of Genesis is a case in point. Joseph merely recounted his dreams about his brothers' sheaves of corn that bowed low to his sheaf, and about the sun, moon and eleven stars that bowed low before him; it was his brothers and father who interpreted them as meaning that Joseph would become king. There are strong implications in this story that dreams of obvious symbolic content were taken to be both prophetic and to reflect the wishes of the dreamer.

In spite of the prevalence of belief in magic, there were some who were able to examine the subject more dispassionately. Aristotle (384–322 BC), with remarkably modern insight, pointed out the connection between dreams and previously experienced external events, and recognized the way in which normal sensory data can be distorted by emotional factors. In the Roman period, Marcus Tullius Cicero (106–43 BC), in his book *De diviniatione*, mounted a scathing attack on the dream superstitions of the age. Six hundred years later, the prophet Muhammad (c. AD 570–632), distressed at the extent to which the lives of the people were being influenced by dream divinations, expressly forbade the practice.

These few voices of common sense did not, however, have much effect on the mass of public superstition, and it was not until well into the 19th century that thinking people began to recognize that dreams were more likely to be a reflection of what was going on in the mind than what might later happen outside the body.

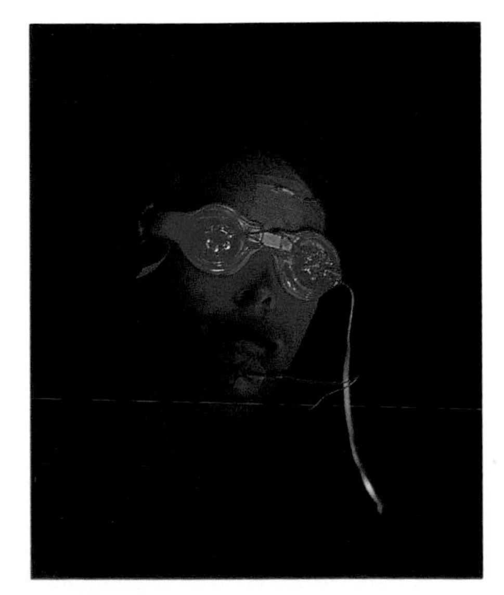

In rare cases adults may continue to suffer night terrors in their sleep, a phenomenon usually confined to childhood. These terrors bring on sudden attacks of severe anxiety, racing pulse and respiration, and the person may wake up screaming. During these terrors, sufferers may experience a sense of suffocation, claustrophobia, or even the conviction that they are about to die. Recording the sleeper's eye movements, as shown here, may uncover the cause. (Gamma)

Parapsychology

Reported instances of the paranormal power of the mind – foretelling the future, mind-reading, telepathy, and so on – are common and arouse endless interest. These reports are so numerous as to suggest either that such things really can happen or that we have an overwhelming desire to believe that they can.

Can such a mass of evidence, albeit anecdotal, be ignored? Many people think not and have been so impressed by it that they have devoted their lives to research into the subject. When it was first coined, the term 'psychical research' gave a strong impression of scientific respectability. But seriousness of purpose does not necessarily preclude credulity, and much that was reported as genuine has, in retrospect, been shown to be fraudulent. The result was that psychical research remained plausible only to those determined to believe at all costs. If parapsychology – the current expression – seems to be retaining its acceptability with ordinary people, this is because those engaged in it have been forced to adopt an attitude of scepticism at least as strict as that of orthodox scientific researchers.

David Hume on Miracles

The eminent Scottish philosopher and historian David Hume (1711–76) had a great deal to say about miracles, and his ideas on the subject have been highly influential. He discussed the subject in his book *An Enquiry Concerning Human Understanding*:

> **No testimony is sufficient to establish a miracle, unless the testimony be of such a kind that its falsehood would be even more miraculous than the fact which it endeavors to establish. When anyone tells me that he saw a dead man restored to life, I immediately consider with myself whether it be more probable that this person should either deceive or be deceived, or that the fact which he relates should really have happened. I weigh the one miracle against the other; and according to the superiority which I discover, I pronounce my decision, and always reject the greater miracle.**

David Hume argued that we can know something only in so far as it can be perceived by the senses. (ME)

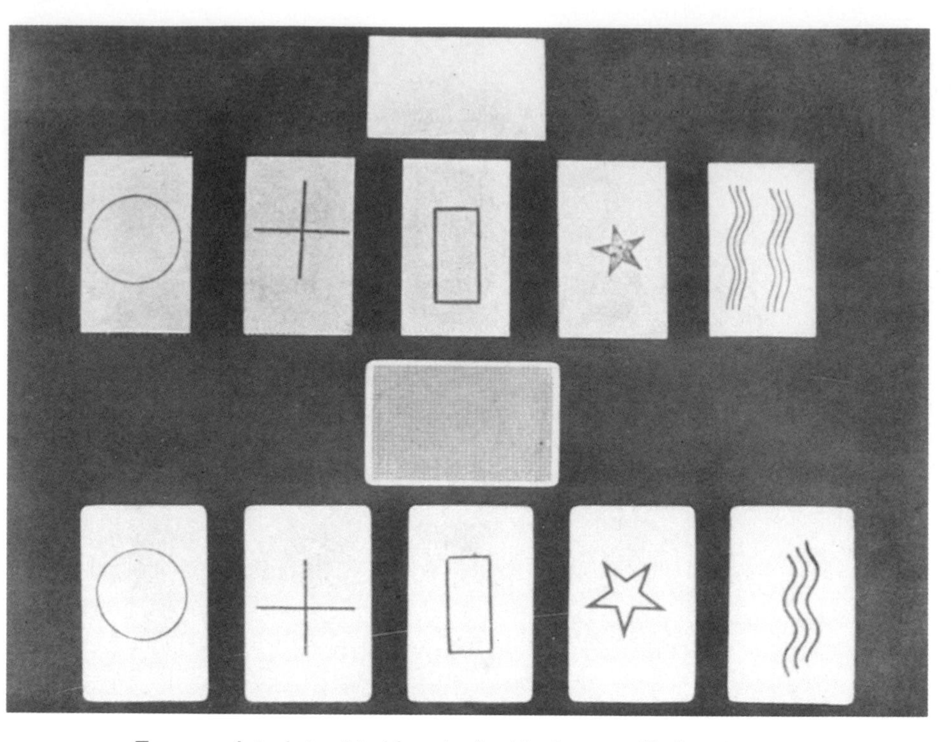

Zener cards in their original form (top), with a later, modified version. (ME)

Modern positivistic science is built on the principle of causality. So, if one steps out of a tenth-floor window, one rightly expects gravity to cause one to fall rather than fly. While we cannot claim that causality is any more than an idea based on previous experience, it is, nevertheless, an idea that we repeatedly trust our lives to. The problem with paranormal 'phenomena' is that, by definition, they are phenomena for which modern science can offer no causal explanation. This does not entirely rule out their existence, but they have not so far been scientifically established – despite the mass of anecdotal evidence.

Spiritualist research

Many 19th-century scientists suffered because of the conflict, real or apparent, between science and religion. As science, especially physiology, advanced and developed along strictly materialistic lines, there seemed to be less room, or even necessity, for the soul or, indeed, for any spiritual reality. This prompted several people to try to apply scientific methods to the investigation and detection of out-of-body spirits (ghosts), or to achieve contact with those who had 'passed over'.

Some very distinguished scientists devoted much time and effort to these investigations. The pragmatic psychologist William James (1842–1910; ➪ p. 163) held that it was quite wrong for science to insist on mechanistic laws in the face of 'divinations, inspirations, demonical possessions, apparitions, trances, ecstasies, miraculous healings and productions of disease and occult powers'. These, he claimed, were facts of experience: 'science, so far as it denies such exceptional occurrences, lies prostrate in the dust for me'.

Finding themselves devoid of psychic powers, the researchers turned to those who claimed they did have such abilities and, not always with appropriate detachment, repeatedly examined the work and claims of these 'mediums'. But a lot of these scientists started with preconceptions and then proceeded to try to prove them. Often the standards of rigour left much to be desired, and some of these workers showed a most unscientific gullibility. Even so, medium after medium was caught out in fraud and trickery. The practice of holding séances was thoroughly discredited and by the end of the 19th century only the most tenacious were willing to continue.

Much the same happened to 'research' into apparitions during the first half of the 20th century. Ironically, the rising standards applied by investigators (such as the Society for Psychical Research in the UK) to research into hauntings tended progressively to deprive them of their *raison d'être*. The notable debunking of the popular British ghost-raiser Harry Price, who had persuaded large numbers of people that Borley Rectory was haunted, did much to bring about the demise of belief in such phenomena.

Rhine and ESP

The American botanist J.B. Rhine (1895–1980) came to believe in the existence of psychic phenomena after hearing a lecture by Sir Arthur Conan Doyle, a convinced spiritualist. In 1930 Rhine and the psychologist William McDougall (1871–1938) set up a parapsychology laboratory at Duke University, North Carolina. Four years later he, Rhine, published *Extra-Sensory Perception*, which was an account of his experiments into telepathy, and in 1937 a popular work entitled *New Frontiers of the Mind*. This caused a sensation and became a best seller, bringing to the attention of the world Rhine's claims that he had proven thought transference scientifically.

Rhine's attitude was scientific. He believed that the supposed paranormal mental powers were an extension of normal psychological activities and were susceptible to statistical analysis. Using packs of 25 'Zener' cards, each card bearing one of five symbols – square, circle, cross, star and wavy lines – Rhine was able to quantify the results of a test in which one person turned the cards over and looked at them, one at a time, while another, who could not see the cards, wrote down his or her impression of which symbol was being observed. In a sufficiently large number of trials, pure chance would dictate that correct answers occur, on average, five times for each run of the pack. Rhine obtained correct results at a significantly higher rate, and immediately claimed that he had proved the existence of extrasensory perception (ESP).

In response to criticisms of his methods, Rhine repeated his trials with each half of the pairs in different buildings and with wholly independent verification and statistical analysis of the findings. Above-average scores, though, occurred less often. Millions who read of his work accepted Rhine's conclusion, and for a time it seemed that ESP was an established scientific fact. Later, Rhine claimed to have proved the existence of *psychokinesis* – the ability to move objects by the power of the mind without physical intervention. This work was conducted along similar lines to the ESP trials, and it was claimed that one could influence dice that were randomly thrown to produce better-than-average scores.

Science does not accept new data without independent verification, so several other psychologists repeated Rhine's tests under similarly rigorous conditions. None was able to find any evidence of above-average results. The response of Rhine's supporters to this was that the demonstration of ESP might depend on the attitude of the experimenter. Those who were sceptical might have a negative influence and only those who believed could demonstrate it. Such an argument did little to convince the scientists. It seemed more likely that, in some way not yet apparent, the powerful desire of the experimenter to prove the hypothesis was leading to unconscious bias or even cheating, later events showing this to be so. After Rhine had retired, his successor, Walter J. Levy, was discovered, in 1974, to be fraudulently increasing the scores in an experiment so as to support ESP.

Today, rigorous parapsychological investigation continues, and much use is made of electronics and automation. Attempts are being made to study the effects of attitude, beliefs, mood and altered states of consciousness on claimed ESP scores. It is said that people with a positive attitude to ESP, and who are relaxed and emotionally healthy, are most likely to produce good scores.

The problem of proof

The great bulk of claims of paranormal phenomena – such as materializations at séances, communications with the dead and levitation – can be dismissed, usually on the grounds of fraud. Few of them can be sustained after trained investigation. Claims by those who use normal scientific methods are not so easily dismissed, but their case has been weakened by the few notorious examples of cheating.

In fairness to those who support parapsychology, it must be said that they do not, and cannot, claim that paranormal phenomena can be repeated at will. If they could, such phenomena would not be paranormal. So it proves nothing to conduct experiments that give negative results. Such paranormal phenomena as have been claimed are rare; the prevalence of outright fraud among parapsychology workers is also rare, but it is possible that there is a connection. It is also possible that unconscious self-deception and fraud could account for all the claimed positive findings.

A good deal of support for belief in the paranormal arises from a well-recognized quirk of human nature. If we want to believe something and a highly unlikely coincidence occurs that supports the belief, we will be highly affected by it. We will not, however, remember the numerous occasions on which such coincidences did *not* occur.

The public appears to have an insatiable appetite for the out-of-the-ordinary, and will read avidly any account of claimed paranormal phenomena, especially if presented in a seemingly scientific manner. By contrast, there is little interest in accounts that disprove the paranormal, possibly because, in the minds of the uncritical, the 'evidence' for the paranormal enormously exceeds the evidence against it. Perhaps this imbalance arises from a human desire to believe that we are more complicated than our physiology suggests.

SEE ALSO

- TOWARDS A SCIENCE OF THE MIND p. 162
- IMAGINATION AND CREATIVITY p. 176
- DREAMS p. 178

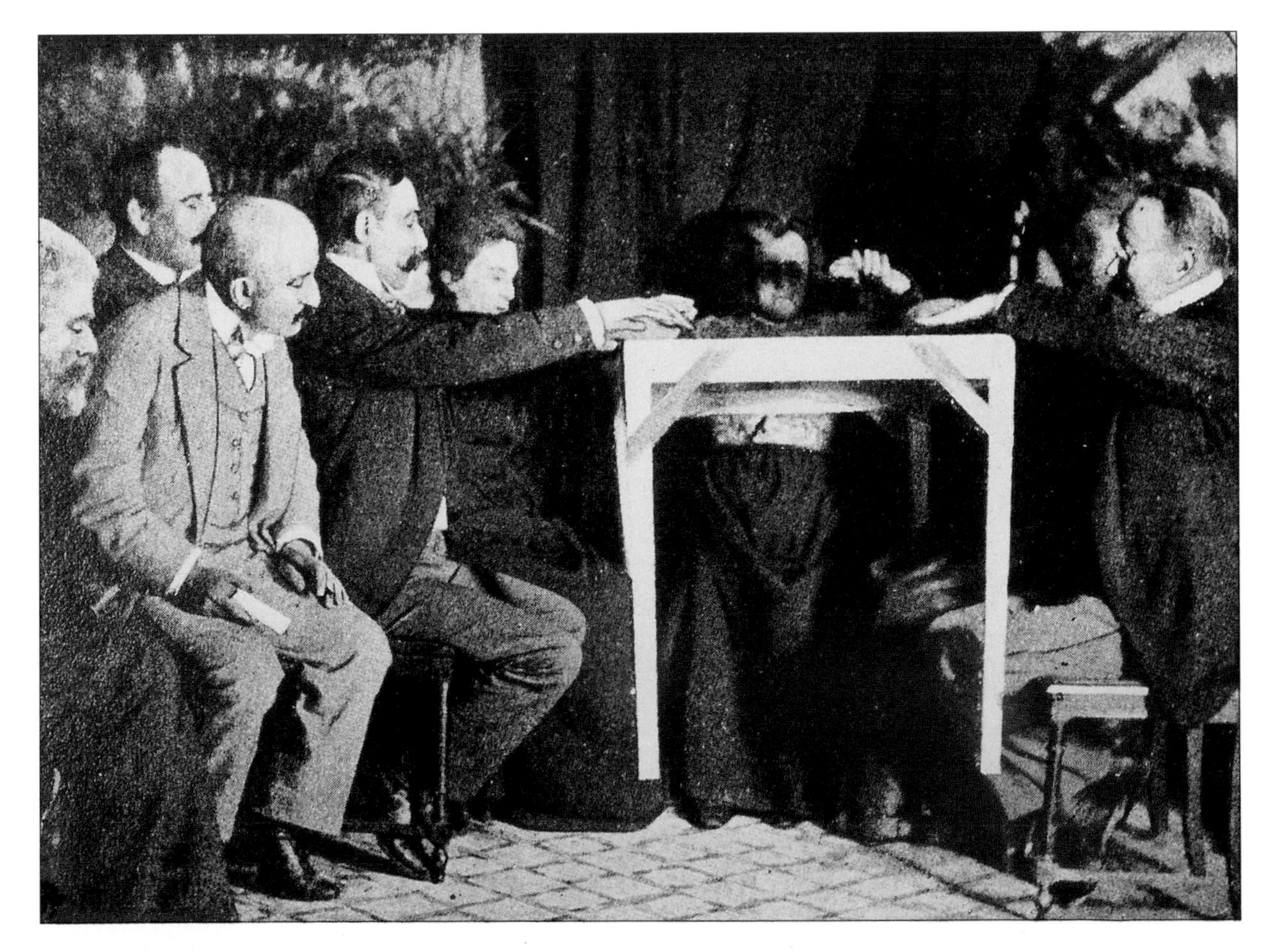

Levitation of a table at a séance in Milan, 1892. (Images)

MENTAL DISORDERS

A History of Mental Illness

The history of human attitudes to mental disorder shows some curious early peaks of enlightenment. These were followed by a re-descent into near savagery and then a gradual restoration of humanitarian views. Real progress in the treatment of the mentally ill, however, did not begin until the 19th century, when doctors started to look at mental illness as an illness.

Witch-hunts in 18th-century America, as depicted here by a later artist, were quite common. Much of the peculiar behaviour of women suffering from mental illness was mistaken for signs of witchcraft. Old and isolated women in particular were suspected of being witches. (Image Select)

All human intercourse involves judgement of the state of the minds of our fellows. We assess, compare and privately criticize. This process is habitual, so much so that even when we know that another person's mind is disordered we tend to continue to apply customary criteria in assessing his or her behaviour. Unless we remind ourselves of the injustice of so doing, we thoughtlessly attribute responsibility and apportion blame. When a mind appears to be so disordered that none of the normal criteria can apply, we become confused and do not know how to respond.

Although evidence of early attitudes is fragmentary we know that some attempts were made, notably by the early Chinese, by people in Siberia and Malaya and by the priests of the early Egyptians, to treat mental disorders humanely. An Egyptian stele (a decoratively carved and inscribed stone slab) in the Bibliothèque Nationale in Paris tells the story of the successful treatment, for 'demonic possession', of a princess of the 20th Dynasty of Pharaohs. Later Greek and Roman attitudes also seem to have been, in some cases, relatively civilized.

Greek and Roman views

Greek mythology abounds in accounts of mental disorder and the Homeric tradition suggests that the early Greek views were somewhat ahead of those of earlier civilizations. Mental disorder, to them, manifested itself in odd and irrational behaviour, which was, they thought, interesting, and not necessarily frightening. Homer's Ulysses, pretending to be mad, yoked a horse and a bull together, ploughed a beach and sowed salt. Madness, the Greeks believed, was due to the action of the gods in taking away the mind. In early Roman times most people viewed mental disorder as a visitation of the gods, and the mentally ill were often refused entrance to the temples, but some were believed to be inspired and were consulted as oracles.

Hippocrates (c. 460–377 BC), the great Greek physician, was thoroughly scornful of the view that madness was caused by malign spirits. He classified mental disorders as *mania* (abnormal excitement), *melancholia* (depression) and *paranoia* – by which he meant what we would now call dementia (progressive loss of mental capacity). Hippocrates attributed all of these to physical causes. He also believed that hysteria (now called 'conversion disorder'), a disorder he wrongly thought was restricted to women, was caused by wandering of the unsatisfied womb and cured by marriage.

Although epilepsy is not a mental disorder, Hippocrates' attitude to this condition illustrates his good sense and humanity. In his book *The Sacred Disease* he wrote: 'It is thus with regard to the disease called sacred: it appears to me to be nowise more divine nor more sacred than other diseases, but has a natural cause from which it originates like other affections.' Hippocrates correctly believed epilepsy to be caused by disordered function of the brain.

The Greek philosopher Plato (c. 427–347 BC) was less enlightened and attributed some forms of mental disorder to divine intervention – a sign from the gods that the afflicted person was being punished for past wrongdoing. Plato's influence was enormous, especially during the early part of the Christian era, and this may have been partly responsible for the horrors that were to follow.

Witchcraft and heresy

Many of the mentally ill who hear voices and have visions commonly identify them as belonging to God or the Devil or spirits. Many of these outbursts were considered heresy or even evidence of witchcraft, leading to torture and death by burning at the stake. The Book of Leviticus plainly states, 'The Lord spoke to Moses and said . . . any man or woman among you who calls up ghosts or spirits shall be put to death. The people shall stone them; their blood shall be upon their own heads.' The widely held principle that attributed divine authority to every word in the Bible served for centuries to justify the killing of millions of people and to warrant and rationalize the deep hostility felt towards millions of others who suffered mental disorder.

From the 15th century onwards, when the Christian Church initiated a formal campaign to 'cleanse' society of those corrupted by the devil, a wave of hysteria over the practice of witchcraft swept across Europe, and many thousands of people, mainly women, were tortured and killed. A notable factor in this persecution was the infamous book the *Malleus Maleficarum* ('Hammer of Witches'). From the Middle Ages until as late as the 19th century, mental disorder was equated with 'possession' by devils. This view was held even by those most advanced in other fields of thought. The 1st-century Roman medical writer Celsus recommended restraint in chains. The great physician and surgeon Ambroise Paré (▷ p. 34), for all his notable contributions to medicine, was convinced that the devil entered into women and made them witches and that, in accordance with the Book of Leviticus, witches should be killed.

Those who dared to suggest otherwise were often persecuted. The Belgian physician Johann Weyer (or Wier) (1516–88) believed that 'witches' were deluded or suffering from hallucinations, and wrote a diatribe against the folly and cruelty of witch trials. He was saved from the fury of the clergy by the protection of his patron and patient the Duke of Jülich. In England, Reginald Scot (1538–99), whose book *The Discoverie of Witchcraft* (1584) was an attempt to expose the absurdities underlying the belief in witchcraft, aroused the anger of King James I. The King had Scot's book burned by the public hangman and wrote *Daemonologie* (1597), damning the opinions of Weyer and Scot.

The belief that mentally disturbed people were witches or possessed by demons provoked much hatred towards them, and these unfortunates were usually treated with great harshness. Many were locked up for years in dark dungeons. Others, if they avoided being burnt at the stake, were tortured and beaten. The authority of the Church was taken to be justification for unimaginable cruelty by men whose motives were at least ambiguous.

By the late 16th century religious intolerance and hardening attitudes on the part of the authorities to those sections of society that they considered disruptive led to the widespread incarceration of people with mental disorders. Formal medical treatment, even of those not suspected of witchcraft, was not renowned for its humanity.

Bedlam

Europe's first asylum for the insane – St Mary of Bethlehem (Bedlam) – was officially established in the 16th century and stayed open well into the 19th century. Official attitudes were explicit: the insane, being deprived of reason, were less than human and were to be treated as beasts. They could be dangerous, it was believed, and were to be closely restrained. They understood physical punishment, however, and could be beaten into submission. They were impervious to heat or cold, and suffered no shame – clothes were therefore unnecessary. It became a popular amusement to visit Bedlam and view the 'human animals'. In the early 18th century the charge was one penny, and in one year 96 000 spectators attended. Institutions such as Bedlam did not offer medical treatment to their inmates. Insanity was considered incurable and incarceration necessary and permanent.

An engraving of Bedlam (Bethlehem Hospital) from Hogarth's *Rake's Progress*. It shows a group of sightseers looking at the patients. The hospital became notorious for its brutal treatment of the insane. The word Bedlam, meaning uproar, is derived from the behaviour in the hospital. (RB)

Philippe Pinel and the humanitarian movement

By the 18th century, however, attempts at treatment, if it can be called that, were being made in some European centres. These were largely at the whim of individual doctors. The German psychiatrist Johann Christian Reil (1759–1813), widely regarded as an advanced and humane practitioner, recommended inflicting hunger and confinement on psychiatric patients, flogging with a cowhide, restraining them with straitjackets, firing canons near them, and throwing them into cold water 'in order to bring them to their senses'.

In 18th-century France, a notable humanitarian advance occurred. The physician Philippe Pinel (1745–1826), at the Bicêtre Hospital, decided that, in some cases, the insane behaviour displayed by his psychiatric patients was due to their being chained. Pinel, at great risk from the authorities, ordered their chains to be removed. He also allowed certain patients the freedom to move freely around the hospital grounds and to enjoy the fresh air. The experiment was a dramatic success. Many – who were not genuinely disordered or who were only depressed – recovered so completely as to be able to be released. Pinel's successor, Jean-Étienne Dominique Esquirol (1772–1840), pursued the same policy and was so influential that he was able to have most of the regulations governing the management of mentally ill people amended.

Similar reforms were effected in England by members of the distinguished Quaker Tuke family. William Tuke's asylum in Yorkshire became famous for its enlightened treatment of psychiatric disorders. In the USA the Massachusetts schoolteacher Dorothea Dix was able to achieve similar reforms by bringing the plight of such patients to the notice of the public. Dorothea Dix also campaigned in Europe with considerable success. By the end of the 19th century a widespread movement towards more humane treatment of those with mental disturbances had been achieved.

Psychiatry in the 19th century

During the 19th century rapid advances in neuroanatomy and pathology began to show that many mental disorders had a counterpart in organic disease. Conditions such as tertiary syphilis (the third and most dangerous stage of the disease), alcoholic dementia, various toxic states and a number of hereditary conditions were so convincingly linked with mental disorder that many doctors began to believe that all such conditions would eventually be shown to be of organic origin. The French neurologist Jean Martin Charcot (1825–93), who carried out brilliant work on disease of the nervous system, favoured an organic basis for all mental disturbance, although the concept has remained unsubstantiated.

At the same time, doctors were beginning to attempt a formal classification of mental disorders. Prominent among those interested in this was the German psychiatrist Emil Kraepelin (1855–1926). Kraepelin collected and studied thousands of psychiatric case histories and accounts of the lives of patients prior to their illnesses. These enabled him to see that mental disorders could be grouped into a comparatively small number of classes with features in common. The first edition of his textbook of psychiatry was published in 1883 and has remained influential ever since.

Although the foremost 19th-century medical thinkers were convinced that psychiatric disorders would be shown to be of organic origin, some of them, including Charcot, did begin to become aware of the importance of psychological factors. Charcot was especially interested in hysteria (⇨ p. 188) and found that so-called hysterical manifestations in women, such as paralysis, could be induced by hypnotism. This suggested that mental disorders were not necessarily of organic origin. He believed, however, that this was possible only in women who were already considered hysterics. In 1887 Charcot, in collaboration with Paul Richer, published a study entitled *Les Démoniaques dans l'art* ('Demonic Possession in Art') in which he clearly showed that the manifestations of 'demonic possession' were typical of the hysteric. Even so, Charcot was still convinced that hysteria had an organic basis, and believed that the ovaries were implicated.

Charcot's former student, Sigmund Freud (1856–1939), however, took an almost wholly psychological view of the origins of mental disturbance. In the course of his investigations into the origins of the neuroses, conducted by encouraging patients to talk freely and disclose the contents of their minds, he developed the school of psychiatric treatment known as psychoanalysis (⇨ p. 164).

SEE ALSO

- IMAGINATION AND CREATIVITY p. 176
- WHAT ARE MENTAL DISORDERS? p. 186
- PERSONALITY AND ITS EXTREMES p. 188
- ANXIETY AND NEUROSES p. 190
- PSYCHOSES p. 192
- TREATING MENTAL DISORDER p. 194

Creativity and Mental Illness

The idea that high achievers are often mentally unstable was given its most familiar form by the poet John Dryden in his diatribe against the politician Lord Bolingbroke in *Absolom and Achitophel*:

> **Great wits are sure to madness near alli'd**
> **And thin partitions do their bounds divide.**

Bolingbroke certainly was not mad, but can we be so sure about William Blake and Vincent Van Gogh? Blake, poet and engraver, claimed that he conversed with the ghost of a flea which featured in one of his drawings. Many of his writings are clearly the result of visionary preoccupations which lie beyond the range of normal experience. The poet Southey certainly thought him insane, but if he was it is difficult to believe he would have had anything worthwhile to say if he had been 'cured'. His genius seems to have been inseparable from his madness.

The same link between creativity and mental ill health is apparent in the life and work of Vincent Van Gogh. Although Van Gogh's work as an artist spanned ten years, most of the pictures that made him famous were painted in 1888–90, when he was either clinically insane or very near to it. Van Gogh's masterpieces date from 1888, starting with his move from Paris to Arles in February and ending with the breakdown in December. His breakdown, including the famous episode in which he cut off part of his left ear, had a typically schizophrenic quality. During these three years he produced 200 canvases of extraordinary originality. Some features of the paintings suggest that his perceptions were enhanced: his suns and stars seem to shimmer and dance, the lights and colours glow and vibrate. Visual experiences of this sort occur in schizophrenia and those taking hallucinogenic drugs. Some people believe that the psychotic distortion of the world around Van Gogh played a central role in pushing him to a creative level that he would not otherwise have achieved.

Schizophrenia is, of course, a destructive illness. Van Gogh's last paintings caricature and debase his style and most critics agree that very little artistically was gained from a psychiatric breakdown. Nijinsky, hailed as the greatest male dancer of all time, never danced again after a schizophrenic breakdown at the age of 29. But it is arguable that mental illness has always been a part of the spectrum of human experience and that the world would be a poorer place if everyone had textbook standards of mental health.

What are Mental Disorders?

There is no complete consensus among psychiatrists about the cause of mental disorders. Many possible factors are involved. The main points of contention are the definition of the mind, the correlation between brain anatomy and mental functioning, the genetic basis of mental disorder, the significance of biochemical changes associated with mental disorder, and the implications of life events.

SEE ALSO

- GENETIC CHARACTERISTICS p. 28
- TOWARDS A SCIENCE OF THE MIND p. 162
- PERSONALITY AND ITS EXTREMES p. 188
- ANXIETY AND THE NEUROSES p. 190
- PSYCHOSES p. 192
- TREATING MENTAL DISORDER p. 194

Various opinions are represented in the medical literature, and the best the layperson can do is to try to take a detached view and understand some reasonably secure principles. Not all people seen by psychiatrists are ill in the medical sense, many of them are simply distressed. This makes it difficult to give an overall definition of mental disorder, but clinically classified disorders come under three broad categories: organic psychiatric syndromes, neuroses and psychoses.

The organic psychiatric syndromes

Organic psychiatric syndromes are mental disorders that result from actual known pathological changes (disease) in the brain or from known brain malfunction caused by poisons or deprivation of essential materials. Any of them can be so severe as to amount to a psychosis (⇨ below). Their common feature is a state of confusion of varying degree, sometimes amounting to delirium. This is manifested by impairment of consciousness (from slight loss of alertness to deep coma), distortion of perception, slowing of thought processes, lack of spontaneous speech, agitation and restlessness, and memory disturbance. The disturbance of memory leads to disorientation in space and time, so that the affected person may readily become lost in previously familiar surroundings and may be unable to state the day, month or year.

Salvador Dali was well known for his flamboyance and eccentricity. Many would have considered his actions and paintings as indications of 'madness'. There is a very thin line between mental disorder and eccentricity and the final definition is often determined by the environment and the community that makes the judgement. (Sygma/J. Bryson).

There are as many causes of the organic syndromes as there are factors that can interfere with the normal functioning of the brain. They include head injury, a drop in the blood supply to the brain from arterial disease, cerebral thrombosis, cerebral haemorrhage, heart failure, brain compression from bleeding within the skull, brain tumours, brain infections (including AIDS), brain abscesses, alcohol and other toxic influences, carbon monoxide poisoning (which reduces the oxygen-carrying capacity of the blood), vitamin B deficiency, hormonal (endocrine) imbalances, and conditions such as kidney and liver failure which result in accumulation of toxic metabolic products in the blood.

Some brain infections, however, cause more specific mental changes. Before the widespread use of antibiotics (before about 1950), untreated tertiary syphilis (the third and final stage) commonly led to the condition known as 'general paralysis of the insane'. This featured progressive dementia with severe tremor, delusions of grandeur, irresponsible behaviour and usually death within a few years.

Neurosis and psychosis

The distinction between neurosis and psychosis is fundamental to an understanding of the current classification of psychiatric conditions. People suffering from neurotic disorders retain a normal perception of external reality; people with psychotic disorders do not. Neurotic people are aware that something is wrong with them; psychotic people, in general, are not. In neurotic people wide areas of perfectly normal mental function remain; psychotic people have more global mental disturbance. The problems of neurotic people can be considered as extremes of common or normal difficulties; in the case of psychotic disorders there is a clear distinction between what is experienced by normal people and the experience of those affected.

People with psychotic disorders lose logical coherence and have no insight into the disorder, which is manifested in hallucinations or delusions, or both. Hallucinations are false perceptions, while delusions are false beliefs. Hallucinations may involve any of the five senses, the commonest being hearing. 'Seeing things' is rare except in delirium tremens, a condition caused by withdrawal from heavy alcohol or drug use. Delusions are most commonly paranoid – false convictions of persecution or of the hostile intention of others. They are nearly always self-directed.

What is Normality?

In the context of mental health, no absolutely clear distinction can be made between what we call normal and abnormal. How can we distinguish what is normal from what is mere eccentricity, or eccentricity from the pathological? This difficulty is especially great in the fields of neurosis and the personality disorders. None of us can be said to be entirely free from neurotic traits – a touch of obsession, the odd phobia, a tendency to become over-anxious or to be easily depressed. These are so common as almost to define normality.

Perhaps the most useful criterion in distinguishing normality from abnormality is the extent to which the neurotic trait interferes with the usual conduct of life. In the case of the personality disorders a useful criterion might be the extent to which the disorder interferes with the lives of others. It is not clear, either, whether we can be sure of the distinction in the case of the psychoses. Psychiatrists rely heavily on evidence of hallucinations and delusions in making a diagnosis of psychosis. But both of these are subjective and cannot be directly recognized by anyone other than the person who claims to experience them. They must therefore be taken on trust by the doctor, and are susceptible to misinterpretation.

Some patterns of what we call psychotic behaviour suggest that the affected person has chosen deliberately to reject the conventional paradigm that society or the family would impose, and has retreated into a 'private world' with its own values. Conversely, 'normality' has sometimes been defined by the State in political terms, which the conscientious have been unable to accept. In many of these cases, dissidence has been equated with madness, and used to justify 'psychiatric' incarceration.

The neuroses (⇨ p. 190) all feature varying degrees of anxiety. Some are manifested by apparently pure anxiety, but certain others can be distinguished by specific features – phobias, depression, somatoform disorders (psychological upsets manifesting as physical symptoms) and obsessive-compulsive behaviour. Anxiety is very unpleasant and is associated with a variety of symptoms such as headache, difficulty in breathing, tightness in the chest, raised pulse rate, muscle ache, dizziness, weakness, difficulty in swallowing, frequency of urination, and so on. There are also psychological features such as loss of concentration, apparent loss of memory, and, in severe cases, fear of death and fear of madness.

Although these symptoms are always associated with, and may even be caused by, known hormonal changes, few doctors believe that the neuroses are organic syndromes. This is partly because no organic brain changes have ever been shown unequivocally to be the cause of any neurosis. The general consensus seems to be that they are of psychological origin and are the result of life experience, especially early influences.

With the exception of the organic syndromes, the psychoses – schizophrenia, manic-depressive psychosis and the paranoid states (⇨ p. 192) – also seem to be unassociated with any demonstrable structural or other changes in the brain. If there are changes, medicine is not yet advanced enough to detect their subtleties, even by the most searching electron microscopic study of brain cells. If the psychoses are to be explained on an organic basis such an explanation will almost certainly be at a biochemical level. Some doctors, however, believe that these psychoses are also the result of earlier life experience.

The importance of heredity in determining the psychoses is difficult to assess. Family patterns of schizophrenia and the result of twin studies do, however, suggest that genetic factors are involved. The current view seems to be that the occurrence of a psychotic disorder is the result of environmental factors operating on a genetically induced tendency. A person may inherit the tendency but may be fortunate enough to avoid the external stress or other factors that precipitate the disorder. Traumas such as divorce bereavement, loss of money, or serious injury might precipitate a psychiatric breakdown.

Neuroses and psychoses have psychological features only. Unlike organic psychiatric disorders they cannot be diagnosed with the use of blood tests, brain scans or x-rays. Diagnosis in these circumstances is more of an interpretative art than a science and it is hardly surprising that psychiatrists often disagree.

It takes time for a psychiatrist to get a proper picture of a patient – their personality, life history and current disturbance. This means that the diagnosis can change as more information comes in. For example, a depressed patient may be so convincing about the reason for his depression (his wife, whom he adores, is divorcing him) that the psychiatrist classifies him as a normal person experiencing abnormal stress. When the relatives arrive and explain that his wife is only leaving him after years of broken promises, hare-brained schemes and failure to face reality, it is apparent that the patient is more properly classed as having a personality disorder (⇨ below). Finally, after he has been on the ward for a few days, the patient confesses to hearing voices telling him that he is God, and the diagnosis has to be changed, yet again, to psychosis.

The classification of personality

Human personality is a very variable quality and the distinction between what is normal and abnormal is hard to make. Some people are outgoing (extroverts), some inward-looking (introverts) but few show either pattern to an extreme degree. Some react in either way, depending on circumstances. Many psychologists claim that we can all be fitted into one or other of the following groups:

Schizoid (aloof, withdrawn, intellectual, austere, defensive).

Obsessional (controlled, orderly, meticulous, rigid, intolerant).

Hysterical (emotionally changeable, dramatic, exaggerated, manipulative).

Sociopathic (conscienceless, unfeeling, unconcerned).

Any of these, taken to extremes, is said to constitute a *personality disorder* (⇨ p. 188). The possession of such a personality can hardly be said to be an illness; it is a life-long pattern dating from childhood or, at the very latest, adolescence. Treatment, in the form of counselling, is seldom successful.

A woman suffering from dementia. The symptoms of dementia range from mild confusion to marked senility. There can also be a personality change for the worse. In most cases dementia is caused by serious loss of functioning brain tissue. About half of the cases of dementia in people over 60 years old are due to Alzheimer's Disease. (SPL/Oscar Burriel)

DEMENTIA

Dementia simply means 'loss of mind'. It can be caused by any of the causes of the organic syndrome (⇨ text), but there are some additional causes. Some common diseases of dementia are listed below.

ALZHEIMER'S DISEASE is the commonest reason for dementia in people over the age of about 60 years old. The cause remains unclear although we do know that there are specific brain changes, such as tangles of fibres within the nerve cells, and severe loss of nerve tissue.

SENILE DEMENTIA is caused in most cases by progressive loss of brain function from multiple areas of brain death. This is a result of reduced blood flow to the brain.

CREUTZFELDT-JAKOB DISEASE is an infectious disease caused by, or associated with, an agent simpler even than any virus, known as a prion protein. This 'organism' can reproduce, resist normal sterilization methods and can be transmitted during surgery.

HUNTINGTON'S CHOREA is a genetic condition with dominant inheritance (⇨ p. 28). Sufferers alternate between excitement and depression.

PICK'S DISEASE is a rare genetic disorder causing atrophy of the frontal and temporal lobes of the brain and mainly featuring loss of speech function.

PARKINSON'S DISEASE is a disease of the central nervous system. Some 40% of sufferers develop dementia, and this often starts with hallucinations.

REPEATED BRAIN TRAUMA FROM BOXING is associated with multiple small bleeds throughout the brain that gradually destroy function.

True dementia is invariably associated with serious loss of functioning brain tissue and has most of the features of the organic brain syndrome, especially memory loss and progressive slowness of thinking, but no clouding of consciousness. In addition, there is often a marked personality change for the worse. The demented person becomes rude, aggressive, tactless, emotionally changeable, insensitive to the feelings of others, dishonest, dirty, unkempt and often unconcernedly incontinent.

Certain reversible organic influences such as hypothyroidism or severe anaemia, and some forms of depression and schizophrenia, can cause a pseudo-dementia that can be corrected. Doctors recognize the importance of not assuming that a person's symptoms are those of dementia simply because that person is old.

Boxing can cause Repeated Brain Trauma. It is associated with multiple small bleeds throughout the brain that gradually destroy it. The mental disorder is therefore organic in that it is caused by brain malfunction. (Allsport)

Personality and its Extremes

Many mental disorders arise as a result of the effect of environmental influences on the human being. Some people have powerful, resilient personalities capable of rising above, and overcoming, the most appalling blows of fate; others have vulnerable personalities that can succumb to what most would consider quite minor stresses.

Baron Münchausen was the subject of a series of exaggerated adventure tales written by R. E. Raspe in the 19th century. His name has come to mean a kind of tall-story teller. The Münchausen syndrome was named after him as it refers to a mental disorder in which the patient deliberately simulates disease to obtain attention. People suffering from the syndrome may study medical textbooks to report a plausible list of symptoms to doctors. They often prefer to undergo surgical treatment and some sufferers will have multiple operation scars. (AR)

COMMON DEFENCE MECHANISMS

During a lifetime there may be many anxious situations that must be overcome. None of us remains for long without being aware of our liability to risk of some kind. Most of these threats – whether physical, emotional, professional, social or financial – are minor and can be coped with easily by the employment of one or other of a range of avoidance strategies called *defence mechanisms*. On some occasions, such as the threat of fatal illness, financial ruin or social disgrace, the defence mechanisms are very important to us as a means of keeping anxiety within reasonable bounds. In such cases the avoidance reaction may cause behaviour that seems surprising unless we are aware of its origin. Some common defence mechanisms are listed below.

RATIONALIZATION

This is one of the commonest and most widely used defence mechanisms. The method is to find a plausible reason for an action or omission so that the true emotional reason is effectively concealed. A man may insist that he simply cannot find time to consult a doctor when, in fact, he is terrified of discovering that he has cancer.

DENIAL

When employing the denial mechanism a person will pretend that the problem does not exist, typified by an attitude of cheerfulness in the face of adversity or danger. Denial of profound emotion is one of the ways of coping with grief.

REPRESSION

Repression is not quite the same as denial, since in repression, something that causes us unpleasant or painful feelings is simply forgotten. This is an active protective process to spare us continual humiliation, regret, pain or discomfiture.

PROJECTION

The painful sense of a personal defect is projected onto another person, or group of people. People may be astonished to be accused, by friends, of faults they are sure they do not possess but with which they are thoroughly familiar in the accuser. The dishonest may, for instance, accuse others of stealing and cheating.

SUBSTITUTION

Emotion can be so strong that it must have an outlet. When the logical outlet – against the cause – is impossible or unwise, the emotion may be directed against something, or someone, else. Anger against fate may be vented on an innocent person. Anger against the boss may be directed against the spouse.

SPLITTING

Some people cope with anxiety by splitting the world into the good and the bad, identifying strongly with, and hoping for the support of, the good group, and blaming the bad group for everything. This may cause racism or religious bigotry.

DISSOCIATION OR CONVERSION

Strong emotion, especially fear, is disposed of by its conversion into a physical symptom. A good example is the man who will 'lose his voice' if asked to speak in public.

SUBLIMATION

Emotional needs that cannot for some reason be gratified in the most direct and obvious way may be satisfied by devoting oneself to some other purpose. Unfulfilled sexual needs, for instance, may be sublimated into exercise and sporting activities.

The term 'personality', as used in psychology, embraces a wide range of human qualities. It is the total of all those mental characteristics that contribute to create a unique human being. It includes intellectual and educational attributes, the emotional disposition and the behavioural tendencies, and incorporates both character and temperament.

Personality types and vulnerability

Many attempts have been made to classify personality into groups in accordance with behaviour traits, but these have not been particularly successful (⇨ p. 200). This is mainly because the great majority of people fall into an indefinable group in which none of the extremes of personality type is particularly marked. There are, however, some who will be generally agreed to be of the outgoing extrovert type, as defined by Jung (⇨ p. 164). Others will be generally recognized as being of the inward-looking introvert type. Many tend slightly in one or other of these directions; most, however, cannot be realistically classified in this way at all. Rarely do we come across people who manifest one or other of these patterns to an extreme degree.

In addition, a small number of people can be fairly categorized as having obsessional, hysterical or sociopathic personalities. The obsessional person is overconscientious, meticulous, orderly and tidy, rigid in habit and hates disruption of the established patterns of behaviour. The hysterical person is demanding, histrionic, attention-seeking, manipulative, emotionally volatile and exaggerates consistently. The sociopath (also called 'psychopath') is conscienceless, overtly selfish, and behaves with no regard for the rights of others.

In general, people with obsessional, hysterical or sociopathic personalities are mentally more vulnerable than average. Psychiatric disorder commonly takes a form that seems to be determined by the personality (the 'premorbid personality'). The obsessional may develop a severe obsessional neurosis, the hysteric a somatoform disorder (an apparently medical condition, such as an inability to move a limb) and the sociopath may display antisocial behaviour, drinking, stealing and fighting (⇨ box).

When personality types are taken to the extreme they become disabling or socially disruptive and must then be regarded as

Dennis Nilsen murdered 15 young men in London between 1978 and 1983. During Nilsen's trial lawyers argued unsuccessfully that he suffered from a personality disorder that stemmed from an unhappy childhood. (Rex)

disorders. None of these disorders is, however, an actual psychotic illness. There is, for instance, a clear distinction between a person with a *schizoid personality disorder* (⇨ box) and a person with schizophrenia (⇨ p. 192).

The determinants of personality

Personality is the result of the interaction between inherited characteristics and an individual's experience of the environment, especially in early childhood. Heredity on its own is now often dismissed by psychologists as relatively unimportant, but studies on twins have shown how strong genetic influences can be. Identical twins brought up apart retain close similarities in personality. Even allowing for coincidence, some cases of separated identical twins demonstrate this dramatically. One paper on the subject, for instance, describes a pair of American male twins who were separated at the age of five weeks and did not meet again for 39 years. When they compared notes they found that both had married and divorced women called Linda, and had remarried women called Betty. Both had sons called James Allen and dogs called Toy. Both drove Chevrolet cars, had served as sheriffs' deputies, had a white bench around a tree in their garden, had similar hobbies and chain-smoked the same brand of cigarettes.

There is no argument that hereditary factors largely determine the bodily characteristics (⇨ p. 34), the pattern of physical and mental development, and, to a lesser extent, the way in which the endocrine (hormonal) system operates. All of these can affect the personality.

Environmental factors have an apparent and obvious effect on the personality. Most people accept that the qualities of the parents and other members of the family can mould and determine the personality of the growing child. Other important environmental factors include the wider social milieu, educational and cultural influences, life experience generally, nutritional standards, and major events such as serious illness. The family influences that are considered to play the most important role in creating healthy and stable personality are:

- Freely expressed love and affection.
- Unequivocally expressed standards of conduct.
- An unshaken sense of physical and emotional security.
- A consistent pattern of attitude and ethics by parents.
- Ample and rich stimulation of all the senses.
- Freedom to express individual responses within the constraints of the rules.

People with psychopathic personalities often – though not always – have had grossly deprived childhoods with little or no show of affection and a pattern of early conditioning that could hardly fail to induce a criminal outlook (⇨ p. 210). However, the hereditary element is perhaps too readily discounted in current sociology: the 'black sheep' of an otherwise happy and successful family is unlikely to be the result of environmental influences alone.

Cat in a Rainbow by Louis Wain (1860–1919). Wain was a notable painter of cats and his earlier paintings are sentimental and cute illustrations. However, after the onset of mental illness he produced increasingly bizarre paintings of the animals looking wild and startled. (BAL)

SEE ALSO

- THE PSYCHOANALYTIC MOVEMENT p. 164
- A HISTORY OF MENTAL ILLNESS p. 184
- WHAT ARE MENTAL DISORDERS? p. 186
- ANXIETY AND THE NEUROSES p. 190
- PSYCHOSES p. 192
- TREATING MENTAL DISORDER p. 194
- CAN PERSONALITY BE ASSESSED? p. 200

PERSONALITY DISORDERS

SCHIZOID PERSONALITY DISORDER

People suffering from a schizoid personality disorder may be severely disabled by being so shut in on themselves as to be socially isolated. However, they have full insight into their condition and do not suffer hallucinations or delusions. Such people are concerned, even preoccupied, with the things of the mind and with material things, almost to the exclusion of interest in other human beings. Although often highly creative, artistic and intelligent, such people may be aloof and withdrawn, cut off from emotional contacts, easily offended and defensive.

OBSESSIONAL PERSONALITY DISORDER

People with obsessional personality disorder can be seriously disabled. They may be unable to start work until they have arranged their desks in a particular way, may be unable to pass a colleague's telephone with a twisted cord without untwisting it; may become acutely uncomfortable and unable to listen, for instance, in a job interview if the interviewer's security badge is at an angle; and may be outraged by the untidiness or disorderliness of others. A degree of obsessiveness can be an asset in many occupations and can add to efficiency, but an obsessional personality disorder often gives rise to conflict between spouses and close associates, and may cause serious trouble in conditions of stress when urgent action may be needed. Such a disorder commonly overlaps with the obsessive-compulsive neurosis (⇨ p. 186) in which anxiety is dealt with by the establishment of rituals. The chief distinction between this personality disorder and an obsessional-compulsive neurosis is that, in the latter condition, the affected person is compelled to behave in a manner he or she recognizes to be irrational. Only about one third of the people who develop this neurosis have a previous obsessional personality.

HYSTERICAL PERSONALITY DISORDER

People with a hysterical personality disorder express their feelings in exaggerated terms. They may be highly manipulative, and may use emotion and drama to achieve their aims. Anger is a prominent feature. Relationships tend to be superficial and usually brief as these people find it very difficult to form close attachments. They commonly complain of illness. Pain is always 'excruciating'; headaches are always 'migraine'; they do not vomit but are 'violently sick'. Some people with hysterical personality disorder simulate apparently serious illnesses by deliberate deception – putting sugar in a urine sample or damaging the skin to produce an unusual rash. Such 'factitious' illness is an attempt to arouse interest and concern. Extreme cases may amount to the *Münchausen syndrome* in which convincing but fabricated case histories are given, usually reproduced from medical textbooks, so that surgery is performed.

SOCIOPATHIC OR PSYCHOPATHIC PERSONALITY DISORDER

The sociopathic or psychopathic personality disorder manifests itself by repeated acts of criminal behaviour, usually dating back to childhood. The people concerned act as if they are quite unaware of the difference between accepted standards of right and wrong and appear indifferent to the effect their actions have on others. There is usually a rationalization (⇨ box) of the conduct but this is superficial and seldom convincing. There is complete insight into the motives for conduct and no question of delusions or hallucinations. Psychopaths are fully aware of the possible or probable consequences of their actions and may be persuaded to amend behaviour, but only to avoid punishment.

Anxiety and the Neuroses

Anxiety is commonplace and has value in increasing our levels of alertness and effort, and preparing us to meet threats of various kinds. In normal people anxiety occurs only when a threat is perceived, and is proportionate to the size of the threat. Neurotic anxiety, however, may occur not only in the absence of a rational stimulus, but is also so severe in relation to the cause that it becomes socially disabling.

Irrational and excessive anxiety is obvious to the sufferer as well as to the outside observer. Such awareness by the sufferer is called *insight*, and the fact that sufferers have insight into their condition is one of the main distinguishing features between the neuroses and psychotic mental disturbance (➪ p. 192).

Anxiety is probably the most common symptom of mental and physical disorder.

Royal Marines during World War I. The appalling conditions experienced by soldiers during the war sometimes led to fugue, manifested by wandering away from the battle scene and amnesia of their previous lives. Soldiers suffering from the disorder were often executed for desertion. (TPS)

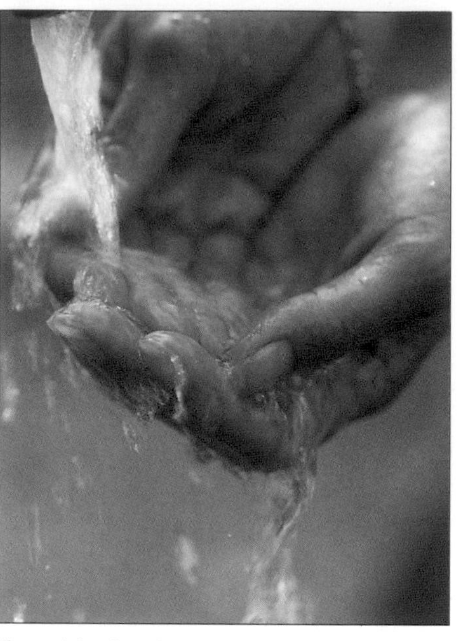

Repetitive handwashing is one of the common manifestations of obsessive-compulsive disorder. A sufferer's compulsion to wash his or her hands is often precipitated by an attack of anxiety. The sufferer is aware that the action is illogical and unnecessary but feels compelled to do it.

(Gamma/Alain Le Bot)

It is a major feature of depressive illness (➪ below), obsessional states and phobic disorders (➪ box). It occurs in dementia (➪ p. 186) and in the toxic confusional states (➪ p. 188) and may also occur in schizophrenia (➪ p. 192). In all these cases, however, it is a secondary reaction to something else. But in many cases of neurotic anxiety it seems to be a disorder in its own right. It has many manifestations – generalized anxiety, phobias, depressions, loss of memory (amnesia), seemingly aimless wandering (*fugues*) and physical symptoms of all kinds, some of them highly dramatic.

Fears and phobias

A *phobia* is an irrational dread of something which does not normally evoke such a reaction. Many different things or situations may evoke the fear, but these tend to fall into well-defined groups (➪ box).

Dissociative reactions

Another possible response to anxiety is the *dissociative reaction*. Here, a person faced with a situation that he or she finds intolerable reacts in such a way that the situation can be ignored. This is often an understandable reaction, and it is arguable whether it should be classified as a disorder. For many people, life is so unpleasant that in many ways it is surprising that dissociative reaction is not more common. Dissociative disorders may take various forms, including amnesia, fugue, depersonalization and multiple personality.

Psychogenic amnesia simply means selective forgetting. Much continues to be remembered and the ability to memorize new material is not affected. It is a defence mechanism (➪ p. 188) against emotional conflict or severe distress. The death of a baby may, for instance, be forgotten. The onset is usually sudden, and the affected person is usually aware that memory loss has occurred. Some seem concerned, but many appear indifferent.

Fugue is an even more extensive reaction to unpleasantness. The affected person wanders away from home and takes on a new identity, often with a new occupation. The underlying cause of fugue is a desire to withdraw from an unacceptable or impossibly painful situation. Whether the response is the result of a deliberate conscious decision or is a wholly unconscious process can probably never be known. The wandering is purposeful and usually lasts for a few days. There is amnesia for the previous life and its associations, and apparently no awareness that the memory has been lost. The affected person usually takes up a quiet, secluded occupation and seems anxious to avoid drawing attention. The distinction between fugue and malingering is therefore hard to make. In the past soldiers suffering fugues in the course of war service have been executed for desertion.

Depersonalization is loss of the reality of one's own existence. The affected person feels detached or estranged from his or her body as if in a dream. The experience is common and does not necessarily imply mental disorder. About 70% of people have had the experience at some time. It can also be caused by organic brain disease, such as tumour, abscess or infection. Neurotic depersonalization, which is rare after the age of 40, may follow an acute anxiety attack. It is frequently associated with anxiety or depression. Another dissociative reaction is the deliberate simulation of psychiatric symptoms (*Ganser's syndrome*). Trance states, daytime somnambulism, and other bizarre behavioural patterns also sometimes occur as dissociative reactions.

Occasionally, a person's response to the intolerable is to develop what appear to be two or more distinct personalities (*multiple personality disorder*), each usually with no knowledge of the other. This reaction is now thought to be due mainly to physical, sexual or psychological abuse in childhood. The affected person switches suddenly and unexpectedly from one personality to the other. In fully established cases, each personality has a normal and quite distinct set of behaviour patterns and a set of relationships with individuals that are distinct from, and unknown to, the friends of the other personality. The different personalities may be of opposite sex, or even of different races. They will usually behave in very different ways. One may be prudish, the other sexually promiscuous, one extrovert, the other introvert.

Many of the cases of multiple personality reported, especially in the 19th century, however, are now believed not to be genuine. On close examination, some were obviously simulated by patients anxious to please doctors known to be fascinated by the condition. Current views on the nature of personality and on

Common Phobias

A person experiencing a phobic reaction shows all the signs of genuine, intense fear. The degree of fear is usually high and quite inappropriate to the level of danger, if any. There may be a facial expression of horror, the hair may stand on end, there may be sweating, trembling, flushing, a fast pulse, palpitations, spontaneous deep breathing, and so on. The affected person usually feels as if he or she is choking and may feel faint, dizzy or nauseated. Affected persons tend to keep their particular phobia constantly in mind and are always on the lookout for the precipitating cause so that it can be avoided. This can be very disabling.

SOCIAL PHOBIAS include a fear of humiliation or embarrassment about performing any particular activity in public, such as eating, speaking, reading or writing. The particular activity can usually be performed comfortably in private.

AGORAPHOBIA is a fear of going out alone into open or crowded places. This is one of the commonest phobias.

ANIMAL PHOBIAS usually arise from an unpleasant incident in childhood and commonly relate to insects or reptiles, but sometimes to domestic animals.

DISEASE PHOBIA usually relates to cancer, heart disease or sexually transmitted diseases.

SPECIFIC PHOBIAS can relate to anything and include abnormal fears of heights, darkness, thunder, the sea, and so on.

the relationship of brain to mind suggest that it is unlikely that one body can genuinely be occupied by more than one personality. It seems likely that, at least in a proportion of cases, the condition is deliberately assumed.

Neurotic depression

Like anxiety, depression is a normal part of life. Sadness affects us all from time to time and is to be expected. The normal response to loss of any kind is mourning, and the greater the loss the greater the sadness. Bereavement generally implies the loss of a person by death, but the same response occurs when there is a loss of anything that is valued. Depression caused in this way is said to be *exogenous*, that is, caused by some external agency. Doctors used to compare this with 'endogenous' depression, as if to imply that in pathological depression the cause was internal. This usage has fallen somewhat out of favour.

Neurotic depression, however, implies sadness that is not caused by any obvious external event or circumstance, or is manifestly out of proportion to such an event. Such depression usually features inappropriate patterns of thought and behaviour. There is anxiety and often obsessive behaviour, hypochondriasis (abnormal and irrational anxiety about one's own health) and psychogenic physical symptoms (symptoms of mental rather than physical origin). Neurotic depression is distinguished from the 'bipolar' mood disorder, the manic-depressive psychosis (⇨ p. 192).

All forms of depression, however, affect people in much the same way. The whole personality and mind, as well as the body, are affected. There is often retardation of thought and movement, apparent tiredness, loss of appetite and weight, sometimes rigidity and immobility of the face, and there may be obvious signs of distress with agitation, restlessness and wringing of the hands. Depressed people have headaches, fluttering in the pit of the stomach, tightness in the chest, palpitations, giddiness, dryness of the mouth, breathlessness and often undue frequency of urination. There is loss of sexual interest and, in women, menstruation often stops.

Clinically depressed people feel unworthy, fit for nothing, hopeless, despairing and lacking in interest. Since most people spend their lives in a quest for happiness, depression must be deemed one of the most serious of human disorders. It is also surprisingly common. Well over one third of all psychiatric consultations are concerned with depression. Fortunately, most cases now respond well to treatment (⇨ p. 194).

Conversion disorder

One of the most dramatic forms of neurosis is *conversion* or *somatization disorder*, commonly known as hysteria. In this condition the mental problem is converted into a physical symptom or sign. The popular concept of hysteria – screaming and laughing and throwing limbs about in a histrionic manner – has little, in fact, to do with the condition. Broadly, hysteria features psychogenic disorders (which are of mental, rather than physical origin) that have specific motives behind them. The symptoms produced may purport to be those of any disease – especially a neurological disorder – but often, for lack of detailed medical knowledge, correspond to no known pathology. Conversion disorder is distinguished from malingering on the grounds that the symptoms seem to be genuine. Conversion disorder may manifest itself as paralysis or severe weakness of limbs or other parts, abnormalities of gait, blindness, deafness, loss of taste or smell, loss of the sense of touch or insensitivity to pain.

Sensory hysteria corresponds to the patient's idea of sensory nerve distribution, rather than to actuality. Common is the 'glove and stocking' sensory loss in the hands and feet, that stops at the wrists and ankles. No nerve disorder can cause such distribution of sensory impairment. The conversion disorder can often be 'explained' on the basis of the conscious or unconscious desires of the affected person. Thus, for example, a person who wants to be an opera singer but is forced to work as a typist might develop paralysis of the fingers.

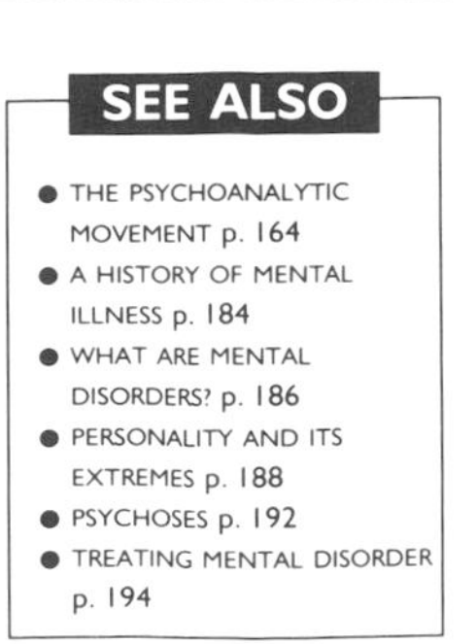

SEE ALSO

- THE PSYCHOANALYTIC MOVEMENT p. 164
- A HISTORY OF MENTAL ILLNESS p. 184
- WHAT ARE MENTAL DISORDERS? p. 186
- PERSONALITY AND ITS EXTREMES p. 188
- PSYCHOSES p. 192
- TREATING MENTAL DISORDER p. 194

Acrophobia, a fear of high places, is one of the most common phobias, which are irrational or excessive fears of particular objects or situations. When elevated to any height an acrophobic will experience intense anxiety and even a panic attack. (Zefa)

Psychoses

All over the world, about one person in every hundred develops a psychotic disorder at some time in his or her life. In the great majority of cases the disorder is schizophrenia. A small proportion suffer from severe mood disorders (manic–depressive psychosis) or a paranoid state. Unless controlled by effective treatment, psychotic disorders usually have a devastating effect on the lives of the sufferers and those close to them.

SEE ALSO

- THE PSYCHOANALYTIC MOVEMENT p. 164
- A HISTORY OF MENTAL ILLNESS p. 184
- PERSONALITY AND ITS EXTREMES p. 188
- TREATING MENTAL DISORDER p. 194

A psychotic disorder – usually equated in the lay mind with 'madness' – is fundamentally different from a neurotic disorder in that affected people are unaware that they are ill. They have no *insight* into the fact that the events they are experiencing are the product of a disordered mind and lack any basis in reality. The majority of psychotic people – unlike sufferers of the neuroses – have *hallucinations* or *delusions* or both. Hallucinations are sensory perceptions without the normal external stimulus such as hearing voices when no one is speaking. They may be of sight (visual), of hearing (auditory), of smell (olfactory), of taste (gustatory) or of touch (haptic or tactile). With the exception of auditory hallucinations, however, all of these are rare in psychotic disorders. (Hallucinations should be distinguished from illusions, which are misinterpretations of normal sensory stimuli.) Psychotic delusions are beliefs that cannot be sustained by reason or that are persisted in even after they have been shown to be absurd.

In *One Flew Over the Cuckoo's Nest* Jack Nicholson plays an inmate of a psychiatric hospital. He is committed to the institution after his criminal behaviour is attributed to mental disorder. However, the film questions the necessity of keeping people with mental disorder under strict supervision and the value of the treatment in such institutions. (Kobal)

The Fairy Feller's Master Stroke by Richard Dadd (1819–87). As a promising young artist Dadd suddenly and inexplicably killed his father. The murder became a famous case in Victorian England. He was probably schizophrenic, and spent the rest of his life in a mental institution. Even those paintings completed before he was committed – such as this one – were bizarre and eccentric. (BAL)

Schizophrenia

The original term for the condition now known as *schizophrenia* was *dementia praecox*. This term has rightly been abandoned because schizophrenia is in no sense a dementia (➪ p. 186). It does, however, as the root *praecox* (premature) implies, nearly always start either in the late teens or early in adult life. It is questionable whether the current term schizophrenia, suggested in 1908 by the Swiss psychologist Eugen Bleuler (1857–1939), is entirely satisfactory either. Literally, it means 'split mind', and many lay people equate the condition with the dissociative disorder, multiple personality (➪ p. 190). This has nothing to do with schizophrenia. Bleuler put the term forward because the disorder seemed to involve a splitting or cleavage between feelings or emotions (affect) and thinking (cognition). Schizophrenia may take one of several forms and is sometimes difficult to classify. Indeed, some psychiatrists believe that it is, in fact, a group of different but related conditions.

Simple schizophrenia has been called the 'skid row' disease because a high proportion of the people we see sleeping on our streets are simple schizophrenics. Many of these are never diagnosed and never treated. Some spend a period under medical care and are then released, only to relapse. The schizophrenic pattern is often slow to develop. A young person begins to lose interest in social relationships, and then in other matters, gradually becoming emotionally indifferent and seemingly irresponsible, careless and lacking in motivation and ambition. He or she is secretive and the ideas become progressively more bizarre. Any career is abandoned, and casual, unskilled jobs are taken when possible, as the need for money becomes pressing. Some will resort to begging for a living. Family and friends are neglected and then forsaken.

Many simple schizophrenics become vagrants, recluses or solitary eccentrics. Female schizophrenics typically drift in and out of prostitution. Schizophrenics of both sexes drink when they can afford it as alcohol relieves their distress in the short term, but makes the symptoms worse in the long run. The disorder affects both mental processes and social functioning, and it has several characteristic features, including:

- Auditory hallucinations in which voices are heard that comment on actions or may even issue orders.
- Hallucinations of voices discussing the sufferer and referring to him or her in the third person.
- Hallucinations of strange bodily sensations, such as electric shocks.
- Delusions in which a normally indifferent stimulus acquires special significance. For instance, the sounding of a car alarm is interpreted as a message to the people watching the sufferer that he or she is approaching.
- The sense of being a passive tool of someone else or of an alien force.
- The conviction that the sufferer's privacy is being breached and that thoughts and actions are being imposed from outside.
- The conviction that the sufferer's thoughts are being interfered with.

In addition, the affected person becomes withdrawn, shut in, isolated and cut off from society. Former friends are avoided and the sufferer spends hours sitting alone.

Not all schizophrenics withdraw in this way, however. Another major class of the disorder is *paranoid* schizophrenia, in which social functioning may be much less affected. In this variant the onset is usually later – between 30 and 40 years of age – and the personality is much less severely affected. The delusional content, however, is strongly paranoid (⇨ below).

Catatonic schizophrenia features periods in which the affected person remains frozen, statue-like, for hours on end. Sometimes the limbs can be passively moved into a new position that will be retained – the *flexibilitas cerea* or wax-like flexibility phenomenon. In this state, there is no response to external stimuli, not even to painful ones. The inertia is, in fact, misleading as catatonic schizophrenics are often experiencing vivid hallucinations. Alternatively, the state of catatonic schizophrenia may have phases in which certain actions are mechanically repeated (*echopraxia*) or words or phrases spoken over and over again (*echolalia*). Catatonia may also pass into a state of frenzied excitement in which violent activity occurs with hurried pacing, a characteristic protrusion of the lips (*schnauzkrampf*), shouting of obscenities, open masturbation or even homicidal or suicidal violence.

Some 15% of simple schizophrenics remain severely disabled all their lives, but in about one third of cases there is only a single episode with full recovery. This outcome is more likely if the previous personality was normal, if the disorder came on suddenly as a result of some obvious precipitating factor such as a bereavement, a change of school, or the use of a hallucinogenic drug, and if it featured conspicuous symptoms and marked change of mood. In just over half of all cases the disorder shows periods of relative normality alternating with relapses. In most of these cases relapses can be prevented by drug treatment (⇨ p. 194) and by family counselling to relieve family tensions.

Paranoia

The current clinical definition of paranoia usually implies a feeling or conviction of persecution. It is a possible feature of other psychoses, such as schizophrenia and the manic-depressive disorder, but may occur as a psychosis in its own right. In all cases, the principal features are the same. Paranoia is also found in the organic brain disorders (⇨ p. 186), especially alcoholism, hallucinatory drug abuse and in solitary elderly people, and it may be a feature of people with a naturally suspicious personality.

Common to most forms of paranoia are sustained delusions that one is being attacked in some way. These delusions are not bizarre and refer to things that might possibly happen, such as being spied upon, poisoned, followed, deceived, deliberately infected, and so on. People with a pure paranoid psychosis are not likely to be convinced that aliens from a UFO are beaming thought rays at them. More probably, they will be convinced that a conspiracy involving everyone in their village has been formed against them, and may cause chaos by sending out anonymous and malicious poison pen letters. The paranoid delusion is systematized – if we were to accept that the delusion was true, the behaviour of the sufferer would be seen to be entirely logical and reasonable. Hallucinations are rare and when they do occur they are not prominent.

In addition to delusions of persecution, a paranoid person often experiences other delusions, often with a sexual content. These include:

- Erotomania – the conviction that some person of high social or other status is in love with the sufferer.
- Morbid jealousy arising from delusions of infidelity by a sexual partner.
- The conviction that the sufferer has a physical defect or disease.
- The conviction that others secretly think the sufferer is homosexual.
- Delusions of grandeur involving a conviction of power, wealth, authority, knowledge or a special relationship to some famous person.

The manic–depressive psychosis

Manic–depressive psychosis is known as a bipolar disorder because the affected person is at one stage profoundly depressed and at another in a state of dangerous elation. The episodes may be months or years apart, and seldom alternate rapidly. Full development of the depressive phase involves low self-esteem, marked retardation of thought and action, and a major risk of suicide. To avoid suicide urgent hospitalization is needed. The manic phase may be less readily recognized at first. The affected person may appear lively, energetic, dynamic, socially brilliant and highly attractive. But it may soon become apparent that the ideas are inconsistent and disoriented, disrespectful of others, seriously indiscreet and unconcerned with truth or reason. The manic person tends to dominate every gathering and talk over others, ignore people's social rights, use obscene language, and make universal sexual advances. In the manic phase, the affected person may go for days without sleep, abusing alcohol or drugs and ending up in a state of exhaustion. Religious mania, with a determination to convert the multitude to some faith or other, is common.

Acute mania may be dangerous both to the sufferer and to others. There is severe irritability and the affected person is easily provoked into violence which may be homicidal. There are wild flights of ideas with incoherent thought, often with an erotic or religious content. *Delirious* mania is the most pronounced stage of the manic phase. The affected person is literally 'raving mad'. Clothes are torn off, furniture is smashed, faeces are smeared on the walls and person, and the sufferer shouts incomprehensibly. Nothing short of an injection of paraldehyde (a powerful sedative) given into whatever part of the body can be accessed will suffice to restore calm.

Insanity as a Legal Defence

In spite of the popular view to the contrary, in Britain the term 'insanity' has no medical or psychological meaning. It is exclusively a legal term meaning that a person is not responsible for his or her actions. In criminal law a person is presumed sane and responsible. Individuals can, however, bring evidence to rebut this presumption and may escape conviction for a crime if they can prove that they were insane at the time.

Legal insanity is defined by the McNaughten rules. These were formulated by judges after the trial in 1843 of Daniel McNaughten, who intended to kill the Prime Minister, Sir Robert Peel, but killed his private secretary Edward Drummond by mistake. McNaughten was suffering under the delusion that the Government was persecuting him. The central point at issue, however, was not whether or not this was true but whether, at the time of the crime, McNaughten was responsible for his actions. The matter went to the House of Lords and the 'McNaughten Rules' were enshrined in law.

According to these rules, defendants must show that they are suffering from a 'defect of reason' arising from a 'disease of the mind', and that as a result of this they did not know what they were doing at the material time, or that, if they did know what they were doing, they did not know that it was wrong. A defendant suffering from an insane delusion is treated as if the delusion were true. If, for example, a man were to cut off another man's head under the insane belief that he was cutting through a loaf of bread, he could be acquitted of murder, since cutting a loaf of bread is a legal act. Similarly, he could be acquitted of murder if he had the insane delusion that he was acting in self-defence, as this is a defence. McNaughten was acquitted on the grounds of insanity, and, although the rules have been repeatedly criticized as inadequate, nothing better has yet been produced.

Homeless in London. Many schizophrenics become vagrants, using alcohol to quieten the voices that they hear. However, in the long run alcohol makes their condition worse. Paranoid outbursts take many of these sufferers to hospital or prison from time to time. (Gamma/Arkel)

Treating Mental Disorder

Until the middle of the 20th century psychiatry was limited to diagnosis, speculation on the causes of mental disorder, and largely unsuccessful attempts at treatment. Little could be done for people with severe psychoses other than restraint and heavy sedation. Severely neurotic people continued to suffer or spent years in supportive but ineffectual psychoanalysis. From about 1950, however, there have been remarkable developments in drug and behaviour therapy.

A group therapy session in progress. Group therapy is often used to treat people suffering emotional disorders that include dysfunctional relationships. A therapist will guide the discussion, which provides a forum for each member to demonstrate his or her particular traits. The aim of the therapy is to give patients insight into their disorder and hopefully modify their behaviour. (SPL/John Greim)

Although there has been no reduction in the incidence of mental disorder, much can now be done to alleviate or cure it. Rational medical treatment depends on an understanding of the causes of disease. Unfortunately, very little is known of the causes and mechanisms of mental disorder, and, until recently, treatment was largely based on 'trial-and-error'. In addition, natural variations in mental responses make it very difficult to be sure that improvement is, in fact, the result of the treatment. Many proposed treatments, hailed as 'breakthroughs' in the past, have sadly been shown to be nothing of the sort. There is no doubt, however, that there has been a genuine revolution in psychiatry following the introduction of effective psychotropic drugs (which affect the mind) in the 1950s.

The treatment of mentally ill patients varies from country to country. Even in the Western world not all people understand the needs of people who suffer from mental ill-health. (Gamma/Eric Bouvet)

Antipsychotic drugs

In 1952 reports began to appear in France of the results of treatment with a new drug known as *chlorpromazine* (Largactil). This drug was derived from the substance phenothiazine during the search for an improved way of destroying intestinal worms. Chlorpromazine was found to have a wide range of useful effects. It quietened the emotions, slowed down hyperactivity, controlled mania without seriously impairing consciousness, and acted in a remarkable way to eliminate abnormal experiences and behaviour in schizophrenia. It had no effect on depression but was useful in treating people whose depression was associated with agitation.

These results were so striking that in a very short time thousands of psychotic patients had been given the drug with excellent results. It is estimated that 100 000 000 patients worldwide have been treated with Largactil since 1953. A drug of such power can hardly be expected to be free from side effects. Chlorpromazine in large doses causes lethargy, and given over a long period it may cause repetitive involuntary movements of the limbs and sometimes of the lips, tongue and mouth. It may also cause an irresistible desire to move about (*akathisia*). It causes flushing and, in elderly people, may cause the body temperature to fall (*hypothermia*) with potentially dangerous consequences unless checked. Even so, the use of chlorpromazine and other associated drugs heralded a new era in psychiatry. Directly as a result of the use of such drugs, the number of people kept in mental institutions dropped markedly, and it proved possible to close many of the old psychiatric hospitals completely.

Antidepressants

The effective use of the phenothiazine derivatives led to intensive pharmacological research into the mode of action of these drugs and a search for drugs that would be useful in other forms of mental disorder. The first antidepressant drugs appeared in the late 1950s. There were two types, the tricyclics and the monoamine oxidase inhibitors.

The tricyclics are structurally related to chlorpromazine, and undoubtedly part of their effect is sedative rather than properly antidepressant. But after a time lag of two to three weeks, tricyclics have a mood-elevating effect that is sufficient to lift half to two thirds of people suffering from depression onto the road to recovery. It should be noted, however, that between one third and one half of people suffering from depression will do as well on dummy pills (the placebo effect). Nevertheless there is a genuine therapeutic element in the drug, and the widespread use of tricyclics has certainly led to a steady decline in the use of ECT (➪ below).

The antidepressant properties of the monoamine oxidase inhibitors (MAOIs) were discovered when it was noticed that certain isoniazid drugs being tested for the treatment of tuberculosis had a stimulant effect on the nervous system. In 1958 a report on the value of the drug iproniazid in cases of severely depressed people started a series of trials which showed an undoubted antidepressant action. But MAOIs have never been as widely used as tricyclics. People taking them have to avoid food rich in amines (cheeses, meat and yeast extracts, and red wines) as failure to do so can have serious consequences. The enzyme that oxidizes amines is inactivated by the drug, allowing them to accumulate and raise the blood pressure to levels that can cause crippling headaches and, in rare instances, a stroke. The risk has made doctors inclined to use tricyclics first and only resort to MAOIs if they fail.

Recently a new class of antidepressants has been developed, the 5HT reuptake inhibitors. They seem to be as effective as tricyclics, and, because they act quicker and lack many of the side effects associated with tricyclics (which include dry mouth, dizziness and, in the susceptible, urinary retention), their use as antidepressants has increased.

Lithium

In manic depressive disorder (➪ p. 192), anti-psychotic drugs are used to treat the manic episodes, and antidepressants are administered for depressive swings. Neither has any effect on the root cause, the instability of mood. Lithium carbonate, however, seems to be able to reduce the amplitude of the fluctuations. It is now routinely prescribed as a prophylactic (preventative drug) for patients with manic depressive illnesses. As in the case of anti-psychotics, the longer the patient stays on lithium, the better the result. To avoid potential danger the level of lithium in the blood has to be checked at regular intervals. Blood levels also have to be carefully monitored in order to avoid complications and side effects. However, some people still cannot tolerate lithium and have to be taken off it. Exceptionally strong manic or depressive surges may still require hospitalization, but, by and large, patients who go on to lithium find it easier to cope with mood swings and require shorter and less frequent spells in hospitals.

Non-drug treatments

Electro-convulsive therapy (ECT) is based on the observation that epileptic seizures can have an antidepressant effect. The essential element in ECT consists of a seizure artificially induced by passing an electric current across the patient's temples. Six to eight treatments are usually required to control severe depression. For many years the treatments have been given under general anaesthesia, so that the patient sees nothing of the procedure, and using a muscle relaxant, without which muscle tendons can tear and bones fracture. In the early days of ECT, in the 1940s, it was sometimes given without an anaesthetic and muscle relaxants were not then available. As a result the treatment became notorious: the convulsions were distressing to watch and bones were sometimes broken during seizures. In many cases too many treatments were given to unsuitable patients, who sometimes suffered memory impairment as a result.

ECT remains the treatment of last resort for severe depression, and many doctors still believe that it should be used, with due care, to treat sufferers who have failed to respond to antidepressants. The number of doctors and patients who believe in maintaining the treatment, however, is dwindling.

A treatment that has been almost entirely discontinued is leucotomy (prefrontal lobotomy), in which a surgeon divides the main tract connecting the frontal lobe with the rest of the brain (➪ p. 76). Popular in the 1940s, leucotomy was effective in reducing the anxiety felt by some severely ill psychotic patients, but had no effect on the psychosis itself. In many cases the operation was quite crudely carried out and unforeseen personality changes occurred, often rendering the patient excessively placid and tranquil. The operation was largely dropped when the new drug therapies were discovered.

Psychotherapy

Formal psychotherapy purported to treat mental disorder by examining the unconscious mind to find the psychological causes, and was widely practised, especially in the USA, until the 1960s. It is a time-consuming and unproven procedure and is no longer believed to have any part to play in the treatment of psychotic disorders (➪ p. 192). Many sufferers from the neuroses (➪ p. 190), however, find the supportive element in psychoanalysis helpful, especially if the analyst is regarded as authoritative. The particular school of analysis, or psychotherapy generally, seems to be less important than the personal qualities of the therapist. Most people cannot afford psychotherapy, so much of it is, by necessity, relatively brief. As such the technique merges into counselling and common sense.

Recently, psychologists have developed behavioural therapies based upon learning theory. These have been shown to have specific value in the treatment of a few kinds of neurotic disorder. The methods are based on the belief that neurotic conditions are the result of inappropriate conditioning or learning. Behaviour therapy has been found to be most valuable in the treatment of phobias (➪ p. 190). It is more successful in treating these states when they are linked to a specific stimulus. Even with specific phobias, however, the relief may be short-lived and the treatment may have to be repeated every few years.

SEE ALSO

- THE PSYCHOANALYTIC MOVEMENT p. 164
- PSYCHOLOGY TODAY p. 166
- A HISTORY OF MENTAL ILLNESS p. 184
- WHAT ARE MENTAL DISORDERS? p. 186
- PERSONALITY AND ITS EXTREMES p. 188
- ANXIETY AND THE NEUROSES p. 190
- PSYCHOSES p. 192

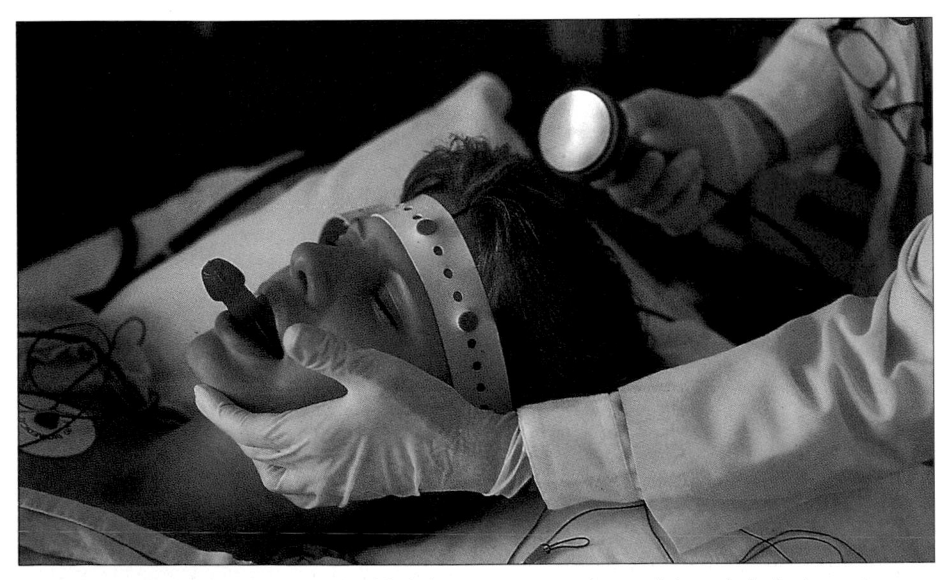

In electroconvulsive therapy a modified alternating current is passed through the brain causing a convulsion in the patient. It is used to treat severe cases of depression and occasionally schizophrenia and mania. It is not known how the treatment works but it is believed to have the same antidepressant effect as an epileptic fit. The effectiveness of the treatment is debated and many hospitals no longer make use of it. (SPL/Will McIntyre)

CARE IN THE COMMUNITY

Even before the development of the new mind-controlling drugs in the early 1950s it had begun to be recognized that many of the functional problems experienced by patients in mental hospitals were the result of long-term institutionalization itself. The development of drugs like chlorpromazine, which could prevent relapses and control the more bizarre manifestations, then precipitated a revolution in thinking on the management of mental disorder. The result was that, throughout the Western world, large numbers of patients who would formerly have been expected to spend many months or years as virtual prisoners in mental institutions were released into the community.

The majority of new cases of mental illness that arise are now dealt with in small psychiatric wings in general hospitals, and patients are released after a comparatively short period of treatment. This development, however, has not been without its critics. Some psychologically oriented psychiatrists have objected to a policy that does nothing more than suppress symptoms and has no effect on the disease. Unfortunately, genuine causal treatments do not exist. In their absence, the pragmatic approach has prevailed.

Treatment in the community is sustained only if patients take their drugs, and major problems arise if patients do not. Compliance rates for all forms of medication are much lower than doctors like to think, and in the case of anti-psychotic drugs, compliance is very poor. As the patients do not consider that they are ill they reject treatment of any kind. Although sufferers can be forced to undergo hospitalization if they are sufficiently disturbed, they are often returned to the community after a short spell where there is no guarantee that medication will be taken. The result may be another breakdown and a second round of treatment. Only when patients have undergone this sequence several times do they accept that what is available is maintenance therapy, not a cure. Up to a third of them never accept the idea of treatment at all.

Although these limitations may seem to put the idea of treatment in the community in some doubt, the results are more reassuring. Even if it should take ten years and five admissions to bring a schizophrenic to the point where he or she accepts the need for maintenance therapy, this is still preferable to a lifelong stay in hospital. Success rates are improving as doctors gain experience in running their community programmes. Many of the nurses who were formerly employed in psychiatric hospitals are now working in clinics tracing non-attenders and persuading them to return to the clinic to continue medication.

Unfortunately many patients in the community do badly, especially those who are homeless. Of the latter, a large proportion end up in hostels or sleeping rough. Studies have shown that many move beyond the reach of medical care and relapse. Fewer than 20% of them will be in permanent full-time work and most will be aware of social difficulties and restricted lives. The death rate from suicide is about twice that of a mentally well group of a comparable age. People suffering from psychoses are not only a possible danger to themselves but also to others. Some fears have been expressed for the safety of the public when potentially violent psychotic people are not taking their medication.

NIKE
prince
MS

OURSELVES AND OTHERS

How We Present Ourselves to Others

Human beings' perception of one another is a central element in social intercourse. This process of perception is surprisingly complex. Presentation involves the display of a wide range of cues or signs by which we are recognized as a particular type of individual.

Commuters on their way to work. The clothes we wear to work will often be similar to those worn by colleagues. The desire to be accepted as part of the working group will encourage us to conform in our dress. In offices across the developed world the shirt, tie and jacket has become standard wear for male office workers. (Spectrum)

The cues by which we are classified include age, sex, body size, skin colour, voice, accent, physical attractiveness and dress. Recognition of them is called *social cognition*. Constellations of cues tend to build up into stereotypes and these help us to make quick decisions about other people. Stereotypical choices often have to be amended, however, when some of the cues fail to fit the pattern.

Few of us are entirely indifferent to the impression we make on others. Our concerns will vary, however, with the circumstances and especially with the importance we attribute to the goodwill of the person concerned. From time to time situations arise in which that goodwill is highly important to us. On these occasions we will be preoccupied with the way we present ourselves and, in the process, strict honesty may go by the board. Many people, who would never dream of telling a lie in the course of an important human interaction, will prevaricate unconsciously in non-verbal ways. Some will even do this deliberately. As Jean-Jacques Rousseau (1712–78) noted in his *Discourse on the Arts and Sciences* (1750), politeness of expression may conceal a ruthless, calculating and determined egoism. In such instances our body language can reveal our true feelings, as it is often involuntary (⇨ p. 206).

Introductory rituals

The early interactions between two people previously unknown to each other are especially interesting. They are rich in the rituals of recognition – purely ceremonial acts and statements designed, not to convey factual information, but to demonstrate attitudes. In most cases, neither party will be particularly interested in the factual content of what is said, much of which, as in talk about the weather, is well known to both. However, each party will be deeply interested in whether the other is willing to perform the ritual – taking turns in conversation, giving way, providing listening cues, and exhibiting positive body language. Quite often, when there is goodwill on both sides, long conversations may be sustained that are entirely constructed of such elements and in which nothing new is said. The willingness to participate in such a ritual is to acknowledge tacitly, or to seem to acknowledge, that the other person is a member of the same social, educational or professional class. In this, as in other encounters, nonverbal statements of this kind speak louder than words. If explicit verbal statements suggesting acceptance are accompanied by the denial of the ritual, it is the latter that will be taken to denote the true state of affairs.

During the introductory process people can offer marks of identity – personal opinions or attitudes – for perusal and approval. These are often proffered in a tentative way and then fortified, amended or even denied in accordance with the response. Such adjustments are commonly made on both sides if there is a mutual wish to confirm that both belong to the same group.

In an introductory situation, we will almost always wish to present ourselves in a favourable light. The other person will, of course, be aware of this and will tend to discount that part of our performance – mainly verbal – that we can readily control. They will pay more attention to the expressive part – body language (⇨ p. 206) – that is thought to be less under direct control. A host will be less impressed, for instance, by our comments on the quality of the cooking than by the gusto with which we appear to be enjoying the food.

Sophisticated people, aware of others' recognition of body language, will try to make theirs conform with the data they wish to convey. They might, for instance, adopt an expression of pleased anticipation well before approaching the front door of a house to which they have reluctantly accepted an invitation. The most subtle of all will, if necessary, adapt body language to convey the desired message while deliberately underplaying verbal expression. This kind of manipulation of the seemingly spontaneous aspects of behaviour may be apparent to a keen observer who will then look for, and probably find, the true motive.

In polite conversation, objections are not raised to statements on matters important to the speaker, but not to others, even if there is tacit (silent) disagreement. For example, when car salesman Eric tells computer programmer Phillip that the design of car interiors is not what it used to be, Martin will not disagree even though he quite likes them as they are. Similarly, Martin's tirade against a certain brand of computer software will be met by a non-committal Eric who occasionally uses the brand in the office. Thus, although no real agreement is reached on the facts, there is clear agreement on whose assertions on what matters are, at least for the present, to be respected.

Interpersonal behaviour

One of the most interesting of the many theories describing how people interact is the dramaturgical theory of the American sociologist and anthropologist Erving Goffman (1922–82). According to this theory, each one of us is at all times playing a part – taking a role in the continuous drama of life. The *persona* (Latin for 'actor's mask') is the conception we have formed of ourselves, the person we would like to be, and often the person we finally become.

Adopting a role often involves dressing for the part. Although sartorial factors are less important than they used to be, they remain critical in many situations,

The Significance of Accent

Many studies have shown that human prejudices about accents are deeply entrenched. Although the effect of accent is probably most acute in Britain, it is found all over the western world. Similar, if less striking, effects occur, for instance, in relation to standard New York versus Bronx or Brooklyn, Parisian versus Provençal or a Hamburg accent versus Bavarian.

In George Bernard Shaw's play *Pygmalion* Professor Higgins declares 'An Englishman's way of speaking absolutely classifies him. The moment he talks he makes some other Englishman despise him'. Most British accents convey information about the speaker's geographic origin or current location. Only one accent does not do this – what linguists call 'received pronunciation' (RP). RP is wholly nonregional but is usually thought to prevail in the south of England, Oxford, Cambridge and independent schools. It is the accent of the Establishment and its use at once conveys to the listener a sense of the possession of status, power and money on the part of the speaker. The Waspish and Ivy League accents in the USA convey similar information.

Accent alone, regardless of what is said, is highly influential. Different accents, in people not well known to the listener, evoke surprisingly different evaluations of the speaker, and the effect is most striking when the voices are heard without the speaker being seen, as on radio. RP speakers are invariably voted highest for intelligence, education, leadership and reliability. They are considered likely to be taller and more physically attractive, and even to have higher standards of personal hygiene, than people with non-RP accents. A close correlation is also automatically drawn between the type of accent and the probable occupation. Repeated demonstration that there is no factual basis in such correlations seems to do little to prevent people from making them.

RP is not, of course, absolutely standard and there are many individual variations among RP speakers. Some are so highly refined as to arouse hostility in many by seeming to imply a sense of infinite superiority. Some have a hint of a regional quality. Such accents, known as 'off-RP', are often also highly regarded. Other accents somehow achieve the flattering feat of implied equality.

In *My Fair Lady* Audrey Hepburn plays a cockney flower girl under the tutelage of arrogant elocutionist Rex Harrison. Her ability to learn how to speak in a refined upper-class voice allows her to join the world of high society. (Kobal)

especially in professions such as the Armed forces, the Law and the Church. The response of others to a recognized form of dress is automatic and often has nothing to do with logic. If, for instance, in a poorly disciplined Army unit, a group of officers dressed in jeans and sweatshirts were to pass a group of private soldiers, the soldiers might ignore them or turn away, even if they knew who they were. But if the officers approached them dressed in uniform they would all spring to attention, salute and remain at attention until the officers were well past.

Role-playing may also involve the adoption of appropriate accent, language, posture, facial expression, gravitas, gestures, and so on. Many of these characteristics are acquired – gradually or quickly – in the course of adopting, or advancing in, a new occupation or profession. The new recruit quickly recognizes that a particular preferred image exists and that conformity to it is desirable. He or she may take as a role model an especially distinguished, successful or popular member of the group, and may copy, in some detail, the superficial characteristics of this person. In some cases, more usefully, the real qualities of the hero may be emulated. Role-playing also involves activity that demonstrates a quality or capability. For many, such activity is part of the job, as it is for professional footballers, snooker players, concert pianists, TV comedians and neurosurgeons, for example. Others, however, may have to dramatize a little. Students wishing to impress lecturers with their attentiveness, fix their gaze steadily upon them and adopt an attitude of rigid attention and interest. Academics, however, are not often impressed by this kind of performance and have other and better ways of judging interest.

When we present ourselves to others, we try to offer an idealized picture – one that manifests to an unattainable degree the officially accredited values of the group we wish to join. At work people often dress in a style which conforms with others in the organization, even if it is against their own tastes. They will also make sure that they keep strong opinions or beliefs to themselves in order to avoid conflict. This idealization reflects the values of the group and we behave in this way even if to do so conflicts with conventional morality. To gain identifications with, and approval of, the group, we may disingenuously project values we do not actually hold.

People who possess certain social characteristics expect to be valued and treated in an appropriate way. Indeed they claim this as a moral right. This is one of the bases on which society is organized. The corollary is that those who explicitly display certain characteristics are morally obliged to be what they claim to be. Those who are not are, whatever their motives, regarded as confidence tricksters out to gain an advantage by cheating.

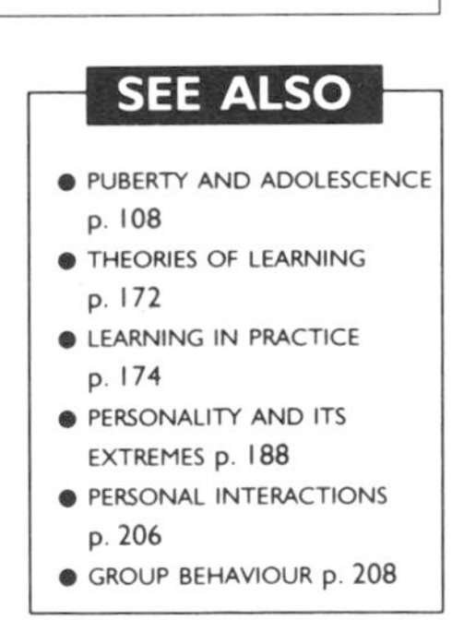

SEE ALSO
- PUBERTY AND ADOLESCENCE p. 108
- THEORIES OF LEARNING p. 172
- LEARNING IN PRACTICE p. 174
- PERSONALITY AND ITS EXTREMES p. 188
- PERSONAL INTERACTIONS p. 206
- GROUP BEHAVIOUR p. 208

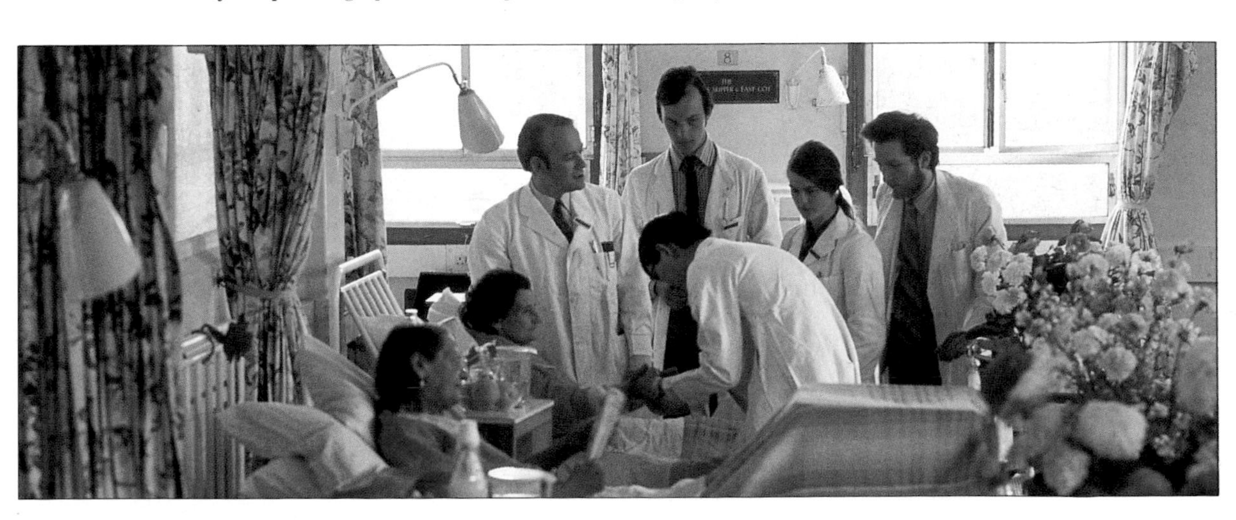

In some professions the clothes we wear are vitally important to the data we convey. A doctor's white coat, for instance, suggests authority, intelligence and competence. (Telegraph)

Can Personality be Assessed?

Personality is an all-embracing term, incorporating all the behavioural and mental characteristics by which an individual human being is recognized (⇨ p. 188). To claim to be able to assess it is to imply that behaviour can be predicted. Unfortunately, there are conflicting views among psychologists on the nature of personality.

Those psychologists who take the view that personality is simply a collection of traits believe that if you get answers to enough questions you will make an adequate assessment. Psychodynamic theorists (those who hold that personality is centrally related to motivation and drive) maintain that useful assessment is impossible without investigation of the unconscious by such means as projective tests like the Rorschach Inkblot Test (⇨ below). Social learning theorists (who emphasize the importance of experience) maintain that no one test can reliably predict how a person will behave in a particular situation.

Personality assessment is obviously of importance if we are to try, for instance, to match up people with the occupations best suited to them. However, little attention has been paid to what temperaments are suitable for particular professions. Many employers ignore the possibility of personality assessment and take the pragmatic view that employees should be judged only on the quality of their work. Some employers, however, especially those who invest large sums in the training of their employees, take a greater interest in personality assessment. Empirical evidence suggests that personality factors cannot safely be ignored in the selection of such people as recruits to the Armed Forces, civil airline pilots, candidates for the senior Civil Service and the diplomatic corps, and senior business executives.

Education authorities in developed countries have traditionally tried to assess at least some aspects of personality – notably the intellect and the capacity to learn – in selecting candidates for higher education. But little if any attempt, for instance, has been made to determine the level of humanitarian qualities in candidates for medical training. Medical students, in general, are selected only on the basis of their past academic records. Some Deans would argue that the development of human qualities is a continuing process and that the outcome is determined as much by later life experience as by the personality at the time of starting training.

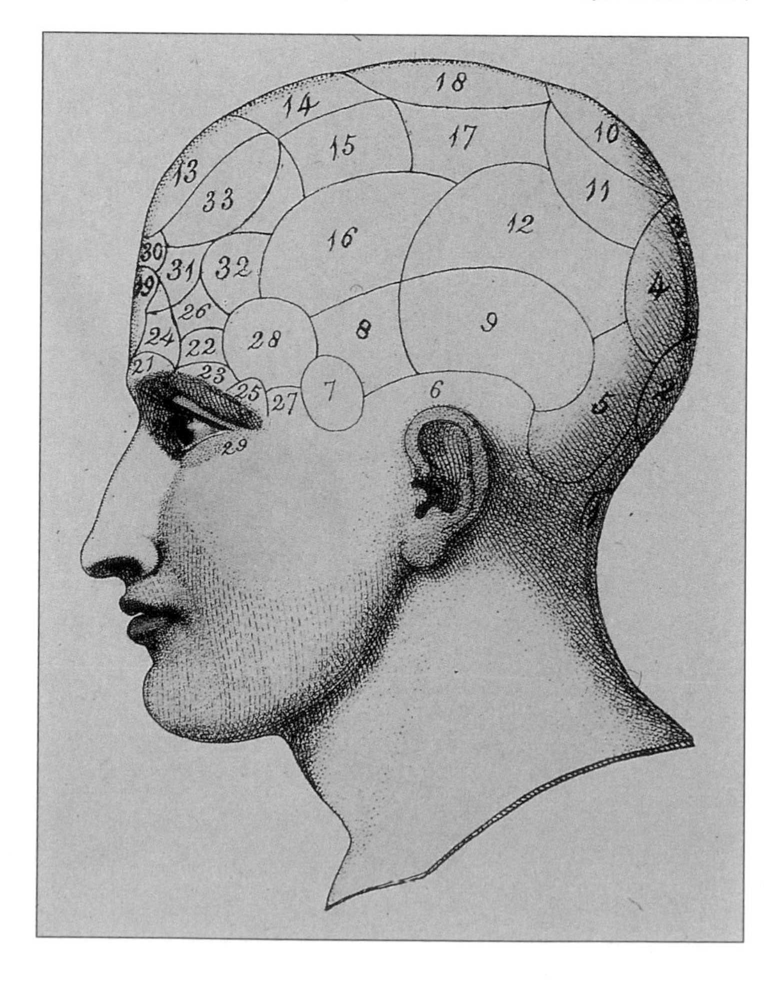

A phrenologist's division of the head. In phrenology the brain was divided up into distinct regions, each responsible for a different faculty or trait. The size of any one region would be indicative of the strength of that faculty or trait. In the diagram the areas marked 10, 18 and 14 indicate 'self-esteem', 'firmness' and 'veneration' respectively. Franz-Joseph Gall and Johann Kaspar Purzheim popularized the study in the 19th century and it was not scientifically discredited until well into the 20th century. (AR)

The trait theory of personality

Folk wisdom has it that people fall into readily recognizable groups by virtue of possessing certain characteristics or 'traits'. Some people are naturally shy, others socially bold; some are reserved and aloof, others outgoing and participatory; some intellectual and scholarly, others unconcerned with the things of the mind; some trusting, others suspicious; and so on. Trait theorists, such as Hans Eysenck, take this seemingly commonsense view and then proceed to an analysis of the way in which the many possible characteristics may be associated. In *Factor analysis* psychologists measure a large number of statistically separate 'factors' or behavioural dimensions that are then used to define personality traits.

Methods of assessment

Interviews by prospective employers or their personnel managers are by far the most widely used method of attempting to assess personality, but are not highly regarded by most psychologists, who consider that such interviews are seldom properly organized and do not serve their nominal purpose. The results of interview testing may be affected by many factors that are not easily controlled, such as the interviewer's manner, and the hopes, fears or expectations of the subject. Complex social, emotional, even sexual interactions between the subject and the interviewer may have largely unpredictable effects. Formally structured interviews conducted by experienced psychologists, often using the same sets of questions, are generally believed to give more valid and reliable assessments than free interviews, particularly if the interview is combined with other forms of assessment, such as personality questionnaires, aptitude tests, or behavioural observations.

Behavioural observation is an important element in personality assessment. This may be done in various ways but will inevitably be selective and the behaviour observed may not be representative. Behaviour may be assessed in the following ways: by the test subject's own descriptions of his or her past history; by observations of actual conduct in real or simulated situations; by facial expression, gestures, posture, body language, dress, personal cleanliness, and so on. In assessments for the British higher Civil Service, for example, candidates are asked to perform in simulated situations. This can include acting as Chairperson of a committee dealing with some contentious issue and having as members some particularly difficult, obstreperous and disorderly people. Several governments have exposed candidates for the Secret Service to simulated inquisition of the type a captured spy might expect.

Intelligence tests (⇨ p. 168) and aptitude and achievement tests are widely used for specific educational and vocational purposes. Vocational tests often concentrate on the nature of the test subjects' interests, as these have been shown to be strongly correlated with suitability for particular employment. It has been found that interests that are well established by late adolescence or early adult life remain permanent in enough cases to allow valid prediction. Formal personality tests take the form either of questionnaires or, in the case of tests conducted by adherents to the psychoanalytic schools, the so-called 'projective' tests. The two best-known of these are the *Rorschach*, in which the test subject is asked to look at folded inkblot patterns and say what the shapes suggest to him or her; and the *Thematic Apperception Test* (TAT), in which the subject is shown drawn or photographed set scenes and is asked to invent appropriate stories to describe the role of the main figure. Such tests are said to work on the basis that an individual 'projects' something of his or her own personality into the visualization or story. Interpretation of these tests calls for more than usual expertise from the psychologist, and members of different schools of psychoanalytic thought may offer different interpretations. Interest

A Rorschach Inkblot Test (left) devised by Hermann Rorschach in 1921. During the test the subject will be asked to describe what he or she is reminded of or sees in 10 inkblots, half in various colours and half in black and white. In theory these perceptual responses were supposed to indicate basic personality traits. However, there is some doubt about the link between perception and personality and therefore the value of the tests is questionable. (SPL)

in projective tests reached a peak in the 1950s but these have since declined in popularity.

The Cattell 16 Personality Factor test

One of the best-known trait theorists is the English psychologist Raymond Cattell (1905–), who, after various academic appointments in the USA, became Research Professor and Head of the Laboratory of Personality Assessment at the University of Illinois. Cattell's 16PF (Personality Factor) scale lists 16 categories, or dimensions, of personality, each of which can be assessed on a scale of 1 to 10 from low score to high. Cattell drew up a questionnaire to be used to quantify these 16 dimensions. This has several hundred multiple-choice questions and the test subject is asked to select from each question the answer that seems to them to describe most accurately their own personalities. The results are then used to construct a personality profile (⇨ illustration). Cattell's test is widely used and, because it is said to be as helpful in selecting a marriage partner as in choosing a career, computer-dating agencies often make use of it.

The Minnesota Multiphasic Personality Inventory (MMPI)

The MMPI test was originally compiled in 1942 as an attempt to systematize the diagnosis of mental disorder, and although it did not prove very useful for this purpose, it was found to give a better assessment of personality than most other tests. It therefore became one of the earliest standard personality assessment methods. The influence of MMPI has been considerable and many subsequent tests have been based on it. Like many of its successors MMPI contains a collection of statements about how a person might behave in certain situations, or how he or she might feel or think in response to them. The test subject is asked to consider whether the statement would apply to him or her and to indicate this by a 'true', 'false' or 'don't know'. The MMPI is a major test containing 550 statements of varied types:

- I like tall women.
- I hear strange voices all the time.
- There are those out there who want to get me.
- When I get bored I like to stir up some excitement.
- My bowel movements are irregular.
- I am in touch with flying saucers.
- My father was a good man.
- I work under a lot of tension.
- I have never done anything dangerous for thrills.
- I seldom have headaches.
- My hands and feet are usually warm enough.

The intention behind some of these statements is clear: the second may detect the auditory hallucinations of the schizophrenic or a simple, literal-minded person; the third may elicit paranoid tendencies; the fourth may reveal a psychopathic personality; the fifth a possible tendency to hypochondriasis; the sixth a frank psychosis; and so on.

One of the most important advances in the design of the MMPI was the attempt to assess the validity of the test. Thus, for instance, people with many more than the usual 'don't know' responses may be deemed to be evasive; those whose answers seemed to show them in an improbably good or improbably bad light could be assumed to be faking for particular purposes; and those who made many corrections to their choice could be considered to be defensive. Unfortunately, the MMPI proved to have been standardized on a sample that was not representative of society generally. It also failed to give particularly consistent results when the same candidates were retested.

A development of the MMPI, and one that greatly benefited from a critical analysis of it, was the California Psychological Inventory (CPI; 1956). Although it adopted over 200 of the statements from the MMPI, those that were selected had no relevance to mental disorder. The CPI was standardized on 13 000 people from many different walks of life and the result provides scores on such personality traits as masculinity, femininity, sociability, responsibility, self-control, flexibility, tolerance, dominance, self-acceptance and achievement orientation. The test has been tried out against numerous peer assessments made by friends of the test subjects and good correlation has been found. A later development, the Personality Research Form (1984), assesses 20 essential 'needs' that were postulated in 1938 as being the basis for human motivation.

Criticisms

An obvious criticism of methods such as the Cattell 16PF test and the MMPI is that the questions asked are likely to give the person taking the test an idea of the desired answers. Since one can hardly expect candidates to be wholly disinterested, some deliberate or unconscious bias in the selection of answers seems likely. The choice of questions in such tests may also be criticized, since they do not necessarily reflect a wide consensus view of what constitutes, or reveals, certain personality attributes. Differences in cultural background between the candidates and those who set the questions can also readily introduce unfairness.

To be valid any psychological measurement technique must measure what it purports to measure. Unfortunately, unlike physical entities or quantities, personality traits cannot be objectively defined. The content of questionnaires may not always be clearly understood by the test subject and the same question may mean different things to different people. The selection of questions is clearly fundamental, but different experts will inevitably differ in their choice. Validation of personality tests is also difficult.

Test reliability can be assessed by noting the consistency of results when the same individual is given different tests at around the same time. Its reliability is also indicated when the results from interviews match the results from tests. Doubts about a test may arise if the same test is carried out on the same person at various times and the results differ. However, there is always the possibility that personality is not a stable entity.

Although widely used, personality tests have not lived up to their promise as important tools in the development and extension of psychology. Perhaps the most cogent criticism is that no test can be sufficiently comprehensive to allow realistic extrapolation from its strictly limited situation to the infinite complexities of real life.

SEE ALSO

- WHAT ARE MENTAL DISORDERS? p. 184
- PERSONALITY AND ITS EXTREMES p. 188
- ANXIETY AND NEUROSES p. 190
- PSYCHOSES p. 192
- PERSONAL INTERACTIONS p. 206
- GROUP BEHAVIOUR p. 208

Sexuality

Sexuality is possessed to a greater or lesser degree by all human beings. Although the biological purpose of sex is perpetuation of the species, sexuality plays a major part in our make-up. It promotes and colours much of our association with others and frequently modifies our behaviour.

The sex-specific behaviour of children – such as little girls playing with dolls (right) – is still believed by many to be innate. However, there is no evidence to support this and a strong indication that powerful cultural stereotyping from a very early age dictates much of their behaviour. (Gamma)

The transsexual singer Leila Umar. A number of transsexuals have undergone careful investigation and assessment, surgical reconstruction, hormone therapy and psychological support to become well-adjusted men or women, able to react as a heterosexual to those whom they deem to be the opposite sex. (Gamma)

Many of our ideas of sexuality are conditioned by our life experience and especially by the sexual attitudes and inclinations of our parents. Newborn babies, although anatomically male or female, are largely gender neutral. The child's sex will, however, determine the attitudes of the parents who will, in general, apply the local cultural stereotypes – usually treating boys in quite a different way from girls. Different parents also subtly impose on their young children considerable differences in sexual attitudes. These are very influential on the later sexuality of the child (➪ below). Sexual attitudes that are determined by the physical differences are harder to quantify (➪ below).

The physical and hormonal basis of sexuality

The physical sexual differences are dependent upon the fertilizing sperms; whether they are carrying an X chromosome, to produce a female, or a Y chromosome, to produce a male (➪ p. 22). If fertilization is with a Y-bearing sperm, the early embryo, sexually undifferentiated during the first two months or so, begins to produce male sex hormones (androgens). These cause the gonads, which would otherwise become ovaries, to become testicles, and the penis and scrotum to develop from the tissue that otherwise becomes the clitoris, vagina and labia (➪ p. 96). If fertilization is with an X-bearing sperm no androgens are produced and the embryo develops into a female.

Androgens have a second major role to play in the differentiation of the sexes after the onset of puberty. A large increase in the production of androgens from the cells lying between the sperm-producing tubules of the testicles causes skeletal and muscular development, male pattern of hair growth, and the maturation of the sexual organs to the adult male pattern (➪ p. 108). Most of these changes can occur in the female, also, if androgens are produced in abnormal quantities by other endocrine organs (➪ p. 78). The virilization that results includes enlargement of the clitoris and a marked anabolic effect. As a corollary, severe underproduction of androgens in the male, as in the case of castration before puberty, results in the adult female body pattern of the eunuch, with wide hips, well-marked breasts, high voices and female distribution of body hair.

Female sex hormones (oestrogens) promote the female secondary sexual characteristics and regulate menstruation and fertility. Oestrogens do not influence the libido of the adult female, although given to a male they reduce his sex drive. Contrary to popular opinion, sex hormones have nothing to do with sexual orientation. Homosexual people do not have abnormal levels of androgens or oestrogens.

Sex-related differences in infancy

New babies of different sex have much in common but also show some sex-specific differences (other than physical ones). These have been assessed by observers unaware of the sex of the babies they were observing. Boys are larger at birth and are more fractious and irritable, and are more difficult to calm than girls. Infant boys are more aggressive than girls and are more prone to anger. They are also less timid and show less fear than girls and seem more independent of maternal protection. Girls react more sensitively to

Sexual Intercourse

Sexual intercourse would appear to be an instinctive activity, although there is anecdotal evidence of people not knowing what to do. During sexual arousal the penis will become erect and the woman's labia, vagina and clitoris engorged with blood. Intercourse takes place when the penis is positioned inside the vagina. Both partners may make thrusting movements which usually cause rising excitement and pleasure to both. Stimulation of the clitoris and penis result in a climax (orgasm), followed by a sense of physical and emotional release and relaxation. Both male and female orgasms involve involuntary rhythmical muscle contractions in the genital area. Within a few minutes of male orgasm the penis returns to its normal size. This is followed by a period of about half an hour during which very few men can achieve a full erection or another orgasm. Women are capable of achieving further orgasms almost immediately.

Most human beings distinguish sexual intercourse from its reproductive function and view it as an activity in its own right, as a source of pleasure and an expression of love. Effective contraception (➪ p. 100) now allows this separation and largely frees people from the concern over the possibility of unwanted pregnancy. A satisfactory sex life is, for most people, an essential component in a contented life. Certainly, the failure to achieve this is often a major source of unhappiness.

touch, smile earlier and more often, respond more readily to human contact and usually talk earlier than boys. They are generally quieter and are more readily frightened by strange situations. These characteristic differences are observed in most cultures and also in all species of mammals.

However, the neurological differences between the male and the female, if any, must be subtle. In line with the general size differences between male and female, the male brain is, on average, larger than the female. Other than this, no structural or intellectual differences can, currently, be demonstrated.

Because parents act differently towards and respond differently to boys and girls, it is almost impossible to determine which characteristics are innate and which are conditioned. Many parents are convinced that baby boys are intrinsically interested in 'male' concerns, such as cars and guns, while girls are interested in 'female' concerns, such as dolls. However, it seems quite plausible that such differences are accounted for by cultural explanations, such as parental provisions, and sometimes rewards, for 'correct' behaviour. By the age of about two, infants normally have an unequivocal sense of being either male or female. This is made more potent by the wide cultural differences in the treatment of boys and girls.

There is no complete consensus of opinion by behavioural scientists on whether sex-specific behaviour is genetically determined or is cultural. The difficulty here arises from the fact that cultural sexual influences are so pervasive and powerful that they tend to submerge and conceal innate influences. It is also clear that all sexually-related attitudes and behaviour can exist in a healthy child of either sex.

Homosexuality

Homosexual activity must be distinguished from homosexual preference, which is what is usually meant by homosexuality. The term derives from the Greek *homo-*, meaning 'the same', not from the Latin *homo*, 'a man'. Many people, of both sexes, experience homosexual interest or engage in homosexual activity at some time in their lives, often during the period of sexual development, but settle in the end for exclusive heterosexuality (from the Greek *hetero-*, meaning 'different'). Some people are attracted to members of both sexes, and are called 'bisexual' (from latin *bi-*, meaning 'two'). Homosexual behaviour is not exclusive to the human being, and has been observed in most animal species.

Homosexual people vary as widely in their sexual behaviour as heterosexual people. Some are promiscuous, others monogamous. Some have a few affairs, some many. Some settle into a life-long homosexual or heterosexual relationship, often, in either case, bringing up children, while others remain celibate. Some openly acknowledge their homosexuality, braving social prejudices, while others conceal it. The origins of homosexuality remain obscure, and, although there is no shortage of theories, there are no definite facts on the matter. Recent evidence suggests a genetic basis for the predisposition to homosexuality.

Social prejudices against homosexuality are still commonplace. With some exceptions, such as that of ancient Greece, homosexuality has always been proscribed. Until well into the 20th century homosexual activity was a criminal offence in almost all western countries and many men suffered long terms of imprisonment. Later, homosexuality came to be thought of as an illness and until as recently as 1973 it was still officially regarded as a mental disorder by the American Psychiatric Association.

During the early 1980s the promiscuous sexual activity among male homosexuals was shown to have been a major factor in the rapid spread of HIV (the human immunodeficiency virus) and rising incidence of AIDS (➪ p. 60). However, there is nothing specific to homosexuality in the nature of the disease.

Gender identity

The gender of a person cannot in every case be established simply by an inspection of the genitalia, or even by checking whether the sex chromosomes are of the male (XY) or the female (XX) configuration. Quite often, babies with ambiguous genitalia are, accidentally or otherwise, reared in a gender opposite to that of the biological sex. This can also happen when a parent's desire for a child of the opposite sex leads them to apply the opposite cultural stereotype. In these cases, the gender in which they are reared nearly always becomes the one they accept and wish to retain. This demonstrates the conditioning power of the environmental influences during the critical early period of life.

Transsexuality

The gender assumed by parents and by the individual during the early years will almost always become definitive. If this later proves to be discordant with the chromosomal sex and with the developing physical appearance, serious conflicts may arise. In some cases a person will have a strong feeling that they belong not to their genetic sex but with the opposite sex – transsexuality. Many transsexuals feel so unhappy with their genetic sex that they are willing to go to any lengths to change it, including 'sex-change' surgery so that the appearance conforms to the perceived gender.

Some researchers have suggested that the phenomenon might arise because of unduly high levels of the sex hormones of the opposite sex before birth, but there has never been any evidence for this. In view of the power of early parental conditioning, it seems highly likely that transsexuality arises from the conscious or unconscious imposition on the infant of the desire of one or both parents for a child of the other sex.

— Sex Research and Therapy —

The Kinsey reports – *Sexual Behaviour in the Human Male* (1948) and *Sexual Behaviour in the Human Female* (1953) – were the first large-scale studies of sexual behaviour. Published by the Institute for Sex Research at Indiana University they were received by a shocked and sometimes disbelieving public. The reports contained the collated findings of researchers who questioned over 10 000 men and women on matters previously considered too private for general mention – frequency of marital and extra-marital sexual intercourse, methods of intercourse, sexual orientation, masturbation, female orgasm, and so on. Among their findings were that nearly all males and about 75% of females masturbate at some time; that 4% of adult men were exclusively homosexual and another 13% were predominantly homosexual; that more than one man in three had had a sexual interaction leading to orgasm with another male; and that the reported rates of female homosexuality were about half of those for men.

Kinsey and his colleagues made a major contribution to the knowledge of an important aspect of human behaviour. Their findings were broadly accepted, and their pioneering work prompted much further valuable research. Perhaps above all, the Kinsey reports helped people to overcome repressive taboos about open discussion on sex.

The gynaecologist William Masters and the psychologist Virginia Johnson first came to public notice when studying the nature of the physiological changes occurring during sexual arousal. Their use of polygraphs (lie detectors) to record these changes, however, revealed a great deal of new information. It became clear, for instance, that sexual dysfunction must be considered in the context of an interaction between both partners and not as a problem affecting one only. Despite the widely accepted criticism that observation alters behaviour, their studies have led to some effective methods of sex therapy and to a more realistic public awareness of the nature of human sexuality. In particular, they helpfully demonstrated the role of anxiety in male sexual failure.

SEE ALSO

- CHROMOSOMES AND CELL DIVISION p. 22
- NATURE VERSUS NURTURE p. 30
- THE CONTROL SYSTEM 4: GLANDS AND HORMONES p. 80
- PASSING ON LIFE p. 96
- PUBERTY AND ADOLESCENCE p. 108

Homosexual behaviour is still considered unacceptable by many people, and it takes considerable courage for homosexuals to display affection in public. (Gamma)

The Emotions

Emotions cannot be properly defined without reference to the physiological changes that accompany them. It is easier to enumerate the emotions – there are scores of them – than to define them, but this does not mean that they are unimportant. For many of us the evocation of certain emotions (such as happiness) is the principal aim in life.

In his book *The Expression of the Emotions in Man and Animals* (1872), Charles Darwin (1809–82) (➪ p. 8) noted the similarity between the bodily reactions of animals and those of humans in similar situations. The similarities were so close as to suggest that something was happening in the animals very similar to what happens in humans. A comparison with other close species shows that the same emotion-producing stimulus produces contraction of the analogous facial muscles. Dogs and humans can snarl, sneer and smile in common and do so in similar circumstances. The similarities of behaviour in the more extreme situations of acute fear or rage are even more impressive.

Darwin also showed that the expressions of emotions were not learned behaviour but were innate. The facial expressions that accompany the various emotions – joy, sadness, surprise, anger, fear, disgust, etc. – are common across all cultures, and in children who are born blind. They are the same even in remote human groups, as has been found by research, for example, into the South Fore tribe of New Guinea, which has had little or no contact with the outside world.

Adrenaline and the emotions

When we examine samples of blood taken from humans and other animals while experiencing rage or fear, we find that there is a substantial rise in the level of the hormone adrenaline (➪ p. 80). A sharp rise in the adrenaline content of the blood makes the heart beat faster, the rate of breathing increase, blood pressure rise, the pupils dilate, the skin turn pale, the hair stand on end, and the muscle ring at the outlet of the stomach contract tightly, causing a characteristic 'butterfly' sensation in the upper abdomen. Calming alpha and beta blocker drugs, which prevent adrenaline causing these reactions in the heart and respiration and in the smooth (involuntary) muscles of the body, largely eliminate the emotional effects of high blood-adrenaline levels. As the effects of adrenaline are so reminiscent of the features of high emotion it is not, perhaps, surprising that some scientists equate such bodily changes – 'visceral reactions' – with the emotions themselves (➪ box).

The limbic system

The most primitive part of the brain is the *limbic system* (➪ p. 76), which is the seat of the emotions. It operates in much the same way in other animals as in human beings. Humans have a highly developed cerebral cortex – the outer layer of the brain – which allows us to associate the emotions with complex thought. Other animals cannot do this and operate largely on an emotional level. Electrical stimulation of parts of the limbic system in animals produces a range of emotional responses such as fear, rage, contentment, sexual arousal, and so on, which occur with no involvement of the cerebral cortex. Stimulation of the limbic system in humans, in the course of brain surgery, produces a similar range of emotions.

The hypothalamus (➪ p. 78) links the brain functions and the endocrine system. All emotional activity is associated with the production of the hormones adrenaline and cortisol from the adrenal glands (➪ p. 78), and with the effects that these hormones have on the body. The stronger the emotion the higher the hormonal levels. A rise in the production of adrenaline and cortisol occurs in states of excitement whether they are associated with fear, anger or even pleasurable anticipation. An injection of adrenaline causes an emotional response, but the nature of this emotion – whether pleasant or unpleasant – is known to be determined by the associated circumstances, although such artificial arousal is more likely to be interpreted as an unpleasant than as a pleasant emotion.

Modern views

It is now generally accepted that every human emotional experience is a synthesis of bodily (visceral) hormonal arousal and conscious intellectual evaluation of the situation. Emotion is no longer regarded as an atavistic characteristic. Emotions are aroused when an incongruity or discrepancy occurs between our expectations and reality. This, for instance, is the basis of most humour and of much literary appreciation, but it is more obviously the case when the emotions aroused are the powerful and fundamental ones of fear, anger, joy and sadness. The recognition of such an inconsistency between expectation and eventuality produces the nonspecific hormonally induced visceral events that bring the matter unequivocally to our notice and help us to remember it. Viewed in this way, the emotions play an important role in our struggle for survival. Recognition and recollection of past dangers, for instance, are greatly enhanced by the powerful memory of the bodily events we experienced at the time.

The emotions also induce action, usually of a self-preserving kind. We may retreat hastily from a source of danger, or we may attack it. The physiological changes that accompany emotion improve our performance in times of stress and danger.

Primary and secondary emotions

Although there are at least 200 words in English that describe different emotions, most psychologists agree that the number of basic, or primary, emotions is much smaller and is probably limited to a list consisting of fear, anger, joy, sorrow, surprise, acquiescence and disgust. Emotions as experienced, however, are rarely pure, and are usually mixtures of the primary emotions. These mixtures may be called secondary emotions. The habitual experience of secondary emotions often characterizes personality (➪ p. 188). In very simple terms, a person who is commonly sad and fearful may be described as an anxious personality; one who experiences sadness and disgust may be thought of as a depressive; one who manifests surprise and joy will be seen as an optimistic extrovert; one who manifests anger and disgust might be said to have a sarcastic personality; and the joyfully acquiescent will be deemed sociable.

WILLIAM JAMES AND THE NATURE OF THE EMOTIONS

In the psychology journal *Mind* in 1884, William James suggested that emotional bodily reactions were not the result of the emotion, but that the mental states were the psychological accompaniments of the physical changes. These, in conjunction with the perception of the events that had caused the visceral reactions, *were* the emotion. James's idea was also promoted enthusiastically by others, notably the Danish anatomist and psychologist C.G. Lange (1834–1900). The James–Lange theory dominated psychological thinking on the nature of the emotions until well into the 20th century.

According to James–Lange theory, we feel fear because we are shaking, our hearts are bounding and our mouths are dry; we feel happy because the facial muscles that elevate the corners of our mouths have contracted; and we feel sad because we are crying.

It is true that there is some evidence that the deliberate assumption of certain facial expressions can produce the kind of physiological changes associated with the emotions that normally evoke these expressions. To some extent a cheerful countenance deliberately assumed can help induce the physiological changes that accompany happiness. Similarly, a chronically miserable expression can precipitate unhappiness. Nevertheless, the James–Lange theory is no longer accepted. Emotion is not simply the awareness of the bodily reactions to the hormones. This was clearly shown by the English physiologist Charles Sherrington (1857–1952) when he demonstrated that emotions were experienced even when all sensory connections to the parts of the body concerned were cut. Paralysed people experience all the emotions, although to a reduced degree. It is also clear that the bodily responses are by no means specific to the kind of emotion experienced, and we now know that the emotion precedes the visceral response by one or two seconds.

SEE ALSO

- CONTROL SYSTEM 2 p. 76
- CONTROL SYSTEM 3 p. 78
- CONTROL SYSTEM 4 p. 80
- PERSONALITY AND ITS EXTREMES p. 188
- ANXIETY AND THE NEUROSES p. 190
- PSYCHOSES p. 192

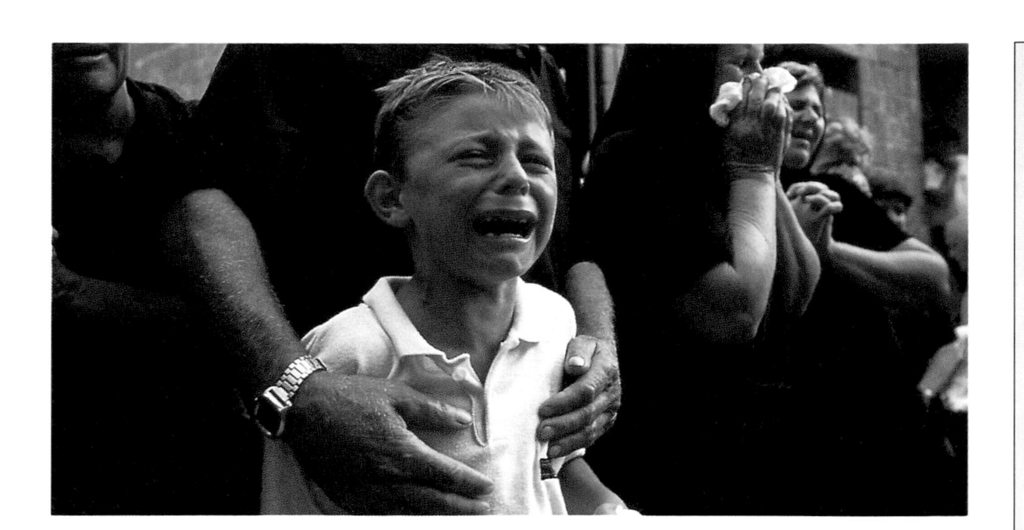

Emotions are rarely pure. They are often mixtures of primary emotions such as joy and surprise, and anger and disgust. Similarly the intensity of emotions will be determined by the circumstances. The above pictures show: a grieving boy in the former Yugoslavia (Colorific!); an angry man during the London Poll Tax riots (Colorific!); a mother and daughter sharing a happy moment (Spectrum); and a depressed woman (TCL).

HUMAN MOTIVATION

The idea of motivation – what makes people do things – is a central psychological question. Many philosophers believed that humans were activated by no more than the desire to achieve the greatest possible level of self-gratification and the maximum avoidance of pain. This is known as the *hedonistic philosophy.* This view, however, neglects the element of *instinct* which clearly has a major influence on behaviour. We show many behavioural patterns that occur without learning and must be regarded as 'built-in' or, in other words, instinctive. Among the most obvious of these are aggressive reactions, sexual behaviour, and the impulse to eat and drink when hungry and thirsty. We differ from other animals, however, in that we are able to exercise a good deal of control over whether or not we give way to our instinctive patterns of behaviour.

Some psychologists believe that motivation is based on homeostatic regulation. When there is an imbalance in psychological or bodily forces the body will act to restore the equilibrium. The *drive theory* of motivation – any impulse to satisfy a desire – is a homeostatic theory. These theories are plausible in the context of physiological upsets like hunger or thirst but are not particularly successful in wider application. Moreover, the theory does not consider the fact that many people choose voluntarily to behave in such a way as to increase homeostatic disturbance when, for instance, they starve themselves, practise chastity, pay to go on roller-coasters, and watch horror films.

Cognitive psychologists (➪ p. 166), in general, interpret motivation as the anticipation of something of value (*positive incentive motivation*). We behave, they suggest, in such a way as to bring about some desired goal. In addition, we can be motivated to behave in such a way as to avoid unpleasant outcomes. This is called *negative incentive motivation.* This theory is, of course, very close to the old hedonistic philosophy.

It is obvious that many of us are motivated by the desire to excel over others in some activity or other. This is called *achievement motivation,* and it appears to be unrelated to instinct or to genetics. It is not present in all cultures. Achievement motivation is a learned phenomenon and can be greatly enhanced by training. It should be remembered, however, that people motivated in this way are well aware that to succeed is pleasant and to fail is painful.

No one theory of motivation has been accepted. The homeostasis theories have fallen from favour and many psychologists now emphasize the external factors determining behaviour.

WHO ARE MORE EMOTIONAL – WOMEN OR MEN?

It is often said, and widely believed, that women are more expressive of emotion than men. The freedom of expression of emotion, however, is very much a cultural matter and varies a good deal from society to society. In contemporary western societies, formal studies have shown that women do, indeed, on average, display the facial expression of emotion more readily and more fully than do men. Trials were conducted in 1972 in which both men and women were covertly watched over closed-circuit television while being shown projected pictures evocative of emotion. The observers, who could not see the pictures, scored significantly higher in identifying the type of picture from the facial expression when watching women than when watching men.

But does this mean that women are more emotional? Not if the emotional response is measured physiologically rather than by facial expression. Among many other physiological responses, emotion produces slight to profuse sweating and this is sensitively reflected in the electrical resistance of the skin. When this and other physiological parameters were used to measure emotional response to various scenes, women were found to respond less markedly than men. So it seems possible that men are actually more emotionally responsive than women but that women show their emotions more. Culturally, of course, western men tend simply to behave in accordance with the standard stereotype of the stoical male because they have learned from an early age to repress the expression of emotion.

Personal Interactions

Human beings are not, by nature, solitary creatures and their personalities cannot be characterized in isolation. In most social interactions, the behaviour of one participant acts as the stimulus for the behaviour of the other. The judgements we make of each other depend very largely on how we behave in different situations and, perhaps most importantly of all, how we are treated.

In his 17th Meditation the poet John Donne (1572–1631) reminds us that 'No man is an Island, entire of it self.' This statement serves to remind us of our social interdependence and of the immensely complex way in which we interact with others. Moreover, our assessment of each other varies as a result of these social interactions. Such interactions constantly change us as we learn, by happy, or more often unhappy, experience; and as we change, so do others' perceptions of us.

The relativity of personal qualities

The relativity of the perception of personal qualities is often comically illustrated in hierarchical situations; the different aspects of the personality of a middle-ranking person are revealed or concealed depending on whether he or she is relating upwards or downwards. Major Y has a superior officer, Colonel X, and a junior, Captain Z. Colonel X sees the Major as a charming and courteous officer who invariably agrees with him, who is flexible, accommodating, energetic and right-thinking. Captain Z has a very different perception of the Major, whom he sees as a difficult, fault-finding martinet of rigid ideas, who never does any work. One might think that the truth about Major Y might be found by talking to his wife. But doubtless, her perception of the Major is determined by aspects of his behaviour conditioned by a different set of criteria. Mrs Y might, for instance, find him selfish, inconsiderate and boring. To his friends, on the other hand, he might be the ideal companion – kind, understanding, ready to make allowances, generous and amusing. The assessment of a spouse or partner, however, is likely to be more realistic than that of a casual acquaintance or professional colleague. Clearly, our appraisals of our fellow human beings are inevitably incomplete.

In any judgement of another person we cannot ignore the personality, tastes, inclinations and opinions of the person or persons making the judgement. To take an extreme example, in Dickens's *Oliver Twist*, Fagin's estimate of the Artful Dodger was based on the criteria that a propensity for thievery and high skills as a pickpocket were virtues of the first order. To Fagin, the Dodger was a pearl beyond price; to the average law-abiding citizen he was a social menace.

Relationship Between Person, Behaviour and Situation

Person

Behaviour

Situation

Verbal Communication

The American physicist, philosopher, mathematician and linguist Charles Sanders Pierce (1839–1914) pointed out that when we speak we do not directly transmit thoughts, only sounds and visual cues. These have to be interpreted by the person to whom we are speaking if communication is to occur. That interpretation will be determined by the listener's past experience, which will necessarily be different from that of the speaker. Pierce suggested that the meaning conveyed was not inherent in the sounds and cues but resulted from the relationship between them and the subject to which the speaker was referring.

Pierce adopted the earlier medical word 'semiotics' (the scientific study of the symptoms of disease), which has since come to mean the science of communication studied through the interpretation of signs and symbols. The term 'semantics' was later adopted for the study of the relationship between signs, such as sounds and visual cues, and meaning. This was elaborated in the important book *The Meaning of Meaning* (1923) by the English linguist Charles K. Ogden (1889–1957) and the English literary critic Ivor A. Richards (1893–1979). Together, these two men also produced 'Basic English', a proposed international language of only 850 words by which all thought could be expressed, albeit often in a clumsy and circumlocutory way. 'Tears' becomes 'eye water'; 'sweat' becomes 'body water'; 'I cannot define it' becomes 'I am unable to give any clear account of it'; and so on.

There is the further question of the stability of personality with circumstance and with time. An assessment made only in the context of limited experience is valid only for the qualities demonstrated in that situation. However warlike his conversation, no reliable evaluation of Major Y's personal courage under fire can be made from his behaviour in the Officers' Mess. Because social interactions can have a profound effect on the personality, either positively or negatively, we should not assume that even apparently basic qualities remain unchanged over a period of time. There are numerous instances of seemingly unpromising youngsters, noted for idleness and lack of interest, who, as a result of contact with an enthusiastic teacher or mentor, have ended up as striking successes in some discipline or profession.

It is a useful analogy to consider these interactions as triangular, in the manner suggested by the American psychologist Albert Bandura in 1978. Each angle represents one of the following – the person, the behaviour and the interactional situation. Each is joined to the other two by two-way arrows (⇨ diagram). A person who behaves in a socially acceptable way will positively affect the interaction and

President Bill Clinton campaigning in the Southern states. Politicians are adept at presenting their best side to the public. Their personal interactions are often orchestrated and will vary, depending on the type of people being addressed. In his shirtsleeves Clinton is in close physical contact with the crowd, stressing his credentials as a politician of the people. (Gamma)

also any consequent behaviour. Unacceptable behaviour will lead to negative consequences for the interaction and for the person. The interaction may affect both the behaviour and the person.

Body language

If by personal communication we simply mean transmission of information, then much of the communication between people is nonverbal. As soon as two people come within visual range of one another, communication starts. Information is conveyed by clothing, physical attitude, facial expression, proximity, body position, tone of voice, rapidity or otherwise of speech and response, body odour, and so on. Much of this information is conveyed without volition or even unconsciously, but many people are well aware of their nonverbal messages, and are also aware that this information is often more convincing than words.

Nonverbal communication is part of the stock-in-trade of salespeople, politicians, preachers, actors and teachers. To the perceptive, most nonverbal communication is obvious and its content and significance can be analysed. But sometimes such information is acquired without conscious recognition of the fact. It may still, however, strongly colour one person's attitude towards another, perhaps producing or reinforcing prejudice.

An important aspect of nonverbal communication has been termed 'body language'. This is the information about a relationship conveyed by conscious or unconscious bodily actions. A man puts his arm around a woman's shoulder in public; she immediately stiffens and goes rigid. A couple sit ever closer together on a park bench. In conversation a man crosses his arms (indicating defiance) and another talks behind his hand (suggesting lies or reluctance). A police suspect under interrogation leans back casually and puts his hands in his pockets. All these interactions convey a great deal of information. Such nonverbal messages commonly express emotions or desires or reveal attitudes. They often qualify, to a greater or lesser degree, the words that are spoken at the same time. Body language frequently demonstrates contradictions between what is said and what is felt.

Obviously, facial expression and movement convey much about the current state of the emotions, but body posture and muscle tension also reveal the emotions. Studies have shown that certain physical actions or reactions have specific meanings. Eye contact is particularly eloquent and regularly signals significant changes in an interaction. Pupil size can also signal interest or excitement, although this mode of communication is usually unconscious.

Personal interactions, involuntary and voluntary, verbal and nonverbal, have become a focus of study for psychologists. Similarly most people have become more aware of the meaning of our personal interactions beyond the spoken word.

GAMES PEOPLE PLAY

In 1964 the psychoanalyst Eric Berne published a book entitled *Games People Play* in which a form of group therapy known as 'transactional analysis' was described. The method proposed to analyse the nature of the interaction between people in terms of a transaction – a kind of deal or agreement between them. Berne recognized that people structure their lives in such a way as to achieve 'payoffs' from which they derive social gratification, or 'strokes'. A certain minimum of such gratification is necessary for reasonable contentment. To achieve this, people consistently play certain games in which they assume one of three alternative personality states: a somewhat irresponsible but fun-loving 'child'; a reasonable and mature 'adult'; or a stern, disciplining 'parent'. Any person, in relationship to another, may adopt, temporarily or in the long term, any one of these roles, as the situation dictates. The three states correspond roughly to Freud's *id*, *ego* and *superego* (➪ p. 164).

Relationships are said to be balanced if appropriate complementary states occur, as, for instance, when an employee accepts the 'adult' role in relation to the 'parent' role of the boss. Unbalanced relationships, in which appropriately reciprocal roles are not adopted, inevitably cause trouble. Examples would be when either a husband or a wife adopts a role other than that of 'adult' to each other, or when an unscientific husband married to a research physicist refuses to adopt, in matters of science, the 'child' role in relation to the wife's 'parent' role. In some marital relationships, stability is best achieved when both partners agree on a fixed interaction. A wife might, for instance, agree to her husband adopting the role of 'parent' to her 'adult' or even 'child', or the other way around. Others play different roles on different occasions, taking turns as 'parent' and 'child'.

Berne postulated four basic states that people experience in their interactions with others. These are: 'I'm OK, you're OK'; 'I'm OK, you're not OK'; 'I'm not OK, you're OK'; 'I'm not OK, you're not OK.' It was important, he believed, to be able to recognize which one was in effect so that an attempt could be made to correct any imbalance. In transactional analysis the function of the therapist is to help the members of the group make the analysis themselves and recognize when they are adopting which particular role. Berne's ideas were popularized when his method was described by Thomas Harris in the book *I'm OK, You're OK* (1969).

BARRIER GESTURES

Any position is an important element in body language. Crossing one or both arms across the chest can create a barrier, to be interpreted according to the facial expression or posture.

The standard arm cross signifies a negative or defensive state of mind – it is commonly seen in situations where people feel insecure. It can also indicate that a listener disagrees with what is being said.

If the arms are crossed and fists are clenched, the attitude is more strongly defensive and negative.

A partial barrier can be formed by folding one arm over the body to grasp the other arm – this often signifies a lack of self-confidence.

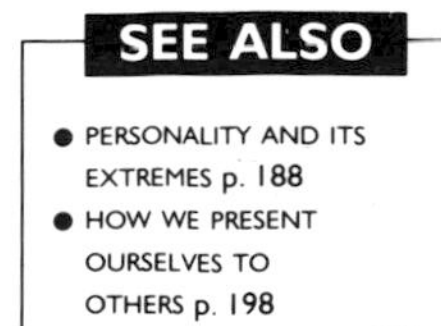

Group Behaviour

Throughout history, human beings have congregated in groups of many kinds. These associations may have many advantages, including mutual protection and psychological support and comfort. Some groups have a brief life span and are soon dissolved, while others – such as the extended family – may last for years or centuries.

SEE ALSO

- THE SPREAD OF HUMANS p. 12
- THE TALKING ANIMAL p. 16
- HOW WE PRESENT OURSELVES TO OTHERS p. 198
- PERSONAL INTERACTIONS p. 206
- THE FAMILY p. 210

The psychology of the group is not simply a multiplication of the psychology of the individual but has features that differ from that of any one of its members. The interaction of people in groups such as families, associations at work, clubs, political parties, restricted societies, athletic teams, and religious organizations may differ in many ways, but they also have a great deal in common.

Human beings in groups

Human beings are remarkably intolerant of solitude and constantly tend to reach out towards some existence larger than their own. Loneliness and mental isolation are to be assuaged only by human association. Individuals will therefore invariably elect to be members of a group of some kind; otherwise, they may suffer a deep sense of incompleteness. One can hardly exaggerate the importance of the group to the individual or the strength of the influence it wields. As the English surgeon, physiologist and philosopher Wilfred Trotter (1872–1939) remarked of man in his influential book *Instincts of the Herd in Peace and War* (1916), 'He is more sensitive to the voice of the herd than to any other influence. It can inhibit or stimulate his thought and conduct. It is the source of his moral codes, of the sanction of his ethics and philosophy. It can endow him with energy, courage and endurance, and can as easily take these away.' It is a remarkable example of how the influence of the group has changed since 1916, certainly in the West, that the quotation should continue: 'It can make him acquiesce in his own punishment and embrace his executioner, submit to poverty, bow to tyranny, and sink without complaint under starvation.'

Sometimes a person will seek to lead the group; more often he or she will look to another member of the group to provide leadership. In selecting or accepting a leader, the group will rarely make a reasoned choice and will nearly always be swayed by the skills of the demagogue rather than by solid merit. The judgement of the group, as a whole, is usually worse – and the intelligence less – than that of the most intelligent members and better than that of the least intelligent, and the group will commonly be influenced by appearances rather than by reality. This is one of many examples of the way in which the characteristics of the group differ from those of the individual members. Another is in the behaviour of the group as a mob (a disorderly or riotous crowd), in which behaviour is nearly always much worse than that of the individual members would be. Referring to the man in the crowd, Trotter said: 'He is subject to the passions of the pack in his mob violence and the passions of the herd in his panics.'

Characteristics of the group

Human groups can be loose or cohesive depending on the degree to which people value membership. In a highly cohesive group the members strongly subscribe to common opinions, are anxious to participate in the group's activities and wish to help it to achieve its goals. Membership of such a group – a political party, for instance – confers a feeling of identity and security and increases the feeling of

Milan football supporters. Many football supporters feel a strong sense of identity with their football teams, following them across the world to watch them play. Although most supporters find an acceptable outlet for their partisanship, a minority allow the psychology of the mob to take over and become violent towards supporters of other teams. (Popperfoto)

personal worth. Some groups, as they mature, develop standards or norms for the opinions, attitudes, values, behaviour and even sometimes dress of the members. These norms are important to the cohesiveness of the group and those who fail to conform to the most important of them are disapproved of as deviants. Conformists are rewarded by acceptance and inclusion; nonconformists may be punished by ostracization.

Norms are seldom statutory, but certain groups, such as private golf clubs, may become so anxious to emphasize them that they will formalize rules and put them in print. The uniformity that results from these cohesive pressures has the positive effect of strengthening the group, but often leads to a sterile rigidity of outlook and to chauvinism in its wider sense.

Inevitably, certain members of well-established groups tend to assume the role of leaders and establish a hierarchical structure. Functions are allocated, subgroups formed and formal channels of communication created. The imposition of such structures can insulate the leadership too much from members perceived to have low status, but who may have valuable contributions to make. The result, especially in large groups such as political parties, may be a diminishing satisfaction for those of low status, a tendency to fragmentation and the formation of splinter groups or factions.

In the wider context of society certain norms – such as using utensils at table, or queueing for attention in a bank rather than fighting for it – become so well established that they are shared by almost everyone. These are then known as 'folkways', and are called 'mores' when they become – by common agreement – essential to society. Anyone who contravenes the mores of a society is deemed at best eccentric, at worse, immoral or wicked.

The psychology of the mob

People in a crowd or mob behave differently from the way they behave when they

A mass wedding of members of the Unification Church (often referred to as Moonies). Group practices are very important to the Church. The mass blessings by Reverend Moon (the head of the Church) and his wife are seen as the most important ritual of the Church. Some Cult churches have been accused of brainwashing individuals into relinquishing their individuality in favour of the group. (Gamma)

are alone. Violence, aggression or hysteria can develop in a group of people who would not exhibit such behaviour on their own. One reason for this is that a person who identifies with the crowd feels a lowered sense of personal responsibility. It seems to him or her impossible that so many other people could be wrong. Most individuals are only too ready to accept the leadership of others, to conform to the obvious majority wish. Secondly, people in an excited mob experience a markedly heightened emotionality, partly because of the observation of the signs of emotion in others and partly because of the release of the normal controlling influences on their behaviour. As Freud would put it, the crowd has taken the place of the superego (➪ p. 164). Thirdly, many people harbour deep feelings of resentment that they would like to release in violence but are restrained from so doing because of fear of sanctions. Membership of the mob confers a sense of anonymity and allows the expression of these feelings. The resulting violence need not be directed at the real cause of the resentment but at any apparent symbol of it. Thus, if a man bitterly resents the fact that he cannot afford a luxurious car, he may, as a member of a mob, vent this resentment by damaging or destroying the first expensive vehicle he encounters.

Mobs have always been recognized as being potentially dangerous. Reason is impotent against the mob, and police units generally aim to implement policies of containment, though, unfortunately they themselves have on occasion been seen to resort to violence.

Another feature of mob psychology is its unthinking partisanship. While even unimaginative individuals can recognize that not all members of an alien community are alike, the mob often does not, and immediately identifies any recognizable member of such a community as an enemy to be attacked. This phenomenon is commonly carried to remarkable extremes. As has become only too apparent in Europe in recent years, football team partisanship can confer an unthinking recklessness on a mob, especially when exacerbated by the effects of alcohol. Supporters of one team need no further justification than this to attack the supporters of another.

The tendency to mob reaction is, of course, greatly aggravated by prevailing social inequality and injustice. As a result, race riots have become a feature of many contemporary urban situations. After the assassination of black civil rights leader Martin Luther King in 1968, race riots occurred in 150 American cities. The Los Angeles race riots of April–May 1992 were precipitated by the acquittal, by an all-white jury, of white police officers who had been seen brutally attacking a black man on television. To the mob, this elicited an automatic condemnation of the justice system and the Establishment, resulting in unfocused and widespread violence. It was pointless to suggest to the rioters that many of those people robbed or injured were also outraged by the police action.

Group Dynamics

The psychological and social forces that operate when people associate together in groups are studied in the social science of *group dynamics*. This term was first used by the German psychologist Kurt Lewin (1890–1947) in the course of a study of group psychology. His studies led to the development of the influential *field theory*, which has inspired a great deal of research in social psychology. Lewin rejected the idea that a person is motivated by instincts and desires and operates in an indifferent environment. Instead he proposed the concept of the 'life space'. The life space of a person is not simply the objective environment but the complex of the person and the environment, together with the way it is perceived and interpreted by that person. It also includes cultural pressures that determine when a particular form of behaviour is appropriate and when it is not.

The interaction of the life spaces of the members of a group produces a field of force containing many different elements, some tending to drive the group in one direction of opinion, some pulling it in another. The resultant of these forces determines the group behaviour. A group might, for instance, contain members who were strongly racially prejudiced and others who saw the need to respect the rights of others, regardless of race. If such a group were involved in a situation in which a threat was perceived from a different racial group, tensions would arise and the outcome in terms of behaviour of the group would be determined by the relative strength of these forces. In any attempt to alter the outcome, little is likely to be achieved by trying to change individual opinion. Since the life space of each member of the group is subjected to the pressures of the group field of force it is this that must be altered if a change is to be achieved.

Lewin was trained in physical science and was much attracted by analogies between the dynamics of the group and physical fields of force, which he represented by topological and other mathematical models. This method – always impressive to the non-mathematician – gave an unwarranted air of precision to his arguments. This, however, is not a subject that can be accurately represented mathematically.

In William Golding's novel *Lord of the Flies* a group of English schoolboys is shipwrecked on a deserted island. Without the restraining influence of adult society, order breaks down. The demagogic Jack wrests control of the group from the more democratic Ralph, and becomes leader of what turns into a savage tribe of boys. (National Film Archive)

The Family

The family is one of the oldest and most familiar of all human institutions and an essential unit and element of social cohesion. On evolutionary and archaeological evidence, including that of the habits of other apes, it may be assumed to have existed from the earliest appearance of the human being.

Because of the long developmental period required for the upbringing of children, family cohesion is prolonged, up to 20 years or more. Nuclear families, in which a father and a mother protect and bring up children separately from other kin in accordance with the prevailing culture, occur in almost all known societies. Families may, however, take different forms, in accordance with different cultures.

Most of us will experience at least two families; that of our parents, the family of origin, and the one we help form when we leave our parents and have children, the family of procreation. These two (or more) families, however, whatever the quality of their relationship, remain linked by blood kinship.

Families are also linked by former generations, although in the West this has little significance, other than for the purposes of inheritance. Technically, lineage implies at least four generations and may be reckoned backwards through the ancestry of the father (patrilineage) or of the mother (matrilineage). In some cultures, especially those of traditional India and China, lineage has been valued even more greatly than membership of a family, and a preoccupation with it is often associated with ancestor worship.

Family beginnings

In order to organize the sexual appetites of males and females most civilizations have formulated rules to govern their pairing. For example, in many past civilizations strong dominant males will have selected mates and kept other males at bay. Female selection of the best provider will have led to the rejection of other males. Such rules will have varied considerably, with little in common except for the prohibition of sex between close relatives.

The taboo against incest ranges across cultures and is almost universal, although the practice of incest has always been common. The origins of the taboo, which is strengthened and perpetuated once established, probably lie in the fact that incest is almost universally found to lead to rivalry and other problems within the family and to have been a cause of group disruption. Incest may also have had the effect of uneconomically increasing the number of dependants. The incest taboo has the incidental advantage of directing sexual interest away from the family and increasing the links between separate families, thereby increasing genetic diversity. Indeed highly inbred populations can become gene pools for hereditary disorders.

Family structures

Once mates are taken from outside the family, the offspring increase the size of the primary family. At the same time complex links are formed with the families of those taken in marriage, so that the secondary family can grow very rapidly.

In most families in the Western world the mated pair remain the central feature (monogamy). But in many other parts of the world the man or woman may take more than one spouse concurrently (polygamy). If a man takes more than one wife, this is called polygyny; if a woman takes more than one husband, it is polyandry.

Polygyny occurs in every part of the world and has been common in societies in which men are outnumbered by women, sometimes because of deaths in battle. It is especially common in Africa, where it is the rule among the Nupe people of Nigeria, the Serer of Senegal and Gambia and the Mende of Sierra Leone and Liberia. Various social customs encourage this, including the rule that a man is obliged to marry his brother's widow or the practice of marrying the sister of a barren wife. Islam permits polygyny with up to four wives, but insists that all should be treated equally. This is still practised by many well-to-do Muslims and can lead to a great increase in the complexity and size of the primary and secondary families. Polygyny was formerly the accepted practice among the Mormon community in America.

Polyandry is much less commmon than polygyny but occurs in some Indian societies. In general, none of the husbands has exclusive sexual access to the wife, but among the Sinhalese, the main ethnic group in Sri Lanka, who practise both polygyny and polyandry, the first husband in a polyandrous marriage has priority and the others have to have his consent to sexual intercourse with the common wife.

The extended family

A family comprising three or more generations living together, namely grandparents, parents and children, is known as an extended family. This usually occurs when children grow up, remain with their parental family and bring their spouses to live in or near the parental home. The extended family may contain a cluster of nuclear families around the original parents. Such a family is not destroyed by the deaths of the oldest members as these are replaced, in turn, by the eldest children. In a patrilineal extended family – the most common type – the male members remain attached when they marry, so the backbone of the family consists of the grandfather, his sons and their families and their sons. Unmarried daughters remain in the extended family until they marry, but then become members of the different patrilineal extended families of their husbands. Less commonly, in matrilineal societies, daughters remain members of their mothers' families after marriage and the husbands and children become part of their kin group.

The extended family is characteristic of agricultural societies and becomes less relevant and less common in highly industrialized societies, mainly because family members need no longer work the land together. Close kinship with a larger number of people who acknowledge family responsibilities clearly has advantages when food and other means of subsistence are hard to obtain. This type of family is rarer among the relatively well-off in industrialized countries, but still occurs among the poor and in some immigrant minority groups that retain traditional family attitudes. In addition to the economic advantages, extended families can provide essential support and comfort to those who suffer financial, social and political disadvantages.

In western societies many people combine the benefits of living in an extended family with those of living in a nuclear family by forming a modified extended family. This is characterized by a nuclear family that maintains social, economic and emotional ties with wider kin. Moreover, it is a family form that seems to be spreading with industrialization and urbanization.

The changing family pattern

In the industrialized world, membership of a family has declined greatly in

THE PSYCHOLOGICAL IMPACT OF FAMILY LIFE

Among the factors that determine the nature of a child, few can be so important as his or her experience of the family. Much of this experience is essential to future psychological well-being. The most important elements of family life include an early maternal or paternal bonding; the establishment of easy and secure relationships with parents and siblings; the development of the child's sense of personal worth, particularly to others; and understanding the advantages of submerging natural selfish impulses in the interests of harmonious relationships. The establishment of the latter can be a very prolonged and sometimes painful operation, but it is essential if later life is to be reasonably contented.

These conditioning elements are far more likely to be found in the family life of most people than in any other environment. They are not, however, to be found in every family. In some cases the origins of sociopathic personality disorder can be traced to a family environment in which acceptable conditioning, of the kind described above, was absent. Sociopaths – also known as psychopaths – have a strong tendency to commit antisocial and sometimes violent acts, and often show an apparent inability to feel guilt. In many cases these characteristics are the consequence of an upbringing by parents whose own unhappy conditioning has deprived them of the ability to understand their children's needs. It would be wrong to imply that family influences are the only factors that engender the sociopath, but they are certainly among the most important.

Although there are family environments which destroy self-esteem, inhibit the individual's capacity to love and to attract love, and force the individual to seek satisfaction at the expense of others, it would be untrue to say that it is always a disadvantage to live in a family. Human beings are gregarious animals and those who are driven away from one family by neglect of their emotional needs or worse are likely to look for, and find, a substitute.

The male-dominated family was regarded as the central institution of 19th-century Western society. Families were typically large, but after 1875 the gradual spread of contraception led to a decline in the birth rate in Europe and North America. (ME)

importance compared with pre-industrial times. Before the 18th century, kinship provided the main forms of social organization, stabilized the association between people and the land, promoted sustained agricultural effort and regulated the inheritance of wealth. For most of us today, with the exception of the latter purpose, kinship has ceased to have practical or economic importance and is now of social and emotional significance only. The phenomenon whereby the head of the family exercised unchallenged authority is now rare. Expectations of automatic obedience, even of respect, cease when the children acquire maturity and leave the nuclear family. Authority, once vested in the *paterfamilias*, has now been assumed by various social, legal, religious and commercial institutions which are more complete and pervasive than they were in pre-industrial societies. Today, the former powers of the patriarch would probably be regarded as something close to tyranny.

The changes to the structure and function of the family are by no means confined to Western industrialized nations. The changes are also occurring in many of the developing countries of the world. In large parts of China the traditional extended family is giving way to nuclear families. Similar changes have occurred in India, where, for instance, child marriage is no longer generally considered acceptable. These changes are a direct result of the declining necessity for the economic support of the extended family and the growth of wage-earning. Invariably the changes have begun among the growing urban middle classes where relative wealth and security allow nuclear families to seek autonomy, self-sufficiency and privacy.

Despite the changes, though, the family, whether nuclear or extended, still plays an important part in the life of most people. It is widely regarded as the most satisfactory structure for bringing up children, and a stable early family life is thought by many to be the necessary basis for future psychological wellbeing. Alternatives to the family do, however, exist. In Israel, for example, a minority believe in the kibbutz ideal in which children are largely brought up communally. A minority in the West also believe in the boarding school system in which children, often from the age of eight, are virtually taken over by an institution for large parts of the year, the parents' influence being almost wholly superseded.

For most adults, the importance of the family is primarily emotional or social. Although family breakdown is common the successful family provides a loving environment; support against a sometimes hostile world; partisanship in the face of opposition; a forum for the safe expression of opinion; and a usually willing audience for the recounting of day-to-day detail that would probably be of little interest to anyone else.

The contemporary family

In spite of ever-rising divorce rates in the West and a greatly increased tendency to live together, marriage remains remarkably popular. The great majority of heterosexual people eventually marry or will enter into some permanent relationship. Many homosexual people also form life-long partnerships. Changing norms of premarital sex seem to have little effect on the prevalence of marriage except possibly to raise the average age of people entering into it (because many cohabit before marriage). For women marriage used to be the main guarantee of economic security and was therefore sought early. This is no longer the case and women – and men – often enter marriage after establishing their careers.

The selection of partners continues to be determined largely by social, educational and economic similarity (homogamy), although marriages between people of different ethnic groups have increased and become more generally accepted. Marriage is still embarked upon for romantic rather than practical reasons and the partners expect to be 'in love'. Sexual infatuation commonly matures into real affection and deep mutual regard. When it does not, the disappointment of romantic expectations may increase the risk of marital breakdown.

Belief in the sanctity, and expectations of the durability, of marriage have changed and many people now candidly acknowledge that they do not necessarily regard marriage as a permanent state. The incidence of divorce in the Western world is high. The rise is generally recognized as a result of the increasing number of spouses who refuse to accept poor marital relations, and the greater ease of obtaining divorce, rather than the result of increasing marital discord and unhappiness. Divorce is not so high in developing countries and in countries with large Roman Catholic populations such as Ireland (where divorce is illegal), Spain, Italy and Mexico. The number of single-parent families in the West has increased dramatically in recent years partly as a result of rising divorce rates.

In the West, many people no longer accept that they should fulfil the traditional roles within the family, in which the male is the provider, active in the economic sphere, and the female is the domestic worker and family nurturer. Today many partners seek to share both the financial and the domestic work, although women still play the major role within the home. Conflicting opportunities and demands may, however, lead to tensions. Equally promising, but geographically incompatible, career opportunities may occur for both partners, and such dilemmas can seldom be resolved without conflict within the relationship. It is still more common, however, for couples to move in response to the male career. As the recognition of sexual equality grows, it seems inevitable that some form of social reorganization must evolve to address the problem.

SEE ALSO

- NATURE VERSUS NURTURE p. 30
- THE CYCLE OF LIFE pp. 94–110
- LEARNING IN PRACTICE p. 174
- PERSONAL INTERACTIONS p. 206
- GROUP BEHAVIOUR p. 208

The nuclear family, which spans two generations, is the characteristic family unit of developed industrialized societies. However, most nuclear families are not isolated but are part of modified extended families in which they maintain social, economic and emotional ties with wider kin. (Images)

abdomen, the belly. The region of the body bounded by the diaphragm above and the pelvis below; ➪ pp. 42–3.

abortion, spontaneous or deliberate ending of a pregnancy before the fetus is capable of independent existence outside the womb; ➪ MISCARRIAGE.

abscess, a collection of pus surrounded by a zone of inflammation and often scar tissue.

absorption, movement of fluids, usually containing dissolved substances, across a membrane in the body. The lining of the intestine is one of the major absorbing membranes in the body; ➪ pp. 44–5.

accent, the speech patterns and pronunciation characteristic of a particular geographic region or social class; ➪ pp. 198–9.

accommodation, the adaptation of the focus of the eye for different distances, by changing the radius of curvature of the internal crystalline lens. In youth the lens is naturally elastic and assumes a shape giving increased light convergence for close viewing when the pull on it by suspensory fibres running from the surrounding circular ciliary muscle is relaxed. The circular muscle ring becomes smaller when it contracts, thereby relaxing the pull; ➪ pp. 84–5.

acetabulum, the cup-shaped hollow in the side of the pelvis with which the head of the thighbone (femur) articulates.

acetylcholine, one of the principal NEUROTRANSMITTERS of the body. Acetylcholine diffuses across the gaps of SYNAPSES so as to propagate nerve impulses.

achievement motivation, the persistent impulse to attain a high standard of performance in any activity.

Achilles tendon, the prominent tendon at the back of the ankle that connects the calf muscles to the heel bone and powerfully extends the ankle in walking and running; ➪ pp. 42–3.

achondroplasia, a genetic defect causing dwarfism; ➪ pp. 28–9.

acid, any compound that can release hydrogen ions when dissolved in water. Acids have a pH (the common logarithm of the reciprocal of the hydrogen ion concentration) of less than 7 and turn blue litmus red. In the body, the main concentration of acid is in the STOMACH.

acne, one of the commonest of persistent skin diseases. Acne mainly affects adolescents and young adults, causing inflammation of the oil-secreting sebaceous glands of the face and upper back and featuring BLACKHEADS, pustules and sometimes scarring.

acoustic nerves, a pair of nerves that connect the ears directly to the brain, carrying impulses concerned with hearing and BALANCE; ➪ pp. 86–7.

acquired characteristics, features of an organism, such as the human being, arising from environmental influences or bodily functioning, rather than from heredity.

acromegaly, a growth disorder featuring marked enlargement of the hands, feet, jaw and other parts, caused by excessive production of growth HORMONE in adult life; ➪ pp. 40–1.

ACTH, adrenocorticotropic HORMONE, the hormone that prompts the ADRENAL GLAND to secrete steroid hormones.

actin, ➪ pp. 42–3.

active transport, the movement of dissolved substances across a membrane in the direction opposite to that of normal diffusion. Active transport requires energy expenditure.

acupuncture, a form of traditional Chinese medical practice now used by a few Western doctors. Fine steel needles are inserted into the skin along the lines of imaginary 'meridians', and twirled or vibrated, for the purpose of 'unblocking' them; ➪ pp. 158–9.

acute, sharp and short. Compare CHRONIC.

Adam's apple, the popular term for the protrusion caused in the neck by the cartilage of the LARYNX, especially in men.

adaptation, the way an organism, such as the human being, adjusts to environmental changes.

addiction, the state of dependence on repeated resort to a drug or activity to achieve an artificial comfort of body and serenity of mind.

Addison's disease, an ADRENAL GLAND disorder in which there is inadequate production of the essential steroid hormones. This leads to great weakness and tiredness, colouring of the skin and severe susceptibility to injury or the effects of surgery.

additive, any substance added to something, especially a food, in order to improve or preserve it. Additives are of economic and nutritional importance but some people may display allergic sensitivity to some of them.

adenine, one of the four bases whose sequence in DNA forms the genetic code; ➪ pp. 24–5.

adenoids, collections of lymphoid tissue at the back of the throat. These have a protective function in early life but may enlarge and obstruct the openings to the EUSTACHIAN TUBES leading to deafness and middle ear infection.

adhesion, the abnormal healing together of raw tissues within the body that have been deprived of their normal 'non-stick' lining (epithelium) by injury or disease; ➪ pp. 62–3.

adipose tissue, fat.

Adler, Alfred, ➪ pp. 164–5.

adolescence, the period from the onset of puberty until full adult physical maturity is reached around the age of 20; ➪ pp. 108–9.

adoption, the bringing of a child into a new family relationship in which the adopters assume the full role of parents.

adrenal glands, a pair of endocrine glands that perch one on top of each kidney. The inner parts of each adrenal secrete adrenaline; the outer parts, each in three layers, secrete a range of steroid hormones, especially cortisol, corticosterone, aldosterone and the male sex hormone (androgen) dehydroepiandrosterone; ➪ pp. 80–1.

adrenaline, one of the hormones of the ADRENAL GLANDS. Adrenaline causes an increase in heart and breathing rate, a rise in blood pressure, a sense of alertness and the emotions characteristic of fright; ⇨ pp. 80–1.

adultery, voluntary sexual intercourse between a married person and a person who is not the legal spouse. For good reasons, adultery has generally been condemned. It has ethical, theological and legal implications. In some cultures adultery still carries a death penalty.

Aegyptopithecus, ⇨ pp. 10–11.

Aesculapius, the Greek and Roman god of healing.

affective disorder, any mental disorder that features abnormal elevation or depression of mood, such as mania or pathological sadness, or any other severe disturbance of the emotions.

affective functions, the brain functions concerned with the emotions, especially those of fear, pleasure, gratification of all kinds, sexuality and jealousy. Affective functions are centred in the most primitive parts of the brain, especially the limbic area which we have in common with many other animals.

afferent nerve, a nerve that carries impulses, and hence information, from the periphery of the body to the central processing unit – the brain.

Afro-Asiatic language family, a group of languages spoken throughout most of the Middle East and in large areas of Africa. The group currently includes Arabic and Hebrew and is also known as Hamito-Semitic.

afterbirth, ⇨ PLACENTA.

after image, a visual illusion of the persistence of a bright image, usually in changed form or colour, for a short time after the removal of the stimulus.

ageing, the gradual accumulation of minor bodily injuries or degenerations, often associated with a gradual decrease in functional capacity, that affects all human beings, to a greater or lesser degree, after middle age; ⇨ pp. 110–11.

ageism, discrimination on the grounds of age.

age roles, certain characteristic patterns of behaviour expected of people of different ages. Age roles, once of central importance, and marked in their transitional stages by formal rites of passage, have become less prominent in modern industrialized society.

agglutination, the sticking together in clumps of small particles of matter such as bacteria so as often to form small visible masses. Antibodies cause agglutination.

aggression, the tendency to attack in an unprovoked manner.

agnosia, a disorder of perception of sensory input resulting from brain damage.

agnosticism, a term coined by Thomas Henry Huxley (⇨ pp. 8–9) in 1869. Agnosticism is the belief that scientific knowledge of religious doctrine is impossible, on the grounds that such data is inherently unknowable. The term is distinguished from atheism, which postulates the non-existence of God.

agoraphobia, fear of open or public places.

ague, an old-fashioned term for a fever with shivering stages, such as malaria.

AIDS, the acquired immune deficiency syndrome, a plague of pandemic proportions currently sweeping the world and decimating some populations. AIDS differs from all previous infectious diseases in that it deprives the victim of the normal resistance to infection and some kinds of CANCER. It is caused by a RETROVIRUS called the human immunodeficiency virus (HIV) that invades certain of the 'T' lymphocytes, the helper cells, killing them and depriving the immune system of an essential component. The incubation period of AIDS – the period from the time of infection to the appearance of the first signs or symptoms of the disease – is anything up to 10 years or longer. The relative loss of immune capacity then leads to a syndrome of effects known as the AIDS-related complex (ARC). AIDS features a very low resistance to infections of all kinds and many micro-organisms that do not normally affect the human body produce serious illness. These are called opportunistic infections and treatment exists for all of them. AIDS victims are also prone to develop a form of multiple blood vessel cancer called Kaposi's sarcoma. This is very rare except in people with AIDS. The volume of medical research devoted to the elucidation of AIDS is unprecedented in medical history. Unfortunately, no cure or effective vaccine has, so far, been developed; ⇨ pp. 60–1.

air, the mixture of gases forming the atmosphere of the earth. It consists of about 78% nitrogen, 21% oxygen, 0.1% argon, 0.03% carbon dioxide and smaller proportions of rare gases and ozone.

albinism, a genetic disorder in which body colour fails to develop normally; ⇨ pp. 28–9.

albumin, a soluble protein present in the blood and important for maintaining its concentration and power of absorbing water from surrounding tissue.

alcohol, one of a range of chemical compounds that contain one or more hydroxyl groups bound to carbon atoms that are not in a benzene ring. Ethyl alcohol, the principal constituent of intoxicating drinks, is produced by the enzymatic fermentation of sugars.

alcoholism, a disease featuring the habitual or repetitive consumption of ALCOHOL in quantities that threaten health, prejudice work performance, disrupt relationships and endanger personal safety and the safety of others. There is controversy as to whether alcoholism is the result of a pre-existing personality defect, which can be temporarily assuaged by alcohol, or the result of excessive drinking.

aldehyde, a product of dehydrogenated (metabolized) ALCOHOL, hence the name. Aldehydes cause most of the toxic effects of overindulgence in alcohol.

Alexander technique, a method of therapy for various physical and psychological disorders said to be caused by faulty posture. The pupil is taught how to break habits of slouching and adopt a new and better bearing, thereby, it is claimed, being relieved of such problems as insomnia, lethargy and chronic ill health; ⇨ pp. 158–9.

alienation, a state of estrangement from, or inability to relate to, other people, concepts, social norms, or even oneself. Alienation, especially of the latter type, may be a feature of psychiatric disorder, but equally it may result from an accurate perception of the social environment.

allantois, part of the early embryo that, on development, helps to form the UMBILICAL CORD and the PLACENTA.

allele, an abbreviation for allelomorph. An allele is any one of a number of alternative forms of a GENE. Originally used to define gene variants such as the ones controlling eye colour, the advent of molecular genetics now permits the identification of allelic genes that differ only by a single base pair in their DNA sequences.

allergy, abnormal sensitivity of the body to contact with a foreign substance including pollens, various chemicals, a few food ingredients, dusts, mites and some metals. The result may be hay fever, skin inflammation such as eczema, skin wheals, asthma, or even general collapse; ➪ pp. 142–3.

alphabet, a fixed list of letters, usually arranged in a particular order, each one being a representation of a spoken sound (phoneme) or group of sounds in a language.

Altaic language family, a subdivision of the Ural-Altaic family, this group encompasses the Mongolian, Turkic and Manchu-Tungus sub-families. The Altaic languages are spoken over a very wide area. Mongolian extends from China as far west as Afghanistan and the lower Volga; Turkic from Anatolia to the Volga basin; and Manchu-Tungus in the north-east Siberian coast as far south as the Amur river.

alternative medicine, ➪ pp. 158–9.

altitude sickness, a dangerous condition characterized by excess fluid (oedema) in the lungs and the brain caused by too rapid ascent to high altitudes where the oxygen tension in the atmosphere is low. Also known as mountain sickness.

altruism, behaviour manifesting unselfish concern for the advantage of others. Much seemingly altruistic behaviour can be shown, on analysis, not to be so, and there are those who hold that altruism is a myth. Most social scientists, however, accept the concept.

alveolus, a small bodily cavity or sac, as in the tiny air-sacs of the lungs or the sockets for the teeth.

Alzheimer's disease, a form of DEMENTIA; ➪ pp. 186–7.

amenorrhoea, absence of menstruation, most commonly caused by pregnancy or excessive physical stress.

Amerindian language family, the group of nearly 1000 languages, many now extinct, that were and are spoken by the indigenous American Indian population in all parts of North and South America. Over 700 Amerindian languages are still spoken; ➪ pp. 14–15.

amine, an organic base formed when one or more of the hydrogen atoms of ammonia is replaced by an organic group; ➪ pp. 26–7.

amino acid, one of the 20 fundamental constituents, or 'building blocks' of PROTEIN – the main structural material of the body; ➪ pp. 26–7.

amnesia, loss of memory. This may be of organic or psychological origin (➪ FUGUE), selective or global, mild or severe. Retrograde amnesia is loss of memory for a period following a trauma such as a head injury. Anterograde amnesia is loss of memory for a period prior to the injury.

amniocentesis, sampling the fluid in the womb by suction through a needle passed through the abdominal wall. This is done to obtain biochemical information about the fetus and to get samples of cellular material from which the chromosomal and genetic constitution of the fetus can be determined and hereditary disorders detected early; ➪ pp. 102–3.

amnion, the inner of the membranes that enclose the fetus in the womb. The amnion secretes AMNIOTIC FLUID.

amniotic fluid, the fluid in which the fetus in the womb floats and, from time to time, swallows; ➪ pp. 102–3.

amoeba, a mobile, single-celled, free-living organism. Many of the free-living cells of the body, especially some of those of the IMMUNE SYSTEM, are described as amoeboid because of this characteristic method of moving and engulfing foreign substances; ➪ pp. 24–5.

amphetamine, a drug causing stimulation of the nervous system and a sense of euphoria. It is commonly abused, leading to side effects, some of which may be serious.

ampulla, a widening in a body tube or gland.

amputation, surgical or traumatic removal of a part of the body.

amylase, an ENZYME that accelerates the breakdown of starch into simple sugars. Amylase is present in saliva; ➪ pp. 44–5

anabolic steroids, male sex hormones (ANDROGENS) that promote muscular growth; ➪ pp. 108–9.

anabolism, building up of the tissues of the body. Compare CATABOLISM. METABOLISM includes both anabolism and catabolism.

anaemia, one of a range of conditions in which the amounts of oxygen-carrying pigment in the BLOOD, haemoglobin, are reduced.

anaerobic, capable of generating biological energy in the absence of oxygen. Only certain very simple organisms, such as bacteria, are capable of anaerobic existence. However, human muscles can, for short periods, generate energy under anaerobic conditions, leading to the accumulation of lactic acid and muscle cramps.

anaesthesia, absence of sensation or awareness in a part or the whole of the body. This may result from disease or may be induced deliberately for surgical or other purposes; ➪ pp. 152–5.

anal canal, ➪ ANUS.

analgesics, drugs or other agencies that relieve pain.

anal sex, COPULATION in which the anal canal is used as a substitute for the vagina.

anal stage, a stage in the psychosexual development, proposed by Sigmund Freud, in which the child's preoccupation is with the anal region and the faeces.

anarchism, a political concept holding that all government is bad and should be dispensed with. Anarchists envisage a society based on individual cooperation.

anatomy, the structure of, or the study of the structure of, the human body; ➪ pp. 4–5.

ancestor worship, religious beliefs and activities relating to the spirits of dead relatives or of people believed to be forebears, especially those deemed to be important. The practice is widespread in many cultures, especially in Africa, Asia and the Pacific area.

androgens, male sex hormones. Androgens bring about sexual maturation in males together with all the secondary sexual characteristics such as increased muscularity and BONE structure. Androgens are anabolic, that is, they promote the build-up of body tissues; ➪ pp. 202–3.

aneurysm, a dangerous swelling on an artery, liable to burst. The common sites for aneurysm formation are on the AORTA – the largest artery in the body – and on the small arteries under the base of the BRAIN that supply the blood to it. The former are usually caused by ATHEROSCLEROSIS, the latter by a congenital weakness of the vessel wall; ➪ pp. 140–1.

angina pectoris, a tight gripping pain in the chest, the symptom of inadequate BLOOD supply to the heart MUSCLE from narrowing of the coronary arteries that supply the heart muscle with blood. This is usually caused by the arterial disease ATHEROSCLEROSIS; ➪ pp. 138–9.

angioplasty, a surgical procedure to restore the full flow of blood in an important artery that has become partially or wholly blocked. This may be done by expanding balloon pressure, by laser or by means of a miniature, high-speed rotating cutter, all of which are introduced by way of a fine catheter.

animism, the belief held by many primitive peoples that a spirit resides within every object, controlling its existence and influencing events in the natural world.

ankle, the joint between the lower part of the leg and the foot. The ankle joint is capable of movement only in the vertical plane; sideways movement of the foot occurs between the bones of the foot itself; ➪ pp. 40–1.

ankylosing spondylitis, a persistent (chronic) inflammatory disease of the spine that progressively leads to fusion and fixation of the vertebrae until the spine becomes solid and incapable of bending or twisting.

anomie, lack of moral principle, whether in an individual or a society.

anorexia nervosa, a serious psychiatric disorder affecting almost exclusively young women. It involves a distorted perception of body size and leads to determined self-starvation. The mortality rate is high; ➪ pp. 98–9.

antacid, a drug that tends to neutralize the ACID in the STOMACH so as to prevent damage to the lining of the stomach or the duodenum from excess acidity.

antagonist, anything that acts in a direction contrary to anything else. Most of the body muscles have antagonist muscles which relax, without unduly slackening, as their agonists contract; ➪ pp. 42–3.

anterior, the anatomical term for the front of, or towards the front of, the body. Compare POSTERIOR.

anthrax, a serious infective disease of skin, lungs or intestine usually acquired from infected animals or animal products. It causes black boils, severe pneumonia or gastroenteritis.

anthropoid ape, any tail-less, human-like primate of the family *Pongidae* with a highly developed brain and long arms. The anthropoid apes include orang-utans, chimpanzees, gorillas and gibbons.

anthropology, the science of humankind, and of human cultural differences, from the earliest times to the present. Anthropology is thus a very wide subject, concerned not simply with the less familiar human groups but with every aspect of mankind in a social context. Increasingly, anthropology overlaps the social sciences, but, at the same time, preserves a certain detachment from concern with the more utilitarian aspects of such studies, as befits one of the basic sciences. Cultural anthropology, or ethnology, is a comparative study of cultural systems and includes concern with early archaeology, RELIGION, myth, political and economic systems and language. Other branches of cultural anthropology include psychological, legal and urban anthropology. The observation, recording and analysis of anthropological data in the course of 'field work' are called ethnography. Physical anthropology is the study of human evolution, including recent diversification of humans. Social anthropology covers the whole field of human beings in their social context.

anthropometry, measurement of the size and shape of the human body for scientific and other purposes; ➪ pp. 4–5.

anthropomorphism, the conceiving of a deity or animal in terms of human characteristics or appearance. Man, because of his experiential limitations, commonly resorts to an anthropomorphic concept of anything transcendental.

antibiotics, drugs that are able to kill or inactivate infective organisms without harming the infected person; ➪ pp. 152–3.

antibodies, special proteins (immunoglobulins) capable of a protective function against infecting organisms and other foreign material introduced into the body; ➪ pp. 58–9.

anticoagulants, drugs that slow BLOOD clotting; ➪ pp. 152–3.

antidepressants, drugs used in the treatment of abnormal and persistent sadness (mood depression); ➪ pp. 194–5.

antidiuretic hormone, a hormone produced by the

PITUITARY GLAND that acts on the KIDNEYS to control water loss from the body. Also known as vasopressin.

antigen, any substance that prompts the IMMUNE SYSTEM to produce a specific ANTIBODY to it; ➪ pp. 58–9.

antisemitism, attitudes, policies and behaviour deliberately directed against the Jews.

antisepsis, the attack on micro-organisms of medical importance usually by means of chemical solutions. Compare ASEPSIS.

anus, the external orifice at the lower end of the intestine through which the waste materials of the bowel (faeces) are periodically discharged; ➪ pp. 64–5.

anxiety state, a condition of persistent fear, dread or trepidation in the absence of the normal causes of these emotions; ➪ pp. 186–7, 190–1.

aorta, the main artery of the body, springing from the left lower chamber of the heart, from which almost all other arteries branch, directly or indirectly; ➪ pp. 50–1.

ape, ➪ ANTHROPOID APE.

aphasia, a speech disorder resulting from brain damage; ➪ pp. 76–7, 82–3 (boxes).

aphrodisiac, a drug or other agency purporting to promote sexual interest or arousal in a person.

apocrine, of a gland that gives off some of its cellular contents as part of its secretion. Apocrine sweat gland secretion produces body odour; the sweat from non-apocrine glands does not.

apoplexy, an old-fashioned term for a STROKE.

appendicitis, ➪ APPENDIX.

appendix, a short, blind-ended, worm-like, vestigial structure hanging from the first part of the large intestine (the caecum). Acute inflammation occurring in a blocked appendix is called appendicitis and carries the risk of rupture and the more serious condition of PERITONITIS; ➪ pp. 44–5.

appetite, desire, whether for food, drink, sex, work or anything else that humans can enjoy. Lack of appetite for food is called anorexia, of which a particularly dangerous kind is ANOREXIA NERVOSA.

aqueous humour, the water that fills the space between the back of the outer lens of the eye (the cornea) and the internal focussing (crystalline) lens behind the iris; ➪ pp. 84–5.

Aquinas, St Thomas, ➪ pp. 70–1.

ARC, ➪ AIDS.

archetypes, a term used by the Swiss psychiatrist and mystical thinker Carl Jung (1875–1961) to characterize some of the features of the 'collective unconscious' he believed common to all mankind. Archetypes were, he believed, inherent tendencies to experience and symbolize the many different and important human situations in particular ways. Jung pointed out that all the great mythological and religious systems display these archetypes in common; ➪ pp. 164–5.

areola, the pink or brownish disc-shaped area of skin surrounding the nipple.

aristocracy, any specially privileged class in a society, especially those deemed to be of 'high birth' or of hereditary title and those who have enjoyed wealth for several generations.

Aristotle, ➪ pp. 4–5, 20–1, 70–1, 150–1, 162–3.

arm, the paired forelimb of the human being. The arm skeleton consists of a single upper arm bone, the humerus, and two lower arm bones, the radius and the ulna. These bones are invested by many longitudinally placed muscles that act to move the shoulder joint, the elbow joint, the wrist joint and the fingers; ➪ pp. 40–1.

armpit, medically known as the axilla, this region of the body is formed by the edges of various arm and shoulder muscles and is lined with hair-bearing skin containing many APOCRINE sweat glands. Under the skin lie several LYMPH NODES and these may enlarge and become prominent in breast cancer or arm infections.

aromatherapy, a form of alternative therapy in which essential oils from herbs, flowers and spices are selected according to the nature of the problem and are massaged into the skin and then inhaled. The procedure is claimed to be especially effective for anxiety and depression but is also used to treat various skin disorders; ➪ pp. 158–9.

arterioles, the smallest arteries; ➪ pp. 50–1.

arteriosclerosis, 'hardening of the arteries' by fibrous thickening and loss of elasticity of the walls; ➪ pp. 138–9.

artery, one of many thick-walled elastic tubes that carry high-pressure blood from the heart to the lungs and to all parts of the body. Compare VEINS, that return blood at low pressure to the heart; ➪ pp. 50–1.

arthritis, inflammation of any joint from any cause whether or not associated with permanent joint damage.

articulation, the physical process involved in speaking. The term is also used for a joint between bones.

artificial heart, a device that attempts, so far with little success, to perform the functions of the human heart.

artificial insemination, any method of achieving fertilization in which spermatozoa are brought into proximity with the ovum by means other that COPULATION; ➪ pp. 98–9.

artificial intelligence, the capacity shown by certain machines or electronic networks to perform some of the mental functions of complex biological systems. Because psychologists cannot agree on a definition of human or animal intelligence, there has been much pointless argument over the nature, or even existence, of artificial intelligence.

artificial organs, a range of prosthetic and other devices intended to replace, supplement or externally perform the function of natural organs such as the kidneys, heart or lungs.

artificial respiration, the maintenance of oxygenation of the blood and tissues by the application of external forces to ensure the regular movement of oxygen or air into the lungs. These forces may be applied manually or mechanically or more often by blowing or pumping air into the lungs via the mouth or a tube passed into the windpipe.

aryan, a fictitious concept in Nazi ideology, referring to a non-Jewish caucasian.

asbestosis, a serious and persistent form of lung scarring (fibrosis) and loss of respiratory function caused by inhalation of asbestos fibres.

ascorbic acid, vitamin C, a substance necessary for the formation of healthy COLLAGEN; ➪ pp. 46–7.

asepsis, an important preliminary to surgery in which live organisms are completely eliminated from instruments, dressings etc., usually by physical means such as boiling or steaming in autoclaves.

aspirin, a painkilling drug with several other valuable properties, including the power to reduce the tendency for blood to form thromboses.

asthma, a disease in which there is abnormal tightening of the muscles in the walls of the air tubes in the lungs. This causes narrowing and great difficulty in breathing; ➪ pp. 138–9, 142–3.

astigmatism, an optical eye defect caused by an aspherical cornea; ➪ pp. 84–5.

atheism, the denial of the existence of God. The atheist is not necessarily irreligious; RELIGIONS such as Buddhism and Taoism feature atheism.

atheroma, a localized disorder of the wall of arteries featuring degenerative changes, deposition of CHOLESTEROL and fibrous tissue and an increase in the number of smooth MUSCLE cells. Atheroma is the main pathological change in ATHEROSCLEROSIS; ➪ pp. 140–1.

atherosclerosis, an artery disease, the commonest cause of death in the Western world; ➪ pp. 110–11, 140–1.

athlete's foot, the popular name for a fungus infection of the outer layer (epidermis) of the skin between the toes.

atlas bone, the top bone in the vertebral column on which the skull rocks when the head is nodding and which rotates on the next bone, the axis, when the head is turning.

ATP, adenosine triphosphate, a chemical substance involved in the release of energy from food fuels in the cell.

atrium, a chamber, specifically one of the two upper chambers of the heart, into which flows the venous blood returning from the body.

atropine, a drug derived from the plant *Atropa belladonna* ('deadly nightshade') that relaxes muscle spasm in the intestines and elsewhere and widely dilates the pupil of the eye.

attention, the direction of some of the channels of sensory input to a restricted area of the environment. Since the totality of possible sources of information in our environment is so great, attention, which selectively directs and concentrates awareness and controls input, is clearly of the first importance. Attention is seldom continuous for long because we are all prone to periods of reverie. The continuity, as well as the effectiveness in promoting memory storage, is determined mainly by the degree of interest in the source of the information. It is no accident that people with a wide range of strong interests tend to have well-stocked minds. Motivation, as towards learning or achieving qualification, is a less powerful stimulus to attention than interest. Fortunately, interest grows with knowledge. Attention can be objectively demonstrated by such methods as electro-encephalography or PET scanning, which show special activity in the parts of the brain most employed at the time.

attraction, sexual, the universal tendency, experienced by all but the simplest of animals, leading to physical approximation which may proceed to COPULATION. In most cases the attraction occurs between species members of opposite sex, but homosexual attraction also occurs.

auditory canal, the skin-lined tube leading from the exterior to the ear drum; ➪ pp. 86–7.

Australopithecus, an extinct human-like genus believed by many to be an evolutionary ancestor of contemporary man. The full title is *Australopithecus africanus* meaning 'southern ape of Africa'; ➪ pp. 10–11.

Austronesian language family, a language group, the Malayo-Polynesian family, whose members are spoken halfway round the world from Easter Island to Madagascar, but particularly in Indonesia, the Philippines, Taiwan, Malaysia and Singapore.

authoritarianism, support for the view that political and other control should be exercised by a small, powerful elite which should be strictly obeyed.

autism, a rare but severe childhood disorder in which there is a gross deficit in social responsiveness and interaction and an apparent lack of interest in, or unawareness of, the environment. Response to sensory stimuli may be bizarre. Such children do not wish to talk and sign language may be helpful.

autocracy, government by a single person. A dictatorshipp.

autoeroticism, sexual arousal of oneself. MASTURBATION.

auto-immune disease, disease caused by a defect of the immune system of the body which results in certain tissues being regarded as 'foreign' and being attacked and damaged; ➪ pp. 60–1.

autonomic nervous system, the involuntary part of the nervous system that controls the action of the heart, the tone of the arteries and the intestines, and other parts containing smooth muscle and glands.

autopsy, ➪ POST-MORTEM EXAMINATION.

autosome, one of the 22 pairs of chromosomes other than the pair of sex chromosomes; ➪ pp. 22–3.

auto-suggestion, a form of therapy in which a person supplies his or her own motivation by repeating certain maxims or rules of conduct.

aversion therapy, a form of behaviour therapy in which attempts are made to eradicate unacceptable patterns of behaviour by closely associating them with pain, distress or punishment.

Avicenna, ⇨ pp. 150–1.

axis bone, ⇨ ATLAS BONE.

axon, the long fibre of a NERVE cell. Nerve cell axons may be by far the longest cell processes in the body; ⇨ pp. 72–3.

AZT, azidothymidine, a drug used to prolong life in people suffering from AIDS. Recently its effectiveness has come into question.

B-cell, a LYMPHOCYTE that produces antibodies; ⇨ pp. 58–9.

bacilli, rod-shaped BACTERIA.

bacteria, single-celled organisms of many kinds and of great biological, medical and economic importance. Bacteria are capable of producing enzymes that can break down complex organic molecules and return their constituents to the cycle of life (putrefaction). A comparatively small number flourish at human body temperature and produce toxins dangerous to the human being; ⇨ pp. 120–1, 134–5.

Baha'i religion, a religion founded in 1863 by Mirza Husayn Ali (Baha Ullah) who is believed to be the latest of a series of divine revelations that include Jesus, Muhammad, Zoroaster and the Buddha. The faith emphasizes the unity and value of all religions.

balance, the ability to maintain the upright posture. Balance is mediated by structures within the inner ears that respond to changes in the position and acceleration of the head and inform the BRAIN of such changes so that compensatory muscular action can be arranged; ⇨ pp. 86–7.

baldness, loss of head hair, usually partial. Male pattern baldness is hereditary but very variable. Local, patchy baldness is called alopecia areata. Female baldness is rare but hair thinning is common, especially after the MENOPAUSE.

balloon angioplasty, ⇨ ANGIOPLASTY.

Banting, Sir Frederick (1891–1941), a Canadian physiologist who, in conjunction with Charles Herbert Best (1899–1978), discovered how DIABETES could be treated with INSULIN and lives saved.

baptism, a Christian religious ceremony involving sprinkling of water on the head or immersion in water and symbolizing cleansing from sin and union with the Church.

barbiturate drug, a class of sedative drugs now largely superseded by the benzodiazepines; ⇨ pp. 152–3.

Barnard, Christiaan (1923–), the South African surgeon who in 1967 performed the first human heart transplant and initiated a procedure that has since become established and successful.

barrier methods, ⇨ pp. 100–1.

barter, the trading of goods or services for other similar benefits rather than for money; ⇨ pp. 232–3.

basal cell, a cell of the single-cell-thick lowest layer of the epidermis of the skin, from which all the more superficial layers are derived; ⇨ pp. 56–7.

basal ganglia, several discrete masses of nerve cells lying among white nerve fibres near the base of the brain. These nuclei are mainly concerned with the control of motor functions; ⇨ pp. 74–5.

base (chemistry), a chemical compound that combines with an acid to form a salt and water. In biology the bases of greatest interest are the four chemical groups adenine, thymine, guanine and cytosine which, arranged in sequences of three, form the genetic code. The base triplets lie along the DNA molecule and each specifies an individual AMINO ACID; ⇨ pp. 24–5.

battered-child syndrome, a specific pattern of injuries seen in young children and recognized as strongly suggesting physical abuse rather than the accidents usually claimed by the parents.

beard, ⇨ FACIAL HAIR.

bed sores, skin, sometimes muscle, ulcers over bony points of pressure in people who are bedridden and who do not move, or are not sufficiently moved.

bed wetting, involuntary, or occasionally voluntary, discharge of urine in bed, often during deep sleep.

bedlam, the slang term for the Bethlehem Royal Hospital in London, the first psychiatric hospital, founded around 1400. Visits to observe the noisy madmen became a popular entertainment and the term soon came to be used for any uproar; ⇨ pp. 184–5.

behaviour, the totality of an individual's conduct, including speech. The term is also used for a particular pattern of conduct; ⇨ pp. 208–9.

behaviour disorder, any pattern of behaviour of which society currently disapproves.

behaviourism, the psychological school that holds that information about the mind can be reliably derived only from observation of behaviour and not from reports of conscious experience; ⇨ pp. 162–3, 166–7.

behaviour therapy, ⇨ pp. 194–5.

belching, the noisy expulsion of gases from the stomach, via the mouth. Persistent belching is often the result of semi-voluntary air swallowing.

Bell's palsy, paralysis of one side of the face or part of one side, resulting from damage to the cranial nerve that supplies all the muscles on that side.

belladonna, ⇨ ATROPINE.

bends, a neurological and joint disorder due to the formation of nitrogen bubbles in divers who have ascended too rapidly from great depths. The bubbles block small blood vessels and can cause pain in the joints ('bends') and sometimes brain damage.

benign, of favourable outlook. A benign tumour is a local growth, from an increase in the number of cells, which has no tendency to invade adjacent tissues or to seed out to remote parts of the body. Benign tumours are commonly enclosed in a definite capsule. They can, however, cause trouble by local pressure effects, especially in confined spaces such as the inside of the skull.

benzodiazepine drugs, a class of sedative drugs that has largely replaced the barbiturates.

beri-beri, a nutritional deficiency disease due to lack of vitamin B_1 (thiamine) and featuring brain, spinal cord and heart damage with memory loss, irritability, painful feet, paralysis and often widespread fluid retention (oedema).

Berkeley, Bishop George, ➪ pp. 70–1.

Bertillon, Alphonse (1853–1914), a French police officer who devised a method of criminal identification by making a series of measurements of the body (anthropometry).

Best, Charles Herbert, ➪ BANTING, SIR FREDERICK.

bestiality, sexual intercourse with an animal.

beta blocker drugs, drugs that interfere with the action of adrenaline at the receptor sites for this hormone in the arteries and the heart.

biceps, the prominent, two-headed MUSCLE of the upper arm that bends the elbow and rotates the forearm outwards; ➪ pp. 42–3.

bigamy, purporting to marry when already legally married. Bigamy is usually committed for purposes of sexual access by deception and some cases may thus be regarded as a kind of rape. It attracts similar sentences.

bile, a secretion of the LIVER into the intestine, necessary for the emulsification and absorption of fats; ➪ pp. 48–9.

bilharzia, a popular name for SCHISTOSOMIASIS.

binary fission, a method of asexual reproduction in which single-celled organisms and body cells elongate and then split into two separate daughter cells.

Binet-Simon test, one of the earliest attempts to quantify intelligence; ➪ pp. 168–9.

biochemistry, the study of the chemical processes, largely under the influence of ENZYMES, occurring in living cells; ➪ pp. 36–7.

bioengineering, the interface between medical science and engineering, concerned with the many ways in which mechanical and electronic engineering can assist doctors in providing artificial aids such as hip and other joints, pacemakers, artificial kidneys, artificial limbs and so on. The term is also used in the context of BIOTECHNOLOGY.

biofeedback, the attempt to control autonomic functions, such as blood pressure and heart beat, by various methods that provide the subject with information about the state of these parameters. The effect of deliberate relaxation may thus be gauged.

biogenesis, the universally accepted principle that all living organisms now originate from other living organisms, in contrast to the theory of spontaneous generation.

biological clock, physiological timing devices, commonly synchronized to the day/night cycle; ➪ pp. 66–7.

biopsy, a tissue sample taken for examination by a pathologist so that a positive diagnosis can be made.

bio-rhythms, ➪ pp. 66–7.

biotechnology, the expanding technology that makes use of micro-organisms for the production of new substances, either by means of the ENZYMES they carry, as in fermentation, or by manipulation of their genetics; ➪ RECOMBINANT DNA.

biotin, one of the B vitamins needed in small quantities. Deficiency leads to fatigue, nausea, depression, muscle pains and skin problems.

birth, the emergence of the fetus from the womb and vagina to take up existence independent of direct connection with the mother; ➪ pp. 104–5.

birth control, the prevention of live birth by any means, such as celibacy, chastity, continence, CONTRACEPTION or ABORTION; ➪ pp. 100–1.

birth mark, a skin blemish present at birth and due either to a defect of pigmentation (brown) or to an abnormality of the skin blood vessels (purplish or red). Some of the latter disappear spontaneously during childhood.

bisexuality, having sexual impulses towards people of both sexes, whether or not manifested in sexual activity.

Black, Sir James Whyte (1924–), the Scottish-born pharmacologist who developed the beta-blocker class of drugs for heart disease and high blood pressure, and the drug cimetidine which has revolutionized the treatment of stomach and duodenal ulcers.

Black Death, a popular term for the plague; ➪ pp. 118–19.

blackhead, a small fatty plug with a darkened oxidised tip, that forms in the ducts of the sebaceous glands of the skin.

bladder, the small sac lying in the midline of the pelvis behind the pubic bone, which acts as a temporary receptacle for urine secreted by the KIDNEYS and conveyed to it by the ureters; ➪ GALL BLADDER pp. 48–9.

blastocoel, the cavity of the BLASTOCYST.

blastocyst, a stage in the development of the embryo at which it consists of little more than a hollow sphere of cells; ➪ pp. 102–3.

bleeding, loss of BLOOD from an artery or a vein. This may be due to injury or to a defect of blood clotting or to disease of blood vessels.

blepharoplasty, a cosmetic operation to remove 'bags' from eyelids.

blindness, total loss of the faculty of vision, usually from major disease or injury to the eyes but sometimes from serious brain damage; ➪ pp. 84–5.

blockade (pharmacological), the use of drugs of similar chemical constitution to natural hormones or neurotransmitters, to occupy the receptor sites of these agents so as to prevent them from acting fully.

blood, a complex fluid, partly liquid, partly cellular, that circulates around the body during life under the pumping action of the heart. The blood contains many substances in solution including all those necessary for the growth and maintenance of body structure, which are derived from digested food. The cellular element consists of the red cells (erythrocytes), containing haemoglobin for oxygen transport, and a range of white cells which are some of the active elements of the IMMUNE SYSTEM. Failure of an adequate blood circulation from any cause is known as shock. Unless rapidly corrected by transfusion or other means shock will often prove fatal; ➪ pp. 34–5, 50–1.

blood clotting, the change in the blood from the fluid to a semi-solid or solid state. Clotting is important as a means of preventing excessive blood loss after injury to blood vessels.

blood groups, ➪ pp. 50–1.

blood poisoning, an informal term for SEPTICAEMIA.

blood pressure, ➪ pp. 50–1.

blood tests, a wide range of investigations commonly performed on blood samples to determine such things as haemoglobin levels, red and white cell counts, and the presence and quantities of such substances as protein, sugar, electrolytes such as sodium, potassium, magnesium, chloride and calcium, cholesterol, urea, hormones, tumour markers, antibiotics, alcohol and lead.

blood transfusion, ➪ pp. 50–1.

blue baby, a baby suffering from a disease, usually congenital heart disease, that results in insufficient oxygenation of the blood and hence a bluish tinge to the skin (cyanosis).

body and soul, a duality that has been accepted, largely without question, since the earliest times. The term soul is, however, barely definable and is taken to be an entity associated with the body but clearly distinguished from it. Various accounts of the properties of the soul have been asserted from time to time by theologians, but such assertions are not of a nature as can be verified; ➪ pp. 70–1.

body language, the communication of information, usually of a personal nature, without the medium of speech, writing or other agreed codes. Body language involves a range of subtle or obvious physical attitudes, expressions, gestures and relative positions. It can, and often does, eloquently reflect current states of mind and attitudes towards others, whether positive or negative. Body language is often at variance with explicit verbal statement and in such cases is often the more reliable indicator; ➪ pp. 14–15, 198–9, 206–7.

Boerhaave, Hermann (1668–1738), Renaissance Dutch physician and polymath who made numerous contributions to scientific medicine especially in the fields of anatomy, postmortem appearances and medical education.

boil, an ABSCESS in the skin arising from a bacterial infection of a hair follicle.

bone, the material from which the skeleton of the body is made. Bone consists of a flexible PROTEIN (COLLAGEN) structure, made rigid by impregnation with calcium and phosphorous salts. Many bones serve a dual purpose by providing a location for the BLOOD-forming tissues (marrow). Living bone is in a constant state of flux with outflow and inflow of calcium and AMINO ACIDS to and from the blood. Bone is susceptible to a range of diseases, the commonest being osteoporosis in which there is loss of both protein and minerals with consequent weakening; ➪ pp. 40–1, 62–3.

botulism, a severe form of food poisoning caused by the toxin of the microorganism *Chlostridium botulinum.* Spores of this organism are sometimes found in improperly preserved meat.

bowels, the intestines or digestive tract; ➪ pp. 44–5.

Bowman, Sir William, ➪ pp. 64–5.

Boyle, Robert, ➪ pp. 150–1.

Bragg, Sir William Lawrence, ➪ pp. 24–5.

brain, the central and most vital organ of the body, to the support and maintenance of which all the rest of the body is dedicated; ➪ pp. 74–5, 76–7.

brain death, the state in which there are no indications of electrical activity in the brain. No recovery is possible from brain death.

brainstem, ➪ pp. 74–5.

brainwashing, an informal term for the attempt to alter beliefs, modes of thought and action by the sustained application of physical or psychological punishment. The intention is to replace the existing mental content with a new ideology.

breast, the mammary gland of the female which produces milk after a pregnancy and is one of the more important secondary sexual characteristics. Dissatisfaction with the size of the breast may lead a woman to seek breast augmentation by a cosmetic operation usually involving a synthetic implant. The female breast is peculiarly prone to BREAST CANCER; ➪ pp. 106–7, 108–9.

breast augmentation, ➪ BREAST.

breastbone, the elongated flat bone, or sternum, with which the front ends of the ribs articulate by way of cartilages so as to allow outward movement in breathing.

breast cancer, ➪ pp. 140–1.

breast reconstruction, ➪ BREAST CANCER.

breathing, ➪ RESPIRATION.

breech presentation, the appearance, immediately before birth, of the buttocks instead of the head; ➪ pp. 104–5.

bridewealth, the payment made to the family of a bride by the groom or his family in order to ratify the marriage. Return of the price may be demanded on divorce. Also known as bride-price.

brittle bone disease, a genetic disorder in which BONE fractures occur on minimal trauma; ➪ pp. 28–9.

Broca's area, ➪ pp. 76–7.

bronchiole, ➪ pp. 52–3.

bronchitis, INFLAMMATION of the lining of the main air tubes in the lungs; ➪ pp. 138–9.

bronchus, ➪ pp. 52–3.

Bronze Age, the transitional stage in human cultural evolution, between the stone age and iron age, when humanity put into widespread use the discovery that melting together (alloying) copper and tin would produce a metal, much harder and more useful than either. This metal, bronze, could be fashioned into effective weapons and tools that provided notable advantage over the stone implements of the neolithic period. The bronze age occurred at different times in different places and in some areas was omitted altogether. Smelting of copper dates from about 3800 BC in Iran and it is assumed that bronze was an accidental by-product; ➪ pp. 12–13.

brucellosis, an infectious disease acquired from sheep, goats or cows and featuring recurrent, but progressively milder, episodes of fever, lethargy and muscle aches.

bruise, the mark made by blood released into the tissues under the skin following an injury or in a BLEEDING disorder.

bubonic plague, ➪ PLAGUE.

Buddhism, one of the major RELIGIONS of the world. Buddhism was founded in northern India around 500 BC by Siddhartha Gautama, the Buddha, who, while on a religious quest, achieved enlightenment under a bodhi tree in Bodh Gaya. He then went about teaching and gathering disciples. All beings, he taught, are inextricably caught up in a cycle of birth and rebirth (samsara) and in suffering caused by their actions (Karma). Suffering is caused by desire, but the chain of suffering can be broken by the practice of the noble Eightfold Path – a system of ethical disciplines and practices that includes training in concentration and meditation and leads to the development of wisdom and enlightenment. After the death of the Buddha various councils of Buddhist monks were held and sectarian differences arose. Later, various thinkers added new ideas, some of which mirrored Western philosophy. There are believed to be some 250 million Buddhists in the world today.

Buffon, Georges, ➪ pp. 4–5.

bug, any insect of the order *Hemiptera* such as the bedbug, the assassin or kissing bug, or the chinch bug, some of which spread disease. Also used informally for infective micro-organisms; ➪ pp. 136–7.

bulimia, a compulsive eating disorder in which bouts of large food intake are followed by attempts to induce vomiting or purgation.

bunion, a tender soft tissue swelling over the joint of the big toe resulting from sideways displacement of the toe and pressure from unsuitable footwear.

bureaucracy, an informal term for the characteristics of the administrative systems of large organizations with power over people. Bureaucracy arises from the human tendency to conduct repetitive tasks in a stereotyped and dehumanized manner.

burn, a heat or caustic chemical injury of the skin or other tissues. The significance depends on the depth. Partial thickness burns may heal; full-thickness burns require skin grafting.

bursitis, inflammation of one of the protective tissue pads that lie over superficial bony points such as the head of the big toe or the kneecap.

Burt, Cyril, ➪ pp. 30–1, 168–9.

bushmen, hunter and gatherer people of South Africa, notable for short stature and prominent BUTTOCKS.

buttocks, the prominent twin muscle masses at the base of the back, behind, each consisting of three flat gluteal muscles, the gluteus maximus, medius and minimus. The buttock muscles pull back the thigh bone in walking and standing; ➪ pp. 42–3.

caecum, the wide first part of the large intestine, lying in the lower right corner of the ABDOMEN and bearing the APPENDIX.

Caesarian section, surgical delivery of a baby through an incision in the front wall of the ABDOMEN; ➪ pp. 104–5.

caffeine, an alkaloid drug with brain stimulant properties that occurs in tea, coffee and cocoa and is mildly addictive.

calcium, ➪ pp. 80–1.

calculus, a stone that may form by crystallization anywhere in the urinary tract, in the gall bladder, in a salivary gland or duct, or elsewhere in the body. Calculi usually cause trouble by obstructing a natural channel.

calf, the curved bulge on the back of the lower leg caused by the muscles that are attached via the ACHILLES TENDON to the heel bone and which, on contracting, extend the ankle.

callus, BLOOD clot containing minerals and BONE-forming cells engaged in the process of healing a bone fracture; ➪ pp. 62–3.

calorie, the amount of heat needed to raise the temperature of one gram of water by 1°C. In dietetics the calorie (often spelt with a capital C) is one thousand times this amount, i.e. the amount required to raise 1kg of water by 1°C.

cancer, a general term for the class of all malignant tumours. Cancers are characterized by abnormal local overgrowth of cells, invasion and destruction of adjacent tissue and spread (metastasis) by lymphatic and BLOOD channels to set up new colonies (metastases) in other parts of the body. Different kinds of cancer characteristically vary widely in their degree of malignancy – the rapidity with which they spread. Some remain limited to the site of origin for months or years while others spread early. In general, the outlook in any case of cancer depends on the degree of malignancy and on how early the tumour is detected. Cancers may be treated by surgical excision, radiotherapy or chemotherapy or by a combination of these. New methods of treatment, by which anti-cancer agents are carried to the malignant cells, are constantly being developed; ➪ pp. 140–1.

candidiasis, ➪ THRUSH.

canine teeth, the four pointed teeth lying immediately to either side of the eight central incisors; ➪ pp. 44–5.

cannabis, ➪ MARIJUANA.

cannibalism, the eating of human flesh. There is now a general taboo against cannibalism which is seldom broken except to save life. Many primitive communities have practised cannibalism both for ritual and utilitarian purposes.

canon law, church law, especially that laid down by an ecclesiastical council.

capillaries, the smallest of all blood vessels which form a network pervading the whole body and through the walls of which oxygen, nutrients and cellular waste products pass; ⇨ pp. 50–1.

capital (economics), the totality of productive assets including the stock of goods and money used as means of production. Capital includes labour, buildings, machinery, land, vehicles of transport and the knowledge and skills of the workforce.

capitalism, an economic system in which the means of production are privately owned and which features ready access to borrowed finance and freedom of enterprise. Capitalism is driven by the desire for wealth and directed by market forces.

capital punishment, punishment by death.

carbohydrates, complex sugars (polysaccharides) that form the main staple of the human diet; ⇨ pp. 36–7.

carbolic acid, phenol, a chemical used as a disinfecting agent.

carbon, an element of central importance in living organisms because all organic molecules contain carbon. Carbon is tetravalent, i.e. it can combine with one, two, three or four other atoms, and can thus form an almost unlimited number of compounds, often by linking with many other carbon atoms to form chains of rings; ⇨ pp. 36–7.

carbon dating, a method of establishing the age of any organic matter. Cosmic radiation causes a small but fixed proportion of the carbon atoms (C^{12}) in atmospheric CARBON DIOXIDE to change to the radioactive form C^{14}. During life this isotope is absorbed into the body in the same proportion. After death it decays with a half-life of 5.57×10^3 years. The time since death can thus be estimated fairly accurately from the ratio of C^{12} to C^{14}.

carbon dioxide, the main product, with water, of the oxidation of sugar fuels to release energy; ⇨ pp. 52–3.

carbuncle, a multi-headed BOIL.

carcinogen, any substance or agency that can cause cancer; ⇨ pp. 140–1.

cardiac arrest, cessation of an effective heart beat, either by complete stoppage of contractions or by the development of an ineffectual fluttering action (fibrillation); ⇨ pp. 138–9.

cardiac massage, an inaccurate term for the process of applying rhythmical constricting pressure to the heart, usually by pressing on the front of the chest wall, so as to maintain the circulation and attempt to save life in the event of heart stoppage; ⇨ pp. 138–9.

cardiac muscle, the tissue from which most of the wall of the heart is made. A unique net-like arrangement of muscle fibres with the power of spontaneous repeated contraction; ⇨ pp. 50 1.

cargo cults, religious, usually revivalist, movements occurring in the South-west Pacific area and featuring the expectation or promise that Western consumer goods will be supplied in cargoes of ships or aircraft as a result of the mediation of ancestral spirits.

carotid, pertaining to one of the two large arteries in the neck which, with the two vertebral arteries, supply the head and BRAIN with BLOOD; ⇨ pp. 50–1, 53–4, 74–5.

carpus, the wrist, or the eight small bones of the wrist. Compare tarsus, which are the analogous bones of the foot.

Cartesian dualism, ⇨ pp. 162–3.

cartilage, ⇨ pp. 40–1.

caste, ⇨ pp. 230–1.

castration, the deliberate removal of the testicles. If this is done before puberty the effect is profound, as masculine features do not developp. After puberty, sterility results but the other effects are comparatively minor.

catabolism, the process in metabolism in which complex organic molecules are broken down to simpler forms with the release of energy. Compare ANABOLISM.

catalepsy, prolonged rigid immobility, sometimes in a bizarre posture, that occurs in some cases of SCHIZOPHRENIA.

catalyst, a chemical substance that accelerates a chemical reaction without itself being changed. ENZYMES are catalysts.

cataract, any permanent opacity in the internal crystalline lens of the eye, usually as a result of degenerative changes in its PROTEIN fibres. Cataract is irreversible and the only remedy for severe visual loss from cataract is to remove the opaque lens. In practice, the lens is carefully ejected from its capsule, all residual opaque material is removed by microsurgery, and a tiny plastic lens is inserted. The results of this operation are excellent; ⇨ pp. 84–5, 110–11, 156–7.

catarrh, inflammation of a mucous membrane, especially that of the nose or throat, featuring the discharge of excessive mucus.

catatonia, the tendency to adopt inappropriate positions or to make inappropriate movements as a feature of psychotic disorder, brain damage or hysteria.

category mistake, a philosophic concept, attributed to the English philosopher Gilbert Ryle, which purports to show that all the speculation, discussion and theorizing on the 'mind–body' problem has been a waste of time and effort; ⇨ pp. 70–1.

catharsis, a psychoanalytic term for the relief from anxiety and tension said to be experienced when repressed unpleasant memories are brought to consciousness.

catheterization, the passage of a narrow tube, usually of soft plastic, into any cavity of the body such as the urinary bladder, any part of the circulatory system, or the heart.

Caucasian, belonging to the light-complexioned or white-skinned groups of humankind.

causality, the agency by which one event leads to another and by which they are connected in a manner more than merely consecutive. The Scottish philosopher David Hume (1711–76) pointed out that no necessary connection can be observed between cause and effect and that the association is merely one of expectation based on past experience.

cauterization, the use of heat or caustic chemicals to destroy diseased tissue or coagulate blood and control bleeding.

celibacy, the state of being unmarried and of avoiding sexual intercourse. Celibacy is mainly associated with the professional clergy of various religions. Roman Catholic priests, for instance, are forbidden marriage by canon law. Celibacy should be distinguished from CHASTITY.

cell, the basic structural entity of the body; ➪ pp. 38–9.

cellulite, a popular term for the dimpled appearance, especially on the thighs, caused by the natural tethering of the skin that is brought into prominence by obesity.

cellulitis, inflammation of the skin; ➪ pp. 130–1.

cellulose, a polysaccharide that is the major constituent of plant cell walls and commonly appears in food but for which we have no digestive ENZYMES. Roughage that is neither digested nor absorbed.

Celtic languages, members of the INDO-EUROPEAN LANGUAGE FAMILY and including Irish and Scottish Gaelic, Welsh and Breton.

centriole, one of the two rod-like bodies in cells forming the poles of the spindles during cell division (MITOSIS).

cephalosporin antibiotic, a range of antibacterial drugs of structure similar to that of the penicillins and of wide application and low toxicity.

cerebellum, ➪ pp. 74–5.

cerebral aneurysm, a small balloon-like swelling in a localized weak area on one of the arteries supplying the brain. Rupture of such an aneurysm is highly dangerous and often fatal.

cerebral cortex, the much-infolded thin outer layer of the BRAIN which, by virtue of its massive development, distinguishes the human brain from that of lower animals. The cortex can be mapped out into functional areas; ➪ pp. 14–15, 76–7.

cerebral haemorrhage, bleeding within the brain.

cerebral hemisphere, ➪ pp. 14–15, 76–7.

cerebral palsy, a form of partial paralysis, present from birth and affecting usually the legs, but sometimes other parts of the body. In a proportion of cases there is also intellectual retardation.

cerebral thrombosis, clotting within an artery supplying part of the brain with blood. This usually causes a STROKE; ➪ pp. 74–5, 138–9.

cerebrospinal fluid, the watery fluid that surrounds and cushions the brain and that fills the spaces (ventricles) within the brain.

cerebrum, the main part of the brain.

ceremonies, ➪ RITUAL.

cervix, the neck of the womb; ➪ pp. 104–5.

Chain, Ernest, ➪ pp. 152–3.

chancre, the primary sore of SYPHILIS.

chancroid, a tropical sexually transmitted disease causing painful genital ulcers and swelling of the groin lymph nodes. Also known as 'soft sore'.

characteristics, ➪ pp. 28–9.

Charcot, Jean-Martin (1825–93), French neurologist who worked at the Salpêtrière in Paris from 1862, made notable observations of a wide range of neurological and hysterical disorders, practised hypnotism, and greatly inspired and influenced SIGMUND FREUD.

chastity, abstention, usually by choice, from sexual activity, generally from a conviction that the state of virginity possesses or confers merit or constitutes a sacrifice to God. Chastity is voluntarily undertaken by monks and nuns.

chauvinism, an unreasonable and offensive degree of expression of partisanship, patriotic sentiment or jingoism. The term derived from the name of a simple-minded French soldier Nicolas Chauvin who was loud in his expression of satisfaction with all things Napoleonic. Later the sense changed to denote undue partiality to any place or social group, and it is now sometimes narrowed to indicate a sense of male gender superiority; ➪ pp. 208–9.

cheekbone, a prominent narrow bony arch known as the zygoma, prone to suffer a disfiguring depressed fracture from direct violence, that requires surgical elevation.

chemical messengers, an alternative term for hormones; ➪ pp. 78–9.

chemoreceptor, a specialized chemical binding site on the surface of a cell to which HORMONES and other substances will attach if they are of the complementary chemical 'shape', thereby effecting a change within the cell.

chemotherapy, ➪ pp. 152–3.

chewing, ➪ pp. 44–5.

Cheyne-Stokes respiration, periods of very shallow breathing alternating with periods of deep breathing. This characteristic pattern of breathing often precedes death.

chickenpox, a usually minor infectious disease of childhood featuring a rash of tiny, itching blisters that crust and form scabs. The virus remains in the nerve cells and may later reactivate to cause SHINGLES in adults. Thus adults with shingles can cause chickenpox in children, but not *vice versa*; ➪ pp. 132–3.

chilblain, raised, red, itchy areas of skin, especially of the fingers and toes, that occur in cold weather.

childbirth, ➪ pp. 104–5.

childhood, ➪ pp. 106–7.

child rearing, ➪ pp. 106–7.

chimpanzee, ➪ pp. 14–15.

chin, the front part of the face below the lower lipp. The chin is formed by the jaw bone (mandible) and the muscles and skin that cover it. It is a very variable feature of the human face, sometimes large and protruding, sometimes small and receding. There is no correlation between chin and character, as is sometimes suggested.

chiropody, a paramedical specialty devoted to the care of the feet and the treatment of minor foot and toenail disorders.

chiropractic, ⇨ pp. 158–9.

chlamydial infection, ⇨ pp. 98–9.

chloroform, a general anaesthetic agent introduced in 1847 that provided an easy and pleasant induction and made anaesthesia respectable and popular. It has long been abandoned, however, because of its dangerous tendency to cause heart stoppage and fatal liver damage.

cholera, a water-borne infection of the intestinal tract that causes damage to the lining of the bowel, severe diarrhoea and massive fluid loss from the body. The resulting dehydration is the usual cause of death; people adequately treated with intravenous fluids nearly always recover; ⇨ pp. 124–5.

choleric type, the personality type characterized by easily aroused anger, supposed by early philosophers to be due to an excess of yellow bile.

cholesterol, a vital ingredient of the body, present in all cell membranes and an essential basic material for the synthesis of the steroid hormones. Large quantities of cholesterol are excreted by the LIVER into the intestine in the bile but most of this is reabsorbed into the BLOOD. Popular concern about cholesterol arises from the knowledge that it is deposited in the atheromatous plaques in the linings of arteries in the dangerous disease of ATHEROSCLEROSIS. The relationship between a high dietary cholesterol intake and the tendency to develop this disease is by no means obvious and there is no real evidence that the two are related. Soluble dietary fibre can bind intestinal cholesterol so that it is not reabsorbed but is eliminated from the body; ⇨ pp. 46–7, 78–9.

Chomsky, Noam, ⇨ pp. 14–15, 174–5.

chorea, ⇨ pp. 28–9, 186–7.

chorionic villi, the early stage of the formation of the placenta. Since both the chorionic villi and the embryo are derived from the same fertilized ovum a sample of the former provides material for genetic studies of the latter. Chorionic villus sampling has become an important method of early pre-natal screening for genetic defects; ⇨ pp. 102–3.

Christianity, one of the principal RELIGIONS of the world, based on the enlightened social and theological teachings of the 1st century Palestinian prophet Jesus of Nazareth. Biographical material and a record of many of the sayings of Jesus are recorded in the four, considerably overlapping, gospels of four of his disciples, Matthew, Mark, Luke and John. These writings, recorded many years after the event, form the main basis for Christian belief. Christianity was widely promulgated, largely by the much-travelled and articulate apostle Paul of Tarsus who, after initially persecuting Christians, had a religious experience and conversion on the Damascus road. Much dogmatic theology and ritual practice have been superimposed on the ideas expressed by Jesus, and the religion has repeatedly been divided by conflicting factions. Currently its main divisions are Catholicism and Protestantism. Recent attempts to unite the Christian churches (ecumenicism) have not been particularly successful.

chromatography, a method of separating and analysing substances in solution or mixtures of gases by selective absorption. Chromatography is widely used in medical diagnosis.

chromosome, ⇨ pp. 20–1, 22–3.

chronic, lasting for a long time, persistent or permanent. Chronicity does not imply severity. Compare ACUTE.

chyme, the semi-fluid mixture of partially digested food and digestive juices that is moved from the stomach to the small intestine for further DIGESTION and ABSORPTION.

cigarette smoking, ⇨ TOBACCO.

cilia, tiny, beating, hair-like structures found on the surface of certain lining cells.

circadian rhythm, ⇨ pp. 66–7.

circulation, ⇨ pp. 50–1.

circumcision, removal of the foreskin of the penis.

cirrhosis, fine, diffuse, fibrous scarring, especially of the liver; ⇨ pp. 48–9.

citric acid cycle, ⇨ KREBS CYCLE.

city state, miniature autonomous kingdom confined to a single city or to a city and its dependents, especially those of ancient Greece, Rome, Sparta and Carthage.

civil disobedience, a deliberate policy of non-violent protest against authority to achieve a political end. Disobedience takes the form of refusal to pay taxes or to obey some of the law.

civil rights, entitlements, privileges and protection concerning person and property, conferred by enlightened governments and guaranteed by law. Civil rights also include security from discriminatory treatment by governments, administrative bodies and individuals.

clairvoyance, the claimed ability to perceive other than by the senses. Most if not all episodes of alleged clairvoyance are either illusory or fraudulent.

clan, a grouping of people connected by common ancestors or by marriage or merely by surname. Clans recognize allegiance to a common, often hereditary, leader.

class, ⇨ pp. 230–1.

clavicle, the collar bone; ⇨ pp. 40–1.

cleft palate, a congenital defect associated with a cleft lip and due to a failure of full fusion of the processes that form the face early in fetal life. The mouth cavity communicates with the nasal cavity causing a severe defect of voice production.

climacteric, ⇨ MENOPAUSE.

climax, ⇨ ORGASM.

clitoris, the small, highly sensitive and erectile organ situated centrally at the front of the vulva and mainly covered by a hood of skin. The clitoris is the female analogue of the penis and, as far as sexual arousal is concerned, serves a similar purpose. Appropriate stimulation can induce an ORGASM; ⇨ pp. 96–7.

clock, biological, ⇨ pp. 66–7.

clone, a perfect copy, or a population of perfect copies of an individual cell or organism of any kind, usually as a result of asexual reproduction. In molecular biology a clone is a genetically engineered replica of a DNA sequence.

clothing, body coverings that serve a protective, heat-insulating, decorative and modesty-preserving function. The clothes a person wears are incorporated to a remarkable degree into other people's perception of that person. It is almost as if the clothes were part of the body and were to be identified with the person. Thus, a person of no great physical attractions may appear attractive if attractively dressed; a short person may appear taller by putting on a tall hat; a smart uniform confers authority; conformity to a particular fash–ion in dress confers acceptance by a group; and so on. This phenomenon has made preoccupation with dress one of the central concerns of a large section of humanity and has spawned a huge industry. It also tells us something about the relative importance of emotion and logic in the assessments most of us make of our fellow men and women.

clotting, ⇨ BLOOD CLOTTING.

club foot, a popular term for one of a variety of congenital foot deformities of which the commonest is known as talipes equino-varus. In this deformity, the ankle is extended and the foot is inverted and curled inwards.

clubbing, an abnormality of the fingers or toes in which the usual concavity at the root of the nail is replaced by a convexity. Clubbing is a feature of various persistent disorders such as congenital heart disease, chest infection, emphysema and bronchiectasis.

cocaine, ⇨ pp. 146–7.

coccus, a common form taken by bacteria. Cocci are roughly spherical and stick together either in strings (streptococci) or bunches (staphylococci).

coccyx, the skeleton of the residual tail, consisting of five tiny vertebrae fused together.

cochlea, the snail-shell-shaped organ in the inner ear that contains the structures by which sound vibrations are analysed and transduced to the form of frequency-modulated NERVE impulses. These are carried by the acoustic nerves to the BRAIN to prompt the sensation of meaningful sound.

codon, any group of three consecutive BASES along the length of a DNA molecule, that code for a particular AMINO ACID.

coenzyme, a non-protein substance that combines with an ENZYME to allow it to act. Most of the B vitamins are coenzymes.

cognition, ⇨ pp. 166–7, 198–9.

cognitive functions, ⇨ COGNITION.

cognitive psychology, the psychology concerned with those aspects of brain activity subserving perception, sensation, reasoning, memory, imagination, ideation and the formation of new concepts; ⇨ pp. 166–7.

coitus, COPULATION or SEXUAL INTERCOURSE.

coitus interruptus, an unreliable method of CONTRACEPTION in which the penis is withdrawn from the vagina before ORGASM occurs.

cold, common, the commonest of all infectious diseases, an inflammation of the lining of the nose and throat, lasting for about a week and causing sore throat, sometimes fever, and nasal congestion and discharge. Colds are caused by one of some 200 different viruses, and their range and ready tendency to mutate makes it very difficult to produce an effective vaccine; ⇨ pp. 132–3.

cold sore, the painful, crusty blister at the junction of the skin mucous membrane of the mouth or nose caused by the *Herpes simplex* virus.

colic, a pain due to stretching of the bowel wall in its efforts to move along an irritant or a partial obstruction.

colitis, inflammation of the COLON.

collagen, the main structural PROTEIN of the body, of great tensile strength and constituting about a quarter of the total body weight. Collagen is the main fibrous element in bone, tendon, cartilage, skin, blood vessels and teeth and is present in nearly all the bodily organs as a 'connective tissue' holding the cells and tissues together; ⇨ pp. 26–7, 36–7.

collar bone, the clavicle, a short, blunt-ended bone that links the upper end of the BREASTBONE with a bony protrusion on the shoulderblade (scapula); ⇨ pp. 40–1.

collective unconscious, an entity, deemed to be a kind of storehouse of ancestral memory, proposed by the Swiss psychiatrist and philosopher Carl Gustav JUNG to explain similarities in symbolism among disparate peoples; ⇨ pp. 164–5.

collectivism, the political concept of the common ownership, by the state or the people, of the means of production. Collectivism is a central tenet of COMMUNISM.

colloid, a state of matter in which finely particulate substances, too small to settle by gravity, are suspended in a fluid medium. Milk, for example, contains colloidal fat.

colon, the main part of the large intestine. The colon starts in the lower right part of the ABDOMEN at the CAECUM, ascends on the right side, swings across to the left side and then descends on the left side. An 'S' shaped part, the sigmoid colon, then moves to the mid line to enter the rectum; ⇨ pp. 44–5.

colostomy, a surgically formed artificial exit for the bowel, usually through the front wall of the ABDOMEN.

colour (skin), the characteristic hue due to the concentration of melanin pigment, the thickness of the skin and the degree to which it is transparent to underlying blood vessels; ⇨ pp. 22–3, 56–7.

colour blindness, ⇨ pp. 84–5.

colour vision, the ability to perceive the world in colour, mediated by the retinal cones which provide a maximal output signal to the fibres of the optic NERVE for one of three spectral wavelengths, thus discriminating between the three primary colours of ambient light; ⇨ pp. 84–5.

coma, a state of deep, unrousable unconsciousness due to a temporary or permanent disorder of the brain.

comedone, ➪ BLACKHEAD.

commune, a group of people or families living together and sharing possessions and duties.

communication, the transmission of information. Human communication is conducted through many different channels, some less obvious than others. These include speech, the written or printed word, body language, recorded data, art, music, drama and film; ➪ pp. 14–15, 206–7.

communism, an idealistic political system conceived on the basis of an unrealistic appraisal of human nature. Communism involved collective ownership of the means of production, central economic planning and a monolithic party system.

comparative anatomy, the study of the similarities and differences between the body structure of different animals. Although external appearances may vary considerably, in many cases the similarities are much greater than the differences. This observation has been one of the principal reasons for the belief that we have evolved from common ancestors.

complementary pairs, the pairs of bases that link together, like the rungs of a ladder, along the length of the DNA molecule. The whole vital process of DNA replication depends on the fact that adenine can only link to thymine and guanine can only link to cytosine; ➪ pp. 24–5.

complex, a psychoanalytic term for an unconscious or partly unconscious wish to behave in accordance with a particular tendency unacceptable to the conscious mind and thus repressed with damaging effect.

compulsive behaviour, the performance of often irrational or pointless acts in spite of a strong wish to the contrary.

computer/brain analogy, ➪ pp. 72–3.

computer intelligence, ➪ ARTIFICIAL INTELLIGENCE.

computerized axial tomography, CT or CAT scanning; ➪ pp. 128–9.

conative functions, the brain functions concerned with needs, drives, desires, purpose and motivation, whether reasonable or unreasonable, voluntary or involuntary.

conception, ➪ pp. 96–7.

concussion, a form of brain damage resulting from physical force and often leading to a brief period of unconsciousness and then an apparently full recovery. Repeated concussion, as sustained by many boxers, leads to obvious and permanent loss of brain function.

conditioned reflex, an involuntary response that has become associated with a stimulus that normally produces the response; ➪ pp. 172–3.

conditioning, an important element in human programming and behaviour. Conditioning is a form of learning in which a particular stimulus will eventually and reliably elicit a particular behavioural response; ➪ pp. 172–3.

condom, a long-established device in the form of a sheath used during sexual intercourse for purposes both of contraception and protection against sexually transmitted (venereal) disease. The traditional male condom is worn on the erect penis; the female condom is a larger device used to line the vagina; ➪ pp. 100–1.

cone, one of the light- and colour-sensitive elements that are concentrated towards the centre of the RETINA. Compare the colour blind rods that are concentrated more peripherally; ➪ pp. 84–5.

cone biopsy, a method of removing a cone-shaped or cylindrical sample of tissue from the inside of the CERVIX of the womb for examination to establish whether CANCER is present and, if so, whether the procedure has removed it. Bleeding and scarring and an increased MISCARRIAGE rate are possible complications.

Confucianism, the ethical system of the Chinese philosopher Confucius (K'ung Fu-tzu) (c. 551–479 BC), that emphasizes moderation, absolute justice, gentlemanly values, virtue and respect for ancestors.

congenital disorder, any bodily abnormality present at the time of birth.

conjunctiva, the thin, transparent membrane that covers the white of the eye and the insides of the eyelids and forms a deep cul-de-sac near the root of each lid; ➪ pp. 84–5.

conjunctivitis, inflammation of the CONJUNCTIVA. Also known as 'pink eye'.

connective tissue, fibrous COLLAGEN structures that hold the organs and tissues of the body together.

conscience, a seemingly innate sense of right and wrong, formerly commonly attributed to divine guidance but now generally recognized as being due to forgotten conditioning early in life.

consciousness, full awareness of self and of one's environment.

Conservatism, a political system popular with the middle classes and based on a general opposition to radical change and the preservation of all that is deemed to be best in established Church and the monarchy, free enterprise, minimal governmental interference with the individual and anti-inflation policies.

constipation, less than normal frequency in the emptying of the rectum, often with difficulty and discomfort, usually as a result of unsuitable diet and of ignoring the call of nature.

consumerism, advocacy of a policy of high spending in order to stimulate the economy.

consumption, an old-fashioned term for TUBERCULOSIS of the lungs.

contact lenses, small, plastic optical devices worn directly on the surface of the corneas to correct refractive errors of the eyes.

contagion, transmission of disease from person to person by direct contact or by way of contaminated material; ➪ pp. 130–1.

contamination, pollution of an object, person or environment with bacteria or material considered to be dirty or impure.

contraception, ➪ pp. 100–1.

contraceptive methods, ➪ pp. 100–1.

contractions, tightening of the muscular wall of the womb so as to expel the fetus in the process of childbirth; ➪ pp. 104–5.

convergent thinking, analytical thinking that follows a set of rules, as in arithmetic, or in which the logical validity of the thought processes is checked and verified. Compare divergent or creative 'lateral' thinking, characterized by unorthodox mental processes but often productive of a number of different and sometimes valuable solutions.

coordination, effective and cooperative action between various parts of an organism, as provided by the CEREBELLUM and other parts of the BRAIN and their connections.

coprophilia, abnormal attraction towards, and interest in, excrement.

coptic, a long-extinct Afro-Asiatic language that used the Greek alphabet.

copulation, the act of performing the sexual function in which the erect penis is moved in the vagina so as to stimulate ORGASM and the emission of SEMEN.

cornea, the external lens of the eye behind which the coloured iris can be seen; ➪ pp. 84–5.

corneal dystrophy, one of a number of disorders of growth or development of the CORNEA.

corneal graft, a surgical procedure to replace a central disc of opacified CORNEA with a clear disc taken from a donated eye.

coronary artery, one of the two arteries whose branches form a kind of crown over the heart (hence the name) and plunge into the ever-moving heart muscle to supply it with the considerable throughput of BLOOD it needs to maintain its constant contractions; ➪ pp. 50–1.

coronary thrombosis, ➪ HEART ATTACK.

corpuscle, an old-fashioned term for a cell suspended in a fluid, especially a red blood cell (erythrocyte) or a white blood cell (leucocyte).

cortex, the outer layer of an organ such as the BRAIN or the ADRENAL GLAND; ➪ pp. 76–7.

Corti, organ of, the mechanism in the cochlea of the inner ear that converts sound vibrations into nerve impulses and provides the brain with the electrical information that is interpreted as meaningful sound.

corticosteroid drugs, drugs that reproduce or simulate the action of the hormones produced by the outer layer (cortex) of the adrenal gland. Such drugs have a powerful anti-inflammatory action and boost the supply of fuel to the body enabling it to cope in an emergency. They can be life-saving.

cortisol, the principal natural steroid produced by the outer layer of the adrenal gland. Hydrocortisone.

cortisone, a commercial preparation that is converted into CORTISOL in the liver; ➪ CORTICOSTEROID DRUGS.

cosmetic surgery, any operation designed to improve appearance.

cot death, the sudden and unexpected death of a baby during sleep for reasons that are not immediately apparent.

cough, sudden expulsion of air from the lungs for the purpose of removing an irritant of any kind from the air passages; ➪ pp. 126–7.

counter-culture, any cultural trend that runs contrary to the prevailing social patterns.

courtship, a period during which the favour of another person is cultivated with the aim of persuading that person to agree to marriage.

crab louse, the crab-shaped human louse *Phthirus pubis* that is usually confined to the region of the pubic hair and is conveyed from person to person during SEXUAL INTERCOURSE.

crack, a form of cocaine; ➪ pp. 146–7.

cramp, a sustained and painful abnormal contraction of a muscle or group of muscles.

cranium, the vault of the skull; ➪ pp. 40–1.

creation myth, ➪ pp. 218–19.

creationism, the belief that the account of the creation of the world contained in the first chapter of the book of Genesis is literally true. The implication, often expressed, is that the scientific account is false.

creativity, ➪ pp. 76–7, 176–7.

cremation, destruction of the dead body by burning.

creole, a fully formed language that has developed from a PIDGIN to become a primary vehicle of communication; ➪ pp. 14–15.

cretinism, a severe form of physical and mental retardation due to gross iodine deficiency before and after birth. Very little iodine is needed but a small quantity is essential for the synthesis of the thyroid hormones that have a profound effect on cell METABOLISM.

Crick, Francis, ➪ pp. 24–5.

criminology, the study of criminal behaviour, of the nature and origins of crime and of crime prevention.

Cro-Magnon man, one of a group of prehistoric, but anatomically modern, humans who lived in what is now the Dordogne and in other parts of France and Italy in the late Pleistocene era, between about 30 000 and 10 000 years ago; ➪ pp. 10–11.

cross-dressing, ➪ TRANSVESTISM.

crossed eyes, an informal term for strabismus or SQUINT.

crossing-over, the important exchange of genetic material between homologous pairs of chromosomes that helps to ensure the genetic diversity of individuals and their differences from their parents; ➪ pp. 20–1, 22–3.

croup, inflammation of the air tubes in young children with swelling causing painful, noisy breathing and a peculiar barking cough; ➪ pp. 126–7, 132–3.

crown (of tooth), the part of the tooth that is visible above the level of the gums and covered with enamel; ➪ pp. 44–5.

crying, the uttering of inarticulate sobbing or wailing sounds, associated with the secretion of tears and often with facial contortion, that expresses the emotion, usually of grief or sadness but sometimes of joy. Crying in babies and infants is prompted by minor distressful stimuli and has value in exercising the respiratory muscles, but may, if excessive, cause severe parental stress.

cryonics, rapid freezing of a human corpse soon after death for the purpose of preservation in the expectation of future medical advances that may make restoration of life possible.

cuddling, ➪ pp. 16–17.

culture, the totality of the common traditions, beliefs, values, activities, artistic expressions, dress, customs, prejudices, superstitions, indeed of all the shared information, of a society.

cuneiform writing, the system of wedge-shaped pictographic characters used in the writing of several ancient peoples such as the Sumerians and Babylonians from about 4000 to 2000 BC.

cunnilingus, sexual stimulation of the female genitalia by the partner's mouth and tongue. Compare FELLATIO.

curare, a powerful alkaloid drug, derived from South American trees of the *Strychnos and Chondrodendron* species, that for centuries has been used as a paralysing arrow poison. It is now widely used by anaesthetists to obtain extreme muscle relaxation for surgical purposes so as to eliminate the need for large doses of central nervous system depressants.

Cuvier, Baron Georges, ➪ pp. 4–5.

cyanide, a dangerous poison that blocks the action of vital cellular enzymes causing headache, convulsions, coma and often rapid death.

cybernetics, the study of the control and communication systems common to machines and animals, including the human being. The study of the analogies between complex feedback control systems and human physiology has been fruitful to both disciplines.

cystic fibrosis, a genetic disorder affecting body glands causing them to produce abnormally thick, sticky mucus which clogs them or tends to obstruct body passages. There are many complications. Recently the gene that has mutated in sufferers has been identified, thus paving the way for GENE REPLACEMENT THERAPY; ➪ pp. 124–5, 148–9.

cystitis, inflammation of the urinary bladder, usually from infection, causing frequency of urination, a burning pain on passing urine and a sense of incompletion afterwards.

cytoplasm, ➪ pp. 38–9.

cytosine, one of the four bases whose sequence in DNA forms the genetic code; ➪ pp. 24–5, 26–7.

dandruff, scaliness of the scalp which, if severe, is probably due to infection with the fungus *Malassezia furfur*. The medical term for dandruff is pityriasis capitis.

Darwin, Charles, ➪ pp. 8–9, 204–5.

deafness, partial or total loss of the faculty of hearing from disorder of any of the structures of the middle or inner ear, of the acoustic nerves, or of the part of the brain concerned with hearing; ➪ pp. 86–7.

death, termination of the vital respiratory and metabolic processes, whether at a cellular, tissue, organ or whole body level; ➪ pp. 112–13.

death wish, Freud's 'thanatos', which, like so many of his concepts, was derived from classical mythology. This idea, conceived late in his career, proposed that responses such as denial and rejection of pleasure or the repeated seeking of extreme danger indicated a general wish or instinct for death; ➪ pp. 164–5.

defecation, the process of emptying the rectum by way of the anal canal; ➪ pp. 64–5.

defence mechanisms, an important set of unconsciously motivated protective behaviours by which we conceal from ourselves matters, including aspects of our own personalities, which are unpalatable, threatening or anxiety-provoking. The process is not always healthy, and some forms of dynamic psychotherapy are based on the elucidation of defence mechanisms; ➪ pp. 188–9.

defibrillation, the use of a sudden electric shock to restore normal rhythm to a heart which has passed into a state of ineffectual and rapidly fatal fluttering.

de Graaf, Regnier, ➪ pp. 94–5.

dehydration, an abnormal and sometimes dangerous state of reduced water content of the body. This occurs either by excessive water loss as in cholera, persistent diarrhoea or excessive sweating, or from inadequate water intake.

déjà vu, ➪ pp. 170–1.

delayed immune response, the response of the immune system to infection, or the introduction of foreign material, that is mediated by T LYMPHOCYTES rather than by ANTIBODIES.

delirium, a mental disturbance featuring restlessness, confusion, disorientation and often fearfulness and indicating some disturbance of brain function.

delirium tremens, DELIRIUM characteristic of alcohol withdrawal, often with HALLUCINATIONS.

deltoid muscle, ➪ pp. 42–3.

delusion, ➪ pp. 192–3.

dementia, ➪ pp. 186–7.

democracy, government by the people through representatives voted into office by a society nominally free to do so.

demotic writing, the hieroglyphic cursive handwritten script used for business and other common purposes in ancient Egypt from about 700 BC until about AD 500.

dendrite, one of the many short processes of NERVE cells that carry impulses towards the cell body. They allow highly complex inter-connections between nerve cells that act as summation 'gates' controlling the movement of impulses; ➪ pp. 72–3.

dengue fever, a tropical disease spread by mosquitoes featuring high fever, severe muscle and joint aches, prostration, lymph node enlargement, a widespread skin rash and persistent weakness.

dentistry, the profession dedicated to the care, preservation and cosmesis of the teeth.

dentition, ➪ pp. 44–5.

deoxyribonucleic acid, ➪ pp. 24–5.

depression, ➪ pp. 190–1.

deprivation, failure to obtain or to be provided with a sufficiency of the material, intellectual or spiritual requirements for normal development and happiness.

dermatitis, any inflammatory disorder of the skin; ➪ ECZEMA.

dermatomyositis, an inflammatory and degenerative disorder affecting both skin and muscle and causing rashes and weakness.

dermis, the true, living skin that lies under the partially dead EPIDERMIS.

Descartes, René, ➪ pp. 70–1, 150–1.

despotism, the arbitrary and tyrannical rule of an absolute monarch or dictator; ➪ DICTATORSHIP.

determinism, the philosophic theory that everything we do is absolutely determined by preceding causes and that free will, in the sense of the ability to make an arbitrary choice, is an illusion.

detoxification, ridding the body of poisons. The term is commonly used to refer to the process of trying to wean addicts off alcohol or drugs.

developmental psychology, ➪ pp. 166–7.

deviance, failure to conform to patterns of behaviour generally taken to be acceptable or normal; ➪ pp. 226–7.

de Vries, Hugo Marie, ➪ pp. 20–1.

Dewey, John, ➪ pp. 162–3.

diabetes, a complete or partial failure in the production of the hormone INSULIN that appears in two forms – insulin dependent diabetes mellitus (IDDM) or Type I, and non-insulin-dependent diabetes (NIDDM) or maturity-onset diabetes (diet-dependent Type II). Type I diabetes characteristically affects children and young adults and is due to destruction of the insulin-producing beta cells of the pancreas by an auto-immune process. There is a large output of heavily sugared urine, great thirst, MUSCLE wasting, an aromatic smell to the breath and, in untreated cases, coma and death. Treatment is by regular doses of insulin adjusted to balance the diet and energy expenditure. Type II diabetes affects overweight middle-aged people whose insulin output is insufficient for their metabolic needs. It can often be controlled by dieting or by drugs that boost the insulin output; ➪ pp. 80–1.

diabetes insipidus, a rare disease caused by a deficiency of a PITUITARY hormone and featuring the output of large volumes of dilute urine with resulting thirst.

diagnosis, ➪ pp. 128–9, 154–5.

dialect, the form a spoken language takes in a particular geographic locality, or in a particular social or occupational groupp. Dialects differ in pronunciation, grammar, syntax and vocabulary.

dialysis, ➪ pp. 156–7.

diaphragm, any flat bodily structure lying between two compartments, especially that forming the floor of the chest and the roof of the abdomen; ➪ pp. 42–3, 52–3.

diarrhoea, intestinal hurry with unduly watery stools and unduly frequent emptying of the bowels.

diastole, the period between contractions of the heart chambers during which blood continues to circulate under the persisting effect of the force of the contraction in SYSTOLE; pp. 194–5.

diazepam, a widely used sedative and tranquillizing drug commonly known as Valium.

dictatorship, a form of government in which power is vested in a single person who is not answerable to law. Much experience of dictatorship has demonstrated the justice of Lord Acton's famous dictum: 'All power corrupts; absolute power corrupts absolutely.'

diencephalon, the central lower part of the brain that contains a number of large masses of nerve cell bodies (grey matter) and includes the HYPOTHALAMUS and the PITUITARY GLAND.

diet, ➪ pp. 46–7.

digestion, ➪ pp. 44–5.

digit, a finger or a toe.

digitalis, a drug that can strengthen the pumping action of the heart and produce a slower and more regular pulse. It is often of value in the management of HEART FAILURE.

Dioscorides, Pedanius, ➪ pp. 100–1.

dioxin, one of a range of highly toxic substances found as contaminants of herbicides and some dyes. Dioxins can cause cancer and CONGENITAL DISORDERS.

diphtheria, a dangerous infectious disease featuring a severe throat inflammation that can lead to airway obstruction and death from asphyxia or can damage the heart.

diplococcus, one of several species of COCCUS that characteristically form, and remain in, pairs. The organism causing GONORRHOEA is a diplococcus.

diploid, pertaining to the number of chromosomes (23 pairs) found in body cells. Compare the haploid number (23 chromosomes) found in sperms and eggs; ➪ pp. 22–3.

dipsomania, a compulsive and usually uncontrolled desire for alcohol.

dirt eating, repeated ingestion of non-nutritious and usually objectionable substances such as soil, coal, clay and paper. Medically known as pica, the practice is rare in adults except in the severely mentally retarded and sometimes in pregnant women. It is common and seldom particularly harmful in young children.

disaccharide, a sugar, such as cane sugar, consisting of two simple (monosaccharide) sugars linked together.

disassociation, a defence mechanism (➪ pp. 188–9) in which any mental or behavioural process is segregated from other mental activity. Sometimes, as in dissociative disorders, an idea is separated from its usual emotional content.

disc, any flat, roughly circular and narrow structure, specifically the intervertebral disc that lies between adjacent vertebrae of the spine to cushion compressive strain.

disease, ⇨ p. 120–1.

dislocation, an abnormal separation of the bearing surfaces of a joint usually with damage to the ligaments holding them together and to the joint capsule.

disorder, any divergence from a state of health, harmony, regularity of pattern, or acceptable uniformity, whether of the human body or of society.

diuresis, an unusually large output of urine.

diuretic drugs, drugs that promote a large output of urine so as to rid the body of unwanted excess accumulated water.

divergent thinking, the kind of imaginative, sometimes eccentric, unstereotyped thinking characteristic of the creative artist rather than of the deductive logician. Compare CONVERGENT THINKING.

diverticulitis, inflammation of small, abnormal outpouchings from the colon (diverticula), causing a clinical condition similar to APPENDICITIS.

divination, prophesying the future through supernatural power, often by using alleged natural signs such as unusual weather or claimed irregularity in the abdominal organs of birds or other animals.

divorce, the dissolution of a marriage by legal process or, in some cultures, by declaration.

DNA, ⇨ pp. 24–5, 26–7.

Domagk, Gerhard, ⇨ pp. 152–3.

dominance, the power of a GENE to express its characteristic over the ALLELE present on the other, paired chromosome. In humans the allele for brown eye colour, for example, is dominant over blue eye colour. Thus an individual who is HETEROZYGOUS for eye colour, having both the blue- and the brown-eyed alleles, will be brown eyed; ⇨ pp. 28–9.

Down's syndrome, a severe genetic disorder formerly known as mongolism. It is caused by the presence of an extra chromosome number 21. Affected people have oval, down-sloping eyes, small ears and a large tongue. There is usually some degree of mental retardation but many people with Down's syndrome are able to undertake simple work; ⇨ pp. 120–1.

dowry, a sum of money or a quantity of goods or property brought by a bride to a marriage often to increase her attractiveness to a potential husband. On marriage, the dowry commonly becomes the property of the husband.

drama therapy, a form of treatment in which insight is gained into emotional problems by literally acting them out under the guidance of a psychotherapist.

Dravidian language family, the language group predominant in southern India, northern Sri Lanka, Pakistan, Afghanistan and some parts of the Far East. It includes such languages as Tamil, Malayalam, Kannada, Telugu, Kurukh and Brahui and is spoken by over 100 million people.

dreams, illusory experiences, mainly visual, that occur in the course of rapid eye movement (REM) sleep; ⇨ pp. 164–5, 178–9.

dress, ⇨ CLOTHING.

dropsy, an old-fashioned and obsolete term for the abnormal accumulation of fluid in the tissues (oedema).

drug, any substance that can be taken as a medication. In informal language the term is usually limited to drugs that have a psychoactive effect, such as the narcotics.

drug abuse, ⇨ pp. 146–7.

druidism, an ancient, pre-Christian religion practised in Britain, Ireland and Gaul (France) and sometimes revived by small groups in modern times.

Dryopithecus, an extinct primate genus whose remains have been found in many areas including Europe, East Africa, India, Pakistan and China. The genus flourished in the Miocene epoch some 7 to 25 million years ago; ⇨ pp. 10–11.

Dubois, Eugene, ⇨ pp. 10–11.

ductless glands, ⇨ pp. 78–9, 80–1.

duodenum, the first short segment of the small intestine into which the STOMACH empties. The duodenum receives bile from the LIVER and digestive ENZYMES from the PANCREAS. The term derived from the dimensions of the viscus – said to be twelve finger-widths long. Because the duodenum receives the full trauma of acid material emerging from the stomach, it is especially prone to ulceration; ⇨ pp. 44–5.

dura mater, the outer and strongest of the three membranes that cover and protect the brain.

dwarfism, abnormal shortness of stature from one of a number of causes such as ACHONDROPLASIA.

dysentery, bowel inflammation featuring DIARRHOEA, the passage of blood and mucus, fever, abdominal pain and sometimes abscesses in the liver.

dyslexia, a neurological disorder, not associated with intellectual deficit, that greatly increases the difficulties of learning to read or to understand what is read. Dyslexics benefit from early diagnosis and specialized teaching.

dyspepsia, any minor symptoms associated with the digestive system, such as pain after eating, a sense of fullness, FLATULENCE or nausea.

dysphagia, difficulty in swallowing, whether from obstruction of the gullet or a neuro-muscular disorder.

dysphasia, any disorder of speech caused by an injury to the brain, usually from STROKE.

ear, ⇨ pp. 86–7.

ear drum, ⇨ pp. 86–7.

ear lobe, the variable-sized flap of skin that hangs at the lower extremity of the external ear (pinna).

The lobe is of no functional value except as a vehicle for decorative rings or pendants.

ear piercing, the widespread human practice of forming a permanent, skin-lined perforation in the lobe of one or both ears. In some cultures heavy weights are suspended from the perforation so that the tissue stretches and the skin proliferates. In most cases the purpose of ear piercing is simply to facilitate the wearing of rings or pendants.

earwax, the secretion of glands in the external ear passage that traps dust, discourages the entry of small insects and, if excessive, may cause DEAFNESS from blockage.

eccentricity, unconventional behaviour.

eccrine glands, glands that secrete externally, as on to the surface of the skin. Sweat glands are eccrine. Compare APOCRINE.

ECG, ➪ ELECTROCARDIOGRAPHY.

echocardiography, a form of ultrasonic scanning used to build up a moving image of the living heart and assist in the DIAGNOSIS of heart disorders.

eclampsia, a dangerous complication of pregnancy in which seizures occur.

ecology, the study of the relationship between living organisms, especially human beings, and their environment. Much contemporary ecological concern relates to the damaging effect humans can have on the natural environment, but this is only one aspect of a much larger subject.

ecstasy, a state of high rapture, bliss or euphoria. The term is also used informally to refer to a stimulant amphetamine derivative drug which is used to promote a strong sense of well-being and excitement but which is not without danger.

ECT, ➪ ELECTROCONVULSIVE THERAPY.

ectoderm, the outermost of the three primitive germ layers of an embryo which develops into skin, BRAIN and nervous system, and the sense organs; ➪ pp. 6–7, 102–3.

ectopic pregnancy, a pregnancy occurring anywhere outside the womb – a highly dangerous circumstance.

eczema, skin inflammation, often due to contact allergy, with scaling, itching, weeping of serum and crusting; ➪ pp. 142–3.

education, the provision of instruction and information, ideally in such a manner as to inculcate in the recipient the desire to continue the process, spontaneously, throughout life. Literally, a 'drawing out', education should be regarded as probably the most important activity in which a human being can engage. As it is a 'programming for life' an enlightened society sees that it is of the highest quality available. It can be a serious mistake to make premature judgements about the educability of an individual as there are plenty of examples of highly successful 'late developers'. However, John Locke insisted that the aims of education, in order of importance, were virtue, wisdom, breeding and learning.

educational psychology, the study of the processes of learning and child development and of the assessment of the efficiency of the processes; ➪ pp. 166–7.

EEG, electroencephalography; ➪ pp. 128–9.

efferent nerve, a nerve carrying impulses away from the brain or spinal cord, usually to cause muscles to contract or glands to secrete.

egalitarianism, the doctrine that seeks to promote human equality – political, social and economic.

ego, a term with a spectrum of meaning, not least because Freud, who used it widely, redefined it from time to time. 'Ego' is the Latin word for 'I' and is deemed to be one's concept of oneself. Freud saw it as only one component of the individual, comprising a complex of functions by which one deals with reality, and contributing to the personality and remaining relatively stable throughout most of mature life.

egoism, the philosophic theory that the motive for all conduct is the promotion of one's own interest rather than the interest of others. This is a seductive hypothesis, easily supported by argument and hard to attack. Although widely rejected it continues to challenge other views.

egomania, abnormal and excessive self-love and regard for all things connected with oneself.

Ehrlich, Paul, ➪ pp. 152–3.

ejaculation, ➪ pp. 96–7.

elation, a sense of personal joy, satisfaction or high spirits, not amounting to ECSTASY.

elbow, the joint between the bone of the upper arm (the humerus) and the two bones of the lower arm (the radius and the ulna). The ulna articulates with the humerus in such a way as to prevent extension much beyond 180°; and the radius articulates so that it can rotate on the humerus thus allowing the hand to be turned freely; ➪ pp. 40–1.

elective surgery, surgery that is not necessary for the preservation of life or health but that may be undertaken to achieve some advantage, such as a cosmetic or functional improvement.

Electra complex, the Freudian concept, complementary to the OEDIPUS COMPLEX, featuring the sexual attraction of a female child for her father.

electrocardiography, an important aid to the diagnosis of heart disorders, in which the tiny electric currents generated by the contraction of the heart muscle are picked up by electrodes, amplified and recorded as a continuous tracing on a moving strip of paper.

electroconvulsive therapy, ➪ pp. 194–5.

electroencephalography, ➪ pp. 128–9.

electrolysis, the passage of an electric current through a conducting solution and its breakdown into components, such as the hydrogen and oxygen of water. Human electrolysis is used to destroy individual hair roots so as to remove unwanted hair.

electron microscope, a compound microscope that uses a beam of electrons instead of light and shaped electromagnetic fields instead of lenses. The resolving power can be great enough to reveal individual molecules.

electrophoresis, a method of separating the dissolved and electrically charged components of a mixture by the application of an electric field. Electrophoresis has many applications including some in medicine.

elements, ⇨ pp. 34–5.

elephantiasis, a disease due to blockage of lymph vessels by microscopic parasitic worms (filaria) so that the drainage of fluid from the tissues is impeded and an extremity, such as a leg or the scrotum, becomes progressively more swollen and enlarged; ⇨ pp. 136–7.

embalming, the preservation of dead bodies by draining out most of the BLOOD and replacing it with disinfecting fluids such as formaldehyde so as to retard the bacterial activity that leads to putrefaction.

embolism, the sudden blockage of an artery by any material, especially a blood clot, brought to the site of blockage in the bloodstream.

embrocation, a stimulating lotion that can be rubbed into the skin to prompt an improved blood flow in the underlying tissue and relieve pain. Embrocations have little real effect and are now rarely used.

embryo, ⇨ pp. 6–7, 102–3.

embryology, ⇨ pp. 102–3.

emetic, any drug or agency that induces vomiting.

emotion, ⇨ pp. 204–5.

empathy, the ability to identify so closely and imaginatively with the feelings of another that they are virtually shared.

emphysema, a lung disease in which the tiny air sacs break down to form larger cavities with a much reduced total surface area for oxygen transfer from the atmosphere to the BLOOD. The effects are essentially those of oxygen deprivation and may be very serious.

enamel, the hard outer layer of the crown of the tooth; ⇨ pp. 44–5.

encephalitis, inflammation of the brain.

encephalopathy, any organic disorder of the brain, especially one of a degenerative nature or toxic origin.

encounter therapy, a method of promoting self-awareness and insight into personal problems by participating in a group licensed to express feelings and attitudes openly and sometimes even physically, under the guidance of a psychotherapist.

endemic, continuous within a local community or area. Usually used to refer to a disease, such as malaria which is endemic in West Africa.

endocrine system, the bodily control system of glands that secrete 'chemical messengers' (HORMONES) directly into the BLOOD; ⇨ pp. 78–9.

endoplasmic reticulum, a net-like membranous structure within the cell that carries the RIBOSOMES which synthesize protein and the site of other metabolic processes.

endorphins, morphine-like substances secreted by the BRAIN under conditions of stress and pain. The brain contains specific receptors for endorphins and it is these that are affected by the drug MORPHINE; ⇨ pp. 54–5.

endoscope, an optical instrument, rigid or flexible, that allows medical diagnosis by direct visualization of the interior of the body. The endoscope is passed in either through a natural orifice or through a small incision in a body wall; ⇨ pp. 128–9.

enema, a method of treating CONSTIPATION by introducing fluid into the rectum through a tube passed in via the anus. This softens the faeces and allows normal evacuation.

energy, the capacity of a body to do work. Energy occurs in several forms – potential as in a compressed spring or a mass in a high position, kinetic as in motion, chemical as in petroleum and nuclear as in the binding forces of the atomic nucleus. Its effect, when manifested, is to bring about a change of some kind. The term is also used metaphorically to refer to human vitality and appetite for exertion or work.

enkephalins, ⇨ pp. 54–5.

enteritis, inflammation of any part of the intestine.

environment, ⇨ pp. 16–17.

environmentalism, the psychological concept that behaviour is mainly determined by a person's environment.

enzyme, a PROTEIN that acts as a catalyst in organic chemical reactions. Enzymes are specific for particular reactions. Most enzymes are named by adding the ending '-ase' to a brief account of the reaction catalysed or the substance broken down, as in *proteinase*. Enzymes are vital in human biochemistry and physiology and thousands are involved; indeed enzyme action is probably the most important single factor in maintaining life and health. Each one is coded by its own gene; ⇨ pp. 36–7.

epidemic, the occurrence of many similar cases of a disease in a population within a short period of time. There are usually at least some years between epidemics, and in these intervals many people lose resistance or are born without it. An epidemic may be caused by such a reduction in immunity, a novel form of the disease, as happens with influenza, or a new disease, as with AIDS; ⇨ pp. 124–5.

epidermis, the outer layer of the skin, in which there is a gradual transition, from living cuboidal cells (the basal cells), through irregularly shaped cells (the prickle cells) to the flattened, dead surface layer of horny cells, packed with the tough PROTEIN keratin. The horny layer is a highly protective barrier, waterproof and resistant to the passage of organisms and many different molecules; ⇨ pp. 56–7.

epididymis, the long, coiled tube lying behind the testicle that connects it to the VAS DEFERENS. Sperms formed in the testicle pass into and through the epididymis taking about twelve days to do so and maturing, the while, to the fully active form; ⇨ pp. 96–7.

epiglottis, the small, leaf-shaped cartilage that lies at the entrance to the voice-box (larynx) and acts as a lid, preventing the entry of food during swallowing; ⇨ pp. 44–5.

epiglottitis, an often dangerous inflammation of the EPIGLOTTIS that may lead to obstruction of the airway from swelling.

epilepsy, the result of an abnormal electrical discharge in the brain, which may give rise to a variety of effects which include major seizures (GRAND MAL), brief 'absence' attacks (PETIT MAL) or a range of unpleasant subjective hallucinatory experiences.

epiphenomenalism, the belief that mental events are solely a consequence of physical events, specifically neural activity, and never the causes of them. Once considered heretical, the view is now widely held by scientists.

episiotomy, ⇨ pp. 104–5.

epistemology, the study of the theory of knowledge, its sources, reliability and limits.

epithelium, the thin, coating layer of all body surfaces except the inner lining of BLOOD and lymph vessels. Epithelium forms a 'non-stick' surface allowing organs and tissues to move against each other without the formation of adhesions. There are various forms of epithelium, which may be single-celled, multi-layered and stratified with the cells becoming ever more flattened towards the surface, or ciliated – covered with fine, wafting, hair-like structures. Epithelium also commonly bears mucus-secreting goblet cells.

erectile tissue, any tissue capable of being engorged with blood so as to become stiffer and firmer, as in the case of the nipple or the PENIS.

erection, ⇨ pp. 96–7.

ergonomics, the study of the relationships of human beings to machines and to their physical environment at work. The increasing application of complex technology has resulted in increasing human discomfort, difficulties and dangers. Ergonomics seeks to solve such problems.

erogenous zones, parts of the body which, when stimulated by touch in a suitable context, are especially liable to arouse sexual interest or desire.

eroticism, having the quality of stimulating sexual interest or desire, whether in human intercourse, art, literature or symbolism.

erythrocyte, ⇨ RED BLOOD CELL.

ethics, moral philosophy. The study of what is meant by right and wrong and of whether these can be distinguished. Practical ethics is the study of right conduct and its relationship to currently accepted behaviour.

ethnicity, pertaining to the characteristics of a human group with racial, linguistic and other features in common.

ethnography, the subdivision of ANTHROPOLOGY concerned with the direct study of cultural groups such as tribes, communities, social classes or any other definable class of people. Ethnography involves fieldwork in which observations are made and recorded, and then a written account and analysis of the data.

ethnology, the branch of ANTHROPOLOGY concerned with races and peoples, with their characteristics, origins and movements and with their relationships to each other.

ethology, the study of the behaviour of animals in their normal habitat.

etiquette, a set of rules for the conduct of human beings in particular social and professional contexts. Conformity to the current etiquette is mandatory if one is to be generally accepted; non-conformists, unless exceptionally valuable in other respects, usually suffer disadvantage or rejection. Some items or codes of etiquette – such as the practice of men giving way to women – have a logical and utilitarian basis and help to regulate social conduct. Some – such as a slavish adherence to precedence on social occasions – confer exclusiveness on a group; some provide entertainment and preoccupation for the rich and idle. But many – such as tilting soup-plates away or leaving the bottom button of a waistcoat unbuttoned – are arbitrary and meaningless, with no greater justification than the desire of one human being to establish some kind of ascendancy over another.

eugenics, the study or practice of trying to improve the human race by encouraging the breeding of those with desired characteristics (positive eugenics) or by discouraging the breeding of those whose characteristics are deemed undesirable (negative eugenics). The concept implies the improbable proposition that there exists some person or institution capable of making such decisions. It also implies possible grave interference with human rights. For these reasons, the principles, which have long been successfully applied to domestic animals, have never been adopted for humans except by despots such as Adolf Hitler.

eukaryote, an organism whose cells have a demarcated nucleus in which the genetic material is carried. All multi-cellular organisms are eukaryotes, but some single cellular organisms are also eukaryotes.

eunuch, a castrated male.

euphoria, a feeling of extreme or exaggerated delight or exhilaration.

Eurasian, of mixed European and Asian descent.

eustachian tube, the air pressure-equalizing link between the middle ear and the back of the nose; ⇨ 86–7.

euthanasia, assisting the suffering and the incurable to die in comfort. Euthanasia may be passive, by the withdrawal of medical support, or active as in 'mercy killing'. The latter is still illegal in most countries, but the debate is gaining momentum.

Eve hypothesis, the suggestion that all mankind is descended from an African woman who lived 200 000 years ago. Support for the hypothesis comes from a study of mitochondrial DNA which is inherited only from the mother. Knowledge of the average rate of the occurrence of mutations allows one to extrapolate backwards in time and estimate the date of the universal maternal ancestor. The hypothesis has been criticized.

evolution, ➪ pp. 6–7, 10–11.

excretion, discharge of waste matter, such as urine or faeces, from the body.

exercise, ➪ pp. 54–5.

exhibitionism, a persistent desire to boast or to display one's abilities or person. The term is also used for a compulsive desire to arouse shock by an unwanted display of one's genitalia.

existentialism, a 20th-century philosophic movement promoted by Jean-Paul Sartre (1905–80) and others which emphasizes human freedom in a meaningless universe; the importance of the personal response to experience and of the willingness to accept personal responsibility for one's life and its outcome.

expectorant, any drug or agency that liquefies and loosens phlegm so that it can more easily be coughed up and spat out.

exposure, sustained lack of adequate shelter from inclement weather conditions, especially cold and damp. Exposure can lead to HYPOTHERMIA and, in some cases, to reduced resistance to infection.

extended family, ➪ pp. 211–12.

extensor, a muscle whose shortening (contraction) causes a joint to straighten. Compare FLEXOR.

extrasensory perception, ➪ pp. 180–1.

extravert, ➪ pp. 164–5.

eye, ➪ pp. 84–5.

eyebrow, the curved bony ridge above each eye and the narrow curved strip of hairs that grow on the ridge. The eyebrows help to keep forehead sweat out of the eyes and their movement and position can eloquently convey emotion such as surprise or interrogation.

eyelashes, the double or triple row of stiff, short hairs growing from the margins of the eyelids and curving outwards. Eyelashes help to keep sweat out of the eyes.

eyestrain, the symptoms of fatigue and ocular discomfort sometimes experienced after prolonged close work.

Eysenck, Hans (1916–), German-born British psychologist noted for his studies into human personality and intelligence and for his savage criticism of psychoanalysis. He aroused controversy when he published his views on racial differences in intelligence in the book *Race, Intelligence and Education* (1971).

Fabricius, Hieronymus, ➪ pp. 34–5.

face, the area of the front of the head that incorporates the complex system of the muscles of expression. These act on the mouth, the eyelids, the nostrils and the eyebrows, causing them to vary widely in shape and position so as to convey great complexity and subtlety of emotion.

face lift, a cosmetic surgical operation in which the facial skin is separated from the underlying tissue, pulled tightly upwards so as to stretch wrinkles flat, and sewn in place. Surplus skin is removed.

facial hair, hair that grows in the beard and moustache areas or elsewhere on the face. This is often welcomed in men but seldom in women who may be driven to various expedients to remove it.

factor VIII, one of the many factors needed for normal BLOOD CLOTTING, the absence of which causes haemophilia.

faeces, ➪ pp. 64–5.

fainting, transient loss of consciousness from a brief period of inadequate blood supply to the brain with recovery contingent on lying down.

faith healing, the attempt to cure organic disease by prayer or religious ritual. Much of the claimed evidence for the efficacy of faith healing is suspect. The psychological effect can be powerful, and unjustified hopes for miraculous cures are commonly aroused.

fallen arches, an informal term for the loss of the longitudinal and/or transverse arch of the foot, resulting in the orthopaedic condition of flat foot (pes planus) and often in persistent aching.

fallopian tube, the uterine tube that extends sideways from each upper and outer corner of the womb (uterus) to the ovaries. The open outer ends of the uterine tubes bear many finger-like processes that drape over the ovaries and help to collect eggs released by the ovaries and direct them into the tube. To be successful, FERTILIZATION must occur in the fallopian tube; ➪ pp. 96–7.

Fallopius, Gabriel, ➪ pp. 34–5, 100–1.

false pregnancy, an illusory or delusory conviction that one is pregnant, often because of a strong desire for a child.

family, ➪ pp. 214–5.

family therapy, a form of psychotherapy in which the whole family of the patient is actively involved in the therapeutic process, usually by guided group discussion.

farmer's lung, a severe allergic lung disease caused by inhalation of dust from mouldy straw, hay or mushroom compost containing the spores of a fungus. There is fever, coughing and breathlessness and repeated attacks can cause lung scarring and eventual death.

fascism, any ultra-rightwing political movement or social attitude featuring AUTHORITARIANISM, CHAUVINISM, intolerance, persecution of disapproved minorities and a fundamental denial of liberty and democratic rights.

fasting, deliberate abstention from eating for certain periods, especially as a form of religious observance.

fat, ➪ pp. 36–7.

fatigue, severe physical or mental tiredness due to physical exertion or psychological disorder.

fatty acids, ➪ pp. 36–7.

fear, ➪ pp. 204–5.

febrile, fevered.

federalism, advocacy of a form of government in which political power is partly centralized and partly delegated to variably autonomous peripheral regions or states.

feelings, emotions; ➪ pp. 204–5.

fellatio, sexual stimulation of the penis by the partner's mouth. Compare CUNNILINGUS.

feminism, the movement dedicated to the advocacy of women's rights.

femur, the thigh bone; ⇨ pp. 40–1.

fertility, ⇨ pp. 98–9.

fertility drug, a drug used to stimulate ovulation so that the chances of conception are increased. Fertility drugs often result in multiple pregnancies (triplets, quads, quins).

fertilization, the first stage in sexual reproduction in which the sperm penetrates the egg and the genetic material from the father joins that from the mother to make up the full complement required for the production of a new human being. Once a single sperm has penetrated the outer layer of the egg, a membrane quickly forms to prevent further access; ⇨ pp. 96–7.

fetal alcohol syndrome, the damaging effects on the growing fetus of maternal alcohol consumption. These include low birth weight, a small head, mental retardation, congenital heart disease and a high fetal death rate.

fetal distress, slowing or irregularity of the fetal heart or the passage of fetal stools, occurring during pregnancy or labour and indicating that the fetus is in danger.

fetishism, ⇨ RELIGION, SEX.

fetus, ⇨ pp. 102–3.

feudalism, a medieval form of European local government, usually based on a castle, in which the lord alloted land to his subjects in return for military service.

fever, a rise in the body temperature above the normal level of 37°C.

fibre, dietary, the indigestible and non-absorbable polysaccharide part of certain diets, particularly those containing vegetables. High fibre intake is valuable as a protection against constipation and possibly some more serious intestinal disorders and as a way of reducing blood cholesterol levels; ⇨ pp. 46–7.

fibrillation, an ineffectual fluttering of any muscle, especially the heart muscle. Fibrillation of the upper heart chambers (atrial fibrillation) causes pulse irregularity; fibrillation of the lower, main, pumping chambers is a form of cardiac arrest and is soon fatal unless reversed.

fibroid, a common benign tumour of muscle and fibrous tissue that may grow, singly or multiply, in the wall of the womb and sometimes cause pain, infertility and heavy menstrual periods.

fibrosis, ⇨ pp. 62–3.

fibrositis, an ineptly named range of minor symptoms, unconnected with inflammation of fibrous tissue and mainly due to muscle or tendon strain but without demonstrable organic abnormality.

fibula, the delicate long bone that lies to the outer side of the main bone of the lower leg (the tibia).

fight or flight response, ⇨ pp. 204–5.

filariasis, a group of parasitic diseases, spread by mosquitoes and other biting flies, and featuring microscopic worms (microfilariae) that inhabit the blood and lymph vessels and settle in the tissues where they may grow into adult worms; ⇨ ELEPHANTIASIS.

finger, the digits of the hand, each consisting of three short bones or phalanges (the thumb has two), and a system of flexor and extensor tendons that respectively bend and straighten them; ⇨ pp. 40–1.

fingerprints, ⇨ pp. 30–1.

Fischer, Emil (1825–1919), German organic chemist whose work on the understanding of amino acids, peptides and proteins was of fundamental importance in the development of organic chemistry and biochemistry. He was awarded the Nobel Prize in Chemistry in 1902.

fistula, an abnormal communication between the interior of the body and the surface of the skin or between two internal organs.

fitness, ⇨ pp. 54–5, 122–3.

fixation, a psychological term indicating a close and persistent attachment to a person or object, of a degree appropriate to an immature stage of development.

flagellum, a long, whip-like process found on a number of microscopic organisms as a means of locomotion in fluid.

flatfoot, ⇨ FALLEN ARCHES.

flatulence, a feeling of fullness in the abdomen associated with a desire to belch or fart.

flatworm, any of the parasitic or free-living members of the phylum *Platyhelminthes* that includes human parasitic flukes and tapeworms.

flea, one of a range of small, wingless, blood-sucking parasites of animals including man. Fleas transmit PLAGUE and TYPHUS.

Fleming, Sir Alexander, ⇨ pp. 152–3.

Flemming, Walther, ⇨ pp. 20–1.

flexor, a muscle which, on contraction, causes a joint to bend. Compare EXTENSOR.

Florey, Howard, ⇨ pp. 152–3.

flu, ⇨ INFLUENZA.

fluoridation, the artificial addition of small quantities of fluorides to drinking water deficient in natural fluorides so as to restore the tooth-protective advantage provided by the element.

foetus, ⇨ FETUS.

folic acid, a B vitamin necessary for the synthesis of DNA and RED BLOOD CELLS. Deficiency can cause pernicious anaemia and other disorders. The vitamin is now being used to help to reduce the incidence of neural tube defects (e.g. spina bifida) in fetuses.

folie à deux, a delusional psychosis in which the delusions of the sufferer are voluntarily accepted by another person for purposes of gaining approval and mutual support.

folk medicine, ⇨ pp. 158–9.

folklore, a collection of legends, sayings, songs, proverbs, riddles, etc., forming an unwritten literature of a people.

follicle, any sac-like cavity or sheath-like depression in the body, especially one that is secretory of hairs (hair follicle) or eggs (Graafian follicle in the ovary).

fontanelle, one of the two gaps left, at the front and the back, between the growing bones of the skull of a young baby, and readily felt through the skin and scalpp. The fontanelles usually close by about 14 months.

food, diet; ➪ pp. 46–7.

food allergy, a rare hypersensitivity to a foodstuff or to a food additive; ➪ pp. 142–3.

food poisoning, intestinal irritation or damage from the toxins of *Salmonella* and other organisms.

foot, the pedal extremity of the vertebrate hind limb, that has, in humans, adapted for little other than walking. The human foot articulates at the ankle and can be forcibly brought into almost straight alignment with the leg by contraction of the powerful muscles of the calf, via the Achilles tendon.

forebrain, the wide front segment of the developing brain in the embryo that eventually forms the paired cerebral hemispheres.

forehead, the part of the front of the head between the hairline and the EYEBROWS. The skeleton of the forehead is the frontal bone that contains the frontal sinuses. This is covered with muscle capable of wrinkling the overlying skin and contributing to the facial expression.

forensic medicine, medicine applied to criminal investigation.

foreplay, the progressively intimate talking, touching, kissing and caressing that precedes COITUS.

foreskin, ➪ PREPUCE.

Fracastoro, Girolamo, ➪ pp. 134–5.

fracture, a break, especially of a bone, which may be internal (simple) or may, more seriously, penetrate the skin (compound). A greenstick fracture involves bending with a break on one side only and in a comminuted fracture the bone is shattered. Fractures may be transverse, oblique or spiral in direction.

Franklin, Rosalind, ➪ pp. 24–5.

freckle, harmless local aggregations of the skin pigment melanin.

free radicals, short-lived but highly active chemical groups thought to be the immediate cause of tissue damage in many disease processes. They attack DNA, PROTEINS and cell membranes. They are said to be promoted by various agencies including radiation, smoking, atmospheric pollutants and inadequate BLOOD supplies and to be capable of being 'mopped up' by vitamins C and E; ➪ pp. 46–7.

free will, the ability to choose our own destiny. Philosophic analysis suggests that free will may be an illusion.

Freud, Anna (1895–1982), Austrian psychiatrist daughter of SIGMUND FREUD who adopted his mantle, his interests and his literary fertility, and was one of the pioneers of child psychiatry and a custodian of the Freud Archives.

Freud, Sigmund, ➪ pp. 164–5.

Freudian slip, a humorous reference to an error of speech that reveals the thoughts the speaker wished to conceal.

Friedreich's ataxia, an inherited degenerative disease of the nervous system causing severe unsteadiness, movement and speech defects, loss of sensation, spinal curvature and heart defects.

Fries, James, ➪ pp. 110–11.

frigidity, a persistent failure to achieve erotic arousal; a pejorative term applied by men to women whom they have failed to arouse.

Fromm, Erich (1900–80), German psychoanalyst, anthropologist and philosopher who was notable for his emphasis on the role of CULTURE, rather than sexuality, in the formation of personality.

frontal bone, the skeletal basis of the FOREHEAD.

frontal lobe, the part of the brain lying immediately behind the forehead. It is concerned with voluntary movement, initiative, spontaneity and strength of personality.

frostbite, superficial death of tissue, usually at the extremities, as a result of freezing.

frottage, a male activity, usually engaged in in densely packed crowds, in which the active male rubs his genitals against a woman's buttock or thigh. Some frotteurs are buttock fetishists (pygophiliacs) but most are sexually inadequate people who cannot achieve a more satisfactory outlet.

fructose, a simple monosaccharide sugar derived from fruit or from the breakdown of the disaccharide cane sugar (sucrose).

fugue, a rare psychological entity that occurs usually as a response to an intolerable situation, such as exposure to danger in warfare or serious domestic unhappiness. The affected person appears to lose his or her memory in a selective manner, wanders off to a new place of residence and adopts a new identity and occupation. Needed information and education are retained. If recovery occurs, there is amnesia for the fugue period.

fundamentalism, any religious movement that denies the claims of science and insists on the literal truth of everything in holy scripture, such as the Bible. Christian fundamentalists, for instance, believe in the creation of the universe in six days, the miracles, the virgin birth, the resurrection and the divinity of Christ.

fungus, ➪ pp. 120–1.

fungus infection, the establishment of one or more colonies of fungus in the EPIDERMIS of the skin or, less commonly, in the interior of the body.

funny bone, a popular name for the elbow end of the ulna forearm bone. The ulnar nerve runs through a groove in this bone very near the surface where it is susceptible to painful injury that may cause a marked physical reaction.

Galen, Claudius, ➪ pp. 34–5.

gall bladder, the small sac lying under the LIVER in which bile, secreted by the liver, is stored and concentrated until released into the intestine under the stimulus of fat in the diet; ➪ pp. 44–5, 48–9.

Gallo, Robert, ➪ pp. 60–1.

gallstones, hard concretions that can develop by crystallization from the bile in the GALL BLADDER and may cause problems by obstructing the bile ducts.

Galton, Francis, ➪ pp. 30–1.

Galvani, Luigi, ➪ pp. 42–3.

games, ➪ PLAY.

gamete, a SPERM or an egg (OVUM).

gamma globulin, a class of ANTIBODIES present in the blood.

ganglion, a collection of nerve cell bodies surrounded by a capsule and usually situated outside the brain or spinal cord. The term is also used for a cystic, fluid-filled, benign swelling on a tendon sheath or joint capsule; ➪ pp. 74–5.

gangrene, death of tissue, usually from an inadequate blood supply. In dry gangrene the affected part will often drop off. Wet gangrene usually causes serious toxic effects from infection. Gas gangrene is a form of ANAEROBIC muscle infection causing muscle death and severe general upset.

Gardner, Allan and Beatrice, ➪ pp. 14–15.

gargoylism, one of a number of rare genetic disorders of mucopolysaccharide metabolism that lead to coarsening of the facial features, excessive hairiness and often mental retardation.

gas gangrene, ➪ GANGRENE.

gastric ulcer, a painful localized loss of tissue of variable depth on the inner lining of the stomach caused by acid and the digestive ENZYME pepsin in an area inadequately protected by stomach mucus.

gastritis, inflammation of the stomach lining from any cause, but often from ALCOHOL or other irritants, stress, blood infection or major burns.

gastroenteritis, inflammation of the linings of the stomach and the intestines, usually from an infection with *Salmonella* or other organisms.

gastroenterology, the medical specialty concerned with the whole digestive system and its various appendages.

gastroscopy, direct visualization of the inside of the stomach for diagnostic purposes. A steerable, flexible, fibre-optic endoscope is used.

gay bowel syndrome, fissuring and ulcerative damage to the ANUS, anal canal and rectum from male homosexual practices such as finger or hand insertion ('fisting').

gender identity, ➪ pp. 202–3.

gene, a sequence of DNA that codes for a single product, usually a protein. Each gene has a unique position on the chromosome, and in humans, who have two copies of each chromosome, each gene is present in the cell twice (one on each chromosome) and has been derived from each parent. If the pair of genes have different DNA sequences their expression may either be a complex interaction between their products, or one of them may be DOMINANT. A gene not only contains the DNA sequence containing the information for the gene product, but also has associated stretches of DNA that control the expression of the gene. In humans, genes occur as very small islands of information amongst vast stretches of apparently meaningless junk, DNA; ➪ pp. 22–3, 24–5, 26–7, 28–9.

gene flow, ➪ pp. 4–5.

gene pool, the totality of genes in a species, from which future generations will derive their DNA.

gene replacement therapy, a recombinant DNA technique in which a defective gene is replaced by a fully functional gene. Although gene replacement therapy will become widely available it is, at present, only at the initial research stage. The moral and ethical dilemmas associated with altering the human GENE POOL should be fully discussed before these techniques become commonplace.

general paralysis of the insane, ➪ pp. 186–7.

generic drug, a drug known under the official name for the basic active substance it contains. Compare trade or brand name.

genetic counselling, ➪ pp. 148–9.

genetic drift, a change in the GENE POOL of a small isolated population that occurs by chance. The long-term effect of genetic drift may be more important than selection in determining the evolutionary fate of small populations.

genetic engineering, a range of techniques by which the genetic makeup of organisms such as bacteria or fungi is deliberately modified to achieve some practical result or desired product.

genetic fingerprinting, a means of specifically identifying an individual from a minute sample of tissue, blood, semen, etc. from that person's body, so long as the sample contains cells and thus a specimen of the GENOME. Each individual, excepting identical twins, has a unique genetic fingerprint.

genetics, ➪ pp. 28–9.

genital stage, the last and most mature of the stages of psychosexual development as described by SIGMUND FREUD. This stage is said to be reached in late adolescence or early adult life.

genital warts, ordinary skin warts affecting the genital area, transmitted by sexual contact and tending to spread profusely in the area.

genitals, the sexual and reproductive organs, both external and internal; ➪ pp. 96–7.

genius, a variously defined term for a person of exceptional achievement or accomplishment, or for the capacity that allows such achievement. Genius is commonly manifested in the arts, literature or invention.

genome, the complete GENE set that is contained in the nucleus of each living cell of the body; ➪ pp. 26–7, 132–3.

genotype, the genetic constitution. Compare phenotype, which is the physical constitution.

geriatric medicine, ➪ pp. 110–11.

germ cell, a reproductive cell or GAMETE.

German measles, a popular name for rubella, a mild infectious disease of childhood causing only a slight rash, lymph node swelling and mild fever.

Rubella during early pregnancy, however, is much more serious and often causes congenital heart disease, CATARACT and other serious abnormalities in the fetus; ➪ pp. 131–2.

Germanic languages, a sub-family of the Indo-European group of languages that includes English, German, Frisian, Dutch, Flemish, Danish, Swedish, Icelandic, Norwegian, Faeroese, Afrikaans and Yiddish; ➪ pp. 16–17.

gerontology, ➪ pp. 110–11.

gestalt, a configuration or pattern that can be perceived in a particular context by noting the relationship between different components, rather than merely the sum of the parts. An entire psychology and a method of psychotherapy have been erected on the basis of this idea.

gestation, the development of an individual from fertilization to birth. The time this takes is called the gestation period and is 40 weeks, plus or minus a few days, in humans.

gesture, significant movements of the hands, arms, shoulders, head or body which can convey information, especially about the emotions or mental attitudes.

gigantism, the condition of abnormally great physical stature. This usually results from an excessive production of PITUITARY growth hormone in childhood, before the growing ends of the long bones have fused.

gingivitis, inflammation of the gums, usually as a result of neglect of toothbrushing.

gland, ➪ pp. 56–7, 66–7, 80–1.

glandular fever, a virus infection commonest in young adults featuring fever, weakness, headache, sore throat, enlargement of lymph nodes and spleen and often followed by prolonged fatigue and recurrences of fever; ➪ pp. 132–3.

glans, the acorn-shaped bulbous tip of the penis which is liberally supplied with sensitive nerve endings; ➪ pp. 96–7.

glaucoma, an eye disorder featuring damage to the origins of the fibres of the OPTIC NERVE as a result of a persistent rise in the pressure of the fluids within the eye. There are several different forms of the disease but the commonest is insidiously free of symptoms often until a late and irremediable stage. Routine screening of pressures can detect the disorder at an early stage and the damage can then be prevented; ➪ pp. 84–5, 110–11, 154–5.

globulin, one of a range of proteins dissolved in the blood that includes the immunoglobulins (antibodies).

glomerulus, ➪ pp. 64–5.

glottis, the vocal cords of the larynx and the space between them.

glucagon, one of the hormones produced by the PANCREAS and having effects opposite to those of INSULIN; ➪ pp. 80–1.

glucose, a sugar, the starting point for the generation of energy in the body; ➪ pp. 36–7, 80–1.

glue ear, a common childhood disorder in which the secretions of the middle ear become viscous and stiff and impede free movement of the ear drum and small auditory bones (ossicles) so that deafness results; ➪ GROMMET TUBE and pp. 86–7.

glue sniffing, the inhalation of one of a number of organic solvents for their intoxicating or hallucinogenic effects – a highly dangerous pursuit.

gluteus maximus, one of the pair of the largest of the three flat muscles that form the buttock and act to extend the hip joint and move the thigh bone (femur) backwards in walking; ➪ pp. 42–3.

glycerol, the sweet-tasting liquid glycerine that forms the 'backbone' to which three fatty acids are attached in the fat molecule (triglyceride); ➪ pp. 46–7.

glycogen, a polymerized form of glucose, stored mainly in the LIVER and the muscles, from which glucose can be released as required. Glycogen has been called 'animal starch'; ➪ pp. 46–7.

glycolysis, the enzymatic process by which the chemical energy of the cell, ATP, is generated from glucose.

gnosticism, the philosophic basis of a religious sect claiming secret knowledge that conferred salvation. Gnostics believed that the spirit was imprisoned in material things, all of which were evil, and could be freed only by the application of the secret ways.

goitre, any swelling of the thyroid gland in the neck, as from overactivity, inflammation, cancer or iodine deficiency; ➪ pp. 80–1.

golfer's elbow, inflammation, from overuse, of the area on the inside of the elbow where several forearm muscle tendons are attached.

Golgi apparatus, ➪ pp. 38–9.

Golgi, Camillo, ➪ pp. 72–3.

gonad, a sex gland, specifically the ovary in the female and the testis in the male.

gonorrhoea, ➪ pp. 98–9.

goose flesh, the skin appearance caused by contraction of the tiny erector pili muscles each of which is attached to a hair follicle. This occurs in cold conditions or as a result of certain emotions.

gout, an acute and painful form of joint inflammation (arthritis) caused by a metabolic disorder that leads to the deposition of urate crystals in and around joints.

government, ➪ pp. 224–5.

Graafian follicle, one of many nests of cells in the ovary that, about once a month, forms a fluid-filled cyst containing a maturing egg (ovum).

grafting, surgical transfer of tissue or organs from one body to another of the same species (homograft) or from one area of a body to another area of the same body (autograft); ➪ also IMMUNE SYSTEM and ORGAN GRAFTING.

grand mal, a major epileptic seizure consisting of a massive overall contraction of most of the muscles of the body (tonic stage) followed by a series of smaller jerky contractions (clonic stage).

Graves' disease, a form of overactivity of the

THYROID GLAND (thyrotoxicosis) that may involve hyperactivity, fast pulse, anxiety, shakiness, palpitations, loss of weight, good appetite, sweaty skin and protruding eyes.

grey matter, those parts of the brain and spinal cord in which there are close accumulations of nerve cell bodies. Compare white matter which consists largely of nerve fibres; ⇨ pp. 74–5, 76–7.

groin, the region of the junction of the thigh and the lower abdomen.

groin strain, any minor stretching of the tendinous attachments of the several muscles that are inserted into the central ligament of the groin and the adjacent bones of the pelvis.

grommet tube, a tiny plastic flanged tube that is inserted into a hole made in the eardrum to provide middle ear drainage in the treatment of GLUE EAR.

group behaviour, ⇨ pp. 208–9.

group therapy, a form of psychotherapy in which up to a dozen patients participate, under the guidance of a psychotherapist, in discussions and analysis of their emotional and other problems. Many different schools of psychotherapy engage in group therapy of different kinds.

growing pains, an informal term for various aches arising, often without apparent cause, in the musculo-skeletal systems of rapidly growing children or adolescents.

growth, the process of bodily enlargement or elongation. The term is usually restricted to the period from conception to early adult life when bone growth is complete. Growth is a complex process whose control is not fully understood. It involves controlled cell reproduction but also the destruction and removal of earlier cells; ⇨ pp. 106–7, 108–9.

growth hormone, ⇨ pp. 78–9, 106–7.

guanine, one of the four bases, the sequence of which along the DNA molecule forms the genetic code; ⇨ pp. 24–5.

guilt, a sense of having transgressed against generally accepted modes of conduct, against an implicit or explicit contract with another person, or against divine command.

gullet, the muscular-walled tube that leads down from the back of the throat to the stomach for the passage of food. The medical term is oesophagus.

gumboil, an ABSCESS in the gum usually arising from an infection at or around the root of a tooth.

gums, the smooth, red, mucous membrane-covered connective tissue that is firmly attached to the bones of the upper and lower jaw and surrounds the necks of the teeth. The medical term is gingiva.

gynaecology, the medical specialty concerned with diseases of the reproductive system of women.

habit, one of many learned responses that lead to regularly repeated behaviour performed with little thought in response to a given stimulus. Habits are strengthened by repetition until they become automatic. Life would be intolerable without habits as they allow us to perform many essential repetitive tasks while devoting our thoughts to other things. Most habits are useful in this way; some, such as systematic working habits or the habit of eating in moderation, are of the greatest value. Many are, however, harmful and may be difficult to break. Behaviour therapy includes some effective techniques for breaking undesired habits.

Haeckel, Ernst (1834–1919), German zoologist and comparative anatomist who was one of the first to point out the similarities between the embryonic development of the individual and its evolutionary history. He also produced an evolutionary and genealogical tree of the animal kingdom.

haematoma, any accumulation of blood anywhere in the body outside the circulation that has partially clotted to form a semi-solid mass. Haematomas are commonly caused by injury but may result from disease of blood vessels or disorders of blood coagulation.

haemoglobin, the PROTEIN-iron molecule that fills the red BLOOD cells (erythrocytes), and forms a loose combination with oxygen in environments of high oxygen tension and releases it in areas of low oxygen tension; ⇨ pp. 52–3.

haemophilia, a genetic disorder of blood clotting of variable severity featuring often severe spontaneous bleeding into joints, skin or other areas; ⇨ pp. 148–9.

haemorrhage, ⇨ BLEEDING.

haemorrhoids, ⇨ PILES.

Hahneman, Christian Friedrich Samuel, ⇨ pp. 158–9.

hair, ⇨ pp. 36–7, 56–7.

halitosis, bad breath. This may be due to dental, nasal, respiratory or digestive problems or to smoking or the ingestion of odorous substances such as garlic.

Haller, Albrecht von (1709–77), Swiss anatomist and physiologist who was the first to recognize the full significance of the nervous system as the central controlling and directing agency of the whole body. He was the founder of experimental PHYSIOLOGY.

hallucination, a sensory perception in the absence of the normal stimulus. Hallucinations may be visual, auditory, tactile, gustatory, olfactory or dimensional and do not necessarily imply mental disorder. They should be distinguished from delusions, which are mistaken ideas; ⇨ pp. 192–3.

hallucinogenic drugs, drugs that induce HALLUCINATION.

hamstring muscles, the three muscles on the back of the thigh whose cord-like tendons (the hamstrings) are conspicuous on the back of the knee. The hamstring muscles bend the knee and straighten the hip joint; ⇨ pp. 42–3.

hand, the gripping organ at the end of the arm. The hand skeleton consists of the five metacarpal bones in the palm and the phalanges of the FINGERS and THUMB. Most finger movement is mediated by forearm muscles, but as well as the many tendons to operate the fingers, the hand also contains small intrinsic muscles that allow the fingers to spread and the thumb to flex.

haploid, the number of CHROMOSOMES in a sperm or an egg. This number is equal to half the number in a normal body cell (the DIPLOID number). At fertilization the diploid number is achieved; ➪ pp. 22–3.

hardening of the arteries, ➪ ARTERIOSCLEROSIS.

hare lip, the deformity of the upper lip caused by an inadequately repaired congenital cleft lipp.

Harvey, William, ➪ pp. 34–5.

hashish, ➪ MARIJUANA.

hay fever, an allergic inflammation of the nose lining from contact with inhaled tree or grass pollens or grains, that causes sneezing, profuse watery discharge, congestion and nasal obstruction, and smarting and watering of the eyes; ➪ pp. 142–3.

head, the upper extremity of the body that contains the BRAIN, carries the major organs of the senses and bears the major agency of emotional expression – the FACE.

headache, pain originating in the scalp or facial muscles or in the blood vessels of the scalpp. The brain is insensitive to pain but, in comparatively rare cases, an expanding mass in the brain can cause headache by stretching the covering membranes. The great majority of headaches are due to sustained muscular tension; ➪ pp. 122–3.

healing, ➪ pp. 62–3.

health, ➪ pp. 122–3.

health education, the inculcation of knowledge, the possession of which can help to promote health and reduce the chances of disease. Health education is concerned with such matters as personal hygiene, cleanliness, exercise of body and mind, good diet, care of the skin and hair, and the avoidance of hazards such as smoking, excessive drinking and the abuse of drugs.

hearing, sense of, ➪ pp. 76–7.

heart, ➪ pp. 50–1.

heart attack, the acute and usually serious disorder occasioned by the obstruction of a branch of one of the coronary arteries supplying the ever-contracting heart MUSCLE (myocardium) with BLOOD. As a result of this, part of the heart muscle dies and, if the victim survives, is replaced with non-contracting scar tissue so that the efficiency of the heart may be reduced. The gravity of a heart attack depends almost entirely on the size of the coronary branch affected. Coronary obstruction is nearly always caused by the clotting of blood on top of a plaque of atheroma – a feature of the arterial disease ATHEROSCLEROSIS. This is called a coronary thrombosis and the effect on the heart muscle is called a myocardial infarction; ➪ pp. 138–9.

heart block, the effect of an interruption of the specialized muscular conducting tissue that coordinates the heart-beat. This is usually due to disease of the coronary arteries. The result may be a severe slowing of the rate of contraction of the lower pumping chambers. An implanted artificial pacemaker may restore a more normal rate and improved bodily function; ➪ pp. 138–9.

heart failure, the state in which the heart MUSCLE is incapable of a sufficiently energetic beat to maintain an adequate BLOOD circulation. Failure may affect either the right or the left side of the heart, or both. The effects may be debility, fatigue, listlessness, low tolerance of exertion, breathlessness especially when lying down, fluid accumulation in the tissues, lungs and abdomen, swelling of the ankles and coldness of the extremities. Treatment can do much to improve the general condition; ➪ pp. 138–9.

heart-lung transplant, an operation in which a donated heart and one or both lungs are inserted into the body after removal of corresponding diseased organs. A heart-lung transplant is technically easier than a heart transplant but carries its own special complications.

heart murmur, a sound caused by turbulent blood flow or vibration of a heart valve that may or may not indicate abnormality. Many heart murmurs are innocent.

heart valve, ➪ pp. 50–1, 138–9.

heartburn, the burning sensation felt behind the BREASTBONE when acid from the stomach regurgitates up into the GULLET.

heat, the form of energy exhibited by bodies by virtue of atomic movement and associated with electromagnetic radiation. Human body heat is maintained by oxidation of glucose and fatty acids and is kept within narrow limits by an efficient regulating mechanism. The term has a number of metaphoric uses including those to describe anger or sexual excitement.

heaven, a concept common to many RELIGIONS and referring to the abode of GOD and of those whose behaviour during life has qualified them for residence there after death.

hedonism, the philosophic and psychological proposition that pleasure, or gratification, is the only ultimate good, and that the pursuit of pleasure is the ultimate motivating force. The concept of 'pleasure' is, of course, susceptible to a variety of definitions.

Heidelberg man, the origin of a fossil jaw found in a quarry at Heidelberg, Germany in 1907. The species is thought likely to be that of *Homo erectus* who flourished some 400 000 years ago; ➪ pp. 10–11.

Heimlich manoeuvre, a method used to try to dislodge an object obstructing the airway and causing asphyxia. Sudden thrusts are made up into the angle of the ribs by a person standing behind the victim and encircling the upper abdomen with the arms and clasping the fists in front. The idea is suddenly to compress the air in the lungs and force out the obstruction.

Helmholtz, Hermann Ludwig Ferdinand von (1821–94), German physiologist and polymath whose researches and studies greatly promoted our understanding of the detailed function of the eye and the ear and the conservation of energy in muscular action. He elucidated colour vision, much of physiological optics and the whole subject of the perception of musical tone and vowel sounds.

helper cells, ⇨ pp. 60–1.

hemiplegia, paralysis of one side of the body.

hemisphere, ⇨ CEREBRAL HEMISPHERE.

hepatitis, inflammation of the liver. This may be caused by a range of different viruses that cause different clinical patterns of varying seriousness; ⇨ pp. 132–3.

herbal medicine, a form of medical treatment using extracts of herbs. Many orthodox and important drugs are derived from herbs, but herbalists concentrate on those not considered by pharmacologists to be of sufficient medical value to exploit. Instances of poisoning by herbal remedies regularly appear in the medical press; ⇨ pp. 158–9.

heredity, the transmission of genes from parent to offspring; ⇨ pp. 20–1, 30–1.

hermaphrodite, a person who possesses both male and female reproductive organs.

hernia, ⇨ RUPTURE.

heroin, an opium alkaloid with powerful pain-killing properties that is much abused because of the EUPHORIA it induces.

herpes simplex, ⇨ COLD SORE.

herpes zoster, ⇨ SHINGLES.

heterosexuality, the common pattern of sexual attraction and inclination between and towards a person of the opposite anatomical sex. Compare HOMOSEXUALITY.

heterozygote, ⇨ pp. 22–3, 28–9.

hiccups, repeated involuntary spasms of the diaphragm that cause sudden inspiration quickly followed by closure of the GLOTTIS; ⇨ pp. 126–7.

hindbrain, the part of the embryonic brain from which the CEREBELLUM and the brainstem, with the nuclei of most of the cranial nerves, developp. Technically known as the rhombencephalon.

Hinduism, one of the handful of great RELIGIONS of the world and the principal religion of India. Unlike most Western religions, Hinduism does not have a formal structure and a narrow body of dogmatic principle. Its development, over the course of some 4000 years, has led to great diversity of doctrine and belief loosely based on respect for a particular way of life, known as Dharma. Great emphasis is placed on the moral principles of honesty, self-control, service, purity, courage and the avoidance of violence. Dharma is considered superior to Kama – the satisfaction of desires – and Artha – the achievement of material prosperity. The highest aim in life is emancipation from, and renunciation of, the things of the world. Another central element is the doctrine of METEMPSYCHOSIS or transmigration of the soul under the influence of one's behaviour (Karma). There are six philosophical systems, one of which is YOGA. Hinduism is also based on the caste system and on the importance of conforming to the duties of one's caste and fulfilling one's role in society.

hip joint, the ball and socket joint between the head of the thigh bone (femur) and the deep hollow (acetabulum) in the side of the bony pelvis; ⇨ pp. 40–1.

hip replacement, a surgical operation in which the head and neck of the thigh bone (femur) are sawn off and a metal ball on a pointed arm is cemented into the hollow of the bone. The plastic cup of the joint is then cemented into the natural hollow in the pelvis and the two parts articulated.

hippocampus, ⇨ pp. 170–1.

Hippocrates, ⇨ pp. 34–5, 94–5, 124–5, 150–1.

Hippocratic oath, ⇨ pp. 150–1.

histamine, a powerful and irritating substance released from MAST CELLS in an allergic reaction; ⇨ pp. 142–3.

histology, the science of microscopic ANATOMY.

HIV, the human immunodeficiency virus, the cause of AIDS.

hives, ⇨ URTICARIA.

Hodgkin's disease, a cancer of lymphatic tissue, especially lymph nodes, that can affect several parts of the body simultaneously. Modern treatment offers a good outlook.

Hohenheim, Theophrastus Bombastus Paracelsus von, ⇨ PARACELSUS.

hole in the heart, an informal term for an abnormal opening anywhere in the central wall that divides the heart into right and left sides.

holistic medicine, medicine practised with due regard to the patient as an entire person with fears, prejudices and attitudes and with a personality determined by his or her cultural background (⇨ pp. 158–9).

homeopathy or homoeopathy, a system of medicine based on the belief that diseases can be cured by giving almost infinitesimally small doses of substances which, in larger quantity, can produce some of the symptoms of the disease. The logic of this is rejected by scientific doctors, but the method is widely practised on an empirical basis. It is, however, dangerous to rely on homeopathy for the treatment of serious or potentially serious disorders; ⇨ pp. 158–9.

homeostasis, the set of physiological mechanisms by which constant conditions are maintained in the body. It is essential for the maintenance of health, and often even life, that the many variable elements (parameters), such as temperature, heart rate, blood acidity, blood levels of many substances, hormone levels and blood pressure should be kept within prescribed limits. There are thousands of such parameters and all must be controlled. Some have natural limits, but in many cases control must be exercised by a process of automatic monitoring and self-regulating information feedback. Homeostatic control is most obviously exercised by the endocrine system under the general supervision of the PITUITARY GLAND, but other systems are extensively involved, especially the nervous system. Transducers of various kinds – chemical, pressure, orientational, positional – convert deviations from the norm into nerve impulse variations that are fed to the brain to effect an appropriate corrective response. Homeostasis is one of the central principles in PHYSIOLOGY.

hominid, any primate of the family *Hominidae*. The hominids include modern humans and their predecessors.

hominoid, resembling humans, or of the *Hominoidiae* superfamily that includes the ANTHROPOID APES and *Homo sapiens*.

Homo erectus, ⇨ pp. 10–11.

Homo habilis, ⇨ pp. 10–11.

Homo sapiens, ⇨ pp. 4–5.

Homo sapiens neanderthalensis, ⇨ pp. 10–11.

Homo sapiens sapiens, ⇨ pp. 10–11.

homologous chromosome, one of a pair of chromosomes with mainly identical features and carrying close or identical copies of each other's genes in corresponding positions (loci).

homology, ⇨ pp. 4–5.

homophobia, a slang term for fear or hatred of homosexual people.

homosexuality, sexual preference for a person of the same anatomical sex; ⇨ pp. 202–3.

homozygous state, ⇨ pp. 22–3, 28–9.

hookworm, a roundworm whose larvae are capable of penetrating the intact skin and whose adult forms are equipped with hooks at one end by which they attach themselves to the lining of the intestine; ⇨ pp. 136–7.

Hopkins, Sir Frederick (1861–1947), English biochemist whose pioneering work on vitamins earned him the Nobel Prize in 1929.

hormone, one of many chemical messengers of diverse type, the effectors of one of the major control systems of the body. Some hormones are small molecules derived from AMINO ACIDS. Others are polypeptides or PROTEINS, steroids derived from cholesterol, or eicosanoids derived from a polyunsaturated fatty acid. They are secreted directly into the BLOOD which carries them to the site of their action where they specifically alter the action of target cells; ⇨ pp. 78–9, 80–1.

hormone replacement therapy (HRT), any medical treatment in which the deficiency of a normal body HORMONE is compensated for by supplementation with the hormone. This is necessary when the endocrine gland normally producing the hormone has had to be removed or has ceased, for any reason, to secrete. Hormones may sometimes be given by mouth and are often delivered from an implant or a skin patch. The commonest form of HRT is for oestrogen deficiency in women after the menopause.

hospice, a hospital for the terminally ill where, by the skilled management of pain and the exercise of emotional, psychological and spiritual support, people are enabled to die peacefully and with dignity.

hospital, an institution dedicated to the care of human beings suffering from medical, surgical or psychiatric disorders or requiring surveillance in, or assistance with, childbirth.

Hounsfield, Godfrey Newbold, ⇨ pp. 128–9.

housemaid's knee, inflammation, from sustained pressure, of the fluid-filled tissue pad (bursa) that lies in front of the kneecap.

human genome project, the attempt to establish the sequence of the entire genetic code of the human being. Although the genome is only about two metres long it consists of a sequence of some three billion consecutive samples of one of four bases. The task of identifying these is appallingly large, but ever more refined automation methods have brought the project within the realms of possibility. The latest developments have advanced the expected date of completion to around the turn of the century.

humanism, the doctrine that proposes that human behaviour should be based on reason, respect for human values and on the worth of the individual rather than on religious dogma.

Hume, David, ⇨ pp. 180–1.

humerus, the bone of the upper arm; ⇨ pp. 40–1.

humour, the possession of, or the capacity to perceive, those things which excite laughter or the desire to laugh. Humour is one of the more mysterious characteristics of the human being and its nature has been endlessly argued. We laugh when we are painlessly surprised; when we perceive foolishness or qualities to which we consider ourselves superior; when we see the pompous deflated, the powerful threatened or the consciously superior mocked. Theories about humour abound, but none of them are entirely convincing.

hunger, the symptoms of abdominal discomfort, pain, contractions of the STOMACH and craving for food induced by a drop in the level of glucose in the blood passing through the HYPOTHALAMUS of the brain.

Hunter brothers, John and William, ⇨ pp. 150–1, 154–5.

hunter-gatherer society, a society that subsists on the animals its members can kill and the edible material they can find growing or can cultivate.

Huntington's chorea, ⇨ pp. 28–9, 186–7.

Huxley, Thomas Henry, ⇨ pp. 8–9.

hydrocele, a normally potential cavity in the SCROTUM that forms a painless collection of fluid around the TESTIS. Usually benign, the condition sometimes indicates cancer.

hydrocephalus, ⇨ WATER ON THE BRAIN.

hydrochloric acid, ⇨ pp. 44–5.

hydrotherapy, the use of water for medical treatment either to sustain body weight in the mobilization of orthopaedic patients or to take advantage of the alleged therapeutic properties of spa water or baths.

hygiene, the study of measures to promote health; ⇨ HEALTH EDUCATION.

hymen, the variable thin membrane of skin that partly occludes the entrance to the VAGINA in a virgin.

hyoid bone, a delicate U-shaped bone in the neck entirely suspended from fine strap-like muscles and forming a base for the movements of the tongue.

hyperactivity, given to excessively vigorous or prolonged exertion and bustle. Hyperactivity is fairly common in children but rare in adults other than those suffering from MANIA.

hyperglycaemia, an abnormally high level of glucose in the blood; one of the salient features of DIABETES.

hypermetropia, a dimensional eye disorder in which the distance from the lens system to the focal plane on the retina is too small. Young people with active accommodation can compensate by employing the same action normally-sighted people use when reading. But as accommodation weakens with age, the hypermetropia becomes manifest and glasses are needed.

hypertension, high blood pressure.

hyperthyroidism, overactivity of the THYROID GLAND with overproduction of thyroid HORMONES. These affect almost every cell in the body causing accelerated function. There is general hyperactivity, fast pulse and respiration, sweating, good appetite but loss of weight, intolerance to warm conditions and sometimes protruding eyes (exophthalmos); ➪ pp. 80–1.

hypertrophy, an increase in cell size causing a part to enlarge, as in the effect of exercise on muscles. Compare with hyperplasia in which the enlargement is due to an increase in the number of cells; ➪ pp. 42–3.

hyperventilation, rapid or unusually deep breathing. This may be physiological as during strenuous exercise or it may be pathological in BRAIN damage, fever, poisoning, HYPERTHYROIDISM or in an acute anxiety state where it may lead to a change in the BLOOD acidity and MUSCLE spasm (tetany).

hypnosis, a state of high receptiveness to suggestion in which the attention is concentrated on the pronouncements of the hypnotist. Most instructions are obeyed mechanically and all modalities of sensation can be experienced as dictated. In spite of its promise, hypnosis has never proved of much practical value.

hypochondria, a neurotic preoccupation with health and often conviction of non-existent illness.

hypoglycaemia, abnormally low levels of sugar (glucose) in the BLOOD. The condition is dangerous; ➪ pp. 54–5.

hypospadias, a congenital abnormality of the penis in which the urine tube (urethra) opens on the under surface, often near the scrotum, instead of at the tip.

hypothalamus, the area of the BRAIN immediately above the stalk of the PITUITARY GLAND on the middle of the undersurface. This area receives wide connections from other parts of the brain and is closely linked to the pituitary both by NERVES and by HORMONES. Among other functions it acts as the link between the neural and the endocrine control mechanisms; ➪ pp. 66–7, 78–9.

hypothermia, a drop below normal in the core body temperature; ➪ pp. 144–5.

hypothyroidism, underactivity of the THYROID GLANDS. There is general slowing of the physical and mental processes and a tendency to inactivity and weight gain; ➪ pp. 80–1.

hypoxia, an inadequacy in the supply of oxygen to a part or to the whole body. Hypoxia, unless very mild, is always serious.

hysterectomy, surgical removal of the womb.

hysteria, ➪ pp. 164–5, 184–5, 190–1.

id, SIGMUND FREUD's notion of that part of our nature concerned with personal, especially sensual, gratification and indifferent to reason or humanity. The id is said to be the source of our psychic energy; ➪ pp. 164–5.

identity, the awareness of one's own nature and personality, especially in relation to a social context. An identity crisis arises when one feels unable to 'identify' with one's social environment; ➪ pp. 198–9.

ideology, the beliefs and aspirations of a group, political party or society that underlie and guide political action and policies.

idiopathic, of unknown cause.

idiot savant, ➪ pp. 168–9.

ileum, the lower of the two parts of the small intestine; ➪ pp. 44–5.

ilium, one of the paired bones of the pelvis.

illusion, a misleading or deceptive appearance of reality. Compare DELUSION, which is a false belief.

imagination and creativity, ➪ pp. 176–7.

imaging techniques, a wide range of methods used in medicine to produce visual images of the interior of the body for diagnostic purposes. They include X-RAYS, contrast X-rays, contrast angiography, digital subtraction angiography, CT and MRI scans, radionuclide scans, PET scans, and various forms of endoscopy.

Imhotep (c. 2980 BC), Egyptian physician, sage and architect of great repute who was identified by the Greeks and Romans with their own god of healing AESCULAPIUS.

immune system, the elaborate set of cellular mechanisms, mediated by LYMPHOCYTES and a variety of scavenging (phagocyte) cells, without which survival in our environment of hostile micro-organisms would be impossible. The cells of the immune system are scattered throughout the body, in the tissues and in the BLOOD and are also concentrated in the lymph nodes and the spleen. Body cells carry surface markers (antigens) by which the immune system can 'recognize' them as self – a process that involves the fitting together of appropriately shaped 'lock and key' molecules. Cells so recognized are respected. Others, such as micro-organisms, cancer cells, or tissue from another individual, are tagged with markers that identify them as foreign to the immune system. Over-activity of the immune system can be unpleasant, as with allergies, or dangerous as in tuberculosis, where repeated attacks by the immune system on a bacillus that is hidden in the long tissue causes extensive cell damage. However, the immune system is vital to our defence against disease, and immuno-comprised individuals rapidly succumb to many diseases; ➪ also ORGAN-GRAFTING and pp. 58–9.

immunization, ➪ pp. 58–9.

immunodeficiency disorders, any disorder caused by a reduction in the efficacy of the system of the body that protects against infection. Immune deficiency may be congenital or acquired and there are several causes of the latter, apart from AIDS.

immunoglobulin, an antibody; ➪ pp. 58–9.

immunology, ➪ pp. 58–9.

immunosuppressive drugs, drugs used to prevent the normal immune system rejection processes from interfering with the acceptance of a grafted organ such as a kidney or a heart.

impetigo, a staphylococcal skin infection causing small crusting blisters and encouraged by neglect of skin cleanliness.

implantation, ➪ pp. 102–3.

impotence, the inability to achieve or sustain a sufficient penile erection to allow normal sexual intercourse. Most cases are not of organic origin and many are relative to the partner.

impulse (nerve), a zone of electrical depolarization that passes along a nerve fibre in both directions and can be transmitted to other nerves or to a muscle or gland to cause, respectively, contraction or secretion.

in vitro fertilization, literally, fertilization 'in glass' i.e. fertilization of an ovum in a glass dish by the artificial application of sperms. Once fertilization is confirmed the ovum can be implanted into the womb; ➪ pp. 98–9.

inbreeding, ➪ pp. 30–1.

incest, sexual intercourse between close blood relatives. The prohibited degrees vary in different societies. The widespread and strong incest taboo is not, as was once thought, based on genetic reasons; ➪ pp. 30–1.

incisor, one of the eight biting front teeth; ➪ pp. 44–5.

incontinence, inability to maintain normal control over the discharge of urine or faeces.

incubation period, the time interval between the acquisition of an infection and the first appearance of signs or symptoms; ➪ pp. 130–1.

incubator, a temperature-controlled chamber used to grow micro-organisms on a suitable culture medium; or to accommodate a premature baby in a safely regulated environment.

indigestion, an informal term for abdominal discomfort felt after eating.

individualism, the state of mind or opinion manifested by the desire to assert one's individuality and independence from other influences.

Indo-European language family, a group of languages spoken by half the world's population and on every continent. The family includes many branches. The Germanic branch includes English, German, Dutch and the Scandinavian languages; the Italic or Romantic branch includes the Latin derivatives – Italian, French, Spanish, Portuguese and Romanian; the Celtic branch includes Irish and Scottish Gaelic, Welsh and Breton; the Indo-Iranian branch includes Hindi, Bengali, Persian; the Baltic branch includes Lithuanian and Latvian; and the Slavic branch includes Russian, Polish, Czech and Bulgarian; ➪ pp. 16–7.

induction of labour, deliberate and artificial initiation of womb contractions so as to bring about delivery of a baby in late pregnancy. This is usually done by means of a drug known as a prostaglandin.

industrial hazards, ➪ pp. 144–5.

industrial psychology, ➪ pp. 166–7.

infancy, ➪ pp. 106–7.

infant mortality, the number of infants per thousand live births who die before the first birthday. Infant mortality is a reliable indication of the quality of local standards of public health and health care. In 1900 the figure in Britain was about 150; today in some areas it is as low as eight.

infarction, the effect of the loss of blood supply to a local area of tissue, with death of a wedge-shaped segment.

infection, ➪ pp. 130–1, 132–3.

infectious mononucleosis, ➪ GLANDULAR FEVER.

inferiority complex, ➪ pp. 164–5.

infertility, ➪ pp. 98–9.

infibulation, female circumcision in which the labia minora and CLITORIS are removed and the edges sometimes sewn together. A dangerous and disfiguring form of circumcision that has no practical value.

inflammation, the response of any body tissue to any kind of injury. Blood vessels dilate, the area becomes swollen, red, hot and painful, cells are damaged and tissues injured. Inflammation is the commonest of all pathological processes and is not wholly destructive in nature. The local increase in blood supply, which is its chief feature, ensures that a good supply of antibodies is brought to the area and that toxic substances produced by bacteria are diluted and swept away. The final stage of inflammation may, if appropriate, be associated with scar formation. The inflammatory reaction is greatly inhibited by the cortico-steroid hormones and by analogous drugs. The use of such drugs can be clinically advantageous, but can also be dangerous; ➪ pp. 62–3, 130–1.

influenza, a highly infectious viral disorder of the respiratory tract causing fever, sore throat, cough, headache, muscle aches and prostration. There are several varieties of varying virulence, some of which may be fatal to the elderly or the infirm; ➪ pp. 132–3.

ingrowing toenail, a painful but inaccurately named condition in which INFLAMMATION and swelling of the soft tissues at the edges of the big toenail overlap the nail.

inhibition, restraint, hindering or prevention. Psychological inhibition implies the existence of a mental condition that impedes the full expression of the personality or the aspirations of the individual.

injury, ➪ pp. 120–1.

ink blot test, ⇨ RORSCHACH TEST.

inoculation, ⇨ pp. 58–9.

insanity, a legal term indicating mental disorder of a degree that interferes with legal responsibility for action.

insomnia, a tendency to lie awake at night when sleep is desired; ⇨ pp. 144–5.

inspiration, the act of breathing in. The term is also used metaphorically to refer to the sense of stimulation of the thoughts or emotions especially in a creative mode.

instinct, one of the many inherited, complex, unlearned patterns of partly predictable behaviour that underlie social activity.

insulin, the pancreatic HORMONE that controls the BLOOD sugar levels. Blood is monitored by the pancreas and if the sugar levels become too high, more insulin is released. This acts to promote storage of sugar in the muscles and the liver. A deficiency of insulin causes DIABETES; ⇨ pp. 80–1, 152–3.

intelligence, a roughly quantifiable or comparative measure of the speed, range and efficiency of the intellectual processes, and especially of the speed with which relationships are perceived; ⇨ pp. 168–9.

intelligence tests, ⇨ pp. 168–9.

intensive care, a continuous form of medical and nursing surveillance and management provided for people whose condition is either critical or may suddenly become so. MONITORING can reveal dangerous changes and the means of combating these is kept immediately to hand.

intercostal, lying between the ribs.

intercourse, sexual, literally and ideally, the totality of mutual sexual stimulation and response between two people, including the psychic and emotional elements as well as the erotic and sensual. In popular usage the term often implies only COPULATION.

interferon, one of a number of antiviral PROTEINS produced by cells that have been invaded by viruses. Interferons are not specific to the original invaders but provide protection against other viruses also and this protection is extended to other cells likely to be invaded; ⇨ pp. 132–3.

intervertebral disc, a fibrous ring (annulus fibrosus) enclosing a small mass of elastic pulpy material (nucleus pulposus) that lies between and cushions adjacent vertebrae in the spinal column; ⇨ pp. 40–1.

intestine, ⇨ pp. 44–5.

intoxication, poisoning. The term is usually applied to the mild poisoning with a drug, such as alcohol, that causes effects ranging from a valued loss of social inhibitions and a temporary EUPHORIA to a dangerous state of coma.

intrauterine contraceptive device, a plastic or metal loop or coil of varying shape that is inserted into the womb and left in place for many months to prevent pregnancy, usually by interfering indirectly with the implantation and survival of fertilized ova; ⇨ pp. 100–1.

intravenous, into or within a vein.

introvert, ⇨ pp. 164–5.

intuition, knowledge apparently acquired without either observation or reasoning. The idea, although romantically attractive, wilts in the presence of modern psychological and physiological ideas. Few experts now believe that anything can come out of the BRAIN that has not previously gone in, in however fragmentary a form. However, the brain has a powerful ability to synthesize conclusions from the most meagre information, and this power may be the basis of intuitive thought.

invasive, of a tumour, having a strong malignant tendency to grow into and damage normal tissue. The antonym, non-invasive, is used to refer to methods of medical treatment or investigation that do not involve a breach of the body's surface.

involuntary muscle, a muscle of the unstriped or smooth variety, as in the walls of the intestines or the arteries, that is stimulated only by nerves of the AUTONOMIC NERVOUS SYSTEM.

involutional melancholia, a severe form of depression, often with paranoid and hypochondriacal elements, that occurs for the first time between the ages of about 45 and 65 years.

IQ, the INTELLIGENCE quotient; ⇨ pp. 168–9.

iridology, medical diagnosis by examination of the iris of the eye and the location of 'clefts' in areas said to represent the various parts of the body. The procedure is rejected by orthodox ophthalmologists as having no scientific basis.

iris, the coloured diaphragm of the eye lying behind the cornea and perforated, centrally, by the pupil through which light passes. Only one pigment – brownish melanin – is involved and the range of colours results from varying concentrations and positioning of this pigment and varying thickness of the iris tissue and the density of its muscle and blood vessels; ⇨ pp. 84–5.

iritis, inflammation of the IRIS.

iron, a metallic element essential for the health of the body as it is necessary for the formation of HAEMOGLOBIN.

iron age, the period following the BRONZE AGE, and starting about 1100 BC, in which the use of iron tools and weapons rapidly spread over much of the inhabited world.

irritable bladder, ⇨ CYSTITIS.

irritable bowel syndrome, a common condition affecting mainly women between 20 and 40 and featuring attacks of lower abdominal pain that are often relieved by defaecation and sometimes a sense of distention. There is morning diarrhoea and often an urge to defaecate after meals and frequently abdominal noises and excessive passage of wind (flatus).

irritation, the condition of being aroused to annoyance or anger.

Islam, one of the five major RELIGIONS of the world, first articulated by the prophet Mohammed in the 7th century AD. It is the dominant religion of the Middle East but is also represented in almost every country in the world, especially in Africa and parts of Asia. Orthodox Muslims make a total commitment in obedience and trust to the one God Allah. In accordance with the prescription of their holy book, the Koran or Qur'an, they do this in the manner of Abraham who did not hesitate to sacrifice his son at God's command. Indeed, Islam is seen by the most orthodox as a return to the religion of Abraham. The religion is comprehensive in its demands on its adherents, whose basic duties include conformity to the five pillars of Islam – profession of faith in God and in his prophet; ritual prayer, five times a day facing Mecca; the giving of alms; abstaining from food and drink during daylight hours throughout the month of Ramadan; and making a pilgrimage to Mecca. On the last day every soul will have to answer for its deeds during life, the details of which have been written down by a recording angel, and will be judged. Sinners are sent to Hell; the righteous to Paradise.

isometrics, schemes of physical exercise involving ISOTONIC tensioning of muscles and their ANTAGONISTS so that contraction and movement are prevented.

isotonic, having equal tension. The term is more often used in medicine to refer to solutions having the same tendency to cause a flow of water across a membrane (osmotic pressure).

itching, a tickling or irritable sensation in the skin, caused by local stimulation of sensory nerve endings, that prompts the desire to scratch; ➪ pp. 136–7, 142–3.

IUD, ➪ INTRAUTERINE CONTRACEPTIVE DEVICE.

Jainism, an ancient Indian religion that has been practised for well over 2000 years and which teaches the sacredness of all forms of life, leading to great caution to avoid injury to anything.

James, William, ➪ pp. 162–3, 180–1, 204–5.

jaundice, ➪ pp. 48–9.

Java man, ➪ pp. 4–5, 10–11.

jaw, the mobile mandible, a U-shaped bone that hinges on the base of the skull in front of the ears, being pulled upwards by powerful muscles running down from the skull. The term is also applied to the fixed maxilla or upper jaw.

jejunum, ➪ pp. 44–5.

Jenner, Edward, ➪ pp. 58–9, 116–17.

jet lag, ➪ pp. 66–7.

jogger's heel, any disorder of the heel, such as blistering, strain of the Achilles tendon, or persistent bony tenderness, that can be attributed to jogging.

jogger's nipple, painful inflammation of the nipple caused by constant rubbing against a garment in the course of recreational running.

joint, ➪ pp. 40–1.

joint replacement, a surgical procedure in which a diseased or damaged joint, especially a hip or knee joint, is cut out and a mechanical prosthesis fitted in its place.

Judaism, the religion of the Jews, the first of the monotheistic religions, based on the Old Testament, especially the first five books (the Pentateuch or Torah), and the laws and interpretations contained in the Talmud. Judaism holds that, although God's providence extends to all, he entered into a special covenant with the ancient Israelites through which they were to be an example to the whole word. Central to the faith is a belief in the coming of the Messiah. Judaism prescribes individual practices and rules concerning prayer, study, the recital of blessings, marital relationships, the preparation of food, diet, observance of the Sabbath, and so on.

judiciary, relating to courts of law and to judges and their activities. Compare LEGISLATURE.

jugular veins, the complex and variable system of neck veins, of which there are six main vessels, that drain the blood from the head and return it to the heart.

Jung, Carl, ➪ pp. 164–5.

junk food, an informal term for highly refined and palatable food with a high saturated fat content that readily provides calories but is deficient in vitamins, minerals and fibre.

justice, the quality of fairness or equity, especially when entrhroned in a constitutional legal system applicable equally to all people in a society.

juvenile delinquency, criminal behaviour by a young person often because of boredom, idleness, lack of social opportunity, shortage of spending money and peer pressures. Poor school performance leads to truancy and resentment of authority and the desire for some form of status and self-expression. Delinquency often follows. Most delinquents, however, learn to conform to more acceptable patterns of behaviour.

Kaposi's sarcoma, a condition featuring multiple tumours of blood vessels, especially of the skin, that commonly affects homosexual males with AIDS; ➪ pp. 118–19.

karma, the idea, common to all Indian religions, that current behaviour influences future beatitude, either in this life or in a future life; ➪ pp. 70–1.

keloid, a thick, conspicuous, overgrown and disfiguring scar due to abnormal healing.

keratin, a tough protein found in nails, hair and the outer layer of the skin (the epidermis).

ketosis, high levels in the blood of substances (ketones) produced when fats are used as fuel because of starvation or because the normal fuel, glucose, cannot be used. Ketosis is a feature of untreated DIABETES. Ketones are acidic and dangerous.

keyhole surgery, surgery performed using an operating ENDOSCOPE passed into the body through a very small incision. Much gynaecological and gall-bladder surgery is performed in this way and the use of this technique is steadily expanding; ➪ pp. 154–5.

Keynsian economics, economic principles recommended by the British writer, lecturer, businessman and economist John Maynard Keynes (1883–1946). Keynes taught that an increase in money supply and low interest rates would stimulate a stagnant economy and reduce unemployment. He also recommended a fiscal policy of spending on public works during recessions and, if necessary, an unbalanced budget to increase the demand for goods and services. These ideas were, at the time, considered revolutionary but were eventually widely accepted.

kidney failure, a late stage in kidney disease in which both have been so damaged that they cannot excrete body waste products fast enough. Life can be saved only by dialysis or a kidney transplant.

kidney stone, a hard, irregular body that forms in the kidney by crystallization of various substances dissolved in the urine, especially if there is DEHYDRATION.

kidney transplant, the insertion of a donated kidney into the ABDOMEN and its connection to a main artery and vein and to the bladder; ➪ TRANSPLANT SURGERY.

kidneys, paired, bean-shaped organs lying on the back wall of the abdomen; ➪ pp. 64–5.

kilocalorie, the amount of heat required to raise the temperature of a kilogram of water by 1°C. This has long been the standard nutritional unit of energy but is now being replaced by the kilojoule. 1 Kcal = 4.2 KJ.

kin, ➪ pp. 214–5.

Kinsey reports, the first of the large-scale studies of human sexual behaviour. *Sexual Behavior in the Human Male* (1948) and *Sexual Behavior in the Human Female* (1953) by Alfred C. Kinsey, Wardell B. Pomeroy and Clyde E. Martin had a marked effect on public and private attitudes to sex and helped to encourage open discussion on the subject.

Kirlian photography, a variable halo, or aura, that appears on a photograph taken of an object, such as the human hand, when it is exposed to high frequency electric currents. This phenomenon, which is probably due to an electrostatic effect that varies with the moisture content, has been claimed to reveal psychic phenomena.

kissing, the widespread human practice of pressing the lips against some part of the body of another person, especially the mouth. Kissing has social as well as sexual functions and these are usually kept apart. Some societies accept public kissing between adult males; others do not.

kiss of life, mouth-to-mouth or mouth-to-nose artificial respiration.

Klein, Melanie (1882–1960), a Viennese psychiatrist and developmental psychologist who pioneered psychoanalysis in children and invented play therapy.

kleptomania, an often uncontrollable impulse to steal things, neither wanted nor needed, in order to obtain relief of emotional tension. Kleptomania is a common defence in theft proceedings, but is a very rare disorder.

knee, the hinge joint between the lower end of the thigh bone (femur) and the upper end of the main lower leg bone (tibia). The knee cap (patella) is a flat bone lying within the strong tendon of the thigh muscles and lies outside the joint; ➪ pp. 40–1.

knock-knee, the condition in which the knees are closer together than the ankles when the legs are straightened so that the knees tend to touch.

knuckle, the popular name for the prominence caused by a finger joint.

Koch, Robert, ➪ pp. 116–17, 152–3.

Krebs cycle, ➪ pp. 46–7.

kwashiorkor, a serious disease of young children caused by gross deficiency of dietary PROTEIN and a high intake of CARBOHYDRATE of low nutritional value. Growth and development are retarded, the abdomen is protuberant, the muscles are wasted and the hair assumes a reddish tinge. The ill effects are usually life-long.

labia, the four lips of the female genitalia, the thin inner labia minora and the thick, outer, hair-covered labia majora. The latter are normally closed and conceal the rest of the GENITALS; ➪ pp. 96–7.

labour (childbirth), the three-stage process of delivering a baby and the afterbirth (PLACENTA) by involuntary contraction of the MUSCLE wall of the womb and voluntary contraction of the muscles of the abdominal wall and diaphragm. The first stage is from the onset of pains to the full widening of the cervix; the second continues until the delivery of the baby; and the third stage ends with the delivery of the placenta; ➪ pp. 104–5 .

labyrinth, the elaborately shaped membranous structure of the inner ear that contains the organs of balance (the vestibule and semi-circular canals) and hearing (the cochlea); ➪ pp. 86–7.

Lacan, Jacques (1901–81), controversial French psychiatrist who applied Freudian ideas to structuralist linguistics and taught that the acquisition of language by a child began the process of repression of emotions and thoughts that could lead to psychiatric disorder.

lacrimal gland, the glandular elements of the BREAST that secrete milk under hormonal and emotional stimuli at the end of pregnancy.

lactation, milk secretion.

lactic acid, an acid formed in painful excess when muscles contract strongly for long periods. The acid is also formed in the VAGINA from the action of bacteria on carbohydrates.

lactose intolerance, a disease caused by shortage of an ENZYME, normally present in the lining of the small intestine, that breaks down the lactose in milk. The accumulating sugar is acted on by gas-forming intestinal bacteria, leading to discomfort, colicky pain and diarrhoea.

Laennec, René (1781–1826), French physician who invented the stethoscope and thereby greatly advanced the science of diagnosis of chest and heart disease by the correlation of sounds with known conditions.

Laing, R.D. (1927–), Scottish psychiatrist, writer and EXISTENTIALIST thinker who came to regard SCHIZOPHRENIA as a deliberate strategy adopted by the sufferer as a means of dealing with an intolerable life situation, often caused by a disturbed family environment.

Lamarck, Jean Baptiste, ➪ pp. 6–7.

Landsteiner, Karl, ➪ pp. 50 1.

Langerhans, islets of, the localized collections of specialized cells in the PANCREAS that secrete INSULIN and GLUCAGON under the influence of changing blood sugar levels, and other hormones.

language, a locally agreed system for the transmission of information by the articulation of sounds or the transcription of symbols; ➪ pp. 14–15, 174–5.

laparoscopy, ⇨ pp. 154–5.

laparotomy, an exploratory operation performed for purposes of diagnosis.

laryngitis, inflammation of the voice-box (larynx), with sore throat, hoarseness and difficulty in speaking but, in most cases, full recovery in a few days, especially if the voice is rested.

larynx, the voice box at the upper end of the windpipe (trachea). The larynx contains the vocal cords and the tiny muscles that control them, causing them to press together so that they can vibrate under the action of expired air to produce voice sounds, or pulling them apart to allow free breathing; ⇨ pp. 42–3, 52–3.

latency, a general term for the interval before a reaction occurs, or for something present but not yet manifest or having an effect. Freud described a latency stage in psychosexual development, between the ages of 6 and puberty, when sexuality remains dormant.

lateral, of, at or towards the side of the body. Unilateral means on one side only; bilateral means on both sides.

laughing gas, a popular term for the anaesthetic gas nitrous oxide.

laughter, the expression of the emotion of amusement by a rapid sequence of repeated single inarticulate syllables together with an upward curving of the corners of the mouth and often closure and watering of the eyes and sometimes convulsive movements of the body.

Lavoisier, Antoine-Laurent, ⇨ pp. 52–3.

laxative, a drug used to promote emptying of the rectum, especially in cases of CONSTIPATION. Laxative drugs are not the ideal way to treat constipation; a high fibre diet and regular habits are better.

lead poisoning, brain and blood damage from the ingestion of lead compounds, usually in small quantities over a long period. Chronic lead poisoning may cause serious and permanent brain damage with loss of memory, intellectual power and coordination; ⇨ pp. 144–5.

Leakey, Louis (1903–72), Mary (1913–) and Richard (1944–), family of British anthropologists who made major contributions to the study of human evolution by the work of Louis and Mary on the human dwelling sites in the Olduvai Gorge, Tanzania, and by the discovery by the son Richard of even earlier hominid fossils in East Turkana, Kenya.

learning, ⇨ pp. 172–3, 174–5.

Leeuwenhoek, Antony van, ⇨ pp. 116–17.

Left, the, the body of people with political opinions tending to the radical, progressive and revolutionary rather than the conservative.

leg, either of the lower limbs, whose skeleton consists of the thigh bone (femur), the knee cap (patella), the shin bone (tibia) with, on the outer side, the delicate fibula, and the tarsal, metatarsal and phalangeal bones of the foot. These bones are secured with ligaments, invested in muscle, blood vessels and nerves, and covered with sheets of tendon (fascia) and skin.

Legionnaires' disease, a form of pneumonia caused by *Legionella* bacteria which can grow in warm, moist places such as air-conditioning towers and spread in air-borne water droplets. The disease features headache, muscle pain, diarrhoea, cough, high fever, mental confusion, and kidney and liver damage. Lung damage may be fatally severe.

Leishmaniasis, a skin or internal infection, endemic especially in the Mediterranean area, caused by a single-celled microscopic parasite spread by sandflies. Skin damage can be disfiguring and internal (visceral) Leishmaniasis (Kala azar) dangerous.

lens implant, ⇨ pp. 156–7.

leprosy, a skin and nerve infection of very low infectivity and long INCUBATION PERIOD that leads to loss of sensation (anaesthesia), destruction of the extremities, facial disfigurement and sometimes blindness. Now more often known as Hansen's disease. The causative organ is closely related to the tuberculosis bacillus; ⇨ pp. 116–17.

leptospirosis, an infection by a spiral-shaped organism (spirochaete) transmitted in the urine of rats or dogs to people in contact with water. The disease features headache, muscle pain and tenderness, conjunctivitis, vomiting, a skin rash, jaundice, liver damage and sometimes MENINGITIS or HEART FAILURE. Also known as Weil's disease.

lesbianism, female homosexuality.

lesion, a term, indispensable to doctors, denoting any injury, wound, infection, or any structural or other abnormality anywhere in the body.

leucotomy, ⇨ LOBOTOMY.

leukaemia, one of a group of blood disorders in which white blood cells reproduce in a disorganized and uncontrolled way, progressively and dangerously displacing the normal constituents of the blood. Leukaemia is a form of CANCER but there are effective treatments for most forms.

leukocyte, an independent white cell of the blood or tissues, such as a LYMPHOCYTE, PHAGOCYTE or MACROPHAGE.

Lévi-Strauss, Claude, ⇨ pp. 212–13.

levitation, ⇨ pp. 180–1.

liberalism, the political philosophy favouring individual freedom, tolerance, progress, breadth of cultural interest and human values.

libertarianism, a philosophy that supports freedom of thought, expression and behaviour and opposes legislation that restricts such freedom. The term is also used for the philosophic belief in the idea of free will.

libido, sexual drive; ⇨ pp. 164–5.

lie detector, a popular terms for the polygraph - an assembly of devices that monitor and record pulse rate, blood pressure, rate and evenness of breathing and the moistness of the skin. As these vary with the state of the emotions, the results can indicate an emotional response to certain questions or statements. They do not, however, indicate lying or guilt and, if the interpretation is unimaginative or insensitive, the test can be very unfair.

life span, ⇨ pp. 110–11.

life style, ➪ pp. 164–5.

ligament, ➪ pp. 40–1.

ligature, any thread-like material tied tightly round a structure – commonly a blood vessel to control bleeding. Ligatures may be absorbable and can be buried.

lightening, the sense of relief felt, near the end of pregnancy, when the head of the fetus descends into the cavity of the pelvis and there is a corresponding reduction in the oppressive volume of the abdominal contents; ➪ pp. 104–5.

limb, an elongated projection from the body adapted to locomotion or manipulation; ➪ ARM, LEG.

limbic system, a centrally placed ring-shaped structure that represents much of the BRAIN of more primitive mammals. The limbic system is concerned with automatic processes such as hunger, thirst, wakefulness and sexual interest and with their emotional concomitants. Limbic disease causes emotional disturbances; ➪ pp. 74–5, 204–5.

Lind, James, ➪ pp. 150–1.

lineage, from descent, a single common ancestor or a group of people who can trace such descent. Matrilineage is descent through the ancestry of the mother; patrilineage through that of the father.

lingua franca, any auxiliary language used between people whose primary languages are different. PIDGIN English is a widely used lingua franca; ➪ CREOLE.

Linnaeus, Carolus, ➪ pp. 6–7.

lipid, a fatty acid, wax, fat or sterol; ➪ GLYCEROL.

lipoma, a benign tumour of fat.

lipoproteins, one of various complexes of fat combined with PROTEIN found in large quantities in the blood. High density lipoproteins protect against arterial disease; low density lipoproteins contain relatively large amounts of cholesterol and are dangerous. A high intake of saturated dietary fats causes a rise in blood low density lipoproteins.

lip reading, an important method of communicating with the deaf. About 10 out of 40 possible distinguishable mouth patterns can be reliably identified, but other facial, bodily and contextual clues to meaning are also provided and a skilled lip reader can discern most of what is said by a familiar and cooperative person.

lips, the muscular and highly mobile structures that surround the mouth and are important in the production of speech and emotional expression and in eating and drinking.

Lister, Joseph, Lord (1827–1912), English surgeon noted for his application of Louis Pasteur's ideas of infective micro-organisms to surgery. Lister's adoption of antisepsis, in 1867, greatly improved the outlook for many operations and revolutionized surgery.

liver, ➪ pp. 48–9.

liver abscess, a walled-off collection of pus in the liver, causing high fever, pain and tenderness in the upper right abdomen and prostration. Liver abscesses are usually secondary to infection elsewhere, such as amoebic dysentery, and must be drained surgically and the cause treated.

liver cancer, malignant disease affecting the liver, usually secondarily to a primary CANCER elsewhere in the body. Primary cancer originating in the liver (hepatic cancer) is rare in Britain.

liver failure, the final stage of liver disease in which the liver can no longer meet the metabolic needs of the body. There is JAUNDICE and an ultimately fatal accumulation of poisonous substances in the blood, unless a LIVER TRANSPLANT can be done.

liver function tests, a group of biochemical investigations that can confirm or deny the presence of liver disease and indicate its nature. They are performed on the blood, the urine and on BIOPSY samples from the liver.

liver transplant, the grafting of a donated liver, or part of a liver, into the body of a person suffering from LIVER FAILURE. A transplanted segment of liver will quickly grow to the required size.

lobe, any well-defined subdivision of an organ such as the brain, the lung, the liver, the thyroid gland or the prostate gland.

lobotomy, surgical removal of a LOBE, especially the prefrontal lobe of the brain.

Locke, John, ➪ pp. 70–1.

locked knee, a knee that has suffered a tear and displacement of an internal cartilage which acts as a wedge to prevent the bent knee from being straightened.

lockjaw, an informal term for uncontrollable spasm of the powerful chewing muscles, usually in the disease tetanus. This clamps the teeth so tightly together that they can barely be separated. The medical term is trismus.

logic, the processes of REASON by which conclusions are correctly drawn from premises, or the analysis of these processes.

loins, the soft tissue of the back, on either side of the spine, between the lowest ribs and the pelvis.

lordosis, exaggeration of the normal forward curvature of the lower part of the spine, usually with abnormal backward curvature of the upper part (kyphosis). The buttocks may appear unduly prominent.

Lorenz, Konrad, ➪ pp. 208–9.

louse, one of a range of human ectoparasites, that includes the head louse, the body louse and the CRAB LOUSE; ➪ NIT and pp. 136–7.

Loyola, Ignatius, ➪ pp. 30–1.

LSD, lysergic acid diethylamide, a hallucinogenic drug sometimes used, unwisely, for recreational purposes. LSD can induce a psychotic delusional state that can last for months; ➪ pp. 146–7.

lumbago, severe and incapacitating pain in the region of the lower back – a symptom rather than a disease, that has many different causes.

lumbar, pertaining to the LOINS and lower back.

lumbar puncture, a method of neurological investigation in which a long needle is passed between two vertebrae of the spine from behind into the fluid-filled space below the lower end of the spinal cord. A sample of cerebrospinal fluid is sucked out with a syringe for laboratory examination.

lumpectomy, a minimal operation for breast cancer in which only the obvious lump is removed and radiation or chemotherapy given; ➪ MASTECTOMY.

lung, one of the paired, elastic, spongy organs that occupy each side of the chest on either side of the heart; ➪ pp. 52–3.

lung abscess, a severe lung infection, often from an inhaled foreign body, that has progressed to tissue destruction and the formation of a pus-filled cavity lined with inflamed tissue.

lung cancer, a term commonly applied to a CANCER of the inner lining of one of the air tubes (bronchial carcinoma), a tumour that accounts for more than half of all male cancer deaths and an increasing proportion of female cancer deaths. The most important cause is smoking. Types differ, but even in the most favourable and with good surgical treatment, less than one third of victims survive for five years; ➪ pp. 138–9, 140–1.

lung function tests, tests that determine the efficiency of the lungs and the respiratory system generally in transferring oxygen from the atmosphere to the blood.

lungs, ➪ pp. 52–3.

lupus erythematosus, an inflammatory disease of connective tissue caused by an AUTO-IMMUNE process and affecting either the skin (discoid lupus) or the internal organs (systemic lupus). The former type causes raised, red bumps on the skin and loss of hair; the latter is usually more serious and may be life-threatening.

lupus vulgaris, a rare tuberculous infection of the skin that can cause much tissue destruction and facial deformity, especially around the nose and the inside of the mouth.

luteinizing hormone, a HORMONE released by the PITUITARY GLAND that causes egg production (ovulation) from the ovary, in the female, and the secretion of the sex hormone testosterone from the testis, in the male; ➪ pp. 78–9.

lycanthropy, the supposed transformation of a human being into a wolf.

Lyme disease, a disease caused by a spirochaete transmitted by the bite of the tick *Ixodes dammini*. Expanding red rings appear on the skin followed by fever, fatigue, headaches, stiff neck, persistent arthritis and enlarged lymph nodes. Some affected people develop nervous system complications; others develop mental illness or profound fatigue and weakness that may last for months or years. Treatment at the ring stage can prevent these complications.

lymph, ➪ pp. 50–1.

lymph node, ➪ pp. 50–1.

lymphatic system, the bodily system responsible for returning to the circulation, fluid in the tissues that has leaked out of the BLOOD. Such fluid, mainly water, is returned via a complex of thin-walled tubes, the lymph vessels, and a series of bean-shaped lymph nodes packed with LYMPHOCYTES. These can usually deal with any infective material that is carried by the fluid. Lymph nodes are often erroneously described as 'glands' but have no glandular function; ➪ pp. 50–1.

lymphocyte, one of a large number of small round white cells, central to the functioning of the IMMUNE SYSTEM. They form two large classes – the 'B' cells and the 'T' cells. The former produce specific antibodies and include a huge range of variants from which the most appropriate can be selected by the system to meet the current antibody need. The latter effect 'cell-mediated immunity' that helps to protect against virus and other infections and CANCER; ➪ pp. 50–1, 58–9, 60–1, 124–5.

lymphoid tissue, ➪ pp. 50–1.

lymphoma, one of a group of CANCERS of lymphoid tissue, especially of the lymph nodes and the spleen, that vary widely in severity and outlook.

lymphosarcoma, the term formerly used for one kind of LYMPHOMA.

Lysenko, Trofim Denisovich, ➪ pp. 6–7.

lysosome, ➪ pp. 38–9.

McDougall, William, ➪ pp. 180–1.

macrophage, one of many large scavenging and ingesting (phagocytic) cells found all over the body and especially in the LIVER, lymph nodes, spleen and bone marrow. Some are stationary, some free-ranging; ➪ pp. 58–9.

magic, the use of RITUAL to attempt to summon supposed supernatural forces. Historically, there has been considerable overlap between religion, magic and science and these have been separated only with great difficulty and, even today, incompletely.

magnetic resonance imaging, ➪ pp. 128–9.

malabsorption, a group of disorders featuring a failure of movement of certain foodstuffs from the small intestine into the bloodstream so that MALNUTRITION results.

maladjustment, a psychological term signifying the inability to function normally in the context of social relationships and cultural standards and, in consequence, to suffer emotional upset.

malaise, a feeling of being unwell. Although a feature of most diseases, it has no specific diagnostic significance.

malaria, one of several types of infection that cause at least a million human deaths each year throughout the world. Malaria is caused by single-celled parasites, mainly *Plasmodium vivox* and *Plasmodium falciparum*, that reproduce in mosquitoes and are then transferred to humans by the mosquito bite: uninfected mosquitoes biting infected humans and thereby being infected by the parasite complete its life cycle. There is periodic fever occurring every one, two or three days, shaking, shivering, severe aches and pains, headache, and in some cases dark urine (blackwater fever) or dangerous brain complications. The condition may recur at intervals for years after the initial infection; ➪ pp. 136–7.

Malebranche, Nicholas, ➪ pp. 70–1.

malignant, dangerous and tending to cause death. The term is usually applied to cancer but is also used to qualify particularly serious forms of other diseases.

malignant melanoma, ➪ MELANOMA.

malingering, the pretence that one is suffering from a disease so as to achieve an advantage; ➪ MUNCHAUSEN'S SYNDROME.

malnutrition, any disorder resulting from an inadequate diet or from failure to absorb or assimilate dietary elements. The term has been extended to cover the effects of a poorly chosen or even excessive diet, either of which can damage health.

Malthus, Thomas Robert, ➪ pp. 8–9.

maltose, a common dietary sugar consisting of two linked molecules of glucose. It is produced when starches are split by digestive ENZYMES.

mammary gland, ➪ BREAST.

mandible, the JAW bone.

mania, an abnormal state of physical and mental hyperactivity with compulsive and sometimes repetitive movement and constant speaking or raving.

manic-depressive psychosis, a 'bipolar' psychiatric disorder of unknown cause featuring severe depressive phases of up to a year, after two to four of which there may be a manic phase in which the affected person is inappropriately euphoric with ever-changing flights of ideas, disordered judgement, grandiose notions and often socially or financially ruinous behaviour; ➪ pp. 192–3.

manipulation, any operation on the body performed with the hands, especially one intended to restore displaced parts to their normal position.

marasmus, severe wasting or emaciation from starvation, especially in infants.

marijuana, the dried leaves, flowers or stems of various species of the hemp grass *Cannabis* whose resin contains a drug that promotes laughter or giggling, an apparent heightening of the senses, especially vision, with distortion of dimensions, and a delusory sense of deep philosophical insight. Persistent heavy users may become apathetic and lose interest in the normal concerns of life (amotivational syndrome).

marital counselling, the process of analysis of relational problems between spouses and the giving of advice on how they may be relieved.

marrow (bone), the fatty connective tissue that fills the hollow interior of bones and which, in the case of the flat bones, also contains the blood-forming tissue of the body; ➪ pp. 40–1.

Marxism, a political philosophy, derived largely from the writings of Karl Marx, which gave rise to the now almost defunct political system of communism. Essentially, Marxism was concerned to right the imbalance of justice between the workers, who create wealth, and the capitalists, who enjoy it.

masochism, a form of sexual activity in which arousal is achieved by suffering pain, humiliation, abasement, or physical or mental abuse.

mast cell, a cell, occurring in large numbers in the skin and respiratory and lymphatic systems, that plays a central part in allergic reactions. It is full of granules of powerfully irritating substances such as histamine, serotonin and heparin which are released to cause symptoms when allergens contact the cells; ➪ pp. 142–3.

mastectomy, surgical removal of the breast, almost always for the treatment of cancer. Radical mastectomy involves the removal of all breast tissue, skin, underlying pectoral muscles and the lymph nodes in the armpit. In simple mastectomy, only the breast tissue is removed; ➪ pp. 140–1.

Masters, William (1915–) and Johnson, Virginia (1925–), a husband and wife team of sexologists and sex therapists whose laboratory investigations of the physiology of human sexual intercourse using volunteer subjects greatly advanced knowledge on the subject. Their book *Human Sexual Response* (1966) became a best-seller.

mastication, chewing.

mastitis, inflammation of the breast, usually from infection acquired through a nipple crack during breast feeding. Early treatment may prevent ABSCESS formation.

mastoid bone, the prominent bony process behind the lower part of the ear. This bone is honeycombed with air cells that communicate with the middle ear and infection can spread to these from a middle ear infection.

masturbation, self-stimulation of the genitals, accompanied by the enactment of erotic fantasies, to produce sexual arousal and often orgasm. This is one of the normal forms of sexual activity and is commonly practised by those deprived of sexual relations with others.

materialism, a general philosophic belief, taking various forms in different areas of thought, that all events involve the interaction or interrelationship of matter, and nothing more.

matriarchy, social organization in which the tribe, community, family or other group is headed by a woman and descent is traced through the female line. Compare PATRIARCHY.

matrilineage ➪ LINEAGE.

maxilla, the upper jaw.

ME, ➪ POSTVIRAL SYNDROME.

measles, an infectious virus disease of childhood acquired by droplet inhalation and featuring fever, cough, running nose, general misery, CONJUNCTIVITIS and an irregular, red, mottled, slightly raised rash which lasts for about a week and then fades.

meatus, any passage or opening in the body.

meconium, thick, greenish-black, sticky stools passed by a baby during the first day or two of life, or sometimes just before birth.

Medawar, Sir Peter (1915–87), British zoologist and immunologist whose studies on the nature of immune rejection paved the way for successful transplant surgery.

medial, situated toward the midline of the body. Compare LATERAL.

median nerve, one of the main nerves of the arm, that activates most of the muscles and provides sensation to part of the hand.

medicine, ➪ pp. 124–5, 150–1, 158–9.

medulla, the inner part of an organ, such as the kidney or the adrenal gland, or of the shaft of long bones. Compare CORTEX; ➪ pp. 74–5.

medulla oblongata, the part of the brain stem lying immediately above the spinal cord in front of the cerebellum, and containing the nuclei of four pairs of nerves, the vital centres for breathing and control of heart-beat and the long motor and sensory tracts running to and from the spinal cord; ➪ pp. 74–5.

megalomania, a delusion of great power, wealth or social status.

meiosis, the important stage in cell division in the production of sperms and ova in which the 23 paired chromosomes (diploid number) separate into two sets of 23 (haploid number) and only one of these sets is transmitted to the next cell. This reduction division is necessary so that at FERTILIZATION the diploid number can be restored; ➪ pp. 22–3.

melancholic type, one of the supposed personality types described in the system of the humours (➪ pp. 34–5) and said by it to be brought about by an excess of black bile.

melanin, ➪ pp. 56–7.

melanoma, a dangerous form of CANCER of the skin and a less dangerous cancer of the eye. Half of the skin melanomas arise from pigmented moles, half arise independently. The characteristics are essentially changes – in size, shape, colour, regularity, protuberance, softening and tendency to crumble. Some become itchy or painful. The outlook depends on the size, especially on the depth, and on the speed of reporting. Treatment is wide surgical excision. Eye melanomas tend to progress slowly and spread late and usually present as a visual field defect; ➪ pp. 140–1.

membrane, cell, the limiting structure of the cell, a specialized sheet of fatty (phospholipid or glycolipid) molecules with embedded proteins. The cell membrane has semi-permeable properties and provides intimate communication with the cell's environment by way of the proteins, and the reception of information in this way commonly leads to the release within the cells of substances known as 'second messengers'; ➪ pp. 38–9.

memory, ➪ pp. 76–7, 170–1.

memory cell, one class of B LYMPHOCYTE that retains information about previous challenge, as by infective organisms, so that ANTIBODIES can be more rapidly produced in response to a subsequent infection with the same agent; ➪ pp. 58–9.

menarche, the time of the onset of the menstrual periods, marking the beginning of the reproductive life of the female; ➪ pp. 16–17.

Mendel, Gregor Johan, ➪ pp. 20–1.

Ménière's disease, a disease of the inner ear caused by an increase in the fluid pressure in the labyrinth and featuring dizziness, a tendency to fall, nausea, hearing loss, a sense of fullness in the head and ringing in the ears (TINNITUS).

meninges, the three layers of membrane that surround the brain and the spinal cord, respectively, from in–out, the pia mater, the arachnoid mater and the dura mater; ➪ pp. 74–5.

meningitis, inflammation of the MENINGES, usually from a virus or other infection. Severity varies from a mild illness to one of grave outlook with headache, stiff neck, fever, weakness or paralysis of muscles, speech disturbances, double vision, epileptic fits and drowsiness that may progress to coma and death. Most patients recover fully, but some have residual nervous system damage.

menopause, the end of MENSTRUATION and of the reproductive period of a woman's life. The menopause usually occurs between the ages of 48 and 54 and is associated with reduced production of oestrogen hormones. This may result in various undesirable effects, such as loss of BONE strength (osteoporosis), thinning and drying of the vaginal wall, hot flushes, night sweats and general irritability. The arrival of the menopause may also have adverse psychological effects depending on how greatly a woman's fertility is valued by her or by others. This is a damagingly narrow view. Happily, many women greet the menopause as a signal of emancipation from the fear of pregnancy and the inconveniences of menstruation, and embark on a new lease of active life; ➪ pp. 110–11.

menorrhagia, abnormally heavy and prolonged menstrual periods.

menstruation, the periodic shedding of the lining of the womb (endometrium); ➪ pp. 96–7.

mental development, ➪ pp. 106–7.

mental retardation, a failure to reach a level of intellectual capacity generally deemed to fall within normal limits.

meritocracy, a system of government by those who have shown themselves most capable of governing effectively or by those whose general achievements have been greatest.

mescalin, a substance with hallucinogenic effects derived from the spineless cactus *Lophophora williamsii*. Mescalin is related to the amphetamines and is also known as peyote.

mesentery, the complex, double-layered folded membrane that encloses the bowels and suspends them from the back wall of the abdomen. Blood and lymphatic vessels run to and from the intestines between the two layers of the mesentery.

mesoderm, the middle of the three layers of the early developing embryo lying between ECTODERM and ENDODERM. Mesoderm develops into bones, muscles, the heart and circulatory system, and most of the reproductive system.

metabolism, the totality of the cellular chemical activity, largely under the influence of ENZYMES, that results in work and growth or repair. The 'building-up' aspects of metabolism are known as anabolic and the 'breaking-down' as catabolic. Metabolism involves the consumption of fuel (glucose and fatty acids), the production of heat and the utilization of many constructional and other biochemical elements provided in the diet, such as AMINO ACIDS, fatty acids, carbohydrates, vitamins, minerals and trace elements. The metabolic rate is increased in certain disorders, such as HYPERTHYROIDISM, and decreased in others. Anabolism can be artificially promoted by the use of certain steroid male sex hormones (ANDROGENS or anabolic steroids).

metacarpal bone, one of the five long bones of the palm of the hand immediately beyond the carpal bones of the wrist. Compare METATARSAL BONE; ➪ pp. 40–1.

metastasis, the spread of any disease, especially cancer, from its original site to another part of the body where the same disease process starts up as a new focus. Metastasis usually occurs via the bloodstream or the lymphatic channels. The disease at the new site is also called a metastasis.

metatarsal bone, one of the five long bones of the foot lying beyond the tarsal bones. Compare METACARPAL BONE; ⇨ pp. 40–1.

microbiology, the science of microscopic organisms such as viruses, bacteria, fungi and protozoa.

micro-organism, any organism of microscopic size.

microphage, ⇨ pp. 58–9.

microsurgery, ⇨ pp. 154–5.

micturition, urination; ⇨ pp. 64–5.

mid-life crisis, a psychological upset, sometimes affecting people in middle age, who feel that their lives have become meaningless and devoid of satisfaction. This may lead to injudicious sacrifice of established achievement and a change of life style.

midbrain, the central part (mesencephalon) of the primitive embryonic neural tube that develops into the structures immediately above and behind the upper part of the brainstem (PONS).

midwifery, ⇨ pp. 94–5.

migraine, a specific form of headache caused by abnormal widening of some of the arteries of the scalp and brain, usually on one side only. This is preceded by partial closure of these arteries causing temporary loss of part of the field of vision or other transient effects of disturbed brain function. Severe migraine features vomiting, prostration and acute intolerance to light.

migration, movement of populations from one area, region or country to another.

milk, the secretion of the female breast of any mammal that is promoted by hormonal changes occurring at the end of pregnancy and maintained by the stimulus of suckling. The production of milk from the breasts (mammary glands) is one of the primary characteristics of a mammal.

mind and body, an association long regarded as a duality of disparate entities, but now becoming recognized by most scientists as aspects of a unity. The prevailing view, today, is that the manifestations of the mind are wholly the result of neurological (mainly BRAIN) activity and that, given sufficiently detailed knowledge of the structure and function of the nervous system, the association between function of the nervous system and corresponding mental capacity will be seen to be complete. Many mental processes involve widespread brain function. Recently, the mind–body problem has been dismissed by some philosophers as a CATEGORY MISTAKE; ⇨ pp. 70–1.

minerals, chemical elements required in the diet to maintain health. Apart from iron, calcium, sodium and potassium only very small amounts are needed. The other necessary minerals are magnesium, copper, selenium, phosphorus, fluorine and zinc; ⇨ pp. 46–7.

miscarriage, spontaneous loss or death of the fetus before the stage of development at which independent existence is possible. As medical technology advances this stage becomes ever earlier; ⇨ pp. 102–3.

mite, one of many small, free-living arachnids of the order *Acarina*, some of which, such as the scabies mite, are parasitic on human beings; ⇨ pp. 136–7.

mitochondria, small organelles found within the cells of all multicellular organisms. Bacteria-like in shape, the mitochondria partition of an area in which the complex metabolic processes that produce biological energy (ATP) from glucose and oxygen (AEROBIC RESPIRATION) occur. The presence of bacterial-like chromosones within mitochondria supports the suggestion that they are the result of an ancient symbiotic association between a microorganism and an eukaryotic cell. MITOCHONDRIAL DNA is inherited only from the mother and, apart from spontaneous mutations, remains unchanged throughout the generations. The quantity of mitochondrial DNA is tiny compared to that in the cell nucleus and is present in complete rings as in bacterial chromosomes. Surprisingly, mitochondrial DNA does not use the same 'universal' genetic code as the nuclear chromosomes but a slightly different code; ⇨ pp. 38–9.

mitochondrial DNA, ⇨ pp. 4–5, 38–9.

mitosis, the normal process of cell division in which a single cell gives rise to two daughter cells each having a genetic composition identical to the mother cell. Compare MEIOSIS; ⇨ pp. 22–3.

mittelschmerz, pain or discomfort in the lower abdomen sometimes felt by some women at the time of ovulation.

mob behaviour, ⇨ pp. 208–9.

molar tooth, one of the 12 back grinding teeth.

mole, a coloured and sometimes hairy BIRTH MARK (naevus) that may be disfiguring but is never dangerous unless, as rarely happens, there is a change in size, shape or colour. Such change should be investigated at once.

monarchy, rule by a king whose succession is usually hereditary, but whose power may vary from absolute to nominal.

monetarism, the economic theory that inflation is caused by an excess of money in a community and that price stability can be achieved by a policy of monetary and budgetary control in a context of free market forces.

mongolism, ⇨ DOWN'S SYNDROME.

mongoloid, belonging to the racial group of mankind characterized by slanting eyes, a short nose, straight black hair and a yellowish complexion.

monitoring, continuous surveillance of people liable to suffer a sudden and dangerous deterioration in health. Checks include pulse rate, temperature, respiration rate, pupil size, ELECTROCARDIOGRAPH, level of consciousness and the concentration of OXYGEN and CARBON DIOXIDE in the blood.

monogamy, marriage to only one spouse at a time. Compare POLYGAMY (more than one spouse at one time), POLYANDRY (more than one husband), POLYGYNY (more than one wife); ⇨ pp. 210–11.

monosaccharide, the simplest form of sugar, classified by the number of carbon atoms in the molecule. The commonest monosaccharide in the body is glucose, which is a hexose, with six carbons; ⇨ pp. 36–7.

monotheism, any system of RELIGION that recognizes only one God.

mons veneris, the rounded eminence or mound of fatty tissue that covers the junction of the pubic bones and is generally more conspicuous in women than in men. Also known as the mons pubis or mount of Venus.

Montagnier, Luc, ➪ pp. 60–1.

mood swings, sudden changes of humour from happiness to misery, or vice versa.

moon face, the hamster-like appearance caused by large doses of corticosteroid drugs or excessive production of the natural adrenal cortical hormone in Cushing's syndrome.

morality, conformity to currently conventional views of acceptable behaviour and character, or to some absolute standard of right or wrong and good or bad, which may or may not be based on divine precept.

Morgagni, Giovanni Battista, ➪ pp. 116–17.

Morgan, Lewis Henry, ➪ pp. 212–13.

Morgan, Thomas Hunt, ➪ pp. 20–1.

morning-after pill, CONTRACEPTIVE or abortive medication effective some hours after coitus; ➪ pp. 100–1.

morning sickness, the vomiting or retching that commonly assails pregnant women from the 6th to the 12th week but seldom affects health.

morpheme, the smallest element of speech that conveys either factual or grammatical information. Compare PHONEME.

morphine, a narcotic drug, one of the principal alkaloids in opium, used for its powerful pain-killing (analgesic) properties.

mortality, the condition of being mortal and hence liable to die. In statistics, the ratio of the total number of deaths from any cause or from a particular cause, in a year, to the number of people in the population; ➪ INFANT MORTALITY.

Morton, William Thomas (1819–68), American dentist who first popularized the use of ether as an anaesthetic for surgical purposes.

mosaicism, the state in which some of the body cells in a person have a different number of chromosomes from the norm or different chromosomes from the majority in the body. The effect varies with the proportion of cells containing abnormal chromosomes or chromosome numbers.

motion sickness, nausea and sometimes vomiting induced by sustained, repetitive, movement of the body as in a boat, car, aircraft, swing, etc. Severe motion sickness features pallor, sweating, salivation, nausea, vomiting, apathy, depression and even loss of the will to live.

motor neuron disease, a rare disease of unknown cause affecting men twice as often as women and featuring gradual and progressive degeneration and destruction of motor nerve cells with resultant paralysis. Intelligence is never affected, but there is no known treatment.

motor system, the part of the nervous system concerned with voluntary movement. The motor system starts on the surface of each hemisphere of the BRAIN where NERVE fibres (axons) arise that pass right down through the substance of the brain, brainstem and spinal cord to end in SYNAPSES at the front of the cord at all levels. There they synapse with the nerve cells whose axons form the peripheral motor nerves that run to the muscles. The motor system is thus exposed to any major disaster affecting the brain and is almost always affected in stroke (the result of cerebral haemorrhage or thrombosis); ➪ PYRAMIDAL SYSTEM.

mountain sickness, a potentially very serious disorder caused by the reduced atmospheric oxygen tensions above about 3000 m (9750 ft). This causes reduced capacity for work, headache, dizziness, insomnia, nausea, vomiting, deep breathing and a rapid pulse. The main danger is from BRAIN swelling and the accumulation of fluid in the lungs. Either may be fatal unless the victim is quickly brought down to a lower altitude.

mouth, the orifice surrounded by the LIPS in the lower part of the face and the cavity behind it. The mobility of the tongue and the lips allows the rapid change in the shape and volume of the mouth cavity necessary for speech articulation and facilitates eating, drinking, chewing and swallowing.

movement, ➪ pp. 76–7.

mucous membrane, the moist, mucus-secreting lining of most body cavities and hollow internal organs such as the mouth, nose, eyelids, intestines and vagina.

mucus, a slimy, jelly-like mucopolysaccharide produced by goblet cells in MUCOUS MEMBRANE. Mucus is protective and lubricant. It prevents self-digestion of the walls of the stomach and intestines, traps particulate matter including smoke in the inhaled air, lubricates swallowing and the transport of the bowel contents, and facilitates sexual intercourse.

multiple births, the production of more than one individual in a single parturition. Twins occur in about one in every 84 births in Britain, and about one in 300 pregnancies results in identical twins. The frequency of multiple births varies considerably from country to country. In some South American countries, twins are comparatively rare – about one in 125. Some African tribes are said to have a twin frequency as high as one in 22. Multiple births may occur as a result of the near-simultaneous release and fertilization of more than one ovum or from the fertilization of a single ovum which then separates into two or more parts, from each of which a new individual develops. In the latter case the genetic constitution is the same in all and they start physically identical. Any physical changes that occur thereafter – such as differences in size – are wholly of environmental cause. Non-identical (dizygotic) twins may be of the same or of different sex and are no more alike than any other two siblings. Each has his or her own PLACENTA and set of membranes. Identical (monozygotic) twins share a placenta and often compete for nourishment so that they are of unequal weight when born. Multiple pregnancies are shorter than single pregnancies. Triplets and higher number births are biologically rare. Triplets have a natural frequency of about one in 7000 births. In recent years they have been much more common as a result of the use of anti-infertility drugs that induce multiple ovulation.

multiple personality, a rare psychiatric dissociative disorder in which a person appears to have two or more distinct and often contrasting personalities at different times, with corresponding differences in behaviour, attitude and outlook. Although sometimes confused with it in the popular mind, this condition is quite distinct from SCHIZOPHRENIA.

multiple sclerosis, a nervous system disease in which the insulating layer of nerve fibres is damaged in an apparently random and patchy manner causing a wide range of neurological defects such as loss of vision, numbness, partial paralysis or incoordination.

Munchausen's syndrome, calculated simulation of carefully studied disease to obtain attention, status, often surgical operations, and free accommodation and board in hospital.

muscle, an organised mass of contractile PROTEIN in the form of long, fine, spindle-shaped cells that shorten when stimulated by a NERVE impulse or tense when stretched. All bodily movement is subserved by muscle. Because of its fine structure voluntary muscle shows microscopic stripes (striated muscle). Involuntary muscle is unstriped (smooth muscle). The heart muscle forms a kind of network of fibres called a syncytium and contracts repeatedly and rhythmically even when deprived of all nerve connections; ➪ pp. 42–3.

muscular dystrophy, one of a group of hereditary degenerative MUSCLE disorders that lead to progressive weakness and disability at various stages in life; ➪ pp. 148–9.

mutagenesis, ➪ MUTATION.

mutation, a change in the sequence of the DNA so that it is transmitted to all subsequent replications of the cell concerned. The change may be a replacement such that, for example, a cytosine is changed to a thymine, or it may be an insertion or deletion of a number of bases. Some mutations are silent and do not affect the gene, others result in an altered gene product. The evolution of a species depends upon the NATURAL SELECTION of those individuals that have beneficial mutations. Most mutations are not beneficial. A mutation that occurs in a body cell, other than a sex cell, may cause a clone of mutated cells and even CANCER, but is not passed on and dies with the death of the person concerned; a mutation in a sperm or ovum will also usually die with the person, but if the affected sperm or ovum participates in a successful fertilization, all the body cells, including the sex cells, of the resulting individual will contain the mutation and it may thus be passed on. Mutations usually involve a single GENE but may affect a major part, or even the whole, of a chromosome; ➪ pp. 8–9, 26–7.

myalgia, muscle pain, especially if persistent and associated with a persistent muscle disorder.

myalgic encephalitis, ➪ POSTVIRAL SYNDROME.

myasthenia gravis, an AUTO-IMMUNE DISEASE that affects nerve stimulation of muscle so that muscles that are in use weaken rapidly. The eyelids droop, there is double vision, difficulty in swallowing and speaking and general weakness of the limbs.

myocardial infarction, ➪ pp. 138–9.

myocarditis, ➪ pp. 140–1.

myopia, ➪ SHORT-SIGHTEDNESS.

mysticism, belief in, or direct experience of, anything that transcends normal understanding, sensory experience or reality.

myth, ➪ pp. 218–9.

NADH, nicotinamide adenine dinucleotide, a substance that accepts hydrogen in the course of the action of many ENZYMES (dehydrogenases) that remove hydrogen from molecules in the course of changing them.

naevus, a growth or mark on the skin present at birth. A birth mark.

nail, ➪ pp. 36–7, 56–7.

nappy rash, skin irritation in the nappy area from ammonia formed from the bacterial breakdown of urea, or from irritating faeces or thrush.

narcissism, exaggerated self-love and a constant need for admiration; ➪ pp. 164–5.

narcolepsy, a disorder of unknown cause in which the sufferer has a marked tendency to fall asleep when relaxed. People with narcolepsy fall asleep many times a day and sleep for minutes or hours.

narcosis, a state of unconsciousness, often caused by drugs, that may vary in depth from light sleep to deep, irreversible coma.

narcotic drugs, drugs which cause sleep and relieve pain but which, taken in excess, may cause coma and death. Most narcotics are derived from opium or are synthetic substances chemically related to MORPHINE.

nasal congestion, swelling of the MUCOUS MEMBRANE lining the nose, usually from a common COLD virus infection or from HAY FEVER.

nasal septum, the thin, partly bony, partly cartilaginous, MUCOUS MEMBRANE-covered central partition that divides the inside of the nose into two passages.

nasopharynx, the space at the back of the nose, above and behind the soft palate.

nation, a large community of people sharing a common geographic area, language and culture and recognized as a political entity.

nationalism, adherence, often chauvinistic, to the independence and separatism of one's own country and to its cultural characteristics.

natural law, the idea that there exists a 'built-in' body of ethical or moral principle common to all mankind and perceived without instruction. Human experience may seem to support the idea, but an equally strong case can be made for the proposition that the illusion of a natural law may arise from forgotten early conditioning and from the notion of justice induced by personal experience of injustice.

natural selection, the Darwinian concept of the basis of evolution. Random but hereditable variations in the characteristics of individuals make them more or less likely to survive to breed in the context of a particular environment. In this way, certain characteristics may become more common in a population and others less common. So species change in a manner that improves their ability to tolerate their natural surroundings. Over a long period this can lead to such major changes that different species form; ➪ pp. 8–9.

nature versus nurture, the perennial argument as to whether heredity or environment is more influential in determining the outcome of any individual's development. It is now apparent that the two are so intimately inter-related in their effects as to be almost inseparable; ➪ pp. 30–1.

naturopathy, a system of folk medicine based on the belief that any disease can be cured by taking a vegetarian diet free from all contaminants and drugs. The assumption that disease is caused only by ingested substances is not accepted by medical science; ➪ pp. 158–9.

navel, the depressed scar in the centre of the abdomen left when the umbilical cord – the connection to the PLACENTA – drops off. The medical term is umbilicus.

Neanderthal man, the sub-human sub-species *Homo sapiens neanderthalensis* that inhabited much of Europe, North Africa and parts of Asia from about 125 000 to 40 000 years ago. The first remains were discovered in 1856 in a cave in the Neander valley near Dusseldorf. Much argument persists as to whether these were a direct ancestor of man or a separate blind-ended branch of evolution; ➪ pp. 10–11.

nebulizer, a type of inhaler used mainly in the control and treatment of ASTHMA.

neck, the flexible conduit that connects the head to the body and contains passage-ways for air, food and drink, major vessels supplying the head and brain with blood, and vital bundles of nerve fibres connecting the brain to all parts of the body. The term is also applied to any narrowing or constriction in a body part.

neck rigidity, difficulty and discomfort in bending the head forward, because of spasm of the neck and spinal muscles, that is an important sign of MENINGITIS.

necrophilia, the desire for sexual intercourse with a dead person.

negroid, belonging to the racial group characterized by black or brown skin, full lips, a short nose, well-marked ridges above the eyes and tightly curled hair.

nematodes, unsegmented worms, many of which are parasites of humans. They include the whipworms, muscle worms, hook worms, threadworms, roundworms and puppy dog worms.

neolithic period, the last phase of the STONE AGE that immediately preceded the BRONZE AGE. In the neolithic period stone implements were ground and polished, animals were domesticated and bred, agriculture was practised and pottery made.

neonatal, pertaining to a new-born baby.

neoplasm, ➪ pp. 120–1, 140–1.

nephritis, inflammation of the kidney.

nephron, ➪ pp. 64–5.

nerve, a bundle of very large numbers of nerve cells or neurons. The nerve cell is a specialized cell of particular excitability, consisting of a cell body, one or numerous short processes called dendrites and a single long fibre called an AXON. Because of the axon, which may extend almost for a metre, some nerve cells are by far the longest cells in the body. Like all body cells, the nerve carries a charge on its outer surface and an opposite charge on the inner side of its limiting cell membrane. The fact of the very long, very fine axon, however, makes this of special significance in the case of the neuron. A local change in the polarity of this charge can be propagated a considerable distance from the cell body (➪ NERVE IMPULSE). This fact subserves all neurological and brain function and all human consciousness. Many nerve tracts (white matter), consisting of large numbers of bundled neurons, occur in the central nervous system. Peripheral nerves are cord-like structures conveying millions of simultaneous nerve impulses. The main peripheral (spinal) nerves connected to the spinal cord are compound nerves, carrying motor (outgoing or efferent) impulses from the brain and cord to the periphery and sensory (incoming or afferent) impulses from the periphery to the cord and brain; ➪ pp. 74–5.

nerve cell, ➪ pp. 74–5.

nerve impulse, a zone of depolarization that passes relatively slowly along NERVE fibres. The membrane of the resting fibre is positively charged on the outside and negatively on the inside. Depolarization involves a local reduction in the charge or even a reversal so that the outside becomes negative and the inside positive. These changes are occasioned by the movement of charged sodium and potassium ions through the membrane. Depolarization is propagated automatically as a nerve impulse in both directions. When a nerve impulse reaches the link (SYNAPSE) with a MUSCLE fibre, gland or other nerve it causes contraction, secretion or the initiation of another nerve impulse, as the case may be. Thus, although the phenomenon is essentially electrical, movement of zones of depolarization occurs at a much slower rate (a few feet per second) than the normal propagation of electricity.

nervous breakdown, an informal term used to describe any emotional or psychiatric disturbance, from an episode of hysterical behaviour to a major psychotic illness.

nervous system, ➪ pp. 74–5, 76–7.

nettle rash, ➪ URTICARIA.

neuralgia, pain caused by a disorder of a sensory nerve.

neurasthenia, a state of apathy, fatigue, depression and loss of energy and motivation, most commonly due to psychological rather than organic causes. The condition is not due to a nerve disorder as the term implies.

neuritis, inflammation of a nerve, from infection, inadequate blood supply, mechanical injury, nutritional deficiency, poisoning, auto-immune disorder or other causes.

neurology, the medical speciality concerned with the nervous system and its disorders.

neuron, neurone, a single cell with a very long, fibre-like extension, called an AXON, and one or many short extensions called DENDRITES. The neuron is the functional unit of the nervous system; the brain, the spinal cord and the peripheral nerves are composed of many millions of them; ➪ pp. 74–5.

neurophysiology, the study of the function of the nervous system.

neurosis, ➪ pp. 190–1.

neurosurgery, the speciality concerned with the treatment of those disorders of the nervous system and its associated structures and blood vessels that can be relieved or cured by surgical intervention.

neurotransmitter, one of a range of substances released from a nerve ending on the arrival of a nerve impulse that stimulates receptors on another nerve or other structure so as to trigger off a response.

niacin, nicotinic acid, one of the B group of vitamins, present in liver, meat, grains and legumes. Deficiency causes PELLAGRA; ➪ pp. 138–9.

nicotine, a highly poisonous alkaloid drug derived from the leaves of the tobacco plants *Nicotiana tabacum* and *Nicotiana rustica*. The very small dose obtained by inhaling the smoke from burning tobacco has a stimulant and mood-elevating effect, increases the heart rate, raises the blood pressure and is addictive. The real danger from smoking comes not from the nicotine, but from the other constituents of tobacco smoke.

nicotinic acid, ➪ NIACIN.

night blindness, defective vision in dim light that affects some people for no apparent reason but is also a feature of vitamin A deficiency and degenerative diseases of the retina especially retinitis pigmentosa.

Nightingale, Florence (1820–1910), English nurse and pioneer of medical reform whose application of the principles of hygiene, sanitation and good nursing in the Crimean War greatly reduced mortality in the sick and wounded and established much improved standards for posterity.

nightmare, a frightening dream occurring during REM sleep (➪ pp. 178–9) and often associated with a traumatic prior event.

night terror, a sudden panic attack occurring during deep non-REM sleep in children of around the age of 5, and causing screaming and a very rapid pulse and respiration rate. Affected children describe a sense of suffocation or bodily confinement in a small space or a conviction of impending death.

Nilo-Saharan language family, a group of African languages spoken in and immediately south of the Sahara desert, in Kenya, Uganda and northern Tanzania, and from Mali in the west to the Nile basin in the east. These languages are characterized by the use of 'click' consonants; ➪ pp. 16–17.

nipple, the conical or cylindrical projection from the breast that is surrounded by a darker AREOLA. The female nipple is larger than that of the male and is perforated by the ducts of the milk-secreting segments of the breast.

nit, a LOUSE egg, usually glued to the shaft of a hair near the skin level.

nitrogen, an inert, colourless and odourless gas constituting about 80% of the atmosphere.

nitrous oxide, a weak anaesthetic and painkilling gas seldom used alone to produce general anaesthesia but commonly used in conjunction with more potent anaesthetic agents.

noble savage, an entirely fictitious and romantic notion, mainly promulgated by Jean Jacques Rousseau in his novel *Emile*, that those who dwell in natural surroundings, and are untouched by civilization, remain innocent and noble. Rousseau believed that the child should not be exposed to the corrupting influence of civilization but should be allowed to develop his or her own natural and inherent virtue.

nociceptor, ➪ pp. 90–1.

nocturnal emission, a 'wet dream' – a spontaneous orgasm in the male with ejaculation, occurring during sleep often at the climax of an erotic dream.

noise, ➪ pp. 144–5.

nomadism, ➪ pp. 12–13.

noninvasive, ➪ INVASIVE.

nonspecific urethritis, a obsolescent term for the sexually transmitted infection of the urethra now known mainly to be caused by the organism *Chlamydia trachomatis*.

non-verbal communication, transmission of information from person to person without the use of words, as by gesture, bodily attitude, expression, exclamation, and so on.

noradrenaline, a powerful NEUROTRANSMITTER released at many SYNAPSES and secreted by the MEDULLA of the ADRENAL GLAND.

nose, the prominence in the centre of the face containing two air passages separated by the NASAL SEPTUM which are the normal and preferred entry route for inspired air. The nose is equipped to warm, moisten and clean incoming air and bears in its roof the nerve endings of the olfactory nerves which subserve the sense of smell; ➪ pp. 88–9.

nostril, one of the paired openings into the NOSE that contains hairs which can trap gross particulate matter in the inhaled air.

nuclear family, ➪ pp. 214–15.

nuclear magnetic resonance, ➪ pp. 128–9.

nucleic acids, ➪ pp. 24–5.

nucleotide, ➪ pp. 24–5.

nucleus, ➪ pp. 38–9.

nursing, the dedicated employment of skills in personal management, of medical knowledge and techniques, and of human sympathy to assist the medical profession to maintain health and fitness, to assist in recovery from illness or injury, to relieve pain or distress or ease the process of dying.

nutrition, ➪ pp. 16–17, 44–5.

nymphomania, a rare condition said to feature an excessive and uncontrollable desire by a woman for copulation.

obesity, the effect of a food intake consistently greater than that needed to meet the energy requirements of the body. Surplus food is converted to fat which is deposited under the skin and in the abdomen. Obesity is a hazard to health, longevity and happiness and should be combatted by establishing new eating habits; ➪ pp. 122–3.

obsessive-compulsive disorders, ➪ pp. 186–7, 188–9.

obstetrics, ➪ pp. 94–5.

occipital lobe, the part of the brain, right at the back, that is concerned with vision; ⇨ pp. 76–7.

occiput, the lower part of the back of the head.

occupational therapy, a method of promoting recovery from physical or mental illness by exercising the body and mind in following selected forms of work.

oculomotor nerves, the pair of the nerves arising directly from the brainstem that activate most of the small muscles that move the eye, raise the upper lid and constrict the pupil.

oedema, excessive accumulation of fluid, mainly water, in the tissue spaces of the body.

Oedipus complex, ⇨ pp. 164–5.

oesophagus, the gullet. A muscular tube down which food passes from the back of the throat to the stomach; ⇨ pp. 44–5.

oestrogens, female sex hormones. These are responsible for bringing about the female secondary sexual characteristics at puberty and maintaining the health and normality of the primary sexual characteristics and the bones in later life; ⇨ pp. 80–1, 202–3.

oestrus, the period of sexual interest and receptivity, accompanied by ovulation, manifested at regular intervals by most female mammals, but not by women.

old age, ⇨ pp. 110–11.

Olduvai Gorge, ⇨ pp. 10–11.

olfactory nerves, the paired nerves of smell that arise directly from the brain and run to the roof of the nose; ⇨ pp. 88–9.

oligarchy, government by a small group of people.

oligopoly, a near-MONOPOLY enjoyed by a small group.

onanism, an obsolescent term for MASTURBATION or COITUS INTERRUPTUS.

oncology, the study of the causes, signs, symptoms and treatment of CANCER.

open heart surgery, surgery that involves entering the heart and, in most cases, requires that the heart should be isolated from the circulation and stopped. In this case the circulation must be maintained by a heart-lung machine.

ophthalmology, the medical and surgical speciality concerned with the eye and its disorders.

ophthalmoscope, an instrument that illuminates the inside of the eye to allow observation of the internal details and disorders.

opiate, a drug derived from, or similar to, the poppy opium alkaloids. The natural body substances known as the enkephalins or endorphins are also opiates or, more correctly, opioids; ⇨ pp. 54–5.

opium, a mixture of narcotic alkaloids found in the juice of the pods of the oriental poppy *Papaver somniferum*, the most important being MORPHINE and codeine.

opportunistic infection, infection by organisms of low virulence that are normally excluded or rapidly destroyed by the IMMUNE SYSTEM. Opportunistic infections are a feature of AIDS and other immune deficiency disorders.

optic nerves, the two nerves which connect the eyes to the brain and subserve vision; ⇨ pp. 84–5, 90–1.

optometry, eye testing and the determination of the lenses needed to correct visual defects.

oral contraceptives, hormone drugs or combination of drugs taken by mouth for the purpose of preventing pregnancy. Oral contraceptives contain an OESTROGEN and/or a PROGESTERONE and act by preventing the ovaries from producing eggs (ova); ⇨ pp. 100–1.

oral sex, sexual stimulation of a partner's genitals with the mouth.

oral stage, the first and most primitive of Freud's proposed stages of psychosexual development, in which the mouth is the focus of the libido and the primary source of satisfaction.

orang utan, one of the PRIMATES, the ANTHROPOID APE *Pongo pygmaeus* found in Borneo and Sumatra; ⇨ pp. 4–5.

orbit, the bony cavern in which the eyeball resides and moves; ⇨ pp. 40–1.

orchitis, inflammation of the TESTIS, usually a result of a mumps infection, and featuring swelling, exquisite tenderness, acute pain and usually high fever. Testicular atrophy may follow.

organ, any part of the body consisting of more than one tissue and performing a particular function.

organ grafting, recent decades have seen a massive increase in the volume and scope of transplant surgery. This has become possible because of advances in understanding of the rejection process that occurs when a foreign tissue enters the body, and of the ways in which this process can be prevented. An *autograft* is a transplant made from one site to another in the same individual. So long as an adequate blood supply is assured at the new site, there are no immunological problems. An *allograft* is a transplant from one individual to another of the same species.

When an organ, such as a kidney, a heart or a liver, from a genetically identical person (an identical twin) is grafted and the arteries and veins are joined up so that the graft has an adequate blood supply, new blood vessels grow across the interface, healing is rapid and the grafted organ is soon functioning fully. There are no problems with rejection and the graft is gradually, and over a long period, replaced by tissue growing into it from the host. But when an organ is grafted from a different individual and no other special action is taken, the result is very different. To begin with, new blood vessels grow across the interface and along the inside of the small vessels of the graft. After about a week, however, the host T cells, which encountered the graft antigens at the time of surgery, have cloned in large numbers and millions of them accumulate at the graft–host interface together with many B cells and macrophages. A violent inflammatory reaction occurs with severe damage to, and even destruction of, the small graft blood vessels. This inflammation spreads into the grafted organ causing swelling, haemorrhages and eventually destruction of the graft. These processes occur because the immune system, exercising its protective role, cannot distinguish a useful but foreign tissue from normally dangerous foreign tissue.

If grafting is to succeed, the immune response must be suppressed. This can be done using a variety of drugs including: corticosteroids, which interfere with interleukin production; cytotoxic drugs and X-rays, which interfere with lymphocyte production; and, in particular, the drug cyclosporin, which prevents the activation of cells that encounter antigen. Any immunosuppressive treatment necessarily involves danger, because resistance to infection is reduced, but drugs like cyclosporin, while highly effective in preventing rejection, have a less general immunosuppressive effect than other drugs.

organelle, any of the microscopic structures within the CELL that might be described as a little organ of the cell.

organic brain syndrome, any psychiatric disorder, whether of reason, personality, mood or perception, caused by abnormalities of BRAIN structure, function or biochemistry. The organic psychoses may feature delirium, dementia, amnesia, hallucinations, delusions, depression, euphoria, apathy, personality changes or anxiety. They are mainly caused by brain damage from physical injury, infection, haemorrhage, lack of oxygen, poisons, drug overdosage, alcohol excess, vitamin or mineral deficiency and various general disorders such as Alzheimer's disease, multiple sclerosis, Huntington's chorea, Parkinson's disease and severe liver or heart disease.

orgasm, the recurrent (clonic) muscular spasms in the genital region that accompany ejaculation and the release of sexual tension at the climax of sexual intercourse or masturbation; ➪ pp. 96–7.

orthodontics, the dental speciality concerned with the correction of irregularities of tooth placement and of the relationship of the upper teeth to the lower (occlusion).

orthopaedics, the branch of surgery concerned with disorders of the skeletal and muscular systems whether from injury or disease.

orthoptics, a speciality ancillary to OPHTHALMOLOGY concerned with the diagnosis and treatment of SQUINT.

ossicle, a tiny bone, specifically one of the three bones in the middle ear that link the ear drum to the COCHLEA of the inner ear; ➪ pp. 86–7.

osteoarthritis, an age- and trauma-related degenerative joint disease involving damage to the bearing surfaces and sometimes the ends of the bones involved in the joint.

osteoblasts, specialized cells that secrete large quantities of the PROTEIN COLLAGEN to build up the scaffolding of new BONE. This is then mineralized with calcium and phosphates; ➪ pp. 40–1.

osteoclasts, large, bone-absorbing cells that secrete ENZYMES to digest bone and release the minerals. Osteoclast activity precedes OSTEOBLAST activity in the repair of bone fractures; ➪ pp. 40–1.

osteomyelitis, persistent infection of bone, including the marrow, and often featuring abscess formation.

osteopathy, a branch of alternative therapy based on the proposition that many bodily ills are the result of displacement of one or more of the many faceted joints of the VERTEBRAL COLUMN. The treatment involves a careful analysis of the state of these joints followed by manipulation. The underlying assumption is not universally accepted by the orthodox medical profession; ➪ pp. 158–9.

osteoporosis, a disorder in which both the PROTEIN structure and the mineralization of BONE become progressively reduced so that the bones weaken and may break on minor trauma. Osteoporosis, and resulting fractures, are especially prevalent in women after the menopause because they are deprived of the bone-protective effect of the sex hormone oestrogen; ➪ pp. 110–11.

otitis externa, inflammation of the skin of the external ear canal or of the external ear (pinna), usually from infection or allergy.

otitis media, inflammation in the middle ear cavity, usually as a result of spread of infection from the nose or throat by way of the EUSTACHIAN TUBE.

otosclerosis, a hereditary ear disease in which one of the tiny, sound-transmitting middle ear bones becomes fused at its articulation with the inner ear, causing progressive deafness.

ova, ➪ OVUM.

ovarian cyst, fluid-filled, usually benign, closed sacs of greatly variable size growing from an OVARY.

ovary, one of the two female organs in which the eggs (ova), from which new individuals can develop after fertilization, are produced; ➪ pp. 96–7.

overbite, protrusion of the upper teeth beyond the line of the lower teeth when the mouth is closed.

ovulation, the monthly release of one or more OVUM from an OVARY about the mid-point of the menstrual cycle. Released ova are wafted into the FALLOPIAN TUBE along which they pass in the course of two or three days. If fertilization and successful implantation are to occur, the ovum must meet an adequate number of sperms in the course of this passage; ➪ pp. 96–7.

ovum, an egg. The female gamete. The human ovum is the largest cell of the body and contains all the genetic material provided by the mother. The ovum contains 23 chromosomes – half the normal number of a human body cell. The other 23 are contributed by the fertilizing spermatozoon; ➪ pp. 96–7.

Owen, Richard, ➪ pp. 4–5.

oxidation, the addition of OXYGEN to an element or a compound.

oxygen, the chemical element most vital for the survival of the human being. Deprivation of oxygen for a few minutes is fatal because the brain is unable to support its activity without a constant supply. Oxygen is used to generate energy by a controlled biochemical process equivalent to burning; ➪ pp. 52–3.

ozone, a harmful gas consisting of three linked oxygen atoms. An ozone layer in the outer atmosphere offers a protective barrier against excess dangerous ultraviolet radiation from the sun; ➪ pp. 16–17.

pacemaker, ➪ pp. 156–7.

pacifism, the conviction that violence is never justified and that one should never, under any circumstances, resort to war.

paediatrics, the medical speciality concerned with the diseases and disorders of childhood and with all aspects of child health and development.

paedophilia, sexual urges towards a prepubertal child by a person over the age of 16 and at least five years older than the child. If acted on, these urges constitute a criminal offence.

pain, a localized sensation causing distress, anxiety and a sense of being punished. Pain is caused by stimulation of specific nerve endings called nociceptors, or by strong stimulation of other sensory nerves and commonly serves as an indication of bodily disorder. The appropriate response to pain is to try to find and remove the cause rather than simply to suppress it with PAINKILLERS.

painkillers, drugs that act either at the site of injury to prevent the release of substances that stimulate PAIN nerve endings, or on the brain to suppress the perception of pain.

palate, the roof of the mouth, formed partly from bone and partly from soft tissue, both covered with MUCOUS MEMBRANE. The soft palate is a small flap of muscle and fibre that can press firmly against the back wall of the throat to seal off the opening to the nose during swallowing.

palliative treatment, treatment that relieves symptoms but does not cure their cause.

palm, the creased front surface of the hand, from the wrist to the bases of the fingers. The skin of the palm is bound down to the underlying bones so as not to protrude when the palm is folded during gripping.

palpitation, abnormal awareness of the heart's action either because of irregularity or rapidity.

palsy, an obsolete term for PARALYSIS.

pancreas, an organ lying behind and under the stomach that secretes digestive enzymes into the intestine and the hormones insulin and glucagon into the BLOOD; ➪ pp. 44–5, 80–1.

pandemic, a world-wide EPIDEMIC.

panic attack, a common feature of a phobia; ➪ pp. 190–1.

pantheism, the religious belief that GOD is everywhere identified with the world and does not possess an independent personality or entity. In the pantheistic view God is the ultimate reality lying behind our imperfect and illusory perception of the universe.

papilla, any small, nipple-like projection.

papilloma, a harmless TUMOUR of skin or mucous membrane in which cells grow outward from a surface around a connective tissue core containing blood vessels. Most papillomas are WARTS.

Pap smear, a popular term for the cervical smear or scrape screening test for cancer originated by the Greek-born American anatomist, George Nicholas Papanicolaou (1884–1962).

Paracelsus, ➪ pp. 152–3.

paracetamol, a PAINKILLER drug also used to reduce fever. A trade name is Panadol.

paradigm, a human being's mental model of the world, which may or may not conform to that of others but is often stereotypical. In the philosophy of science, a general conception of the nature of scientific operation within which a particular scientific activity is undertaken. Paradigms are, of their nature, persistent and hard to change. Major advances in science – such as, for instance, the realization of the concept of the quantum – involve painful paradigmic shifts which some people, notably the older workers, are unable to make.

paralysis, loss of the power of movement of any part of the body as a result of damage to peripheral nerves or spinal nerve tracts or to the motor parts of the brain. Hemiplegia is paralysis of one side of the body; paraplegia is paralysis of the legs and lower parts of the body; quadriplegia is paralysis of all four limbs; ➪ pp. 126–7.

paranoia, a mental disorder featuring especially delusions of persecution, but in which the personality remains relatively intact; ➪ pp. 192–3.

paraplegia, ➪ PARALYSIS.

parapsychology, the study of mental phenomena that do not conform to everyday experience. Interest in parapsychology is fuelled by the innate longing of the human being for magic and miracles. The more rigorous the study of such matters, however, the less evidence for them appears. The history of science has been a long and painful struggle to escape from the realms of magical thinking and superstition and many scientists are concerned at the possible dangers of conferring a kind of respectability and plausibility on matters which they consider to be without scientific basis; ➪ pp. 180–1.

parasite, an organism that lives on, and at the expense of, another organism; ➪ pp. 136–7.

parathyroid gland, one of four small glands lying in the thyroid gland, that secrete a HORMONE that controls the BLOOD levels of calcium by regulating its release from bones; ➪ pp. 80–1.

paratyphoid fever, a mild form of TYPHOID FEVER.

Paré, Ambroise, ➪ pp. 34–5, 154–5.

parenting, the process of caring for, nurturing and bringing up a child.

parietal, pertaining to the wall or outer surface of a part of the body.

Parkinson's disease, a brain disorder affecting mainly men and involving progressive loss of muscular coordination and often severe tremor. The outlook varies greatly. The later the onset the less severe it is likely to become. Attempts have been made to treat the disorder by implanting cells from the fetal mid-brain in the hope of replacing missing dopamine-forming tissue. It is still too early to judge the results; ➪ pp. 186–7.

parliamentary system, the political system of a DEMOCRACY that features a public assembly of people's elected representatives and appointed ministers regularly meeting to consider new legislation, amend or repeal old, and discuss matters of national interest and concern.

parotid gland, the main salivary gland, one of which is situated in each cheek just in front of the ear.

Parsiism, the RELIGION, customs and practices of the descendants of the Zoroastrians, most of whom live in and around Bombay in a closed community. Much ritual, sacrifice and the observance of the strict rules of ZOROASTRIANISM surround every aspect of the life of the Parsee.

partial reinforcement, the process of intermittently re-strengthening a CONDITIONED REFLEX by repetition of the association between the unconditioned and the conditioned stimuli.

parturition, childbirth; ➪ pp. 104–5.

passive transport, the movement of dissolved material through a biological membrane in the direction of fluid flow and without the expenditure of energy. Compare ACTIVE TRANSPORT.

Pasteur, Louis, ➪ pp. 116–17.

pasteurization, a method of destroying bacteria and other micro-organisms in milk and other liquid foods by rapid heating and then cooling.

pastoralism, a way of life dependent on the grazing of livestock.

patella, the knee cap; ➪ pp. 40–1.

patenting, acquiring the rights to exploit an invention for profit with protection against others wishing to do the same.

paternity testing, tests performed to confirm or deny the claim that a man is the father of a particular child. Simple blood-grouping tests can confirm the fact that the man is *not* the father; and the more complex GENETIC FINGERPRINTING can confirm paternity.

pathogen, any agent, especially a micro-organism, that causes disease; ➪ pp. 134–5.

pathology, ➪ pp. 116–17.

patois, a regional dialect of a language.

patriarchy, family rule by a hereditary male head, with descent through the male line. Compare MATRIARCHY.

patrilineage, ➪ LINEAGE.

Pauling, Linus, ➪ pp. 24–5.

Pavlov, Ivan Petrovich, ➪ pp. 172–3.

pectoral, pertaining to the chest wall in the breast region.

pederasty, anal intercourse, especially with a boy.

pediculosis, any kind of louse infestation.

Peking man, early humans, of the species *Homo erectus* who lived in northern China some 500 000 years ago. A mass of skeletal remains of 44 people jumbled with thousands of broken and burned animal bones and many stone tools were found at Chou K'ou Tien, near Beijing, in the decade following 1927; ➪ pp. 4–5, 10–11.

pellagra, a disease, caused by deficiency of the B vitamin NIACIN, featuring skin cracking and blistering, diarrhoea and, in severe cases, delirium and DEMENTIA.

pelvis, the bony girdle at the lower end of the spine with which the legs articulate. The term is also used for any funnel-shaped body structure such as the pelvis of the kidney.

penetrance, genetic, the frequency with which an ALLELE causes its effect. This may be influenced by environmental factors or by other genes; ➪ pp. 28–9.

penicillin, ➪ pp. 152–3.

penis, the male organ of copulation, which also contains the outflow tube for the urine. The turgidity necessary for copulation occurs when the organ becomes temporarily suffused with BLOOD under pressure. This occurs as a reflex response to sexual stimuli; ➪ pp. 96–7.

penis envy, the Freudian concept that all women have a repressed wish to have a penis and resent the fact that they are incomplete men. The idea is no longer taken seriously.

pepsin, a digestive enzyme that splits PROTEIN into simpler components; ➪ pp. 44–5.

peptic ulcer, a localized area on the lining of the STOMACH, DUODENUM or GULLET which has been eroded and partly digested by stomach acid and digestive ENZYME to expose the underlying muscle layer.

peptide, a chemical compound formed by the linkage between the amino and carboxyl groups of adjacent AMINO ACIDS. This allows them to form long and elaborate chains as polypeptides and PROTEINS; ➪ pp. 26–7.

perception, the operation of any of the sense organs. The term is often also taken, in a metaphorical sense, to include the mental activity that follows the receipt of sense data; ➪ pp. 76–7, 84–5.

percussion, tapping of the chest or the abdomen to detect, by the quality of the sound produced, whether the underlying area is air-filled, fluid-filled or solid.

pericarditis, INFLAMMATION of the double-layered fibrous bag that surrounds the heart; ➪ pp. 140–1.

pericardium, ➪ pp. 140–1.

perinatal, pertaining to the period immediately before and after birth.

perineum, the part of the floor of the PELVIS that lies between the anus and the CLITORIS in females and the anus and the scrotum in males.

period, ➪ MENSTRUATION.

periosteum, the membrane that surrounds and closely invests any bone, assisting in its nutrition and in its repair after a FRACTURE.

peristalsis, a coordinated squeezing action of the MUSCLE layers of the wall of the intestine and other body tubes that ensures the onward movement of the contents; ➪ pp. 44–5.

peritoneum, ➪ pp. 98–9.

peritonitis, a dangerous INFLAMMATION, usually from infection, of the sterile membrane that invests or covers the organs of the abdomen; ➪ pp. 98–9.

peritonsillar abscess, ➪ QUINSY.

personality, ➪ pp. 76–7, 188–9.

personality tests, ➪ pp. 200–1.

perspiration, ➪ SWEAT.

pertussis, ➪ WHOOPING COUGH.

pessary, a solid but readily melting vehicle for medication that is placed in the vagina to treat infections or act as a contraceptive spermicide. The same term is used for a plastic or rubber ring placed in the vagina to support the womb or other pelvic organs.

petit mal, ➪ EPILEPSY.

phagocyte, one of the cells capable of engulfing and digesting organic material, especially material that has been 'labelled' as foreign by the IMMUNE SYSTEM. There are two main classes of phagocytes

– the giant macrophages and the small polymorphs; ➪ pp. 58–9, 124–5.

phalanges, the bones of the fingers and toes.

phallic stage, the stage in Freud's theory of psychosexual development, following the ANAL STAGE, in which interest is centred on the penis or clitoris. The phallic stage is said to be followed by the Oedipal stage.

phallus, ➪ PENIS.

phantom limb, the strong illusion that a limb which has been amputated is still present. This is due to stimulation of the ends of the cut nerves in the stump producing impulses that can only be interpreted as coming from the original limb. Successful use of a prosthetic device is often facillitated by the cohesion of the missing limb.

pharmaceutics, ➪ pp. 152–3.

pharmacology, ➪ pp. 152–3.

pharyngitis, inflammation of the throat, usually from a virus infection, with pain, discomfort on swallowing and often enlarged LYMPH NODES in the neck. Commonly known as 'sore throat'.

pharynx, the throat; ➪ pp. 44–5.

phenotype, the observable appearance of an organism. Compare GENOTYPE.

phenylketonuria, a genetic disorder due to the absence of an enzyme needed to convert the dietary AMINO ACID phenylalanine to tyrosine. Toxic products accumulate in the body causing BRAIN damage and other effects. These can be prevented by taking a diet free from the amino acid; ➪ pp. 28–9, 30–1.

pheromone, an odorous body secretion that affects the behaviour of others. Pheromones are much less important in humans than in other animal species.

phlebitis, ➪ VEIN.

phlegmatic type, the personality group, according to the ancient theory of the humours (➪ pp. 34–5) caused by a preponderance of phlegm (MUCUS) and featuring stolid lack of emotional expression.

phobia, inappropriate or irrational fear of an object or situation, of a degree that interferes with normal living. Phobias may take many forms but most commonly involve fear of closed spaces (claustrophobia) or fear of public places (agoraphobia). Phobias are most effectively treated by behaviour therapy; ➪ pp. 190–1.

phoneme, one of the many sounds in speech that distinguish the meaning of one word from another, as in the case of 'b' and 'w' which distinguish, for instance, 'bed' and 'wed' in English.

phonetics, the branch of linguistics concerned with the study of the speech sounds (PHONEMES) of a language and their classification and representation.

phosphorus, an essential mineral whose salts or esters are incorporated into bones and teeth and are important in stabilizing the acidity of the blood and other tissue fluids and in the energy reactions in the cells.

photophobia, undue intolerance to light whether from ALBINISM or acquired disease.

photosynthesis, the vital process in nature by which plants are able to use the energy in sunlight to form organic compounds, such as sugars, from atmospheric carbon dioxide and water. This fixation of carbon is the ultimate process by which all living things are directly or indirectly nourished; ➪ pp. 8–9.

phylogeny, the sequence of events in the evolution of a species; ➪ pp. 6–7.

physiology, the study of the function of the body, as distinct from its structure (anatomy). Physiological studies, however, presuppose a knowledge of anatomy and often the two are so closely inter-related as to be inseparable; ➪ pp. 36–7.

Piaget, Jean, ➪ pp. 106–7, 166–7, 174–5.

pidgin, a limited form of language arising for use between unintegrated social groups with restricted intercourse, whose natural languages are different. Grammar is non-inflexional and highly simplified. When a pidgin is adopted by children and becomes the primary language of a community it becomes a CREOLE; ➪ pp. 14–15.

Pierce, Charles Sanders, ➪ pp. 206–7.

pigeon toes, a popular term for the largely cosmetic defect in which the legs or feet are rotated inwards.

pigment, the body colouring substance melanin.

piles, varicose veins in the walls of the canal of the anus, internally or externally and of varying size. Large internal piles can protrude. The medical term is haemorrhoids.

Pill, the, ➪ ORAL CONTRACEPTIVES; ➪ pp. 100–1.

pimple, a small skin ABSCESS commonly occurring in adolescents and young adults suffering from ACNE.

Pincus, Gregory, ➪ pp. 100–1.

pineal gland, a tiny, cone-shaped structure situated deep within the brain, which secretes the hormone melatonin, mostly at night, and which may be important in synchronizing CIRCADIAN RHYTHMS.

Pinel, Philippe, ➪ pp. 184–5.

pituitary gland, the central controlling endocrine gland that hangs on a short stalk from the middle of the underside of the BRAIN. The pituitary HORMONES promote growth and lactation, control the body's water balance and prompt all the other ENDOCRINE organs to secrete their own hormones. The BLOOD level of these hormones is monitored by the pituitary and its output of stimulating hormones adjusted accordingly. The pituitary is under the control of the HYPOTHALAMUS; ➪ pp. 78–9.

placebo, a drug or treatment given to satisfy, please or impress the patient rather than to bring about any direct bodily change. The effect of a placebo can be remarkable as is shown by the fact that dummy tablets believed by a subject to be painkillers can promote the production of endorphins by the BRAIN. Phenomena of this kind can complicate trials of new drugs and sometimes confer on a drug or form of treatment a quite undeserved reputation.

placenta, the structure that grows on the inner wall of the womb throughout pregnancy and in which the maternal BLOOD is brought into close contact with that of the fetus, but without mixing. Oxygen, nutrients and other substances, including many drugs, are thus able to pass to the fetus from the mother, and waste substances, especially carbon dioxide and urea, are able to pass out. When the placenta is delivered, after the baby is born, it is sometimes called the afterbirth; ➪ pp. 102–3, 104–5.

plague, ➪ pp. 118–19.

plaque (arterial), a deposition of cholesterol, degenerate muscle and other materials under the lining of an artery, as in the common disease of ATHEROSCLEROSIS, that causes narrowing and may lead to blockage.

plaque (dental), a mixture of food debris, dried saliva and bacteria deposited around the necks of neglected teeth and liable to lead to tooth decay.

plasma, the fluid in which the red and white blood cells are suspended, or blood from which all cells have been removed. Plasma contains proteins, carbohydrates, fats and electrolytes and will coagulate if exposed to air.

plasma membrane, ➪ MEMBRANE, CELL.

plastic or reconstructive surgery, any surgical procedure performed to try to restore normal appearance and function by repair or reconstruction after injury, disease or congenital malformation.

platelet, one of millions of fragments of large cells called megakaryocytes that are present in the blood and are essential for normal blood clotting. Deficiency is known as thrombocytopenia; ➪ pp. 40–1.

Plato, ➪ pp. 14–15, 184–5.

play, a usually non-constructive, voluntary activity engaged in for amusement, diversion, entertainment or competition. Play is commonly a make-believe activity natural to children and is their principal mode of self-expression and a principal element in psycho-social development. In adults play tends to become institutionalized as a part of the culture and is often engaged in with all the seriousness shown by children in their play.

play therapy, a method of child psychological treatment pioneered by the Austrian psychoanalyst Melanie Klein in which observation of children at play reveals patterns of emotional and interactional disturbance; and appropriately directed play allows the therapeutic release of bottled-up emotions.

pleasure, any enjoyable or agreeable emotion or sensation, to the pursuit of which most people devote their lives.

pleasure principle, Freud's notion of the tendency to seek immediate gratification of instinctual desires and to avoid pain – a primitive id reaction that is gradually modified by the more mature ego function – the reality principle.

pleura, the thin, double-layered membrane that separates the lungs from the inside of the chest wall.

pleurisy, inflammation of the PLEURA, usually from virus infection. Pain, which becomes acute and stabbing on breathing in, is a central feature.

pluralism, a sociological theory that proposes that society is best served by the existence of several discrete but interdependent competitive and changing power groups, none of which is able to dominate.

PMT, ➪ PREMENSTRUAL TENSION.

pneumoconiosis, any one of a number of industrial lung diseases caused by long-term inhalation of irritating mineral dusts; ➪ pp. 16–17.

pneumonia, inflammation of the air sacs of the lungs, usually from infection but also from contact with inhaled irritant or toxic material. There is fever, chills, shortness of breath, cough with greenish or blood-stained sputum and often PLEURISY; ➪ pp. 116–17, 144–5.

pneumothorax, air in the normally closed space between the two layers of the PLEURA which expands the pleural cavity and collapses the underlying lung.

poison, any substance which can, in small amounts, damage or kill living organisms. Many of the most poisonous substances act by interfering with cell enzyme systems.

poliomyelitis, a viral infectious disease spread by fecal contamination that affects the motor cells of the spinal cord and brainstem and may, after a short period of fever, sore throat and headache, cause severe PARALYSIS; pp. 132–3.

pollution, the undesirable and potentially dangerous dispersal of substances capable of causing harm directly or indirectly to living things or to their environment. Polluting substances are commonly waste-products or by-products of industrial activity; ➪ pp. 144–5.

polyandry, marriage to more than one man, simultaneously.

polygamy, marriage to more than one spouse, simultaneously.

polygraph, ➪ LIE DETECTOR.

polygyny, marriage to more than one woman, simultaneously.

polysaccharide, a polymer consisting of many simple sugars linked together. Some polysaccharides, such as starch, are readily digested by humans; ➪ pp. 36–7.

polytheism, the belief in the existence of many gods.

polyunsaturated fats, ➪ UNSATURATED FATS.

pons, the middle part of the brainstem lying immediately above the MEDULLA OBLONGATA and containing the nuclei for some of the cranial nerves that move the eyes, those for balance and hearing and those that supply sensation to the face.

population, ➪ pp. 100–1.

population control, the hope that the catastrophe that currently seems to threaten the world by the present explosion of population can somehow be averted by bringing about a falling birthrate.

pore, any tiny opening, especially in the skin, through which sweat or sebaceous secretion (sebum) passes to the surface. Most sebaceous pores are also hair follicles; ➪ pp. 56–7.

porphyria, any one of several inherited enzyme deficiency disorders in which porphyrins, formed in the course of HAEMOGLOBIN synthesis, accumulate in the body causing a wide range of ill effects.

These include hypersensitivity to sunlight, skin blistering, abdominal pain, vomiting, constipation, paralysis and psychotic disorders.

port-wine stain, an often disfiguring flat, purple-red birthmark caused by a non-malignant tumour of small skin blood vessels.

positivism, a school of philosophy that rejects value judgements, metaphysics and theology and holds that the only path to reliable knowledge is that of scientific observation and experiment.

positron emission tomography, ⇨ pp. 72–3.

posterior, pertaining to the back of the body or of a part of the body. Compare ANTERIOR.

postmortem examination, an autopsy or pathological examination of a dead body in order to discover the cause of death or advance medical knowledge.

postnatal depression, maternal depression after childbirth that may vary from mild sadness to a severe suicidal or infanticidal psychotic depression.

postpartum haemorrhage, excessive and sometimes dangerous blood loss immediately after childbirth usually from the site to which the PLACENTA was attached.

post-traumatic stress disorder, a persistent anxiety reaction with insomnia, nightmares, guilt, irritability, isolation and loss of concentration, caused by a major personal trauma such as severe injury, assault, rape or involvement in or witnessing of a major disaster.

posture, the relationship of different parts of the body to each other and to the vertical. Faulty posture may affect health as well as appearance.

postviral syndrome, a disorder often inaccurately called myalgic encephalitis (ME) that may follow a virus infection and features severe FATIGUE, limitation of activity, muscle aching and emotional disturbance. The condition is not caused by brain inflammation as the name implies and there is controversy as to whether it is of organic origin but there is no questioning the distress and disablement of the sufferers.

potassium, a metallic element essential for the normal functioning of nerves and muscles. The amount in the blood must be kept within strict limits if health is to be preserved; ⇨ pp. 46–7.

potato nose, a popular term for the condition of rhinophyma in which overgrowth of sebaceous tissue and blood vessels causes a bulbous deformity of the nose.

potency, the ability of a man to obtain and sustain a penile erection and so perform SEXUAL INTERCOURSE. The term is also used to describe the strength of a drug based on its ability to cause bodily change.

poverty, lack of adequate resources to obtain food, shelter, clothing or a standard of living commensurate with the average in a society.

pragmatism, action determined by the need to respond to immediate necessity or to achieve a particular practical result, rather than by established policy or dogma. The term is also used for the philosophic principle that the truth and meaning of an idea is entirely relative to its practical outcome.

predestination, the theological notion that the ultimate fate of a person's soul is determined by GOD prior to that person's life and will take effect regardless of that person's behaviour during life. The main exponent of the idea, St Augustine, taught that a person can behave well only by God's grace and that this grace is vouchsafed or withheld by God without regard to human merit. These ideas have been incorporated into much subsequent religious dogma and are held by many to this day.

pre-eclampsia, ⇨ pp. 104–5.

pregnancy, ⇨ pp. 102–3, 104–5.

prejudice, the maintenance of an adverse opinion about a person or class of persons in spite of evidence to the contrary. This is a common characteristic of the human being and is linked with the habit of arguing illogically from the particular to the general and the tendency to unreasoning CHAUVINISM.

Premack, David, ⇨ pp. 14–15.

premature ejaculation, a male orgasm occurring before, at the time of, or immediately after vaginal penetration, thus depriving both partners of satisfaction. This is usually due to inexperience or over-excitement.

premenstrual tension, any of a range of symptoms, such as irritability, fatigue, depression, headache, backache and a sense of fullness, related to hormone changes experienced during the few days prior to the onset of menstruation.

premolar, the four pairs of permanent grinding teeth situated on either side of each jaw behind the canines and in front of the molars; ⇨ pp. 44–5.

prepuce, the foreskin of the penis. This is the part removed in circumcision, whether for ritual or medical reasons; ⇨ pp. 96–7.

presbyacusis, progressive loss of hearing as a function of age. Presbyacusis results from loss of hair cells in the cochlea of the inner ear and is more likely to be due to the sum of a lifetime's acoustic trauma and other damaging effects than simply to age; ⇨ pp. 86–7.

presbyopia, progressive loss of the power of focussing the eyes on close objects as a function of age. Presbyopia results from loss of elasticity of the internal crystalline lenses of the eye so that they are no longer capable of relaxing into the more spherical shape needed for close viewing. The cause is unknown; ⇨ pp. 84–5, 110–11.

presidential system, the political system of a republic in which the head of state, who is also often the chief executive, is known as the president.

pressure points, places where arteries can readily be compressed between the fingers and an underlying bone so as to control arterial bleeding.

pressure sores, ⇨ BED SORES.

priapism, a condition of persistent, often painful, erection of the penis without sexual interest that calls for urgent treatment by withdrawal of blood to avoid permanent injury.

prickly heat, an irritating skin disorder caused by excessive sweating.

Priestley, Joseph, ⇨ pp. 52–3.

primary teeth, the first set of 20 teeth that begin to appear, usually around the age of 6 months. Also known as milk teeth.

primates, ➪ pp. 4–5.

prion, a protein that appears to be capable of replicating itself and of resisting normal methods of medical sterilization. It is associated with, and possibly the cause of, the range of spongiform encephalopathies that includes (BSE) bovine and feline spongiform encephalopathy, scrapie in sheep, and kuru, Creutzfeldt-Jakob disease and the Gertsmann-Straussler syndrome in people; ➪ pp. 132–3.

private enterprise, business or other commercial or economic enterprise undertaken under private, rather than public ownership.

process (anatomy), any outgrowth or projection from a part, especially a bone.

progesterone, ➪ pp. 100–1.

prognathism, abnormal protrusion of either jaw, especially the lower.

prognosis, ➪ pp. 128–9.

prokaryotes, the large class of organisms, such as bacteria, with no defined nucleus; ➪ pp. 134–5.

prolactin, the PITUITARY GLAND HORMONE that stimulates the growth of the breasts in either sex and promotes and maintains the flow of milk at the end of pregnancy; ➪ pp. 78–9.

prolapse, displacement of a body part or tissue into an abnormal position, especially of the downward movement of the womb or rectum, or the backward movement of the pulp of an intervertebral disc.

proletariat, that part of society that owns little capital or property and has to work for a living. The working class, which, according to Karl Marx and others, is exploited by the capital-owning class.

promiscuity, a loose term for sexual excess – a common pattern of behaviour in which sex is valued for its variety and immediate gratifications rather than as one of the important bases for a long-term relationship. Some psychologists equate promiscuity with social immaturity; others hold it to be unrelated. Promiscuity has always carried penalties of some kind, either imposed by society or in the form of sexually transmitted diseases such as syphilis, gonorrhoea, herpes and chlamydial infections. In the context of the spread of AIDS, these have become much more significant.

pronunciation, received, a form of phonation, cultivated in many public schools and believed by many members of the Establishment to be correct and desirable and to promote a sense of identity with the ruling classes.

propaganda, a term derived from an institution of the Roman Catholic Church concerned with missionary work. It is now taken to mean any system of sustained and organised dissemination of information, whether true or false, for the purposes of modifying popular opinion in favour of a government or movement or against an enemy. Once seen for what it is, propaganda is often counter-productive.

prostaglandins, ➪ pp. 104–5.

prostate gland, the chestnut-sized and shaped solid gland that surrounds the start of the urine exit tube (the urethra) on the underside of the bladder. The gland produces alkaline secretions that help to dilute the sludge-like mass of spermatozoa issuing from the vas deferens and so improve sperm motility. The gland is subject to an increase in size with age that often causes urinary output problems in elderly men. It is also a common site of CANCER; ➪ pp. 96–7.

prosthesis, any artificial replacement for a part of the body.

prostitution, sale of sex, most commonly by women. This may be a part-time or full-time private enterprise or one organized on a small or large scale by pimps, brothel-keepers or call-girl ring organizers. In general, the lot of the prostitute is not a happy one and most of the girls involved are driven by economic necessity into an unpleasant and often dangerous trade. Many of them, through inadequacy of one kind or another, are unable to sustain more conventional employment. The legal status of prostitution varies considerably from country to country and even within a country. Prostitution is legal, for instance, in Nevada, but illegal in other American states. Most male prostitutes offer services to other men, but a few (gigolos) cater for women.

protein, one of many Different types of large molecule each consisting of up to thousands of AMINO ACIDS linked together by peptide bonds. Proteins may be insoluble and fibrous, as in COLLAGEN – the main structural material of the body – or soluble. Soluble proteins include the ENZYMES, many HORMONES, antibodies, albumin and haemoglobin. Many proteins are contractile - able to sHorten and thicken under suitable stimulation. MUSCLE is such a protein and, together with collagen, makes up a large part of the bulk of the body. Chromosomes carry GENES that code for all the proteins of an individual, most of which are enzymes. Protein is formed by cell organelles called ribosomes, using a pattern of amino acid sequence conveyed from the DNA by RNA; ➪ pp. 36–7, 38–9, 42–3, 80–1.

Protestantism, a major religious movement of CHRISTIANITY, promoted from about 1517 mainly by Martin Luther in Germany, which rejected the claim of the Roman Catholic Church that it was the only channel of divine authority. The movement led to the Reformation during which the church was split into Catholic and Protestant divisions. The latter soon broke into many schismatic groups such as the Lutherans, the Calvinists, the Anglicans and the Anabaptist and Mennonite sects. Most of south and southeast Europe and all of Russia remained Catholic but much of central Europe and Scandinavia became Protestant. The hundred years from about 1550 saw a succession of religious wars. In thc 18th and 19th centuries both sides were attempting to proselytize, especially in Africa, where Protestant gains were considerable. South America remained largely Catholic.

protoplasm, the internal substance of the living cell.

protozoa, ➪ pp. 120–1.

pruritus, ITCHING.

FACTFINDER

psittacosis, a severe influenza-like disease acquired by inhaling the dust of bird droppings containing a *Chlamydia* organism.

psoas muscle, the two-part muscle that runs from the front of the lower spine to the pelvis and to the front of the top of the thigh bone (femur) and that acts to flex the hip joint.

psoriasis, a skin disease in which dull red or salmon-pink, oval, thickened, scaly patches occur anywhere on the body but especially around the elbows, Knees, lower back and scalpp.

psyche, the mind, as opposed to the body.

psychiatry, ➪ pp. 186–7, 192–3.

psychic phenomena, ➪ pp. 180–1.

psychoanalysis, a method of treatment for mental disorder proposed by SIGMUND FREUD and based on his theories of the mind. The method involves months or years of largely on-interactional free association and, although patients often become dependent on the process, there is little evidence that it has real therapeutic value. Classical Freudian analysis assumed many of the features of a RELIGION and, like religion, was 'by schisms rent asunder'. There are now many schools; ➪ pp. 164–5.

psychogenic, of symptoms or disorders that appear to arise from mental rather than physical causes.

psycholinguistics, the study of the mental processes involved in the production and comprehension of speech and in the acquisition of language.

psychology, ➪ pp. 166–7.

psychometrics, ➪ pp. 166–7.

psychopathology, ➪ pp. 186–7.

psychopathy, any systematic behaviour pattern with strong anti-social elements.

psychosis, ➪ pp. 186–7, 192–3.

psychosomatic disorder, ➪ pp. 190–1.

psychosurgery, ➪ pp. 194–5.

psychotherapy, any form of treatment for mental disorder. Conventionally, the term excludes drug or physical methods of treatment such as ECT, and usually implies an interaction between patients and therapist either on a one-to-one basis or in groups. There are many different schools of psychotherapy; ➪ pp. 194–5.

psychotropic drugs, drugs that affect the state of the mind in any way.

ptosis, drooping of the upper eyelid, whether CONGENITAL or acquired.

puberty, the period of development during which the child is transformed into a sexually capable adolescent; ➪ pp. 108–9.

pubic bone, the paired bones of the pelvis that meet and join behind the pubic mound.

pubic hair, the area of springy hair that covers the MONS VENERIS.

public health, the general state of physical and mental wellbeing of a populace; and the branch of medicine concerned with prevention of disease and the promotion of health in populations; ➪ pp. 124–5.

pudenda, the external genitalia.

puerperal sepsis, ➪ pp. 94–6.

puerperium, the period following childbirth when the WOMB and VAGINA are returning to their normal state.

pulmonary, pertaining to the lungs.

pulse, the rhythmically repeated expansion of an ARTERY, at a rate of between about 50 and 80 beats per minute, caused by the pumping force of the heart.

pupil, the circular opening in the centre of the iris of the eye which becomes smaller in bright light and widens in dim light; ➪ pp. 84–5.

purgative, a strong laxative drug.

pus, a yellow or greenish-yellow viscous fluid consisting of dead white blood cells, bacteria, partly destroyed tissue and protein, usually formed at the site of bacterial infetion but sometimes as a result of inflammation from other causes; ➪ pp. 58–9, 130–1.

putrefaction, ➪ pp. 116–17.

pyramidal system, a general term for the large motor tracts that originate in the motor cortex of the BRAIN and descend, as inverted pyramids, through the substance of the brain and brainstem to end at various levels in the spinal cord. The pyramidal tracts on each side mAinly cross over to the other side in the brainstem. This is why a stroke on the right side of the brain causes left-sided paralysis, and vice versa.

pyromania, the uncontrollable impulse to start fires or to set fire to property.

Pythagoras, ➪ pp. 116–17.

Q fever, a pneumonia-like illness caused by organisms found in the excreta and birth products of farm animals, featuring high fever, headache, aches and pains, cough and pain in the chest.

quadriceps muscle, the powerful group of four muscles on the front aspect of the thigh. The quadriceps are inserted into a thick tendon, containing the kneecap (patella), that passes across the front of the knee to be inserted into the upper end of the front of the main lower leg bone (the tibia). When the quadriceps are contracted the knee is straightened; ➪ pp. 42–3.

quadriplegia, paralysis of the muscles of both arms, both legs and of the trunk, either from high spinal cord damage or neurological disease.

quarantine, isolation of a person who has been exposed to an infectious disease.

quinine, the first effective drug for the prevention and treatment of MALARIA but now largely replaced by more recent drugs; ➪ pp. 152–3.

quincy, an abscess between the TONSIL and the underlying wall of the throat, occurring as a complication of TONSILLITIS.

rabies, a serious infection of the nervous system acquired from the bite of a rabid animal and causing fever, headache, neck stiffness, disorientation, violent spasms of the throat and diaphragm, gagging, choking, panic, hallucinations, coma and death. Both vAccination and human antirabies globulin given after a bite can prevent the disease. Also known as hydrophobia.

race, ➪ pp. 4–5, 12–13.

radial nerve, one of the main nerves of the arm and hand activating the forearm muscles to straighten the flexed wrist and also providing sensation to the back of the forearm and hand.

radiation sickness, the effects of large whole-body doses of ionizing radiation, including nausea and vomiting, diarrhoea with blood in the stools, bleeding into the skin, severe ANAEMIA, hair loss and sterility; ➪ pp. 144–5.

radiation therapy, ➪ RADIOTHERAPY.

radiography, the use of X-rays to produce photographic images to assist in diagnosis. Compare RADIOLOGY.

radiology, the medical speciality concerned with the interpretation of X-rays in medical diagnosis and the use of X-rays in treatment; ➪ RADIOTHERAPY.

radiopaque, relatively impermeable to the passage of X-rays. Radiopaque substances are used to outline hollow structures such as the intestines (barium meal), the urinary system (urography) and the arteries (angiography) in medical diagnosis.

radiotherapy, ➪ pp. 156–7.

radius, one of the two forearm bones, that articulate on the thumb side.

Ramapithecus, ➪ pp. 10–11.

rape, sexual intercourse with a woman who does not, at the time, consent to it, or who is asleep or unconscious; or sexual intrcourse obtained by fraud or threats or by the use of physical force, or with a woman incapable of understanding what she is consenting to.

rash, any at least moderately extensive inflammatory skin eruption.

rationalism, a general term for the group of philosophic schools that reject received or authoritarian wisdom and dogmatic RELIGION and hold that knowledge is to be obtained only from observation and the application of logic to data so derived. Rationalism does not necessarily exclude religious beliefs, but tends to do so.

reason, the faculty by which new information is derived from old, judgement exercised and argument pursued.

receptor, any sensory nerve ending capable of receiving stimuli or any structure on a cell membrane that can bind a HORMONE or a NEUROTRANSMITTER or other information-conveying substance; ➪ pp. 75–6.

recessive, of inheritance of a characteristic carried by an ALLELE on one of the paired chromosomes that is not expressed because of the DOMINANCE of the allele or the other chromosome. Those parents with brown eyes, but each carrying recessive blue eye alleles, may have blue-eyed children if each parent donates the blue-eyed allele to their children; ➪ pp. 28–9.

recombinant DNA, DNA containing segments from different individuals or species, produced by laboratory manipulation.

recombination, the process by which segments of paired chromosomes, each from a different parent, are exchanged immediately after fertilization of the ovum. In this way each individual chromosome in the offspring contains segments from each parent.

reconstructive surgery, ➪ PLASTIC SURGERY.

recovery position, the prone lying position, with the head turned to one side and the hip joint on the same side well flexed for stability, into which unconscious patients are placed so as to avoid lethal inhalation of vomit or obstruction to the airway by the tongue.

rectum, the lowest, and most distensible, part of the large intestine lying immediately above the anal canal. When bowel contents pass into the rectum the desire to defaecate is felt; ➪ pp. 42–3, 44–5.

red blood cells, the haemoglobin-filled oxygen-carrying, biconcave disc free cells that circulate in their billions in the blood; ➪ pp. 40–1, 50–1.

referred pain, pain felt in a place other than the site of the causal disorder. Shooting pains in the left arm, as a signal of heart attack, are an example of a referred pain.

reflex, an automatic, predictable and involuntary response to a stimulus; ➪ pp. 76–7.

reflexology, a form of alternative therapy based on the hypothesis that the body contains channels of 'life force' similar to the meridians of ACUPUNCTURE. These terminate in the feet and practitioners of the method claim to be able to detect blockage in the channels by feeling the feet and toes. Massage and manipulation of the feet are then performed to unblock the channels; ➪ pp. 158–9.

refraction, the determination of the optical errors of the eyes so that appropriate correcting spectacles can be prescribed.

regression, a psychoanalytic term indicating a return to a more primitive stage of psychosexual development; or a psychological term meaning a temporary return to a less mature form of thinking in the process of learning how to manage new complexity.

reinforcement (learning), ➪ pp. 172–3.

relapsing fever, a SPIROCHAETE infection transmitted by louse or tick bite that causes successive recurrences of bouts of fever, shivering, headache, muscle aches, vomiting and PHOTOPHOBIA.

relationships, associations or connections of any kind between human beings, whether by genetics, marriage, friendship, enmity, profession, business or any other common activity or interest; ➪ pp. 206–7.

relaxation, deliberate release of involuntary muscle tension, usually for the purpose of calming anxiety or relieving the symptoms (such as headache) commonly associated with such tension.

religion, an almost indefinable entity covering a wide field of human activity and experience and encompassing doctrinal, historical, literary, devotional, experiential, behavioural and transcendental elements. In general, religion is concerned with humanity's responses to its concept of the characteristics of God. Religion may be formalized in dogma or entirely free and individual. It may be a matter of indifference or a matter of the most central importance. Its influence may be beneficial or malign.

REM sleep, ➪ pp. 178–9.

remission, a marked reduction in the severity of a disease, or its temporary disappearance.

renal, pertaining to the KIDNEY.

repair, the automatic process of cellular extension and reproduction that follows injury so as to restore, so far as possible, the original state of the body; ➪ pp. 62–3.

repetitive strain injury, a movement disorder, affecting people engaged in any frequently repeated activity, such as keyboard operators, musicians, packers, machine operators, manual cleaners and others, that causes cramp-like stiffness and sometimes the inability to continue in the associated occupation.

replication (DNA), ➪ pp. 24–5.

repression, one of the psychological defence mechanisms by which thoughts and emotions associated with anxiety are excluded from consciousness; ➪ pp. 164–5.

reproduction, ➪ pp. 96–7, 102–3, 104–5.

respiration, ➪ pp. 52–3.

respiratory function tests, ➪ LUNG FUNCTION TESTS.

respiratory system, the major system of the body concerned with the oxygenation of the blood; ➪ pp. 52–3.

response, any action, such as a movement or a complex sequence of behaviour, that follows automatically, involuntarily and sometimes predictably on receipt of a stimulus.

resuscitation, the restoration of acceptable function in a person whose heart action, blood pressure or body oxygenation had dropped to critical levels.

retina, the neural membrane that lines the inside of the eye and that responds to images formed upon it by the lens system by producing a complex of frequency-modulated and bit-mapped signals that pass along the optic NERVES and subsequent nerve tracts to the back of the BRAIN. The retina is not simply a passive transducer of light-sensitive RODS AND CONES, but performs computing functions on the impulses produced by these so as greatly to extend the range of light sensitivities and to increase the perceived quality of the image; ➪ pp. 84–5.

retirement, the period after the permanent cessation of work.

retrovirus, a virus, such as the HIV, that has only one copy of an RNA GENOME but carries an enzyme, reverse transcriptase, that enables it to create a complementary DNA copy; ➪ pp. 118–19, 132–3.

Rhazes, (c. 860–923) Baghdad physician, follower of Hippocrates and Galen, whose many medical writings made him the most celebrated and probably the greatest physician of his day, and whose influence on the development of medical science was considerable.

rhesus incompatibility, the condition in which a rhesus negative mother bears a rhesus positive fetus from a rhesus positive father. The mother produces antibodies against the fetal RED BLOOD CELLS which are broken down, often with serious consequences; ➪ BLOOD GROUPS.

rheumatic fever, an acute illness involving fever and a fleeting joint disorder and often serious damage to the heart valves and sometimes the nervous system.

rheumatism, a term applied to any joint disorder involving pain and stiffness as well as to RHEUMATOID ARTHRITIS and OSTEOARTHRITIS.

rheumatoid arthritis, a general disease that may cause progressive joint deformity, joint destruction and disability, especially of the small joints of the fingers and hands.

Rhine, Joseph Banks, ➪ pp. 180–1.

rhinitis, inflammation of the MUCOUS MEMBRANE lining of the nose; ➪ pp. 88–9.

rhythm method, the unreliable method of CONTRACEPTION in which an attempt is made to predict the time of ovulation (about 14 days before the onset of the next menstrual period) and to avoid coitus immediately before and during this time; ➪ pp. 101–2.

rib, ➪ pp. 40–1.

riboflavin, vitamin B_2. Deficiency causes reddening of the lips, cracking at the corners of the mouth and eye problems.

ribonucleic acid, ➪ pp. 26–7.

ribosome, a cell organ (organelle) made of RNA and PROTEIN and present in countless numbers in most cells. Ribosomes may be free or attached to the endoplasmic reticulum. They are the sites of protein synthesis, which they effect by moving along messenger RNA chains, picking up AMINO ACIDS from the cell fluid in the right order, in accordance with the code provided by the base triplets, and linking them together to form polypeptides and proteins; ➪ pp. 38–9.

rickets, a disease of childhood due to a deficiency of vitamin D and hence calcium. The growing bones are softened and become distorted, usually permanently.

Right, the, the conservative, often anti-SOCIALIST and sometimes totalitarian political division of society. The sense is that of opposition to the 'LEFT' rather than to the 'wrong'.

rigor mortis, the stiffening of muscles which usually starts 3 to 4 hours after death as a result of the coagulation of muscle by accumulating lactic acid.

ringworm, ➪ TINEA.

rite, ➪ pp. 216–17.

rite of passage, ➪ pp. 216–17.

ritual, ➪ pp. 198–9.

Rivers, William Halse, ➪ pp. 212–13.

RNA, ➪ RIBONUCLEIC ACID.

robotics, the branch of technology concerned with the development of machines capable of performing complex tasks of a kind normally limited to humans. Robotic machines of limited function controlled by computer have now become commonplace in the manufacturing industries, but the expected development of anthropomorphic or humanoid robots, in the manner predicted by the writer Karel Čapek in his 1921 novel *Rossum's Universal Robots*, has not been fulfilled in any but a trivial sense.

rods and cones, the light-sensitive elements of the RETINA. The rods are highly sensitive to light intensity but not to colour and are concentrated towards the periphery of the retina; the colour-sensitive cones are in highest concentration centrally. Any particular cone exhibits maximum sensitivity at a wavelength equal to that of one of three colours – red, green or blue. This provides a basis for colour vision by differential simulation; ➪ pp. 84–5.

Roman catholicism, the branch of CHRISTIANITY holding that the Pope is the successor of the apostle Peter and is thus appointed by God, and is authorized to make infallible pronouncements on matters of faith and morals. Catholic practices include the mandatory sacraments of baptism, confirmation, attendance at the Mass (Eucharist), penance and the anointing of the sick, and the practice, at least once a year, of confession of sins to a priest. The Mass is the supreme sacrament which must be entered into in a proper state of mind.

Romance languages, those languages derived from Latin, including Italian, French, Spanish, Portuguese and Romanian.

Röntgen, Wilhelm Konrad, ➪ pp. 128–9.

root-canal treatment, drilling out the whole of the diseased pulp of a tooth and sealing the cavity.

Rorschach test, ➪ pp. 176–7.

Ross, Surgeon Major Ronald, ➪ pp. 136–7.

roundworms, NEMATODE parasites of the intestine acquired by fecal contamination of food or finger; ➪ pp. 136–7.

RU486, an abortion pill; ➪ pp. 100–1.

rubella, ➪ GERMAN MEASLES.

rupture, a popular term for a hernia – a protrusion of an organ, especially a loop of bowel, through a natural or abnormal opening in the body such as the inner layers of the wall of the abdomen. Hernias commonly pass down into the SCROTUM or cause a bulge in the groin or at the navel.

Ryle, Gilbert, ➪ pp. 70–1.

sacral region, the area of the SACRUM.

sacroiliac joint, the ligamentous junctions between the edges of the SACRUM and the two outer bones of the pelvis (iliac bones).

sacrum, the large, triangular, wedge-like bone that forms the centre of the back of the PELVIS.

sadism, taking pleasure, often of a sexual nature, in the infliction or witnessing of cruelty.

sadomasochism, a sexual deviation in which arousal is achieved by inflicting or receiving pain or abuse of various kinds.

SADS, ➪ SEASONAL AFFECTIVE DISORDER SYNDROME.

safe period, ➪ pp. 100–1.

safe sex, measures, such as the use of CONDOMS and non-penetrative sex, taken to reduce the risk of acquiring sexually transmitted disease, especially AIDS.

saliva, the lubricant and starch-splitting watery fluid secreted into the mouth by the SALIVARY GLANDS.

salivary glands, six glands that surround the mouth and secrete a watery fluid (saliva) which aids the mastication and swallowing of food and contains an ENZYME maltase that can break down the disaccharide sugar maltose to sweet and easily absorbed monosaccharides; ➪ pp. 44–5.

Salmonella, a genus of bacteria containing hundreds of species, all of which can cause food poisoning and some of which cause TYPHOID FEVER; ➪ pp. 134–5.

salt, any substance, such as sodium chloride or common salt, that dissociates in solution into ions of opposite charge.

Sanctorius of Padua, ➪ pp. 150–1.

sanguine type, of a cheerful and optimistic personality, or of a ruddy complexion. One of the personality types once believed by the adherents of the humoral theory to be due to a preponderance of the influence of blood over phlegm and yellow and black bile.

sarcoma, a cancer of connective tissue such as bone, muscle, cartilage, fat, fibrous tissue and blood vessels; ➪ pp. 140–1.

saturated fats, ➪ UNSATURATED FATS.

satyriasis, an extreme, persistent and sometimes uncontrollable male craving for copulation.

scab, a protective crust formed when exuding serum mixes with pus and dead skin and then clots; ➪ pp. 62–3.

scabies, an infestation with the mite parasite *Sarcoptes scabei* which burrows into the skin to feed on dead epidermal scales and lay eggs.

scalp, the layers covering the vault of the skull and consisting of a thin sheet of muscle (epicranium), a layer of connective tissue richly supplied with blood vessels, and the skin.

scanning techniques, ➪ pp. 128–9.

scapula, the shoulder blade.

scar, fibrous COLLAGEN repair tissue that firmly binds together the edges of wounds if reasonably approximated, or gradually fills the gap between edges kept apart; ➪ pp. 62–3.

scarlatina, a mild attack of SCARLET FEVER.

scarlet fever, a comparatively rare streptococcal infection, usually of children, that causes fever, sore throat, headache and a rash of thousands of tiny red spots that spreads to cover the whole body. Complications include RHEUMATIC FEVER and kidney inflammation (nephritis).

scepticism, the tendency to question established dogma and belief and to refuse to accept received wisdom simply on assertion. The philosophic and scientific stance that rejects authoritative claims to certainty.

schistosomiasis, ➪ pp. 136–7.

schizophrenia, the most common major psychiatric disorder; ➪ pp. 192–3.

Schwann, Theodor (1810–82), German physiologist who, among many important medical advances, extended the CELL theory from plants to animals, discovered the STOMACH enzyme PEPSIN and showed that bacteria were responsible for putrefaction.

sciatic nerve, the main nerve of the leg that arises from several of the lower spinal nerves and runs down through the BUTTOCK to activate the leg muscles and provide sensation to the skin.

sciatica, pain in the lower back and leg, usually from pressure on the spinal roots of the main nerve of the leg (the sciatic nerve) by pulpy material squeezed from an INTERVERTEBRAL DISC.

sclera, ➪ WHITE OF EYE.

sclerosis, hardening of tissues, usually from deposition of SCAR tissue, following INFLAMMATION.

screening, examination of apparently healthy people to detect disease at an early stage.

scrofula, TUBERCULOSIS of the lymph nodes of the neck, usually from milk from infected cows, now very rare in developed countries.

scrotum, the bag of skin and muscle that contains the testicles and the lower part of the spermatic cord.

scurvy, an uncommon disease, brought about by a deficiency of vitamin C, that causes spontaneous bleeding into the gums, joints, muscles and skin – the effects of defective COLLAGEN production; ➪ pp. 26–7.

seasickness, ➪ MOTION SICKNESS.

seasonal affective disorder syndrome, a condition affecting people in the winter months, in which the mood is said to be depressed by inadequate exposure to light.

sebaceous glands, ➪ pp. 56–7.

seborrhea, excessive production of the oily secretion of the sebaceous glands of the skin (SEBUM), causing a greasy scalp and predisposing to ACNE.

sebum, ➪ pp. 56–7.

secondary sexual characteristics, ➪ pp. 108–9.

security object, any soft object, such as a baby blanket, a former night garment or a teddy bear, that brings comfort and a sense of security to a young child.

sedative drugs, a large group of drugs that includes sleeping drugs, tranquillizers, antianxiety drugs, antipsychotic drugs and some antidepressant drugs.

self, sense of, awareness of one's own identity as an individual and of one's bodily constitution – an awareness that may or may not be accurate and that commonly differs in many particulars from the perception of others.

self-mutilation, acts of destruction of parts of one's own body, such as amputation of fingers, limbs or genitals, gouging out of eyes, and so on, chiefly brought about by psychiatric disorder.

semantics, the study of meaning, of the effectiveness with which thought is translated into language, and of the relationship between words and symbols and meaning; ➪ pp. 206–7.

semen, ➪ SEMINAL FLUID.

semicircular canals, the three fluid-filled, looped channels of the inner ear, each lying in a plane at right angles to the others, which detect changes in the orientation of the head and pass this information to the BRAIN; ➪ pp. 86–7.

seminal fluid, the sticky, opalescent fluid ejaculated during the male orgasm that provides a vehicle for the SPERMATOZOA and is mainly secreted by the SEMINAL VESICLES.

seminal vesicles, two short, narrow sacs, about 5 cm long, lying between the bladder and the rectum and opening through the PROSTATE gland into the urethra. The seminal vesicles do not store sperms. They do, however, contribute a fluid that makes up most of the volume of the ejaculated seminal fluid.

seminiferous tubules, the long, much convoluted tubules of the testicle within which body cells are converted into spermatozoa; ➪ pp. 96–7.

semiotics, the study of signs, including words, symbols, gestures and body language, and of their cardinal role in conveying information. Semiotic studies suggest that meaning, although it may often seem self-evident, is always the result of social conventions. Cultures can be analysed in terms of a series of sign systems. One difficulty, perhaps responsible for a certain vagueness in discussion of the subject, is that the experts have never been able to reach full agreement on the exact definition of the central terms 'sign', 'symbol' and 'signal'.

Semitic-Hamitic language family, ➪ AFRO-ASIATIC LANGUAGE FAMILY.

Semmelweiss, Ignaz Philipp, ➪ pp. 94–5.

senile dementia, ➪ DEMENTIA, ALZHEIMER'S DISEASE.

senility, ➪ pp. 110–11.

sensation, the subjective experience during the act or process of obtaining information (using the word in its strict, all-embracing, sense) through stimulation of any of the receptive structures of the nervous system. These include the eyes, ears, nose, tongue, and the range of receptors for touch, pain, temperature and vibration situated in the skin and in some of the internal organs; ➪ pp. 76–7.

senses, the channels by which information about the environment or the state of the body enters the mind – specifically vision, hearing, smell, taste, touch, body and joint position, sense, and the senses of pain, temperature, pin-prick and vibration; ➪ pp. 82–91.

sensory cortex, ➪ pp. 76–7.

sensory deprivation, the loss, usually as a result of deliberate research action, of input from as many as possible of the sense organs. Research subjects might be in a perfectly silent sealed chamber, floating in water at exact skin temperature, and effectively blindfolded. The results of such research – severe distress, disorientation, hallucination and other mental disturbances – make it clear that rich sensory input is necessary for the maintenance of normal BRAIN function; ➪ pp. 74–5.

sensory nerve ending, a receptor for touch, pain, pressure, temperature, pinprick and vibration situated at the end of a sensory nerve, that converts these physical events into nerve impulses that are then carried to the brain; ➪ pp. 75–6, 90–1.

separation anxiety, excessive levels of concern experienced by children when separated from parents or minders, or threatened with such separation.

sepsis, the presence in the body of infection-producing micro-organisms or their toxins.

septic shock, a serious disorder in which the blood vessels relax so completely, under the influence of bacterial toxin, that the volume of the blood is insufficient to fill them and maintain the circulation.

septicaemia, the presence of dangerous organisms in the blood. Popularly known as blood poisoning; ➪ pp. 152–3.

septum, any thin dividing wall or partition in the body, such as the NASAL SEPTUM.

serotonin, a powerful HORMONE and NEUROTRANSMITTER concerned with the control of mood and levels of consciousness.

serum, the clear, straw-coloured fluid left when the red blood cells and the clotting proteins are removed from the blood.

sex, a term with a wide spectrum of meanings that include the genetically determined gender; the condition of being male or female; the impulse to copulate; the act of copulation; and the genitalia.

sex change, major plastic surgery and hormonal treatment to alter, so far as is possible or desirable, the external manifestations of gender.

sex chromosomes, ➪ pp. 22–9.

sex determination, ➪ pp. 22–3.

sex hormones, a considerable range of steroids produced by the testes and ovaries and by the outer layer of the adrenal glands. Male sex HORMONES are called ANDROGENS and are, in general, anabolic. Female sex hormones include oestrogens, which are feminizing, and progesterone which is essential for the menstrual cycle and the maintenance of pregnancy; ➪ pp. 202–3.

sex-linked, ➪ pp. 28–9.

sexism, discrimination, especially against women, on the basis of sex. The conscious, unconscious or merely habitual expression of traditional male domination in spoken or written language or in behaviour.

sex therapy, counselling and advice designed to overcome sexual problems such as IMPOTENCE, PREMATURE EJACULATION, failure of ORGASM and VAGINISMUS.

sexual abuse, imposing on any person, especially a minor, sexual attention likely to cause physical or psychological harm.

sexual deviation, sexual practices generally disapproved of, such as SADISM, SADOMASOCHISM, PAEDOPHILIA, FROTTAGE, VOYEURISM, obscene telephone calling, and so on.

sexual dysfunction, those disorders that may be treated by SEX THERAPY.

sexual intercourse, ➪ pp. 96–7, 202–3.

sexuality, ➪ pp. 202–3.

sexually transmitted diseases, a range of diseases transmitted, usually through the genital mucous membrane, during sexual intercourse. The STDs include AIDS, candidiasis, chancroid, chlamydial infections, crab lice, *Gardnerella vaginalis* infections, genital herpes, genital warts, gonorrhoea, lymphogranuloma venereum, nonspecific urethritis, syphilis, trichomoniasis and yaws; ➪ pp. 118–19, 130–1.

shamanism, an Asian RELIGION holding that the world is populated by good and evil spirits whose activities can be controlled only by specialists in ritual and trance (shamans) able to communicate with them.

shame, a distressing emotion involving a strong sense of having transgressed against a social or moral code. Shame is always relative to current mores or to the upbringing of the person concerned.

shell shock, ➪ POST-TRAUMATIC STRESS DISORDER.

Sherrington, Sir Charles, ➪ pp. 204–5.

shin, the hard front part of the leg below the knee and above the ankle where the front edge and part of the front surface of the TIBIA lie immediately under the skin.

shingles, a nerve disease caused by reactivation of a previous CHICKENPOX infection that leads to a rash of tingling blisters on the skin of the flank or forehead and often long-persistent pain afterwards. Also known as herpes zoster.

Shinto, the indigenous polytheistic RELIGION of Japan which is without a historic founder or a body of sacred scripture. The practice of Shinto involves ritual purification, worship at shrines, silent prayer, making of offerings, and observation of festivals and important occasions.

shock, in medicine, shock has nothing to do with emotional disturbance. It is a specific surgical condition featuring a substantial loss of circulating BLOOD volume, either from haemorrhage or from displacements of water from the blood into the tissues or on to burned surfaces. The reduction in blood volume may deprive the BRAIN and heart MUSCLE of adequate oxygenation and may be rapidly fatal unless corrected by urgent transfusion. Shock can also be caused by any agency that results in widespread dilatation of arteries, or by virulent bacterial toxins that increase the permeability of blood vessels (toxic shock).

short-sightedness, a dimensional eye anomaly in which the distance from the lens system to the focal plane on the retina is too great. Light rays from near objects, which are more divergent than rays from distant objects, may come to a sharp focus on the retina so that the object is seen clearly; rays from distant objects, however, cannot be focussed on the retina. Vision can sometimes be improved by 'stopping down' – narrowing the lids, but clear distance vision is adequately achieved only by the use of concave diverging lenses in glasses or contact lenses; ➪ pp. 84–5.

shoulder, the region of the junction of the arm with the body where the upper arm bone (humerus) articulates with the shoulder-blade (scapula) to form a joint heavily covered with muscles, most of which are concerned with raising the arm in different planes and rotating it.

shoulder blade, the flat bone, or scapula, lying on the back of the upper ribs, that provides a socket for the articulation of the upper arm bone (humerus).

Siamese twins, identical twins who failed to separate completely after the first division of the egg and remain partially joined together at birth.

sibling rivalry, strong competition or feelings of resentment between siblings, especially between an older child and a new baby. Sibling rivalry may persist throughout life.

sick building syndrome, a range of symptoms of unknown cause, attributed to work in a modern office building, and including headache, dryness and itching of the eyes and nose, sore throat and fatigue.

sickle cell anaemia, one of the major haemoglobin disorders; ➪ pp. 28–9.

sighing, the emission of a long and audible exhalation as an expression of various emotions such as weariness, vexation, relief or despair.

sight, sense of, a principal sensory modality whose operation, so far as its optical and neurological elements go, is fully understood, but whose translation from the neurological to the sensational remains one of the mysteries of science. Most people enjoy binocular vision in which simultaneous perception with the two eyes provides the BRAIN with two slightly dissimilar but overlapping images viewed from separate positions. The fusion of these gives rise to the illusion of solidity (stereopsis). Double vision (diplopia) is experienced by people capable of simultaneous perception when the two images fall on areas of the two RETINAS other than those that normally correspond. This is usually the result of a defect of action of the muscles that move one of the eyes, often from nerve dysfunction, or of the fusional reflexes mediated by the brain; ➪ pp. 84–5.

sign language, ➪ pp. 14–15.

Sikhism, a monotheistic Indian RELIGION founded in the 16th century in the Punjab to combine Hindu and Muslim elements into a single creed, thereby attracting the enmity of both. Sikhism emphasizes the brotherhood of man and rejects the CASTE system. Its adherents take a vow not to cut their hair or to smoke or drink alcohol, and wear a comb in their hair and a steel bracelet on the right wrist. They wear soldier's shorts and carry a sword.

silicosis, lung damage with loss of function from widespread scarring caused by long-term inhalation of silica-containing dusts.

Simpson, Sir James (1811–70), Scottish obstetrician and pioneer in the use of general anaesthetics, who discovered the use of chloroform and championed it against Establishment resistance until it was used by Queen Victoria in 1853, when it became generally accepted.

single-parent family, ➪ pp. 214–15.

Sino-Tibetan language family, the family of some hundreds of languages spoken by Chinese in their homeland and wherever they may be throughout the world, and by many of the inhabitants of Tibet, Burma and northeast India.

sinuses, the cavities, situated around the nose, in the bones of the skull, explicitly the frontal sinuses in the forehead bone, the maxillary antrums in the upper jaw, the ethmoidal air cells in the bones between the upper parts of the nose and the eye sockets (orbits) and the sphenoidal sinuses in the central body of the bone forming the back wall of the nose and part of the base of the skull. These paranasal sinuses are all lined with mucous membrane and all drain into the nose. They lighten the skull and confer resonance to the voice, greatly altering its quality and timbre. Inflammation of the linings is known as sinusitis; ➪ pp. 88–9.

sinusitis, ➪ SINUSES.

skeletal muscle, ➪ pp. 42–3.

skeleton, the bony framework of some 206 bones, mostly paired, by which the rest of the body is supported. Variability in the skeleton is common. Most people have 12 pairs of ribs, but many have 13. People with Down's syndrome often have only 11 pairs. We are born with many more bones because certain of the adult bones are formed from the fusion of more than one bone. The ends of the long limb bones, for instance, are separated from the shafts to allow growth at the EPIPHYSEAL PLATES until early adult life. By about age 25 all bones have fused. The axial skeleton consists of the skull and VERTEBRAL COLUMN. The shoulder girdle consists of the collar bones (clavicles) and the shoulder blades (scapulas); the pelvic girdle consists of the triangular sacrum centrally and the two innominate bones, each consisting of the fused pubic, iliac and ischial bones (on which we sit). The upper arm bone (the humerus) articulates with the scapula and the upper leg bone (the femur) with a hollow in the side of the innominate bones. The arm and leg bones are analogues of each other. The forearm has two bones (the radius and ulna) and the lower leg has two bones (the stout tibia and the delicate fibula). The bones of the wrist, hand and fingers correspond to those of the ankle, foot and toes – respectively carpals, metacarpals and phalanges in the hand and tarsals, metatarsals and phalanges in the foot; ➪ pp. 40–1.

skin, ➪ pp. 56–7.

skin graft, the surgical transfer of an area of skin from one part of the body to another so as to cover denuded areas or improve appearance.

Skinner, B.F., ➪ pp. 162–3.

skull, ➪ pp. 40–1.

slang, an informal, often colourful, vocabulary and mainly limited to special groups of all kinds. Slang is universally employed and, although most of it dies out, those elements that have been found generally useful will often be generally adopted and become an established part of the language. The notion of slang exists only when attempts are made to formalize and fix the 'correct' elements of a language. Prior to this tendency neologism was accepted as a natural part of the evolution of language.

slavery, the process of making another human being into a chattel without rights, so that he or she can be exploited in any way – personal, commercial or sexual – and sold at will. Slavery has been a feature of the human condition since the earliest times and continues today in many parts of the world. It has been justified by a variety of rationalizations, especially that in which the slave is said to belong to an inferior class to that of the master and to be insensitive to suffering. By any worthwhile standards, a community that practises or tolerates slavery must be regarded as socially underdeveloped.

Slavic languages, ➪ INDO-EUROPEAN LANGUAGE FAMILY.

sleep, the nightly condition of reduced consciousness and METABOLISM that occupies about one-third of the average person's life; ➪ pp. 178–9.

sleep deprivation, actively preventing a person from sleeping by forcible arousal whenever sleep is threatened. This causes frequent 'microsleeps' and loss of attention and registration followed by hallucinations and a paranoid psychological breakdown.

sleep talking, a common activity, occurring usually during non-REM sleep (➪ pp. 178–89) and of which there is seldom any waking recollection. Speech varies from meaningless mumbles or monosyllables to clear, articulate and lengthy statements often with pauses as if being one side of a conversation. Secrets are hardly ever revealed during sleep talking.

sleep terror, apparently harmless episodes in which a sleeping child suddenly wakes, starts to scream, gives every indication of severe agitation, seems unable to recognize faces or surroundings, and then returns to sleep and has no subsequent memory of the event.

sleeping pills, drugs, such as the benzodiazepines and sometimes the barbiturates, used to try to re-establish sleep patterns and so overcome insomnia.

sleeping sickness, the parasitic tropical disease African trypanosomiasis; the term is also used for the rare brain disorder encephalitis lethargica.

sleepwalking, a state of partial sleep mainly in children, harmless barring accidents, in which the child gets out of bed and wanders about for a few minutes.

slipped disc, an informal and misleading term for the squeezing out of the pulpy centre of an intervertebral DISC through a degenerative defect in its outer fibrous ring. The resultant pressure on nearby nerve roots or on the spinal cord causes great pain and disability.

smallpox, a disease with the unique distinction of having been eradicated. Smallpox was an often fatal virus disease spread mainly by droplet infection and causing fever, headache, muscle pain and a rash of raised pustules that left deep pitted scars; ➪ pp. 124–5.

smegma, a cheesy-white, foul-smelling sebaceous gland secretion occurring under the foreskin of uncircumcised males with poor personal hygiene.

smell, sense of, ➪ pp. 88–9.

smelling salts, volatile crystals of ammonium carbonate or other substances once widely used as a stimulant to be sniffed on occasions of feeling faint.

smiling, facial expression of the emotion of pleasure, happiness, amusement or friendliness involving an upturning of the corners of the mouth, exposure of the upper teeth and a narrowing of the eyelids.

smoking, the dangerous practice of inhaling tobacco smoke that results in the absorption of many substances, some of which are capable of causing cancer and other serious diseases; ➪ pp. 138–9.

smooth muscle, muscle which performs involuntary functions such as the movements of the bowels and the tightening of the arteries and the air tubes of the lungs. Unlike voluntary muscle, smooth muscle does not display a striped appearance on microscopic examination; ➪ pp. 42–3.

sneezing, a sudden reflex and involuntary ejection of air through the mouth and nose as a result of irritation of the nose lining.

sniffing, quick inhalation through the nose, often to prevent nasal secretions from dripping out, but sometimes as a purposeless habit or to express disdain.

snoring, a noise, heard only by others, caused by vibration of the soft palate in people sleeping on their backs with their mouths open.

Snow, John, ➪ pp. 28–9.

social conditioning, the CONDITIONING to behave in a manner acceptable to society that is applied, almost without our awareness, in the course of participation in social activity.

Social Darwinism, the view, supported by HERBERT SPENCER and others, that the evolution of societies is a natural process of competition for resources and that the fittest have a moral justification in seeking to survive.

social security, the provision of relief of human want from public resources as a legal entitlement, in the form of pensions for the elderly, benefit for the unemployed and the disabled, free medical care, economic assistance with the bringing up of families and free social services.

socialism, a political system based on the proposition that social justice and equality, progress, freedom and happiness can be imposed on society by administrative means. Socialism is inherently opposed to *laissez-faire* capitalism which it would replace by public ownership. The greatest social experiment was that of COMMUNISM.

sociopathy, an antisocial personality disorder.

Socrates, ➪ pp. 150–1.

sodium, a highly reactive metallic element present almost exclusively in the form of its many SALTS, especially common salt (sodium chloride), which is an essential constituent of the body; ➪ pp. 46–7.

sodomy, anal copulation or copulation with an animal. Also known as buggery.

solipsism, the philosophic notion that the universe exists only in the perception of the person holding this view. This extreme position is seldom seriously proposed, however, except to illustrate the general point that none of us can be sure that this is not so.

solvent abuse, ➪ GLUE SNIFFING.

soma, the body, as distinct from the germ cells (eggs and sperm), or, sometimes, as distinct from the mind.

somatic, ➪ SOMA.

somatic nervous system, that part of the nervous system concerned with volition, sensation and consciousness, as distinct from the unconscious AUTONOMIC NERVOUS SYSTEM.

somatotype, the physical build of a person, sometimes claimed without good evidence, to give an indication of the personality type.

somnambulism, ➪ SLEEPWALKING.

spastic paralysis, paralysis, of varying degree, of both lower limbs with muscle spasm, due to disorders of brain, spinal cord, spinal nerve roots or peripheral nerves, or sometimes of genetic origin.

species, ➪ pp. 4–5.

speech, ➪ pp. 76–7, 82–3, 106–7.

sperm, ➪ SPERMATOZOA.

spermatozoa, the male germ cells that carry the genetic contribution from the father and unite with ova to initiate new individuals; ➪ pp. 96–7.

spermicides, substances that can kill sperms and can be used as CONTRACEPTIVE preparations.

sphenoid bone, the bat-like central bone of the base of the skull.

sphincter, any ring-like muscular structure that surrounds a tube or orifice of the body and is capable of exercising control on material passing through it.

sphygmomanometer, ➪ pp. 128–9.

spider naevus, ➪ NAEVUS.

spina bifida, failure of normal development of the rear part of one or more of the VERTEBRAE of the spine with varying degrees of protrusion of the nerve tissue and its coverings (MENINGES) and resulting neurological defects of varying severity, such as lower body paralysis and incontinence.

spinal cord, ➪ pp. 72–3, 74–5.

spinal nerves, ➪ pp. 90–1.

spinal tap, ➪ LUMBAR PUNCTURE.

spine, ➪ pp. 40–1.

spiritualism, ➪ pp. 180–1.

spirochaete, a highly motile, spirally shaped bacterium of a class that includes the causal agents of SYPHILIS, LEPTOSPIROSIS (Weil's disease) and LYME DISEASE; ➪ pp. 134–5.

spleen, a solid LYMPHATIC organ situated high on the left side of the abdomen just under the DIAPHRAGM. The spleen is packed with LYMPHOCYTES and red blood cells and acts as a source of immune cells and ANTIBODIES and as a blood filter; ➪ pp. 58–9, 60–1.

split personality, ➪ MULTIPLE PERSONALITY.

spondylosis, any disorder of the spine (vertebral column).

spouse abuse, physical or mental cruelty to a wife or husband by husband or wife.

sprain, stretching or slight tearing of a joint ligament or of the fibres of a joint capsule.

Sprenger, Jacob, ➪ pp. 94–5.

sprue, a disorder in which nutrients are poorly absorbed from the intestine.

sputum, mucus from the MUCOUS MEMBRANE lining of the air passages, often mixed with PUS or blood, and coughed up. Also known as phlegm.

squint, the condition in which only one eye is directed at the object of regard, the other being directed either too far inwards (convergent squint), too far outwards (divergent squint), or upwards or downwards (vertical squint).

stapes, the innermost of the chain of three tiny bones of the middle ear (auditory ossicles) that transfer vibrations from the eardrum to the cochlea of the inner ear.

starch, a complex polysaccharide of linked glucose molecules. Most natural starches consist of various mixtures of amylose, containing from 200 to 500 glucoses, and amylopectin containing 20 cross-linked glucose molecules. Starch is most abundant in rice, cereals and potatoes and constitutes about 70 per cent of the world's food; ➪ pp. 36–7.

starvation, ➪ pp. 46–7.

status, a position in any social, professional, business or other hierarchy that brings esteem and privilege.

stem cell, a large progenitor CELL present in the bone marrow and other blood-forming tissues from which various series of white cells of the immune system and their clones develop.

stenosis, narrowing of a duct, orifice or tube.

sterility, freedom from infective micro-organisms or incapable of reproduction.

sterilization (bacterial), the process by which anything, especially a surgical instrument or dressing, is rendered free from living micro-organisms.

sterilization (contraceptive), ➪ pp. 100–1.

sternum, the breast bone; ➪ pp. 40–1.

steroids, members of the class of 17 carbon atom fat-soluble organic compounds that include the sex hormones, the hormones of the CORTEX of the ADRENAL GLAND, the progestogens, the bile salts and a large range of synthetic drugs; ➪ pp. 80–1.

stethoscope, ➪ pp. 128–9.

stillbirth, birth of a dead baby.

stimulants, drugs or other agencies that increase the rate of activity of bodily organs or systems.

stimulus, ➪ pp. 166–7.

Stokes-Adams syndrome, a condition featuring repeated brief episodes of loss of consciousness from extreme slowing or even brief stoppages of the heartbeat with resulting inadequate blood supply to the brain.

stomach, the distensible bag at the lower end of the gullet (oesophagus) into which swallowed food passes. The stomach secretes a strong inorganic acid, hydrochloric acid, which helps to break down gross food lumps, and a powerful PROTEIN-splitting ENZYME, pepsin. When food processing has proceeded sufficiently, it is passed on to the duodenum and the small intestine for further biochemical action and absorption; ➪ pp. 44–5.

stomach pump, an informal term for a wide-bore tube that is passed into the STOMACH in cases of poisoning and through which water is poured to dilute and wash out the poison.

stomach ulcer, ➪ GASTRIC ULCER.

stomatitis, inflammation or ulceration of the mouth.

stool, ➪ FAECES.

Stopes, Marie, ➪ pp. 100–1.

stork bite, a popular term for the small, harmless, benign tumours of blood vessels that often form pinkish skin blemishes around the eyes and on the back of the neck in new-born babies and usually disappear in the first year of life.

strabismus, ➪ SQUINT.

strangulation, constriction, usually by compression, of any passage or tube in the body.

strawberry mark, a bright red, raised, harmless tumour of skin blood vessels that appears within the first few weeks of life and grows rapidly but eventually disappears.

strawberry nose, ➪ POTATO NOSE.

streptococcus, ➪ pp. 134–5.

streptomycin, ➪ pp. 152–3.

stress, any trauma – physical, social or psychological – that threatens a person's wellbeing and produces a demonstrable, often defensive, response that may be distressing and occasionally dangerous, but is sometimes beneficial. Stress responses include secretion of ADRENALINE and STEROIDS, raised heart rate and blood pressure, increased muscle tension and raised blood sugar; ➪ POST-TRAUMATIC STRESS DISORDER.

stress fracture, a hair-like break in a bone, especially of the foot or leg, caused by repetitive or prolonged trauma as in marching, running or ballet dancing.

stress ulcer, an acute GASTRIC ULCER that develops as a result of SHOCK, or severe illness or injury, especially burns.

stretch marks, areas of skin atrophy forming broad, purplish, shiny or whitish lines on the abdomen, thighs or breasts most commonly of pregnant women. Also known as striae.

striae, ➪ STRETCH MARKS.

stroke, ➪ pp. 14–15, 74–5, 138–9.

stupor, a state of reduced consciousness, less deep than coma, from which the affected person can be briefly aroused only by a painful stimulus.

stuttering, a speech impediment common in childhood featuring hesitancy and repetition of consonants, especially those at the beginning of words. Stuttering usually ceases before adult life.

stye, a small ABSCESS around the root of an eyelash.

subclinical, so mild or early as to produce no symptoms or signs.

subconscious, pertaining to mental processes that occur without conscious awareness; and to the large information database (memory) of which only a small part is in consciousness at any time. The term is used by psychoanalysts to indicate the level of the mind through which information passes on its way to full consciousness from the unconscious mind; ➪ pp. 164–5.

subculture, a discrete and separate culture within a national culture displayed by a comparatively small group and manifesting behaviour patterns, beliefs and values that differ from, and often conflict with, those of the main body.

subcutaneous, immediately under the skin. Hypodermic.

sublimation, the directing of potentially damaging or dangerous instinctual drives into socially acceptable activities.

subliminal perception, the reception of stimuli, often complex or verbal and usually visual, that are presented for such a short time as to be barely noticed or unnoticed. Such stimuli can, however, influence behaviour and present potential opportunities for abuse. The conclusion, now verified by physiological research, that consciousness and information transmission may involve different systems that can operate independently, has major implications for psychology and philosophy and has aroused much controversy.

sucrose, the disaccharide carbohydrate cane or beet sugar which is broken down, on digestion, to the monosaccharides glucose and fructose.

sudden infant death syndrome, the sudden death of an apparently well baby, especially one put down to sleep in the face-down (prone) position. In many cases no cause is found even on detailed post-mortem examination. Also known as cot death.

suffocation, oxygen deprivation by physical obstruction to the passage of air into the lungs.

sugar, any of the group of simple water-soluble mono-, di- or oligosaccharide carbohydrates such as sucrose, glucose, fructose, lactose and maltose; ➪ pp. 36–7.

suicide, deliberate self killing. In the developed countries the suicide rate in the 15 to 24 age group has been rising sharply in recent decades. Suicide is the most common form of death in those suffering from severe depression and is often deemed to be a consequence of mental disorder, whether or not there are other indications of this. There are cases, however, in which suicide has clearly been adopted as a reasonable means of escape from an intolerable social, medical, interpersonal or economic situation. Suicides have often made this plain before the event or in letters left.

sulphonamide, one of a large group of antibacterial drugs now largely superseded by the antibiotics except for a few purposes such as the treatment of urinary tract infections.

sunburn, ➪ pp. 144–5.

sunstroke, ➪ pp. 144–5.

superego, a Freudian concept involving an imaginary mental component that exercises a supervisory, not to say censorious, function over the ego, directing it into the paths of righteousness and responsibility; ➪ pp. 164–5.

superficial, near the surface.

superiority complex, an exaggerated and unrealistic conviction that one is better than others often held by people with an underlying sense of inferiority.

suppository, a small block of cocoa butter or gelatin, mixed with a drug, that is placed in the rectum or vagina where the drug is released when the vehicle melts at body temperature.

suppuration, the production or discharge of PUS.

surgery, the diagnosis and treatment of disease, injury or deformity by physical means, especially by the use of instruments; ➪ pp. 154–5.

surrogacy, a term that has developed the special meaning of the bearing of a child for another, infertile, woman, often for gain. The surrogate mother may be impregnated with sperm from the husband and agrees to surrender the child to the couple after its birth. For very good reasons, the practice is illegal in Britain.

suturing, surgical stitching of a wound or incision.

swallowing, ➪ pp. 44–5.

sweat, the salty secretion of the tiny, coiled tubular skin glands that open either directly on to the surface of the skin or into hair follicles so as to cool the body by evaporation; ⇨ APOCRINE glands, ECCRINE GLANDS and pp. 56–7.

swimmer's ear, ⇨ OTITIS EXTERNA.

Sydenham, Thomas, ⇨ pp. 150–1.

symbiosis, an association, for mutual benefit, between two or more organisms, often of different species.

symbol, any object, graphic, person or idea that can, usually by general agreement, be taken to represent something else. Symbols are important elements in art and literature and were deemed by Freud, Jung and others to have major psychological implications, especially in the significance of the content of dreams. They are of great interest to anthropologists.

sympathetic nervous system, the part of the involuntary (autonomic) nervous system which acts to constrict blood vessels in the skin and intestines, widen muscle blood vessels, speed the heart, dilate the pupils, widen the bronchial air tubes, relax the bladder and reduce bowel activity; ⇨ pp. 74–5.

symptom, any subjective perception of bodily change that suggests defect or malfunction. Objective indications of disorder are called signs.

synapse, a microscopic anatomical structure forming a junction or link between NERVE cells or between nerve cells and muscles or glands. The synapse features a gap across which neurotransmitter substances diffuse in response to a nerve impulse. Neurotransmitters stimulate receptors that initiate new nerve impulses, MUSCLE contraction or gland secretion. A single nerve cell may have tens of thousands of synapses on its surface, some excitatory, some inhibitory. The summation of the signals from those that are active will determine whether or not that nerve will fire. Such a mechanism provides the necessary unit for the construction of a highly elaborate parallel computing system. Synapses are of central importance in neurophysiology and the study of their action and potentiation has done much to explain BRAIN function and to elucidate how we learn, how we remember, how we acquire our unique personal attributes, and so on.

syndicalism, revolutionary seizure of the means of production by syndicates of workers.

syndrome, any unique combination of symptoms or signs, however apparently unrelated, that forms a distinct clinical entity.

synovial fluid, a clear, sticky lubricating fluid secreted by the membrane (the synovial membrane) that lines the capsule and all internal structures of a joint except the bearing surfaces; ⇨ pp. 40–1.

synovitis, inflammation of the synovial membrane of a joint; ⇨ SYNOVIAL FLUID.

syntax, the way in which MORPHEMES and words are put together to form meaningful phrases or sentences. The highly complex branch of linguistics concerned with the study of the structure of sentences.

syphilis, a sexually or congenitally transmitted disease, caused by a SPIROCHAETE, that produces a painless primary sore (chancre), lymph node enlargement, skin rashes, and, if untreated, potentially serious late damage to the major arteries, the brain and the nervous system including general paralysis of the insane (⇨ pp. 186–7) and TABES DORSALIS.

systole, the short period during which the pumping chambers of the heart are contracting and driving blood through the circulation. Compare DIASTOLE, which is the period between contractions.

T-cell, ⇨ pp. 58–9.

tabes dorsalis, degeneration of the sensory nerve columns of the spinal cord from late SYPHILIS, causing stabbing pains in the legs, unsteadiness and a characteristic stamping gait with the feet wide apart. Also known as locomotor ataxia.

taboo, ⇨ pp. 212–13.

tachycardia, unduly rapid heart rate.

talus bone, the large bone of the foot that rests on top of the heel bone (calcaneus) and articulates with the TIBIA and FIBULA at the ankle joint.

tapeworm, a ribbon-like colony of joined flatworms that grows from a common head (scolex) with hooks or suckers by which it is attached to the lining of the intestine. Mature segments, each of which contains both male and female reproductive organs, are passed in the stools after being fertilized; ⇨ pp. 136–7.

target cells, abnormal red blood cells, resembling targets, that occur in liver disease and in haemoglobin disorders.

tarsal bone, the seven small bones of the foot corresponding to the eight carpal bones of the wrist; ⇨ CARPUS.

tartar, a hard crust of chalky material (dental calculus) deposited in food and bacterial debris, left around the necks of neglected teeth, and leading to decay and gum disease.

taste, sense of, ⇨ pp. 88–9.

taste bud, ⇨ pp. 88–9.

tattooing, deliberate or accidental introduction of coloured material into the deeper layers of the skin.

TB, ⇨ TUBERCULOSIS.

TCA cycle, ⇨ KREBS CYCLE.

teaching, the imparting of knowledge or skills or causing to learn by example and inspiration; ⇨ EDUCATION.

tears, the salty secretion of the lacrimal glands that forms an important film over the surface of the CORNEA that is essential for vision and comfort.

technology, the application of science to a practical purpose, especially for an industrial or commercial purpose.

teeth, ⇨ TOOTH.

teething, ⇨ pp. 44–5.

temperature regulation, the maintenance of body temperature within narrow limits by a process of monitoring, with reflex shivering if the temperature drops and skin flushing and sweating if it rises. An important part of HOMEOSTASIS.

temple, the area on each side of the forehead above and behind the outer ends of the eyebrows and around the region of the hairline.

temporal, pertaining to the temples.

tendinitis, inflammation of a TENDON, usually from injury.

tendon, a strong band or strap of COLLAGEN fibres, often within a sheath, that joins muscle to bone or cartilage; ⇨ pp. 42–3.

tennis elbow, ⇨ GOLFER'S ELBOW.

TENS, transcutaneous electrical nerve stimulation – a method of treating CHRONIC pain by passing small, harmless electric currents into the spinal cord or nerves through electrodes applied to the skin.

tension, anxiety reflected in muscle contraction that often leads to symptoms, such as headaches.

terminal care, care of the dying.

termination of pregnancy, ⇨ ABORTION.

test-tube baby, ⇨ IN VITRO FERTILIZATION.

testicle, ⇨ TESTIS.

testis, one of the two male gonads, located in the SCROTUM, which produce sperms (spermatozoa) and sex hormones such as TESTOSTERONE.

testosterone, one of the chief male sex hormones (ANDROGENS); ⇨ pp. 80–1, 108–9.

tetanus, a serious infection of the nervous system caused by an organism that enters the body through a penetrating wound, especially if deep, and produces a powerful toxin which causes muscles to contract violently and may lead to death from exhaustion or asphyxia. Tetanus is easily prevented by immunization, which must be repeated at regular intervals through life. Compare TETANY.

tetany, the effects of a low BLOOD level of calcium. This interferes with the normal functioning of peripheral NERVES, causing increased excitability and a strong tendency for certain muscles, especially those that move the fingers and toes, to go into spasm. Compare TETANUS.

thalamus, one of two large masses of grey matter lying on either side of the midline in the base of the brain that act as the main sensory collecting and coordinating centre of the body; ⇨ pp. 74–5.

thalassaemia, one of several hereditable HAEMOGLOBIN abnormalities.

therapy, medical treatment of any condition by any means.

thermometer, a device for measuring and registering body temperature.

thiamin, thiamine, vitamin B_1. Deficiency causes BERI-BERI.

thigh, the part of the human leg between the GROIN and the knee, consisting of a stout bone, the femur, surrounded by a mass of muscle, sheathed in tendon and skin, with various large blood vessels and nerves.

thinking, ⇨ pp. 82–3, 108–9.

thirst, the strong desire to drink with dryness of the mouth arising from water shortage (dehydration) and concentration of the blood.

thoracic, pertaining to the THORAX.

thorax, the chest or upper part of the trunk between the neck and the abdomen; ⇨ pp. 40–1.

threadworms, small, harmless, white parasitic worms about 1 cm long, blunt at one end and with a fine, hair-like point at the other, which infect about one child in five at any one time.

throat, ⇨ pp. 44–5.

thrombosis, BLOOD CLOTTING within a closed blood vessel, usually because of disease of the vessel wall, especially ATHEROSCLEROSIS. Thrombosis is commonly the final event leading to the total obstruction of an artery, often with serious effects on heart (heart attack), brain (stroke) or limb (gangrene); ⇨ pp. 138–9.

thrush, infection with *Candida* fungus, usually in the mouth, VAGINA or skin, causing itching, soreness and white patches like soft cheese, with raw-looking inflamed areas in between.

thumb, the first digit of the hand, having only two bones (phalanges) and thus being shorter than the other digits, which have three, and being opposable to the other digits.

thymine, one of the four bases whose sequence in DNA forms the genetic code; ⇨ pp. 24–5, 26–7.

thymus, a small, flat organ, present behind the breast bone in childhood, and responsible for the processing of primitive LYMPHOCYTES of the IMMUNE SYSTEM so that they differentiate into T-cells; ⇨ pp. 60–1.

thyroid gland, ⇨ pp. 80–1.

thyroid hormones, ⇨ pp. 80–1.

tibia, the stouter of the two bones of the lower leg, partnered by the delicate fibula.

tic, a repetitive, twitching or jerking movement of any part of the face or body that occurs at irregular intervals mainly in children as a means of releasing emotional tension.

tic douloureux, an extremely severe, stabbing neuralgic pain affecting one side of the face in episodes of sudden onset lasting for seconds or minutes. Also known as trigeminal neuralgia.

tick, a small, eight-legged, blood-sucking ectoparasite of humans that can transmit several diseases including Q FEVER, RELAPSING FEVER and LYME DISEASE; ⇨ pp. 136 7.

tickling, touching the body of another in a teasing or exciting manner so as to arouse laughter or promote physical evasion.

tinea, infection of the skin by one of several fungi that feed on the dead outer layer (epidermis), affecting especially the groin (tines cruris) and the skin between the toes (tinea pedis or ATHLETE'S FOOT), causing intense itching and a raised scaly rash that may form circles ('ringworm'). Tinea of the hair or nails is very persistent.

tinnitus, any hissing, ringing, clicking or whistling sound originating in the inner ear or head and perceptible by the person concerned.

tissue, any collection of joined cells and their connections that perform a particular function, such as bone, muscle, nerve, epithelium, fat, fibrous tissue or elastic tissue.

tissue typing, identification of the particular chemical groups (histocompatibility antigens) specific to the individual present on the surface of all body cells, as a preliminary, for instance, to organ grafting.

toe, a digit of the foot, having two bones (phalanges) in the case of the great toe (first digit) and three in the case of all the others.

toilet-training, the process of teaching a child to exercise control and to use a potty or toilet for defecation and urination.

tolerance, one of the more admirable qualities of the human being often acquired, however, only after a lifetime's experience of the effects of intolerance. There are, however, philosophic difficulties. Should one, for instance, tolerate intolerance?

tomography, X-ray or other examination performed so as to produce an image in the form of a slice of the body or part of it. CT and MRI scanning are forms of tomography; ➪ pp. 128–9.

tone, muscle, the normal slight degree of tension maintained in a muscle when not actively contracting.

tongue, the highly mobile and flexible muscular organ that forms part of the floor of the mouth and carries specialized nerve endings called taste buds; ➪ pp. 88–9.

tongue-tie, a rare condition in which the partition under the tongue (frenulum) is so tight as to limit movement and affect speech.

tonic, a PLACEBO prescribed by doctors when no active treatment is required.

tonsillitis, inflammation of the TONSILS from infection, featuring sore throat, swelling, redness and sometimes spots of pus and enlargement and tenderness of the lymph nodes in the neck; ➪ pp. 116–17.

tonsils, oval masses of lymphoid tissue on either side of the back of the throat near the soft PALATE that have a protective role in childhood.

tool use, the ability to use an artificial aid to assist in the performance of a task. Found among some animals, tool use is especially common among the PRIMATES; ➪ pp. 16–17.

tooth, ➪ pp. 44–5.

tooth abscess, inflammation and tissue destruction, with an enclosed collection of pus, around the tip of the root of a tooth, with pain and swelling of the adjacent gum; ➪ ROOT-CANAL TREATMENT.

tooth decay, small areas of local destruction of tooth enamel and the underlying dentine with cavity formation and access of infection to the pulp, often leading to TOOTH ABSCESS. Also known as dental caries.

totipotency, the ability of a cell, specifically a fertilized egg, to differentiate to form any kind of tissue.

touch, sense of, the ability to discern contact with any part of the body that is subserved by a rich distribution of sensory nerve endings in the skin, especially profuse in the fingertips; ➪ pp. 90–1.

tourniquet, a band placed around a limb and tightened so as to compress blood vessels and prevent blood flow. Tourniquets left in place for more than half an hour or so may cause GANGRENE of the whole limb.

toxaemia of pregnancy, ➪ ECLAMPSIA.

toxin, a poison, often of high virulence, produced by a living organism. Bacterial disease is largely the result of poisoning by the toxins they produce. Some of these act locally; others diffuse out and act remotely; ➪ pp. 120–1.

trachea, the short windpipe that extends downwards from the LARYNX to the point of bifurcation at which the bronchi start; ➪ pp. 52–3.

tracheostomy, a surgical operation to make an opening into the windpipe, below the site of an obstruction – usually in the larynx – to allow breathing and save life.

trachoma, an eyelid and corneal infection with a *Chlamydia* organism that can readily be cured but that blinds millions of people in underdeveloped countries every year.

tract, a group of linked organs that form a passage along which liquids, solids or gases can move, such as the digestive tract, the urinary tract or the respiratory tract. The term is also used for a bundle of nerve fibres with a common function.

traction, exerting a sustained pull, especially on a part of the body to maintain proper alignment, as in the treatment of fractures.

trait, any inheritable characteristic.

tranquillizer drugs, ➪ pp. 152–3.

transcription, the production of a copy of RNA from DNA under the action of an enzyme known as RNA polymerase. Reverse transcriptase is an enzyme found in RETROVIRUSES, such as HIV, that synthesize DNA from an RNA template.

transference, the psychological process in which emotions felt towards one person are transferred to another, especially to one in a dominant or authoritative position, such as a psychotherapist. Freud believed transference essential to success in psychoanalysis.

transfusion, the restoration of the circulating volume of the blood by running donated blood or blood products into a vein, in cases of SHOCK, severe blood loss or ANAEMIA.

transient ischaemic attacks, episodes lasting for less than 24 hours and usually only a few seconds or minutes, in which any function of the brain, such as vision, speech, movement or sensation, is temporarily disturbed. Transient ischaemic attacks give serious warning of incipient STROKE and should never be neglected.

translocation, a form of chromosome mutation in which parts of different chromosomes may be exchanged or parts of two chromosomes joined. As long as all the usual chromosome material is present in normal amounts, translocation need cause no ill effects on the body.

transplant surgery, GRAFTING of donated organs or tissues into the body (homograft) with suppression by drugs or other means of the otherwise inevitable immune system rejection process. The term is also used for the surgical transfer of tissue from one site to another in the same person (autograft). Transplant surgery is currently most successful with kidneys and this has become routine. Heart transplantation and heart and lung transplantation are also becoming common operations. Liver transplantation is less common but many successful transplants have now been done. Vascularised bone transplants are being performed to treat ununited fractures. Pancreas transplants for diabetics are sometimes justified.

transsexualism, a person's unshakeable conviction that his or her true gender is the opposite of that represented by his or her anatomical body. If, after lengthy medical and psychiatric investigation, a person is deemed to be a genuine transsexual, his or her wish for sex-change surgery may sometimes be gratified.

transvestism, male desire to wear women's clothing, often to promote sexual arousal. Transvestism has little to do with TRANSSEXUALISM or homosexuality; ➪ pp. 202–3.

trapezius muscle, a large, triangular muscle that extends from the lower part of the back of the skull (occiput) almost to the lumbar region and to the outer tip of the collar bone (clavicle), on each side. This muscle braces the shoulder blade and rotates it outwards when the arm is raised; ➪ pp. 42–3.

trauma, injury of any kind.

travel sickness, ➪ MOTION SICKNESS.

tremor, a rhythmical oscillation of any part of the body, especially the hands, the head, the jaw or the tongue, either as an unexplained phenomenon or as a feature of one of many diseases or toxic effects.

trepanning, ➪ pp. 150–1.

trephine, ➪ pp. 150–1.

tribalism, the state of existing as a TRIBE, or the organization, beliefs and culture of a tribe.

tribe, a system of social organization, especially of an unindustrialized people, with a common territory, language, name, culture or ancestry.

triceps muscle, the three-headed MUSCLE on the back of the upper arm that straightens the elbow; ➪ pp. 42–3.

trigeminal neuralgia, ➪ TIC DOULOUREUX.

triplet (DNA), ➪ pp. 24–5.

tropical diseases, the group of diseases common in underdeveloped countries that are encouraged by poor standards of hygiene and sanitation, food and water contamination, the presence of insect VECTORS, a heavy human parasite load, nutritional inadequacies and poor living conditions; ➪ pp. 136–7.

tropical ulcer, a localized area of skin loss, sometimes with destruction of underlying tissue from one of a number of tropical infections such as LEISHMANIASIS, MALNUTRITION, yaws or cutaneous DIPHTHERIA.

Trotter, Wilfred, ➪ pp. 208–9.

truss, a belt and pad appliance used to exert pressure over the orifice of a RUPTURE (hernia) to prevent protrusion of the bowel.

tuberculosis, infection of the lungs, lymph nodes, skin, bones or other organs with the tubercle bacillus. Lung tuberculosis is the most common and, having been largely eliminated in the West, is now becoming more common again because of AIDS. It causes fever, weakness, loss of appetite and weight, night sweats and persistent cough often with blood-streaked sputum. Tuberculosis elsewhere causes local tissue destruction often with discharge of pus through the skin. Worldwide, tuberculosis still kills more people than any other infectious disease, and the recent emergence of multiple-drug-resistant strains is a cause for great concern; ➪ pp. 116–17.

tumour, ➪ pp. 140–1.

tunnel vision, severe narrowing of the peripheral VISUAL FIELDS from any of a number of causes such as retinal or optic nerve disease, brain damage from stroke or other causes, GLAUCOMA, PITUITARY GLAND tumour, head injury or MULTIPLE SCLEROSIS.

turgid, swollen and congested.

twin, ➪ MULTIPLE BIRTHS.

tympanic membrane, ➪ EAR DRUM.

typhoid fever, an intestinal and general infection by a *Salmonella* organism acquired in faecally contaminated food or water and causing high fever, diarrhoea, a rash of small, raised red spots (rose spots), liver and spleen enlargement and sometimes bowel perforation and PERITONITIS; ➪ pp. 134–5.

typhus, one of a number of infectious diseases caused by Rickettsial organisms, transmitted by lice, ticks or mites, and featuring fever, shivering, headache, general aches, cough, CONSTIPATION, a mottled rash, delirium, prostration, HEART FAILURE, stupor and sometimes coma and death; ➪ pp. 136–7.

typing, a procedure done to establish the grouping or classification of blood or tissues, primarily to avoid the problems of rejection during transplantation or transfusion.

ulcer, any local loss of surface covering (EPITHELIUM) and sometimes of deeper tissue in the skin or a MUCOUS MEMBRANE; ➪ BEDSORES, GASTRIC ULCER, TROPICAL ULCER, ULCERATIVE COLITIS.

ulceration, ➪ ULCER.

ulcerative colitis, a disease of the large intestine (colon) in which the lining becomes inflamed, swollen and covered with ULCERS, some of which may perforate leading to PERITONITIS. There is frequent diarrhoea, with blood, mucus and pus in the stools.

ulna, one of the two forearm bones, that on the little finger side, around which the RADIUS bone rotates when the hand is turned.

ultrasound scanning, body imaging using echoes of high frequency (ultrasonic) sound waves which can be used to generate small electric currents that can be represented as points of light on a screen to build up an image; ➪ pp. 128–9.

ultra-violet light effects, sunburn, damage to skin COLLAGEN, leading to wrinkling, rodent ulcers, squamous cell cancers, malignant melanomas, chronic eye discomfort and extension of the membrane over the white of the eye (conjunctiva) on to the cornea (pterygium).

umbilical cord, the knotted, irregular, rope-like structure that joins the PLACENTA to the site of the future navel on the abdomen of the fetus. The umbilical cord is the nutritional, hormonal and immunological link with the mother. It contains two arteries and a vein and is an extension of the fetal circulation, through which the placenta BLOOD is pumped by the fetal heart; ⇨ pp. 102–3.

umbilicus, ⇨ NAVEL.

unconscious, the, a term with more than one definition. In general, it is taken to be that division of the mental process that proceeds without immediate awareness of the fact on the part of the possessor. Clearly, this is an essential arrangement. Were the whole content of the mind (or memory) in consciousness at all times, we would be overwhelmed, so some form of selectivity is necessary. Access to the unconscious component is of variable difficulty. Ease of access seems to depend largely on associative links with the currently conscious part, such as mnemonics, and with the length of time since last the same information was accessed. Freudian psychoanalysts have their own definition. To them, the unconscious mind, or at least a part of it, is a domain into which are repressed those ID functions too unpalatable to be constantly presented to consciousness. The unresolved conflicts within such material are, they claim, the source of all our psychological troubles; ⇨ pp. 164–5.

undescended testis, a testis that has remained in the abdomen – where the testes develop – or in the groin canal by which they normally descend to the SCROTUM. Undescended testes are sterile and are prone to cancer later in life.

unsaturated fats, fats containing fatty acids whose carbon atoms are linked by double or triple valence bonds. Saturated compounds have only single bonds. Unsaturated fats are liquid at room temperatures; saturated fats are solid. A diet low in saturated fats is desirable.

urea, ⇨ pp. 64–5.

ureter, ⇨ pp. 64–5

urethra, ⇨ pp. 64–5.

uric acid, a waste product derived from two of the bases – the purines adenine and guanine – that form the genetic code in DNA. Abnormalities of purine METABOLISM lead to an excess of uric acid in the body and GOUT.

urinary calculus, a stone that forms within the kidney or the urine collecting system, usually because of highly concentrated urine, infection or metabolic disease.

urinary system, ⇨ pp. 64–5.

urine, ⇨ pp. 64–5.

urology, the study of the kidneys and the urine drainage system and their disorders.

urticaria, an allergic skin condition featuring raised, pink, itchy areas surrounded by pale skin, that may occur after contact with plants, drugs or foodstuffs, after insect bites or jelly fish stings or on exposure to sunlight or cold. Also known informally as nettle rash or hives; ⇨ pp. 142–3.

uterus, the womb. The muscular organ in which the embryo and fetus grow to a stage of maturity allowing independent existence. As the fetus grows, the uterus stretches; ⇨ pp. 96–7, 104–5.

utopia, a mythical place where all social problems have been solved and man's nature has been perfected.

uvula, the soft, MUCOUS MEMBRANE-covered, central, mobile process of the soft palate that can be seen hanging down from the roof of the mouth at the back of the palate.

vaccination, the process of inducing immunological protection against an infectious disease, usually by the injection of a killed or weakened version of the disease organism. Strictly speaking this term should be confined to the process once widely used to confer immunity against smallpox, originally by introducing cowpox material under the skin (Latin *vacca*, a cow). Smallpox has been eradicated and the term is now used for any process of active IMMUNIZATION.

vacuole, a small, spherical space in the cytoplasm of a cell, sometimes surrounded by a membrane, in which cell products are stored or carried to the surface for excretion.

vagina, the sheath-like organ that receives the penis during copulation and that acts as the birth passage during parturition; ⇨ pp. 64–5, 96–7.

vaginal discharge, the release from the VAGINA of more than the normal small amount of clear or creamy fluid. Profuse discharge may be due to a retained tampon or to THRUSH or *Trichomonas* infection.

vaginismus, a seemingly involuntary physical resistance, by a woman, to any attempt at sexual intercourse or gynaecological examination.

vaginitis, inflammation of the VAGINA from any cause. Also known as colpitis.

vagus nerves, a pair of cranial nerves that arise directly from the brainstem and pass down the neck to supply and control the muscles of the throat, gullet, larynx, air tubes (bronchi), heart, stomach and intestines, as part of the AUTONOMIC NERVOUS SYSTEM.

valve, any structure that allows movement in one predetermined direction only, as in the heart, the veins, the lymphatics and the urethra.

varicella, ⇨ CHICKENPOX.

varicocele, a distortion in the network of veins that surround the testis, especially on the left side, so as to form an irregular swelling in the scrotum and sometimes reduce fertility.

varicose veins, enlargement, twisting and distortion of veins, especially in the legs, the anal canal (piles), at the lower end of the gullet (oesophageal varices) or in the scrotum (VARICOCELE). Leg varicosities are due to defects of the vein valves.

variola, ⇨ SMALLPOX.

vas deferens, one of two narrow tubes up which a sludge of mucus and spermatozoa is carried by their constant production in the testicles. This sludge is diluted at the upper ends of the vasa by fluid from the seminal vesicles and the prostate gland. An effective method of male contraception involves cutting and separating the two vasa deferentes (vasectomy) so that the route of the sperms from the testicles is removed.

vasectomy, ⇨ VAS DEFERENS.

vector, any organism, such as an insect, capable of transmitting an infectious disease from one person to another; ⇨ pp. 130–1. In molecular biology a small independently replicating chromosome into which genes can be placed by recombinant techniques.

vegetarianism, the personal policy of excluding meat and sometimes other animal products, such as eggs and milk, from the diet.

vegetative state, life without consciousness, will or the possibility of any form of communication with others, as a result of damage to the 'higher' centres of the brain.

vein, a thin-walled blood vessel that carries low pressure blood back to the heart from the body organs and tissues. Compare ARTERY; ➪ pp. 50–1.

vena cava, one of the two largest veins in the body, through which oxygen-depleted blood from the tissues returns to the right side of the heart to be pumped to the lungs; ➪ pp. 50–1.

venereal diseases, ➪ SEXUALLY TRANSMITTED DISEASES.

ventilator, a mechanical device used to maintain artificial respiration in people unable to breathe for themselves. A ventilator is essentially an air pump that inflates the lungs rhythmically by way of a tube inserted in the windpipe, either through the mouth or nose or through a hole in the neck (tracheostomy); ➪ pp. 156–7.

ventral, pertaining to the front of the body. Compare dorsal.

ventricle, ➪ pp. 50–1.

ventricular fibrillation, a form of CARDIAC ARREST; ➪ FIBRILLATION.

venule, a tiny vein; ➪ pp. 50–1.

vernix, the layer of greasy material, mainly of SEBUM mixed with skin scales and hair, with which a new-born baby is covered.

verruca, ➪ WART.

vertebrae, ➪ pp. 40–1.

vertebral column, the column of 24 individual bones (vertebrae) which, together with the sacrum, form the backbone, the main central structural member of the axial skeleton, providing attachment for the ribs and the pelvis. A hole behind the body of each vertebra lines up with its fellows to form a canal, the spinal canal, which accommodates the spinal cord. The spinal NERVES, leaving and entering the cord, pass out between the vertebrae. The vertebrae are held together by pad-like intervertebral discs and by tough longitudinal ligaments that run down the fronts and backs of the vertebral bodies. The side processes of the neck (cervical) vertebrae are also perforated to accommodate, on each side, a vertebral artery – one of the four arteries that supply the BRAIN.

vertigo, the illusory sense that the environment or the body are rotating, often with a tendency for the sufferer to fall. The many possible causes include MOTION SICKNESS, ALCOHOL, fear of heights, anxiety, drugs, disorders of the balancing mechanisms in the inner ears, MULTIPLE SCLEROSIS or brain tumour.

Vesalius, Andreas, ➪ pp. 34–5.

vesicle, any small blister or pouch.

villus, one of the millions of tiny, finger-like processes found mainly in the intestine, where they greatly increase the absorptive surface area, but also in the PLACENTA, from an early stage, and on the tongue and choroid plexuses of the BRAIN; ➪ pp. 44–5.

virginity, the state of never having engaged in sexual intercourse. Like most matters connected with the sexual function and hence with the emotions, virginity has acquired a special social, marital, economic and even theological significance. Male-dominated societies greatly value virginity and often persecute women who lose it in an unauthorized manner. Non-virgins may be regarded as 'spoiled goods' with greatly lowered value. Such an attitude springs partly from male chauvinism and pride in exclusive possession and partly from a theological notion that virginity can be equated with 'purity'. Sustained virginity can also be regarded as a sacrifice to God, as in voluntary life-long chastity. The miraculous virginal fertilization of Mary the mother of Christ (the immaculate conception) – a manifestation of the same notion – is no longer universally maintained by informed theologians who are aware that parthenogenesis can produce only females.

virility, the quality of male strength, vigour, energy and especially sexuality.

virulence, the degree of the capacity of any infective organism to cause disease and to injure or kill a susceptible host.

virus, ➪ pp. 120–1, 132–3.

vision, ➪ SIGHT, SENSE OF .

visual fields, the whole of the area over which some form of visual perception is possible while the eyes of the subject are directed straight ahead.

vitamins, substances, often co-enzymes, needed for the maintenance of health. In general, the fat-soluble group (A,D,E and K) are obtained from food containing animal and vegetable fats and fish, and the water-soluble group (C and the largely subdivided B group) are obtained from fruit and vegetable matter. Vitamins take part in many important biochemical reactions in the body and are needed in very small quantities. Some are stored long-term in the body, some are not. Excessive intake of certain vitamins, especially A and D, can produce serious disorders. The idea that large doses of vitamins can promote health is generally untrue, but ➪ FREE RADICALS; also ➪ pp. 46–7.

vitreous humour, the delicate, transparent gel, consisting almost entirely of water, that fills the cavity of the eye between the back of the crystalline lens and the RETINA.

vocal cords, ➪ pp. 82–3.

voice box, ➪ pp. 82–3.

vomiting, involuntary expulsion of the STOMACH contents by sudden pressure on the stomach from contraction of the abdominal wall and a downward movement of the DIAPHRAGM.

voyeurism, the practice of secretly watching people undressing or engaging in sexual intercourse, so as to obtain sexual stimulation frequently leading to MASTURBATION.

Vries, Hugo de (1848–1935), Dutch botanist and geneticist who developed a system of genetics,essentially along Mendelian lines, but independently of Mendel. His most important work is *The Mutation Theory* (1901–3).

vulva, the female external genitalia – the mons pubis, the LABIA, the area between the labia minora, and the entrance to the VAGINA; ➪ pp. 96–7.

waist, the usually narrowed part of the trunk of the body below the bottom of the ribs and above the hips.

Wallace, Alfred Russel, ➪ pp. 8–9, 180–1.

wart, a benign, localized, but sometimes disfiguring skin excrescence caused by the human papilloma virus which stimulates overgrowth of the cell layer at the base of the EPIDERMIS. Warts on different parts of the body take different forms because of differences in skin thickness, but all are essentially the same. The medical term is verruca; ➪ pp. 56–7.

waste products, the unwanted and useless residuum of the metabolic processes of the body that are disposed of largely in the urine, and the undigestible and unabsorbable residue of the diet that is disposed of in the faeces; ➪ pp. 64–5.

water, hydrogen oxide, a compound essential for life and constituting about 70% of the body weight. Half the water is within cells, half in the tissue fluid and blood. Intake is via the mouth; losses occur in the urine, in evaporation from the skin, in the expired air and in the faeces. Restricting water intake is dangerous especially in hot conditions.

water on the brain, an informal term for abnormal volumes of CEREBROSPINAL FLUID within, and around, the brain, usually from obstruction of flow to the site of reabsorption. In babies, the head becomes greatly enlarged. Also known as hydrocephalus.

water on the knee, an informal term for an abnormal accumulation of SYNOVIAL FLUID within the knee joint (effusion) from inflammation caused by injury or disease.

Watson, James Dewey, ➪ pp. 24–5.

Watson, John, ➪ pp. 30–1, 162–3.

wax, the secretion of the ceruminous glands in the skin of the outer ear canal, that traps dust and deters small insects; ➪ pp. 86–7.

weaning, substitution of solid foods for milk in an infant's diet.

webbing, flaps of skin between the fingers or toes, or occasionally at the sides of the neck.

weight, the mass of the body, determined primarily by genetic influences and secondarily by the balance between calorie intake and energy expenditure in exercise. Unexplained weight loss is a sign of possible serious disease that should never be ignored.

Weil's disease, ➪ LEPTOSPIROSIS.

whiplash injury, a neck injury resulting from sudden accelerative or decelerative forces to the body so that the neck bends acutely and muscles, tendons or even bones may be injured.

white blood cells, all the CELLS that circulate freely in the blood other than the RED BLOOD CELLS. White blood cells are components of the immune system. The medical term is leucocytes; ➪ pp. 40–1, 50–1.

white matter, the parts of the brain and spinal cord consisting of nerve fibres rather than aggregates of nerve cell bodies; ➪ pp. 76–7.

white of eye, the tough opaque outer coat or sclera, part of which is visible between the lids through the transparent CONJUNCTIVA.

whitlow, infection of the pulp of the tip of a finger or toe, usually from a deep prick. There is dull pain, redness, swelling, throbbing and tenderness which may proceed to ABSCESS formation. Also known as a felon.

WHO, ➪ WORLD HEALTH ORGANIZATION.

whooping cough, an often seriously exhausting infectious disease of early childhood featuring episodes of a succession of rapidly repeated coughs followed by a characteristic long, whooping inspiration and often vomiting. The disease may last for weeks and is often complicated by OTITIS MEDIA, PNEUMONIA, lung collapse or seizures from lack of oxygen to the brain. Immunization is effective in preventing the disease. Also known as pertussis.

wife-beating, ➪ SPOUSE ABUSE.

Wilberforce, Samuel, Bishop of Oxford, ➪ pp. 8–9.

Wilkins, Maurice, ➪ pp. 24–5.

windpipe, ➪ TRACHEA.

Winson, Jonathan, ➪ pp. 178–9.

wisdom tooth, a third molar tooth, the rearmost of the set, that often does not erupt until early adult life, hence the name.

witches, ➪ pp. 184–5.

witches' milk, a brief period of milk production from the breasts of newborn babies of either sex due to transfer to the fetus before birth of small quantities of the HORMONE prolactin present in the mother's blood.

withdrawal symptoms, the effects of denial of a drug of addiction, such as heroin, on a person who is physically dependent on it. Heroin withdrawal symptoms include restlessness, depression, yawning, abdominal pain, vomiting, diarrhoea, loss of appetite, sweating and gooseflesh ('cold turkey') and a persistent craving for the drug. Those from other narcotic drugs are similar but less intense.

womb, the hollow muscular organ in the female, continuous with the upper end of the VAGINA, in the inner lining of which the fertilized ovum implants and the PLACENTA develops and in which the fetus grows, in the course of 40 weeks, to a stage of maturity at which it can survive independently; ➪ pp. 102–3, 108–9.

work ethic, a system of personal behaviour motivated by belief in the merits of hard work, personal discipline, thrift and the value of the individual. The work ethic is the basis of modern capitalism.

worms, ➪ NEMATODES, ROUNDWORMS, TAPEWORMS and THREADWORMS.

wound, ➪ pp. 62–3.

wound healing, a remarkable process that involves a preliminary scavenging by the IMMUNE SYSTEM followed by the production of much new COLLAGEN PROTEIN, by cells called fibroblasts, to bind the wound edges together, and finally a process of collagen maturation and shrinkage, in which the scar becomes less conspicuous; ➪ pp. 62–3.

wrist, joint between the hand and the arm consisting of eight carpal bones, arranged in two rows, that articulate with the forearm bones and with the metacarpal bones of the palm, and are crossed by many tendons that connect the forearm muscles to the fingers and thumb, and by arteries and nerves.

wristdrop, paralysis of the muscles that extend the wrist and fingers.

writing, a means of communicating information by agreed symbols or marks scribed in sequence on a surface, the symbols to represent either things, ideas or concepts (ideograms) or to represent sounds in spoken language. Writing permits the passage of ideas across the centuries and was a major landmark in human development.

Wundt, Wilhelm Max, ➪ pp. 162–3.

X-chromosome, ➪ pp. 22–3.

X-linked disorders, ➪ pp. 28–9.

X-rays, ➪ pp. 128–9.

Y-chromosome, ➪ pp. 22–3.

yawning, an involuntary, deep and often infectious inspiration, accompanied by the desire to open the mouth widely, and occurring most often when the subject is bored or sleepy. Yawning widens the air sacs of the lungs, helps to improve the return of blood to the heart, reduces blood stagnation and increases its oxygen content.

yellow fever, an acute viral infectious tropical disease transmitted between humans by mosquito bite and causing high fever, severe headache, muscle pain, nose bleeds, vomiting of blood, liver damage and jaundice, often followed by delirium, coma and death; ➪ pp. 130–1, 132–3.

yoga, one of six systems of Indian philosophy, of which hatha yoga is most often practised in the West. This is a physical preparation for spiritual development and involves a series of poses (asanas) by which flexibility and control of the body and relaxation and peace of mind may be obtained.

Young, Thomas, ➪ pp. 84–5.

Z-plasty, a surgical procedure for relieving skin tension or releasing contracture by making a Z-shaped incision, freeing the skin and transposing the resulting V-shaped flaps.

zinc, a metallic element required in very small quantities for health. Deficiency, which is rare, results in immune system disorders, skin atrophy, poor wound healing, loss of appetite, diarrhoea, apathy and loss of hair; ➪ pp. 46–7.

Zollinger-Ellison syndrome, severe ULCERS of the stomach and duodenum from excessive stomach acid production under the influence of a HORMONE produced by a tumour of the pancreas.

zoonosis, any one of the many diseases of animals that can be contracted by humans. Compare diseases transmitted from person to person by animal vectors. The zoonoses include anthrax, brucellosis, glanders, leptospirosis, plague, psittacosis, rabies and toxoplasmosis; ➪ pp. 144–5.

Zoroastrianism, a RELIGION founded by Zoroaster (Zarathusthra) (c. 628–551 BC), as a reformation of the ancient polytheistic Persian religion. Zoroastrianism flourished during the Achaemenid, Parthian and Sassanian empires and declined after the Muslim conquest of Persia in the 7th century AD. The modern descendants are the Parsees of Bombay.

zygoma, the cheek bone.

zygote, an egg (ovum) that has been fertilized but has not yet begun to divide. A zygote contains the complete genetic code for a new individual and is DIPLOID; ➪ pp. 96–7.

FACTFINDER

PICTURE ACKNOWLEDGEMENTS

The publishers would like to thank the following for permission to reproduce the pictures in this book, which are individually credited by the abbreviations listed below:

Aerofilms	Aerofilms
AKG	Archiv für Kunst und Geschichte
Allsport	Allsport
AR	Ann Ronan
BAL	Bridgeman Art Library
BC	Bruce Coleman Ltd
CD	Chemical Design Ltd., Oxford
Colorific!	Colorific!
EPG	E.P. Goldschmidt & Co. Ltd
Explorer	Explorer
FAPL	Fine Art Photographic Library
Gamma	Gamma
Giraudon	Giraudon
IB	Image Bank
Images	Images Colour Library
Image Select	Image Select
JW	John Walmsley
Keystone	Keystone
Kobal	The Kobal Collection
KP	Klaus Paysan
LSI	Life Science Images
MD	Muscular Dystrophy Group
ME	Mary Evans
Microscopix	Microscopix
National Film Archive	National Film Archive
NMSB	National Medical Slide Bank
PN	Peter Newark's Historical Pictures
Popperfoto	Popperfoto
RB	Ron Boardman
Rex	Rex Features
Spectrum	Spectrum Photo Library
SPL	Science Photo Library
Sygma	L'Illustration/Sygma
TCL	TCL Stock Directory
Telegraph	Telegraph Colour Library
TPS	The Photo Source Ltd
UNHCR	United Nations High Commissioner for Refugees
UPI/Bettman	UPI/Bettman Newsphotos
Wellcome	The Wellcome Institute Library
WHO	World Health Organization
Zefa	Zefa Stockmarket